W0042258

Progress in Nitrogen Ceramics

NATO ASI Series

Advanced Science Institutes Series

A Series presenting the results of activities sponsored by the NATO Science Committee, which aims at the dissemination of advanced scientific and technological knowledge, with a view to strengthening links between scientific communities.

The Series is published by an international board of publishers in conjunction with the NATO Scientific Affairs Division

A	**Life Sciences**	Plenum Publishing Corporation
B	**Physics**	London and New York
C	**Mathematical and Physical Sciences**	D. Reidel Publishing Company Dordrecht and Boston
D	**Behavioural and Social Sciences**	Martinus Nijhoff Publishers Boston/The Hague/Dordrecht/Lancaster
E	**Applied Sciences**	
F	**Computer and Systems Sciences**	Springer-Verlag Heidelberg/Berlin/New York
G	**Ecological Sciences**	

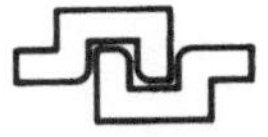

Series E: Applied Sciences – **No. 65**

Progress in Nitrogen Ceramics

edited by

F.L. Riley

Houldsworth School of Applied Science
The University of Leeds
Leeds, Yorksh., UK

1983 **Martinus Nijhoff Publishers**
A member of the Kluwer Academic Publishers Group
Boston / The Hague / Dordrecht / Lancaster
Published in cooperation with NATO Scientific Affairs Division

Proceedings of the NATO Advanced Study Institute on Nitrogen Ceramics, University of Sussex, Falmer, UK, July 27 - August 7, 1981

Library of Congress Cataloging in Publication Data

Main entry under title:

Progress in nitrogen ceramics.

(NATO ASI series. Series E, Applied sciences ; no. 65)
Proceedings of the 2nd NATO Advanced Study Institute on Nitrogen Ceramics, University of Sussex, Falmer, U.K., July 27-Aug. 7, 1981.
"Published in cooperation with NATO Scientific Affairs Division."
Includes indexes.
1. Silicon nitride--Congresses. 2. Ceramics--Congresses. I. Riley, F. L. II. NATO Advanced Study Institute on Nitrogen Ceramics (2nd : 1981 : University of Sussex) III. North Atlantic Treaty Organization. Scientific Affairs Division. IV. Series: NATO advanced science institutes series. Series E, Applied sciences ; no. 65.
TP245.N8P7 1983 666 83-4052

ISBN-13:978-94-009-6853-0 *e-ISBN-13:978-94-009-6851-6*
DOI: 10.1007/978-94-009-6851-6

Distributors for the United States and Canada: Kluwer Boston, Inc., 190 Old Derby Street, Hingham, MA 02043, USA

Distributors for all other countries: Kluwer Academic Publishers Group, Distribution Center, P.O. Box 322, 3300 AH Dordrecht, The Netherlands

PREFACE

The first NATO Advanced Study Institute on Nitrogen Ceramics held in 1976 at Canterbury came at a particularly significant moment in the development of this subject. The five-year period, 1971-75, had been an especially fruitful one in very many respects for work in the areas of covalent materials in general, and of the nitrides in particular. The Institute was therefore able to capture fully the spirit of excitement and adventure engendered by the outputs of numerous national research programmes, as well as those of many smaller research groups, concerning ceramics potentially suitable for applications in a high temperature engineering context. It reflected accurately the state of knowledge with respect to the basic science, the powder technology, and the properties of materials based on silicon nitride and associated systems. The Proceedings of the Institute thus provided a good record for workers already in the field, and a useful textbook for newcomers to the subject of nitrogen ceramics.

The Canterbury Advanced Study Institute had a valuable educational and social function in bringing together for two weeks a large proportion of those workers most closely involved at that time with the nitrogen ceramics. The atmosphere of this meeting, providing both intensive discussions and informal contacts, made a lasting impression on the participants, and inevitably the question was raised of whether, and when, a second Advanced Study Institute might be held on this subject. There was no unique answer to the second of these questions, but an interval of four to five years, in the light of developments, seemed a not unreasonable one, after which a sound case might be made to the Scientific Affairs Division of NATO for necessary financial support. As part of this case, when it was presented in 1980, attention was drawn to the continuing expansion of interest in silicon nitride materials, and in particular in the relatively new areas of sintered silicon nitride and sintered reaction bonded silicon nitride, topics which, as the progress chart in the Concluding Remarks section of 'Nitrogen Ceramics' indicates, in 1976 were in their infancy. Work in the area of 'Sialon' systems had also brought about a considerable expansion of knowledge in breadth and in depth of these complex systems, and their essential importance in connection with the production and properties of the sintered and hot-pressed silicon nitrides was now clear. The distinction between 'silicon nitride' and 'sialon' materials was indeed in certain respects becoming less meaningful. The crystallization of the intergranular glass phase, known to be present but barely detectable at grain interfaces in 1976, and recognized as the key to obtaining high strengths at high temperature, had received attention and by now formed the basis for the development of a range of commercial materials. In presenting the case, attention was also drawn to the practical achievements seen in the expanding number of ceramic gas turbine and diesel engine research programmes in the U.S.A. and in Europe, and the intense interest being shown in high grade ceramic materials in Japan. At the same time, it was recognized by the planning group that the last five years had also seen big steps in the development of other groups of high strength ceramic materials for engineering applications. High grade sintered silicon carbides in particular were now serious contenders, together with the self-bonded or siliconized forms, as gas turbine and diesel engine components. The possibility that the subject of a second Advanced Study Institute might be broadened to include silicon carbide was, however, rejected on the grounds that it would not then be possible, even in the space of a two-week meeting, to treat adequately either set of materials. Nonetheless, in recognition of the fact that the silicon carbides were becoming increasingly mature materials and, in some senses, competitors to the nitrides, it seemed very appropriate that a formal discussion session should be devoted to this subject, so that recent developments in the area of the silicon carbides could be examined, and respective properties compared.

The Institute programme again was a very full one with 15 invited review contributions and some 60 shorter presentations on specific topics. The texts of the majority of these presentations appear in this volume, together with edited selected contributions from the very lively discussion sessions which followed the presentations, and of which teams of reporters, and the speakers, attempted to keep a record. Also included are the prepared texts of the four major contributions to the session on 'National Research Programmes', which will be of interest as a rare statement of the extent of international interest at a high level in nitrogen ceramic materials. As in 1976, the original intention of the planning group was that all types of ceramic nitrides should receive attention during the Institute. As a glance at the 'Contents' section will show, however, the fact is that with only one or two notable exceptions, in the ceramic nitrides area interest is still firmly focussed on silicon nitride and the 'sialon' families of materials.

The success of this second Institute rested largely on the uniformly very high quality of the teaching and research review contributions, and of the ensuing discussions, and it is the participants themselves therefore who must take the credit. We were, however, indebted to the members of the international planning group, who by their efforts helped to ensure that those participants best able to represent the many active groups in the field were indeed present at Sussex. Essential financial support for the Institute was provided by the Scientific Affairs Division of NATO, together with generous contributions from the European Research Office in London (USARDSG-UK), the Office of Naval Research, Arlington, Virginia, and the National Science Foundation, Washington D.C. In the context of the preliminary planning of the Institute, the valuable guidance and help of Dr. Mario di Lullo and Dr. Tilo Kester of the Scientific Affairs Division of NATO is gratefully acknowledged.

The Institute has also to acknowledge with gratitude the behind-the-scenes activities of many members of Staff of the Universities of Sussex and Leeds. In particular, the efficient organisational work of Mrs. S. Michael and Mrs. J. Hutchings of the Sussex Conference Office, and Miss Gillian Horsfall, seconded to the Institute Office from the Leeds Secretarial Office, should be singled out for mention. Equally important for the smooth running of the technical sessions and refreshment breaks was the assistance provided by the technical staff of the Sussex Department of Physics. Messrs. Lu and Oswald of the University of Leeds Department of Ceramics, and Mr. Spacie from the Crystallography Laboratory at the University of Newcastle-upon-Tyne, provided flawless service in the projection room. Finally, the Editor has to acknowledge once again his deep indebtedness to Miss Susan H. Toon, ably assisted by Mrs. Susan Richards, in the Leeds Department of Ceramics, for their calm efficiency in dealing with a very large volume of correspondence and paper-work, both before and after the period of the Institute, and in the typing of many sections of this Proceedings.

F.L. Riley
Editor

CONTENTS

(R) indicates an invited review lecture

SECTION D

STRUCTURAL DEVELOPMENT

PARTICIPANTS

Advisory Committee:

M. Billy	University of Limoges, France
W. Bunk	DFVLR, Köln, Germany
K.H. Jack	University of Newcastle-upon-Tyne, U.K.
R.N. Katz	AMMRC, Watertown, U.S.A.
F.F. Lange	Rockwell International, U.S.A.
F.L. Riley	University of Leeds, U.K. (Director)
P. Vincenzini	IRTEC, Faenza, Italy.

Local Organization:

Mrs. S. Michael and Mrs. J. Hutchings, University of Sussex.
Miss G.L. Horsfall, University of Leeds.

Chairmen and Discussion Reporting Teams:

Session		
July 27th	1. P. Popper	D.P. Thompson, T.Y. Tien
	2. J. Lang	P.E.D. Morgan, G.E. Gazza
July 28th	1. P.E.D. Morgan	K.H. Jack, C. Greskovich
	2. H. Hausner	S. Hampshire, P. Popper
July 29th	1. K.H. Jack	M. Kizilyalli, J. Briggs
	2. A. Mocellin	J.W. McCauley, H. Knoch
July 30th	1. C. Greskovich	F.L. Riley, R.N. Katz
	2. D.P. Thompson	H. Hausner, W. Schmidt
July 31st	1. D.R. Clarke	A. Mocellin, R. Pompe
	2. M. Billy	D. Marshall, J.R.G. Evans
	3. J. Weiss	R.J. Fields, G. Wötting
August 3rd	1. J. Lang	T.M. Shaw, J.A. Mangels
	2. T.Y. Tien	J. Heinrich, J. Desmaison
August 4th	1. D. Marshall	J. Cotton, C.S. Furtado
	2. F. Thümmler	J. Kirk, K. Kriz
	3. A.G. Evans	K. Goebbels, H. Knoch
August 5th	1. J.A. Mangels	H.T. Larker, G. Schwier
	2. G. Ziegler	D.W. Richerson, K. Suzuki
August 6th	1. E. Gugel	R.J. Lumby, C.L. Quackenbush
	2. R.N. Katz	
August 7th	1. F.L. Riley.	

Amundin, Dr. C.O. National Defence Research Institute, S-10450 Stockholm, Sweden.

Babini, Dr. G.N. IRTEC, Via Granarolo 8, 48018 Faenza, Italy.

Billy, Prof. M. Université de Limoges, Laboratoire de Céramiques Nouvelles, 123 avenue Albert Thomas, 87060 Limoges, France.

Boch, Prof. P. Ecole Nationale Supérieure de Céramique Industrielle, 47-73 rue Albert Thomas, 87065 Limoges Cedex, France.

Boyer, Dr. S.M. Standard Telecommunication Laboratories, London Road, Harlow, Essex, CM17 9NA, U.K.

Briggs, Dr. J. Morgan Thermic Ltd., Technological Centre, Bewdley Road, Stourport on Severn, DY13 8QR, U.K.

Bunk, Prof. W. DFVLR, Postfach 90 60 58, 5000 Köln 90, W.Germany.

Clarke, Dr. D.R. Rockwell International Science Center, 1049 Camino dos Rios, P.O. Box 1085, California 91360, U.S.A.

Cotton, Mr. J. British Ceramic Research Association, Queens Road, Penkhull, Stoke-on-Trent, U.K.

Desmaison, Dr. J. Université de Limoges, Laboratoire de Céramiques Nouvelles, 123 avenue Albert Thomas, 87060 Limoges, France.

Elias, Dr. A.M. Faculdade de Ciencias de Lisboa, Laboratorio de Quimica, University of Lisbon, 1294 Lisbon Codex, Portugal.

Epicier, Mr. T. Bât. 502, INSA, 69621 Villeurbanne Cedex, France.

Evans, Prof. A.G. Department of Materials Science, University of California, Berkeley, California 94720, U.S.A.

Evans, Dr. J.R.G. Department of Ceramics, Houldsworth School of Applied Science, University of Leeds, Leeds LS2 9JT, U.K.

Faber, Miss K.T. Department of Materials Science and Mineral Engineering, Hearst Mining Building, University of California, Berkeley, California 94720, U.S.A.

Falk, Miss L. Department of Physics, Chalmers University of Technology, Gothenberg, Sweden 41296.

Fields, Dr. R.J. National Bureau of Standards, Washington D.C. 20234, U.S.A.

Furtado, Prof. C.S. Departamento de Fisica, Universidade de Coimbra, 3000 Coimbra, Portugal.

Gazza, Dr. G.E. Materials Branch, Army Materials and Ceramics Research Center, Watertown, Mass. 02172, U.S.A.

Göebbels, Dr. K. Fraunhofer-Institut für Zerstörungsfreie Prüfverfahren, Universität, Geb. 37, D-6600 Saarbrucken 11, W. Germany.

Goursat, Dr. P. Université de Limoges, Laboratoire de Céramiques Nouvelles, 123 avenue Albert Thomas, 87060 Limoges, France.

Greskovich, Dr. C. General Electric Company, Corporate Research and Development Center, P.O. Box 8, Schenectady, N.Y. 12301, U.S.A.

Gugel, Prof. E. Annawerk GmbH, Ceranox Division, D-8633 Roedental, Postfach 44, W. Germany.

Hakulinen, Mr. M. Saab-Scania, Scania Division, Department BLMM, S-15187 Södertälje, Sweden.

Hampshire, Dr. S. College of Engineering and Science, National Institute for Higher Education, Limerick, Eire.
Hattori, Mr. Y. Research & Development, The Nippon Tokushu Togyo Kaisha Ltd., 14-18 Takatsuji-cho, Mizuho-cho, Nagoya, Japan.
Hausner, Prof. H. Institut für Nichtmetallische Werkstoffe, Englische Strasse 20, 1000 Berlin 12, Germany.
Heinrich, Dr. J. Rosenthal Aktiengesellschaft, Institut für Werkstofftechnik, Wittelsbacherstrasse 49, 8672 Selb, W. Germany.
Hohnke, Mrs. H. 4228 E Engineering Bldg., University of Michigan, Ann Arbor, Michigan 48018, U.S.A.
Hunt, Mr. R.A. Lucas Industries Ltd., Group Research Centre, Shirley, Solihull, B90 4JJ, U.K.
Jack, Prof. K.H. Department of Metallurgy and Engineering Materials, The University, Newcastle upon Tyne, NE1 7RU, U.K.
Katz, Dr. R.N. Materials Branch, Army Materials and Ceramics Research Center, Watertown, Mass. 02172, U.S.A.
Kirk, Dr. J. Morgan Thermic Ltd., Technological Centre, Bewdley Road, Stourport on Severn, DY13 8QR, U.K.
Kizilyalli, Dr. H.M. Department of Physics, Middle East Technical University, Ankara, Turkey.
Kizilyalli, Dr. M. Department of Chemistry, Middle East Technical University, Ankara, Turkey.
Knoch, Dr. H. Elektroschmelzwerk Kempten GmbH, Postfach 1526, 8960 Kempten, W. Germany.
Kriz, Mr. K. Institut für Werkstoffwissenschaften I, Universität Erlangen-Nürnberg, Martensstrasse 5, D-8520 Erlangen, W. Germany.
Lal, Dr. M. Department of Metallurgy & Materials Engineering, Lehigh University, Bethlehem, Pa. 18015, U.S.A.
Lang, Prof. J. Université de Rennes, UER Structure et Proprietes de la Matière, Laboratoire de Chimie Minéràle C, Avenue du Général Leclerc, 35031 Rennes Cedex, France.
Lange, Dr. F.F. Rockwell International Science Center, 1049 Camino dos Rios, P.O. Box 1085, Thousand Oaks, California 91360, U.S.A.
Larker, Dr. H.T. ASEA AB, S-91500 Robertsfors, Sweden.
Larsen, Dr. D.C. IIT Research Institute, Materials Technology Division, 10 West 35 Street, Chicago, Illinois 60616, U.S.A.
Lauritzen, Dr. O. Universitetet i Trondheim, Norges Tekniske Høgskole, Institutt für Silikat-OG Hoytemperaturkjemi, Sem Saelandsvei 12, N-7034, Trondheim-NTH, Norway.
Lecompte, Dr. J-P. Ecole Nationale Supérieure des Mines de Saint-Etienne, 158 Cours Fauriel, 42023 Saint-Etienne Cedex, France.
Lenhart, Mr. A. Institut für Werkstoffwissenschaften, Lehrstuhl III, Universität Erlangen-Nürnberg, Martensstrasse 5, 8520 Erlangen, W. Germany.
Lu, Mr. H.Y. Department of Ceramics, Houldsworth School of Applied Science, The University, Leeds LS2 9JT, U.K.
Lumby, Dr. R.J. Lucas Industries Ltd., Group Research Centre, Shirley, Solihull, B90 4JJ, U.K.

McCauley, Dr. J.W. Ceramics Research Division, Army Materials and Mechanics Research Centre, Watertown, Mass. 02172, U.S.A.
Mangels, Mr. J.A. Ford Motor Company, Turbine Development Dept., 20000 Rotunda Drive, Dearborn, Michigan 48121, U.S.A.
Marshall, Dr. D. Department of Materials Science & Mineral Engineering, Hearst Mining Building, University of California, Berkeley, California, 94720, U.S.A.
Matsuo, Mr. Y. 2808 Iwasaki, Komoki-Si, Aichi Ken 485, Japan.
Maunder, Mr. M.J. de F. Department of Industry, Abell House, John Islip Street, London SW1P 4LN, U.K.
Mocellin, Prof. A. Laboratoire de Céramique, Ecole Polytechnique Fédérale de Lausanne, 34 ch. de Bellerive, CH-1007 Lausanne, Switzerland.
Morgan, Dr. P.E.D. Rockwell International Science Center, 1049 Camino dos Rios, P.O. Box 1085, Thousand Oaks, California 91360, U.S.A.
Mori, Mr. M. Technical Department, Refractories Division, Toshiba Ceramics Co. Ltd., 1 Minamifuji, Ogakie, Kariya-shi, Aichi-Ken, Japan 448.
Mustel, Mr. W. Centre des Matériaux, Ministère de l'Industrie, Ecole Nationale Supérieure des Mines de Paris, B.P. 87, 91003 Evry Cedex, France.
Nicol, Dr. A.W. Department of Minerals Engineering, University of Birmingham, Birmingham B15 2TT, U.K.
Nilsson, Mr. S. Department of Physics, Chalmers University of Technology, Gothenberg, Sweden 41296.
Oda, Mr. I. Research & Development Laboratory, NGK Insulators Ltd., 2-56 Suda, Mizuho, Nagoya 467, Japan.
Orange, Dr. G. Bat. 502, INSA, 69261 Villeurbanne Cedex, France.
Ortali, Dr. P.L. IRTEC, Via Granarolo 8, 48018 Faenza, Italy.
Oswald, Mr. J.R. Department of Ceramics, Houldsworth School of Applied Science, University of Leeds, Leeds LS2 9JT, U.K.
Pompe, Dr. R. Swedish Institute for Silikate Research, Gibraltargatan 5J, S-412 58 Göteborg, Sweden.
Popper, Dr. P. British Ceramic Research Association, Queens Road, Penkhull, Stoke-on-Trent, U.K.
Porz, Mr. F. Institut für Werkstoffkunde II, Universität Karlsruhe (TH), Kaiserstrasse 12, D-7500 Karlsruhe 1, W. Germany.
Quackenbush, Dr. C.L. Precision Materials Technology Department, GTE Laboratories Inc., 40 Sylvan Road, Waltham, Mass. 02254, U.S.A.
Ratcliff, Mr. N.A. NRDC, Kingsgate House, 66/74 Victoria Street, London SW1E 6SL, U.K.
Richerson, Mr. D.W. Materials Engineering, Engineering Sciences, Garrett Turbine Engine Co., 111 S. 34th Street, P.O.Box 5217, Phoenix, Arizona, 85010, U.S.A.
Riley, Dr. F.L. Department of Ceramics, Houldsworth School of Applied Science, University of Leeds, Leeds LS2 9JT, U.K.
Rothwarf, Dr. F. Chief, Materials Branch, European Research Office, United States Army, 223 Old Marylebone Road, London NW1 5TH, UK.

Schmidt, Dr. W. Hutschenreuther AG, Central Laboratory, D-8672 Selb, W. Germany.

Schrimpf, Dipl.Ing. C. Lehrstuhl für Glas und Keramik, Institut Für Steine und Erden, Technische Universität Clausthal, Zehntnerstrasse 2A, D-3392 Clausthal-Zellerfeld, W. Germany.

Schwier, Dr. G. Hermann C. Starck Berlin, Postfach 2540, D-3380 Goslar 1, W. Germany.

Shaw, Dr. T.M. Department of Materials Science & Engineering, College of Engineering, Cornell University, Ithaca, N.Y. 14853, U.S.A.

Siebels, Mr. J.E. Volkswagenwerk AG, E/Zentrallabor-Metallurgie, 3180 Wolfsburg, W. Germany.

Spacie, Mr. C.J. Crystallography Laboratory, University of Newcastle-upon-Tyne, Newcastle-upon-Tyne NE1 7RU, U.K.

Suzuki, Mr. K. Asahi Glass Co. Ltd., Research Laboratory, Hazawa-Cho, Kanagawa-Ku, Yokohama, Japan.

Taunt, Dr. R.J. Smiths Industries Ltd., Putney Division, KLG Works, Putney Vale, London SW15 3DY, U.K.

Thompson, Dr. D.P. Crystallography Laboratory, The University, Newcastle-upon-Tyne NE1 7RU, U.K.

Thümmler, Prof. F. Institut für Werkstoffkunde, Universität Karlsruhe (TH), Postfach 6380, D-7500 Karlsruhe 1, W.Germany.

Tien, Prof. T.Y. Department of Materials Science & Metallurgical Engineering, University of Michigan, Ann Arbor, Michigan 48103, U.S.A.

Weiss, Dr. J. Aluminium-Hütte Rheinfelden GmbH, D-7888 Rheinfelden 1, Postfach 1140, W. Germany.

Winder, Mr. S. Ceramics Research Group, Department of Physics, University of Warwick, Coventry CV4 7AL, U.K.

Wötting, Mr. G. DFVLR, Institüt für Werkstoff-Forschung, Postfach 90 60 58, D-5000 Köln 90, W. Germany.

Yeter, Miss B. Asagi Ayranci, Menevis Sok 41/10, Ankara, Turkey.

Ziegler, Dr. G. DFVLR, Postfach 90 60 58, D-5000 Köln 1, W. Germany.

Group Photograph

GENERAL INTRODUCTION

NITROGEN CERAMICS 1976-1981

R. Nathan Katz

Army Materials and Mechanics Research Center (AMMRC)
Watertown, MA 02172, U.S.A.

I. INTRODUCTION

It is both a great pleasure and a special honor to have been asked by Dr. Riley to present the introductory lecture for this second NATO Advanced Study Institute (ASI) on Nitrogen Ceramics. I must add that it is also a rather awesome responsibility, as the introductory lecture should, ideally, provide the perspective from which one can fit what is to follow during the course of the Institute into the main streams of development in nitrogen ceramics, in particular, and into the broader field of high performance ceramics, in general.

The proceedings of the first NATO ASI on Nitrogen Ceramics (1), and in particular Paul Popper's comprehensive introductory paper (2) and Prof. Richard Brook's concluding remarks (3), provide a clear definition of the state-of-the-art in nitrogen ceramics as of August 1976. What I wish to do here is to provide you with my perspective as to what major developments have occurred since then, together with some feeling for where the field may be heading. Before doing this, however, one important observation needs to be made. Namely, that this is not really a conference on Nitrogen Ceramics. It is, rather, a conference on silicon nitride and related materials. On looking at the preliminary agenda at least 80% of the papers to be given at this conference deal principally with silicon nitride, SiAlON's, or related glasses. Of the remaining papers only four will not deal explicitly with phases containing Si and N. Therefore, when I use nitrogen ceramics in the balance of this lecture, I will be referring to silicon nitride and related ceramics such as "SiAlON's" or oxynitride glasses. I will not deal with materials such as BN, AlN or TiN.

Riley, F.L. (ed.) Progress in Nitrogen Ceramics

From my perspective the major advances in the field of nitrogen ceramics during the past five years have been in the areas of processing science and technology, microstructure and property development and in demonstrations of the feasibility of various applications. Looking at the proceedings of the first ASI (1), one is struck by the fact that we were largely dealing with either hot-pressed or reaction-bonded silicon nitrides (HPSN and RBSN, respectively), and the crystal chemistry of "SiAlON" phases. Sintered silicon nitride (SSN) was just emerging as a feasible concept and there was only one, very brief, paper on nitrogen glasses. By contrast, at this ASI we will see that the emphasis has shifted to sintering and to various types of hot isostatic pressing (HIP). There is more emphasis on (and I would anticipate more sophistication in) microstructural analysis and control. There are a variety of papers on nitrogen-containing glasses, and the development of a new single-phase oxynitride, AlON. As a result of the focus provided by hardware demonstration programs, the variety and realism of property measurement has also significantly increased.

Application feasibility demonstrations for high performance nitrogen (and non-nitrogen) ceramics has also seen very considerable progress since 1976 (4-6). The principal areas for nitrogen ceramics in engineering are likely to be in the heat engine, industrial heat exchanger, cutting tool, bearing and radar window applications. Radar windows aside, each of these applications can help reduce energy or strategic materials consumption. With the shock of the "second" oil crisis in the wake of the Iranian Revolution and the Iraqi-Iranian war, the driving force for energy conservation is accelerating. Similarly, a growing awareness of the need to be less dependent on critical minerals is emerging in the US and other highly industrialized nations. The combination of increasing societal need coupled with successful hardware demonstrations are helping to maintain government funding and industrial interest in nitrogen ceramics.

The emergence of new non-nitride high performance ceramics such as sintered α-SiC have provided a stimulus to silicon nitride producers to find more cost-effective fabrication routes for silicon nitride. Thus, the emergence of alternative materials has played an important role in the dynamics of silicon nitride development during the past few years.

With the above introduction we will now take a more detailed look at some of the key innovations in processing, microstructure control, new and emerging materials, applications demonstrations, and institutional factors which have occurred since the last ASI.

II. PROCESSING ADVANCES

A. Sintered Silicon Nitride (SSN)

The need for a fully dense silicon nitride, which can be readily formed to near net shape by mass production processes, has long been recognized as a key technological requirement if silicon nitride-based ceramics are to become true engineering materials. Conventional sintering was the obvious route to achieve these goals. However early attempts at conventional sintering of silicon nitride yielded material of only ≃90% of theoretical density (TD) (7). The basic problem in sintering silicon nitride is that the compound dissociates at temperatures high enough to achieve the necessary atomic mobility for sintering to occur. Under conventional pressureless sintering conditions the rates of dissociation versus sintering (densification) are such that high weight losses and low TD's result. In 1976 Greskovich et al. (8), Priest et al. (9) and Mitomo (10) independently discovered that a N_2 overpressure would help to suppress the dissociation reactions and permit sintering to near full density, with little weight loss.

Greskovich and his co-workers have carried out basic thermodynamic studies of the sintering of silicon nitride (8,11). They have found that the following conditions are necessary or helpful:

(1) Use ultrafine powders (≃ 0.5 μm). This increases both the thermodynamic driving force for sintering and reduces diffusion distances (and hence times).

(2) Use sufficiently high nitrogen pressure, to keep the system to the right of the solid-liquid coexistence boundary shown in Figure 1 (11).

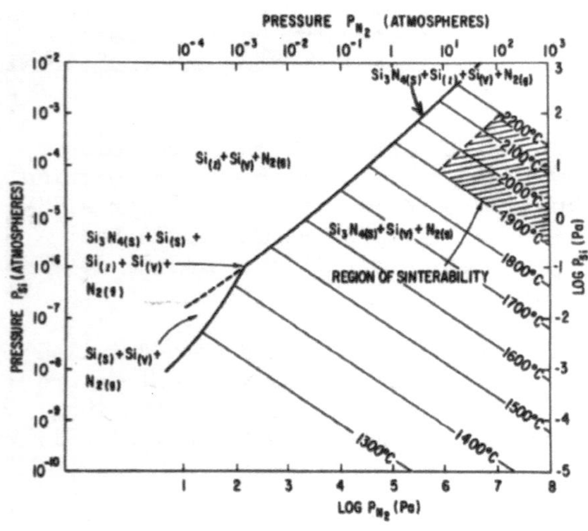

Figure 1. Si vapor pressure in equilibrium with silicon nitride as a function of nitrogen temperature and pressure (Ref. 13).

(3) Prevent silicon vapor loss from the system. This condition is automatically met if one sinters in a closed system.

(4) Have some oxygen present in the system. Using non-oxide additives ($SiBeN_2$), it was learned that a minimum level of oxygen is required to attain sintering of Si_3N_4. For the $SiBeN_2$ additive O_2 levels 2.5% were required (12).

Initially this last point was rather surprising. After all, for years silicon nitride researchers have been assuming that minimizing or eliminating the oxygen in the starting powder would be beneficial. However, since to sinter silicon nitride one starts with powder which goes through a solution precipitation reaction (via an oxynitride glass phase) which finally results in elongated β grains and a boundary phase, it is obviously necessary to have some oxygen available to form the oxynitride glass. This process is the same as the densification/microstructure development process for HPSN. Thus, research groups working on SSN have been able to utilize much of the experience gained from hot pressing studies to optimize their materials.

Even with the above understanding of the sintering process, it is still not possible, in all cases, to attain TD. This is particularly true if one is interested in using an additive such as $SiBeN_2$. This leads to our second processing innovation, the two-step reactive gas pressure sintering process.

B. Two-Step Reactive Gas Pressure Sintering

During the course of the sintering studies carried out by Greskovich and his co-workers at GE (13), it was discovered that Si_3N_4 sintered in N with a $SiBeN_2$ additive to closed porosity (i.e. 92-95% TD) at 2.0 MPa and 1900°C could be further densified by increasing the pressure to ~7.0 to 8.0 MPa, in situ. Densities of 99.6% TD with weight loss of only ~1% have been achieved in this manner. In parallel work by Gazza in our laboratory, using Y_2O_3/Al_2O_3 additives, the two-step process comprises sintering to the closed pore stage at 0.1 MPa, then increasing the pressure to 2.0 MPa at 1780°C. Again densities >99% TD with low weight loss are attained. Figure 2 shows a schematic of this process. It is important to note that compositions which do not yield closed porosity on the first sintering step will not attain TD or near TD on the second, higher pressure step (14). This procedure amounts to "cladless HIP'ing". It has three distinct potential advantages over cladded HIP techniques: (1) it is containerless, (2) sintering and HIP'ing are carried out in the same furnace (less handling) and (3) the maximum pressures are 80-100 atmospheres as opposed to 2000 atmospheres in conventional HIP'ing.

It is appropriate to introduce a word of caution at this point. The process is still in an early stage of development, not well

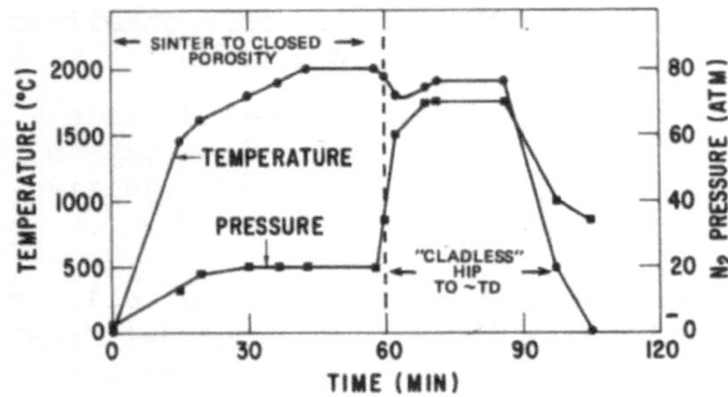

Figure 2. Typical time-temperature pressure diagram for the two-step reactive gas pressure sintering of Si_3N_4 with $SiBeN_2$ additive (Ref.13).

understood, and may change substantially over the next year or so as more is learned about it. While high densities, good microstructures and strengths have been demonstrated in bend bars (as will be discribed in Gazza's paper at this meeting), uniform and controlled fabrication of complex shaped parts has yet to be satisfactorily demonstrated. However, conventional "cladded" HIP of silicon nitride has demonstrated considerable shape-forming capability in the past several years. Therefore, we turn next to this area of processing.

C. Conventional Cladded HIP'ing

The very fact that I titled this section "Conventional Cladded HIP'ing" in itself shows how far this community has progressed in the past five years. HIP'ing of silicon nitride is anything but conventional! In conventional cladded HIP processes a powder preform can be encapsulated in an easily deformable can, or a powder can be placed in a shaped can whose controlled deformation under pressure imposes its shape on unformed powder. In the HIP'ing of silicon nitride, a shaped preform with a glass or tantalum clad (envelope) is used. For HIP'ing of silicon nitride, pressures of 200 MPa and temperatures of 1700-1725 C are generally required. Larker et al. (15), at ASEA in Sweden, have pioneered HIP'ed silicon nitride. Their initial announcement of the process in 1977 indicated that properties similar to HPSN, but isotropic as opposed to anisotropic, could be attained. Most importantly they showed that HIP'ed material could have a significantly higher Weibull modulus, m, than HPSN. The ASEA group is continuing to develop advanced shaping techniques using a glass cladding route. By use of this technique they have recently reported the very impressive achievement of producing a fully bladed silicon nitride axial turbine wheel as part of a joint program with United Turbine (16,17).

In the U.S., Wills and his co-workers are using a tantalum cladding approach. Wills et al. have reported processing and property data for some compositions HIP'ed in this way (18). For Si_3N_4 plus 5% Y_2O_3 additive, full density could not be achieved under uniaxial hot pressing (40 MPa pressure) whereas under HIP conditions (201 MPa), full density was achieved. The 1400 C MOR strength of this material was 449 MPa, which is a very respectable value for a HPSN. The creep rate at 1400 C, in air, was an order of magnitude lower than for any previously reported HPSN or HIP'ed silicon nitride. A particularly interesting inference made by Wills is that in HIP'ing of Si_3N_4 + 5% Y_2O_3 the transformation may proceed via a solid state process, rather than via a liquid solution-reprecipitation process.

While HIP'ing of silicon nitride has made remarkable progress in the past four years in developing material with outstanding properties and producing useful shapes to close tolerances, the process does have some negative factors. Cladding and uncladding steps represent added cost and potential for flaw introduction. While it has not yet appeared as a problem, the possibility of reaction between the cladding material and the nitride exists. Lastly, because of the very high pressures used, the cost of HIP equipment is relatively high. Because of these negative factors in conventional cladded HIP'ing, I, and others, have for some time been advocates of "cladless" HIP'ing. It is important to reiterate that the cladless methods, while having great future promise, have not yet demonstrated either the shaping capability or the high temperature properties that cladded HIP'ing has already achieved.

D. Sintered RBSN

While SSN and HIP'ed or "cladless" HIP of SSN are dramatic and exceedingly important advances, they leave certain issues outstanding. Sinterable silicon nitride powders are not available in large quantities at low prices. Sinterable powders are very fine and thus hard to handle. Green preforms made from such fine powders are typically low density and as a consequence sintering shrinkage tends to be rather high. All of these problems could be overcome if one could sinter RBSN to full, or near full, density. Thus, one of the most significant events of the past five years was first announced by Giachello and Popper (19) who, in a joint program of the Fiat Research Center and the British Ceramic Research Association, demonstrated that it is possible to post-sinter a reaction-bonded silicon nitride preform to ~98% theoretical density, with increased strength and oxidation resistance. Mangels and Tennenhouse (20) at Ford Motor Company have independently followed a similar line of research and have, in fact, fabricated components of sintered reaction-bonded silicon nitride. With this development one could start with a sintering preform that would yield only 6 to 8% linear shrinkage, as opposed to 18 to 20% linear shrinkage for sintered components. It is also possible that sintered silicon nitride bodies of more than 95% theoretical density may be used as preforms for hot isostatic pressing. Such a development would be a major breakthrough toward attaining high-reliability, affordable, high-performance components such as turbocharger rotors, diesel pistons, valve train components and perhaps even bearings and tool bits.

E. Improved Silicon Nitride Powders

As indicated above, one of the driving forces in developing sinterable RBSN was the limited amount of sinterable powders available to the materials developer. The increasing availability of high quality silicon nitride powders by GTE in the US, by H.C. Stark in Germany, and most recently by Toshiba in Japan, has been a major factor in the development of SSN. Several papers at this meeting will deal in depth with the preparation of such powders. One important area for future work is to reduce the costs of sinterable silicon nitride powders. Unless this can be accomplished, SSN will not be competitive with sintered RBSN or α-SiC for many applications.

We turn now from processing, per se, to the related area of microstructure and property development.

III. MICROSTRUCTURE AND PROPERTY DEVELOPMENT

At the previous ASI, most of the focus on microstructure was on the structure of the grain boundary phase and how it affected the high temperature properties of HPSN's. Indeed, we were privileged to have had David Clarke present one of his first papers on grain boundary structure by lattice imaging TEM (21). This powerful tool was later to confirm much of what had only been inferred to as the nature of grain boundary phases in HPSN.

With the focus on grain boundary engineering (22), as a point of departure in 1976 we have returned to a concern with total microstructure. Total microstructure includes concern for both the grain boundary phases, as well as the development and control of grain size and morphology, as illustrated in Figure 3.

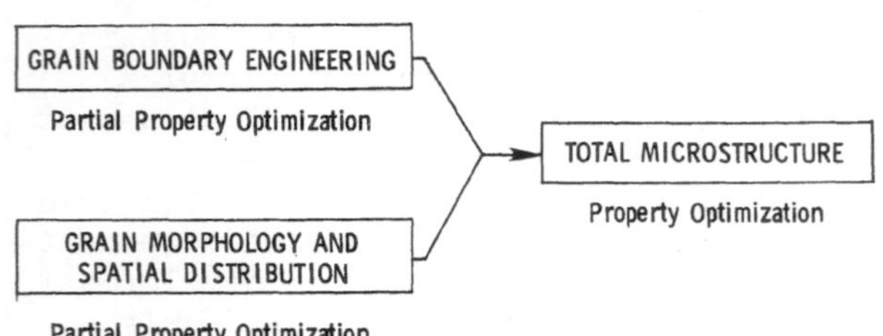

Figure 3. Microstructure development and control strategy for HPSN and SSN.

A. Progress in Grain Boundary Engineering

To distinguish the emphasis on the grain boundary from the more usual interest in total microstructure, a "grain boundary engineering" (GBE) approach was adopted in our laboratory (23) while Tsuge and co-workers at Toshiba focused on "grain boundary crystallization" (GBC). This approach has been particularly successful in developing hot-pressed Si_3N_4 with Y_2O_3 as a densification aid. Gazza (24) and others (25) demonstrated that Y_2O_3 additions provide higher strength at both room and elevated temperatures (to 1400°C) as well as better creep and oxidation resistance than obtained with silicon nitride containing MgO. However this material has been plagued with an intermediate-temperature (~1000 C) oxidation problem, although it appears that postfabrication heat treatment (26) and proper attention to composition and phase equilibria, as discussed by Lange (25), can alleviate this problem.

If one adds Al_2O_3 to the Y_2O_3 addition, outstanding properties at temperatures as high as 1200 C can be attained. For example, in 1978 Tsuge and Nishida (27) reported strengths as high as 965 MPa (140,000 psi) at 1200°C with a $Y_2O_3 + Al_2O_3$ additive, using the grain boundary crystallization approach. Figure 4, showing the general trend of their data, illustrates the power of the GBE/GBC strategy. Through this focus on the grain boundary, the high-temperature strength of HPSN has been increased by an order of magnitude in less than 10 years!

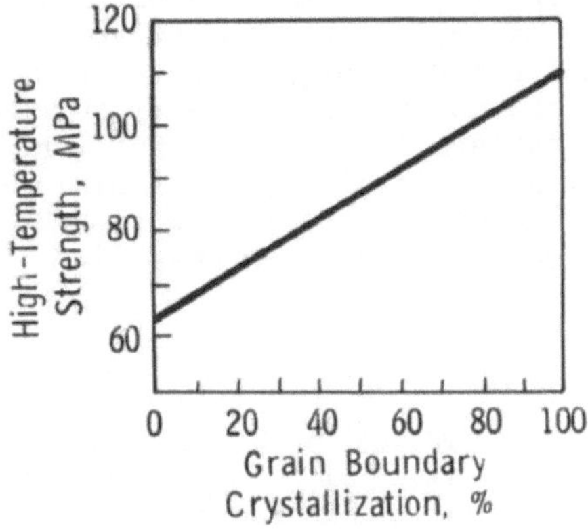

Figure 4. Strength at 1200 C as a function of percent grain boundary crystallization (after Ref. 27).

Of course, properties other than strength are required in high temperature structural applications, for example, creep and oxidation resistance. Recent work by Gazza at AMMRC and by Quackenbush, Smith and co-workers at GTE (which will be discussed in their papers at this meeting) indicate that if one eliminates the Al_2O_3, and uses the correct starting silicon nitride powders, the high temperature (~1300 C) creep and oxidation resistance are significantly increased.

The current goal of all GBE/GBC research is to devise a grain boundary composition which will simultaneously optimize high-temperature strength, creep and oxidation resistance, and, of course, do this reproducibly in actual components. The GBE/GBC approach applied to the development of HPSN is also now being exploited by many groups working on SSN. I am sure that this will be evident in many of the papers which will be given at this meeting on SSN. While GBE/GBC has and will continue to improve HPSN and SSN, one can further improve or control the properties of these materials by understanding and manipulating the development of grain size and morphology. Thus we turn to our next area of concern.

B. Microstructure Development

While Lange early identified the importance of an elongated β grain morphology to attain high strength in HPSN (28), little was done prior to the 1977-1979 time-frame to study the effect of controlling the β-phase morphology to optimize the properties of HPSN. Work by Knoch and Gazza (29-32) demonstrated that, while the interlocking growth of β prisms during the α-β phase transformation produces improved strength and fracture toughness, these properties will degrade if grain growth is permitted to occur. Knoch and Gazza demonstrated that the development of fine grain, high aspect ratio, uniform microstructure in HPSN is primarily a function of the character of the starting powder (i.e. particle size, purity, and α-β phase ratio) and hot-pressing parameters (i.e. time, temperature, and applied pressure). Gazza (Ref. 14 and his paper at this meeting) has shown that starting powder exerts a similar influence on morphology and strength in SSN. Variations in sintering and/or "cladless" HIP times, temperatures and nitrogen overpressures have not yet been studied. However, since microstructure/property relations are expected to follow the same pattern for HPSN and SSN (and the limited data available bear out that expectation), such sintering variables are certain to have a pronounced effect. How pronounced an effect can be inferred from Figure 5 taken from Ref. 31.

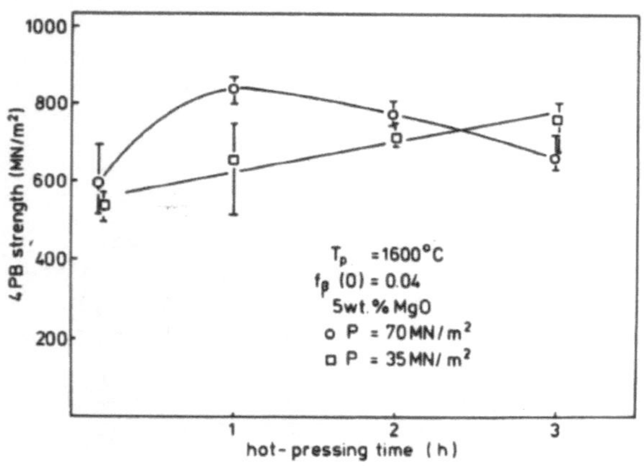

Figure 5. Strength as a function of hot-pressing time for two applied pressures (Ref. 31).

Here we see that for HPSN (with 5% MgO), strength peaks in 1 hour at hot-pressing temperature if one hot presses with 70 MPa pressure. On the other hand, if one hot presses with only half that pressure, 35 MPa, the strength apparently has not peaks after 3 hours. One may look at strength (and other property development) in fully dense silicon nitrides, whether hot pressed, sintered, or presumably HIP'ed, as a competition between the α-β transformation and the coarsening of the resulting β. For optimum strength, grain coarsening should be avoided. Keeping the grain morphology as fine and with as high an aspect ratio as possible will also (for a given volume of grain boundary liquid) keep the grain boundary phase as thin as possible. This will also contribute to strength retention at high temperature.

Although further insights into microstructure development will be forthcoming in many of the papers to be presented at this ASI, this type of research is still in its relative infancy. As work on total microstructure development and its interrelationship with processing parameters matures, I am confident that we shall see dramatic improvements in the properties and reliability of fully dense silicon nitrides.

IV. "NEW" MATERIALS

For the purposes of this paper, "new" materials are those nitride, oxynitride, or even non-nitride materials which have emerged in the past 5 or 6 years. Among these are oxynitride glasses, cubic aluminum oxynitride spinel (AlON), sintered α-SiC and transformation-toughened zirconia. While the latter two are definitely not nitrogen ceramics, several of their potential applications may well influence the directions which nitrogen ceramics development might take over the next few years. Therefore a few comments about them are quite in order at this ASI.

A. Oxynitride Glasses

Oxynitride glasses have been known for some time and were briefly discussed at the last ASI by Jack (33). Nitrogen-containing glasses are intriguing because they increase hardness, softening temperature, viscosity, refractive index, electrical resistivity and decrease thermal expansion. As the result of both the stimulation provided by Jack's work, and the interesting technological promise afforded by the possibility of higher temperature, more thermal shock resistant, and more electrically insulating glasses, work on these materials has accelerated in the past few years. In addition to Jack's efforts, Loehman in the U.S. (34), Makishima et al. in Japan (35), Rabinovich and Abramovici in Israel (36), and Messier in our laboratory (37) are currently involved in oxynitride glass research. It is interesting to observe that processing in an overpressure of nitrogen offers advantages in O-N glass formation, as it does in the case of sintering. Makishima et al. (35) prepared a La-Si-O-N glass under 3 MPa (30 atm) of nitrogen. By this means they were able to obtain a glass containing in excess of 18 atomic % N, the highest concentration that I am aware of.

In looking over the abstracts for the presentations to be made at this ASI on O-N glasses, I wasn't sure if I should be gratified or disappointed! Clearly, as a leading proponent of grain boundary engineering, the fact that most of the O-N glass papers at this conference are oriented in that direction gives me a great deal of gratification. On the other hand, I feel that we are missing something important by not also focusing on the potential of O-N glasses as bulk materials in their own right. The potential for property modification and tailoring afforded by such materials is very large and the payoff, particularly in the optical applications area, may be very great.

B. AlON

McCauley and Corbin (38,39) have shown that it is possible to reactively sinter Al_2O_3 + 27-40 mol % AlN to form a single-phase cubic oxynitride spinel. The resultant material has a unique combination of properties. While possessing mechanical properties very similar to Al_2O_3, of similar grain size, AlON is optically isotropic, i.e. the polycrystalline material can be made transparent, and it has about 20% lower thermal expansion than Al_2O_3. Among the potential applications of AlON are multi-mode missile guidance domes and windows, lamp envelopes, and tool bits. The best optical quality AlON produced to date has been fabricated by Gentilman and his co-workers at Raytheon (40). Dr. McCauley will be presenting an in-depth review of phase equilibria and properties for AlON in his contribution to the conference.

C. Alternative Materials - SiC and Zirconia

Developments in high performance silicon nitride and silicon carbide science and technology have mutually stimulated each other since the early 1970s. The reasons for this are rather straightforward. Materials scientists and technologists working on the various ceramic engine programs have been in close communication and have rapidly seen where a development in one material type could be applied to the other. Similarly those interested in sintering of covalent materials have borrowed from each other. Thus, much of the interest in sintering of silicon nitride came about after Prochazka (41) sintered β-SiC in 1973. After all, if one "unsinterable" covalent compound could be sintered, why couldn't others? Therefore, it should come as no surprise that when sintered α-SiC made from abundant and relatively inexpensive (when compared to sinterable silicon nitride powders) Acheson process powders was developed, this would effect the development of silicon nitride.

Sintered α-SiC, as developed by researchers at the Carborundum Co. (42) has outstanding high temperature strength and creep resistance, it is fabricable into many complex shapes, and it has a low coefficient of friction against steel. For these reasons, among others, sintered α-SiC parts have emerged, since 1975, as major candidates for many automotive ceramic components from turbine wheels to valve lifters. In our talk on "Applications and US Engine Programs" later in the meeting, it will become evident just how far α-SiC has progressed in the few years since its invention.

Similarly, transformation-toughened zirconia, pioneered by Garvie and co-workers at CSIRO in Australia (43), is an outstanding achievement in materials science. This material combines high strength, very high toughness (for a ceramic) and good friction properties. It is currently a leading candidate for various components in adiabatic diesel engine technology, and is being commercially exploited for drawing and extension dies. Its low thermal conductivity and close match of modulus and thermal expansion with steel facilitate joining by shrink fitting.

The key point I wish to make here is that while from 1971 to 1976 it may have seemed that silicon nitride was a panacea, this was because effective materials competition was largely lacking. Today there is competition from alternative materials, and developers of silicon nitrides will have to very carefully consider which specific high performance applications are best matched to their unique properties and processing capabilities. Hot-pressed, sintered and HIP'ed silicon nitride all possess outstanding strength, low friction coefficients, high toughness, unsurpassed erosion resistance, and unique dielectric properties. Future applications will tend to optimize around these strong points. Similarly, to effectively compete with sintered SiC, costs will have to be reduced. Thus, the importance of sintered RBSN (with or without post HIP processing) will be enhanced.

V. DESIGN WITH BRITTLE MATERIALS

The main thrust of materials development for nitrogen ceramics are for structural, highly stressed applications. Thus, it is central to their ultimate utilization that systems designers accept nitrogen, and other high performance ceramics, as materials with which they can confidently design. During the past 5 years much has occurred in a variety of demonstration programs to help to develop such confidence. Table 1 lists several significant milestones in ceramic feasibility demonstrations during this period. I will be discussing most of these in more detail in our later paper. For now let me just cite the importance of the successful flight of the space shuttle, both in building awareness of ceramic technology in the public at large, and in building confidence in the use of ceramics in non-traditional roles.

Table 1

MAJOR SUCCESSES IN BRITTLE MATERIALS DESIGN 1976 ⟶ 1981		
PROGRAM	DURABILITY (hr)	COMMENTS
1. DARPA/FORD	200+	STATIC HARDWARE AND ROTOR, RIG DEMOS
2. MERADCOM/SOLAR	200+	FULL PERFORMANCE DEMO OF HYBRID ALL CERAMIC NOZZLE IN 10-kW TURBO GENERATOR
3. TACOM/CUMMINS	250+	FULL PERFORMANCE DEMO OF CERAMIC CONFIGURED SINGLE CYLINDER ADIABATIC DIESEL
4. DOE/DDA	1500+	ENGINE TEST OF VANES IN TRUCK ENGINE, INCLUDES 100 hr OF ROAD AND TEST TRACK TRIALS
5. DARPA/GARRETT	15+	DEMONSTRATED ENHANCED PERFORMANCE (INC. POWER, DEC. SFC) OF INSERTED CERAMIC BLADES IN A METAL HUB
6. NASA SPACE SHUTTLE	NA+	SUCCESSFUL FLIGHT TESTING AND USE OF CERAMIC THERMAL PROTECTION SYSTEM

\+ = PROGRAM GOAL ATTAINED AND PARTS SURVIVED

Partly because of the use of ceramics in the Shuttle Thermal Protection System, but also for many other reasons, brittle materials design is quite important to NASA. It is important enough that NASA has sponsored a "Brittle Materials Design" curriculum at the University of Washington for the past 4 years. This interdisciplinary program brings together faculty and students from the Ceramic, Metallurgical, Aeronautics and Astronautics, Civil, Electrical, and Mechanical Engineering Departments, as well as Physics, into a unified multi-course curriculum aimed at turning out students who are familiar with brittle materials design and who can function effectively in a multi-disciplinary design team. Aside from courses, the students are organized into design teams and given practical problems to solve, many of which are provided by industry. In one such study, students were presented with a design problem involving an A.C. spark plug, zirconia automotive exhaust gas sensor (44). The students developed a successful proof-test based, quality control procedure and verified that the existing design was conservative. This program has also produced the first available text on brittle materials design (45). I think that it is a milestone in the maturation of any science or technology when major universities start offering programs in it. Thus, academically, as well as in practice, the last 5 years have seen significant gains in the art of brittle material design.

VI. FUTURE DIRECTIONS AND NEEDS

The future of nitrogen ceramics will most probably follow three lines of development:

a) Continued progress in areas presently recognized as promising. Several of these are highlighted in Table 2.

b) Addressing needs which are presently not being addressed with sufficient effort, or in some cases not being addressed at all. Table 3 highlights several of these areas.

c) Increased emphasis on non-silicon oxynitride based materials. The papers given on AlN and AlON fall into this category.

Table 2

AREAS OF BIG FUTURE PAYOFF IN N_2 CERAMICS

- USE OF N_2 OVER PRESSURES IN:
  - SINTERING
  - CLADLESS HIP
  - O-N GLASS PREPARATION
- USE OF THE SINTERED RBSN ROUTE
  POSSIBLY COUPLED WITH CLADLESS HIP
- FURTHER DEVELOPMENT OF O-N GLASSES
- FURTHER DEVELOPMENT OF O-N CRYSTALLINE PHASES SUCH AS SiAlON AND AlON
- GBE/BGC AND GBE/GBC COUPLED WITH MICROSTRUCTURE DEVELOPMENT
- NOVEL PROCESSING OF
  POWDERS (FOR SINTERING)
  POWDER AND ADDITIVE (i.e., FLUIDIZED BED MIXING)

Table 3

SOME NEEDS IN N_2 CERAMICS

- THERE IS STILL MINIMAL DIFFUSION DATA
- THERE NEEDS TO BE BETTER AND QUICKER COUPLING BETWEEN MICROSTRUCTURE STUDY METHODS SUCH AS LATTICE IMAGING TEM AND PROCESSING/MICROSTRUCTURE DEVELOPMENT STUDIES
- JOINING AND ATTACHMENT
- RELIABILITY AND LIFE PREDICTION
- AT LEAST ONE MAJOR COMMERCIAL SUCCESS FOR Si_3N_4 IS URGENTLY REQUIRED TO KEEP THE MANUFACTURER INTERESTED

Perhaps the most critical issue is not technical in the narrow sense of the word. Namely, viable commercial applications for nitrogen and oxynitride ceramics must be developed. While specialized applications may keep a certain amount of government funding available, only commercial applications will keep private capital interested. And without private capital we will not have an industrial capability. It is, therefore, encouraging that firms such as Joseph Lucas Research Laboratory (46) and Rosenthal Technik (47) are currently well along in such commercial applications. However, much more needs to be done in this crucial area.

VII. CONCLUSIONS

The past five years have been considerable progress in sintering, post-sintering processing, and microstructure/property development in silicon nitride based ceramics. New materials such as sintered α-SiC and transformation-toughened zirconia have developed to the point where they provide systems designers with options other than silicon nitride based materials for high temperature, highly stressed structural applications. Successful demonstrations of a number of ceramic materials in engines and other high temperature structures are continuing to provide encouragement to the design community. What these developments would seem to point to, for the near future, is a dynamic highly competitive environment from which reliable, affordable high performance ceramics of many types will find their "niche" in high temperature engineering systems. Nitrogen ceramics will participate fully in this development, and I predict that scientifically and technologically the next five years in nitrogen ceramics will be as exciting as the past five.

If the experience of the past can be used to predict the future, I would expect that the discussions and interchange at this ASI will greatly contribute to the realization of the above projection. With this thought in mind, I want to thank you for your kind attention and wish all of us a most productive and stimulating two weeks.

ACKNOWLEDGMENT

It is a pleasure to acknowledge many of my colleagues whose ideas have greatly influenced my thinking on nitrogen ceramics over the past decade. They include: A.E. Gorum, E.M. Lenoe, G.E. Gazza, D.R. Messier, H.F. Priest, J.W. McCauley, G.D. Quinn, W. Croft, A.F. McLean, E.A. Fisher, J. Mangels, F.F. Lange, C. Greskovich, H. Knoch, D.J. Godfrey, M.W. Lindley, K.H. Jack, R.J. Lumby, F.L. Riley, R.M. Spriggs, T. Vasilos, J.T. Smith, C.L. Quackenbush, E. Gugel, P. Popper, A. Giachello, R. Kossowsky, R. Bratton, M. Torti, R. Phoenix, D. Richerson, K. Styhr, N. Parr, N. Corbin, T. Rockett, and many, many others.

REFERENCES

1. Nitrogen Ceramics, Ed. F.L. Riley, Noordhoff International Publishing, Leyden (1977).
2. P. Popper, ibid. pp. 3-22.
3. R.J. Brook, ibid. pp. 671-674.
4. Ceramics for High Performance Applications II, eds. J.J. Burke, E.M. Lenoe and R.N. Katz, Brook Hill Publishing Co., Chestnut Hill, MA. (1978).
5. Keramische Komponenten fur Fahrzeug-Gasturbinen, W. Bunk and M. Bohmer, eds., Springer-Verlag, Berlin (1978).
6. R.N. Katz, Science, 208, 841-847, 23 May 1980.
7. G.R. Terwilliger and F.F. Lange, J. Mater. Sci. 10, 1169 (1975).
8. C. Greskovich, S. Prochazka and J.H. Rosolowski, ref. 1, pp. 351-358.
9. H.F. Priest, G.L. Priest and G.E. Gazza, J. Am. Cer. Soc. 60 81 (1977).
10. M. Mitomo, J. Mater. Sci. 11 1103-1107 (1976).
11. S. Prochazka and C. Greskovich, "Development of a Sintering Process for High Performance Si_3N_4", AMMRC TR 78-32, July 1978.
12. R.N. Katz, G.E. Gazza and C. Greskovich. "Sintered Si_3N_4", 5th International Automotive Propulsion Systems Symposium DOE Conf-800419, April 1980.
13. C. Greskovich and J.A. Palm, "Development of High Performance Sintered Si_3N_4, AMMRC TR 80-46, September 1980.
14. G.E. Gazza and R.N. Katz, to be published in Communications Amer. Cer. Soc., October 1981.
15. H. Larker, J. Adlerborn and H. Bohman, SAE Technical Paper No. 770335 (1977).
16. H.J. Larker, "HIP Silicon Nitride", in Ceramics for Turbine Engine Applications, AGARD-CP-276, March 1980, Chapt. 18.
17. "Ceramic Turbine Wheels for Automotive Turbines - a Technical Breakthrough", ASEA Journal 55 75 (1980).
18. R.R. Wills, M.C. Brockway, J.G. McCoy and D.E. Niesz, Ceramic Engineering and Science Proceedings, 1 (7-8), 534 (1980).
19. A. Giachello and P. Popper, paper presented at the 4th International Meeting on Modern Ceramic Technologies, St. Vincent, Italy, 28 May - 1 June 1979.
20. J.A. Mangels and G.J. Tennenhouse, Am. Cer. Soc. Bull. 58, 834 (1979).
21. D.R. Clarke, Ref. 1, pp. 433-440.
22. R.N. Katz, and G.E. Gazza, Ref. 1, pp. 417-432.
23. R.N. Katz, in Materials Technology 1976, AIP Conference Proceedings No. 32, American Institute of Physics (1976).
24. G.E. Gazza, Am. Cer. Soc. Bull. 54 778 (1975).
25. R.J. Bratton, C.A. Anderson and F.F. Lange in (2), pp.805-825.
26. G.E. Gazza, H. Knoch and G.D. Quinn, Am. Cer. Soc. Bull. 57 1059 (1978).
27. A. Tsuge and K. Nishida, ibid., p.424.

28. F.F. Lange, J. Am. Cer. Soc. 56 518 (1973).
29. H. Knoch and G. Ziegler, Ber. Deut. Keram. Ges. 55 242 (1978).
30. H. Knoch, G.E. Gazza and R.N. Katz, Energy and Ceramics, Elsevier.
31. H. Knoch and G.E. Gazza, Ceramurgia Int. 6 51 (1980).
32. G.E. Gazza, R.N. Katz and H. Knoch, "Factors Influencing the Quality of Fully Dense Silicon Nitride", presented at the 6th Army Technology Conference: Ceramics for High Performance Applications - III, Orcas Is., WA, July 1979 (to be published by Plenum Press).
33. K.H. Jack, see Ref. 1, pp. 257-262.
34. R.E. Loehman, J. Am. Cer. Soc. 67 491-493 (1979).
35. A. Makishima et al., Yogyo-Kyokai-Shi, 88 701-702 (1980).
36. E.M. Rabinovich and R. Abramovici, unpublished research, Israel Ceramic and Silicate Institute.
37. D.R. Messier, unpublished research, AMMRC.
38. J.W. McCauley and N.D. Corbin, J. Am. Cer. Soc. 67 476-479 (1979).
39. J.W. McCauley and N.D. Corbin, U.S. Patent 4,241,000, Dec. 23 1980.
40. R. Gentilman, unpublished research, Raytheon Co.
41. S. Prochazka, in Ceramics for High Performance Applications, J.J. Burke et al. Eds., Brook Hill, Chestnut Hill, MA. (1974), pp. 239-252.
42. E.H. Kraft and J.A. Coppola, Ref. 4, pp. 1023-1037.
43. R.C. Garvie, R.H. Hannink and R.T. Pascoe, Nature, 258 703-704 (1975).
44. F.D. Gac et al., Ceramic Engineering and Science Proceedings, 1 (7-8) 593-608 (1980).
45. Design and Brittle Materials, Ed. J.I. Moeller, A.S. Kobayashi and W.D. Scott, University of Washington (1979).
46. P. Wright, "New Tools Speed Aero-Engine Output", London Times, Wed. 17 June 1981.
47. S.R. Schindler and A. Krauth, Ber. DKG, 58 (2) 75-84 (1981).

DISCUSSION

Jack: The major successes given for 1976-81 were all (except the 'Space Shuttle') engine applications. The original ARPA specification was for an engine running at 1370°C but this has not been achieved. Too much emphasis has been placed on the ceramic gas turbine and not enough on 'bread-and-butter' applications. Probably 'debased' nitrogen ceramics (like 'debased' aluminas) have bulk uses as refractories.

Katz: The original DARPA programme had as its goal the demonstration of 'Brittle Materials Design' capability. It was believed (correctly) that demonstrating this design capability in a dramatic application would encourage designers to take ceramics seriously as structural materials. We believed that the demonstration of uncooled ceramic hot flow path components in a gas turbine (at temperatures beyond the capabilities of superalloys) would provide the dramatic demonstration of high performance ceramic potential needed to 'change the consciousness' of the design community. The temperature of 1370°C was a goal chosen to assure a level of performance from ceramics far in excess of what could be attained for metals. The engine itself was only a vehicle for this demonstration, and not an end in itself, for that particular programme, from the DARPA point of view. Nonetheless all major stationary ceramic components were run for 200h (175h at 1055°C and 25h at 1370°C) in an engine test rig. A ceramic rotor (HPSN hub/RBSN blades) was run for 200h at 50,000 rpm (100% design speed) at 1200°C turbine inlet temperature. Additionally one rotor was run successfully in an engine for ~10h at 40,000-50,000 rpm at temperatures as high as 1400°C. I fully concur that "lower" technology "bread-and-butter" near-term applications are needed.

McCauley: I perceive a need for quantum increases in toughness to increase reliability etc. Do you think that nitrides based on ceramic-matrix composites and not monolithic materials will result in the required properties?

Katz: Improvements in processing science and technology may be incremental (based on what we already know) but the results of these improvements may lead to more than incremental properties, and certainly to major improvements in reliability. Quantum increases are very hard to predict, even for the case of ceramic-matrix composites. I do believe, however, that ceramic composites will have a major role in the future of nitride oxide ceramics, in particular N-O glass matrix composites with graphite or SiC. It may be that Si_3N_4 fibres could be produced and made into woven structures.

Section A

CRYSTAL CHEMISTRY

SILICATE STRUCTURES AND ATOMIC SUBSTITUTION

J. Lang

Laboratoire de Chimie Minérale C - L.A. 254,
Université de Rennes I, 35042 Rennes Cedex, France.

ABSTRACT. The problems of substitutions in silicate chemistry concern often the cations. Because of the development of the nitrogen ceramics, the substitution of oxygen by different elements and especially by nitrogen has to be studied. First a review of the different structural types of the silicates and their classification is given, pointing out some structures which are of importance for the new ceramic materials. The oxygen-nitrogen substitution is studied in relation with the known compounds but their number is rather small and the number of known structures even smaller. It is possible however to study the relations between the structures of oxynitrides and nitrides related to silicon nitride and those of the silicates. Some aspects are examined for these two series of compounds. The other substitutions of oxygen and the substitution of Si by other elements than aluminium are also briefly reviewed.

1. INTRODUCTION

Silicon is the second element in the order of abundance on the earth (27.72%). The silicates are very important for the mineralogists because of the great number of them which constitute the earth's crust and for the chemist because of the number of artificial compounds, the technological applicability of which is always growing. The silicon chemistry has been the basis of glasses, cements, silicones and today has a large place in the field of electronics and of the new refractories related to Si_3N_4.

The structures of silicates have been investigated soon after the discovery of X-ray diffraction and new structural types are still discovered either in minerals or in new artificial compounds.

Riley, F.L. (ed.) Progress in Nitrogen Ceramics

The problem of their classification is important and has to be solved in a simple and accurate manner. Many classifications were proposed, three of them in the last two decades. We shall use the one proposed and improved by F. Liebau (1,2). Kostov's classification (3) has been developed for geoscientists and takes into account some morphological properties of the silicates. Zoltai's classification is a geometrical one (4). It uses the degree of sharing of tetrahedra regardless of which element occupies the tetrahedra. Thus it can be used for any inorganic compound with tetrahedral groups but is not specifically intended for the silicates.

2. CHEMICAL BONDS OF SILICON

If we compare the energies of the bonds formed by C and Si with different elements, we can see that the values for the carbon with almost all elements are very close explaining the formation of hydrocarbon chains and of organic compounds. On the contrary, silicon forms strong bonds with F and O only. Thus the easy oxidation of silicon compounds except fluorides can be explained and also the stability of the Si-O bond for which some characteristic mean values, obtained from silicates, are given here. With its outer shell $3s^1 3p^3 3d^0$, silicon forms ordinarily 4 σ-bonds corresponding to a tetrahedral environment. However if the ligands have lone pairs and a high electronegativity, they can give interactions with the d-orbitals of the silicon atom and bring some π-bonding character or form an octahedron sp^3d^2. This possibility is growing with the electronegativity of the ligands. With fluorine Si forms easily $SiF_6^=$ octahedra, with oxygen some examples of the SiO_6 group are known; with ligands of lower electronegativity: N (3.07), Cl (2.83), S (2.44) only tetrahedra can be obtained.

	F	O	Cl	H	N	C	Si
C	485	356	339	414	305	347	318 / kJ
Si	565	452	381	314	335	318	222 / kJ

Table 1. Bond energies of C and Si.

[4]	d(Si-O) = 162 pm (mean value) d(O-O) = 264 pm	d Si-O(n. br) = 154 pm d Si-O(br.) = 170 pm
[6]	d(Si-O) = 178 pm	d (O-O) = 254 pm

Table 2. Interatomic distances in silicon polyhedra.

The two SiO_x polyhedra correspond approximately to the amphoteric character of the silica. With strong acids like $H_4P_2O_7$ and H_2SO_4, SiO_2 behaves as a basic oxide and gives compounds such as SiP_2O_7 in which Si is 6-coordinated. With bases SiO_2 is an acid anhydride and gives anionic groups in which SiO_6 octahedra are very rare. The silicates are almost exclusively formed with SiO_4 groups and their structures, like those of the different forms of silica, depend on the relative arrangement of these tetrahedra. According to Pauling's third rule (5) the sharing of elements between two polyhedra has a strong influence on the interatomic distances and on the repulsive forces between two silicon atoms. The stability of silicate anions decreases strongly if, instead of corners, edges or faces are shared. It results that with the exception of fibrous silica (6) analogous to SiS_2, only corners are shared between SiO_4 tetrahedra in the structures of silicates and silica. Thus the oxygen atoms in the anions are of two types: the shared atoms which are bonded to two silicon atoms and forms a bridge between them, the other ones which form an ionic bond with cations and are more highly charged.

Many papers have been published on the problem of Si-O bonds in the silicates to explain the observed differences in the bond lengths and in the Si-O-Si bond angles. Cruickshank (7,8) has explained the short distances with a d-pπ bond, the importance of which increases with increasing electropositivity of the cations. This explanation does not seem necessary to Gibbs and co-workers (9,10). O'Keeffe and Hyde (11,12) have studied the problem on the basis of the Si....Si interactions and the silicate geometry would be largely determined by Si....Si and Si-O distances. They propose a "one-angle radius" for Si, r = 153 pm. Baur (13) has investigated the data published for 314 tetrahedra from 155 structures and applied a regression analysis. The Si-O bond length is a function for example of the mean coordination number of the O atoms (NC) in a tetrahedron of the number of bridging O per tetrahedron (CNM), and the following formula has been proposed to obtain a mean Si-O distance,

$$[\text{Si-O}]\ \text{mean} = 161.5 - 0.47\ \text{NC} + 0.54\ \text{CNM}$$

which can be used for predictive purposes in computer simulation. Baur neglects the role of cations which had been considered previously (14). As a rule the tetrahedra in a silicate structure are never perfectly regular but more or less distorted, and from the published data it can be established that for a tetrahedron the shorter Si-O bond forms the larger bond angles with the 3 other bonds in the tetrahedron.

3. STRUCTURES OF SILICATES

The structure of a silicate is determined by the arrangement of the tetrahedra according to the number of the shared corners. It results in more and more complicated anions characterised by an increasing dimensionality. The grouping of a finite number of tetrahedra gives a finite anion, i.e. infinite in zero dimension (d=o). The grouping of an infinite number can give infinite chain anions (d=1) which together with the previous ones, are the fundamental forms with which all other structures can be built up: layers (d=2) and tridimensional structures (1).

The chains are characterised by their periodicity p, namely the number of tetrahedra in the pattern. By joining several fundamental anions we obtain a composed one, the multiplicity m of which is given by the number of fundamental forms. A general classification can be given corresponding to the known types.

form \ m	1	2	3	4	5
tetrahedra	+	+	+	-	-
chains	+	+	+	+	+
rings	+	+	-	-	-
layers	+	+	-	-	-
frameworks	+				

4. TYPES OF SILICATES

4.1 Nesosilicates

The simplest of them are built up of isolated tetrahedra SiO_4^{4-} and can be considered as a close packing of oxygen atoms in which the silicon occupies tetrahedral positions and the cations are distributed in between in tetrahedral (Be, Zn, Li, Al) or in octahedral (Mg, Fe, Mn, Al...) positions. An example of the first case is the phenacite Be_2SiO_4 which has the same structure as Si_3N_4. Whatever the cation, the structures are identical and the cell parameters increase with the cationic radius while the density and the hardness of the compound decrease. Complete series of solid solutions can be obtained with suitable ionic radii. On the other hand, Mn^{++} can replace Zn^{++} in willemite but only for 1/5.

The manganese silicate Mn_2SiO_4 corresponds to octahedral positions occupied by cations as in olivine $(Mg,Fe)_2SiO_4$ (Fig. 1). MO_6 octahedra have 2 edges (M_1) or 1 edge (M_2) in common with SiO_4 groups. These compounds are all the harder and have the higher melting point as the cation is smaller and the structure denser, e.g. melt. pt. Fe_2SiO_4: 1205°C; Mg_2SiO_4: 1890°C. The calcium can replace Mg or Fe in the position 1 but up to 50% only. On the other hand $CaMgSiO_4$ and $CaFeSiO_4$ give a complete series of solid

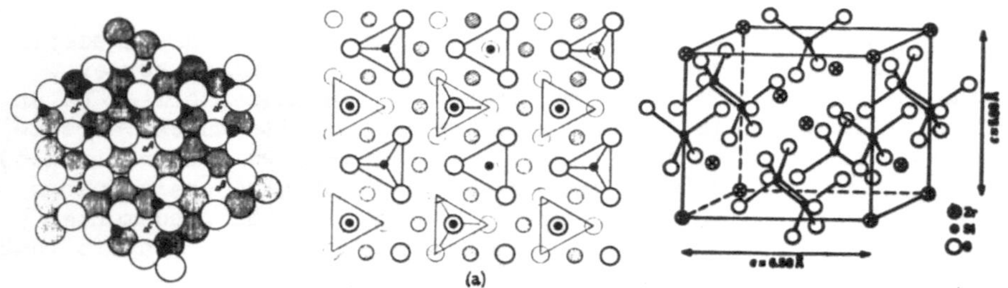

Fig. 1. Structure of phenacite (left), olivine and zircon (right).

solutions. For Ca_2SiO_4, the cell is distorted and the packing of O atoms less dense than olivine.

The garnets $M_3^{II}M_2^{III}(SiO_4)_3$ are very important because of their physical properties and form a large group with M^{II}=Mg, Ca, Fe, Mn...M^{III} = Al, Cr, Fe, Y... The M^{III} cations are to be found in octahedra, every vertex of which is in common with a SiO_4 group. The zircon structure is represented on Fig. 1. The SiO_4 groups are noticeably distorted and the cations are 8-coordinated. Some of these structures could be of interest to explain those of ternary nitrides.

4.2 Sorosilicates

The linear association of several SiO_4 tetrahedra gives isolated anions, the multiplicity of which is equal to 2 or 3 only. In thortveitite the $Si_2O_7^{=}$ anions are associated with octahedral coordinated Sc (Fig. 2a) and in barysilite with Mn and Pb: $MnPb_8(Si_2O_7)_3$.

The melilite group is of interest. Akermanite, $Ca_2MgSi_2O_7$ (Fig. 2b) and gehlenite, $Ca_2Al(AlSi)O_7$, form solid solutions by means of the following substitution.

$$Mg^{++} + Si^{4+} = 2Al^{+++}$$

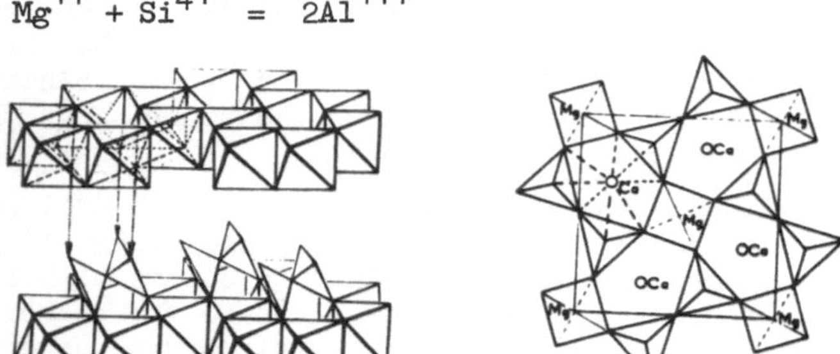

Fig. 2. The structure of thortveitite (left) and melilite (right).

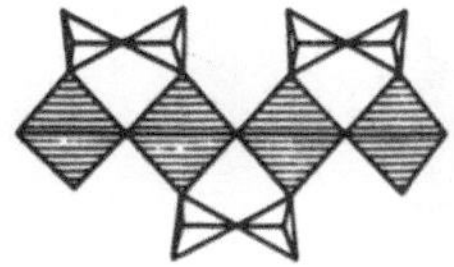

Fig. 3. Calcium octahedra and Si_2O_7 anion in cuspidine.

Nitrogen equivalents are known for them as well as for cuspidine $Ca_4Si_2O_7F_2$ in which CaO_6 octahedra share opposed edges while two neighbour octahedra have two polar vertices in common with a $Si_2O_7^{=}$ group (Fig. 3).

4.3 Inosilicates

Many SiO_4 groups sharing two of their vertices give a unidimensional structure noted $1/_\infty \, |Si_xO_{3x}|^{2x-}$ by Liebau (1). $1/_\infty$ means that we have an infinite one dimensional structure. These simple chains lined up along the c axis are linked by electrostatic interactions with the cations and give elongated crystals with a fibrous texture.

The simplest chain with a periodicity p = 1 is not known for silicates but for germanates like $CuGeO_3$ (Fig. 4a). For p = 2 we have the chain of pyroxene minerals such as enstatite, $Mg_2Si_2O_6$, for which the period corresponds to a length l = 5.2 Å, i.e. twice the diameter of oxygen atoms (Fig. 4b). Another example is the case of diopside represented in Fig. 5. A problem occurs with the bonds between the silicate macroanions and the chains of cationic polyhedra. The bonds can be established if both chains fit their own geometry and the dimensions of their polyhedra. According to

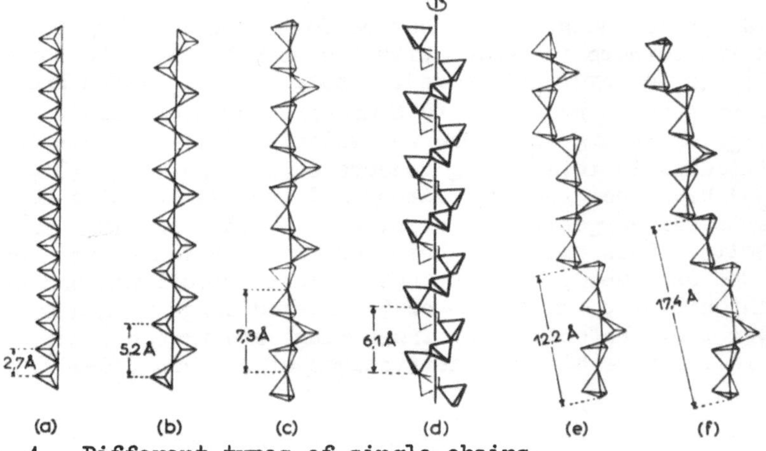

Fig. 4. Different types of single chains.

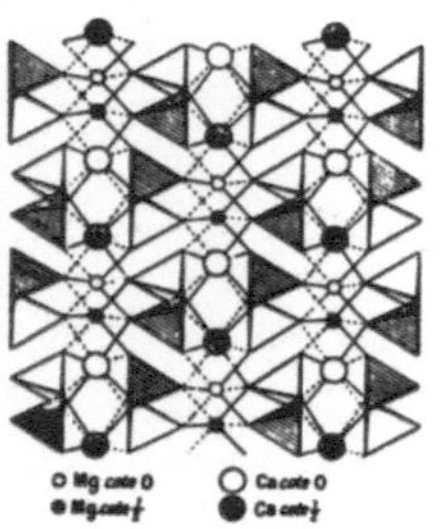

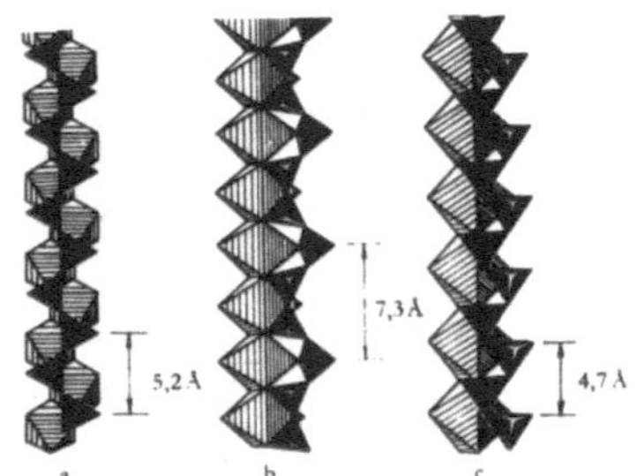

Fig. 5. Diopside structure viewed from the end of the chains (left). Chains in enstatite, wollastonite, and $Ba_2Si_2O_6$.

the cation the anionic chain curls up in a more or less tight helix, its threads not being related to the length of the pattern (Figs. 4 and 5b).

When two simple chains are linked by putting in common free vertices of SiO_4 tetrahedra, we obtain a ribbon or double chain with the same periodicity as the simple ones. If q tetrahedra among the p of the pattern link the chains, the anion can be written:

$$m\ 1/\infty|^{p}\ Si_{2p}O_{6p-q}|^{(4p-2q)-} \quad \text{or} \quad 2\ 1/\infty|Si_{2p}O_{6p-q}|^{(4p-2q)-} \quad \text{with } 1 < q < p$$

the multiplicity, dimensionality and period of the structure being taken into account (1). With $p = 2$ as in the amphibole tremolite $Ca_2Mg_5Si_8O_{22}(OH)_2$ we have $q = 1$. The value of m can be as high as 5 in some ribbons and these linear structures are very important in silicate chemistry. They do not seem to be so important in the case of nitrogen compounds where they would be very rare.

4.4 Cyclosilicates

With a finite chain, a ring can be formed where every tetrahedron has two corners in common with its neighbours. The formula is $c|SiO_3^{=}|_p$ (c for cyclosilicates). The commonest anions are the 3-membered ring and especially the 6-membered ones of beryl, $Be_3Al_2Si_6O_{18}$, or of high pressure wollastonite (15). The question of cyclosilicates is interesting because some large rings (Si_8O_{24}) and (Si_9O_{27}) have been recently discovered (16,17) and also because Bockris has suggested the existence of multiple rings in silicate melts. This point has been confirmed by Vallet and Rossin who admit the existence, besides double rings, of cylindrical or "barrel" anions formed of 3-, 4- or 5-membered rings with a multiplicity over two. Unfortunately these rings are probably not important in the pure nitrided derivatives of silicon (18-21).

4.5 Phyllosilicates

This group is probably the least interesting for the nitrogen ceramist but it includes many of the technologically most important silicates like clays. The tetrahedra are joined by three of their corners and their basis is in the same plane. The sheets can be considered as the result of single chain junction with a high multiplicity or as formed of linked rings. The whole structure, in the case of clays, talc, micas and so on, results from a sort of graft between the rings observed in the anion and those which can be found in some structures of metallic hydroxides such as $Mg(OH)_2$ and $Al(OH)_3$. Additional cations or layers have to be introduced if an Al-Si substitution occurs. The rings in the sheet can be identical (6-membered) or alternate (4- and 8-membered) with the apexes of the tetrahedra pointing out on one or both sides of the layer.

4.6 Tectosilicates

They are all aluminosilicates because the sharing of four corners of the SiO_4 groups gives the structure of a silica. A silicate can be formed with cations only if an Al-Si substitution occurs in the framework necessitating the transfer of electrons from a metallic element. Thus the O/Al+Si ratio is always equal to 2. In almost all cases the framework can be obtained by linking single chains and is noted $3/\infty\,|Si_{4-x}Al_xO_8|^{x-}$ with x = 1 or 2. These silicates can be divided into 3 main groups: feldspars, zeolites, ultramarine. The first group is probably the only interesting one for our purpose because the open structure of zeolites and the complicated structure of ultramarine could probably be obtained only with low nitrogen content. Moreover the already known structures of ternary nitrides are of a closed type.

4.6.1 Feldspars. They play an important role in the formation of rocks. Most of them are solid solutions found in the ternary system $KAlSi_3O_8$ - $NaAlSi_3O_8$ - $CaAl_2Si_2O_8$ and the isomorphic replacements allow for a large number of phases. Contrary to this the small cations like Cr, Fe, Mn are practically non-existent in these silicates. The framework can be considered as built up of rings joined together to form a rather open network (Fig. 6a). The cations are to be found in the holes of the structure and, due to the varied sizes of the cations, the lattices of potash and soda-lime feldspars are somewhat different.

4.6.2 Zeolites. Their framework is much more open than those of the feldspars. Thus the water molecules are loosely bound in the structure and can be removed or taken up again very easily. Many other gaseous substances can be absorbed and the compound can act as a molecular sieve. The cations are exchangeable with those of an appropriate salt solution. These interesting properties are

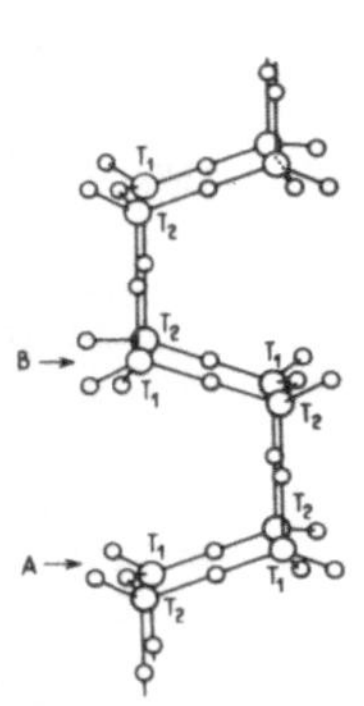

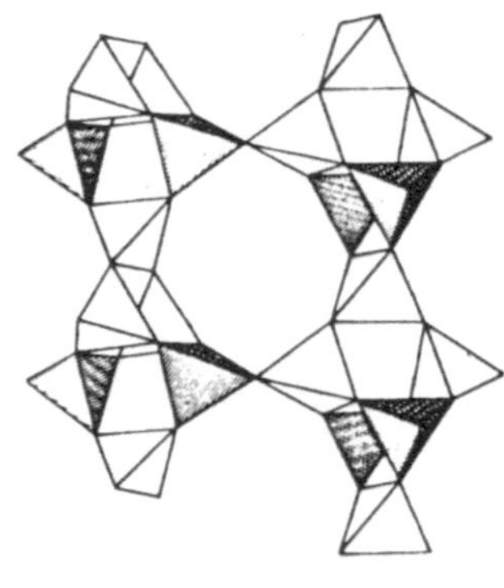

Fig. 6. The network in a feldspar, the chain in a fibrous zeolite.

due to a characteristic feature: the existence of tunnels in the open structure. These tunnels are parallel to

a) one line giving fibrous material such as natrolite, $Na_2Al_2Si_3O_{10}$. $2H_2O$. The chains in the latter are not of a simple but of a branched type (1), Fig. 6b.

b) two lines giving a lamellar structure such as that of clinoptilolite $(Ca, Na)_2Al_2Si_7O_{18}.6H_2O$.

c) three lines. The basket-like structural element is a polyhedron resulting from the truncation of the corners of an octahedron by a cube (Fig. 7). Its surface consists of 4- and 6-membered rings at the vertices of which the Si or Al atoms are situated. If the 4-membered rings are connected together by O-bridges we obtain the cubic structure of the artificial zeolite Linde A in which a large cavity is found at B (truncated cubooctahedron) (Fig. 7). If the hexagonal rings are connected by bridges forming hexagonal prisms we obtain the structure of faujasite (Fig. 8) in which the truncated octahedra build up a diamond-like structure. Polyhedra of less symmetrical types can be found, for example in chabazite.

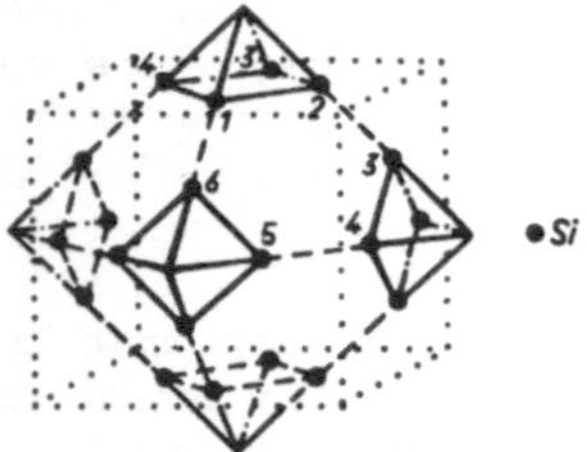

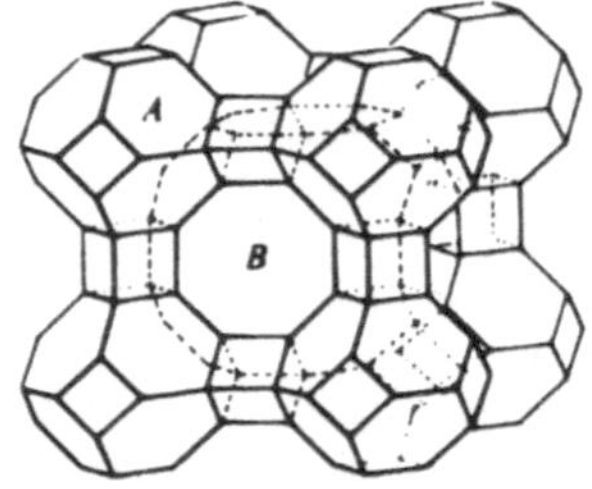

Fig. 7. The truncated octahedron (left) and the zeolite Linde A right.

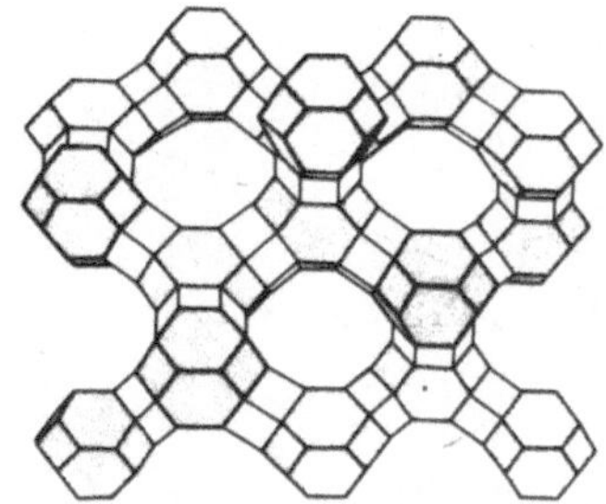

Fig. 8. The structure of faujasite.

4.6.3 Ultramarine. Here the truncated octahedra are linked to each other and the framework also contains electronegative elements such as Cl in sodalite and S in ultramarine, $Na_8Al_6Si_6O_{24}S_2$, which like cations, are to be found in the cavities of the structure. The famous colour of these silicates presents many variations due to the many possible substitutions. A particular one occurs in helvite in which half the silicon atoms are replaced by beryllium: $(Mn,Fe)_8Be_6Si_6O_{24}S_2$.

4.7 Silicates with octahedrally coordinated Silicon

In these compounds with Si-O-E bonds, the octahedral coordination of Si is favoured by higher electronegativity of E as well as by pressure. These compounds are of minor importance because of their very small number. Single octahedra are found in thaumasite, $Ca_3Si(OH)_6(SO_4)(CO_3).12H_2O$ (22) and in various forms of SiP_2O_7 (23) only. The condensation of octahedra has been found only in the high pressure structure of hollandite (24) and in stishovite (25).

4.8 Some features about Silicate Anions

Whatever the type of silicate, anions built up of more than one anionic type are very rare and tend to have the highest dimensionality for a given Si/O ratio. For a long time this ratio was believed to be the means to know that type of structure. This is wrong as many examples are known of different structures corresponding to a given value of this ratio.

The periodicity of silicate anions was also investigated, especially in the case of chains. The pyroxenes with $p = 2n + 1$ contain cations that are octahedrally coordinated by O and the chains have to be adjusted to the strip of joined octahedra as seen above and indicated on Fig. 9. If p is even cations are not, as a rule, octahedrally coordinated. A regression analysis (26) has shown that the higher the degree of stretching of chains, the lower is the electronegativity of cations, the lower their valency, the

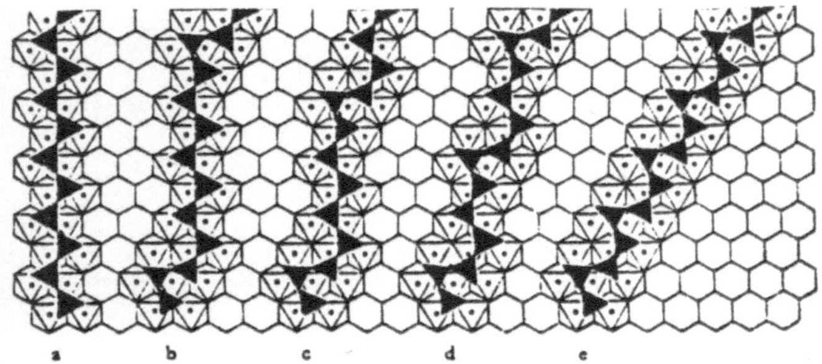

Fig. 9. Adjustment of chains with cationic polyhedra for enstatite, ferrosilite, pyroxferroite, rhodonite and wollastonite.

larger their radius, and the lower the value of p. In the case of phyllosilicates the tensions, due to dimensional differences between anionic and cationic sheets, are reduced by curving or rolling up the layers or even breaking them up to laths with Si tetrahedra pointing alternatively to both sides of the sheet.

Dent Glasser has investigated the problem of binary silicates which are topologically possible and do not exist (27). By plotting them on a graph representing the inverse of charge density versus the composition or the number of oxygen atoms per charge unit in a tetrahedron, we can see that the compounds are all situated in an angular field (Fig. 10). Using the concept of optical basicity (28), which is related to the average electron donor power of oxygen atoms, it appears that compounds with an optical basicity inferior to 0.54 or higher than 0.8 do not form. The basicity of O

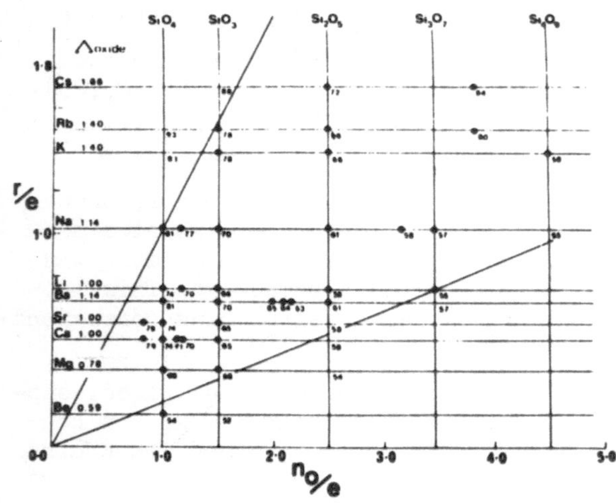

Fig. 10. r^+/e versus number of O per charge unit diagram for binary silicates.

atoms λ has been calculated for the different groups of silicates characterised by the connectivity of their tetrahedra (number of shared corners).

The different values of λ are given in the table.

	C=0	1	2	3	4
Average	0.74	0.70	0.65	0.58	0.48
Terminal	0.74	0.72	0.68	0.63	-
Bridging	-	0.63	0.59	0.55	0.48

Assuming there is some partial neutralisation of respective basicities of oxygen in the anion, the author explains why branched anions are rare. For hydrated silicates, the calculation shows that for a given group λ has a limit value beyond which the possible structures become unstable, e.g. $2ZnO.SiO_2.H_2O$ has the $Zn_2SiO_4.H_2O$ structure, and not the $Zn_2(HSiO_4)OH$ one.

5. SUBSTITUTED SILICATES

The question of oxygen substitution by electronegative elements in silicates has been enlarged recently but only group V- and VI-elements seem to be suitable.

5.1 Oxygen-Sulfur Substitution

The O-S substitution allows for thiocompounds which are analogous by their formula and structure to silicates and to one form of silica. The sulfide SiS_2 is known with a tetragonal chain structure in which each tetrahedra shares two opposite edges with its neighbours. By reaction of appropriate sulfides or of sulfur on silicides, many thiosilicates can be obtained in which SiS_4 tetrahedra are joined in different ways: isolated anions composed of single or multiple tetrahedra, chains, and so on (29-31), but these compounds are of little interest for ceramists.

5.2 Oxygen-Nitrogen Substitution

The Si-N bond is much rarer than the Si-O bond in nature, the only natural product of this type being sinoite, the silicon oxynitride found in meteorites of the condrite group (32). This bond has a high degree of covalency and Si_3N_4 is classified among nitrides as a covalent one. With nitrogen the silicon forms 4 σ-bonds but in some cases as in $N(SiH_3)_3$ and Si_3N_4 where Si and N atoms are coplanar a π-bond character is proposed. The data found in the literature are numerous for molecular compounds where Si and N atoms are linked to organic radicals (33); they are not so numerous for solid state compounds and ceramics, the structure of many of them remaining unknown.

The Si-N mean bond length varies from 172 pm in Si_2N_2O to 175.8 pm in LiSiON. As in SiO_4 tetrahedra, the electropositive element is not situated at the centre of the polyhedron but slightly shifted. This question is still a matter of discussion and many theories or viewpoints have been developed and were briefly reviewed recently (34).

The substitution can occur in two ways:

on a valency basis three oxygen are replaced by two nitrogen atoms:

$$3O^{=} = 2N^{\equiv}$$

on a joined substitution basis, the difference in anionic charge being balanced by a higher cationic one:

$$O^{=} + M^{x+} = N^{3-} + M^{(x+1)+}$$

The preparation work is done by reacting mixtures of oxides and/or nitrides, or by treating a suitable solid compound or mixture, either with a nitrogen or an ammonia stream.

The result of such substitutions is to form new patterns for structures building. While almost all silicates are based on an arrangement of SiO_4 tetrahedra, the nitrided compounds can be formed with them and also with other tetrahedra such as: (a) (SiO_3N); (b) (SiO_2N_2); (c) $(SiON_3)$; (d) (SiN_4). All of them have been encountered in at least one compound.

The structures of the two varieties of Si_3N_4 are well known now. Each silicon is tetrahedrally surrounded by 4N atoms and each N atom is common to 3 tetrahedra. Some π-bond character occurs in the compound. In Si_2N_2O the silicon atoms are bonded to 3N and 1O atoms. Each N atom is common to 3 tetrahedra and each O atom bridges two tetrahedra. These are arranged in such a manner that the whole structure can be considered as formed of $SiON_3$ groups linked together by the N atoms of their basis which lies in a corrugated plane. The oxygen apexes of the tetrahedra point on both sides of this Si-N sheet. They join them together with infinite multiplicity. For other nitrogen compounds, because of their small number, no particular classification can be established and their structures have to be reviewed according to the classification of silicates seen above. However a crude division can be made between nitrides and oxynitrides. The latter have a rather low N-content and single crystal structure determinations have not been achieved. It is thus very difficult, except in some cases, to know how nitrogen is distributed on crystallographic positions. Sometimes all that can be demonstrated is isotypism with a known silicate structure.

In the case of compounds with isolated anions, many examples are known. In olivine structure the joined substitution:

$$M^{++} + O^{=} = M^{+++} + N^{3-}$$

allows for the replacement of one oxygen giving $M^{II}M^{III}SiO_3N$ with type a tetrahedra. This structure is found with small cations inserted in octahedral sites. With large cations such as Ln, the close packing of anions is distorted and the K_2SO_4-β structure is found as in $LnEuSiO_3N$. The tetrahedra are linked together by two 9- and 10-coordinated cations. A series of compounds has been obtained for Ln = La bis Gd (35). One can wonder if a second substitution would be possible to produce $Ln_2SiO_2N_2$?

In the system Ln_2O_3-Si_3N_4, oxynitrides related to akermanite, $Ca_2MgSi_2O_7$, have been obtained (36-38) for a large number of lanthanides through a complicated substitution:

$$3\ II + 4(-II) = 2\ III + IV + 4(-III)$$

$Ln_2Si(Si_2O_3N_4)$ includes a substituted Si_2O_7 group: $Si_2O_3N_4^{10-}$ linked by silicon in the tetrahedral Mg position. The tetrahedral layers thus obtained are joined by Ln in the Ca position (square antiprism) (Fig. 11). This structure was confirmed by 1 MV-high resolution-microscopic observation (39). The anion could be ON_2Si-O-$SiON_2$. Hulliger wondered if less substituted structures could be obtained such as $Ln_2MgSi_2O_5N_2$ or $Ln_2AlAl_2O_6N$ for example, which seems unlikely (40).

In the cuspidine group the nitrogen compound $Ln_4Si_2O_7N_2$ can be related to $Ln_4Al_2O_9$ or to $Ca_4Si_2O_7F_2$. For Ln = La the parameters are twice those of cuspidine (36,37,41). It would be interesting to know whether the nitrogen is located as a bridging atom between the two tetrahedra or not, or whether it is in the place of the F ion. Large single crystals of $La_4Si_2O_7N_2$ have been obtained by the floating zone method (42).

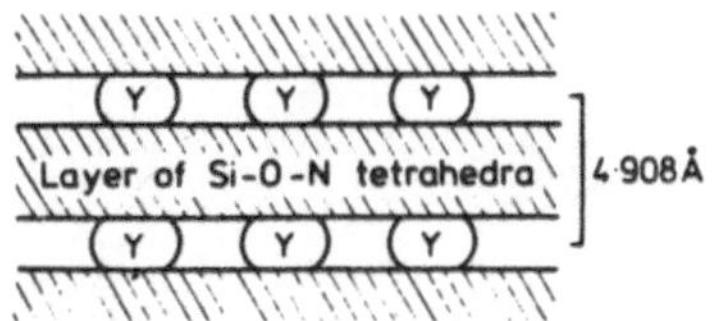

Fig. 11. Layered structure of $Y_2Si(Si_2O_3N_4)$.

Rings are found in the $LnSiO_2N$ which has a pseudo wollastonite structure (43). Rare earth ions are coordinated by 6O +2N, oxygen by 3Ln + 1Si, and N by 2Ln +2Si. It was shown that N is a bridging atom in the six-membered rings and O a terminal one. There are common problems about the stacking of layers along the $\vec{c}$ axis for both the nitrogen compounds and α-$CaSiO_3$.

The structure of apatite, $Ca_{10}(PO_4)_3F_2$, has been extensively studied in the case of phosphatic minerals (45) as well as in the case of silicates (46). The framework is made of SiO_4 tetrahedra linked together by cationic polyhedra corresponding to two crystallographic positions: (a) coordination 6 + 3 in a triangular prism; (b) coordination 7 in a pentagonal pyramid (Fig. 12). Fluoride anions are in a particular position on a 6-axis where they can be replaced by many other anions: X, OH, O, S, CO_3. As a rule this structure offers one the largest varieties of substitution (47). The nitrogen apatites $Ln_{10}Si_6O_{24}N_2$ obtained for many lanthanides and Y have the same structure but with a lower symmetry: sp. gr. $P6_3$ instead of $P6_3/m$, resulting in three independent crystallographic positions for cations. Solid solutions were obtained with different cations (48).

The structural determination (49) has shown that the nitrogen is not located in the place of F ions where $O^=$ can be found. Nitrogen replaces oxygen to form the coordination polyhedron of Si and also of the 6 + 3 coordinated lanthanide. However F positions are occupied, for more than two nitrogen in the cell, there would be an O-N substitution. It was a problem to increase the N content by means of joined substitutions:

$$Ln^{III+} + xO^= = T^{(III+x)+} + xN^{3-}$$

T being a transition element. Compounds with Ti, Ge and V were obtained. In this latter case the maximum N content was 4.7 at/cell, i.e. 18 at % O + N atoms in $Sm_{8.65}V_{1.35}Si_6N_{4.7}O_{21.3}$ (50). By means of both joined substitutions $3O^= = 2N^{3-} + \square$ and $2O^= = N^{3-} + F^-$, no higher N content was obtained. It seems unlikely that defective apatites, well known for oxyapatites, can be obtained in the case of nitrogen ones. In the second case the only compound

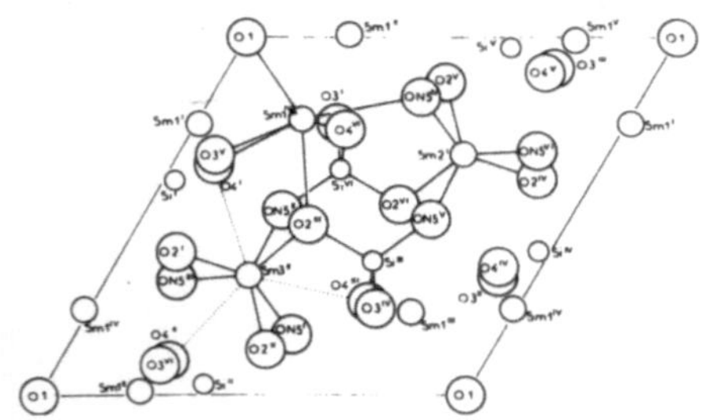

Fig. 12. Projection of the apatite structure.

obtained had a low N content: $Ln_9MnSi_6O_{23}N_2F$ with probably the F atom in one of its particular positions. Further investigations in this field would be needed.

In the case of LiSiON, the problem of nitrogen positions has been solved by using neutron diffraction (51). The structure is related to wurtzite hexagonal packing of O, N atoms (55). Two possible arrangements for cations are given on Fig. 13 and the structure corresponds to the second. Si is surrounded by 3N + O, Li by 3O + N, O by 3Li + Si and N by 3Si + Li. $SiON_3$ tetrahedra form layers parallel to xz plane through common use of their three nitrogen. On the sides of the layers the oxygen corners point out. This structure can be related to that of Si_2N_2O but here, the link between the Si-N layers is formed by O atoms not directly, however, but through $Li(ON)_4$ distorted tetrahedra. In the rather regular $SiON_3$ tetrahedra, the silicon atom is shifted aside. Thus one of the three Si-N bonds is shorter than the two others as this has also been shown for β-Si_3N_4 (59). An important group of oxynitride compounds is formed by sialons (56). These phases result from joining $Si(ON)_4$ and $Al(ON)_4$ tetrahedra and their structure is a tridimensional one. However the order between O and N remains unknown. This group is reviewed in another paper and is left aside here.

For purely nitrided ternary derivatives of silicon, different types of structure can be found or be likely though many of them are not determined. In the systems M_3N_2-Si_3N_4 several phases are obtained, their formulae varying with the ratio of the two nitrides from M_7SiN_6 for Be or M_4SiN_4 for Ca and Mg to $MSiN_2$ (52,53,57). In the Be-Si-N system the structure of most of these phases has been explained by polytypism resulting from intergrowths of $BeSiN_2$ and Be_3N_2 layers (53,54). $BeSiN_2$, as other $MSiN_2$ compounds with tetrahedrally coordinated cations (Mg, Zn, Mn) has a structure related to wurtzite hexagonal close packing of nitrogen atoms in which tetrahedral positions are regularly occupied by Si and M (Fig. 14). Thus each SiN_4 tetrahedron shares its four corners building up a rather close framework (58). M_4SiN_4 with M = Be, Mg, Ca, are typical examples of structures with isolated tetrahedra, the coordination of the metal probably giving very different structural

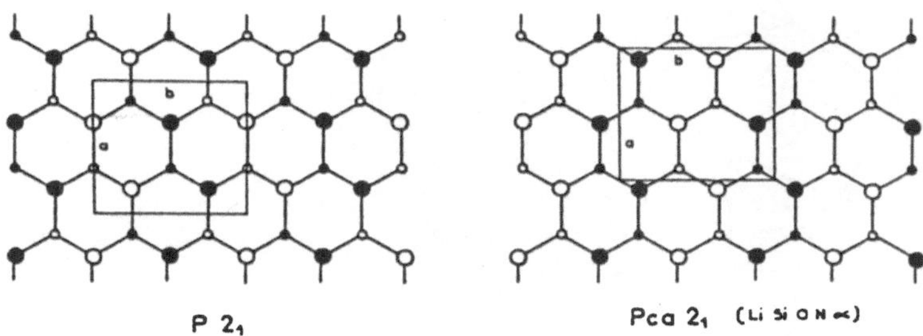

Fig. 13. The possible order for cations in LiSiON.

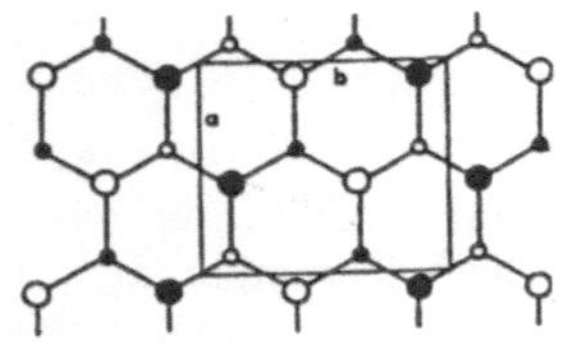

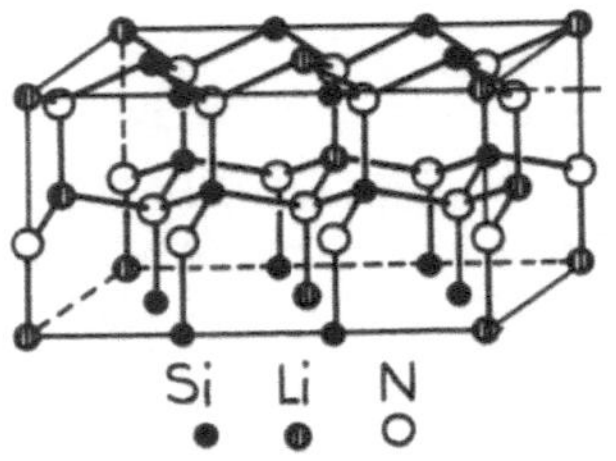

Fig. 14. Order of Si and M in $MSiN_2$, structure of $LiSi_2N_3$.

types. $M_5Si_2N_6$ (57) may be a possible chain or ring structure with SiN_4 tetrahedra linked by two of their corners.

$LiSi_2N_3$ in which two types of coordination for N atoms are found, has also a tridimensional structure related to hexagonal close packing of N atoms (Fig. 14). All tetrahedra share their corners, surrounding the positions occupied by Li.

There has been a recent publication on a very interesting structure, that of $LaSi_3N_5$ (60). This nitride has a pattern of three SiN_4 tetrahedra. They are linked to form 5-membered rings linked together to produce a rather open network (Fig. 15). Between the pentagonal holes the Ln-ions are centrally located. One of the interesting features of this compound is that 2/5 of N atoms are coordinated to 3 Si, the others being surrounded by 2Si + 2La. An N atom coordinated by 3Si is a rare example in a ternary compound though quite common in Si_3N_4. We can hope that the discovery of new nitrides will allow for the discovery of new interesting structures.

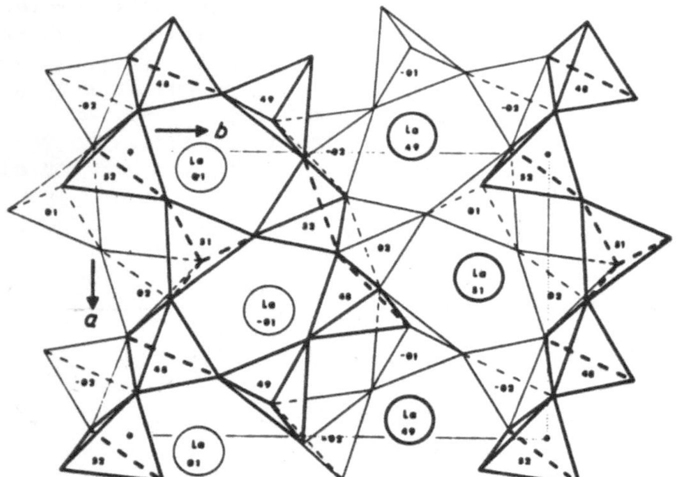

Fig. 15. $LaSi_3N_5$ structure projected on the (001) plane (after 60).

5.3 Oxygen-Halogen Substitutions

They are known for silica with X = F, Cl or Br, but no example seems to exist for silicates. On account of the electronegativity of the halogens and of their bond energy with silicon, chlorine and fluorine only seem to be suitable in two possible ways.

- substitution on a valency basis - 2 halogens replace 1 oxygen

$$2\,X^- = O^=$$

- joined substitution - the oxygen-halogen replacement being accompanied by the replacement of a cation by a less charged one, for example:

$$O^= + M^{x+} = F^- + M^{(x-1)+}$$

One of the most important features in this substitution is the following: a bridging bond $\equiv$ Si-O-, the basis of the existence of the macromolecular silicate anions, is replaced by a terminal bond $\equiv \mathrm{Si}-\underline{\overline{X}}|$. The lone pairs of the halogen are only able to form coordinative bonds of low energy as is the case in molecular compounds such as Al_2Cl_6. Therefore the O-X substitution is a limiting factor for the dimensionality of anions and, in the end, zero-dimensioned tetrahalogenides are obtained.

Moreover on account of its stability, the Si-O bond is easily attacked by fluorine and by chlorine under given energetic conditions: high temperature, presence of a reductor like carbon, use of halogen under molecular form or liberated by an appropriate vector. In these conditions the Si-X bonds form easily but it becomes difficult to control the reaction. Thus the latter pursues itself and brings about the destruction of the silicate structure forming molecular compounds or mixed phases. Here are some examples of reactions.

$$SiO_2 + 2BeCl_2 \longrightarrow SiCl_4 + 2BeO$$

$$3SiO_2 + 2Al_2F_6 \longrightarrow 3SiF_4 + 2Al_2O_3$$

HF is a good solvent of silica which it depolymerizes.

In some cases in silicates there may appear a replacement of oxygen or OH groups by F. Thus the oxyapatite $Ca_2Ln_8Si_6O_{26}$ and the fluoro derivative $Ca_4Ln_6Si_6O_{24}F_2$ are related compounds, but the replacement of oxygen occurs for those atoms which are in a particular position and do not form an Si-O-Si bridge. The silicate skeleton remains unchanged.

Finally we can note that the best known of silicon oxyhalides Si_2OCl_6 the intermediary between SiO_2 and $SiCl_4$ is not directly prepared from silica.

6. SILICON SUBSTITUTIONS

Many elements can replace silicon in its tetrahedral positions, the most important being aluminium. This substitution occurs at the most for 50 at per cent as in micas, feldspars and so on. For a long time it was believed that it was done randomly. According to Pauling's rule the polyhedra around small coordination numbered cations tend to fail to share elements. If, however, elements are shared and, if there exists an alternative structure with higher coordination numbers, the latter will always be more stable. This rule applies especially if the radius ratio approaches the lower limit of stability of the polyhedron. Thus such condensations which are well known in the case of PO_4 or SO_4 tetrahedra do not occur with AlO_4. An oxygen atom can be a bridge between two Al atoms only if one of them has a coordination number of 5 or 6 and thus no two Al atoms can occupy the centre of neighbour tetrahedra. The structures of aluminosilicate anions and especially those with 50% substituted Si atoms are built up of alternate SiO_4 and AlO_4 tetrahedra (61). The analytical data confirm this conclusion (62).

As a rule this sort of substitution increases from neso- (d=0) to tectosilicates (d=3) and the greater the Si-Al order, the lower the temperature.

The substitution of silicon by other elements is much less pronounced in nature (1) but many "metallosilicates" have been prepared with Be (63), Zn (64), Ga (65), Ge (66) for example. The problem with these compounds is to know whether these metals are cations in tetrahedral positions or elements associated with Si to build up the anionic group.

REFERENCES

1. F. Liebau. Handbook of Geochemistry, ed. K.H. Wedepohl, Vol. II/3, Chap. 14A, Springer Verlag (Berlin), 1972.
2. F. Liebau. Amer. Mineral., 63, 918-923 (1978).
3. I. Kostov. Geochem. Mineral. Petrol., 1, 5-41 (1975).
4. T. Zoltai. Amer. Mineral., 45, 960-973 (1960).
5. L. Pauling. J. Amer. Chem. Soc., 51, 1010-1026 (1929).
6. A. Weiss and A. Weiss. Z. Anorg. Allg. Chem., 276, 95 (1954).
7. O.W.J. Cruickshank. J. Chem. Soc., 5486-5504 (1961).
8. W.S. McDonald and D.W.J. Cruickshank. Acta Cryst., 22, 37-43 (1967); Z. Kristall., 124, 180-191 (1967).
9. S.J. Louisnathan and G.V. Gibbs. Mat. Res. Bull., 7, 1281 (1972).
10. G.E. Brown and G.V. Gibbs. Amer. Mineral., 55, 1587-1607 (1970).
11. M. O'Keeffe and B.G. Hyde. Acta Cryst., B34, 27-32 (1978).
12. M. O'Keeffe. Acta Cryst., A35, 776-779 (1979).
13. W.H. Baur. Acta Cryst., B34, 1751-1756 (1978).
14. W. Noll. Angew. Chem. Internat. Edit., 2, 73-80 (1963).

15. F.J. Trojer. Z. Krist., 127, 291 (1968); 130, 185 (1969).
16. A.A. Khan and W.H. Baur. Science, 173, 916 (1971).
17. G. Guiseppetti, F. Mazzi and C. Tadini. Tschermaks Mineral. Petrog. Mitt. 16, 105 (1971).
18. J. O'M. Bockris and D. Lowe. Proc. Royal Soc. 48, 536 (1954).
19. J. O'M. Bockris, J.A. Kitchener and J.D. Mackenzie. Trans. Farad. Soc. 51, 1734 (1955).
20. J.W. Tomlinson, M.S.R. Heynes and J.O'M. Bockris. Trans. Farad. Soc. 54, 1822-1833 (1958).
21. R. Rossin. Thesis-Docteur Ingenieur Rennes (1963).
R. Rossin, J. Bersan and G. Urbain. Rev. Haut. Temp. Refract. 1, 159-170 (1964).
22. R.A. Edge and H.F.W. Taylor. Acta Cryst. B27, 594 (1971).
23. G. Bissert and F. Liebau. Acta Cryst. B26, 233 (1970).
F. Liebau and K.F. Hesse. Z. Krist. 133, 213 (1971).
24. A.E. Ringwood, A.F. Reid and A.D. Wadsley. Acta Cryst. 23, 1093 (1967).
25. S.M. Stishov and N.V. Belov. Dok. Akad. Nauk. SSSR, 143, 951 (1962).
26. F. Liebau and I. Pallas. Z. Krist. (1980).
27. L.S. Dent Glasser. Z. Krist. 149, 291-305 (1979).
28. J.A. Duffy and A.D. Ingram. J. Am. Chem. Soc. 93, 6448-6454 (1971); J. Inorg. Nucl. Chem. 37, 1203-1206 (1975); J. Non-Cryst. Sol. 21, 373-410 (1976).
29. A. Weiss and G. Rocktäschel. Z. Anorg. Allg. Chem., 307, 1-6 (1960).
30. B. Krebs and J. Mandt. Z. Anorg. Allg. Chem. 388, 193-206 (1972).
31. M. Ribes, J. Olivier-Fourcade, E. Philippot and M. Maurin. J. Sol. St. Chem. 8, 195-205 (1973).
32. F. Wlotzka. Handbook of Geochemistry, ed. K.H. Wedepohl, Vol. II, Chap. 7C,D,E,F, Springer Verlag (Berlin), 1974.
33. U. Wannagat. Biochemistry of silicon and related problems, ed. G. Benz and I. Lindquist, Plenum, 1978, pp. 77-90.
34. Y. Laurent, F.F. Grekov, J. David and J. Guyader. Ann. Chim. Fr. 5, 647-655 (1980).
35. R. Marchand. C.R. Acad. Sci., C 283, 281-283 (1976).
36. R.R. Wills, R.W. Stewart, J.A. Cunningham and J.M. Wimmer. J. Mat. Sci. 11, 749-759 (1976).
37. R. Marchand, A. Jayaweera, P. Verdier and J. Lang. C.R. Acad. Sci., C 283, 675-677 (1976).
38. K.H. Jack. J. Mater. Sci. 11, 1135-1158 (1976).
39. S. Horiuchi and M. Mitomo. J. Mater. Sci. 14, 2543-2546 (1979).
40. F. Hulliger. Handbook on the Physics and Chemistry of Rare Earths, ed. K.A. Gschneidner Jr. and L. Eyring. North Holland (Amsterdam), 1979, pp. 227-229.
41. P.E.D. Morgan. J. Amer. Ceram. Soc. 59, 86 (1976).
42. N. Ii, M. Mitomo and Z. Inoue. J. Mater. Sci. 15, 1691-1695 (1980).

43. P.E.D. Morgan, P.J. Carroll and F.F. Lange. Mat. Res. Bull., 12, 251-260 (1977).
44. J. Gaude, J. Guyader and J. Lang. C.R. Acad. Sci., C 280, 883 (1975).
45. D. McConnell. Apatite, its Crystal Chemistry. Springer Verlag (New York), 1973.
46. J. Ito. Amer. Mineral., 53, 890-907 (1968).
47. J. Lang, R. Marchand, C. Hamon, P. L'Haridon and J. Guyader. Bull. Soc. Fr. Mineral Cristal., 98, 284-288 (1975).
48. C. Hamon, R. Marchand, M. Maunaye, J. Gaude and J. Guyader. Rev. Chim. Mine., 12, 259-267 (1975).
49. J. Gaude, P. L'Haridon, R. Marchand and Y. Laurent. Bull. Soc. Fr. Mineral. Cristal. 98, 214-217 (1975).
50. J. Guyader, F.F. Grekov, R. Marchand and J. Lang. Rev. Chim. Min. 15, 431-438 (1978).
51. Y. Laurent, J. Guyader and G. Roult. Acta Cryst. B37, 911-913 (1981).
52. I.C. Huseby, H.L. Lukas and G. Petzow. J. Am. Ceram. Soc., 58, 377-380 (1975).
53. T.M. Shaw. Thesis, Univ. of California, Berkeley (1977). T.M. Shaw and G. Thomas. J. Sol. St. Chem. 33, 63-82 (1980).
54. D.P. Thompson. J. Mat. Sci. 11, 1377 (1976).
55. Y. Laurent, F.F. Grekov, J. David and J. Guyader. Ann. Chim. Fr., 5, 647-655 (1980).
56. K.H. Jack. J. Mater. Sci. 11, 1135-1158 (1976).
57. Y. Laurent. Rev. Chim. Min. 5, 1019-1050 (1968).
58. J. Lang. Nitrogen Ceramics, ed. F.L. Riley, Noordhoff (Leyden), 1977, pp. 90-94.
59. R. Grun. Acta Cryst. B35, 800-804 (1979).
60. Z. Inoue, M. Mitomo and N. Ni. J. Mater. Sci. 15, 2915-2920 (1980).
61. W. Loewenstein. Amer. Mineral., 39, 92-96 (1954).
62. W.A. Deer, R.A. Howie and J. Zussman. Rock forming minerals, Vols. 2,3,4, Longmans (London), 1962-1963.
63. A.A. Goryachev and O.S. Ignatiev. Zh. Neorg. Khim. 15, 1614-1617 (1970).
64. K.F. Hesse, F. Liebau, H. Böhm, P.H. Ribbe and M.W. Phillips. Acta Cryst. B33, 1333-1337 (1977).
65. M.A. Piontkovskaya, G.S. Shameko and I.E. Neimark. Izv. Akad. Nauk. SSSR, Neorg. Mat., 6, 1151 (1970).
66. K.H. Jost, H. Wolf and E. Thilo. Z. Anorg. Allg. Chem., 353, 42 (1967).

THE CHARACTERIZATION OF α'-SIALONS AND THE α-β RELATIONSHIPS IN SIALONS AND SILICON NITRIDES

K.H. Jack

Wolfson Research Group for High-Strength Materials,
Crystallography Laboratory,
The University of Newcastle upon Tyne, UK.

ABSTRACT. The preparation and characterization of sialons with structures based on α-Si_3N_4 extends the science and technology of "ceramic alloying" and also suggests possible relationships between the α and β structures in both sialons and silicon nitrides. The α' structure occurs in M-Si-Al-O-N systems and is derived from the $Si_{12}N_{16}$ unit cell by partial replacement of Si^{4+} by Al^{3+}. Valency compensation is by modifier cations (Li, Ca, Y and all the rare-earth elements except La and Ce) occupying the interstices of the (Si,Al)-N network. Where a modifier oxide is used (e.g. CaO), some O may also replace N, and because there are only two available interstitial sites per unit cell, the α'-phases have the general composition $M_x(Si,Al)_{12}(O,N)_{16}$ where $x \ngtr 2$. Minimum and maximum observed values of x are $Y_{0.3}$ and $Ca_{1.6}$. The limit of replacement of nitrogen by oxygen is probably not more than one atom per unit cell e.g. $CaSi_9Al_3ON_{15}$.

The transformations $\alpha' \rightleftharpoons \beta'$ occur by chemical reactions. By analogy, the relationships between α' and β' support earlier proposals that α-silicon nitride is a defect structure with a range of composition that can accommodate small amounts of oxygen. This accounts for the observed variation in density of α, the relatively wide variation in unit-cell dimensions, the marked differences in properties between α produced by CVD and by SiO-N_2 interaction, and also for the thermodynamics of the Si-O-N system.

1. INTRODUCTION

Our first report (1) of Si-Al-O-N ceramics stated that α-silicon nitride structures expanded by about 3% had been obtained by reaction of $LiSi_2N_3$ with Al_2O_3. Subsequent preparation (2) of

Riley, F.L. (ed.) Progress in Nitrogen Ceramics

α'-lithium sialons showed a variation of cell dimensions when different proportions of $LiAlO_2$ and Si_3N_4 were reacted together, but other phases were always present and the product never contained more than ~30% of the α'-material. A claim by Mitomo (3) of α' solid solutions of Si_3N_4-Al_2O_3 and/or Si_3N_4-Y_2O_3 prompted reports from Newcastle of the preparation and complete characterization of pure Li, Ca and Y α'-sialons (4,5) the proposed compositions for which have been confirmed by Grand et al. (6).

2. PREPARATION AND CRYSTAL STRUCTURE

2.1 The α and β Structures

The "idealised" silicon nitride structures can be described as a stacking of Si-N layers in either an ABAB.... (β) or an ABCD.... (α) sequence as shown in Fig. 1. This gives, in the hexagonal β unit cell containing Si_6N_8, long continuous channels running parallel with the c-direction and centred at 2/3, 1/3. In α, the c-glide plane that relates the layers CD with AB replaces the continuous channels of β by large, closed interstices at 2/3, 1/3, 3/8 and 1/3, 2/3, 7/8. Thus, in the hexagonal unit cell containing, ideally, $Si_{12}N_{16}$ there are two sites large enough to accommodate other atoms or ions. Note that although the Si-N layers in the actual β structure are almost identical with the "ideal" configurations (see Fig. 2), those of α are distorted and nitrogen atoms at heights approximately 3/8 and 7/8 are pulled in towards the centres of the two respective interstices.

2.2 α' Preparation

The α'-sialons are prepared by heating appropriate mixtures of nitrides (e.g. $0.5Ca_3N_2$:$3Si_3N_4$:3AlN) or nitrides plus oxides

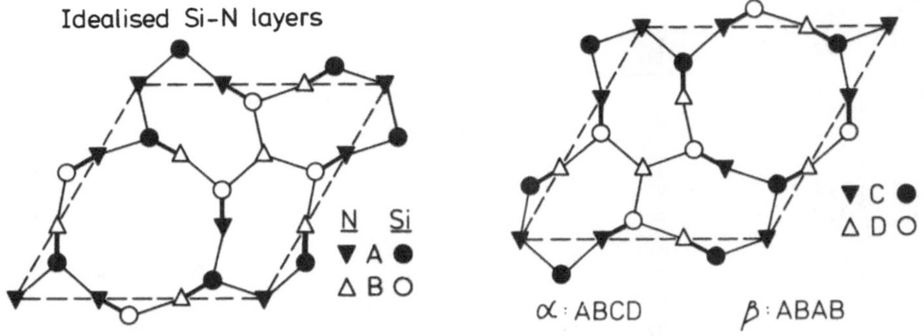

Fig. 1 Idealised Si-N layers in α and β silicon nitrides

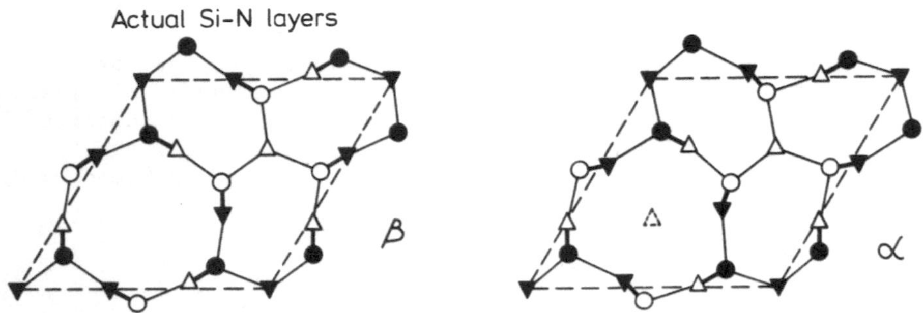

Fig. 2 Actual Si-N layers in β and α silicon nitrides.

(e.g. $CaO:3Si_3N_4:3AlN$) without pressure at 1,750°C for 15 minutes in one atmosphere of molecular nitrogen or argon. Weight losses are negligible and the compositions of the initial mix, calculated from the proportions of powder constituents with allowance for surface oxide on the nitrides, is in good agreement, as shown in Table 1, with direct microanalysis of the product using a Camebax electron probe at AERE, Harwell. It seemed that the Li, Ca or Y of the α'-sialons must occupy some of the two interstitial sites per unit cell to give compositions $M_x(Si,Al)_{12}(O,N)_{16}$ where $x \not> 2$.

La and Ce do not give α'-phases but compositions $0.25Re_2O_3:3Si_3N_4:4AlN$ all gave more than 90% α' with unreacted AlN after 15 minutes at 1,750°C when rare earths of even atomic number elements 60-70 (i.e. Nd, Sm, Gd, Dy, Er and Yb) were used. It is reasonable to assume that the odd-numbered elements behave similarly and that La and Ce do not react because their atomic or ionic radii exceed a critical value.

Table 1. α'-Sialon compositions (atoms per unit cell)

	Ca	Si	Al	O	N	Y	Si	Al	O	N
I-m	0.5	10.6	1.4	1.3	15.0	0.3	10.0	2.0	1.3	14.8
a	0.5	10.5	1.6	0.7	14.8	0.5	10.0	2.0	0.8	15.3
I-m	1.1	9.1	2.9	1.8	14.5	0.5	8.3	3.7	1.5	14.3
a	0.9	9.2	2.8	1.4	14.6	0.6	9.2	2.8	1.1	14.9

I-m: initial mix

a: Camebax electron probe analysis at AERE, Harwell

Table 2. Unit-cell dimensions (Å) and densities (g cm^{-3}) of α'-sialons

	a	c	c/a	d_o	d_c
β-Si_3N_4	7.61	2.91	0.765/2	3.192	3.192
α-"Si_3N_4"	7.76	5.62	0.724	3.16*	3.183
$LiSi_{10}Al_2ON_{15}$	7.83	5.67	0.724	3.12	3.14
$Ca_{0.5}Si_{10.5}Al_{1.5}O_{0.5}N_{15.5}$	7.82	5.68	0.727	3.16	3.20
$Ca_{0.8}Si_{9.2}Al_{2.8}O_{1.2}N_{14.8}$	7.86	5.71	0.727	3.19	3.26
$Y_{0.4}Si_{10}Al_2O_{0.8}N_{15.2}$	7.81	5.69	0.729	3.23	3.25
$Y_{0.6}Si_{9.2}Al_{2.8}O_{1.1}N_{14.9}$	7.83	5.71	0.729	3.28	3.36

d_o, observed density; d_c, calculated density

* a range of values 3.167 - 3.171 g cm^{-3} was observed by WILD et al. (7) for α-needles produced by the SiO-N_2 reaction.

2.3 The α' Structure

Table 2 compares the cell dimensions and densities of typical α'-sialons with those of α and β silicon nitrides. The densities calculated from the cell dimensions and the proposed cell contents are in good agreement with the observed densities. Although the diffraction patterns for α and α' are similar, there are small but distinct differences in intensities that are shown schematically in Fig. 3 for a composition $CaSi_9Al_3ON_{15}$. Comparison of (i) and (ii) shows the expected good agreement between the observed intensities (α_o) for α-Si_3N_4 and those calculated (α_c) from its known structure. In Fig. 3(iii) the intensities

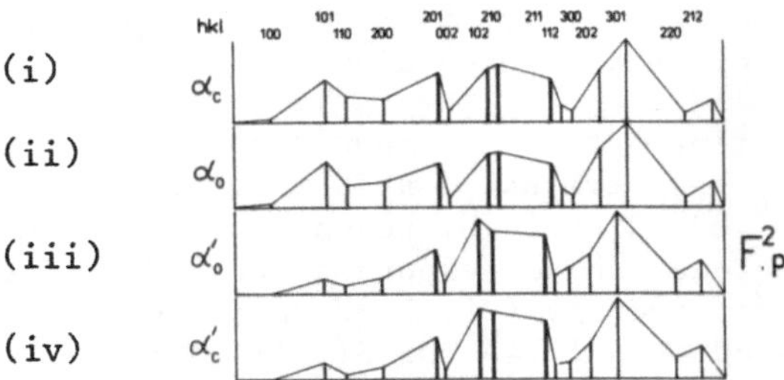

Fig. 3 Observed and calculated diffracted intensities (F^2.p values) for α-Si_3N_4 and α'-sialons. Thick ordinates are half-scale.

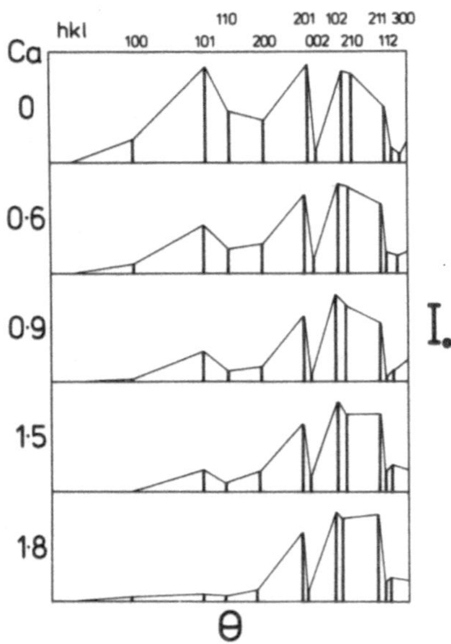

Fig. 4 Intensities of X-ray reflexions (Cu$K\alpha_1$) for α'-Ca-sialons with varying Ca content

observed for α'-Ca-sialon (α'_o) are modified from those of α; e.g. reflexions 100, 101 and 1$\bar{1}$0 are weakened while 102 and 210 are stronger and reversed relative to each other. A complete structure refinement (8) accounts for these changes and shows that each of the two interstitial sites in the unit cell contains, on average, one-half of a Ca atom; compare Figs. 3(iii) and 3(iv). More recently, four α'-Ca-sialons with 0.6, 0.9, 1.5 and 1.8 Ca atoms per unit cell have been compared. Their observed intensities are plotted against θ in Fig. 4 with those of α-$Si_{12}N_{16}$. There are systematic intensity changes due to the increasing contribution of Ca to the structure amplitude F. Thus, the 100 intensity decreases to zero at 1.5 Ca and then re-appears at 1.8 Ca as F passes through zero and changes sign.

The structure refinement of a composition $Ca_{1.83}Si_{8.34}Al_{3.66}N_{16}$ was terminated at R = 0.06. Each Ca is coordinated by seven N atoms at an average distance of 2.53Å whereas the corresponding Ca-N distance in $CaSi_9Al_3ON_{15}$ is 2.59Å; average (Si,Al)-N distances are respectively 1.79 and 1.76Å compared with 1.74Å for Si-N in α-Si_3N_4. With increasing Ca and Al there is, as might be expected, an overall expansion and an increase in the average (Si,Al)-N distance but, at the same time, the increased bond strength between Ca and its ligands shortens the average Ca-N distance.

The α'-structure is derived from α-$Si_{12}N_{16}$ by partial replacement of Si^{4+} with Al^{3+}, and valency compensation is effected by "modifying" cations such as Li^+. Ca^{2+} amd Y^{3+} occupying the interstices of the (Si,Al)-N network. The materials are similar to the "stuffed" derivatives of quartz in which Al^{3+} replaces Si^{4+} and positive valency deficiencies are compensated by "stuffing" cations like Li^+ and Mg^{2+} into interstitial sites. When α' is synthesised entirely from nitrides the product should contain no oxygen and valency compensation is due solely to the introduction of the modifier cations. Because there are only two sites per unit cell for these, the limiting compositions for α'-nitrides might be expected to be $Ca_2Si_8Al_4N_{16}$ and $Y_2Si_6Al_6N_{16}$. These limits have not been achieved, possibly because it is difficult to avoid surface oxide on the nitrides. Where a modifier oxide is used, oxygen replaces nitrogen but the extent to which this can occur and still retain the α'-structure is probably not more than one oxygen atom per unit cell; attempts to prepare $Ca_2Si_6Al_6O_2N_{14}$ have not been successful.

Typical X-ray photographs of Ca, Li and Y α'-sialons are shown in Fig. 5 from which it is clear that exactly the same product is obtained by starting with β-Si_3N_4 as with α-Si_3N_4. Unlike the β'-sialons, $Si_{6-z}Al_zO_zN_{8-z}$, where the replacement without structure change is Si-N by Al-O, the replacement in α' is largely Si-N by Al-N. With bond lengths Si-N$\sim$1.74Å, Al-O $\sim$ 1.75Å and Al-N $\sim$ 1.87Å, the relative increases in unit-cell dimensions for $\alpha \rightarrow \alpha'$ are much greater than for $\beta \rightarrow \beta'$. For a general composition

$$M_xSi_{12-(m+n)}Al_{(m+n)}O_nN_{16-n}$$

m(Al-N) replace m(Si-N) and n(Al-O) replace n(Si-N). Fig. 6

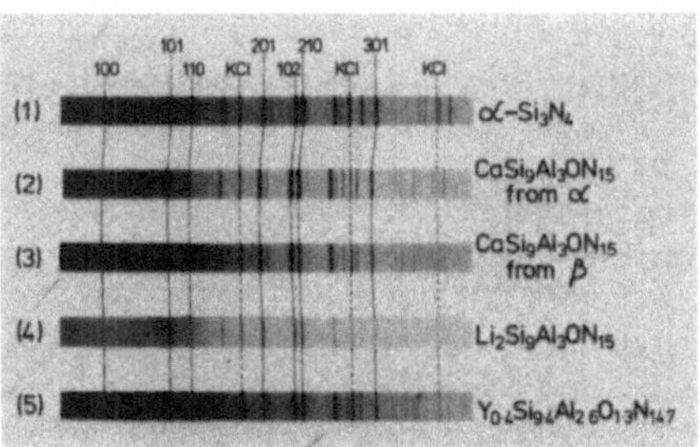

Fig. 5 X-ray photographs of typical Ca, Li and Y α'-sialons.

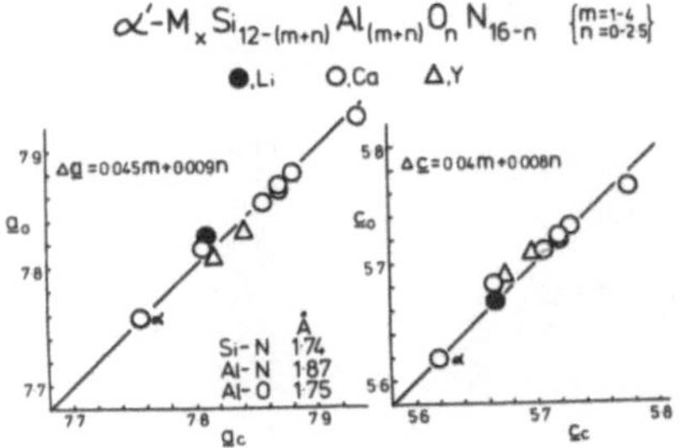

Fig. 6 Observed and calculated unit-cell dimensions for α'-sialons

shows that the cell dimensions fit reasonably with the relationships

$$\Delta \underline{a}(\text{Å}) = 0.045m + 0.009n \qquad .... (1)$$

$$\Delta \underline{c}(\text{Å}) = 0.04m + 0.008n \qquad .... (2)$$

suggesting that the dimensional increases for replacement of (Si-N) by (Al-N) is about five times that for replacement by (Al-O).

3. PROPERTIES OF α'-SIALONS

3.1 Chemical and Thermal

The phase relationships between α' and β' sialons are shown in Fig. 7. As might be expected from these, and as shown by the X-ray photographs of Fig. 8, α' reacts with Al_2O_3 at 1750°C to give β', and β' reacts with AlN plus an appropriate moderator nitride (e.g. Ca_3N_2) to give α'. Equations corresponding to these reactions are:

$$\underset{\alpha'}{Y_{0.4}Si_{9.4}Al_{2.6}O_{1.3}N_{14.7}} + (2.7)Al_2O_3$$

$$\longrightarrow \underset{\beta',\ z = 2.8}{(2.85)Si_{3.2}Al_{2.8}O_{2.8}N_{5.2}} + (0.2)Y_2Si_2O_7 \qquad .... (3)$$

$$\underset{\beta',\ z = 1.2}{(1.45)Si_{4.8}Al_{1.2}O_{1.2}N_{6.8}} + (0.56)Ca_3N_2 + (3.34)AlN$$

$$\longrightarrow \underset{\alpha'}{Ca_{1.2}Si_{7.0}Al_{5.0}O_{1.7}N_{14.3}} \qquad .... (4)$$

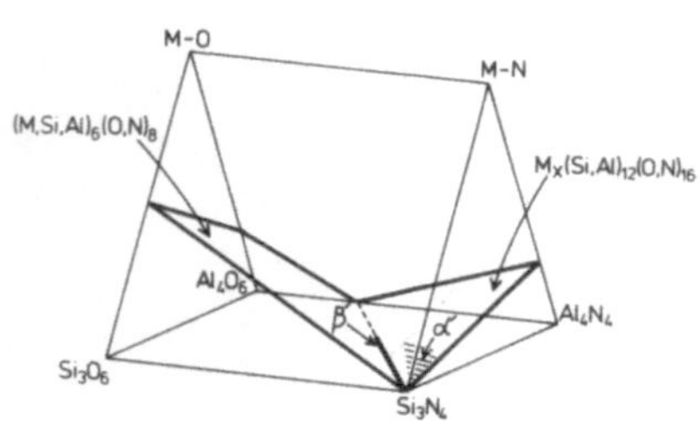

Fig. 7 Phase relationships between α' and β' sialons

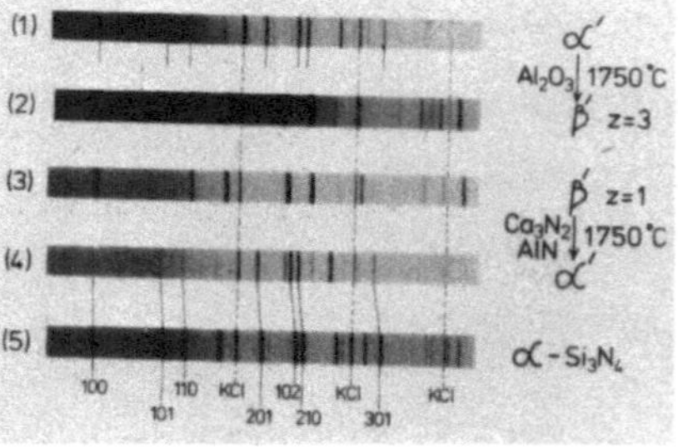

Fig. 8 X-ray photographs showing $\alpha' \rightleftharpoons \beta'$ transformations

Just as β'-sialons can be produced by nitriding mixtures of oxides and carbon with molecular nitrogen, so also can the α'-phases. Using carbon and the mixed oxides finely dispersed by a sol-gel processing route, α'-Ca-sialons have been prepared in this way at temperatures as low as 1400°C.

In inert (argon), nitriding or carburizing atmospheres α'-sialons remain unchanged up to 1750°C. As well as this thermal stability, the thermal expansion is linear over the range 0-1250°C and is almost isotropic with the following low values for the coefficients:

$$\alpha_{\underline{a}} = 3.3 \times 10^{-6}/^{o}C; \quad \alpha_{\underline{c}} = 3.5 \times 10^{-6}/^{o}C$$

3.2 Technological Applications

The ease of preparation, the thermal and chemical stability and the low coefficient of thermal expansion suggest that α' sialons might make useful engineering ceramics. Mitomo et al. (9) have recently prepared fully dense and 98% theoretical density α'-Y-sialon of composition $Y_{0.5}Si_{9.5}Al_{2.5}O_{1.0}N_{14.9}$ by respectively hot pressing and pressureless sintering powder mixtures of Si_3N_4, AlN and Y_2O_3 at 1750°C. These had respective bend strengths of 650 and 450 MN/m^2 at room temperature and that of the hot-pressed material was maintained up to 1000°C. Thermal shock resistance was slightly better than that of hot-pressed β'-sialon. Although this is a preliminary report, the results are promising.

α'-sialons have a potential advantage in that the additive, e.g. Ca, Y or rare earth, that is necessary to provide a high-temperature liquid for densification can subsequently be incorporated into the sialon structure. The phase relationships shown in Fig. 9 suggest the possibility of producing $\alpha' + \beta'$ composites and here again it might be feasible to avoid intergranular glass by a post-preparative heat-treatment that incorporates the oxide additive into the α'.

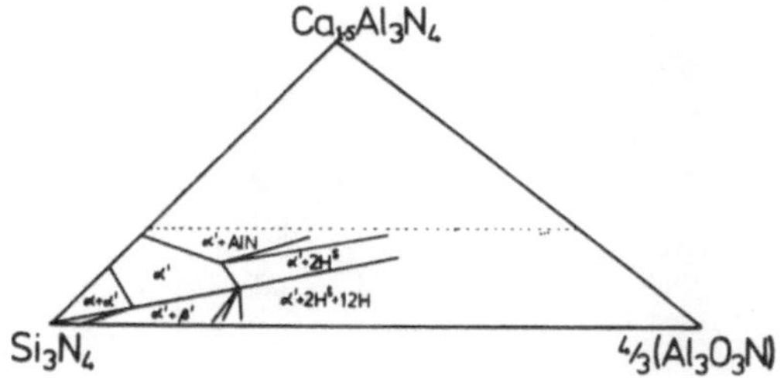

Fig. 9 Tentative phase relationships in part of the Ca-Si-Al-O-N system

3.3 SiC-AlN Solid Solutions

When α'-$CaSi_9Al_3ON_{15}$ is reduced with carbon at 1800°C in nitrogen, the single-phase product has a wurtzite-type 2H structure with unit-cell dimensions corresponding to 3SiC.AlN. It was found unnecessary to use the pre-formed α'-sialon; mixtures of Si_3N_4, AlN and CaO reacted with carbon in the same way and by varying the mix gave a complete series of 2H solid solutions the unit-cell dimensions of which vary smoothly between those of the end members; see Figs. 10 and 11. The CaO is eventually reduced and volatilised as Ca vapour + CO but its addition provides a transient oxynitride liquid in which the reactants are soluble.

Other methods have been used (10) to produce these AlN-SiC solutions which also dissolve other wurtzite-type nitrides, oxy-nitrides and carbo-oxynitrides. The field of sialons has therefore been extended to include M-Si-C-Al-O-N materials.

4. α AND β SILICON NITRIDES

All preparative work on α'-sialons suggests a miscibility gap between α-Si_3N_4 and the α'-phase; see Fig. 9. The α' compositions closest to Si_3N_4 are

$$Ca_{0.5}Si_{10.5}Al_{1.5}O_{0.5}N_{15.5} \quad \ldots. \text{ (i)}$$

and

$$Y_{0.3}Si_{10.5}Al_{1.5}O_{0.5}N_{15.5} \quad \ldots. \text{ (ii)}$$

To stabilise the structure, the equivalent of not less than half a cationic valency ($Ca_{0.25}$ or $Y_{0.16}$) is required in each of the two interstices.

It is suggested that this is also the requirement for α-silicon nitride. Depending on whether or not oxygen is

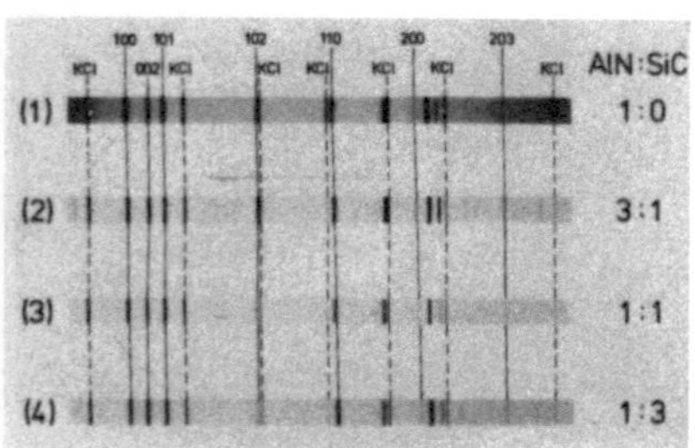

Fig. 10 X-ray photographs of 2H solid solutions AlN-SiC

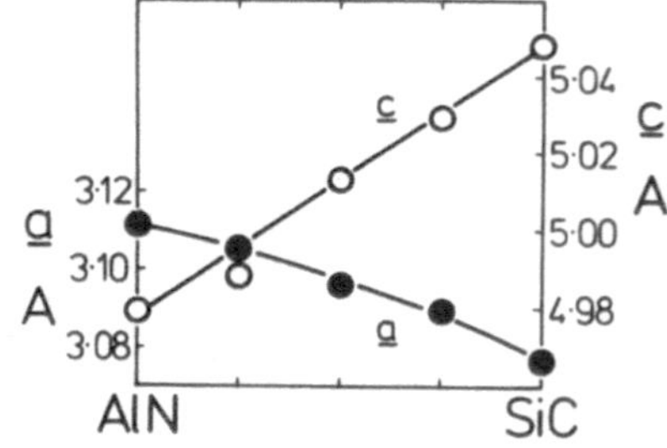

Fig. 11 Unit-cell dimensions of 2H solid solutions AlN-SiC

available for valency compensation, the corresponding two compositions are

$$Si^{3+}_{0.3}\ Si^{4+}_{11.8}\ N^{3-}_{16} \quad \text{.... (iii)}$$

and

$$Si^{3+}_{0.3}\ Si^{4+}_{11.6}\ O^{2-}_{0.5}\ N^{3-}_{15.5} \quad \text{.... (iv)}$$

Si^{4+} is not large enough to be accommodated in the structural interstices and so a lower valency Si^{2+} or Si^{3+} is assumed; the observed length of ~2.2Å for the resultant Si-N bond seems reasonable for Si^{3+}. The partial occupation of the interstitial sites accounts for the structural distortion around them in α as well as in α'. Further, the calculated density for the composition containing oxygen (iv) is in agreement with the density range observed for α needles (7) produced by reaction of silicon monoxide with nitrogen, 3.167 - 3.171 g cm^{-3}, and which cannot be explained by the composition Si_3N_4. Fig. 12 summarises the relationship between α and α' and the differences between non-oxygen and oxygen-containing α-silicon nitride compositions.

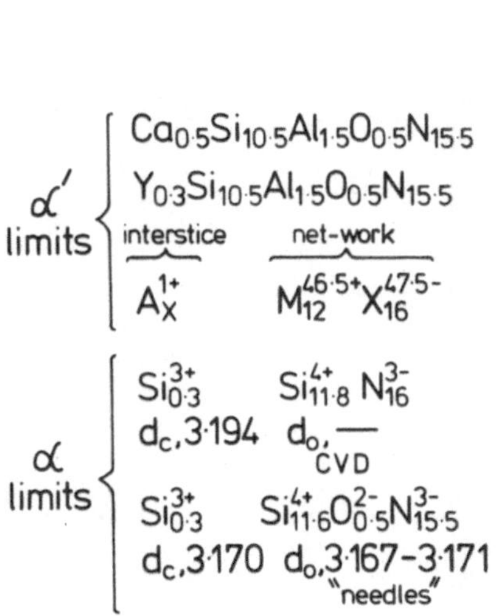

Fig. 12 Comparison of α and α' phases

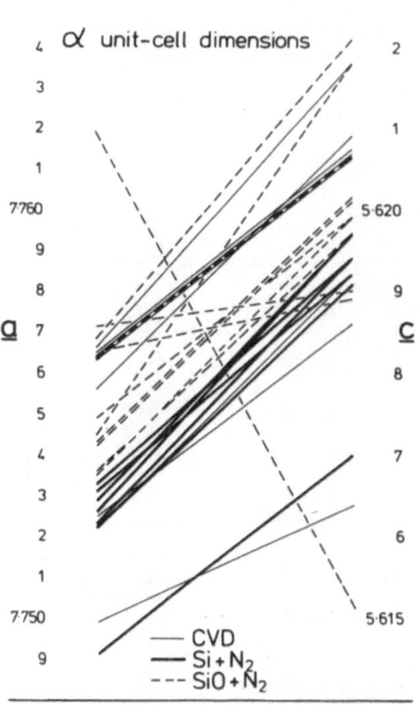

Fig. 13 Unit-cell dimensions of different α-silicon nitrides

4.1 Unit-Cell Dimensions

There is no doubt that α-"Si_3N_4" shows a relatively wide variation in unit-cell dimensions that must be due to a variation in composition. Fig. 13 shows values for 26 different specimens prepared by (i) reacting Si with N_2; (ii) the reaction of SiO with N_2; and (iii) chemical vapour deposition from silicon halides. The same precise method was used in each case by the same investigator (Dr. D.P. Thompson) and the variation in dimensions ($\underline{a}$ = 7.7491 - 7.7572; $\underline{c}$ 5.6164 - 5.6221Å) is well outside the experimental error (± 0.0005Å). The one anomalous value ($\underline{a}$ = 7.7619; $\underline{c}$ = 5.6151Å) is excluded and must be due to dissolved impurity but otherwise the axial ratios $\underline{c}/\underline{a}$ are reasonably constant. In general, reaction (ii) gives higher values than (i) but different specimens of CVD cover the whole range.

4.2 Oxygen-Containing α-"Si_3N_4"

There is also no doubt that α-"Si_3N_4" can accommodate oxygen even though it now seems not to be essential for the stability of the structure. The precipitation of silicon nitrides by nitriding Fe-Si alloys at low temperatures, 500-720°C, showed not only that α and β were not, as previously supposed, merely low and high temperature forms (11) but that pure β was formed at very low oxygen potentials and pure α at higher ones; see Fig. 14.

Furthermore, the extensive thermodynamic investigations of

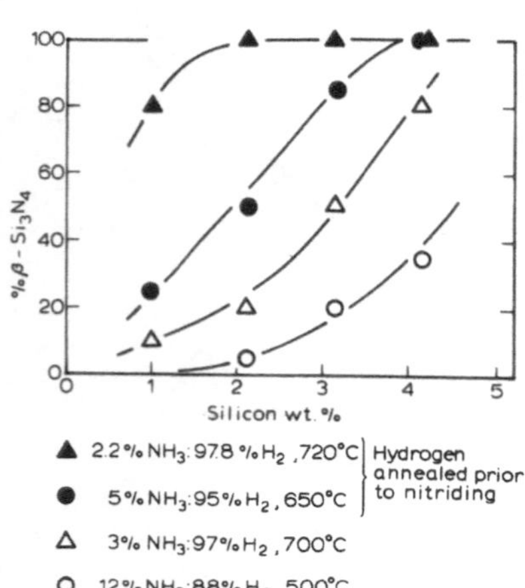

Fig. 14 The precipitation of α & β silicon nitrides from Fe-Si alloys

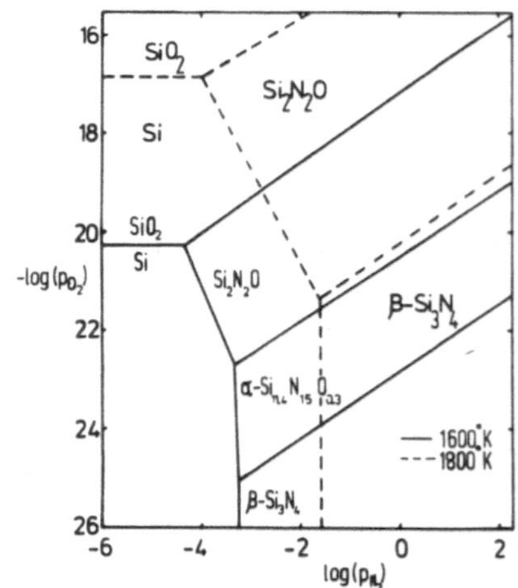

Fig. 15 Thermochemical diagram for the Si-O-N system

the Si-O-N system (12, 13, 14) cannot be ignored. The discrepancy between the work at Newcastle (13) and Trondheim (14) was completely resolved (15) when it was realised that different values for the activity coefficients of silicon in iron had been used in calculating free energies. All results are completely consistent and mutually supporting and can be summarised by the thermochemical diagram of Fig. 15. This shows that α (containing oxygen) becomes unstable with respect to β-Si_3N_4 + Si_2N_2O at $\sim$1400°C and $p_{O_2} \sim 10^{-20}$ atm.

It should be noted that α-silicon nitride formed in absence of oxygen, e.g. CVD silicon nitride of composition (iii), is not necessarily unstable with respect to β until temperatures much higher than 1400°C.

4.3 The α/β Silicon Nitride Question

Morgan (16) has suggested that seeding with pre-formed α or β is the main determiner of the modification that is produced by nitriding silicon, but this does not indicate whether the two crystalline forms - if they are both Si_3N_4 - are monotropic or enantiotropic. Even if equilibrium seldom exists in nitriding systems it is useful to know the directions of driving forces. Morgan also emphasises that α and β are soluble in liquid silicon - a feature implied by the earlier thermodynamic investigations - but the coexistence of two solid phases with liquid silicon and nitrogen gas makes the two-component Si-N system invariant. Except at one specific temperature and pressure, one phase must disappear unless another component (e.g. oxygen) is introduced.

4.4 CVD Silicon Nitride

Silicon nitride deposited on graphite by reaction of silicon halides with ammonia at 1200 - 1500°C is amorphous or crystalline and varies in colour (white, brown, purple, black); the crystalline deposits are invariably α. Oxygen contents of CVDα are often lower than required by composition (iv) and the dense material is remarkably stable. Thus, although Messier and Riley (17) observed some transformation to β with addition of MgO at 1600°C, Hampshire (18) observed none even at 1800°C.

The transformation $\alpha \rightleftharpoons \beta$ is a reconstructive one involving the breaking and reforming of Si-N bonds and requiring either an intermediate vapour phase or a solvent. Dense CVDα is regarded as the purest silicon nitride and if it contains no oxygen or other impurities that might provide liquid, it might be expected to be highly resistant to transformation.

Recently, Dr. P. Korgul at Newcastle has examined white and black CVDα supplied by Dr. F. Galasso (United Technologies

Research Center) and prepared by deposition on graphite from SiF_4 at 1500°C (19). Even the "white" silicon nitride contains an appreciable density of inclusions when examined by electron microscopy; Fig. 16. Black CVD contains a higher concentration which at higher magnifications (see Fig. 17) are shown to be thin discs approximately 250Å average diameter parallel with (00.1) planes and about $2d_{(00.1)}$ = 11.3Å thick. In addition, black CVD contains needle-shaped bubbles lying in (hk.0) planes and about $2d_{(10.0)}$ = 13Å in diameter; see Fig. 18.

The disc-shaped precipitates produce large matrix strains. Heat-treatment of white CVDα in nitrogen in a graphite furnace at 1725°C causes coalescence within each disc and it is then possible to show that the precipitates are amorphous. After 1h at 1850°C, about 20% conversion to β takes place and examination of the β crystals shows that they contain no precipitates or other inclusions. It seems likely that the amorphous precipitates in CVDα are silica and that at 1850°C $\alpha \rightarrow \beta$ transformation occurs by solution and recrystallisation from a small amount of silica liquid.

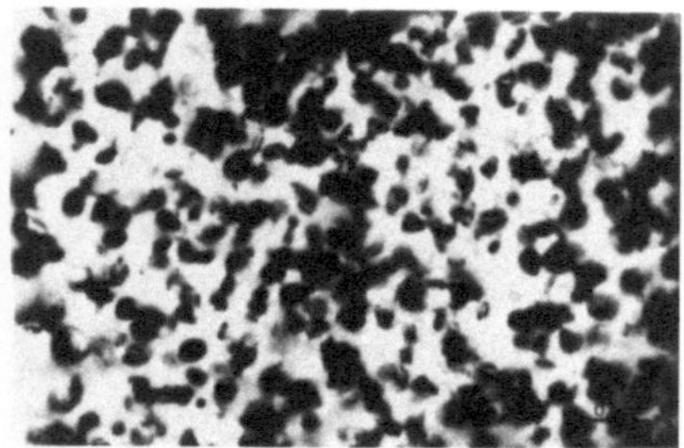

Fig. 16 White CVDα silicon nitride (x 50,000)

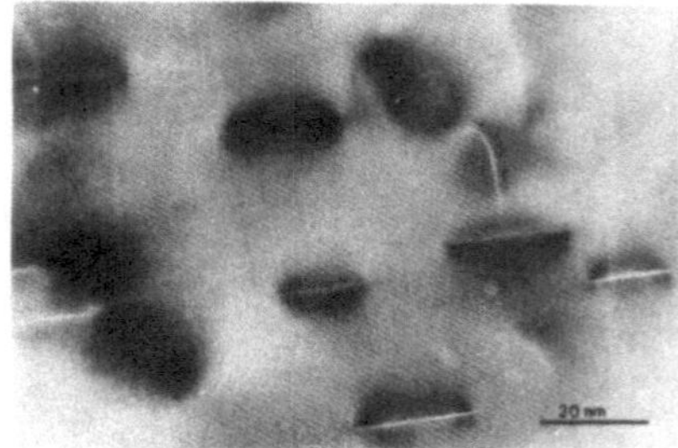

Fig. 17 Black CVDα (x 400,000) showing disc-shaped precipitates on (00.1)

Fig. 18 Black CVDα (x 400,000) showing needle-bubbles on (hk.0)

4.5 The Control of α:β Phase Composition

By analogy with α', it is suggested that lower-valency Si occupying interstitial sites is necessary for the existence of the α structure. At temperatures below 1400°C this is possible by valency compensation with oxygen.

It has further been suggested by Dr. H. Priest, AMMRC, Watertown (20) that control of phase composition to produce a high α content is based on the presence of divalent Si compounds. H_2, O_2 or H_2O vapour in the nitriding gas that are known to promote α formation will produce divalent Si in the form of SiO, while in the formation of CVDα at 1500°C the gas contains at least 50% SiF_2 or $SiCl_2$.

It seems possible that this suggestion by Dr. Priest can be reconciled with the structural occurrence in α of Si^{2+} or Si^{3+}.

REFERENCES

1. Jack, K.H. and Wilson, W.I. *Nature Phys. Sci. 238* (1972) 28.

2. Jama, S.A.B., Thompson, D.P. and Jack, K.H. in P. Popper ed. *Special Ceramics 6* (Stoke-on-Trent, B.C.R.A., 1975), p. 299.

3. Mitomo, M. *Yogyo-Kyokai-Shi 85* (1977) 50.

4. Hampshire, S., Park, H.K., Thompson, D.P. and Jack, K.H. *Nature 274* (1978) 880.

5. Park, H.K., Thompson, D.P. and Jack, K.H. *Science of Ceramics 10* (1980) 251.

6. Grand, G., Demit, J., Ruste, J. and Torre, J.P. *J. Mat. Sci. 14* (1979) 1749.

7. Wild, S., Grieveson, P. and Jack, K.H. in P. Popper ed. *Special Ceramics 5* (Stoke-on-Trent, B.C.R.A., 1972), p. 385.

8. Thompson, D.P. and Patience, M. Unpublished research at Newcastle.

9. Mitomo, M., Tanaka, H., Muramatsu, K., Ii, N. and Fujii, Y. Private communication; to be published in *J. Mat. Sci.*

10. Cutler, I.B., Miller, P.D., Rafaniello, W., Park, H.K., Thompson, D.P. and Jack, K.H. *Nature 275* (1978) 434.

11. Roberts, W., Grieveson, P. and Jack, K.H. *J. Iron & Steel Inst.* (1972) 931.

12. Wild, S., Grieveson, P. and Jack, K.H. in P. Popper ed. *Special Ceramics 5* (Stoke-on-Trent, B.C.R.A., 1972) p. 271.

13. Colquhoun, I., Wild, S., Grieveson, P. and Jack, K.H. *Proc. Brit. Ceram. Soc. 22* (1973) 207.

14. Blegen, K. in P. Popper ed. *Special Ceramics 6* (Stoke-on-Trent, B.C.R.A., 1975) p. 223.

15. Hendry, A. in F.L. Riley ed. *Nitrogen Ceramics* (Leyden, Noordhoff, 1977) p. 183.

16. Morgan, P.E.D. *J. Mat. Sci. 15* (1980) 791.

17. Messier, D.R. and Riley, F.L. in F.L. Riley ed. *Nitrogen Ceramics* (Leyden, Noordhoff, 1977) p. 141.

18. Hampshire, S. Ph.D. Thesis, University of Newcastle upon Tyne (1980).

19. Galasso, F., Kuntz, U. and Croft, W.J. *J. Amer. Ceram. Soc. 55* (1972) 431.

20. Priest, H. Private communication, unpublished.

DISCUSSION

Lange: What is the lower limit of Si^{2+} in the α-Si_3N_4 structure?

Jack: In the α'-Si_4N_4 structure there is 1^+ valency per unit cell, i.e. 0.5 Ca^{2+} per unit cell or 0.25 Ca^{2+} per interstice. Similarly, 0.3 Y^{3+} per unit or 0.16 Y^{3+} per interstice. In the α-Si_3N_4 structure this means, by analogy, 0.25 Si^{2+}, or 0.16 Si^{3+3} per interstice.

Morgan: It seems to me very plausible that Si^{2+} can enter and stabilize the growing crystal. But at room temperature the Si^{2+} might convert to another species.

Jack: Even so, the structure stabilized at high temperature would be maintained, because of the reconstructive nature of the transformation. Your suggestion implies that it might be extremely difficult to prove the existence or effect of Si^{2+}.

Clarke: How do you know that the platelets are amorphous SiO_2?

Jack: On heat treatment the platelets coalesce and within the coalesced region there is no change in contrast on tilting. It is not possible to be unequivocal, and it is extremely difficult to be sure that these inclusions are amorphous SiO_2.

THE STRUCTURAL CHARACTERISATION OF SIALON POLYTYPOIDS

D.P. Thompson, P. Korgul and A. Hendry

Wolfson Research Group for High-Strength Materials,
Crystallography Laboratory,
The University of Newcastle upon Tyne, UK.

ABSTRACT. The six polytypoid phases in the Si-Al-O-N system (8H, 15R, 12H, 21R, 27R, 2Hδ) form a series of structurally similar compounds, analogues of which occur in most metal sialon systems. The structures are determined by their metal:non-metal atom-ratio M:X which for the above series is of the type mM:(m + 1)X where m has the values 4, 5, 6, 7, 9 and 11 respectively. Beryllium, magnesium, scandium and possibly other metal cations can be incorporated into these structures provided that charge balance is preserved and that the overall M:X ratio is retained.

Compounds with an M:X ratio greater than unity have identical structures but with metal and non-metal atoms reversed; such antitypes are observed in the Be-Si-O-N, Al-N-C and Al-Si-C systems. In the Mg-Si-Al-O-N system, two other series of compounds occur; Mg_4N_2O and Mg_4SiN_4 have layer-structures related to Mg_3N_2, and three phases 6H, 14H and 8H form a polytypoid series at MgO-rich compositions in the 6M:7X, 7M:8X and 8M:9X planes respectively.

The structures of the sialon polytypoids have been determined by X-ray powder methods and consist of a wurtzite sequence of tetrahedra broken up by single layers of octahedra inserted every mth layer. Half way between these layers, metal atoms are shared between two adjacent layers of tetrahedra to give an MX_2 unit which reduces the M:X ratio below unity. In the 6H, 14H, 8H series, octahedra occur as double layers in an otherwise similar arrangement. Direct observation of the structures by high resolution electron microscopy gives results in excellent agreement with those obtained by calculation from the X-ray structures.

Riley, F.L. (ed.) Progress in Nitrogen Ceramics

1. INTRODUCTION

The six structurally similar phases which occur near the aluminium nitride corner of the Si_3N_4-SiO_2-Al_2O_3-AlN system and extend into other metal sialon systems have been variously described in the literature. Early work at Newcastle assigned the letters P, Y, M, Q, T, R to these phases (Jack (1)) whilst Gauckler et al. (2) called them X_4, X_2, X_5, X_6, X_7 (X_7 = T and R); after structural characterisation (see Jack (3)) they were described as "polytypes" with Ramsdell symbols 8H, 15R, 12H, 21R, 27R, $2H^{\delta}$. In the strictest sense, polytypes are layer structures which, within very narrow limits, preserve a constant chemical composition and differ only in the way the layers are stacked together. The repeat distances within each layer are identical for every member of the series but perpendicular to the layer the structure repeats after different multiples of the interlayer separation. Common examples of polytypic materials are silicon carbide, zinc sulphide and cadmium iodide (see, for example, Verma & Krishna (4)). In the sialon "polytypes", both the composition and the unit cell dimensions show a systematic variation with structure type and the International Union of Crystallography has assigned the name "polytypoids" to such series (5). They are quite common in inorganic chemistry, other examples being β-alumina structures (6), barium ferrites (7) and several sulphide systems (8).

In addition to the Si-Al-O-N polytypoids mentioned above, there are other series of layer structures which occur in sialon systems. In this paper, the structures of all these compounds are described and the relationship between structure and composition is discussed.

2. CHARACTERISATION

The first sialon layer structures to be characterised were those in the Si_3N_4-SiO_2-Al_2O_3-AlN system (Figure 1). The unit cells are hexagonal or rhombohedral with $\underline{a}$, 3Å and $\underline{c}$, large and widely varying (see Table 1). In polytype structures, the number of layers per

Table 1. Unit cell dimensions for the sialon polytypoids

Type	M:X	No. of layers per block	$\underline{a}$	$\underline{c}$	$\underline{c/n}$
8H	4:5	4	2.988	23.02	2.88
15R	5:6	5	3.010	41.81	2.79
12H	6:7	6	3.029	32.91	2.74
21R	7:8	7	3.048	57.19	2.72
27R	9:10	9	3.059	71.98	2.67
$2H^{\delta}$	11:12	11	3.079	5.30	2.65
2H	1:1	1	3.114	4.986	2.49

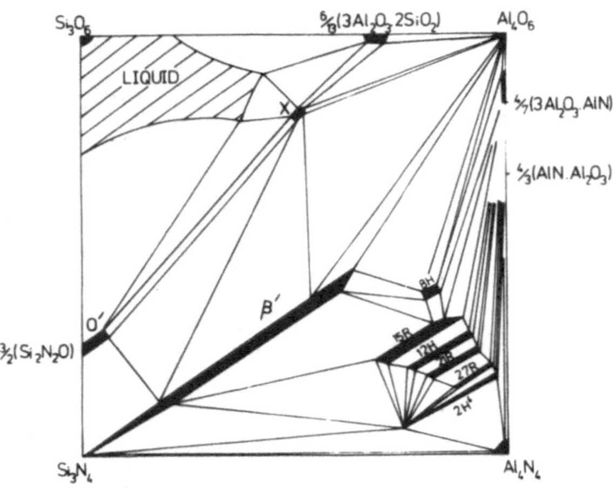

Fig. 1 Phase relationships in the Si_3N_4-SiO_2-Al_2O_3-AlN system at 1700°C

c repeat distance (and hence the Ramsdell symbol) is obtained from the l value of the first strong 00l reflection, the d spacing of which gives the interlayer spacing, c/n. Table 1 shows that this decreases systematically with increasing AlN content. Absent reflections in the X-ray patterns of the hexagonal phases are due to a c glide plane of symmetry which relates the upper and lower halves of the unit cell; in a similar way the rhombohedral phases have absent reflections corresponding to the unit cell being split up into three symmetry-related blocks of structure. The number of layers per block in each phase can therefore be evaluated from the Ramsdell symbol and Table 1 shows that this increases by one from one member of the series to the next. Note that a 16H sialon which should occur between 21R and 27R has not been observed.

From their positions in the phase diagram, the sialon polytypoid phases must either accommodate additional non-metal atoms or be deficient in metal atoms relative to an MX wurtzite composition. Thus the five layers in each block of 15R must consist either of four MX plus one MX_2 layers (five metal atom layers) or three MX plus one MX_2 layers (five non-metal atom layers). Density measurements show conclusively that the first alternative is correct. The fundamental block of structure in each member of the series has the composition M_mX_{m+1}, where m takes the values, 4, 5, 6, 7, 9 and 11 as shown in Table 1. Note that preparative evidence alone is not sufficient to determine the composition. At the high temperatures needed for the preparation of these compounds, weight losses due to silicon monoxide and nitrogen are sufficient

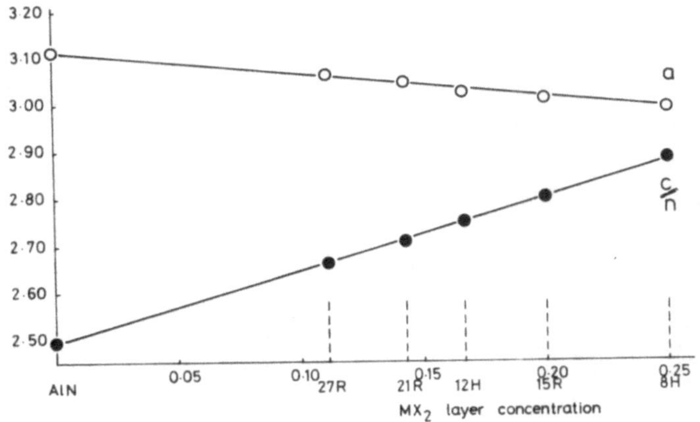

Fig. 2 Graph showing the variation of a and c/n with the proportion of MX_2 layers in sialon polytypoid phases

to move the overall composition from one phase to the next.

The additional non-metal atom in the MX_2 layer increases the thickness of this layer so that the average interlayer spacing c/n increases with frequency of MX_2 layers as shown in Figure 2. The spacing corresponding to zero MX_2 layers, 2.49Å, agrees exactly with the interlayer separation in aluminium nitride. From the slope of the graph, the MX_2 layer is 1.5Å thicker than an MX layer, or alternatively 1.0Å thinner than two MX layers. This latter way of expressing the result is more useful because, as discussed in later sections, the structures are more satisfactorily described in terms of layers of non-metal atoms with occasional metal atoms omitted.

The X-ray pattern of the $2H^{\delta}$ phase has only the lines of aluminium nitride (2H), with 001 and hk0 reflections sharp and hkl reflections broadened. Electron micrographs show extensive faulting on the basal plane. The c/n value corresponds to a 33R structure but the absence of additional reflections and the broadening of the lines show that MX_2 layers are incorporated randomly with an average separation of eleven layers. Compositions richer in AlN than $2H^{\delta}$ occur as two-phase mixtures of AlN and $2H^{\delta}$ rather than as longer-period polytypes.

At higher temperatures the range of homogeneity of the polytypoids increases and extends to the Al_2O_3-AlN join. Collongues et al. (9) reported an "X" phase which indexes primarily as 21R and Gauckler (10) gave a phase diagram for the Al_2O_3-AlN system which included both 21R and 27R compounds. Sakai (11) prepared a

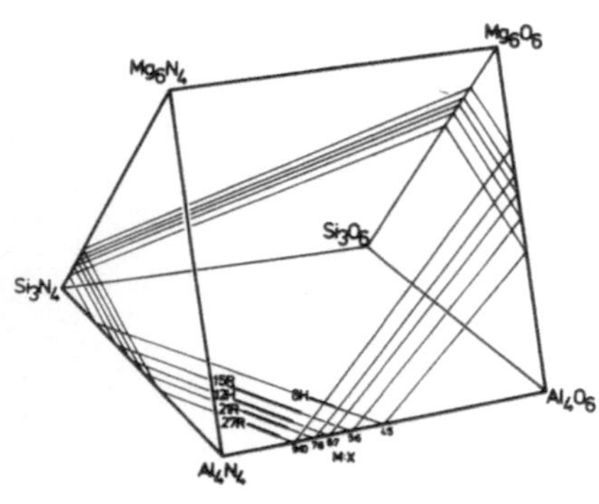

Fig. 3 Planes of constant M:X ratio in the magnesium sialon system

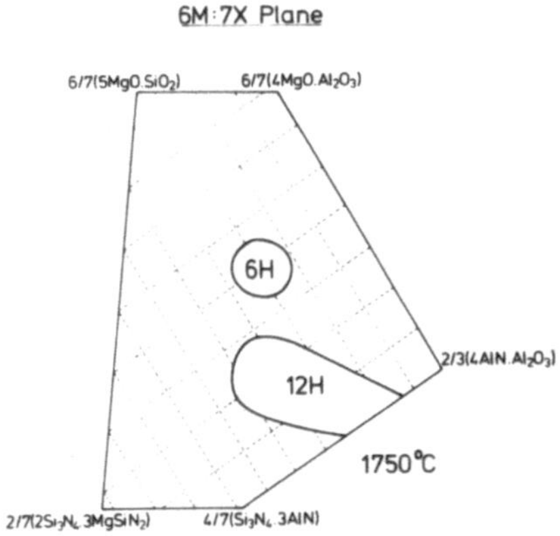

Fig. 4 Phase relationships in the 6M:7X plane of the magnesium sialon system

16H aluminium oxynitride, the first known occurrence of this structure and Bartram & Slack (12) reported 27R and 20H compounds. The most comprehensive work on phase relationships in this system is that of McCaulay (13).

Other cations can be incorporated into polytypoid phases provided the overall metal:non-metal ratio is retained and electrical neutrality is preserved. The additional cation extends

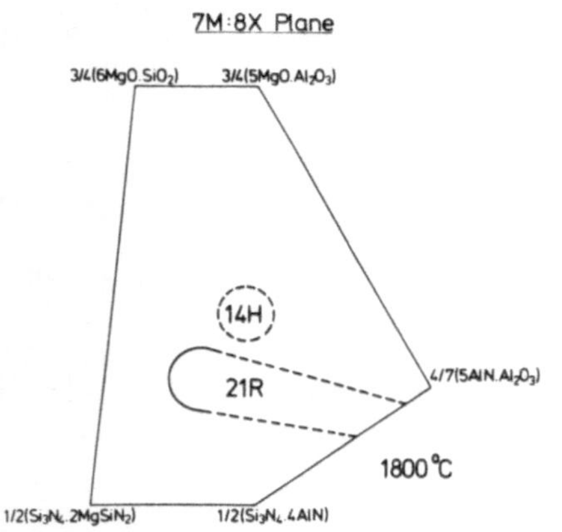

Fig. 5 Phase relationships in the 7M:8X plane of the magnesium sialon system

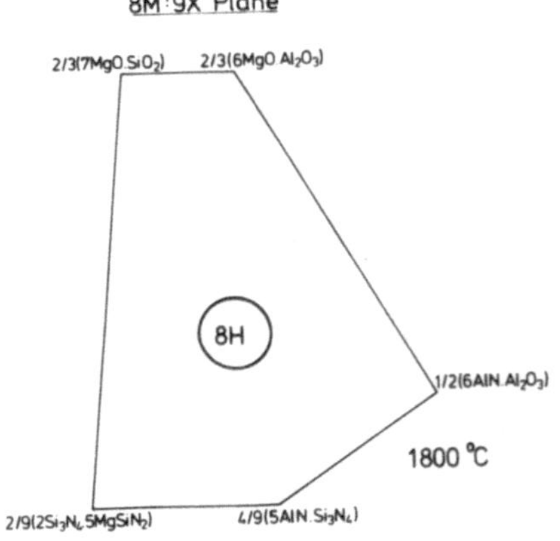

Fig. 6 Phase relationships in the 8M:9X plane of the magnesium sialon system

Table 2. Unit cell dimensions for 6H, 14H and 8H magnesium sialons

Type	M:X	a	c
6H	6:7	3.061	16.33
14H	7:8	3.070	37.52
8H	8:9	3.083	21.15

the range of homogeneity from a line to a plane of constant M:X ratio (see Figure 3). Figures 4 and 5 show the extent of 12H and 21R formation in the magnesium sialon system; the other polytypoid phases accommodate less magnesium (see Buang (14)). Beryllium, lithium (15) and scandium (16) can also be incorporated into these structures. Figures 4, 5 and 6 show a different series of polytypoid phases occurring at more magnesium- and oxygen-rich compositions in the magnesium sialon system. They are described by the Ramsdell symbols 6H (6M:7X), 14H (7M:8X) and 8H (8M:9X); unit cell dimensions are given in Table 2.

Aluminium carbonitrides Al_5C_3N, $Al_6C_3N_2$, $Al_7C_3N_3$ and $Al_8C_3N_4$ (17, 18, 19) have structures similar to 8H, 15R and 12H sialons but with metal and non-metal atoms reversed. The two aluminium silicon carbides Al_4SiC_4 and $Al_4Si_2C_5$ (20) are also similar (see Table 3). Phase relationships in the Be_3N_2-Si_3N_4-SiO_2-BeO system (Figure 7) were studied by Huseby et al. (21), Thompson (22) and Thompson & Gauckler (23) who showed that all the sialon polytypoid structures plus a 9R of composition M_4X_3 and 4H β-beryllium nitride (M_3X_2) occurred with very extensive ranges of homogeneity along lines of M:X ratio greater than unity. The

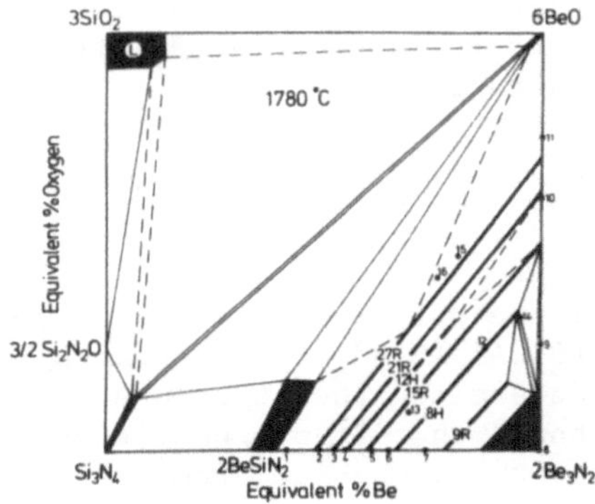

Fig. 7 Phase relationships in the Si_3N_4-SiO_2-BeO-Be_3N_2 system at 1780°C.

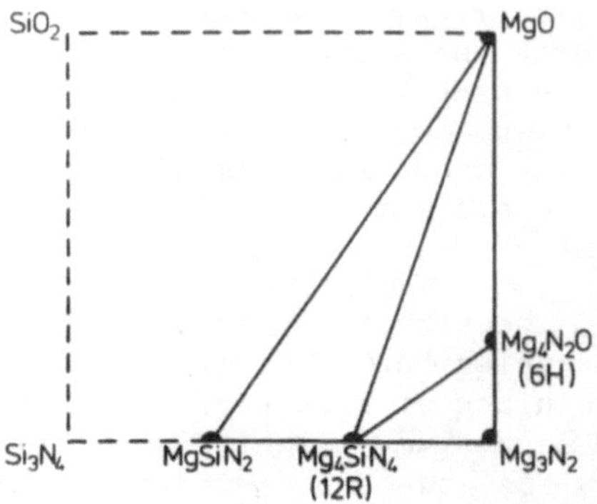

Fig. 8 Phase relationships in the MgO-$MgSiN_2$-Mg_3N_2 part of the Mg-Si-O-N system

Table 3. Unit cell dimensions of Al-C-N and Al-Si-C phases (after (18, 19, 20))

Compound	Type	M:X	a	c
Al_5C_3N	8H	5:4	3.281	21.67
$Al_6C_3N_2$	15R	6:5	3.248	40.03
$Al_7C_3N_3$	12H	7:6	3.226	31.70
$Al_8C_3N_4$	21R	8:7	3.211	55.08
Al_4SiC_4	8H	5:4	3.277	21.68
$Al_4Si_2C_5$	15R	6:5	3.251	40.11

Mg-Si-O-N system is completely different (Figure 8) and only two polytypoid phases occur; Mg_4N_2O has a 6H structure and Mg_4SiN_4 a 12R structure.

3. CRYSTAL STRUCTURES

Structure determinations have been carried out on several phases using X-ray powder methods and the results compared with direct observation of the structures by high resolution electron microscopy.

The structures of 8H and 15R sialons were reported by Thompson (24) (see Figure 9) and consist of layers of $(Si,Al)(O,N)_4$ tetrahedra modified by two additional structural features. In each block of structure one layer of aluminium atoms occupies 6-fold coordinated sites and half way between these layers metal atoms are shared between two adjacent tetrahedral sites. In 8H only one layer of metal atoms is shared but in 15R the symmetry requires two layers to be shared. This arrangement provides the MX_2 layer which reduces the overall M:X ratio below unity and also inverts the sequence of tetrahedra so that it is back in register at the next layer of octahedra. Both structural features are necessary for the structure to occur, but the fundamental reason for the deviation from aluminium nitride stacking is the increased oxygen content which allows 6-fold coordination of aluminium by oxygen.

The structures of other members of the series follow the same principles but incorporate additional $(Si,Al)(O,N)_4$ tetrahedra in each block of structure. Buang (14) determined the structure of 12H magnesium sialon (Figure 11) and showed that it was consistent with the scheme observed in 8H and 15R; the magnesium atoms occupy both 4-fold and 6-fold coordinated sites.

No X-ray structure determinations have been carried out on antitypic Be-Si-O-N compounds. Clarke, Shaw & Thompson (25) interpreted lattice images of 15R and 12H beryllium silicon nitrides

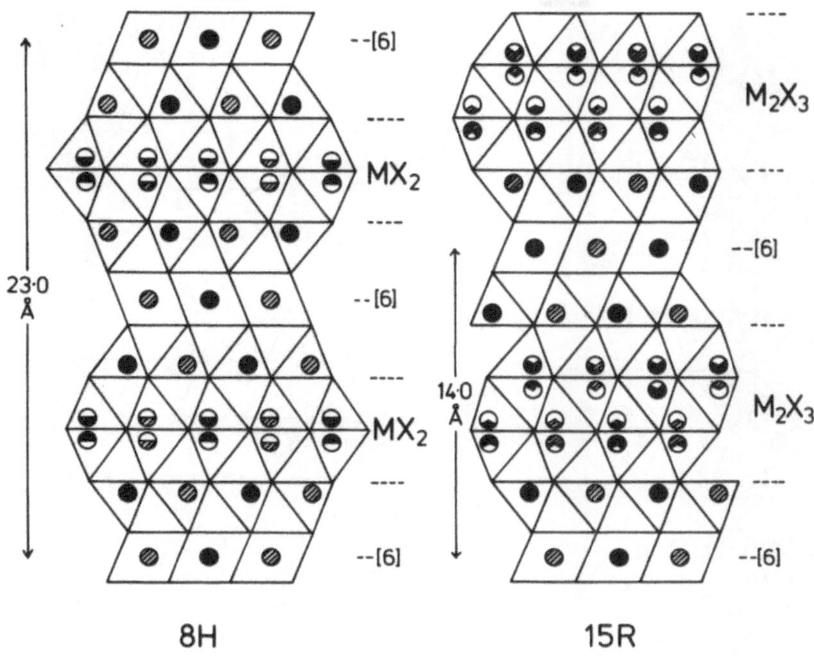

Fig. 9 The atomic arrangement in 8H and 15R sialons. The structures are projected on (110) with metal atoms shown as circles. Shaded atoms are at height ½ above the plane of the paper

in terms of three blocks of five and two blocks of six layers respectively and Shaw & Thomas (26) proposed that the structures of all the compounds in the series were made up of different numbers of $BeSiN_2$ and Be_3N_2 units joined together. Further work is needed to confirm this. In 6H Mg_4N_2O (Figure 10) the non-metals are stacked in a predominantly cubic close-packed sequence with all the magnesium atoms in 4-fold coordination and a well-defined M_2X layer. Note that the non-metal atoms associated with this layer are in distorted 5-fold and 7-fold coordination and therefore antitypes based on this structure are not observed. A diagram of the expected structure of 12R Mg_4SiN_4 is shown in Figure 10 assuming that it incorporates additional layers of SiN_4 tetrahedra into each block of the 6H structure; this still requires confirmation by X-ray diffraction.

The stacking sequence in 6H magnesium sialon is similar to that in the sialon polytypoid series but two adjacent layers of octahedra are incorporated into each block of structure. In Figure 11

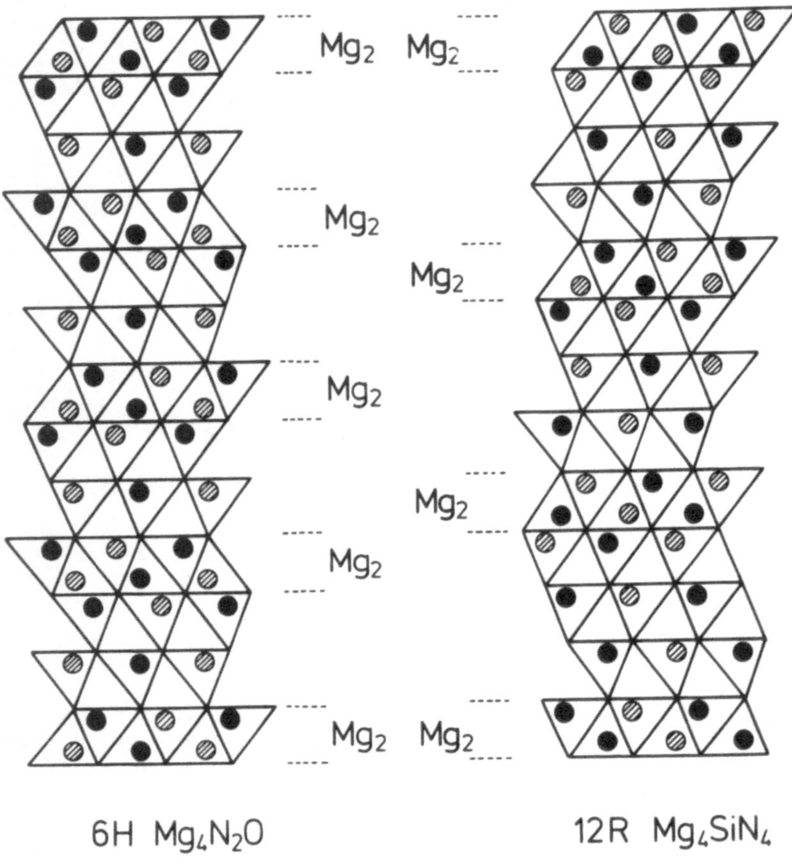

Fig. 10 The atomic arrangement in 6H magnesium oxynitride and 12R magnesium silicon nitride. Symbols as for Figure 9.

two unit cells of 6H are compared with one cell of 12H magnesium sialon (Buang (14)) which contains only one layer of octahedra per block. Again, metal atoms are shared between adjacent tetrahedra half way between layers of octahedra. The double layers of octahedra arise because the composition of 6H is richer in magnesium and oxygen than 12H and magnesium prefers to be 6 - fold coordinated by oxygen. The structures of 14H and 8H, which occur in the same series as 6H, have not been determined but by analogy with other series will be similar to 6H but with one and two additional layers of tetrahedra per block respectively.

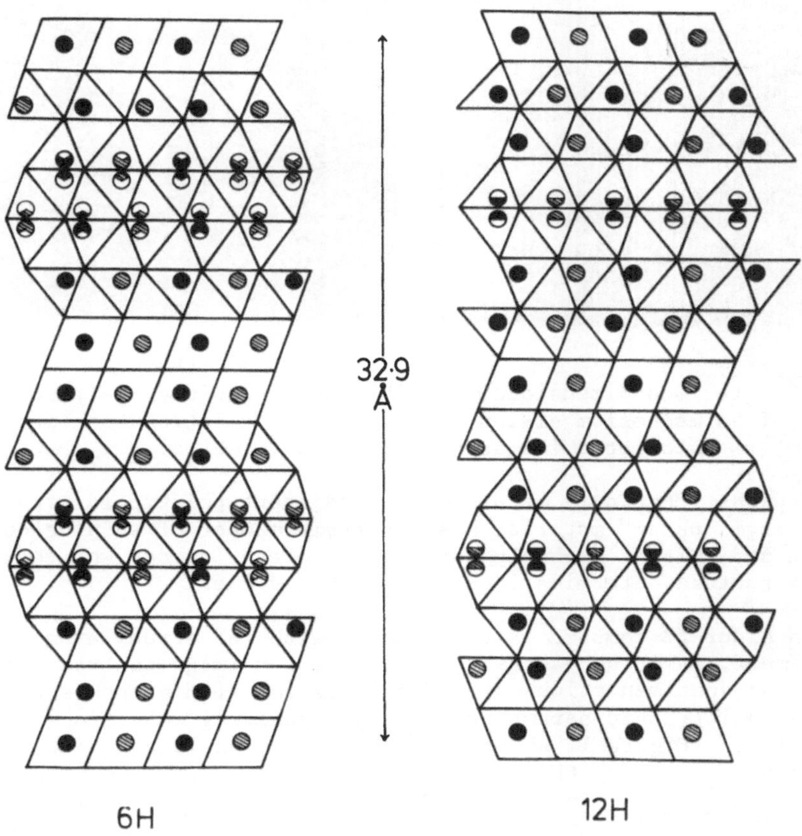

Fig. 11 The atomic arrangement in 6H and 12H magnesium sialons. Symbols as for Figure 9.

4. LATTICE IMAGING

The structures discussed in the previous section were all determined by X-ray powder methods because single crystals were not available. The R indices after refinement were never lower than 6% and were sometimes higher than 10%. Similar high R indices were also found for β-beryllium nitride (27) and the aluminium carbonitrides (18, 19). In such circumstances it is desirable to obtain independent confirmation of the structures and lattice (or structure) imaging is a useful technique for this purpose. Lattice imaging can be used to observe fine-scale structural defects and Clarke, Shaw & Thompson (25), Clarke & Shaw (28) and Hendry & Johnson (29) have used this method to show that

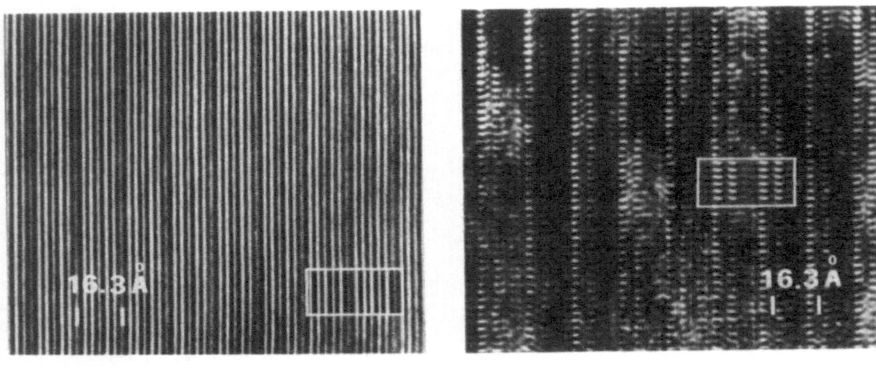

(a) (b)

Fig. 12 (a) One-dimensional and (b) two-dimensional lattice images of 6H magnesium sialon taken with the electron beam parallel to [010]. Calculated images are outlined in white.

intergrowths frequently occur between adjacent polytypoid structures. Even "single-phase" material has a certain number of such defect regions and this is probably the most important reason why the X-ray structure determinations give high R values.

In order to compare lattice images with the actual atomic arrangement, it is necessary to calculate the image contrast for a particular specimen thickness and orientation from a knowledge of the crystallographic parameters and the operating conditions of the microscope. Figure 12 shows one- and two-dimensional observed and calculated images for 6H magnesium sialon taken with the beam parallel to [010]. In both cases the agreement is excellent. Figure 13 shows similar comparisons for 12H magnesium sialon and

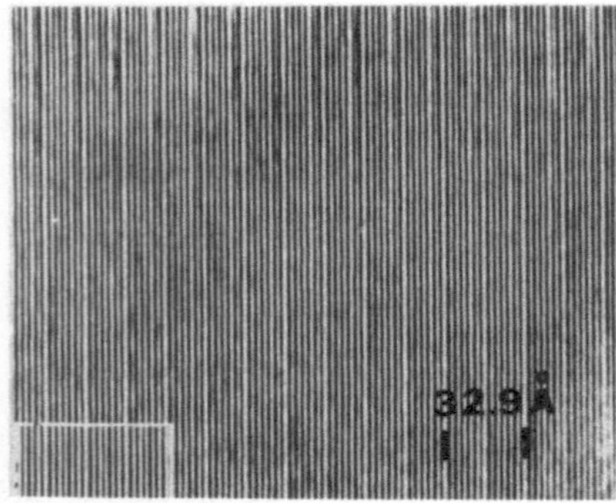

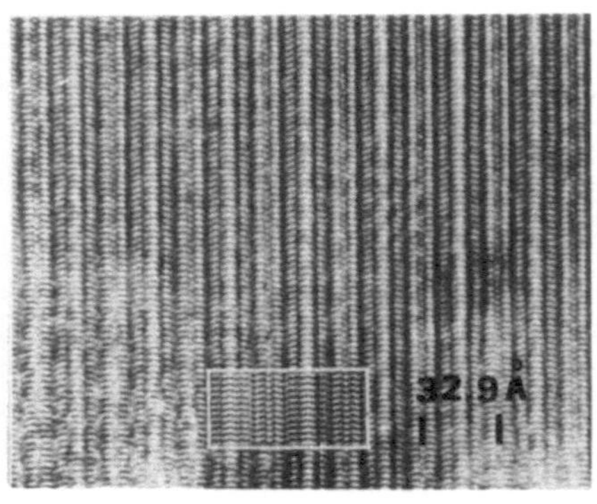

(a) (b)

Fig. 13 (a) One-dimensional and (b) two-dimensional lattice images of 12H magnesium sialon taken with the electron beam parallel to [010] Calculated images are outlined in white.

again the agreement between observed and calculated images supports the correctness of the X-ray structure. The technique is not sufficiently sensitive to determine metal ordering or local variations in the occupation scheme in the MX_2 layer since the scattering factors of magnesium, aluminium and silicon for electrons are all similar and furthermore the specimen thickness corresponds to several unit cells superimposed on top of one another and this averages out variations from cell to cell.

5. CONCLUSION

Figure 14 summarises the polytypoid series so far observed in sialon systems. A combination of X-ray methods and high resolution electron microscopy has been used to characterise these materials and their structures are now established. No detailed evaluation of the ceramic properties of any of the polytypoid phases has yet been attempted. Further work is needed in this area before their usefulness as engineering ceramics can be determined.

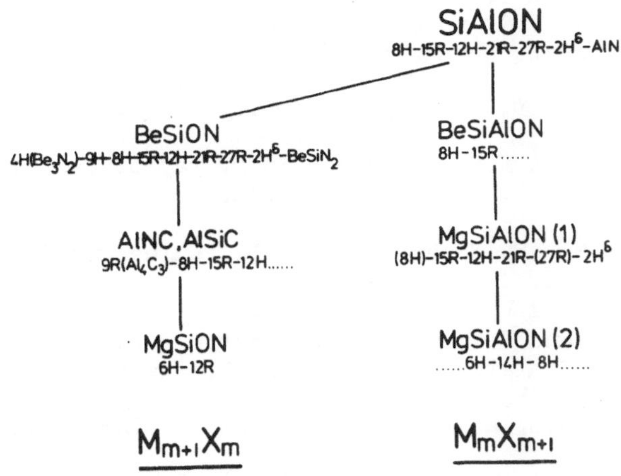

Fig. 14 Summary of the range of sialon polytypoid structures

ACKNOWLEDGEMENTS

We wish to thank Professor K.H. Jack for advice and encouragement and numerous members of the Wolfson Research Group whose work has been included in this paper. Particularly significant contributions have been made by Dr. G. Singh, Dr. K.B. Buang, Dr. P.H.A. Roebuck and Miss K. Liddell.

REFERENCES

1. Jack, K.H. Trans. Brit. Ceram. Soc. 72 (1973) 376.
2. Gauckler, L.J., H.L. Lukas and G. Petzow. J. Amer. Ceram. Soc. 58 (1975) 346.
3. Jack, K.H. J. Mat. Sci. 11 (1976) 1135.
4. Verma, A.R. and P. Krishna. Polymorphism and polytypism in crystals, New York, Wiley 1966.
5. Report of the International Mineralogical Association and the International Union of Crystallography Joint Committee on Nomenclature, Acta Cryst. A33 (1977) 681.
6. Dyson, D.J. and W. Johnson. Trans. J. Brit. Ceram. Soc. 72 (1973) 49.
7. Kohn, J.A. and D.W. Eckart. Zeit. Krist. 119 (1964) 454.
8. Kato, K., N. Morimoto and Y. Gyobu. Acta Cryst. B31 (1975) 2757.
9. Collongues, R., J.C. Gilles, A.M. Lejus, M. Perez y Jorba and D. Michel. Mat. Res. Bull. 2 (1967) 837.
10. Gauckler, L.J. Ph.D. Thesis, University of Stuttgart (1976).
11. Sakai, T. Yogyo-Kyokai-Shi 86 (1978) 125.
12. Bartram, S.F. and G.A. Slack. Acta Cryst. B35 (1979) 2281.
13. McCauley, J.W. This volume, p.111.
14. Buang, K.B. Ph.D. Thesis, University of Newcastle upon Tyne (1979).
15. Jama, S.A.B. Ph.D. Thesis, University of Newcastle upon Tyne (1975).
16. Dodsworth, J. and D.P. Thompson, in D. Taylor and P. Popper ed. Special Ceramics 7 (Proc. Brit. Ceram. Soc., December 1980) (1981) 51.
17. Stackelberg, M., E. Schnorrenberg, D. Paulus and K.F. Spiess. Z. Phys. Chem. A. 175 (1935) 127.
18. Jeffrey, G.A. and V.Y. Wu. Acta Cryst. 16 (1963) 559.
19. Jeffrey, G.A. and V.Y. Wu. Acta Cryst. 20 (1966) 538.
20. Inoue, Z., Y. Inomata, H. Tanaka and H. Kawabata. J. Mat. Sci. 15 (1980) 575.
21. Huseby, J.C., H.L. Lukas and G. Petzow. J. Amer. Ceram. Soc. 58 (1975) 377.
22. Thompson, D.P. J. Mat. Sci. 11 (1976) 1377.
23. Thompson, D.P. and L.J. Gauckler. J. Amer. Ceram. Soc. 60 (1977) 470.
24. Thompson, D.P. NATO Advanced Study Institute on 'Nitrogen Ceramics', ed. F.L. Riley (1977) 131.
25. Clarke, D.R., T.M. Shaw and D.P. Thompson. J. Mat. Sci. 13 (1978) 217.
26. Shaw, T.M. and G. Thomas. J. Sol. State Chem. 33 (1980) 63.
27. Hall, D., G.E. Gurr and G.A. Jeffrey. Z. anorg. allg. Chem. 369 (1969) 108.
28. Clarke, D.R. and T.M. Shaw. Mat. Sci. Res. 11 (1978) 589.
29. Johnson, P.M. and A. Hendry. J. Mat. Sci. 14 (1979) 2439.

DISCUSSION

Jack: Dr. Thompson should have emphasized that it is not sufficient to compare one calculated lattice image with the observed image. It is necessary to calculate different images for different experimental conditions in the microscope. If the calculated images (for a presumed structure) agree with the observed images for 2 or 3 or ...n more different conditions, then the likelihood of the presumed structure increases to the nth power. The amount of effort required to do this makes one wonder whether the results obtained are worth it!

Clarke: You showed a tie-line between Mg_4SiN_4 and Mg_4N_2O in Fig.8. Is there a range of solid solution between the 12R and 6H polytypoids, exhibited perhaps as a series of intergrowths?

Thompson: I would expect both structures to have a slight mutual solubility but the major part of the line is a region of 2-phase equilibrium. The interlayer spacings in Mg_4N_2O and Mg_4SiN_4 are significantly different.

McCauley: Why does not AlN itself form polytypes?

Thompson: AlN has a very stable wurtzite compound which has never been made in 3C form. Polytype formers are always those which occur with equal readiness between both 2H and 3C varieties. The reason why 2H AlN is so stable is not known but must be due to the electron distribution round the atoms. It is worth noting that all other covalent nitrides with equal numbers of metals and non-metals have wurtzite and not zincblende (3C) structures.

Clarke: In examining Al-Mg-Si-N-O polytypoids in TEM it is seen that in the vicinity of the grain boundaries there is often an alteration in the polytypoid lattice spacing. There is also a corresponding change in elemental composition as determined by X-ray microanalysis. This suggests that the regions at the grain boundaries are accommodating excess solute, and further implies that the polytypoid structures are sufficiently flexible to accommodate certain impurities, rather than rejecting these to form an intergranular phase.

Thompson: This is an interesting observation. A wide range of [6]-fold and [4]-fold coordinated cations can probably be accommodated in polytypoid structures.

Section B

THERMODYNAMICS

CALCULATION OF PHASE EQUILIBRIA IN SYSTEMS BASED ON Si_3N_4

J. Weiss, H.L. Lukas and G. Petzow

Max-Planck-Institut für Metallforschung,
Institut für Werkstoffwissenschaften,
Pulvermetallurgisches Laboratorium
Heisenbergstraße 5
7000 Stuttgart-80, West-Germany

INTRODUCTION

The preparation and application of Si_3N_4 ceramics requires the knowledge of phase equilibria . Since within these phase equilibria the vapour atmosphere plays an important role and since some of the condensed components tend to reach equilibrium in a very sluggish manner the use of thermodynamic calculations seems promising in guidance and completion of experimental phase equilibria studies. This paper shall conclude the present state of the art that has been derived and explained in several consecutive papers (1-4).

1 MATHEMATICAL PROCEDURE WITHIN THE CALCULATIONS

In order to explain the procedure of this calculations one at first needs to mention the mathematical background. This has to be outlined here in a short manner, the more interested reader will be refered to a number of special papers on this particular topic.

1.1 Analytical Description of the Gibbs Free Energy

At first the Gibbs free energy of an arbitrary phase with several sublattices and a solubility range - quite commonly found in ionic phases - needs to be described as a function of temperature and composition. For the temperature dependence a function of the kind

Riley, F.L. (ed.) Progress in Nitrogen Ceramics

$$G_k^{is} = a-bT+c(T-\ln T) - \frac{d}{2}T^2 - \frac{e}{2T} - \frac{f}{6}T^3 \quad (1)$$

is used (5-7). Other authors use the same or similar expressions (8,9). Next the composition parameters need to be added for the analytical description. In the formula

$$G^i = \sum_s a^s \sum_k y_k \; (G_k^{is} + RT\ln y_k) + \sum_1 A_1 y_p y_q \; v_p^{m-1} v_q^{n-1}$$

$$+ \sum_{11} B_{11} y_p y_q y_v w_p^{m-1} w_q^{n-1} w_r^{o-1} + \ldots \quad (2)$$

G^i = molar Gibbs free energy of phase i

a^s = lattice site fraction of sublattice s in phase i

y_k = concentration of species k, referred to sublattice sites

G_k^{is} = Gibbs free energy of formation of species k on sublattice s

v_p = $y_p - (1-y_p-y_q)/2$ $\quad v_q = (1-y_p-y_q)/2$

w_p = $y_y - (1-y_p-y_q)/3$ $\quad w_q = y_q - \ldots \; w_v = y_r - \ldots$

A_1 = binary polynomial coefficient

B_{11} = ternary polynomial coefficient

The summation over s and k represents the free energy of ideal mixing. The species to be mixed may be molecules or associates in addition to the elements. The summation over 1 represents the binary polynomials, the 11 summation the ternary ones etc. To each term 1 and 11 the indices p,q,r and the exponents m,n,o are separately given. These polynomials are the generalised Muggianu equations after M. Hillert (10). Hillert compares in this paper the Muggianu formalism with the Kohler formalism and states the first one to be preferable. In figure 1 it is shown, which binary value is used to construct the ternary one in both formalisms. In equation (2) the quantities G_k^{is}, A_1 and B_{11} are represented as functions of the temperature according to equation (1). The total Gibbs free energy for a certain n phase mixture is expressed by a sum over $n^i \cdot G^i$, where n^i is the amount of phase i. The quantities of equation (2) are subject to three types of subsidiary conditions:

The sum of the concentrations in each sublattice is unity.

The sum over all j atoms in all species is the total amount of element j.

The sum over all ionic charges in each phase is zero.

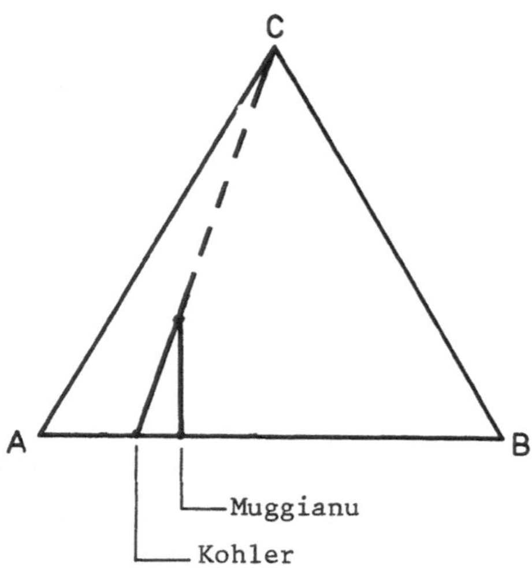

Fig. 1. Graphical comparison of the expression of a ternary composition on one of three binaries.

From these conditions M. Hillert (11, 12) derived a new formulation of the equilibrium conditions that have advantages for computer application compared to the well known equilibrium conditions after Gibbs.

1.2 Strategies for Calculation of Phase Relations (= Strategies for Solution of Equilibrium Conditions)

Based on the discussed formalisms, we wrote a program (13) which calculates after giving a condition for each degree of freedom the equilibrium between a specified number of phases. The strategy to calculate whole phase diagrams using such single equilibria may be divided into two general cases:

1) The first possibility is to calculate for a certain pressure and temperature the chemical composition of each phase in a definite n-phase-field, disregarding, if it is stable or metastable. By calculating the various phase fields the equilibria within a whole system may be obtained and stable parts be distinguished from metastable ones. This strategy of solution is used in the Lukas program for condensed systems (6,7) and in the newer PML program for systems including the vapour phase (4, 13) as well as in the programs of L. Kaufman (14).

2) The second possibility exists in providing a fixed chemical overall composition, pressure and temperature. Here the question to be resolved is, which phases with which compositions and amounts are present under these conditions. This strategy is used in the Eriksson program (1-3, 9) as well as in other programs (15-17). If the conditions are stepwizely altered, the equilibria of a whole system may be calculated likewize as shown by Lin et al. (18), however, in many cases the effort is expected to be higher by this strategy.

1.3 Possibilities of Graphical Representation

The solutions of these calculations may be most simply be represented graphically by diagrams. For the representation of up to three component systems under constant pressure the two dimensions allow a universal representation in a projected temperature concentration diagram. For higher order systems however this possibility reduces to certain temperature-concentration sections or isothermal sections. For unique representation of a whole system the reaction scheme is the only possibility. Single results obtained by the second strategy are best shown by phase amount diagrams where amount and composition of phases are drawn versus the temperature.

2. DATA EVALUTAION

The data for the calculations are taken from thermodynamic tables as far as possible. Due to lack of specific heat data for most phases or species in all calculations the Gibbs free energy was only described as a linear function of temperature: $G = a-bT$. This simplification however normally does not alter the accuracy much in the high temperature range considered here.

For condensed solution phases activity data are normally available for metallic systems only. For semistoichiometric phases like ßss, ideal mixing behavior may be expected from structure (two sublattices) and chemistry, further supported by specific heat data (19, 20). Other phases like the oxide rich liquid do not show such behavior, therefore excess terms have to be estimated. The estimation is done by using experimental diagrams of binary systems. From a miscibility gap the regular term can be calculated after the formula:

$$A_1 = 2RT_c \qquad (3).$$

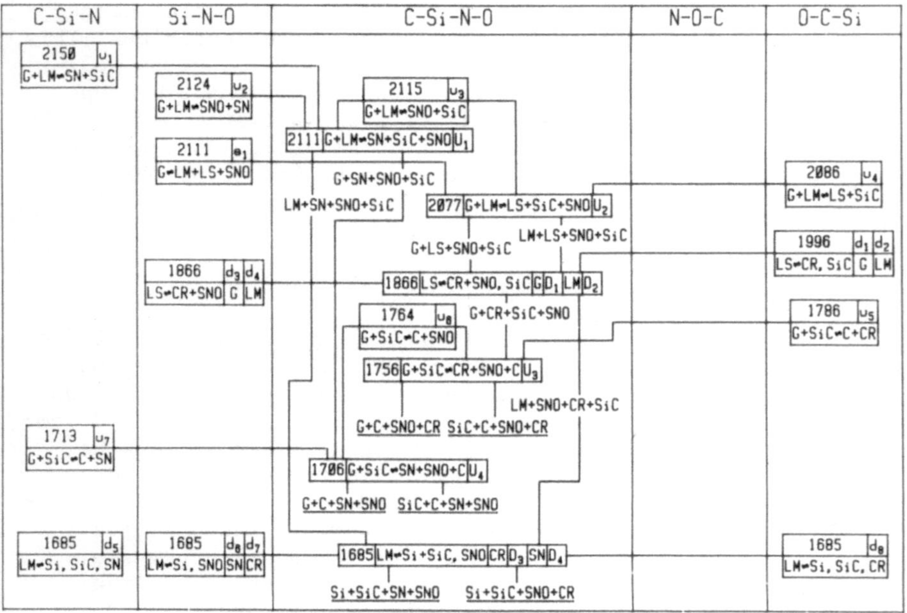

Fig. 2. Reaction scheme of the systems C-Si-N-O (For notation see table 1).

For several quasi binaries this simple approach gave satisfying agreement between experimental and calculated equilibria (13). Only for the quasibinary $MgO-SiO_2$, which shows a distinct deviation from simple regular behavior, higher order polynomial terms had to be introduced (21).

For several phases data were estimated on the basis of known phase equilibria and thermodynamic rules (1-3). Solution phases of less interest like X_2 or spinel in the SiAlON system (2, 3) were thereby in a first approach treated as stoichiometric phases.

3. RESULTS AND DISCUSSION

Figure 2 shows the reaction scheme of the system C-Si-N-O. A T-c section of this system is shown in figure 3 for composi-

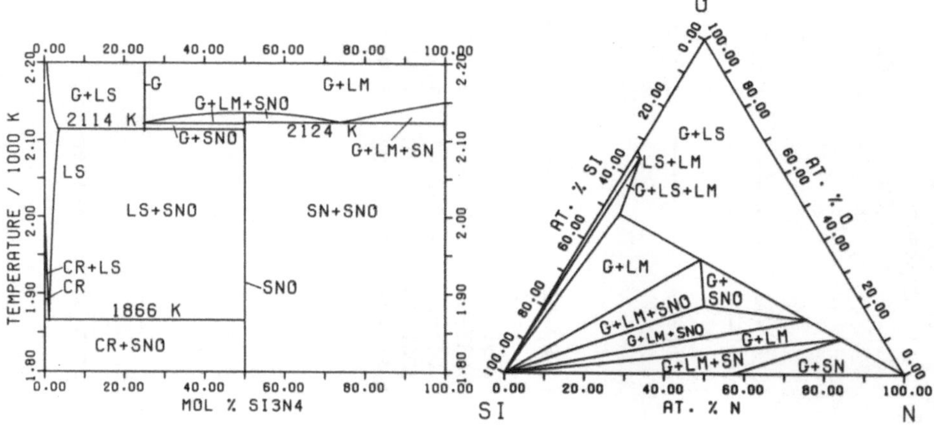

Fig. 3. T-c section SiO_2-Si_3N_4 of the ternary system Si-N-O.

Fig. 4. Isothermal section of the system Si-N-O at 2130 K.

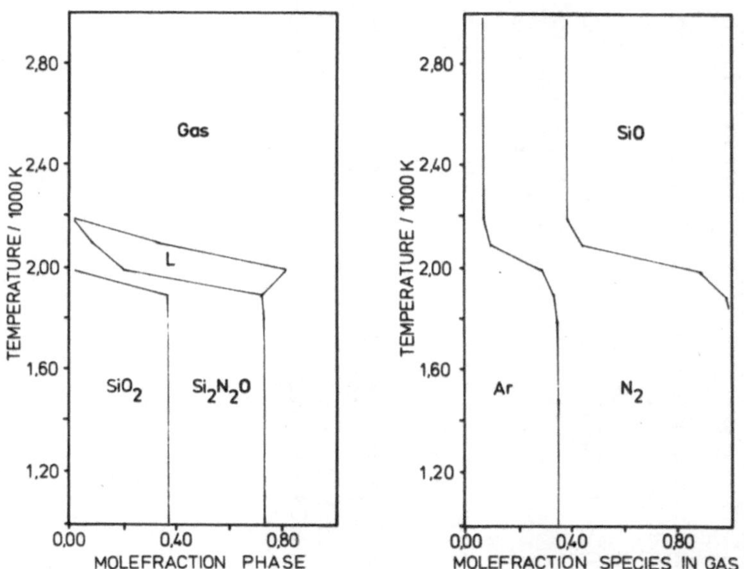

Fig. 5. Phase amount diagram for a composition of Si_2N_2O+SiO_2+N_2.

Equilibrium (Phases)	Temp. K	Equilibrium Partial Pressure of SiO	N_2	Si	CO	O_2
	System Si - N - O					
Gas + LM = Si_2N_2O (max.)	2137	7.50E-01	2.50E-01	1.26E-04	--------	1.50E-06
Gas + LM = Si_2N_2O + Si_3N_4	2124	2.60E-01	7.39E-01	1.63E-04	--------	8.57E-15
Gas = LM + LS + Si_2N_2O	2114	8.06E-01	1.94E-01	1.47E-04	--------	7.47E-14
LS = SiO_2 + Si_2N_2O, Gas	1866	1.50E-02	9.85E-01	2.35E-07	--------	2.25E-15
	System C - Si - N - O					
G+LM = Si_3N_4 +SiC (max.)	2115	3.22E-01	5.00E-01	1.48E-04	1.78E-01	1.23E-14
G+LM = Si_2N_2O+SiC+Si_3N_4	2111	2.32E-01	6.41E-01	1.42E-04	1.26E-01	6.12E-15
G+LM = Si_2N_2O+SiC+LS	2077	6.01E-01	1.20E-01	9.89E-05	2.79E-01	3.00E-14
LS = SiO_2+Si_2N_2O, G, SiC	1866	2.28E-02	2.80E-01	5.43E-07	6.97E-01	9.73E-16
G+SiC = Si_2N_2O+C (max.)	1764	2.49E-03	5.00E-01	3.61E-08	4.98E-01	4.11E-17
G+SiC = Si_2N_2O+C+SiO_2	1756	3.45E-03	2.92E-01	3.04E-08	2.91E-10	7.63E-17
G+SiC = Si_2N_2O+C+Si_3N_4	1706	2.99E-04	9.29E-01	1.11E-08	7.11E-02	4.93E-19

<u>Notation:</u>

G=gas, LM=liquid metal, LS=salt liquid, C=(graphite), Si=Si (solid), CR=SiO_2 (cristobalite), SN=ßSi_3N_4, SNO=Si_2N_2O, SiC=SiC.

Table I: Partial pressure ratios for the non variant phase equilibria of the system CSiNO under 1 bar pressure.

tions of SiO_2-Si_3N_4. By inclusion of the vapour phase the system SiO_2-Si_3N_4 may no longer be treated as a quasi binary system since the composition of the vapour phase is not limited to compositions within this quasi binary. Figure 4 shows an isothermal section of the system Si-N-O. Figure 5 shows a phase amount diagram for a composition of Si_2N_2O+SiO_2+N_2 and underscores the limited thermal stability of the liquid formed by SiO_2 and Si_2N_2O. In Si_3N_4 and SiC based systems the vapour composition exerts a dominant influence on the condensed phase equilibria. As an example for this, the partial pressures present at the nonvariant equilibria of the system C-Si-N-O are listed in table 1.

As a further example of the calculated results an isothermal section of the SiAlON system at 2100 and 1900 K is shown in figure 6. This figure is drawn from at least one calculation by the Eriksson program for each three phase field. An example is shown in figure 5 (2, 3). Calculations with the PML program, that will yield the reaction scheme of this system, are presently carried out.

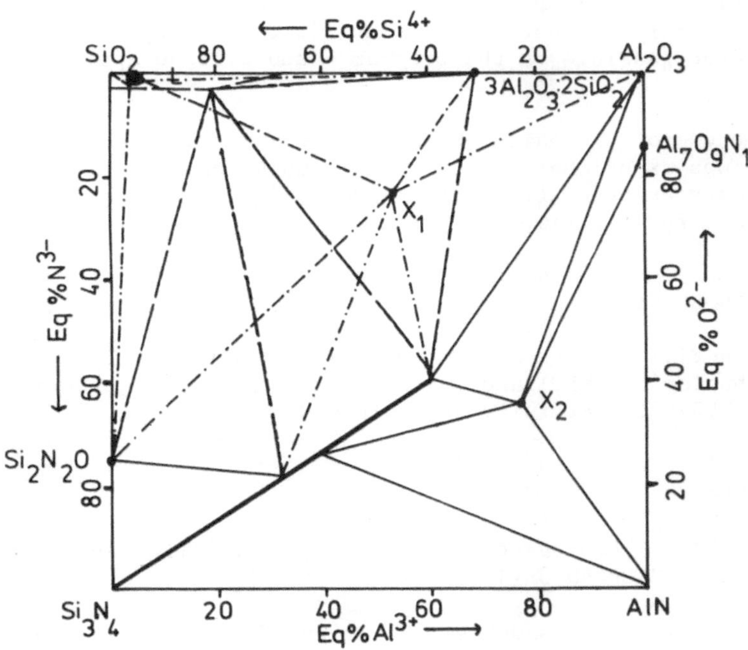

Fig. 6. Calculated isothermal phase equilibria in the SiAlON system.

4. POSSIBILITIES AND LIMITS OF THERMODYNAMIC CALCULATIONS

In the previous section phase equilibria of two Si_3N_4 based systems have been introduced. For the C-Si-N-O system no conclusive experimental data exist yet (22, 23). For the SiAlON system several isothermal experimental studies are known (24, 25). The first result shows that calculations allow extrapolations from known systems into unknown higher order systems. This applies even more, since in the C-Si-N-O system the thermodynamic data are well known except for the single condensed solution phase, the silicate liquid, the agreement with the - unknown - real equilibria should therefore be pretty close.

On the other hand, for the SiAlON system lack of data necessitated several simplifications by either leaving out phases of less importance or treating them as stoichiometric: The comparison with the known experimental data is therefore somewhat limited. For the nitrogen solubility in the liquid phase another fact needs to be pointed out:

The usual experiments are carried out in the reducing atmosphere of a graphite resistance furnace, ergo the oxygen and consecutivly the SiO partial pressures are fixed. The calculation however so far does not 'know' such fixed activity values - which are however possible by the program - and will be applied in the future.

The use of the calculations may be summarized as follows: It combines experimental data to conclusive phase equilibria, allows easy extrapolation into higher order systems and gives guidelines for experimental work.

ACKNOWLEDGEMENT

Financial support by the Bundesministerium für Forschung und Technologie (BMFT) and by the Deutsche Forschungsgemeinschaft (DFG) is gratefully acknowledged.

REFERENCES

1. P. Dörner, L.J. Gauckler, H. Krieg, H.L. Lukas and G. Petzow; CALPHAD 3 (1979) 241.

2. P. Dörner, L.J. Gauckler, H. Krieg, H.L. Lukas, G. Petzow and J. Weiss; J. Mat. Sci. 16 (1981) 935.

3. J. Weiss and G. Petzow; Final Report, NTS 1006 43-Bundesministerium für Forschung und Technologie (1980).

4. J. Weiss, H.L. Lukas, J. Lorenz, G. Petzow and H. Krieg; to be published in CALPHAD 5 (1981).

5. O. Kubaschewski, E.L.L. Evans and C.B. Alcock; "Metallurgical Thermochemistry", Fourth Edition, Pergamon Press, Oxford - New York (1976).

6. B. Zimmermann; "Rechnerische und experimentelle Optimierung von binären und ternären Systemen aus Ag, Bi, Pb und Tl", Ph.D. thesis, University of Stuttgart (1976).

7. E. Th. Henig, H.L. Lukas and G. Petzow; Project Meeting CALPHAD VII, Stuttgart, Germany (1978).

8. I. Barin, O. Knacke, O. Kubaschewski; "Thermochemical Properties of Inorganic Substances", Springer Verlag Berlin - New York (1977).

9. G. Eriksson; Chemica Scripta 8 (1975) 100.

10. M. Hillert; CALPHAD 4 (1980) 1.

11. M. Hillert; Report TRI TA-MAC-0161, Materials Center, Royal Institute of Technology, Stockholm, Sweden (1979).

12. M. Hillert; in Proc. of the Int. Symp. 'Thermodynamics of Alloys', Delft (1980), ed. by A.R. Miedema and G.W. Rathenau, North Holland publishing comp. Amsterdam (1980).

13. H.L. Lukas, H. Krieg, G. Petzow and J. Weiss; in preparation for CALPHAD.

14. L. Kaufmann and H. Bernstein; "Computer Calculation of Phase Diagrams", Refractory Materials Vol. 4, Academic Press, New York (1970).

15. J.F. Counsell, E.B. Lees and P.J. Spencer; Metal Sci. J. 5 (1971) 210.

16. A.D. Pelton and W.T. Thompson; Progr. Solid State Chem. 10 (1975) 119.

17. P.L. Lin, A.D. Pelton, C.W. Bale and W.T. Thompson; CHALPHAD 4 (1980) 47.

18. P.L. Lin, A.D. Pelton and C.W. Bale; J. Am. Ceram. Soc. 62 (1979) 414.

19. H.D. Nüssler and O. Kubaschweski; Trans. J. Brit. Ceram. Soc. 79 (1980) 98.

20. M. Kuyiyama, Y. Inomata, T. Kujima and Y. Hasegawa; Bull. Am. Ceram. Soc. 57 (1978) 1119.

21. R. Müller, J. Weiss, H.L. Lukas, G. Petzow and T.Y. Tien; submitted to CALPHAD.

22. W.A. Krivsky and R. Schuhmann; Trans. Met. AIME 221 (1961) 898.

23. E. Gugel, P. Ettmayer und A. Schmidt; Ber. Dt. Keram. Ges. 45 (1968) 395.

24. L.J. Gauckler, H.L. Lukas and G. Petzow; J. Am. Ceram. Soc. 58 (1975) 346.

25. I.K. Naik. L.J. Gauckler and T.Y. Tien; J. Am. Ceram. Soc. 61 (1978) 332.

26. N.L. Bowen and O.A. Anderson; Am. Journ. Sci. 37 (1914) 488, see Fig. 266 in Phase Diagrams for Ceramists, ed. by M.K. Reser, Am. Ceram. Soc. Columbus (Ohio) (1964).

27. Ya. Ol'shanskii, I. DAN SSSR 76 (1951) 93.

DISCUSSION

Jack: This is an amazingly powerful tool, but if you use data from an existing phase diagram to regenerate another diagram this is working in a circle. Moreover small differences in thermodynamic data make big differences, in general, to the derived diagram. How predictive then is your procedure? (The X_1-phase composition assumed is wrong!)

Weiss: The calculations in general will always be as reliable as the data used. In the particular case of using an assumed composition for X_1-phase previously published data were used, the normal procedure until better data become available.

Tien: The calculation normally cannot prove whether experimental data are correct or not. There are many estimations made in calculation of the Al-Si-N-O system. The correctness of the X_1-phase composition will not make too much difference.

PHASE EQUILIBRIUM STUDIES IN Si_3N_4 - METAL OXIDES SYSTEMS

T. Y. Tien, G. Petzow, L. J. Gauckler and J. Weiss

The University of Michigan, Ann Arbor, MI 48109 U.S.A.
Max-Planck-Institut für Metallforschung, Stuttgart, B.R.D.

This paper is not intended to be a review of phase equilibrium studies in Si_3N_4 - metal oxides systems. This presentation includes only phase diagrams established at The University of Michigan and the Max-Planck-Institut für Metallforschung. The systems to be discussed are: Si,Al/N,O; Si,Be/N,O; Si,Mg/N,O; Si,Y/N,O; Si,Ca/N,O; Si,Th/N,O; Si,Zr/N,O; Si,Al,Be/N,O; Si,Al, Mg/N,O; Si,Al,Y/N,O; Si,Al,Ca/N,O; Si,Al,Zr/N,O and the investigators involved in these two institutes are, in alphabetical order: L. J. Gauckler, H. G. Hohnke, I. C. Huseby, H. L. Lukas, R. Müller, I. K. Naik, S. D. Nunn, G. Petzow, W. Y. Sun, T. Y. Tien, and J. Weiss.

In 1971, Oyama and Kamigaito (1) first reported that Al_2O_3 enters Si_3N_4 lattice forming single phase solid solution. Since then, many research workers have studied the phase relationships in the Si_3N_4 - metal oxide systems. In that period, Oyama in Japan (1) and Jack in England (2) published many phase diagrams of the Si_3N_4 - metal oxides systems.

Gauckler, Lukas and Petzow (3) were first to realize that the Si_3N_4 - metal oxides systems should be treated as reciprocal salts systems; therefore, these systems should be presented in square planar diagrams. The first diagram of this type was published in 1975 by these authors for the system Si,Al/N,O as shown in Fig. 1 which is 1780° isothermal section of the system Si_3N_4-SiO_2-AlN-Al_2O_3. Gauckler showed that the extent of the β-Si_3N_4 solid solution was restricted only to the region having a metal to non-metal ratio of 3:4 and existence of lattice defects is not likely. The existence of AlN polytypes was also reported. These phases have also definite stoichiometric ratios.

Riley, F.L. (ed.) Progress in Nitrogen Ceramics

The compound X_1 and compositions near SiO_2 were melted at temperatures lower than 1780°C. Therefore, the results were presented by dashed lines in this region.

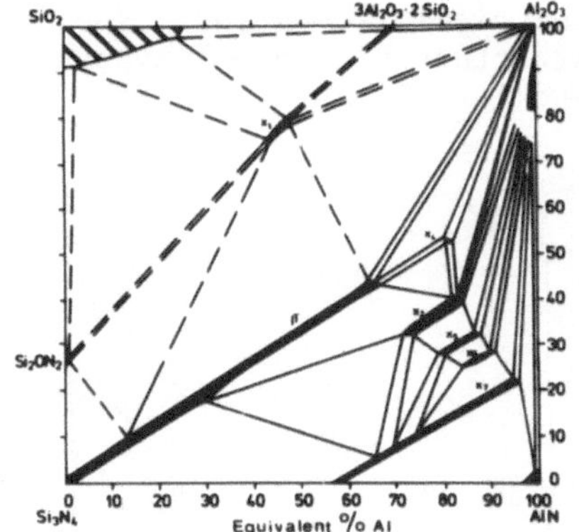

Fig. 1: The system Si_3N_4-SiO_2-AlN-Al_2O_3, 1780°C isotherm (3).

The solid-liquid equilibria at 1750°C were determined by Naik (4). The β-Si_3N_4 solid solution-liquid tie lines were determined experimentally. Samples were sintered at 1750°C and quenched. The lattice parameters of the β phase were measured and the results were used to determine the position of the lower ends of the tie lines. The relative amounts of the β phase were compared with standards and the length of the tie lines were determined by the use of the level rule. The results were used to construct part of the diagram in Fig. 2A. Specimens were then annealed at 1400°C and the results are given in Fig. 2B as the sub-solidus equilibrium of the system Si_3N_4-SiO_2-AlN-Al_2O_3.

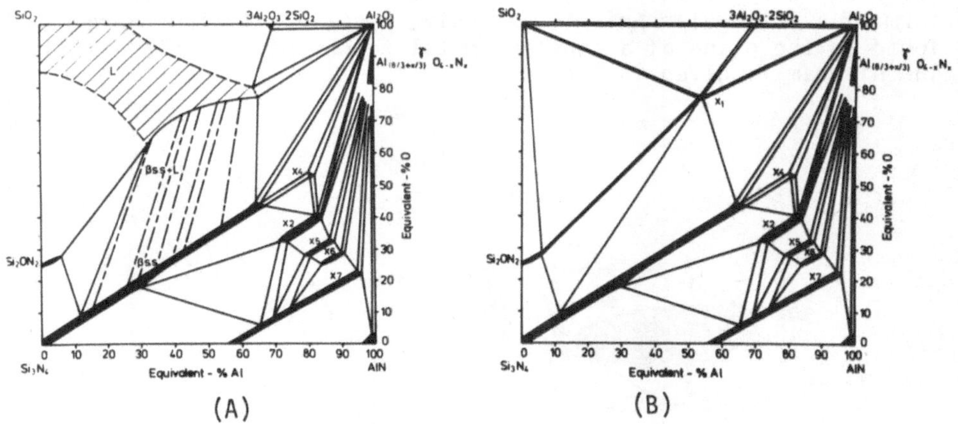

Fig. 2: The system Si_3N_4-SiO_2-AlN-Al_2O_3

A) 1750°C isotherm, solid-liquid tie lines are shown.
B) Sub-solidus.

The system Si_3N_4-$SiO_2$$Be_3N_2$-BeO was investigated by Huseby and was published in 1975 (5). The β-Si_3N_4 solid solutions were also found to have a metal to non-metal ratio of 3:4 similar to that of the solid solutions in the system Si,Al/N,O. Polytypes were also found to have definite stoichiometric ratios. The diagram is given in Fig. 3A. The structure of the polytypes were later identified by Thompson and Gauckler (6) and the diagram was revised as shown in Fig. 3B.

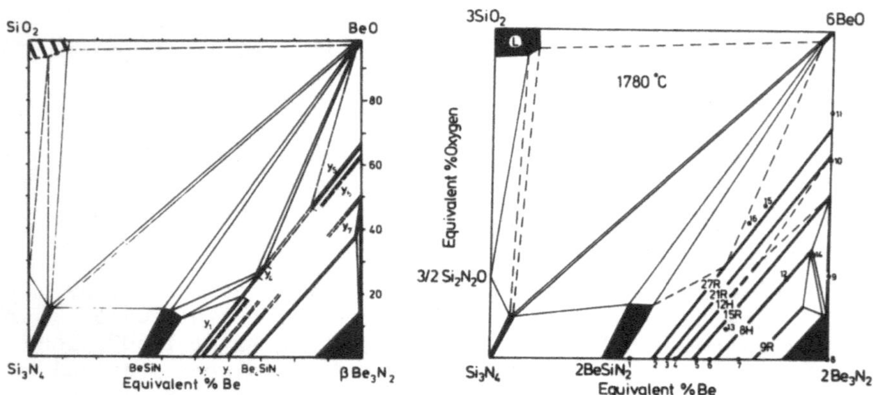

Fig. 3: The system Si_3N_4-SiO_2-Be_3N_2-BeO
A) 1780°C isotherm (5).
B) Shows the polytypes (6).

The pseudo-quaternary system Si_3N_4-SiO_2-AlN-Al_2O_3-Be_3N_2-BeO was studied by Gauckler (7). This system was presented in the form of a triangular prism. Extensive β-solid solutions were found in the plane of a constant metal to non-metal ratio of 3:4. The diagram is given in Fig. 4.

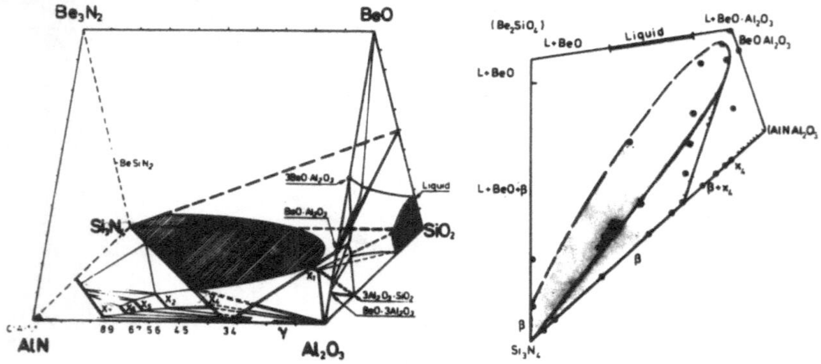

Fig. 4: The system Si_3N_4-SiO_2-AlN-Al_2O_3-Be_3N_2-BeO. The plane having a metal to non-metal ratio of 3:4 is drawn in Fig. 4B. Samples were prepared by hot pressing at 1780°C. (7)

There are three versions of the phase diagram of the system Si,Mg/N,O in the literature as shown in Fig. 5. In the region bounded by Si_3N_4-Mg_2SiO_2-MgO-$MgSiN_2$, Inomata et al (8) and Jack (9) reported the existence of the join Mg_2SiO_4-$MgSiN_2$ while Lange (10) reported that Si_3N_4-MgO form binary join when specimens were hot pressed at 1750°C. Müller (11) studied this system and showed that both the Si_3N_4-MgO and the Mg_2SiO_4-$MgSiN_2$ joins (which cross each other) can be observed depending on firing conditions.

Müller reported that 1550°C fired samples with low weight loss showed the existence of the join Si_3N_4-MgO, the join Mg_2SiO_4-$MgSiN_2$ was also observed when the weight loss was high. Therefore, it is felt that this system should not be treated as condensed system.

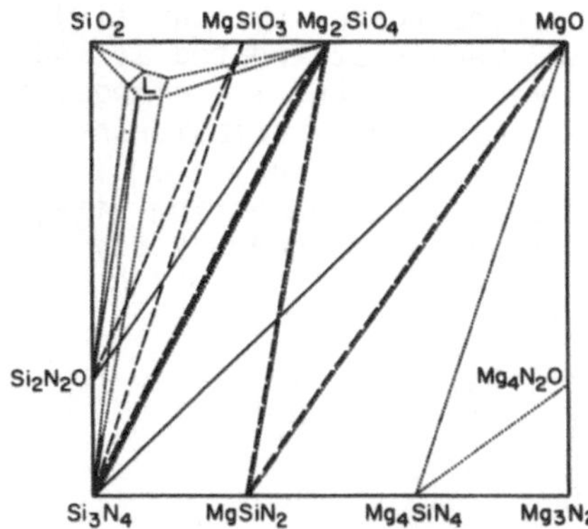

Fig. 5: The system Si_3N_4-SiO_2-Mg_3N_2-MgO shows the disagreement among different authors.

________	Inomata (8)
--------	Jack (9)
_.__.__	Lange (10)

The disagreement in the region Si_3N_4-Mg_2SiO_4-SiO_2 could be caused by sluggishness of crystallization of the liquid. When a refractory compound with high vapor pressure, such as Si_3N_4, react with low melting, high viscosity liquid, equilibrium would be very difficult to reach.

From the size and charge considerations, magnesium should behave similar to Be^{2+}, i.e., Mg^{2+} should enter β-Si_3N_4 lattice forming single phase solid solution (12). The formation of single phase β-Si_3N_4 solid solutions containing Mg^{2+} and Al^{3+} has been reported by Jack (13) in the system Si,Al,Mg/N,O. Weiss (14) has studied the plane of 3:4 in this system and the results are given in Fig. 6. These results showed that single phase materials can only be found in the Mg^{2+} free compositions (15), i.e., in the system Si_3N_4-SiO_2-AlN-Al_2O_3 (3).

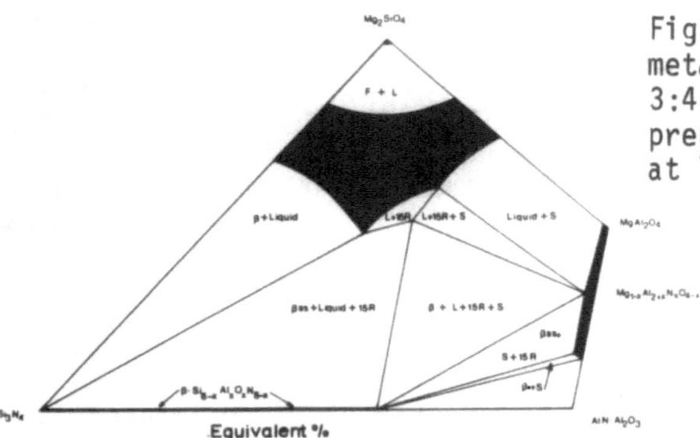

Fig. 6: The plane having a metal to non-metal ratio of 3:4. The specimens were prepared by hot pressing at 1750°C (14).

The sub-solidus phase relationships in the system Si_3N_4-SiO_2-AlN-Al_2O_3-Mg_3N_2-MgO have been investigated by Nunn (16). Specimens were reacted at 1600°C and were heat treated at 1400°C for 24 hours. An unknown compound was found but its composition has not been identified. The sub-solidus phase relationships in terms of compatibility tetrahedra are given in Table I. It should be noted that there is no binary join connecting Si_3N_4 with high melting compound in the pseudo-quaternary system.

TABLE I

A. Quarternary Compatibility Relationships in the System Si_3N_4-SiO_2-AlN-Al_2O_3-MgO

X_1-Cordierite-Mullite-SiO_2
X_1-Cordierite-Spinel-Sapphirine
X_1-Cordierite-Sapphirine-Mullite
X_1-Sapphirine-Mullite-Spinel
X_1-Cordierite-SiO_2-Si_2N_2O
X_1-Mullite-Spinel-Al_2O_3
X_1-Cordierite-Si_2N_2O-Si_3N_4
X_1-Spinel-(β-30)-(β-60)
Cordierite-Enstatite-Forsterite-Si_2N_2O
Cordierite-Enstatite-SiO_2-Si_2N_2O
X_1-Spinel-Al_2O_3-(β-60)
X*-Cordierite-Spinel-X_1
X*-X_1-Spinel-(β-30)
X*-X_1-Cordierite-(β-30)
X*-Cordierite-Si_3N_4-(β-30)
X*-Cordierite-Si_3N_4-Si_2N_2O
X*-Forsterite-Cordierite-Spinel
X*-Cordierite-Forsterite-Si_2N_2O
X*-Forsterite-Si_3N_4-Si_2N_2O
X*-X_2-Spinel-(β-30)

TABLE I (continued)

where β-30 = β-$Si_{6-x}Al_xN_{8-x}O_x$ solid solution x = 2.18
β-60 = β-$Si_{6-x}Al_xN_{8-x}O_x$ solid solution x = 4.0

B. X-ray Pattern of the Unknown Compound X*

d	I/I	d	I/I
5.14	5	1.89	20
3.57	100	1.85	7
3.52	97	1.77	15
2.95	10	1.73	22
2.90	34	1.505	8
2.78	8	1.497	11
2.46	36	1.48	5
2.32	10	1.43	5
2.29	15	1.42	20
2.03	16	1.39	7
2.00	7	1.37	14

Dr. D. P. Thompson suggested that this compound X* has a structure of petelite with a formula Mg_2SiAlO_4N.

The system Si_3N_4-SiO_2-Y_2O_3 has been studied by Hohnke (17). The diagrams are given in Fig. 7. Four ternary compounds were confirmed and two low melting eutectic compositions were located by DTA. The lowest melting composition melts at 1480°C.

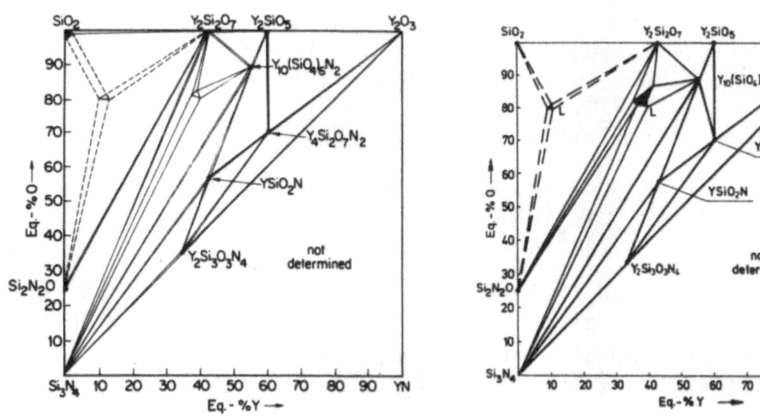

Fig. 7: The system Si_3N_4-SiO_2-Y_2O_3 (17)
A) The 1500°C isotherm.
B) The 1550°C isotherm.

Many authors (18) have shown that yttrium oxide and aluminum oxide, when used together, were efficient sintering aid for Si_3N_4. Naik (19) worked out the sub-solidus phase relationships in part of the system Si_3N_4-SiO_2-AlN-Al_2O_3-Y_2O_3 and the results were reported in the form of compatibility tetrahedra in Table II. It was felt that useful materials might be found in the compatibility tetrahedron Si_3N_4-(AlN:Al_2O_3)-$Y_3Al_5O_{12}$-$Y_2Si_2O_7$; therefore, this sub-system is being studied at present and some of the results are reported in a separate paper (20).

TABLE II

Quaternary Compatibility Relationships in the System Si_3N_4-SiO_2-AlN-Al_2O_3-Y_2O_3

Si_3N_4-(β-60)-$Y_2Si_2O_7$-garnet
X_1-Mullite-Al_2O_3-$Y_2Si_2O_7$
X_1-Mullite-$Y_2Si_2O_7$-SiO_2
(β-60)-$Y_2Si_2O_7$-X_1(β-10)
(β-60)-$Y_2Si_2O_7$-X_1-garnet
(β-60)-Al_2O_3-X_1-garnet
X_1-SiO_2-$Y_2Si_2O_7$-Si_2N_2O s.s.
X_1-$Y_2Si_2O_7$-Si_2N_2O s.s. - (β-10)
Y_2SiO_5-Si_2N_2O s.s. - Si_2N_2O-SiO_2
Si_3N_4-garnet-$Y_2Si_2O_7$-apatite
Si_3N_4-apatite-garnet-$YSiNO_2$
Si_3N_4-$YSiNO_2$-$Y_2Si_3N_4O_3$-garnet
$Y_2Si_3N_4O_3$-$YSiNO_2$-$Y_2Al_2O_9$-garnet
$YSiNO_2$-garnet-$Y_4Al_2O_9$-apatite
apatite-garnet-$Y_2Si_2O_7$-Y_2SiO_5
apatite-garnet-$Y_2Si_2O_7$-$Y_4Al_2O_9$
$Y_2Si_3N_4O_3$-$YSiNO_2$-$Y_4Si_2N_2O_7$-$Y_4Al_2O_9$
$YSiNO_2$-apatite-$Y_4Si_2N_2O_7$-$Y_4Al_2O_9$
$Y_4Si_2N_2O_7$-apatite-$Y_4Al_2O_7$-Y_2O_3
apatite-$Y_4Al_2O_9$-Y_2SiO_5-Y_2O_3

where: β-10 = β-$Si_{6-x}Al_xN_{8-x}O_x$ solid solution, x=0.77
apatite = $Y_{10}(SiO_4)_6N_2$
garnet = $Y_3Al_5O_{12}$

The system Si_3N_4-SiO_2-Ca_3N_2-CaO was studied by Weiss (21) and the results are given in Fig. 8. Si_3N_4 form binary joins with all of the CaO-SiO_2 binary compounds. One ternary compound was identified to have a composition $Ca_2Si_3N_2O_5$.

The system Si_3N_4-SiO_2-AlN-Al_2O_3-CaO is being studied at the present time by Sun (22) and the following compatability relationships have been established and are shown in Table III.

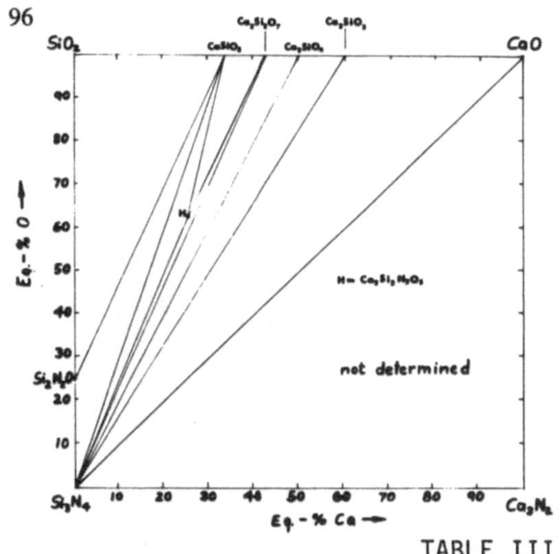

Fig. 8: The system Si_3N_4-SiO_2-CaO: sub-solidus (21).

TABLE III

Quarternary Compatibility Relationships in Part of the System Si_3N_4-SiO_2-AlN-Al_2O_3-Ca_3N_2-CaO (incomplete)

SiO_2-Anothite-Mullite-X_1
Mullite-Anothite-Al_2O_3-X_1
Al_2O_3-Anothite-CaO·$6Al_2O_3$-X_1
CaO·$6Al_2O_3$-Anothite-Gehlenite-X_1
SiO_2-Anothite-$CaSiO_3$-Si_2N_2O
$CaSiO_3$-Anothite-Gehlenite-Si_2N_2O
Si_3N_4-Si_2N_2O-X_1-Anothite
Si_3N_4-X_1-Anothite-(β-20)

where: Anothite = CaO·Al_2O_3-$2SiO_2$
Gehlenite = 2CaO·Al_2O_3·SiO_2
β-20 = β-$Si_{6-x}Al_xN_{8-x}O_x$ solid solution, x = 1.5

The system Si_3N_4-SiO_2-Th_3N_4-ThO_2 was studied by Hohnke (23). The diagram is given in Fig. 9. It should be noted that the compound ThO_2 has the highest melting point among oxides (3220°C) (24). Therefore, the compositions in the triangle Si_3N_4-Si_2N_2O-ThO_2 should be of interest as high temperature material.

The system Si_3N_4-SiO_2-ZrO_2-ArN was studied by Weiss (25). This system cannot be treated as a reciprocal salt system because Zr in ZrN and ZrO_2 have different valence states (26). When mixtures of Si_3N_4 and ZrO_2, and SiO_2 and ZrN were heated in a furnace with flowing nitrogen, the weight losses were very high and the only solid phase remaining was ZrN. The reaction products of these mixtures during heat treatment are given in Fig. 10. However, when specimens were hot pressed and weight losses were kept

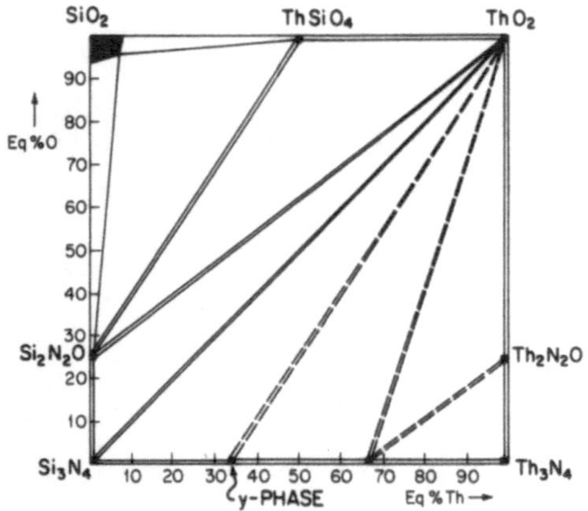

Fig. 9: The system Si_3N_4-SiO_2-Th_3N_4-ThO_2(23). The composition of the new phases are Si_2ThN_4 and $SiTh_2N_4$, respectively.

very low, the sub-solidus phase relationships can be presented as shown in Fig. 11 in a form of compatability tetrahedra.

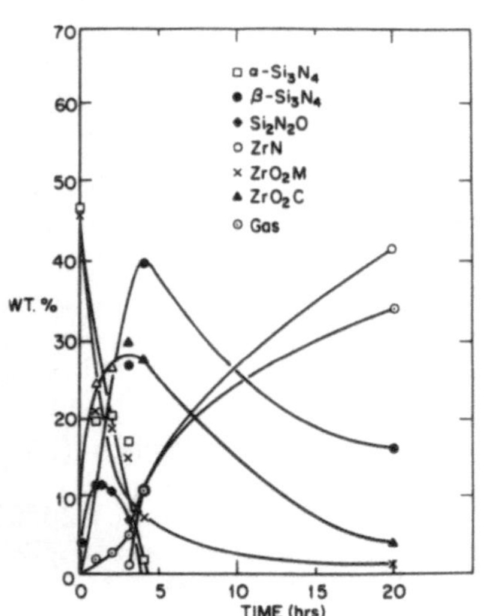

Fig. 10: Phase present when Si_3N_4 and ZrO_2 mixtures were sintered at 1600°C.

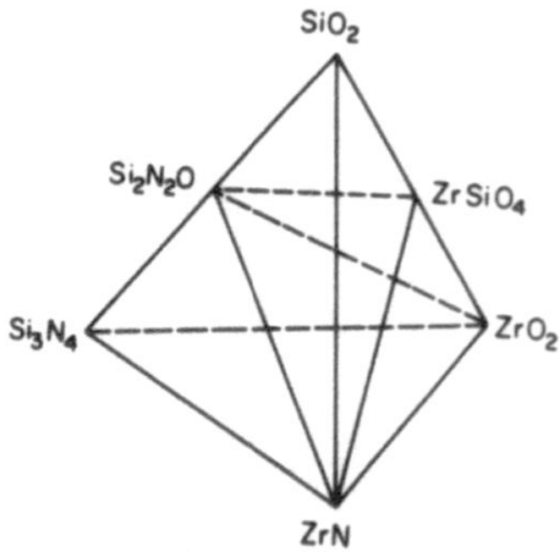

Fig. 11: The sub-solidus of the system Si_3N_4-SiO_2-ZrN-ZrO_2, the specimens were hot pressed at temperatures between 1600-1900°C.

ZrO_2 and ZrN were added to compositions in the system Si_3N_4-SiO_2-AlN-Al_2O_3 and were hot pressed at 1750°C (27). The compatible binary joins between the zirconium compounds are shown as dashed lines in Fig. 12.

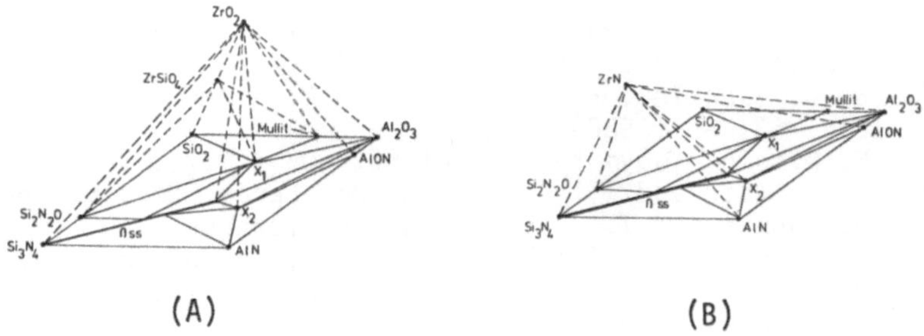

Fig. 12: The system Si_3N_4-SiO_2-AlN-Al_2O_3-ZrN-ZrO_2.
A) ZrO_2-Si,Al/N,O relationships.
B) ZrN-Si,Al/N,O relationships
The samples were hot pressed at 1780°C.

REFERENCES

1. Y. Oyama and O. Kamigiato, Japan J. Appl. Phys. 10, 1637 (1971).
2. K. H. Jack and W. I. Wilson, Nature, Physical Science, 283, 28 (1972).
3. Gauckler, H. L. Lukas and J. Petzow, J. Am. Ceram. Soc. 58, 346 (1975).
4. I. K. Naik, L. J. Gauckler and T. Y. Tien, J. Am. Ceram. Soc. 61, 332 (1978).
5. I. C. Huseby, H. L. Lukas and G. Petzow, J. Am. Ceram. Soc. 58, 377 (1975).
6. D. P. Thompson and L. J. Gauckler, J. Am. Ceram. Soc. 60, 470 (1977).
7. L. J. Gauckler, "The System Si_3N_4-SiO_2-AlN-Al_2O_3-Be_3N_2 Doctoral Thesis, University of Stuttgart, (1975).
8. Y. Inomata, Y. Hasegawa and T. Matsujama, Yogyo-Kyokai-Shi, 85, 29 (1977).
9. K. H. Jack, Processing of Crystalline Ceramics, edited by Hayne Palimour III, R. F. Davis and T. M. Hare, Plenum Press, New York (1978), p. 561.
10. F. F. Lange, J. Am. Ceram. Soc. 61, 53 (1978).
11. R. Muller, "The System Si,Mg/N,O", Doctoral Thesis, University of Stuttgart, 1981.
12. L. J. Gauckler, H. L. Lukas and T. Y. Tien, Mat. Res. Bull. 11, 503, (1976).
13. M. H. Jack, Nitrogen Ceramics, Edited by F. L. Riley, Noordhoff International, Reading, Mass. (1977), p. 109.
14. J. Weiss, Diplomarbeit, University of Stuttgart, (1977).
15. L. J. Gauckler, J. Weiss and T. Y. Tien, J. Am. Ceram. Soc. 61, 397 (1978).

16. S. D. Nunn, H. Hohnke, L. J. Gauckler and T. Y. Tien, Am. Ceram. Soc. Bull. 57, 321 (1978).
17. L. J. Gauckler, H. Hohnke and T. Y. Tien, J. Am. Ceram. Soc. 63, 35 (1980).
18. M. Mitomo, Yogyo-Kyokai-Shi 85, 408 (1977).
19. I. K. Naik and T. Y. Tien, J. Am. Ceram. Soc. 62, 642 (1979).
20. H. Hohnke and T. Y. Tien, This conference.
21. J. Weiss, T. Y. Tien and G. Petzow, unfinished work.
22. W. Y. Sun, T. S. Yen and T. Y. Tien, unfinished work.
23. H. Hohnke, L. J. Gauckler, J. Schneider and T. Y. Tien, Am. Ceram. Soc. Bull. 58, 885 (1979).
24. S. Lungu, J. Nucl. Mater. 19, 157 (1966).
25. J. Weiss, L. J. Gauckler and T. Y. Tien, J. Am. Ceram. Soc. 62, 632 (1979).
26. L. J. Gauckler and G. Petzow, Nitrogen Ceramics, Edited by F. L. Riley, Noordhoff International, Reading, Mass. (1977), p. 41.
27. J. Weiss, L. J. Gauckler, H. L. Lukas, G. Petzow and T. Y. Tien, submitted to J. Mat. Sci.

SOLID-LIQUID REACTIONS IN PART OF THE SYSTEM Si,Al,Y/N,O

H. Hohnke and T. Y. Tien

Materials and Metallurgical Engineering, The University of Michigan, Ann Arbor, MI 48109

Phase relationships in the system Si_3N_4-SiO_2-AlN-Al_2O_3-Y_2O_3 have been studied. Twenty-one compatibility tetrahedra were established in this system at sub-solidus temperatures. In the later stage of the investigation, efforts have been concentrated on the studies of the sub-quaternary system Si_3N_4-Al_2O_3:AlN-$Y_2Al_5O_{12}$-$Y_2Si_2O_7$.

1550^0C isothermal sections of the ternary systems Si_3N_4-Al_2O_3:AlN-$Y_3Al_5O_{12}$ and Si_3N_4-Al_2O_3:AlN-$Y_2Si_2O_7$ were studied. Si_3N_4 ceramics in both systems were hot pressed. Most of the specimens showed only Si_3N_4 solid solution as the crystalline phase after quenching. Grain boundary glassy phase was observed by high resolution microscopy. This grain boundary glassy phase can be crystallized after low temperature heat treatment.

INTRODUCTION

Many authors (1,2) have shown that Y_2O_3 and Al_2O_3 are effective sintering aids for forming Si_3N_4 ceramics when used together. Unfortunately, these ceramics lose their strength at moderate temperatures. It is believed that the presence of either low melting eutectics or glassy phase at the grain boundaries and triple points are responsible for these inferior properties (3).

In general, in a quasi-quaternary system such as Si_3N_4-SiO_2-AlN-Al_2O_3-YN-Y_2O_3, the lowest melting composition should be one of the quaternary eutectics. The ternary and binary eutectic compositions should melt at a temperature higher than that of the adjacent quaternary eutectics.

Riley, F.L. (ed.) Progress in Nitrogen Ceramics

Therefore, in order to determine the optimum amount of Y_2O_3 and Al_2O_3 to be used as sintering aid to produce Si_3N_4 ceramics, the system Si_3N_4-SiO_2-AlN-Al_2O_3-YN-Y_2O_3 was studied.

The sub-solidus phase relationships in the system Si_3N_4-SiO_2-AlN-Al_2O_3-Y_2O_3 has been studied and published by Naik, et al. (4). Among all of the sub-quaternary systems, the Si_3N_4-AlN:Al_2O_3-$Y_3Al_5O_{12}$-$Y_2Si_2O_7$ tetrahedron is of special interest because it is felt that potential high temperature ceramics could be found in this range. Therefore, the solid liquid reactions were studied. The graphic presentation of the subsystem is given in Fig. 1.

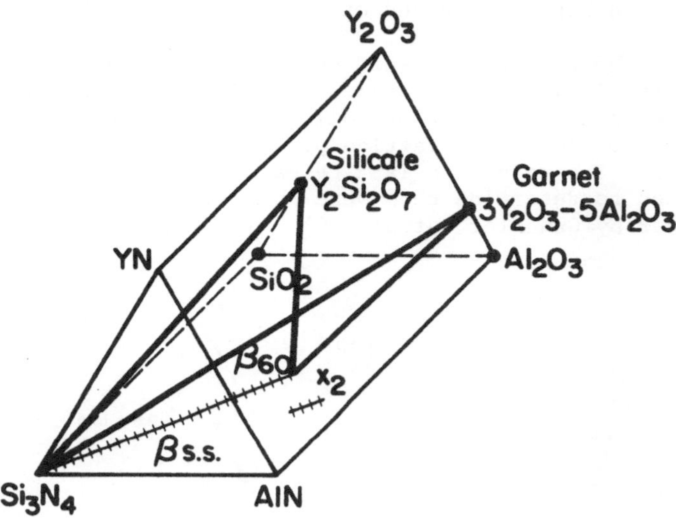

Fig. 1: The system Si,Al,Y/N,O. The compatability tetrahedron Si_3N_4-AlN:Al_2O_3-$Y_3Al_5O_{12}$-$Y_2Si_2O_7$ are shown in the diagram.

EXPERIMENTS AND RESULTS

Si_3N_4, AlN, Al_2O_3 and Y_2O_3 were used as starting materials. Batches of about 100 grams of powders were mixed under methanol in an attritor mill for four hours. The mixtures were then dried and granulated. Samples were sintered in a graphite resistance heating furnace under a mild flow of nitrogen. The experimental arrangement for sintering can be found in papers by Boskovic, et al. (5). Only samples with weight loss lower than 2% were analyzed and the results were used for this report.

Solid-Liquid Equilibria

The system Si_3N_4-AlN:Al_2O_3-$Y_3Al_5O_{12}$

Specimens were sintered at 1550°C for one hour and cooled at a rate of 200°C/min. Phases present in the "as fired" specimens were identified with XRD and the results were used to construct the diagram as shown in Fig. 2. There are three separate phase fields as shown in the diagram.

A. β-Si_3N_4 solid solution and liquid: "As fired" specimens in this region contained β-Si_3N_4 solid solution as the only crystalline phase. The presence of glasses in these specimens was apparent from the high background of their XRD patterns.

The tie lines shown in Fig. 2 were determined as follows: The aluminum content in the β-Si_3N_4 solid solution determines one end of the tie line and the amount of solid and liquid determine the length of the tie line. The lattice parameters of the β-phases were then measured. The reference lattice parameter values and the procedures used to determine them were the same as those described in the paper by Gauckler, et al. (6). The lattice para-

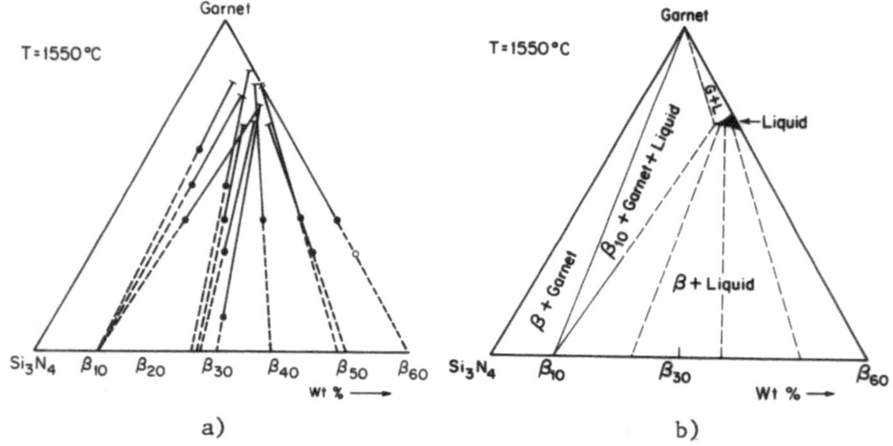

Fig. 2: The system Si_3N_4-(β-60)-$Y_3Al_5O_{12}$.

a) Experimental data: The compositions studied are shown as circles. The straight lines passing through these points are experimentally determined tie lines. (The subscripts under β indicate aluminum concentration in β-Si_3N_4 solid solution in equivalent percent).

b) 1550°C isotherm.

meters of the β-Si_{6-x}-$Al_xN_{8-x}O_x$ solid solution as a function of aluminum content is given in Fig. 3.

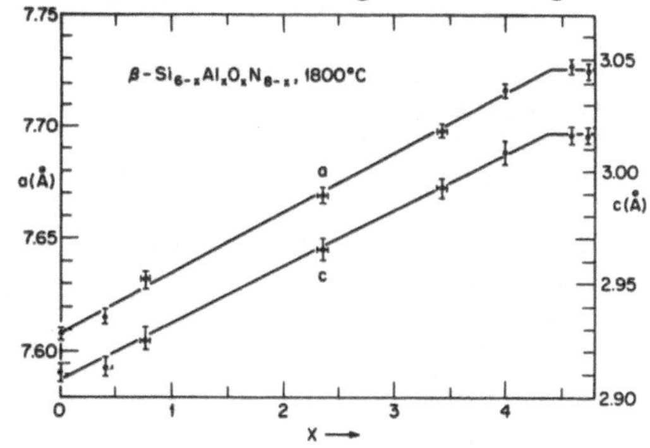

Fig. 3: Composition dependence of lattice parameters of β-Si_3N_4 solid solution containing $AlN{:}Al_2O_3$.

The lengths of the tie lines were determined by the use of the lever rule. Quantitative x-ray powder diffractometry was used to determine the amount of β-phase in the specimens. Calibration standards were made by mixing different amounts of β-solid solution powder with $Y_3Al_2O_{12}$ powder as fillers. Peak heights were compared and the results were used to construct the tie lines in Fig. 2.

B. Three-phase field: β-solid solution, $Y_3Al_5O_{12}$ and liquids: Compositions fired at 1550°C within this triangle showed a unique lattice parameter value which is identical to that of the β-Si_3N_4 solid solution with 10 eq. % Al (β-10). The other apex of the triangle should be the composition of the liquid. This composition point was determined by the tie line with a β-phase whose lattice parameter has the unique value of (β-10), yet, no garnet could be detected by x-ray diffraction.

Hence, the 1550°C isothermal section of the sub-ternary system Si_3N_4-$AlN{:}Al_2O_3$-$Y_3Al_5O_{12}$ can be estimated as given in Fig. 2. It should be noted that liquid is not present in the triangle Si_3N_4-(β-10)-garnet at 1550°C.

The System Si_3N_4-$AlN{:}Al_2O_3$-$Y_2Si_2O_7$

The same procedure as described in the proceeding section was used to determine the 1550°C isothermal section of this system. The results are used to construct the tie lines in Fig. 4. The melting behavior of the binary β-Si_3N_4 solid solution with 60 eq. % Al (β-60) and $Y_2Si_2O_7$ was determined by Müller (8) as shown in Fig. 5. The eutectic in the binary joins Si_3N_4-$Y_2Si_2O_7$ was reported in an earlier paper (9). When these data are used, the 1550°C isotherm can be estimated as shown. The discrepancy between these results can be explained by the large experimental uncertainty of the present experiments.

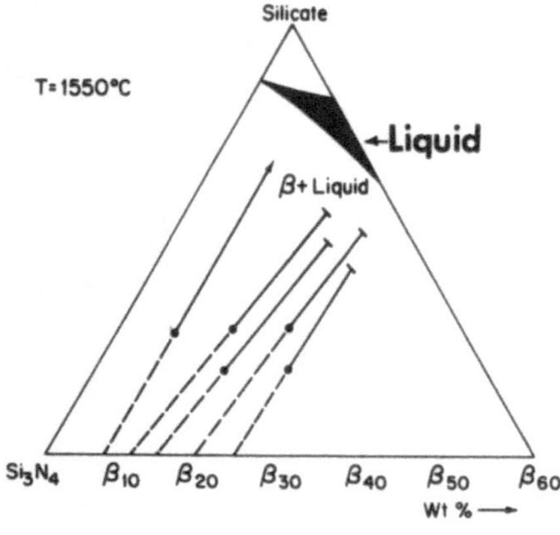

Fig. 4:
The system Si_3N_4-(β-60)-$Y_2Si_2O_7$. The compositions studied are shown as circles. The straight lines passing through these points are experimentally determined tie lines.

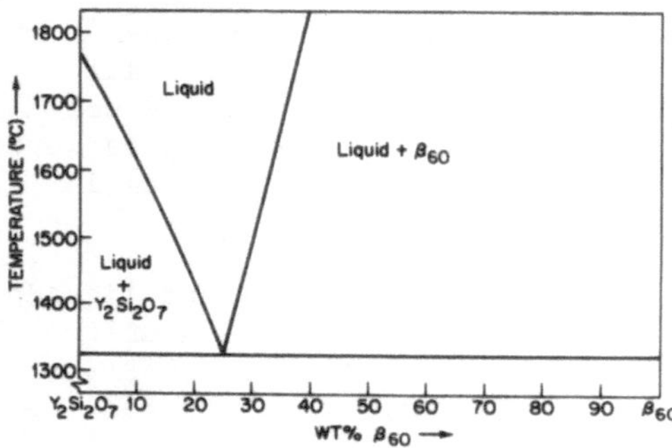

Fig. 5:
The system (β-60)-$Y_2Si_2O_7$.

Crystallization of Grain Boundary Glasses

In most of the silicon nitride--metal oxide systems, the liquids which act as sintering aids contain very high concentrations of SiO_2. These high SiO_2 containing glasses are very viscous and difficult to crystallize. In some systems, the liquids contain lower concentrations of SiO_2 and, hence, are less viscous. We believe that these glasses will devitrify more easily, particularly when their primary phase is not SiO_2 but rather a metal-silicate compound. The triangle Si_3N_4-(β-60)-$Y_3Al_5O_{12}$-$Y_2Si_2O_7$ is an example of this type. The effect of nucleation on crystallization has also been studied. Platinum was used for this purpose.

Compositions in the triangle β-Si_3N_4-(β-10)-$Y_3Al_5O_{12}$ in Fig. 2 contain no liquid at 1550°C. At 1650°C, the liquid region in this ternary system should be larger than that at the 1550°C. The 1650°C isotherm has been estimated as shown by the dotted line in Fig. 6. According to this diagram, at 1650°C composition B in Fig. 6 should contain two phases: β-phase plus liquid under equilibrium conditions. It should become two solid phases: β-phase plus $Y_3Al_5O_{12}$, when annealed at 1400°C for a long time.

Composition B was prepared by mixing Si_3N_4, AlN and Al_2O_3. The mixtures were hot pressed at 1650°C and cooled at a rate of 200°C/min. The cooled specimens were then heat treated at lower temperatures for different lengths of time.

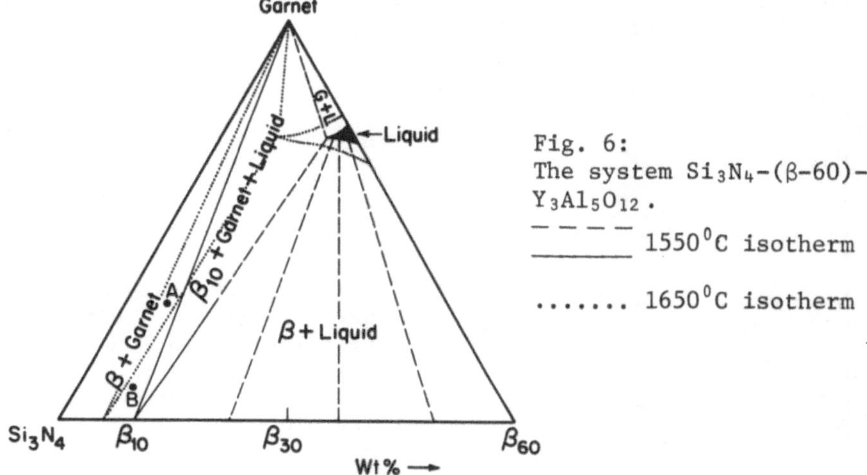

Fig. 6:
The system Si_3N_4-(β-60)-$Y_3Al_5O_{12}$.
– – – –
________ 1550°C isotherm

.... 1650°C isotherm

The hot pressed samples showed β-Si_3N_4 solid solution as the only crystalline phase by XRD. The existence of an amorphous halo due to the glassy phase was also apparent in the x-ray patterns. Upon heating at lower temperatures, the phase $Y_3Al_5O_{12}$ crystallized. The amount of the $Y_3Al_5O_{12}$ was measured by the peak height of the (321) and (422) line and the results are given in Fig. 7. The crystallization of additional β-phase cannot be detected by x-ray diffraction line intensity change because the additional amount of the β-phase was too small to be observed. However, the crystallization of the β-phase can be observed by the appearance of an additional diffraction peak of the β-phase with a d-value corresponding to $2\theta = 32.7^0$ (β_{50}).

The "as quenched" β-Si_3N_4 solid solution has a lattice parameter of 7.631Å. One extra diffraction peak appeared after annealing. This line was identified as the (101) line of the β-Si_3N_4 solid solution with a lattice parameter larger than that of the "as quenched" samples. After prolonged annealing, this extra peak disappeared. The lattice parameter of the β-phase changed with time, as shown in Fig. 8. The change of lattice parameter cannot be interpreted as homogenization of the solid solution because it should have increased instead of decreased. However, it was evident that there were compositional changes in the β-phase. This change should occur when β-Si_3N_4 solid solution crystallizes from the glass.

A second batch of the same composition was prepared. A platinum chloride solution (10% by weight) was added to this mixture to yield compositions containing 0.01 wt. % and 0.05 wt. % platinum as a nucleating agent. These samples received the same heat treatment as those samples without Pt. The as "hot pressed" samples contained β-phase only. Annealing results showed that the presence of Pt did have an effect on the garnet crystallization. The peak heights of garnet for specimens with and without Pt (annealed at different

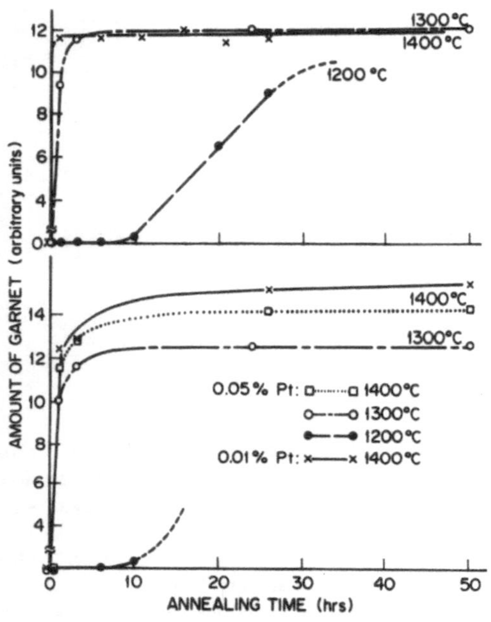

Fig. 7: Relative intensity of garnet ($Y_3Al_5O_{12}$) diffraction lines as a function of annealing time at different constant temperatures.

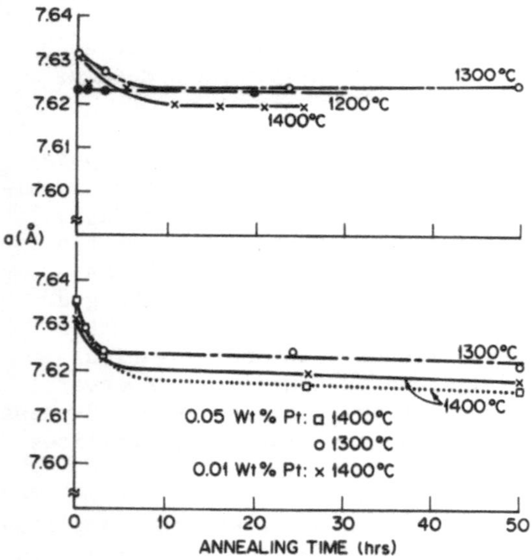

Fig. 8: Lattice parameter of the β-Si_3N_4 solid solution changes as a function of annealing time and temperature.

temperatures) are plotted in Fig. 7 as a function of annealing time. The intensity of the garnet peaks is higher for samples containing Pt as a nucleating agent. However, further work is needed to clarify the effect of Pt. It should be emphasized that when the specimens are heated at a high temperature under lower oxygen pressure, free silicon will form and the specimen will bloat severely. Whenever free silicon is present, Pt and Si will form a low melting eutectic, and the platinum will not be effective as a nucleating agent. The extra diffraction peak at $2\theta = 32.67^0$ has not been observed for samples containing Pt. This phenomenon is not understood.

Crystallization of the liquid (or glass) was evident from the above experiments; however, it is uncertain whether the amorphous phase was crystallized completely. Therefore, high resolution TEM was used to directly observe thin sections of the same specimens. Two microphotographs are given in Fig. 9. It is clear that the amount of glassy phase is reduced after long annealing. How much the improvement of the high temperature mechanical properties is influenced by the remaining small amounts of glassy material needs to be determined by high temperature bend strength measurements.

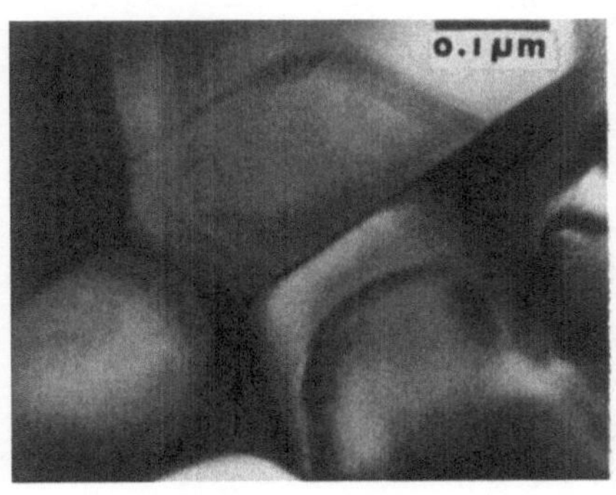

a)

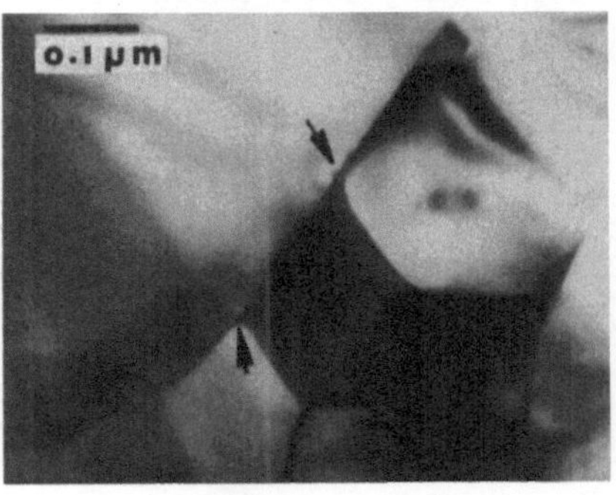

b)

Fig. 9: Transmission electron micrograph of the specimen in the system Si_3N_4-(β-60)-$Y_3Al_5O_{12}$.

a) Hot pressed at 1650^0C quenched. This specimen contains β-Si_3N_4 solid solution and glass. The glassy phase is indicated by the arrows.

b) The same specimen annealed at 1400^0C for 50 hours. The specimen shows β-solid solution, garnet grains (dark) and very small amounts of glass in grain boundaries and triple points.

ACKNOWLEDGMENTS

The authors are grateful to Dr. L.J. Gauckler and Dr. Ruhle for helpful discussions. This research was sponsored by the U.S. Department of Energy, Office of Energy Sciences.

REFERENCES

1. M.H. Lewis, A.R. Bhatti, R.J. Lumby and B. North. J. Mater. Sci. 15, 438 (1980).
2. J.T. Smith and C.L. Quackenbush. Proc. International Symp. "Factors in Densification and Sintering of Oxide and Non-oxide Ceramics", eds. S. Saito and S. Somiya, p.426, 1979, Tokyo Institute of Technology, Japan.
3. R.N. Katz and G. Gazza. "Nitrogen Ceramics", ed. F.L. Riley, Noordhoff (Leyden), 1977, p.417.
4. I.K. Naik and T.Y. Tien. J. Am. Ceram. Soc. 62, 643 (1979).
5. S. Boskovic, L.J. Gauckler, T.Y. Tien and G. Petzow. Powder Met. Int. 9, 185 (1977).
6. L.J. Gauckler, J. Weiss and T.Y. Tien. J. Am. Ceram. Soc. 61, 397 (1978).
7. I.K. Naik, L.J. Gauckler and T.Y. Tien. J. Am. Ceram. Soc. 61, 332 (1978).
8. R. Muller. Diplomarbeit, University of Stuttgart, 1979.
9. L.J. Gauckler, H. Hohnke and T.Y. Tien. J. Am. Ceram. Soc. 63, 35 (1980).

DISCUSSION

Popper: Can you distinguish between β'-sialon and yttrium aluminium garnet under the microscope? What was their crystallite size?

Tien: They can be identified by SEM and TEM, and by optical microscopy if large. The mean grain size was about 0.1 to 0.5 μm for samples densified at 1650°C and heat treated at 1350°C.

Clarke: How did you add your platinum nucleation agent, and did the Pt react with any of the phases? I find that Pt added as chloride reacts to form PtSi and leaves Cl in the intergranular non-crystalline phase.

Tien: Chloroplatinic acid in methanol, with ammonia precipitation of the Pt. The mixture was dried and heated at 500-600°C to try to remove NH_4Cl. We did not analyze for Cl or Pt, but the effect on crystallization was apparent.

Lange: Volatilization causes phase compositional changes and hinders phase equilibrium work. Hot pressing is a better technique than placing powders in a closed furnace.

Tien: We always used powder beds in BN lined graphite crucibles for sintering work; some hot-pressing was done. Samples with weight losses > 2% were rejected.

Greskovich: It would be desirable to extend phase equilibrium studies to > 1800°C because sintering is now being carried out in this region to develop high final densities.

Jack: Some general comments: There is a Mg-β'-sialon solid solution on the 3M:4X plane but unfortunately it is unstable above 1000°C. In the Y_2O_3-sialon system the yttria provides an Al-Si-Y-N-O liquid required for densification, and which can then react by post-preparative heat treatment to give a compatible intergranular crystalline phase with Y-β'-sialon. The Al-Si-Y-N-O glass reacts with β' to give yttrium aluminium garnet (YAG) and the Si and O from the glass go into the β' to reduce its z-value slightly, and so reduce its unit-cell dimensions. We have looked carefully at the possibility of N dissolving in YAG and find that the solubility is very small, with a limit at about $Y_3Al_{4.6}Si_{0.4}O_{11.6}N_{0.4}$ i.e. ~ 2 atom %.

HIGH TEMPERATURE REACTIONS AND MICROSTRUCTURES IN THE Al_2O_3-AlN SYSTEM

James W. McCauley and Normand D. Corbin

Ceramics Research Division,
Army Materials and Mechanics Research Center,
Watertown, Massachusetts 02172, U.S.A.

1. INTRODUCTION

The potential utilization of nitride and oxynitride ceramics in high and low temperature structural engineering systems has focused much interest on these materials. These new efforts have revealed many problems concerned with the formation, stability and sintering of single or multi-phase materials based on compositions in these systems. One of the keys to producing materials with controlled microstructures and optimized properties is an understanding of the phase equilibria. In particular, the phase diagram should include liquid and vapour phases if they occur.

Soon after the discovery of the sialon [β'-$(Si,Al)_3(O,N)_4$] solid solution by Jack and Wilson (1) and Oyama (2), much work centered on studies in the Si_3N_4-AlN-Al_2O_3-SiO_2 quaternary system. Behaviour diagrams and representations in terms of equivalents were used to describe the apparent phase relationships in these systems. However, behaviour diagrams are typically not equilibrium diagrams, and "equivalents" assumes constant valence ionic behaviour of components and phases. A careful review and analysis of these considerations suggested that a systematic study of the Al_2O_3-AlN system could result in much-needed information for the fabrication of silicon-aluminium oxynitride materials and provide a model for other oxynitride systems. Much work has already been carried out in this system by Long and Foster (3), Adams et al. (4), Lejus (5), and Michel (6), and more recently by Kieffer et al. (7), Gauckler and Petzow (8) and Sakai (9).

Riley, F.L. (ed.) Progress in Nitrogen Ceramics

2. PHASE EQUILIBRIA

The experimental method used to determine the phase equilibria has already been described by McCauley and Corbin (10). Careful microstructural analysis coupled with other characterization techniques formed the basis for our deductions. Figure 1 is a new phase equilibrium diagram for the pseudo-binary Al_2O_3-AlN composition joined in one atmosphere of flowing nitrogen. Much of the

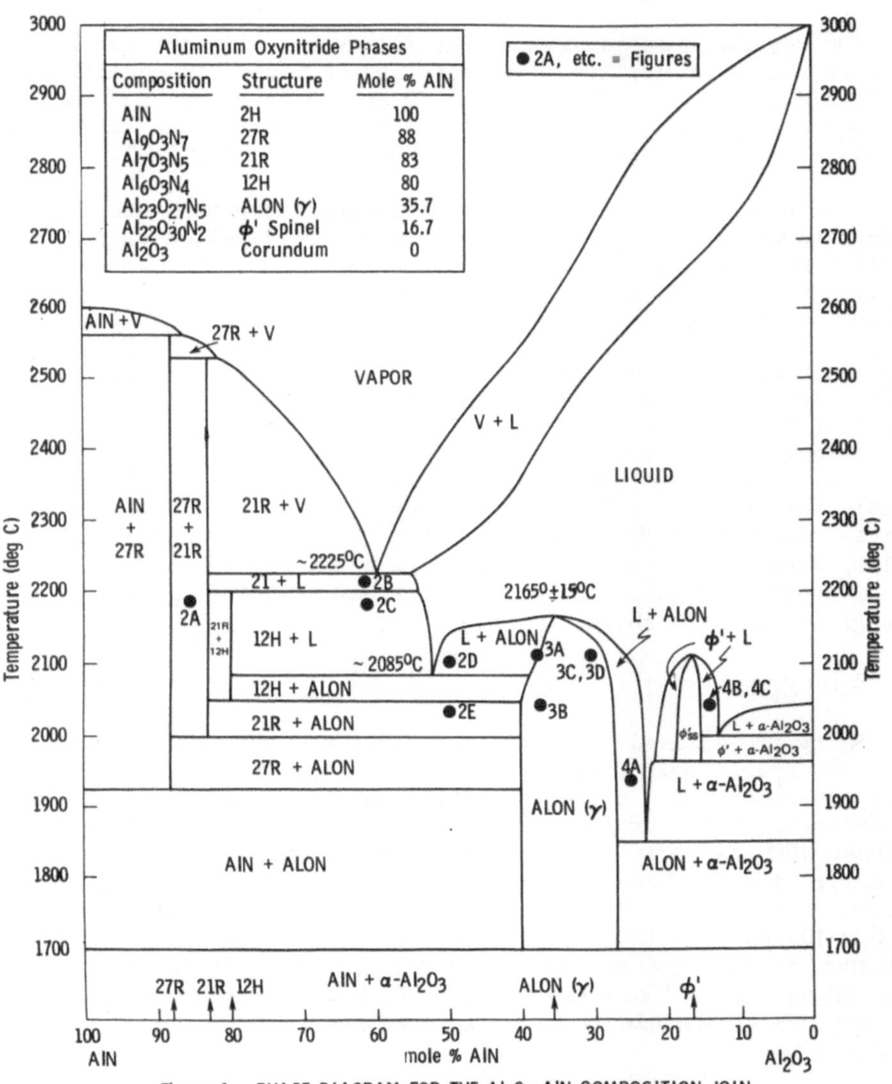

Figure 1. PHASE DIAGRAM FOR THE Al_2O_3-AlN COMPOSITION JOIN IN ONE ATMOSPHERE OF FLOWING NITROGEN.

reasoning involved in the representation emerged from interpretation of Zernike's (11) example of a binary system with one intermediate compound between a subliming (i.e. AlN) and a melting (i.e. Al_2O_3) end member. (See for example pp. 189-203 in Ref. 11.) In fact, the various peculiarities of the diagram might be better understood if it is visualized as an isobaric section through a binary P-T-X diagram. This diagram represents a logical description of the phase relationships as deduced from currently available data, but clearly requires further refinement.

The key features consist of the following: (1) the incorporation of solid-vapour, liquid-vapour, and liquid-solid equilibria; (2) liquid-solid eutectics at about 50 and 25 mole percent AlN; (3) composition and temperature stability limits for the AlN polytype-like oxynitride phases, AlON spinel (γ) and ϕ' (ζ-Al_2O_3).

In order to more clearly relate various microstructures to the phase diagram, all of the photomicrographs are labelled on Figure 1. Figure 2(a) illustrates a 27R-21R assemblage, which showed very little vapour loss. The specimens shown in Figures 2(b) and 2(c), however, showed appreciable weight loss due to the formation of liquid. Further, the liquid that forms solidifies into AlON+12H upon cooling. Figures 2(d), 2(e) and 2(f) illustrate assemblages to the right of the 50 mole percent eutectic, demonstrating the effect of liquid on the micro- and macrostructure. Representative microstructures and a translucent disc of AlON are illustrated in Figure 3. Figure 4(a) depicts the appearance of a liquid near 25 mole percent AlN at 1925°C. Figures 4(b) and 4(c) illustrate typical microstructure of the ϕ' phase (6), which is a ζ-Al_2O_3 or $LiAl_5O_8$ distorted spinel phase (3). Note the cross-hatched extinction effects in Figure 4(c) suggesting complex twinning, defect or intergrowth relationships.

3. PROPERTIES OF AlON

Using the derived phase diagram, AlON was pressureless sintered into nearly pore free, single phase material. This material can be described as a nitrogen stabilized aluminium oxide spinel. The following basic properties have been determined: Knoop (100) hardness 1800 kg/mm^2; elastic modulus of 3.3×10^5 MPa (47.3×10^6 psi); 4-point bend strength of 306 MPa (44.4×10^3 psi) at room temperature, 267 MPa (38.7×10^3 psi) at 1000°C and 190 MPa (27.6×10^3 psi) at 1200°C; and an average thermal expansion (α) from room temperature to about 1000°C of 7×10^{-6} C^{o-1}. This material transmits in the visible, near IR, microwave and millimeter wave regions of the electromagnetic spectrum.

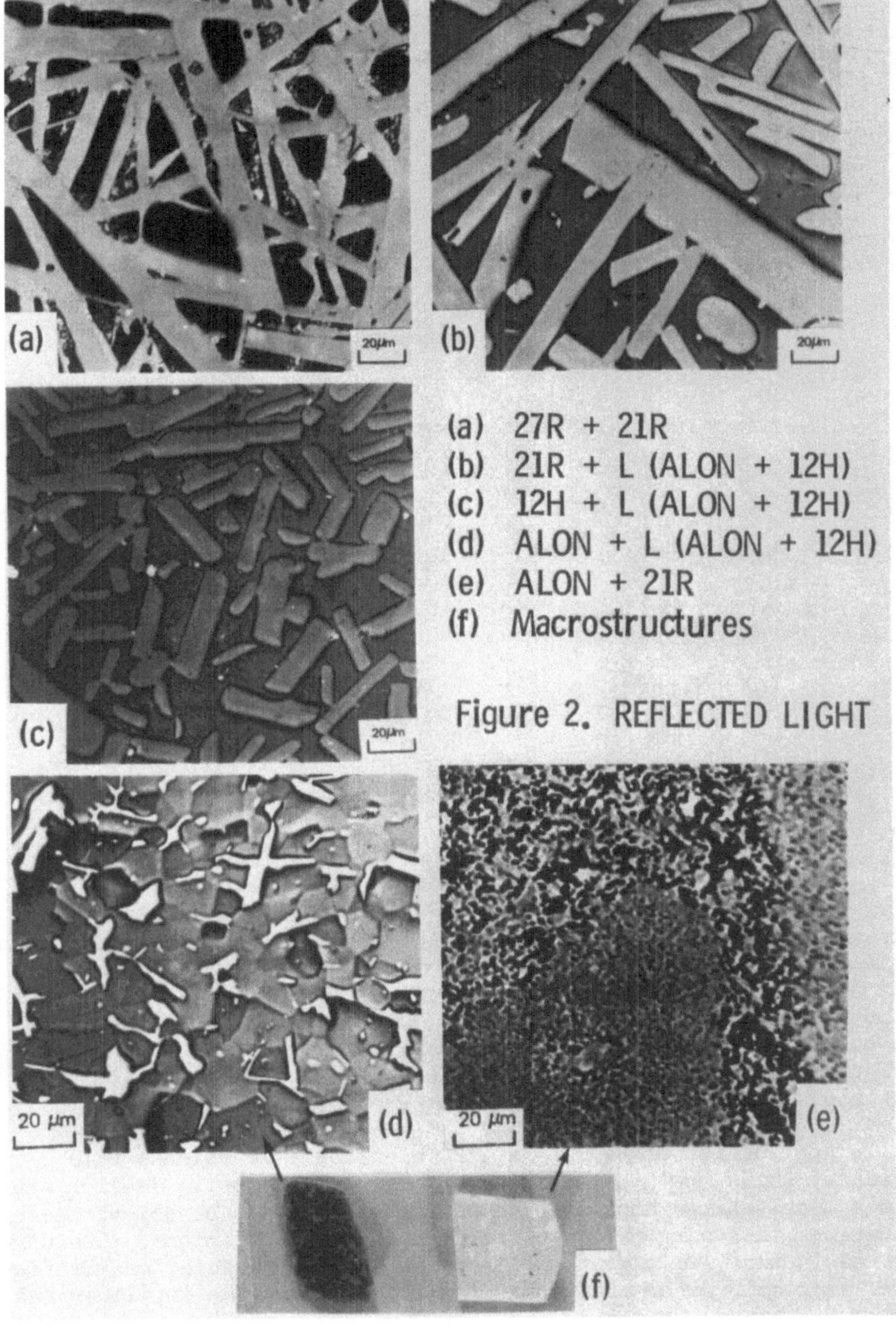

(a) 27R + 21R
(b) 21R + L (ALON + 12H)
(c) 12H + L (ALON + 12H)
(d) ALON + L (ALON + 12H)
(e) ALON + 21R
(f) Macrostructures

Figure 2. REFLECTED LIGHT

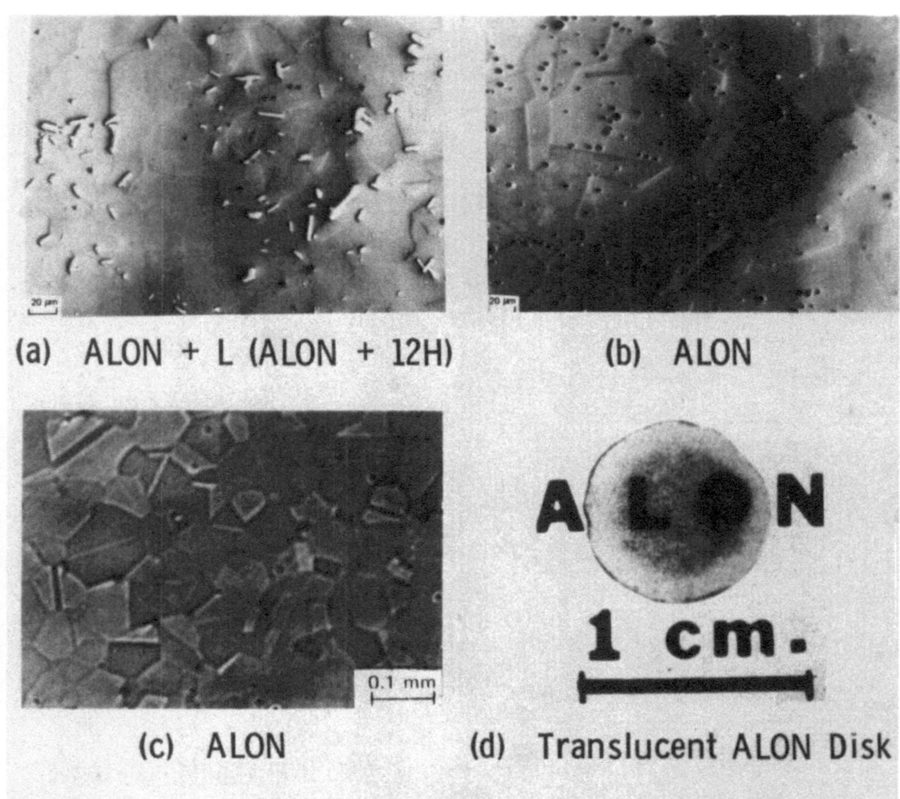

(a) ALON + L (ALON + 12H) (b) ALON

(c) ALON (d) Translucent ALON Disk

Figure 3. ALON MICROSTRUCTURES (REFLECTED LIGHT)

4. CRYSTAL CHEMISTRY

Traditional crystal chemistry normally involves cation-anion relationships, where complex cation replacements and substitutions are considered in an anion framework of fixed chemistry. However, in the Al_2O_3-AlN system, the cation does not change, while oxygen and nitrogen are varied. Further, AlN is primarily covalently bonded, whereas aluminium oxide has a greater ionic character. Figure 5 is an attempt to summarize all of these relationships as related to composition in the Al_2O_3-AlN system.

Application of Pauling's (12) electrostatic valence rule clearly shows that the addition of nitrogen anions to α-Al_2O_3 causes a local charge imbalance on nitrogen which can be alleviated by changing the coordination of Al from six to four anions. The converse is true for oxygen additions to AlN. Therefore, as the figure suggests, minor additions of nitrogen or oxygen to either end

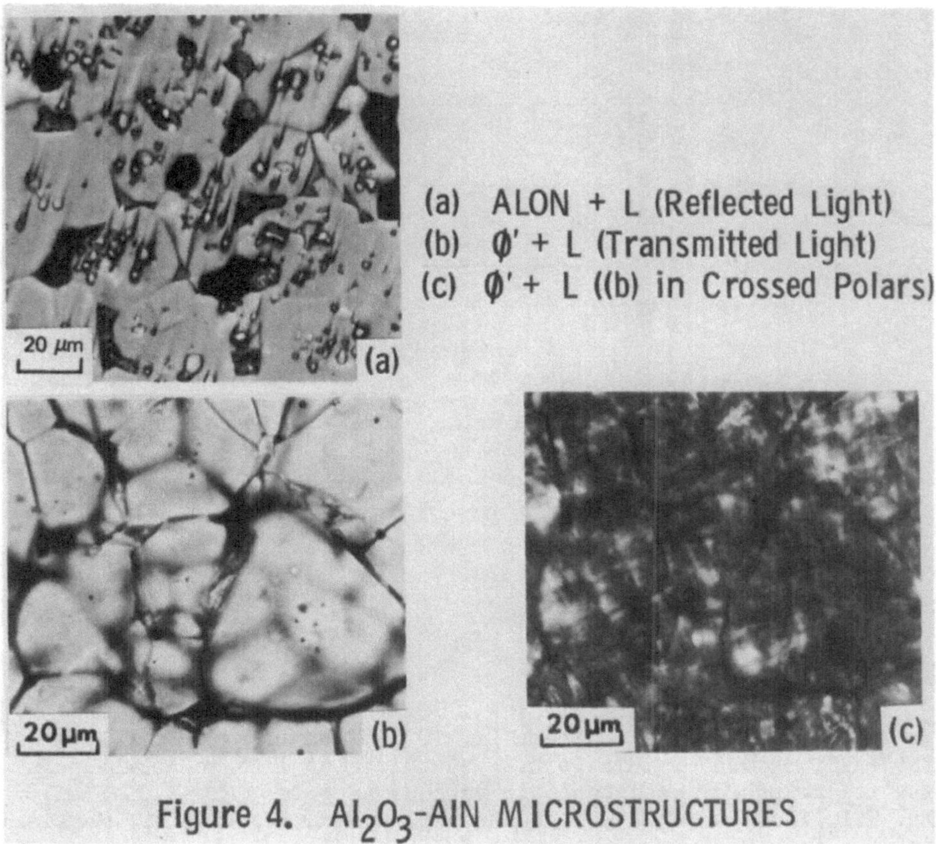

Figure 4. Al_2O_3-AlN MICROSTRUCTURES

member seems to result in a variety of modulated structures based on either AlN or spinel. It is also interesting to note that whereas AlN itself does not seem to form polytypic structures, small oxygen additions result in several polytype-like structures.

A model, assuming a constant anion spinel framework (13), has been successfully used to calculate both the AlON (x=5) and the ø' (x=2) phases:

$$Al_{64+x/3}\,O_{32-x}\,N_x.$$

The last important consideration concerns the bonding in this system. The change from predominantly covalent AlN to increasing ionic characteristics in α-Al_2O_3 seems to cause a change from solid-vapour equilibria to solid-liquid equilibria. This results in extremely complex equilibria in a transition zone which is near 50 mole percent AlN.

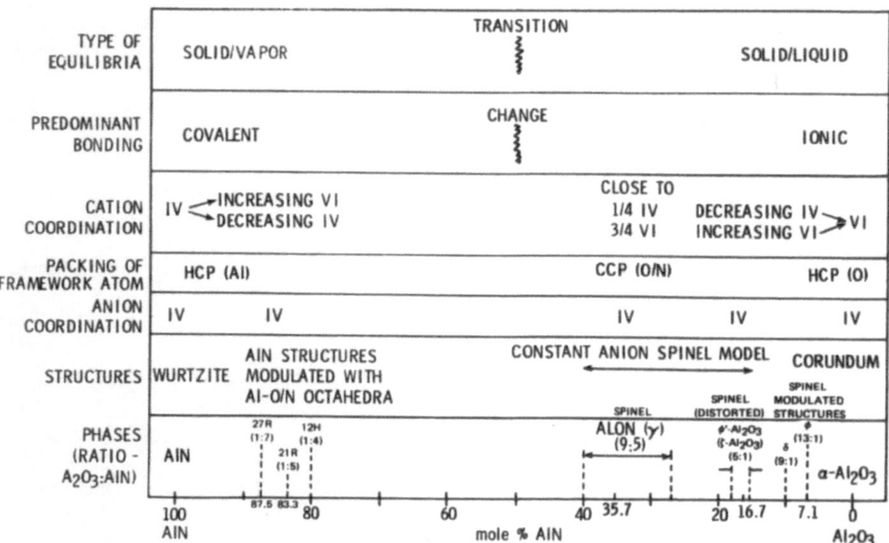

Figure 5. RELATION OF CRYSTAL CHEMISTRY TO COMPOSITION AND PHASE EQUILIBRIUM IN THE AlN-Al_2O_3 SYSTEM

ACKNOWLEDGMENTS

The authors wish to thank Dr. D.J. Viechnicki for his help in the early stages of this work. We would also like to acknowledge the following for assistance in various aspects: Mr. A.J. Zani, Mr. T.P. Sheridan, Mr. G.D. Quinn, Dr. R.N. Katz, Dr. D.R. Messier and Mr. G.E. Gazza.

The authors would also like to acknowledge the help of Professor K.H. Jack with the X-ray analysis.

REFERENCES

1. K.H. Jack and W.I. Wilson. "Ceramics based on the Si-Al-O-N and related systems". Nature Phys. Sci. (London) (1972), 238, 28-29.
2. Y. Oyama. "Solid solution in the ternary system Si_3N_4-AlN-Al_2O_3". Jap. J. App. Phys. (1972) 11, 760-761.
3. G. Long and L.M. Foster. "Crystal phases in the system Al_2O_3-AlN". J. Am. Ceram. Soc. (1961) 44, 255-258.
4. I. Adams, T.R. AuCoin and G.A. Wolf. "Luminescence in the system Al_2O_3-AlN". J. Electrochem. Soc. (1962) 109, 1050-1054.
5. A. Lejus. "Sur la formation a haute température de spinelles non stoechiometriques et de phases dérivées dans plusieurs systèmes d'oxydes a base d'alumina et dans le système alumine-

nitrure d'aluminum". Rev. Hautes Temper. et Refract. (1964) 1, 53-95.

6. D. Michel. "Contribution a l'étude de phénomènes d'ordonnancement de défauts dans les monocristaux de matériaux réfractaires a base d'alumine et de zircone". Rev. Int. Hautes Temper. et Refract. (1972) 9, 225-242.
7. R. Kieffer, W. Wruss and B. Willer. "Propriétés physiques et mecaniques de ceramiques $AlN-Al_2O_3$ obtenues par compression à chaud". Rev. Int. Hautes Temper. et Réfract. (1976) 13, 97-107.
8. L.J. Gauckler and G. Petzow. "Representation of multi-component silicon nitride based systems". F.L. Riley (ed), Nitrogen Ceramics, Noordhoff, Leyden (1977), 41-62.
9. T. Sakai. "Hot pressing of the $AlN-Al_2O_3$ system". Yogyo-Kyokai-Shi (1978) 86, 125-130.
10. J.W. McCauley and N.D. Corbin. "Phase relations and reaction sintering of transparent cubic aluminium oxynitride spinel (AlON)". J. Am. Ceram. Soc. (1979) 62, 476-479.
11. J. Zernike. Chemical Phase Theory, Kluwers Pub. Co. Ltd., Deventer, The Netherlands (1955).
12. L. Pauling. The Nature of the Chemical Bond, Cornell University Press, Ithaca, N.Y. (1960).
13. J.W. McCauley. "A simple model for aluminium oxynitride spinels", J. Am. Ceram. Soc. (1978) 61, 372-373.

DISCUSSION

Popper: Could AlON be considered for high pressure sodium lamps?

McCauley: Preliminary tests for sodium attack suggest that AlON is as good as and possibly a little better than α-Al_2O_3.

Lange: Why is the oxidation at 1300°C linear, and parabolic at 1200°C?

McCauley: Above 1300°C α-Al_2O_3 forms; below this temperature spinel-like nitride oxide phases may form, which are more compatible and more protective.

Billy: We have studied γ-AlON oxidation at Limoges (Materials Chemistry 1 (1976) 1317). The oxidation scale is another nitrogen-containing spinel (γ'-phase), stable to ~1150°C. It then decomposes to α-Al_2O_3.

Section C

FORMATION PROCESSES

SILICON NITRIDATION

F.L. Riley

Department of Ceramics, University of Leeds,
Leeds LS2 9JT, U.K.

1. INTRODUCTION

The primary object of this review is to outline the current state of understanding of the silicon nitridation reaction, and of the formation of reaction-bonded silicon nitride. Attention will be drawn in particular to developments of the last five years since the first NATO Advanced Study Institute (1).

Silicon nitride is a material of great interest for three main reasons:

i) the versatility and established applications of the reaction-bonded form;

ii) in the reaction-bonded form, it is the precursor of a fully dense, post-sintered, material, now attracting increasing attention;

iii) in powder form it is the starting point for the production of dense sintered and hot-pressed silicon nitrides, and the widening range of 'sialon' systems.

To these can be added the CVD, and other pyrolytic forms, produced from volatile, or organic, silicon and nitrogen containing species. This review will, however, be more concerned with the silicon-nitrogen reaction, in the particular context of the reaction-bonding process.

2. THE Si-N-O SYSTEM

The reaction between silicon and nitrogen is a thermodynamically spontaneous process under normal nitrogen partial pressures (0.1 MPa) and becomes kinetically significant at temperatures above

Riley, F.L. (ed.) Progress in Nitrogen Ceramics

1100°C. At temperatures of practical interest the reaction tends to be slow, and for useful bulk production rates high surface area silicon powders (typically $< 10\,\mu m$) must be used.

$$3Si_{(c)} + 2N_{2(g)} \rightarrow Si_3N_{4(c)}; \quad \Delta G^{\ominus}_{1643} = -205 \text{ kJ mol}^{-1} \qquad [1]$$

Reaction [1] is strongly exothermic with $\Delta H^{\ominus} \sim -733$ kJ mol^{-1}. The density of silicon is 2329 kg m^{-3} and that of silicon nitride is $\sim$ 3185 kg m^{-3}, so that a volume expansion of 21.7% occurs during nitride formation. On the basis of the Pilling-Bedworth rule, coherent nitride films would not be expected to form on a silicon surface by normal interdiffusion processes. Moreover at normal nitriding temperatures (1200-1450°C) silicon nitride shows no noticeable plasticity, and as the overall compact volume change during the nitridation of silicon powders is essentially zero, it is clear that considerable internal rearrangement of product material must occur within the pre-existing void spaces of the compact. Solid state diffusion coefficients for Si in pure α- and β-phase silicon nitride are not known with certainty, but are certainly very low. Published values for N show a surprisingly large difference in the pre-exponential terms for the two phases, suggesting that extrinsic effects were dominant (2). Silicon has, however, appreciable volatility in the temperature range of interest:

$$Si_{(c)} \longrightarrow Si_{(g)}; \quad \Delta G^{\ominus}_{1630} = -211 \text{ kJ mol}^{-1} \qquad [2]$$

and a major material transport process must be that of evaporation from exposed silicon surfaces followed by condensation at surfaces of lower silicon chemical potential; in effect the reaction [1]. Silicon mobility in the vapour phase within a powder compact is furthermore likely to be enhanced through the formation of silicon monoxide through reactions involving the inevitable traces of oxygen and water vapour in the nitriding atmosphere:

$$Si_{(c)} + H_2O_{(g)} \longrightarrow SiO_{(g)} + H_2; \quad \Delta G^{\ominus}_{1643} = -82 \text{ kJ mol}^{-1} \qquad [3]$$

$$Si_{(c)} + \tfrac{1}{2} O_{2(g)} \longrightarrow SiO_{(g)}; \quad \Delta G^{o}_{1643} = -238 \text{ kJ mol}^{-1} \qquad [4]$$

Both reactions are effectively quantitative, although the maximum silicon monoxide partial pressure able to exist within the compact is controlled by the dissociation equilibrium:

$$2SiO_{(g)} \longrightarrow Si_{(c)} + SiO_{2(c)}; \quad \Delta G^{\ominus}_{1643} = -143 \text{ kJ mol}^{-1} \qquad [5]$$

giving a value for p_{SiO} in the region of 500 Pa at 1643 K. This is several orders of magnitude higher than that for p_{Si} ($\sim$ 20 mPa). These facts form the basis for the generally contentious issue of whether the primary nitridation process is represented by reaction [1] or [6](3):

$$3SiO_{(g)} + 2N_{2(g)} \rightarrow Si_3N_{4(c)} + \frac{3}{2}O_{2(g)} ; \quad \Delta G^{\ominus}_{1643} = +545 \text{ kJ mol}^{-1} \quad [6]$$

Reaction [6] becomes viable if a convenient sink for oxygen exists, and silicon itself is adequate in this respect. In the void spaces in the silicon powder compact oxygen fluxes between source and sink are likely to be high. At 0.1 MPa pressure, mean free paths are of the order of 1 µm, which is probably larger than the mean interfacial separation distance between growing silicon nitride and unnitrided silicon. Sticking coefficients assume kinetic importance therefore, but are unknown for this system.

Mass transport rates in theory present no problems. On the basis of the standard Langmuir evaporation rate equation:

$$m = \left[\frac{M}{2\pi RT}\right]^{\frac{1}{2}} . p \quad [7]$$

transport rates even for silicon vapour are orders of magnitude higher than those required to account for integrated nitride formation rates, even though the most appropriate value for p in a non-equilibrium environment might be difficult to establish. Initial instantaneous nitridation rates on the other hand are closer to those expected from the Langmuir evaporation rate. Volatilisation from free silicon surfaces should not in itself be a rate controlling process therefore. Surface diffusion of silicon from an exposed silicon surface to nitride growth sites is also feasible, and could account for some of the broad features of the powder nitridation process. However, surface diffusion coefficients for silicon under nitrogen are unknown.

3. NITRIDATION KINETICS

A number of problem areas have been clearly defined, each of which has received detailed attention with varying degrees of success in the development of satisfactory models. These may be enumerated as:

i) initiation of nitride formation, in the presence of a protective barrier film of native silicon dioxide at a silicon-nitrogen interface (4);

ii) attainment of adequately fast reaction rates, while holding the temperature below the melting point of silicon, or of silicon-metal eutectic temperatures, if impurities are present. This is of importance in the earlier stages of the reaction, and it is likely that low temperature nitridation (<1410°C) is to be preferred to a higher temperature reaction. The exothermic nature of the nitridation process is of importance in this context, particularly so for the nitridation of large volumes of silicon powder (5);

iii) the attainment of an adequate degree of nitridation with higher purity silicon powders, since normal commercial powders contain localised impurities responsible for critical flaws in fully nitrided material (impurities can, on the other hand, be tolerated if they are distributed homogeneously, and the addition of finely dispersed iron, or other 'catalysts', is standard practice)(6,7,8);

iv) the development of microstructures containing 'bridging' nitride as opposed to 'non-bridging' nitride, which is expected to be of importance for strength development in the reaction-bonded material (9).

3.1 The silicon dioxide film, and reaction initiation

Silicon, in common with almost all metals, is oxidized spontaneously at room temperature and the best estimates indicate that a 3 nm thick film of oxide is normal (4). This film appears to inhibit access of nitrogen to the silicon surface, and, perhaps more importantly, the volatilisation of silicon. There is evidence that some nitrogen diffusion through the film is possible, with formation of underlying films of barrier nitride at the silicon/oxide interface (10,11). Nitridation can therefore only occur to an appreciable extent when the oxide film has been removed or disrupted. Various techniques are available, used single or in combination, and all appear to have a 'catalytic' action on the nitridation process, generally most apparent in the earlier stages.

i) Vacuum pretreatment at 1000-1200°C[4]:

$$SiO_{2(c)} \rightarrow SiO_{(g)} + \tfrac{1}{2} O_{2(g)} \quad [8]$$

This reaction, in theory, requires that the oxygen partial pressure be less than 1 mPa, not an unreasonably low value.

ii) Hydrogen pretreatment[12], ensuring low oxygen partial pressures and SiO_2 loss through the direct reaction:

$$SiO_{2(c)} + H_{2(g)} \rightarrow SiO_{(g)} + H_2O_{(g)} \quad [9]$$

iii) By addition of (non-volatile) metallic fluorides[12], most probably through the reaction:

$$2MF_{2(c)} + SiO_{2(c)} \rightarrow SiF_{4(g)} + 2MO_{(c)}; \quad (\Delta G^{\ominus}_{1600} \text{ for M=Mg; } 88 \text{ kJ mol}^{-1}) \quad [10]$$

iv) By transition metal additions[14,15]; the mechanism in these cases is not unequivocably established, but there is strong evidence that a form of 'devitrification' and disruption of the silicon dioxide occurs, allowing rapid loss through the reaction:

$$Si_{(c)} + SiO_{2(c)} \rightarrow 2SiO_{(g)}; \quad \Delta G^{\ominus}_{1643} = 143 \text{ kJ mol}^{-1} \quad [11]$$

Many 'catalyst' systems for silicon nitridation almost certainly come into the classes (iii) and (iv). The use of hydrogen at the 5-10% level in nitrogen is standard practice (7), and claims have been made for a synergistic effect of iron and hydrogen in combination (16).

The complete loss of SiO (or SiF_4) through these reactions is only possible in a flowing system because of the relatively large volumes of vapour generated at the low partial pressures and high temperatures. An alternative sink for SiO under nitrogen and in the presence of silicon is the silicon nitride oxide-forming reaction:

$$Si_{(c)} + SiO_{(g)} + N_{2(g)} \rightarrow Si_2N_2O_{(c)}; \quad \Delta G^{\ominus}_{1673} = -400 \text{ kJ mol}^{-1} \qquad [12]$$

which has been shown to occur through the intermediate formation of SiO (17). This reaction would be similarly favoured by low oxygen partial pressures, and agents facilitating the exposure of silicon surfaces. Small traces of Si_2N_2O in normal silicon nitride would be undetectable, although its presence has been reported, presumably where more serious oxygen contamination has occurred(18).

3.2 Nitridation kinetics and mechanisms

The problem with trying to assess literature data amassed in the last five years is the fact that several factors interact and need to be taken into account in most experimental systems - the purity of the starting materials and gaseous atmosphere, particle size and size distribution, and temperature control, probably being the most important. These become more prominent when quantitative comparisons are attempted, particularly with powder systems. Certain qualitative conclusions can be drawn with a fair degree of confidence, however.

3.2.1 Single crystal nitridation

Meaningful quantitative measurements on silicon single crystal nitridations have proved almost impossible because of the problem of adequately excluding oxygen and water vapour (19), and other gas phase impurities such as metal oxide vapours. Qualitative observations of the developing nitride microstructure have, however, been most informative in providing a very useful guide to processes likely to be occurring in powder systems. It is clear that nitride nucleation occurs only at points on the silicon surface, and that nitride growth occurs predominantly at these sites (4,20). Nitrogen partial pressure is an important variable in controlling the nuclei density and therefore the morphological form of the nitride. Evaporation of silicon from exposed points occurs, with the development of extensive pore and cavity formation at the silicon-nitride interface. These pores may subsequently partially become filled by nitride. The product nitride acts as an effective barrier to both vapour and solid state diffusion, and little nitride growth at a silicon-silicon nitride interface can be expected. Eventually the reaction effectively ceases as access points for silicon vapour become sealed off. All nitride formed in the absence of liquid phases can strictly be termed 'CVD' nitride, in the sense that it is formed from gaseous species interacting at a solid surface, which also acts as an essential heat sink in the highly exothermic process:

$$3Si_{(g)} + 2N_{2(g)} \rightarrow Si_3N_{4(c)}; \quad \Delta H^{\ominus} = -2066 \text{ kJ mol}^{-1} \qquad [13]$$

3.2.2 Liquid silicon nitridation

Only a small number of studies have been reported on the nitridation of liquid silicon (21,22). It seems clear that nitrogen is adequately soluble in liquid silicon (estimated solubility 0.02 a/o at 1410°C (23)) and that, once nucleated, single crystals of silicon nitride grow unimpeded into the liquid matrix. In the absence of nuclei, growth initiation appears to be more difficult. Characteristic growths are always observed when silicon powder compacts are taken above a eutectic temperature, or the melting point of silicon itself, during powder compact nitridations, and account for the denser regions of silicon nitride reported in the earlier literature. These growths would not be expected to form interconnected nitride developments in powder nitridations, and they would not therefore make the major contribution to the strength of RBSN materials. This type of nitride appears to be normally β -phase, and the extent of its development has been correlated with the volume of liquid eutectic likely to be present in transition-metal doped silicon powders (8). Quantitative measurements of the rate of formation of nitride crystals in silicon under nitrogen appear to indicate a parabolic rate law, consistent with an increasing length of diffusion path from the silicon-nitrogen interface to the silicon nitride surface (22).

3.2.3 Powder compact nitridations

(i) Condensed phase effects:

Very many attempts have been made to understand the processes taking place within a silicon powder compact during exposure at high temperature to nitrogen. The difficulties of interpreting kinetic data are compounded, in many cases, by absence of complete information regarding the experimental conditions. Silicon powders are almost always contaminated, by iron or by tungsten carbide and cobalt, from milling operations. 'Pure' powders may be obtained by crushing and milling semi-conductor grade material in iron mills, followed by prolonged leaching with acid to remove contaminant iron, but it is very difficult to attain iron levels of less than 100 ppm, an experimentally significant quantity (4).

Numerous attempts to determine rate laws have been made as a route to understanding the nature of the rate-controlling processes (24,25,26,27). In view of the complexity of the nitride growth process as revealed by microscopic studies of single crystal silicon slices, and of the likely involvement of small percentages of liquid phase materials in normal systems, no single rate law is likely to describe uniformly all nitridation reactions. From the practical standpoint of optimising formation conditions for reaction bonded silicon nitride, with an accent on obtaining reproducibility, a semi-quantitative understanding of the function of the variables and of the temperature dependences of the processes becomes more important. However, it is of at least academic interest that three rate laws have been shown to describe fairly closely different stages of the nitridation of high purity silicon powders, and these can be correlated with a reasonable idealised model for the reaction.

The initial stage of the process of nitridation under nitrogen in certain cases can be regarded as an 'induction' period, and may be identified with the time taken for the protective silicon dioxide film on each particle to evaporate. At reduced nitrogen partial pressures in cases where the prior removal of the oxide film has occurred, a linear reaction rate may initially be seen (4). This stage presumably corresponds to the relatively unhindered evaporation of silicon from exposed surfaces. Subsequent kinetics are described reasonably well by versions of the 'pore closure' model (4,26), originally developed to account for metal oxidation where oxygen gains access to a metal surface through high conductivity fissures or pores, which progressively become sealed with development of product (28). The two forms of this law are derived on the assumption that 'pores' either do, or do not, assist in the closure of neighbouring 'pores'. Intuition suggests that the mutual-pore closure model would be more appropriate in view of the vapour phase CVD process operating and the free movement of material within the confines of the voids in a powder compact. 'Pure' forms of the corresponding rate law are unlikely to be found, although the semi-logarithmic treatment of nitridation data can in some instances provide a useful fit to data points over considerable extents of reaction (26,29). For comparative purposes between different systems, values for instantaneous nitridation rates at specified extents of conversion, or reaction time, are often more informative than the integrated 'total extent of nitridation' figures commonly used.

(ii) Gas phase variables:

Nitrogen. The effect of varying nitrogen partial pressure on the reaction rate has been established for the initial, linear, stage of nitridation, by a simple comparison of linear rate constant with nitrogen partial pressure (4). Preliminary data suggest that this is true also of the later stages (26). The picture is clouded by the fact that the nitrogen partial pressure may also change significantly the morphology of the product, and at longer nitridation times nitridation rates have been observed to show an inverse relationship with nitrogen pressure when this is much below 0.1 MPa(4).

Water and oxygen. These gases are impossible to exclude from high-temperature furnaces, and may be assumed to exist in most nitriding gases at the 10-100 Pa level, unless rigorous precautions are taken (30). Remarkably large concentrations of water vapour or oxygen in the long term have surprisingly little effect on the nitridation reaction and the nature of the product (31). The reason is presumably the self-gettering action of the silicon powder compact, and the oxidation of the outer regions of the compact to SiO or SiO_2, removing quantitatively both oxygen and water, and establishing a corresponding partial pressure of SiO (with a maximum value in the region of 500 Pa) and hydrogen in the inner regions of the compact. Nitridation there takes place in effectively constant, low partial pressure of oxygen, and SiO, together with hydrogen derived from the water vapour (26). The larger scale deliberate use of such buffer material has been demonstrated by the nitridation of silicon 'in air', using protective silicon nitride powder beds (32).

Hydrogen. Hydrogen is of interest in the contexts of the reported actions of small percentages of water vapour, and of the well-known accelerating action of the gas itself (29,33,34). Many commercial nitriding atmospheres employ 5-10 v/o hydrogen in nitrogen. The precise nature of the acceleratory function of the hydrogen has been harder to identify. It is clear that one function is to aid initiation of the nitridation reaction by assisting evaporation of surface silica (12). However, examination of subsequent instantaneous nitridation rates (or rate constants when a rate equation can be obtained) shows a longer lasting function of the hydrogen. One model proposed (3) is based on the supposition that an important potential nitride generating reaction is the nitridation of silicon monoxide (reaction [6]), and that the hydrogen in acting as a sink for oxygen supports a higher flux of oxygens (as water vapour) away from nitride growth sites. There is difficulty, however, in visualising gas fluxes within a mean free path dimension. The effectiveness of hydrogen in acting as an accelerator reaches a limit at the 50 to 100 kPa level, whereas the kinetics implied by reaction [6] would require that

$$d[Si_3N_4]/dt \propto p_{H_2} \qquad [14]$$

without a predetermined limit. An apparent induction period has been found to be inversely proportional to p_{H_2}, also with a limiting effectiveness for p_{H_2} around 50 to 100 kPa (26). It is possible that the two features are related, and that the subsequent nitridation rates are in effect determined by processes occurring during the short induction period (perhaps within 10^2-10^3 seconds). These might be, for example, the formation of a thin silicon nitride barrier film at the silicon-silicon dioxide interface, or of silicon nitride oxide films. This aspect requires more detailed study, not least because of the association of hydrogen additions with strength improvements in the reaction bonded silicon nitride (35).

Gas flow rate. Significant differences between 'static' and 'flow' systems have been suggested as being important for the nitridation process, and also the quality of the product, with 'static' conditions being favoured (9,34). This argument concerns essentially the importance of the SiO nitridation reaction for the development of 'bridging' silicon nitride, and it is supposed that under flow conditions the SiO is more readily purged from the system. It is equally possible, however, given the realities of nitridation furnace atmospheres, to equate a static system with a relatively high water vapour, and hence hydrogen, internal partial pressure. Such strength differences may again, therefore, be similar to those observed with the deliberate use of hydrogen and be related to identical nitride development processes.

(iii) Temperature:

Nitridation temperature is an important variable, for the expected reasons of the Arrhenius relationship, and the exponential dependence of equilibrium constants, vapour pressures, and reaction rates on temperature. It assumes added significance, however, in view of the strongly exothermic nature of silicon nitridation, and of the necessity for avoiding premature melting of the silicon, before sufficient nitride skin has developed on each silicon particle to prevent agglomeration of liquified grains with the formation of effectively un-nitridable large volumes of liquid silicon. The indications are that low temperature 'CVD' nitride is 'better' material and is therefore to be preferred to higher temperature 'VLS' nitride (35). Important development work has been carried out on controlled nitridation-temperature conditions in which the rate of temperature rise is programmed with the rate and extent of nitride formation. In this way maximum nitridation rates in theory can be achieved, consistent with optimisation of the homogeneity and quality of the product (7,36,37).

Further important development work has been the extension of attempts to mathematically model reaction and temperature profiles within silicon powder compacts of simplified geometry. The importance of temperature (and nitrogen partial pressure) gradients within powder compacts of moderate size has been indicated, and relationships between temperature and reaction profiles established (38,39). While such conclusions are qualitatively clear, an accurate knowledge of reaction activation enthalpies, and of the variation of reaction rates with time, is required if more quantitative information is to be derived. Such data are not available at present with a satisfactory degree of reliability. This work on the mathematical modelling of the reaction system deserves to be continued.

At much higher temperatures (>1450°C) microstructural changes occur under nitridation conditions, in that Ostwald ripening and grain growth in the nitride occurs, the result of the dissolution of fine scale nitride in liquid silicon, and recrystallization as large grain material. Nitridation completed under such conditions can therefore yield a completely different microstructure (40). The nitridation of liquid silicon has already been referred to, and on a small scale presents no kinetic difficulties.

The use of helium as an inert nitrogen diluent is an interesting development of the controlled nitridation of silicon (41). Its function presumably depends on its high thermal conductivity, and the greater efficiency of heat transfer from the exothermic reaction zone, providing more homogeneous and controllable nitridation conditions.

(iv) Presintering of silicon powder compacts:

Silicon powder compacts may be pre-sintered under argon at temperatures in the region of 1200°C in order to impart strength and to permit easier machining of compacts prior to full nitridation. Pre-sintering is likely to have at least three consequences for the succeeding nitridation:

(a) Some silicon dioxide will be lost, if the oxygen partial pressure at this stage is sufficiently low. Nitridation rates should therefore initially be faster (4).

(b) Loss of surface area due to fine particle sintering (29,42) with an effect opposite to that of (a).

(c) The redistribution of particulate impurities able to form eutectics at sintering temperature. The effect of this redistribution is likely to depend on the size of the original particle, but significant changes in microstructure of the sintered silicon powder compact are observed (6), and which are carried through into the final silicon nitride.

4. MICROSTRUCTURE DEVELOPMENT IN REACTION BONDED SILICON NITRIDE

The microstructure of the 'green' silicon powder compact has an important influence on the microstructure of the product reaction bonded silicon nitride. It is clear, however, that two processes contribute further to the development of the silicon nitride microstructure. Firstly, the 'CVD' production of silicon nitride which can result in well-formed, inter-connected crystals of silicon nitride, likely to be of considerable strength. Secondly, a 'VLS' process at higher temperatures, resulting in the complete nitridation of residual silicon. The strength of the porous product must depend on:

(i) pore size, and grain size distribution - functions to a large extent of the starting silicon particle size and size distribution (43,44);

(ii) the extent of development of interconnecting 'CVD' silicon nitride;

(iii) the presence of impurity inclusions in the silicon powder leading to gross defect formation.

The first and third variables lie outside the nitridation topic. The second appears to be related to the presence of hydrogen in the system, though the exact mechanism of operation is not known.

5. FURTHER ASPECTS

5.1 The 'α/β question'

The phase nature of the product nitride has not been treated in this survey. The thermodynamic arguments have been dealt with elsewhere in detail (45). Experimental evidence from silicon nitridation in general is that the formation of β-phase material is favoured by the presence of liquid phases, though solid phase nucleation is clearly essential for silicon nitride production at all, and nitride may tend to grow epitaxially (46,47). Linear relationships have been observed between the amount of liquid eutectic likely to be present, and the total yield of β-phase material (8), but detailed kinetic studies are required. It has been considered possible that the formation of α- and β-phase may be controlled at will by controlled seeding, given suitable background conditions (46,47).

5.2 Porosity infilling

Filling the interconnected porosity in reaction bonded silicon nitride is a separate question which will be discussed elsewhere in this volume (48). Attempts have been made to fill pore space with pyrolytically formed silicon nitride, with some evidence for slight strength improvement in initially weak material (49). It is clearly necessary for pore infilling to approach 100% for useful improvements in strength to result, and liquid silicon, or metal-silicon eutectics, are obvious candidate materials. This has in the past appeared to present difficulties, because of an apparent poor wettability of silicon nitride by silicon (50), and also by the decomposition of silicon nitride at the high temperature required for adequate reduction in the viscosity of the liquid phase. Recent work has, however, demonstrated that voids can be completely filled, though an assessment of the properties of these materials is not yet available (40).

6. CONCLUSIONS

It is clear that the technology of reaction-bonded silicon nitride production is still ahead of fundamental understanding of the underlying processes and principles. Profitable areas for study remain as:

(i) the full analysis of the kinetic function of additive gases such as hydrogen and helium, working under high purity, fully characterized, conditions;

(ii) controlled schedule nitridations and the function of 'inert' gases in these systems;

(iii) the mathematical modelling of nitriding silicon powder compacts;

(iv) post-nitridation treatments to improve the homogeneity of the material, and in particular to reduce the void volume.

The last five years have seen significant improvements in the density and quality attainable in reaction bonded silicon nitride. There has also been some improvement in understanding of the effects of several variables, but full understanding remains elusive.

REFERENCES

1. Nitrogen Ceramics, Ed. F.L. Riley, NATO ASI Applied Science Series E No. 23, Noordhoff, Leyden, 1977.
2. K. Kijima and S. Shirasaki. J. Chem. Phys. 65 2668, 1976.
3. A.J. Moulson. J. Mater. Sci. 14 1017, 1979.
4. A. Atkinson and A.J. Moulson. Science of Ceramics 8, B. Ceram. Soc. 1976, p.111.
5. J.A. Mangels. See reference (1) p.569.
6. P. Arundale and A.J. Moulson. J. Mater. Sci. 12 2138, 1977.
7. J.A. Mangels. J. Mater. Sci. 15 2132, 1980.
8. S.M. Boyer and A.J. Moulson. J. Mater. Sci. 13 1637, 1978.
9. B.F. Jones, K.C. Pitman and M.W. Lindley. J. Mater. Sci. 12 563, 1977.
10. B.H. Vromen. Appl. Phys. Letters 27 152, 1975.
11. E. Kooi, J.G. van Lierop and J.A. Appels. J. Electrochem. Soc. 123 1117, 1976.
12. D. Campos-Loriz and F.L. Riley. J. Mater. Sci. 14 1007, 1979.
13. D. Campos-Loriz, S.P. Howlett, F.L. Riley and F. Yusaf. J. Mater. Sci. 14 2325, 1979.
14. S.M. Boyer, D. Sang and A.J. Moulson. See reference (1) p.297.
15. W.A. Fate and M.E. Milberg. J. Amer. Ceram. Soc. 61 531, 1978.
16. W.M. Dawson and A.J. Moulson. J. Mater. Sci. 13 2289, 1978.
17. H. Suzuki and T. Hosaka. Yogyo-Kyokai-Shi 75 111, 1967.
18. S.S. Lin. J. Amer. Ceram. Soc. 60 78, 1977.
19. J.W. Evans and S.K. Chatterji. J. Phys. Chem. 62 1064, 1958.
20. A. Atkinson, A.J. Moulson and E.W. Roberts. J. Amer. Ceram. Soc. 59 285, 1976.
21. W. Kaiser and C.D. Thurmond. J. Appl. Phys. 30 427, 1959.
22. S.K. Biswas and J. Mukerji. High Temperatures-High Pressures, 12 81, 1980.
23. Constitution of Binary Alloys, First Supplement, Ed. R.P. Elliott, McGraw-Hill, New York, 1965.
24. F.L. Riley. See reference (1), p.273.
25. P. Longland and A.J. Moulson. J. Mater. Sci. 13 2279, 1978.
26. H. Dervisbegovic and F.L. Riley. J. Mater. Sci. 14 1945, 1979.
27. M.I. Mendelson. J. Mater. Sci. 14 1752, 1979.

28. U.R. Evans. "The Corrosion and Oxidation of Metals - Scientific Principles and Practical Applications", Arnold, London, 1960, p.834.
29. P. Popper and S.N. Ruddlesden. Trans. Brit. Ceram. Soc. 61, 603, 1960.
30. J.E. Still and H.J. Cluley. The Analyst, 1, 97, 1972.
31. D.P. Elias and M.W. Lindley. J. Mater. Sci. 11 1278, 1976.
32. A. Giachello. Science of Ceramics 10, Ed. H. Hausner (Deutsche Keramische Gesellschaft, Welden, 1980) p.377.
33. N.L. Parr, R. Sands, P.L. Pratt, E.R.W. May, C.R. Shakespeare and D.S. Thompson. Powder Met. 8 152, 1961.
34. M.W. Lindley, D.P. Elias, B.F. Jones and K.C. Pitman. J. Mater. Sci. 14 70, 1979.
35. J.A. Mangels. J. Amer. Ceram. Soc. 58 353, 1975.
36. J.A. Mangels. Bull. Amer. Ceram. Soc. 55 395, 1976.
37. D.R. Messier and P. Wong. Bull. Amer.Ceram. Soc. 57 525, 1978.
38. A. Atkinson and A.D. Evans. Trans. Brit. Ceram. Soc. 73 43, 1974.
39. G.S. Hughes, C. McGreavy and J.H. Merkin. J. Mater. Sci. 15 2345, 1980.
40. F.L. Riley and F. Yusaf. To be published.
41. J.A. Mangels. This volume, p.136.
42. N.J. Shaw and A.H. Heuer. To be published, Acta Met.
43. J. Heinrich and H. Hausner, in 'Energy and Ceramics', Ed. P. Vincenzini, Elsevier, Amsterdam, 1980, p.780.
44. S.C. Danforth, H.M. Jennings and M.H. Richman. J. Mater. Sci. 13 1590, 1978.
45. K.H. Jack. This volume, p.45.
46. P.E.D. Morgan. J. Mater. Sci. 14 791, 1980.
47. D. Campos-Loriz and F.L. Riley. J. Mater. Sci. 15 2385, 1980.
48. W. Schmidt. This volume, p.447.
49. K.S. Mazdiyasni, R. West and L.D. David. J. Amer. Ceram. Soc. 61 504, 1978.
50. J.C. Swartz. J. Amer. Ceram. Soc. 59 272, 1976.

THE EFFECT OF SILICON PARTICLE SIZE ON THE NITRIDING BEHAVIOUR OF REACTION BONDED Si_3N_4 COMPACTS

J.A. Mangels

Ceramic Materials Department, Ford Motor Company,
Dearborn, Michigan 48121, U.S.A.

ABSTRACT. Silicon compacts, with green densities of 1680 and 1750 kgm^{-3}, were injection molded using a variety of silicon powders. These compacts were nitrided under identical conditions. The nitriding results show that powders with high surface areas and fine particle sizes are more difficult to nitride than powders with low surface areas and larger particle sizes. The nitriding rate of the compacts was shown to affect the nitriding behaviour. Lower nitriding rates resulted in more complete nitridation.

INTRODUCTION

Much has been written on the nitriding of silicon compacts and factors affecting the microstructure and properties of reaction bonded Si_3N_4 (RBSN) compacts. Moulson (1) in his excellent review article has summarised this work. However the literature contains little information in the area of silicon particle size distributions and their effect on nitridation and strength. Messier and Wong (2) observed that the strength of RBSN could be improved if the maximum particle size of the silicon was reduced from 44 µm to under 30 µm. This observation is consistent with the work of Evans and Davidge (3) who found that the maximum pore size in RBSN was related to the maximum particle size of the silicon used in the fabrication of the material. It is also generally accepted that if the silicon particle size is large, longer nitriding times are required to completely react the silicon to Si_3N_4.

Partially funded DOE/NASA under Contract DEN 3-20.

Riley, F.L. (ed.) Progress in Nitrogen Ceramics

compact, having an intergranular appearance (Fig. 3a). However the coarser particle size compacts had a different form of unreacted silicon (Fig. 3b). As with the previous results, the unreacted silicon was uniformly distributed and present as isolated unreacted particles.

Subtle changes in the nitrogen demand nitriding cycle control parameters could effect changes in the nitriding rates of the silicon compacts (4). Figure 4a shows that by reducing the nitriding rate from 35g Si/hr to 28g Si/hr, in effect increasing the nitriding time from 130 to 160 hours, improves the nitriding behaviour of the compacts. The effect of different powders is also illustrated in Figure 4a. Figure 4b shows the structure of a 2830 kgm^{-3} compact produced using B powder distribution, nitrided at a nitriding rate of 25g Si/hr. The structure is uniform with no evidence of unreacted silicon.

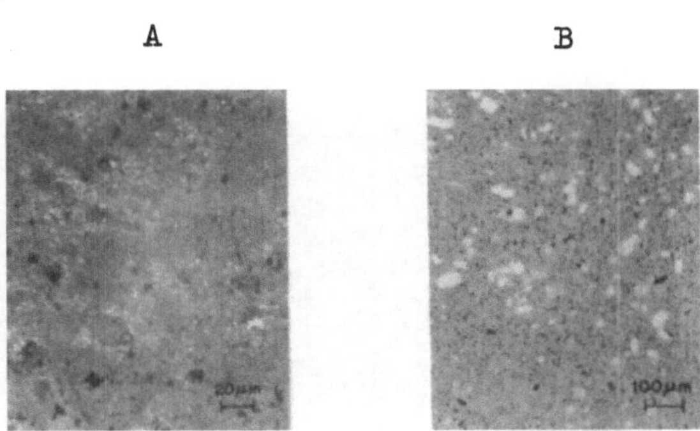

Fig. 3. Microstructure of 2800 kgm^{-3} compacts showing A) intergranular distribution of unreacted silicon (II powder) and B) isolated particles of unreacted silicon (U powder).

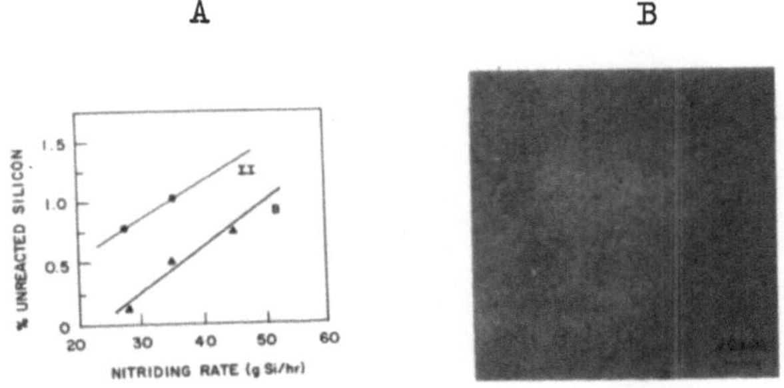

Fig. 4. A) The effect of nitriding rate on the amount of unreacted silicon in 2800 kgm^{-3} compacts and B) the microstructure of a 2830 kgm^{-3} compact nitrided at 25g Si/hr (B powder).

Table 1

Si Powder Characterisation and Nitriding Behaviour

Powder Code	Surface Area m^2kg^{-1}	Rosin-Rammler Parameters a	Rosin-Rammler Parameters b	Nitriding Results (wt% Si) 1680 kgm^{-3}	Nitriding Results (wt% Si) 1750 kgm^{-3}
A	4500	7.4	1.06	0	0.55
B	5000	5.9	1.03	0	0.60
U	3400	13.0	0.91	Trace	0.04
X	2200	10.6	1.02	-	0.3
Y	3400	7.1	1.00	-	0.4
Z	6200	4.4	1.00	-	0.3
HH	6700	5.0	1.00	-	0.73
II	7200	4.7	1.02	0	0.9
JJ	5200	5.9	1.12	-	0.4

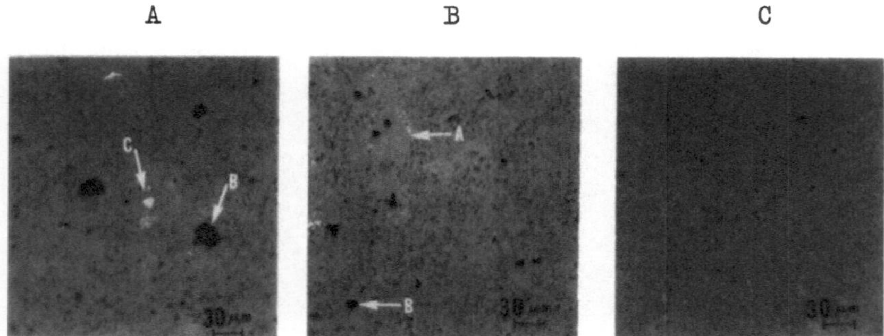

Fig. 1. Microstructure of nitrided 2700 kgm^{-3} compacts. A) Powder U, B) Powder A, C) Powder II.

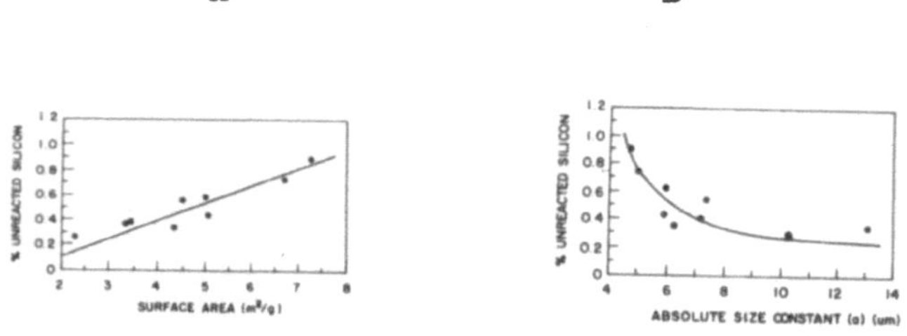

Fig. 2. A) Effect of surface area, and B) particle size on the amount of unreacted silicon present in 2800 kgm^{-3} compacts.

EXPERIMENTAL

A number of silicon powder distributions were generated by dry ball milling silicon powder in various ways (4). The surface area was measured using a single point BET Argon absorption technique, while the particle size distribution was determined using an X-ray sedimentation technique. The particle size distributors were quantified through the Rosin-Rammler expression (5):

$$Wr = 100 \exp - (D/a)^b$$

where Wr = cumulative weight percent retained, D = mean particle diameter, a = absolute size constant (particle size at the 38.6 wt% point) and b = dispersion constant, a measure of the distribution shape.

These powders were each injection molded into 3.2 x 6.4 x 38 mm test specimens with green densities of 1680 and 1750 kgm^{-3} (corresponding to nitrided densities of 2700 and 2800 kgm^{-3} respectively). In every case, 2½ wt% Fe_2O_3 was included as a nitriding aid. These compacts were nitrided using a nitrogen demand nitriding cycle and an atmosphere composed of $H_2/He/N_2$ (4,6).

A number of techniques were employed to evaluate the nitriding results. X-ray diffraction and metallography were used to quantitatively and qualitatively detect the amount of unreacted silicon present in the test samples. X-ray diffraction could accurately detect silicon in silicon nitride at the 0.1% level.

RESULTS

Table 1 summarises both the powder characterisation and the nitriding results. The powders had surface areas ranging from 2200 to 7200 m^2kg^{-1} and particle sizes (expressed in terms of the absolute size constant) ranging from 4.4 to 13 μm.

The nitriding results showed that all powders evaluated, except for the very coarse (U), were completely nitrided at the 1680 kgm^{-3} green density level. However there were distinct differences in microstructure as shown in Figure 1. The coarse powders (U) contained uniformly dispersed regions of unreacted silicon particles and pore sizes of up to 30 μm. The intermediate size powders (A) exhibited a uniform structure with pore sizes ranging from 3 to 15 μm. The finest powders (II) exhibited the most uniform structures, with the finest pore sizes (8 μm maximum).

The nitriding results at the 1750 kgm^{-3} green density level show no compact to be completely nitrided. Relationships were found between the silicon surface area and particle size and the nitriding behaviour, Fig. 2. These results indicate that more complete nitridation can be achieved if coarser powders with lower surface areas are used. Microstructural examination of the samples indicated that unreacted silicon was located at the centre of the

SUMMARY

The properties of the silicon powder were found to greatly affect the nitriding behaviour of RBSN compacts. Powders with a large particle size and a low surface area nitride easier than powders with a finer particle size and a larger surface area. However if the particle size becomes too large (> 10 µm), particles of unreacted silicon will be present in the RBSN microstructure. Nitriding rate was also found to play an important factor in nitriding high density RBSN compacts, with lower nitriding rates resulting in more complete nitridation.

REFERENCES

1. A.J. Moulson, "Reaction Bonded Silicon Nitride: its Formation and Properties", J. Mater. Sci. 14 (1979) 1-17-1051.
2. D.R. Messier and P. Wong, "Kinetics of Formation and Mechanical Properties of Reaction Sintered Si_3N_4", Ceramics for High Performance Applications, ed. Burke, Gorum and Katz, Brook-Hill Pub. Co. (1974), p.181.
3. A.G. Evans and R.W. Davidge, "The Strength and Oxidation Resistance of Reaction Sintered Si_3N_4", J. Mater. Sci. 5 (1970) p.134.
4. J.A. Mangels, "Development of a High Density, Moldable Reaction Bonded Si_3N_4", DOE/NASA Contract DEN 3-20 Final Report, to be issued in 1981.
5. A.F. Taggart, Handbook of Mineral Dressing, John Wiley & Sons Inc., New York (1947), pp.19-145.
6. J.A. Mangels, "The Effect of Rate Controlled Nitriding and Nitriding Atmospheres on the Formation of Reaction Bonded Silicon Nitride", accepted for publication in the Bulletin of the American Ceramic Society.

DISCUSSION

Ziegler: Did you find any relation between the proportions of α and β-phase and the silicon particle size?

Mangels: All samples produced ~75-85 w/o α-Si_3N_4 independent of particle size.

Billy: The fact that the proportion of unreacted silicon increases with decreasing silicon particle size proves that the overall kinetics are governed by internal nitridation through a pore-closure mechanism, and as a consequence the diffusion of nitrogen through the nitride scale is not rate determining.

Riley: Have you given thought to the mechanism of the function of helium? One possible action is to carry heat away from the nitride growth points, assisting the exothermic process.

Mangels: We have only qualitatively considered the effect of He. This entire area requires extensive study to explain the beneficial effects observed.

AN EXPERIMENTAL PLAN FOR THE NITRIDATION OF Si+Al COMPACTS

W. MUSTEL and D. BROUSSAUD

Ecole des Mines de Paris, Centre des Materiaux
B.P. 87, 91003 Evry Cedex, France.

ABSTRACT. This work attempts to approach the nitridation process for silicon-6^{w}/o aluminium powder compacts using an experimental plan called the "Latin square". This plan brings out the respective influence of each parameter of the process through a limited number of experiments. Here, we took into account the following independent parameters:

- nitriding temperature
- nitriding time
- nitrogen flow in the furnace
- chemical nature of the Si starting powder
- density of the compacts before nitridation.

By considering the nitridation rate to be an independent variable function of the five process parameters, thus neglecting any interaction between them, it was shown that two parameters had a major influence: nitriding temperature, and the chemical nature of the starting powder.

1. INTRODUCTION

Silicon nitride materials have been the subject of extensive research for several years. Because pure silicon nitride has so far proved not sinterable without the use of additives (MgO, Y_2O_3), the reaction bonding process appears as a useful way of producing such materials (1).

This reaction involves the nitridation of shapes of compacted silicon powder to obtain silicon nitride pieces. Since it appears that the mechanical properties of these materials are closely controlled by the porosity always present in the final bodies, most investigations are being focused on how the microstructure is affected by the fabrication route (2).

Riley, F.L. (ed.) Progress in Nitrogen Ceramics

The aim of this work is to determine the respective influence of each parameter of the process on the quality of the product using a limited number of runs. An experimental plan called the "Latin square" (3) has been used in this respect. Five parameters have been "a priori" selected:

- the nitriding temperature
- the nitriding time
- the nitrogen flow in the furnace
- the chemical nature of the silicon starting powder (impurity level)
- the green density of the compact.

2. THE LATIN SQUARE EXPERIMENTAL PLAN

Whenever a reaction system depending on p parameters being likely to have n values is investigated, it would be necessary to carry out n^p runs in theory to get the individual influence of each parameter on the reaction system. Assuming a graph can be drawn using five points, i.e. n equal to five, 125 runs would be necessary for three parameters and 625 for four parameters, and so on.

The "Latin square" is an experimental plan allowing the study of a reaction system dependent upon a number of parameters below or equal to six. The fundamental hypothesis of this plan is to consider the reaction system P as an independent-variable function which is expressed as follows:

$$P = f(X_1) + g(X_2) + h(X_3) + i(X_4) + j(X_5) + k(X_6)$$

Consequently the Latin square does not make allowance for any possible interaction between parameters. Its main characteristic is to select among the $5^6 = 15625$ theoretical possibilities 25 runs where each value of a parameter is associated with each value of the other parameters, once and only once. The selected arrangement of values is shown in Table I.

As a rule, the six available parameters are not all taken as physical parameters: on the one hand, a reaction system depending on more than three parameters is not commonly met; on the other hand, the "free" parameters can be used to estimate the Latin square errors which may issue from the hypothesis and from experiments. The error calculation gives access to the standard deviation of the functions, which is used to fit the curves and the standard deviation of the Latin square calculated from the "free" parameters which may translate: scatter of measured values; influence of an aleatory parameter, or of an omitted parameter; interaction between parameters.

3. APPLICATION OF THE LATIN SQUARE TO THE NITRIDATION OF SILICON

Five parameters have been arbitrarily selected among all the parameters which are likely to influence the nitriding process (4). These are listed above in Section 1.

Meas. N°	X 1	X 2	X 3	X 4	X 5	X 6
Meas. N°1	1	1	3	2	1	4
Meas. N°2	1	2	2	1	5	3
Meas. N°3	1	3	1	5	4	2
Meas. N°4	1	4	5	4	3	1
Meas. N°5	1	5	4	3	2	5
Meas. N°6	2	1	4	1	3	2
Meas. N°7	2	2	3	5	2	1
Meas. N°8	2	3	2	4	1	5
Meas. N°9	2	4	1	3	5	4
Meas. N°10	2	5	5	2	4	3
Meas. N°11	3	1	5	5	5	5
Meas. N°12	3	2	4	4	4	4
Meas. N°13	3	3	3	3	3	3
Meas. N°14	3	4	2	2	2	2
Meas. N°15	3	5	1	1	1	1
Meas. N°16	4	1	1	4	2	3
Meas. N°17	4	2	5	3	1	2
Meas. N°18	4	3	4	2	5	1
Meas. N°19	4	4	3	1	4	5
Meas. N°20	4	5	2	5	3	4
Meas. N°21	5	1	2	3	4	1
Meas. N°22	5	2	1	2	3	5
Meas. N°23	5	3	5	1	2	4
Meas. N°24	5	4	4	5	1	3
Meas. N°25	5	5	3	4	5	2

Table I: Selected arrangement of the parameter values.

For maximum efficiency it is necessary to explore a wide range of values for each parameter. The respective ranges are stated in Table II; they have been chosen according to realistic industrial conditions.

3.1 Experimental

Si+6^{w}/o Al powders. In order to study the effect of the "chemical nature" of the silicon starting powder, five powder mixes have been used. Their characteristics are shown in Table III. They all exhibit a specific area (Blaine) around and they were obtained by milling the different batches shown in Table III. Aluminium (6^{w}/o) has been added to the silicon powders to improve the oxidation resistance of the final product, as explained in a previous publication (5).

Compacts. Five green densities were produced by pressing the Si+Al mixes (the two highest densities by isostatic pressing compacts (29x29x3 mm and 29x29x10 mm) were thus prepared).

Nitridations. All the samples were heated in the same muffle furnace shown in Fig. 1. The chronological order of the runs has been fixed using a table of numbers at random. The firing cycle

Parameters	Parameter values 1	2	3	4	5
X_1 = Chemical nature of the Si starting powder	1	2	3	4	5
X_2 = Green density (Mgm^{-3})	1.49	1.53	1.65	1.68	1.71
X_3 = Nitriding temperature (°C)	1250	1280	1310	1340	1370
X_4 = Nitriding time (hrs)	30	40	50	60	70
X_5 = Nitrogen flow in the furnace (cm^3/mn)	8	25	35	50	85

Table II: Ranges of values for each selected parameter.

Powder number	Origin	Specific area (cm^2g^{-1})	Impurity content (Emission spectrometry) ppm Fe	Mg	Mn	Ni	Cr	Ag
1	50% Starck Lot 5 + 50% Starck Lot 2	8678	-	0.05	0.05	0.05	0.05	-
2	100% Starck Lot 5	8628	-	0.05	-	0.05	0.05	0.05
3	100% Metaux speciaux Lot 2	7620	-	0.05	0.05	0.05	0.05	0.05
4	50% Starck Lot 5 + 50% Starck Lot 6	7651	x	0.05	0.05	0.05	0.05	-
5	100% Starck Lot 6	8184	x	0.05	0.1	0.05	0.05	0.05

x = 'saturated', compared with the 0.1% reference sample.

Table III: Silicon powder characteristics.

involves three steps: first, heating to the selected temperature; second holding this temperature for the selected time; and third, cooling. Three compacts per run were placed in the furnace: two 29x29x3 mm and one 29x29x10 mm, so that the Latin square is doubled with the two thicknesses taken into account.

Characterisation of the samples. The weight gain of the specimens after nitridation proved to be one of the most revealing parameters in the Latin square. However, another parameter taken into account was the final density.

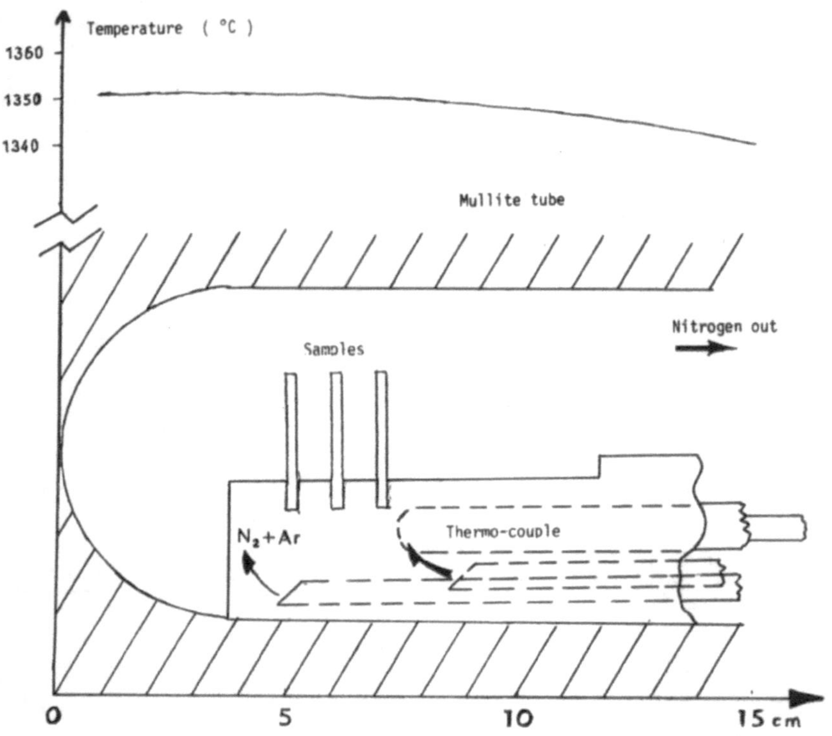

Figure 1. Nitriding Installation.

3.2 Results

All the data of the Latin square were treated using a computer programme. Results are shown in Table IV and on Figs. 2 and 3. The value F in Table IV is a dimensionless number which is used to establish the respective influence of each physical parameter taken into account.

The compared respective results show a major influence of the two following parameters: the nitriding temperature, and the chemical nature of the silicon starting powder. It is also seen that the thicker the specimen, the greater the influence on the weight gain of the samples. It seems that the optimum nitriding temperature should be around 1340°C. Considering the influence of the level of impurities of the silicon starting powder, we have checked the fact that a high level of impurities enhances the nitriding rate. The influence of the nitriding time comes out to be very low in the range 30-70 hours, as well as the influence of the nitrogen flow and green density.

It should be noted that all these results are valid as regards the weight gain and the final density. Further work will allow a better characterization making allowance for the mechanical properties of the nitrided bodies which will be taken into account as Latin square parameters.

		Related weight gain 3mm thick specimens	10mm thick specimens	Final Density 10mm thick specimens
Mean value of Latin square		1.460	1.460	2.224
"F" Parameters	Chemical nature of the powder	7.43	13.82	0.559
	Green density	1.323	3.09	0.2
	Nitriding temp.	11.146	23.46	2.168
	Nitriding time	1.334	2.74	0.813
	Nitrogen flow in the furnace	0.665	0.82	0.586
Latin square standard deviation value		0.057	0.042	0.25
Functions standard deviation		0.026	0.019	0.112
Scattering of measurements		3.9%	2.9%	11.2%

Table IV. Results of the Latin Square.

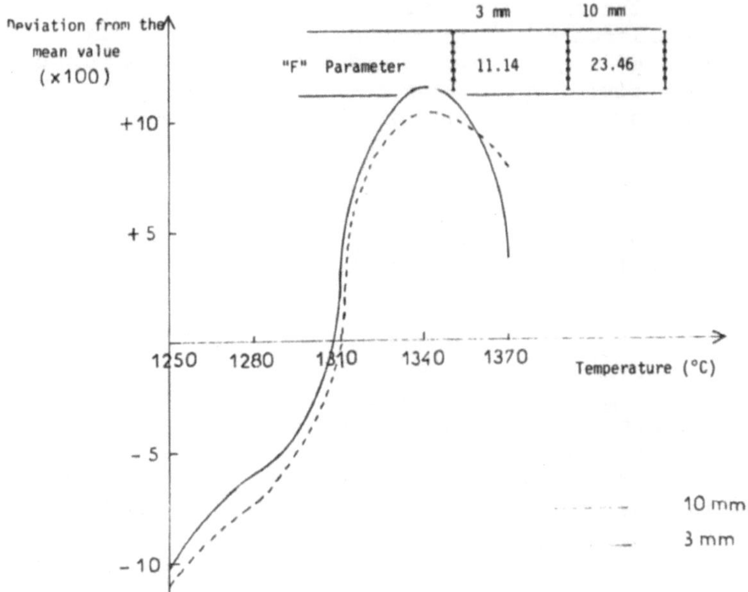

Figure 2. Influence of temperature.

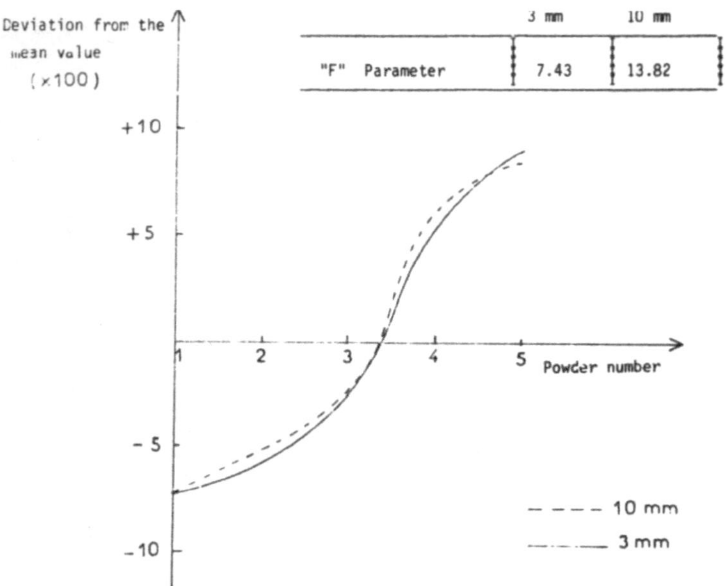

Figure 3. Influence of the chemical nature of the silicon.

4. CONCLUSION

The "Latin square" experimental plan permitted identification of the most influential parameters in the nitriding of silicon compacts. These are: the nitriding temperature, and the level of impurities of the silicon starting powder. Consequently, it should be possible for a given powder and green density to optimize the nitridation firing cycle with regard to temperature and time only, considering the influence of the other parameters as negligible.

5. ACKNOWLEDGMENTS

The authors are grateful to Mr. Minjolle, Head of the Research and Development Department of CERAVER, for his help and for supporting this study.

REFERENCES

1. F.L. Riley, "Nitrogen Ceramics", edited by F.L. Riley, Noordhoff Int. Publishing, Leyden, The Netherlands (1977) 265-288.
2. B.F. Jones and M.W. Lindley. J. Mater. Sci. 10 (1975) 967-972.
3. The Design and Analysis of Industrial Experiments, edited by O.L. Davies; Oliver and Boyd (London 1960).
4. M. Auclair, W. Mustel and J.P. Torre. Science of Ceramics 11 (in press).

5. P. Parlier and J.P. Torre. Demande de brevet Francais No. 7900 268 du 5 janvier 1979.

DISCUSSION

Jack: How confident can you be that there is no interaction between the five process parameters? What are the differences in "chemical nature" of the silicon powders 1, 2 and 3?

Mustel: I am not sure that there is no interaction between the parameters. Although the Latin Square Standard deviation (which reveals the errors coming from the hypothesis and the experiments) is low, some interactions exist. With regard to the analysis, the five silicon starting powders do not seem to be very different.

Mangels: Is there a way of using a Latin Square analysis to determine the interaction of variables?

Mustel: The Latin Square method does not give directly interactions between variables. The only information obtained is whether the interactions are strong or weak. In this case they seem to be weak. A different experimental plan ("2^n") can be used to estimate the interactions.

PREPARATION OF SILICON NITRIDE POWDER FROM SILICA

M. Mori, H. Inoue*, and T. Ochiai**

Refractories Div., Toshiba Ceramics Co., Ltd., Kariya, Japan.
*Research and Development Center, Toshiba Corp., Kawasaki, Japan.
**Metal Products Div., Toshiba Corp., Yokohama, Japan.

ABSTRACT. Although thermodynamic data predict the difficulty in advancing the reaction, $3SiO_2(S) + 6C(S) + 2N_2(G) = Si_3N_4(S) + 6CO(G)$, in a closed system, it has been proved that heating fine powder mixtures of silica and carbon in nitrogen gas flow results in sufficient formation of silicon nitride powder, without detectable coexisting silicon carbide. Apparent activation energy (≃163 kcal/mol) on the reaction approximates to the value of ΔHr (≃160 kcal/mol at 1700 K) calculated on the elementary reaction, $SiO_2(S) + C(S) = SiO(G) + CO(G)$, which therefore seems to be a rate-determining one. The observation of fibrous silicon nitride growing out into gaps among aggregated powder lumps and that of weight losses of samples larger than expected suggest that SiO gas should be generated as an intermediate product. Silicon nitride powder prepared from silica through the improved reaction process possesses excellent characteristics such as high α-phase content, homogeneous shape and size, and very low content of metallic impurities.

1. INTRODUCTION

Because of the commercially wide availability of pure and fine powder of silica and carbon, it is considered very desirable to prepare silicon nitride powder of high purity through the carbothermal reduction of silica followed by nitridation [1,2,3]. Some experiments have proved the feasibility of industrial production of silicon nitride powder with a high α:β ratio by means of the heat-treatment of fine powder mixtures of silica and carbon in nitrogen gas flow.

Riley, F.L. (ed.) Progress in Nitrogen Ceramics

2. EXPERIMENTAL SYNTHESIS OF SILICON NITRIDE FROM SILICA

Very fine powders of amorphous silica and carbon(Lamp Black) were dry mixed in the weight ratio 1:1 and 1:3, i.e. the molar ratio 1:5 and 1:15, with excessive proportions of carbon in relation to the theoretical weight ratio 1:0.4 of the reaction:

$$3SiO_2(S) + 6C(S) + 2N_2(G) = Si_3N_4(S) + 6CO(G) \qquad (1)$$

Heat treatments of a 5 mm deep powder bed of the mixtures were performed in nitrogen gas flow of 2 cm/sec, according to a given program up to 1500°C. The reaction products with residual carbon were decarburized at 700°C in air.

Fig. 1 shows the X-ray diffraction patterns of the products, and Fig. 2 shows the nitrogen contents of the products analyzed by acid-base titration after alkali fusion. Formation of silicon nitride high in α-phase content was achieved effectively at 1500°C, and the nitrogen contents exceeded 37 wt.%. Silicon carbide could not be detected by X-ray diffraction. Scanning electron micrographs of the products are given in Fig. 3, showing that aggregated fine granules decrease and fibrous silicon nitride products increase with the proceeding of heat treatments. Fine granules of silicon nitride were also likely to be produced. Fig. 4 shows the weight changes of samples caused by the heat treatments, which were defined by comparing the weight of the product after decarburization with that of the silica portion of the starting mixtures. The weight losses attained larger percentages than the expected theoretical value of 22.2% cal-

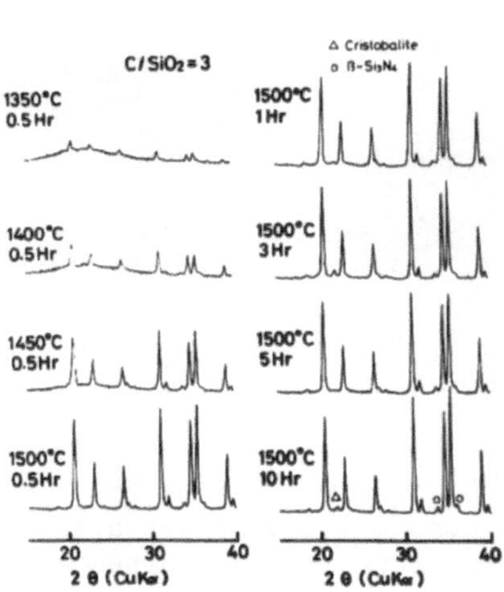

Fig. 1. X-ray diffraction patterns of products.

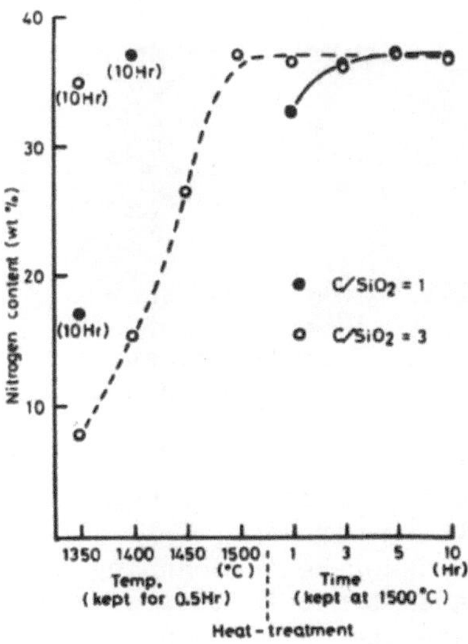

Fig. 2. Relation between nitrogen contents of products & heat treatments.

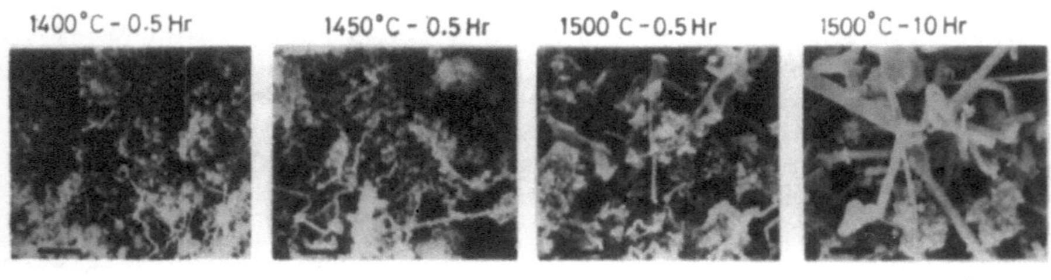

Fig. 3. Scanning electron micrographs of products after decarburization (silica:carbon wt. ratio = 1:3).

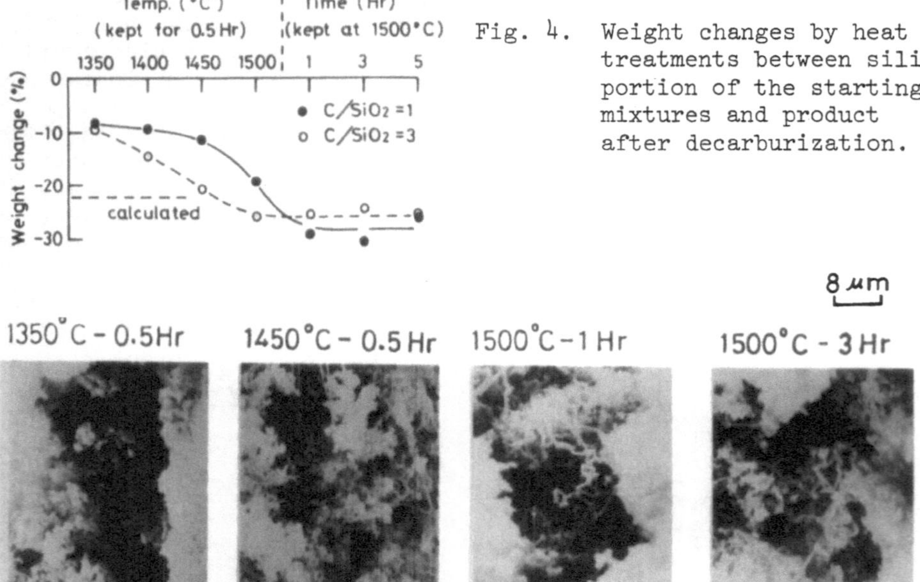

Fig. 4. Weight changes by heat treatments between silica portion of the starting mixtures and product after decarburization.

Fig. 5. Scanning electron micrographs of as-heat-treated products before decarburization (silica:carbon wt. ratio = 1:3).

culated according to equation $3SiO_2 \rightarrow Si_3N_4$. This seems to be due to the generation of volatile silicon monoxide and its escape from the powder bed. Scanning electron micrographs of the as-heat-treated products before decarburization, as shown in Fig. 5, indicate that silicon nitride fibers grew out into gaps among aggregated powder lumps and became thickened in the coarse of the reaction. This also suggests the presence of silicon monoxide gas a an intermediate product.

3. RATE OF REACTION AND SOME THERMODYNAMICAL CONSIDERATIONS

An experiment on the rate of reaction for the formation of silicon nitride from silica was carried out by analyzing the nitrogen contents of the products obtained from various heat treatments of a 7 mm thick powder bed of silica and carbon mixtures in nitrogen gas flow of 1.6 cm/sec, the reaction temperatures being at about 50°C intervals between 1300 and 1455°C. The molar ratio of silica: carbon was fixed to 1:20 (weight ratio 1:4) with a much larger carbon proportion than the theoretical ratio, so that the extent of reaction should be related only to the quantities of the components of interest. And raw materials of high purity were selected to minimize the influence of impurities, especially of iron.

The results of the nitrogen analyses of the products are plotted in Fig. 6 against reaction time at respective reaction temperatures. The reaction was found to proceed even at 1300°C. Chemical kinetics were applied to these data by calculating conversion rates from the nitrogen contents and adopting Jander's rate equation

$$\{1 - (1 - \alpha)^{1/3}\}^2 = kt$$

where α is conversion rate, k is rate constant and t is time, only for rough estimation. An Arrhenius plot was made for reasonable rate constants, as shown in Fig. 7, and the apparent activation energy Q was derived to be about 163 kcal/mol.

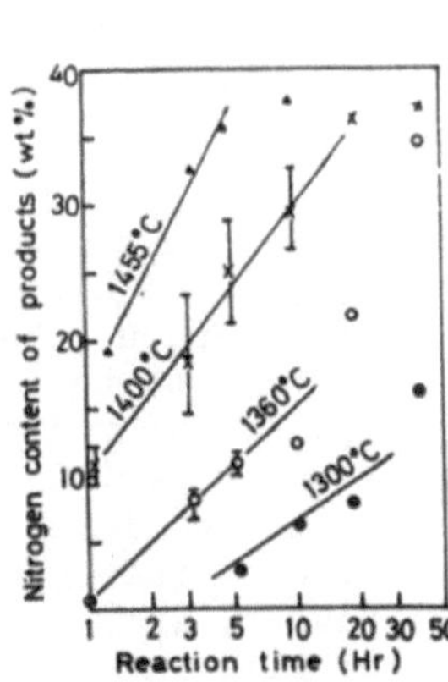

Fig. 6. Nitrogen contents of products vs. reaction time at respective reaction temperatures.

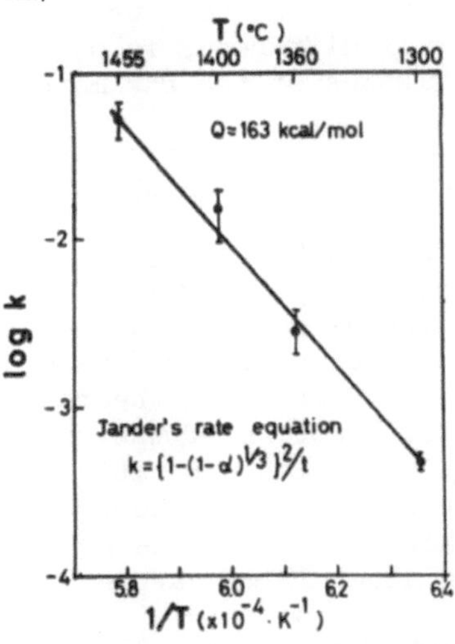

Fig. 7. Arrhenius plot of the obtained data on the rate of reaction.

Thermodynamic data for some reaction formulas related to the production of silicon nitride from silica were examined, based on JANAF Thermochemical Tables [4]. The overall reaction (1) is endothermic with $\Delta Hr \simeq 302.8$ kcal/mol at 1700 K. However, the small values of its equilibrium constants, e.g. 3.51×10^{-3} at 1700 K and 5.04×10^{-1} at 1800 K, indicate the difficulty of advancing this reaction in a closed system. Therefore, it can be considered essential to perform the reaction in nitrogen gas flow with the removal of evolved carbon monoxide out of the system, using a larger amount of nitrogen gas than required for the reaction. The observed apparent activation energy lies close to the enthalpy change $\Delta Hr \simeq 160$ kcal/mol at 1700 K of the elementary reaction:

$$SiO_2(S) + C(S) = SiO(G) + CO(G) \quad (2)$$

which may therefore be a rate-determining one. The equilibrium constants of the reaction:

$$3SiC(S) + 2N_2(G) = Si_3N_4(S) + 3C(S) \quad (3)$$

are calculated to be 1.36 at 1700 K and 0.182 at 1800 K, then it is roughly likely that at 1 atm. N_2 ambient pressure, SiC rather than Si_3N_4 will be produced at around 1500°C or 1773 K. However, SiC was not identified in the above-mentioned experiments at 1300 to 1500°C, where the starting mixtures had excess carbon even with respect to the SiC formation reaction:

$$SiO_2(S) + 3C(S) = SiC(S) + 2CO(G) \quad (4)$$

contrary to the conclusions of the literature [3]. The explanation for this difference is obscure, but seems to be related to the characteristics of the raw materials and their powder beds.

4. SOME PROPERTIES OF PREPARED SILICON NITRIDE POWDER

Although the products of the above-mentioned experiments presented mainly the shape of fibers, as shown in Figs. 3 and 8 (a) by SEM, the granular powder of silicon nitride with homogeneous shape and size of particles, as shown in Fig. 8 (b), has come to be prepared, by improving the reaction process. The typical X-ray diffraction pattern of the granular powder is presented in Fig. 9, indicating a high $\alpha:\beta$ ratio. The narrow range of particle diameter distribution can be noticed in Fig. 10. Table 1 offers the typical characteristics of the granular silicon nitride powder, which contains about 1% of carbon but a small amount of metallic impurities.

TABLE 1. Typical characteristics of silicon nitride powder.
PHASE: 95% α-phase, MEAN PARTICLE DIAMETER (FSSS): 1.2 μm,
CHEMICAL ANALYSIS: Si 59.1%, N 37.5%, O 1.8-2.2%, C 1.05%,
Fe 0.003%, Ca 0.008%, Al 0.002%, Mg 0.002%

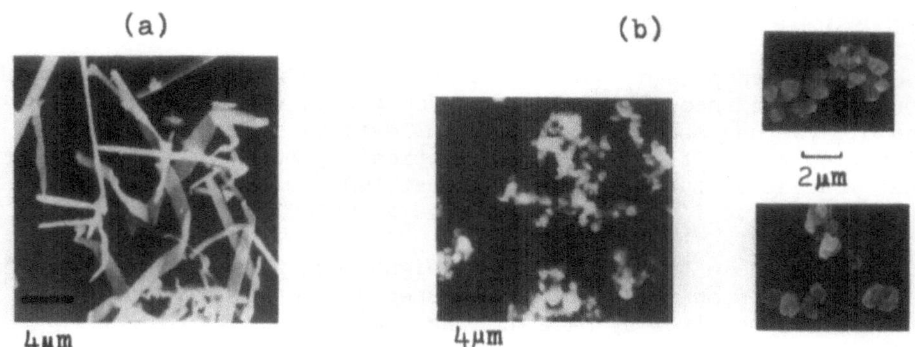

Fig. 8. Scanning electron micrographs of (a) fibrous products and (b) granular products.

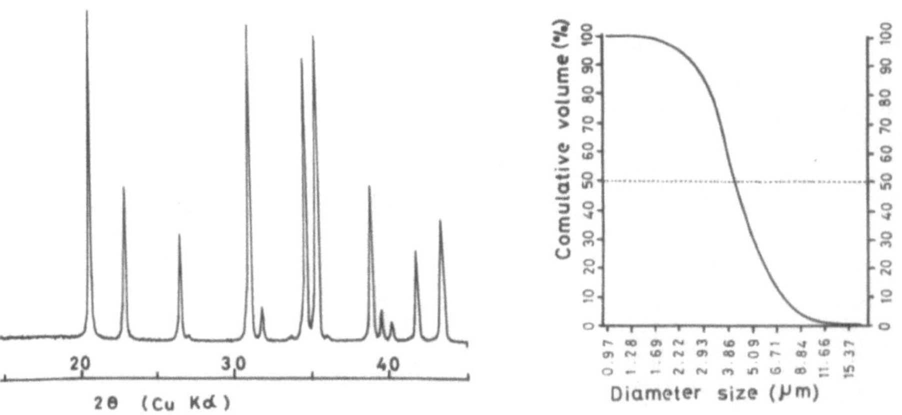

Fig. 9. Typical X-ray diffraction patterns of granular silicon nitride powder.

Fig. 10. Typical particle diameter distribution of granular silicon nitride powder (by ELZONE PARTICLE COUNTER, Particle Data Inc.)

REFERENCES

1. A. Hendry and K. H. Jack, 'Special Ceramics 6', Edited by P. Popper. Stoke-on-Trent, B. Ceram. R. A., 199-208(1975).
2. K. Komeya and H. Inoue, J. Mat. Sci., 10, 1243-1246(1975).
3. J. G. Lee and I. B. Cutler, 'Nitrogen Ceramics-The Proceedings of the 1976 NATO ASI', Edited by F. L. Riley. 175-181(1977).
4. D. R. Stull and H.Prophet, JANAF Thermochemical Tables, 2nd Edition, U. S. Bureau of Stds. Report NSRDS-NBS37, June, 1971.

DISCUSSION

Jack: I am surprised that the weight loss on nitriding is not much more than the theoretical. How do you get an intimate mix of SiO_2 and C? (We use sol-gel processes followed by spray-drying. We also find the N_2 flow rate to be critical.) It is also surprising that you get virtually no SiC, even at 1600°C. Lee and Cutler found SiC above 1450°C and we find it > 1500°C.

Mori: I suppose that our smaller weight loss is due to the characteristics of our raw materials, especially of carbon. Raw materials were mixed by ball milling. We did not get SiC at temperatures between 1300°C and 1500°C. At 1550°C SiC was the dominant product. The effects of nitrogen gas flow rate are shown in the table below. Lee and Cutler used carbon black (Raven 450).

Effects of N_2 Gas Flow Rate on Yields, α-Phase Contents and Nitrogen Contents of Products.

EXPERIMENTAL CONDITIONS:

silica:carbon = 1:2 (wt. ratio)
heat treatment at 1400°C for 5 hours.

N_2 Gas Flow Rate (l/hr)	Yield (%)	α-Phase Content (%)	Nitrogen Content (%)
300	87.5	93.0	36.9
150	89.0	94.0	36.6
100	93.3	95.0	36.2
50	93.0	95.3	36.0
25	93.6	96.1	36.4
12.5	94.3	96.2	37.0

Mangels: Have you incorporated sintering additives into your powder through your processing?

Mori: Yes.

Thompson: How do the properties of hot-pressed and sintered materials prepared from your silicon nitride powder compare with those prepared from nitrided silicon powders? Does the β-content of your powder vary with processing variables?

Mori: In the Toshiba sintering methods our powder is superior to others. The β-content does vary, and we have obtained almost 100% α-phase in some experiments. The present commercial powder contains a few % β-phase.

Briggs: How do you burn out residual carbon without oxidising the si_3N_4?

Mori: At 700°C in a glass container. There is no significant oxidation.

Gazza: Is the 1% residual C intentional, or are you trying to reduce it?

Mori: We would like to have lower levels.

Popper: What is the compactability of the granular and fibrous powders?

Mori: The green density can be 1.7 to 1.8 Mg m^{-3}.

ON THE PREPARATION OF FINE SILICON NITRIDE POWDERS

G. Schwier

Hermann C. Starck Berlin, Werk Goslar.

ABSTRACT. Silicon nitride can be prepared by various reactions. Silicon nitride powders produced by three different methods are compared, and production-specific powder characteristics are explained. The production method chosen by Messrs. Hermann C. Starck is commented on in terms of its powder quality-determining steps, with particular regard to the formation of sinter-active, fine particle size, high-purity α-silicon nitride.

1. INTRODUCTION

The special requirements for silicon nitride components of a gas turbine within the German BMFT programme "Ceramic components for vehicular gas turbines" have since 1975 led to the development of high α-phase silicon nitride powders. Particularly for pressureless sintering, sinter-active powders are required in order to achieve good densification with low additive levels. These sinter-active powders of fine particle size and with narrow particle size distribution are now available and are discussed in this presentation.

2. PREPARATION METHODS FOR SILICON NITRIDE POWDERS

Silicon nitride powders can be manufactured by a number of different techniques. Of major importance are the following methods, already in use for production on a technical scale:

2.1 Nitridation of Silicon Powder

The most common process for the production of silicon nitride is the direct reaction of the elements (1,2,3):

$$3Si + 2N_2 = Si_3N_4.$$

Riley, F.L. (ed.) Progress in Nitrogen Ceramics

Gas mixtures such as H_2+H_2, or ammonia, can also be used. Below 1450°C predominantly α-phase silicon nitride is produced which, however, has to be milled to achieve fine powders.

2.2 Carbothermic Reduction of Silica

In the presence of carbon and nitrogen, fine grain silica reacts according to the equation (4,5):

$$3SiO_2 + 6C + 2N_2 = Si_3N_4 + 6CO$$

Using very fine silica a fine α-silicon nitride powder can be produced directly. Since excess carbon black has to be used, some free carbon remains in the nitride powder. By annealing these powders in air this carbon can at least be partially oxidized, but so is the silicon nitride. As a consequence this powder type seems to have some problems associated with the carbon and/or oxygen content.

2.3 Vapour Phase Reactions

Silicon nitride can be prepared by reaction of gaseous silicon compounds such as silicon tetrachloride or silane, with ammonia (6,7,8):

$$3SiCl_4 + 4NH_3 = Si_3N_4 + 12HCl$$

$$3SiH_4 + 4NH_3 = Si_3N_4 + 6H_2$$

Using correct reaction conditions very fine amorphous silicon nitride powder can be produced directly. Corrosion problems exist with $SiCl_4$ and HCl; high materials costs with SiH_4. These silicon nitride powders are calcined mostly for purification, in the course of which a certain extent of crystallization from amorphous to α-Si_3N_4 takes place.

3. COMPARISON OF DIFFERENT TYPES OF SILICON NITRIDE POWDERS

Silicon nitride powders produced by different techniques possess specific powder characteristics. Fig. 1 provides a comparison of five silicon nitride powders prepared by the three methods listed above. Powder type SN 402 is directly produced from the vapour phase reaction, whereas the type SN 502 is derived from it by calcination.

All five powders in Fig. 1 are of high purity with respect to metallic impurities. Differences are to be seen in the amounts of non-metallic elements, namely of silica, silicon carbide or carbon phases. Toshiba Si_3N_4 powder seems to contain higher C and O contents, due to the preparation method. The purest powder in this respect is the calcined vapour phase powder SN 502.

Regarding the crystallographic state of the silicon nitrides, only slight differences exist depending on the method of preparation.

1. Preparation method		Reaction of $SiO_2 \cdot N_2$ in the presence of C	Nitridation of Si-metal powder		Vapor-phase reaction of $SiCl_4 \cdot NH_3$	
Type of reaction		solid-gas	solid-gas		gas-gas	
2. Example. Manufacturer		Toshiba	H C Starck		GTE Sylvania	
Powder grade		n a	H 1	LC 12	SN 402	SN 502
3.1 Chemical purity						
Σ metallic impurities	%	0,1	0,1	0,1	0,2	0,1
Σ non-metallic impurities	%	4,1	1,7	1,7	4,6	1,1
3.2 Physical state						
α-Phase Si_3N_4 (calc.)	%	88	92	94	-	≈56
β-Phase Si_3N_4	%	5	4	3	-	3
amorphous Si_3N_4	%	-	-	-	92	≈39
SiO_2	%	5,6	2,4	3,0	7,5	1,9
Σ Si_3N_4 (calc. from Phases)	%	93,2	96,3	96,6	91,9	98,0
Σ Si_3N_4 (calc. from N-Anal.)	%	92,4	95,6	95,5	90,3	97,4
3.3 Powder state						
specific surface (BET)	m^2/g	5	9	23	≈11	4
grain size (SEM-foto)	μm	0,4-1,5 (4)	0,1-3 (5)	0,1-1 (3)	0,1-1,5	0,2-2 (10)
apparent density	g/cm^3	0,20	0,37	0,40	0,18	0,10
tape density	g/cm^3	0,43	0,64	0,87	0,26	0,26
morphology of powder		partly aggregated	discret	discret	discret	aggregated
particle (shape)		uniform globular	globular to irregular	uniform globular	uniform globular	globular and whiskers

Fig. 1. Characteristics of three types of α-Si_3N_4 powders.

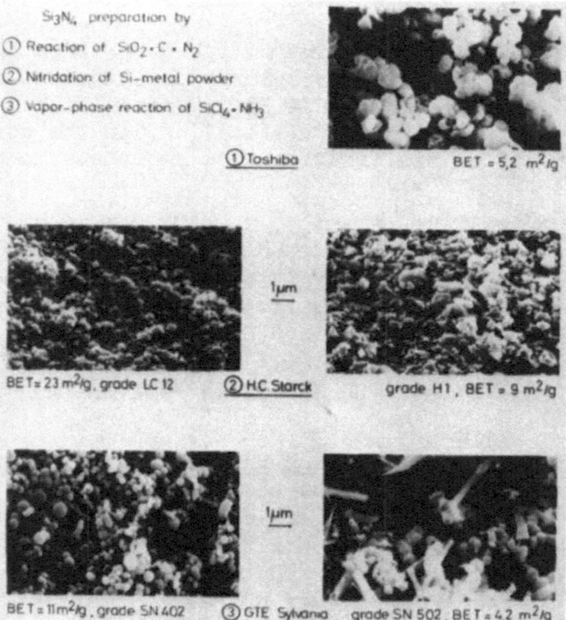

Fig. 2. Silicon nitride powders formed by 3 preparative methods.

All five powders are prepared under conditions (temperature being important) where only little, or no, β-phase is developed. Either α-silicon nitride predominates or, in the case of the vapour phase powders, amorphous silicon nitride.

Essential and preparation-specific differences occur in powder morphology and related powder data (see section 2.3 and Fig. 1). Disregarding the particle size, the three types of powders possess different particle shape and degree of powder aggregation. Aggregated powders with low powder densities require a more intensive milling/mixing during processing to provide a homogeneous mixture with the additives and to achieve not too low a green density giving high shrinkage during sintering. Silicon nitride powders prepared by milling (grade H 1, LC 12 in Fig. 1) consist of discrete particles and therefore possess the highest powder densities. However, powders produced by vapour phase reaction are of comparatively low density. In the case of powder type SN 502 the reason is the high degree of aggregation, together with the existence of Si_3N_4 whiskers (Fig. 2). The Toshiba powder is also partly aggregated.

Scanning electron micrographs (Fig. 2) of the five silicon nitride powders show the different powder morphologies, the particle shapes, and powder aggregation. The Toshiba powder, Starck powder grade LC 12 and the Sylvania SN 402 powder are rather uniform in particle shape and size distribution. The H 1 and SN 502 powders show a wider distribution in this respect; in the case of SN 502 there are globular particles and whiskers.

Silicon nitride whiskers are also developed during nitridation of silicon metal powders. Fig. 5 shows, for example, some whiskers in the silicon nitride raw material and in the lightly milled powder, grade H 2.

4. SELECTION CRITERIA FOR A POWDER PREPARATION METHOD

High-quality silicon nitride ceramics, made by hot pressing or sintering silicon nitride powders, require high purity and fine α-phase powders. Generally powder quality is judged by the sintering result. Chemical and physical powder characteristics which are determined by the powder preparation method are largely responsible for properties of the final product. Fig. 3 shows some common interdependencies between powder parameters and processing with respect to final product quality.

5. PREPARATION OF α-SILICON NITRIDE FROM SILICON

The production of high purity and fine α-silicon nitride powders of definite and reproducible quality chosen by Starck is based on the steps shown in Fig. 4. Nitridation of silicon powder is carried out using ammonia, or nitrogen/hydrogen gas mixtures. This preparative method has proven to be flexible for the production of very different powder qualities. The raw silicon nitride formed in the nitridation process already consists of a great deal of fine

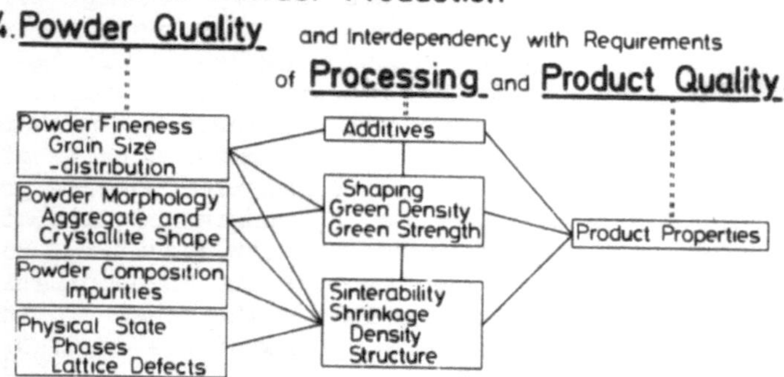

Fig. 3. Selection criteria for powder preparation methods.

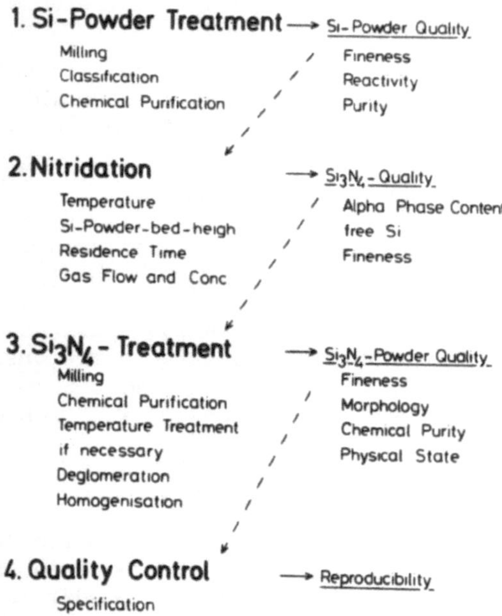

Fig. 4. α-silicon nitride powder production.

Fig. 5. α-silicon nitride, H-grades.

silicon nitride whiskers with a specific surface area in the region of 2 - 3 m^2/g (Fig. 5).

From this crushed raw α-silicon nitride material a powder of about 4 m^2/g is produced by dry milling, or a finer powder of about 9 m^2/g by wet milling, i.e. powder grades H 2 and H 1 respectively. Intensive dry milling would result in different grain size distribution compared with wet milling, while the average particle size and specific surface area are similar. In dry milled powder, coarse particles of 20 μm are present, which are supposed to be sinter-inactive centres. With wet milling, very fine powders of up to 25 m^2/g with narrow particle size distributions can be produced.

Fig. 6 shows some characteristics of three powders. Particle size distributions were determined using an Alpine 400 SZF sedimentation centrifuge.

A summary of powder characteristics of standard Starck silicon nitride powders is shown in Fig. 7. The LC powders also possess a low carbon content.

Fig. 6. α-Si_3N_4 powders, grades H 1, LC 10, LC 12.

Powder Grade	Fineness		Phases			Impurities				
	Specific Surface Area (BET) m²/g	⌀ Grain-Size (FSSS) µm	Beta-Si_3N_4 %	Free Si %	calcul. SiO_2 %	O %	C %	Fe %	Al %	Ca %
H 2	4	1,4	4	0,2	2,4	1,3	0,4	0,08	0,15	0,04
H 1	9	0,7	4	< 0,1	2,4	1,3	0,4	0,04	0,10	0,03
LC 1	9	0,7	3	< 0,1	2,4	1,3	0,1	0,04	0,10	0,03
LC 10	15	0,5	3	< 0,1	2,8	1,5	0,1	0,03	0,08	0,02
LC 12	23	0,4	3	< 0,1	3,2	1,7	0,1	0,02	0,06	0,01

Fig. 7. α-Si_3N_4 powders for hot pressing and sintering (typical analyses).

X-ray diffraction phase analysis is carried out during and after production. Fig. 8 shows the α-Si_3N_4 powder grade H 1. Some characteristic peaks of β-Si_3N_4, free silicon and silica in a range of proportions are shown for comparison.

Finally attention is drawn to quality-determining steps of powder production, which are assumed to be responsible for the sinter-active behaviour of the finest powders. A high α-phase ($>95\%$) content is desirable to ensure beneficial transformation into the β form during sintering, leading to densification and the formation of an interlocked needle structure with high strength. Moreover, due to the intensive milling, mechanical 'activated' surfaces are assumed to be generated, acting as an additional driving force during sintering. Also as a result of the milling, suitable particle shapes and particle size distributions of discrete particles are formed, resulting in a high powder density. With these powders relatively high green densities with low shrinkage can be achieved. The finest powder grade LC 12 provides the highest powder densities because of its' uniformity of particle size (Fig. 9).

In Fig. 9 some processing possibilities and the densification method preferably applied are listed for the powder grades produced.

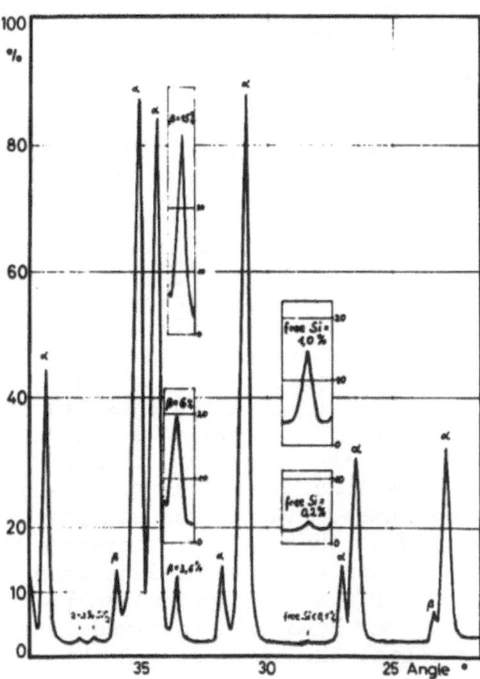

Fig. 8. Phase analysis of α-silicon nitride powder.

Alpha - Si_3N_4 powder grade	fineness BET m^2/g	tape density g/cm^3	Processing-comments advantage	problems	method preferably applied
H 2	4	0,9	high green density	coarse particles need of intens.mill.	HP
H 1	9	0,6	low oxygen, rel.good high-temp. strength		SSN, (HP)
LC 1	9	0,5		some coarse particles	HP, (SSN)
LC 10	15	0,6	good sinterability		SSN
LC 12	23	0,9	high green density, rel. few additives for SSN and good sinterability		SSN

Fig. 9. On the use of α-silicon nitride powders.

To improve the sinterability, even with low amounts of sinter additives, and to achieve fine-grained sintering structures, the finest powder grades are preferred. The silica content is of special importance in respect to the formation of the glassy phase together with the kind and amount of sinter additives. Increasing the silica content of the fine starting silicon nitride powders is easily achieved by oxidation (9). Low silica contents in fine silicon nitride powders are only possible to a certain degree. During wet chemical purification of the powders, even with hydrogen fluoride-containing acids, some hydrolysis takes place at the surface of each particle.

Fig. 10 shows the actual silica contents (as determined by high temperature vacuum fusion and gas chromatography) in our silicon nitride powders, and the silica layer thickness around the nitride particles. The calculation of the silica layer is based on the assumptions that the silicon nitride particles are uniform and spherical in shape, and the total oxygen is due to the silica layer around these spheres. Corresponding to the average particle size measuring method applied - BET or FSSS - silica layers can be calculated to 0.5 or 3 nm. Regarding the dimension of an SiO_2 unit cell as the lower limit for the thickness of a monomolecular SiO_2 layer, about 0.5 nm can be assumed. The silica contents actually analysed almost correspond to that value. Therefore with the purification method applied, silica contents of less than 1.8% for powders with 25 m^2/g specific surface area are unlikely, though in the laboratory 2.3% SiO_2 may have been achieved. Silicon oxynitride is not essentially present, or detected, and therefore is estimated to be less than 0.5%, corresponding to less than an error of 0.15% in the silica estimation.

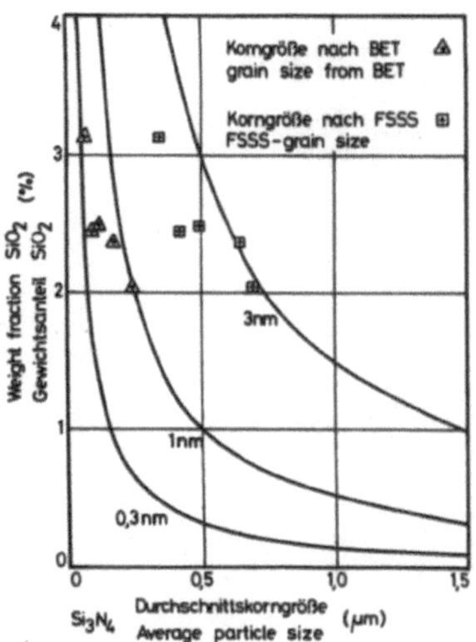

Fig. 10. Silica content and layer thickness of α-silicon nitride powders.

REFERENCES

1. D.R. Glasson and S.A.A. Jayaweera, "Formation and Reactivity of Nitrides", J. Appl. Chem. 18 (3) 65-77 (1968).
2. U.S. Patent 3,937,792, Feb. 1976, "Manufacturing silicon nitride powder", R.J. Lumby (Lucas Industries Ltd.).
3. H. Haag, W.D. Glaeser and B. Krismer, "Preparation and characterisation of silicon nitride powders", in 'Nitrogen Ceramics', ed. F.L. Riley, NATO ASI Applied Science Series E No. 23, Noordhoff, Leyden, 1977, p.315.
4. K. Komeya and H. Inoue, "Synthesis of the α form of silicon nitride from silica", J. Mater. Sci. 10, 1244-46 (1975).
5. U.S. Patent 4,117,095, Sept. 26, 1976, "Method of making alpha type silicon nitride powder", K. Komeya (Toshiba Co.).
6. U.S. Patent 4,073,875, Feb. 14, 1978, "High density high strength Si_3N_4 ceramics prepared by pressureless sintering of partly crystalline, partly amorphous Si_3N_4 powder", S.T. Buljan and P.E. Stermer (GTE Sylvania Inc.).
7. K.S. Mazdiyasni and C.M. Cooke, "Synthesis, characterization and consolidation of Si_3N_4 obtained from ammonolysis of $SiCl_4$", J. Amer. Ceram. Soc. 56 (12), 628-33 (1973).
8. S. Prochaska and C. Greskovich, "Synthesis and characterization of a pure silicon nitride powder", Amer. Ceram. Soc. Bull. 57 (6), 579-81, 86 (1978).
9. C. Greskovich and J.A. Palm, "Controlling the oxygen content of Si_3N_4 powders", Amer. Ceram. Soc. Bull. 59 (11), 1155-56 (1980).

Section D

STRUCTURAL DEVELOPMENT

LIQUID PHASE SINTERING

J. Weiss and W.A. Kaysser

Max-Planck-Institut für Metallforschung,
Institut für Werkstoffwissenschaften,
Pulvermetallurgisches Laboratorium
Heisenbergstraße 5
7000 Stuttgart-80, West-Germany

INTRODUCTION

Liquid phase sintering is a process widely used for production of ceramic and metal parts. The technical empiricisms for application have been known for many years, nevertheless the mechanisms involved in liquid phase sintering are still not completely understood. This paper will try to combine the classical basic ideas on the liquid phase sintering mechanisms with the more recent experimental observations of the last few years. A general statement upon the effective densification phenomena, rearrangement and centre to centre approach of particles will be succeeded by a more detailed description of the mechanisms causing both phenomena. In a final part the contribution of these mechanisms on the densification phenomena of Si_3N_4 ceramics will be discussed, in conscience of the high complexity of Si_3N_4 ceramic systems.

DENSIFICATION PHENOMENA

During liquid phase sintering of some systems just after melting one of the components, a rapid densification occurs by the movement of the solid particles from the initial positions towards a final arrangement of higher degree of space filling. From the classical systematic experiments of Lenel and coworkers (1, 2) this rearrangement was concluded to be the first of three distinct shrinkage stages: rearrangement, solution-reprecipita-

Riley, F.L. (ed.) Progress in Nitrogen Ceramics

tion and skeleton sintering. Actually, rearrangement is an effective densification phenomenon during the whole shrinkage period (3) which even occurs during skeleton and solid state sintering (4). The densification rate due to rearrangement depends on the mobility of the solid particles (5). The "rearrangement stage" was deduced from those experimental cases, where the particles had a high mobility just when the liquid occured (2, 6). This primary rearrangement becomes less or disappears when the initial particle mobility is decreased. Figure 1 shows schematically some effects which may lower the initial mobility as surface roughness (b) (6), particle size (c) and shape (d) distribution, as well as solid contacts (e) formed by cold welding during compaction or by neck formation during heating. If prior solution-reprecipitation processes (as surface smoothening, or the dissolution of small particles between large ones) increase the particle mobility, secondary rearrangement may lead to further densification (7, 8). In the case of secondary rearrangement the rearrangement kinetics may usually depend on the kinetics of the foregoing solution-reprecipitation processes.

From rearrangement we may discern the centre to centre approach of adjacent particles. This second shrinkage phenomenon is based on a change of the particle geometry in the contact region and may for example occur by contact flattening (9, 10) or by the dissolution of small particles and the reprecipitation on large ones.

It is trivial to note, that both densification phenomena are caused by a decrease of the total energy of the vapour-liquid, liquid-solid and solid-vapour interfaces.*

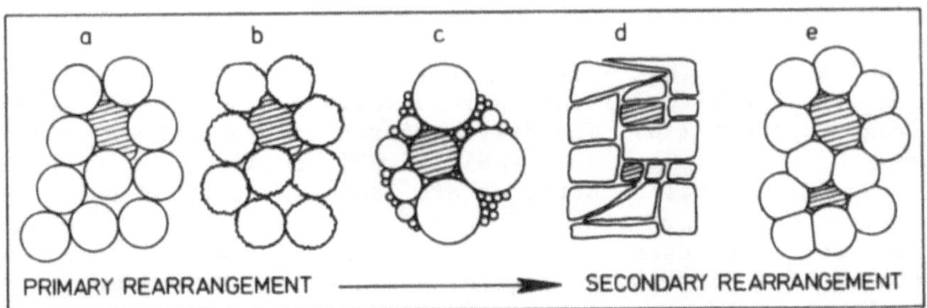

Fig. 1a to e. Factors leading from primary to secondary rearrangement (see text).

* For special cases also swelling may be caused by a decrease of the total interface energy (9, 11, 12).

PRIMARY PARTICLE REARRANGEMENT

The parameters involved in primary rearrangement of large particles have been studied systematically by using large Cu coated W spheres and mixtures of Cu and W spheres. (A summary of this work is given in (13)). The components in this system show practically no solubility into each other (14) thus solution-reprecipitation processes were excluded. The parameters increasing densification of the Cu coated spheres by rearrangement are the wetting angle, the volume fraction of the liquid, and the velocity by which the melting front passed the specimens. Direct observations of the rearrangement process in a hot stage SEM revealed (15), that only one or a very few particles move simultaneously in the region of the migrating melting front and that no further rearrangement occurs outside the melting front. In the mixtures of Cu and W spheres during early rearrangement stage solid particles were drawn into the liquid Cu droplets leading to formation of clusters consisting of 5 to 20 solid particles. Due to the agglomeration of particles in liquid rich clusters larger pores of different size were formed in the sample (16). The later effect explains the strong influence of the initial distribution of the low melting phase and of the pores. These observations were treated with the liquid bridge approach, where it is assumed that the movement of a solid particle is determined by the capillary forces provided by one or a few liquid bridges with its intimate neighbours (17, 18, 19). By this mechanism the particle will migrate towards the neighbour where the first liquid bridge occurs, or towards the direction of the majority of bridges. Thus the particles locally move towards the initially denser regions and a structure consisting from regions where particles are packed to high density and adjacent enlarged pores should occur, as experimentally verified.

The capillary forces which are provided by these enlarged pores are too weak to initiate a remarkable pore closure by further rearrangement (16). The driving force to close these pores becomes, however, larger when their size is smaller, as is the case when smaller solid particles are used. In addition smaller particles show less retarding forces for further movement because of lower forces necessary for liquid bridge disruption, lower friction forces between the particles and thus a higher mobility. For small particles the closure of residual pores becomes the most important and effective process. In an early treatment (20), pore closure was treated on viscous flow principles, whereby the mixture of particles with melt was considered to be a liquid of high viscosity. Beyond that no systematic investigation upon rearrangement during liquid phase sintering of small particles is known to the authors yet. Most principles of primary particle rearrangement are valid for secondary rearrangement too.

SOME IMPORTANT DISSOLUTION-REPRECIPITATION PROCESSES

As already noted in the first chapter, various mechanisms based on dissolution-reprecipitation processes may provide shrinkage by secondary particle rearrangement and centre to centre approach. During two decades contact flattening was assumed to be the only mechanism, but we will see that other mechanisms may give a higher contribution to shrinkage in most cases. Kingery (21) has suggested a theoretical model for the contact flattening mechanism on the basis of which he derived equations for the rate of densification. He assumed that the pressure of pores produces high stresses in the liquid films separating the adjacent particles in the contact regions. Due to the local high stresses the solubility in the contact regions is higher and material is dissolved, transported away and reprecipitated in other regions with lower stress and solubility. The time laws for densification by this mechanism were often experimentally observed, but the mechanism never could be verified to occur by microstructural evidence.

From recent liquid phase sintering experiments it was concluded that a shape accomodated Ostwald ripening provides the essential changes in particle shape and size for densification and that contact flattening plays an unimportant role (7, 22). As fig. 2 presents schematically for a mixture of large and small particles, small particles are dissolved and precipitated on the large spheres due to the differences in specific surface energy of curved surfaces (Ostwald ripening). Precipitation during which the sphericity of the large particles is maintained can only lead to densification if accompanied by rearrangement of the growing spheres. In general, densification by the movement of the large

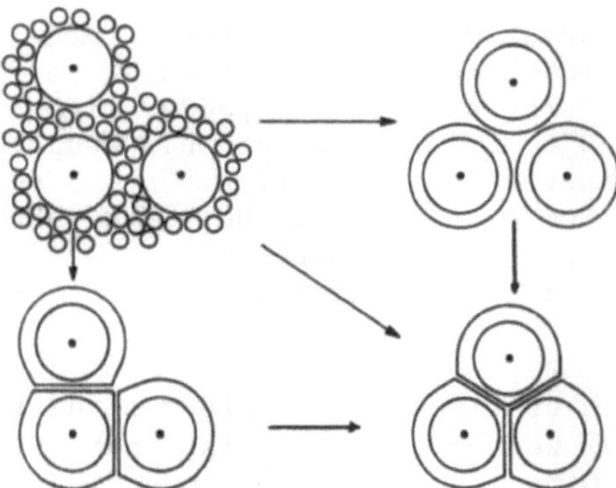

Fig. 2: Shrinkage provided by rearrangement due to shape accomodated grain growth (22).

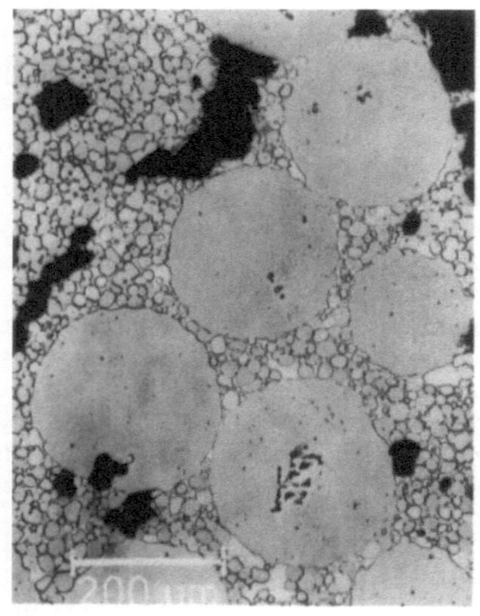

Fig. 3a: 3 min

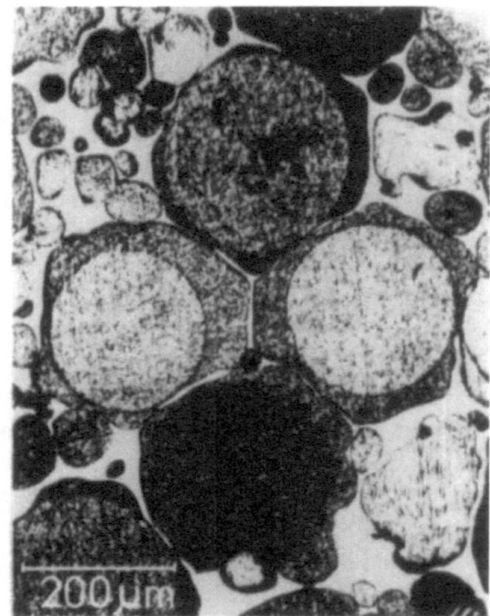

Fig. 3b: 120 min

Fig. 3a and b: Shrinkage provided by rearrangement due to shape accomodated grain growth during liquid phase sintering of W-Ni at 1943K (8).

particles is pronounced if the material reprecipitating on the particles is preferentially deposited away from the areas where the grains are in close contact (shape accomodation).

The latter mechanism was demonstrated in model experiments in the W-Ni system (8) in which mixtures of equal portions of 200-250 µm diameter W spheres and 10 µm W powder with 4 wt.% Ni powder were sintered at 1670°C. After 3 min at this temperature the large spheres remained essentially intact (fig. 3a). After 120 min the porosity decreased from 10 to 2% and the large spheres had grown by reprecipitation; etching with Murakami's solution clearly differentiated the original W spheres and the reprecipitated material (fig. 3b).

For systems with continuous particle size distribution it is still not unequivocal whether contact flattening or shape accommodated Ostwald ripening gives the major contribution to shrinkage.

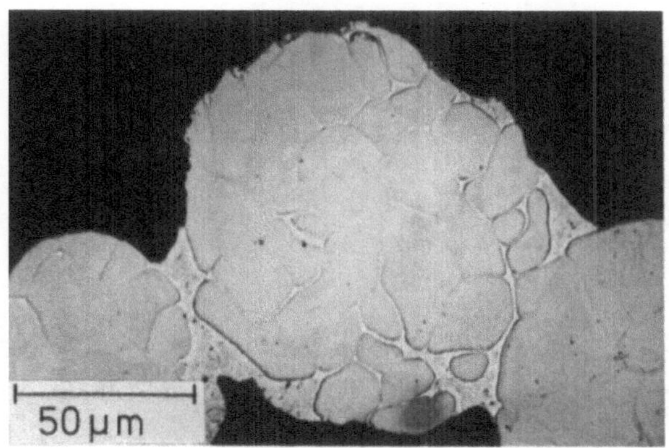

Fig. 4: Microstructure of Fe-10 wt-% Cu after sintering at 1438K for t_s=8 min showing penetration of melt along grain boundary of Fe spheres.

Whereas grain growth is an effect common for all systems the attack on grain boundaries of the solid particles by the melt and the subsequent particle disintegration is a liquid sintering mechanism which is only observed for systems which satisfy the condition $2\gamma_{sl} < \gamma_{ss}$ (γ_{sl} = spec. energy solid-liquid, γ_{ss} = grain boundary energy). Most extensively studied is the Fe-Cu system (12) in which Fe spheres with 10 vol.% Cu spheres, both of 100 µm diameter, were sintered at 1438K for times up to 60 min. After annealing for 3 min above melting point of Cu, all contact regions between the Fe particles were filled with liquid. After 8 min melt had penetrated along the grain boundaries into the interior of the Fe particles as a thin liquid layer (fig. 4). Thereby the tip of the advancing melt takes iron into solution which is transported through the liquid and reprecipitated on the particle surface.

If grain boundary attack along all grain boundaries of a particle is complete, particle disintegration may occur, resulting in loss of original particle shape, as shown in Figure 5a and 5b for polycrystalline Al_2O_3-spheres liquid phase sintered with glassy phase. The disintegration of the Al_2O_3-particles leads to an increased mobility of single grains and enables secondary rearrangement and considerable shrinkage to occur (23, 24).

In the Al_2O_3-glass system it was possible to separate the shrinkage due to secondary rearrangement after particle disintegration from that due to primary rearrangement (24). Figure 24 shows the dimensional change of slightly compacted Al_2O_3-glass samples sintered at different temperatures. Up to 1673K shape

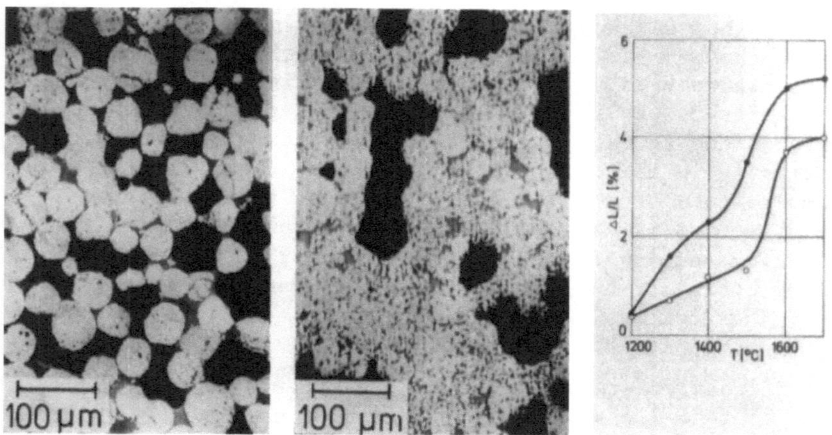

(after 1h, o after 1min)

Fig. 5a: 1h/1673 K Fig. 5b: 1h/1973 K Fig. 5c: Shrinkage

Fig. 5a to c: Microstructural changes and shrinkage during liquid phase sintering of Al_2O_3-alkali borate glass (24).

and size of the polycrystalline Al_2O_3 spheres are maintained and only primary rearrangement occurs. Above this temperature grain boundary attack and particle disintegration occur and the secondary rearrangement rapidly leads to considerable additional shrinkage (shown in fig. 5c and schematically in fig. 6).

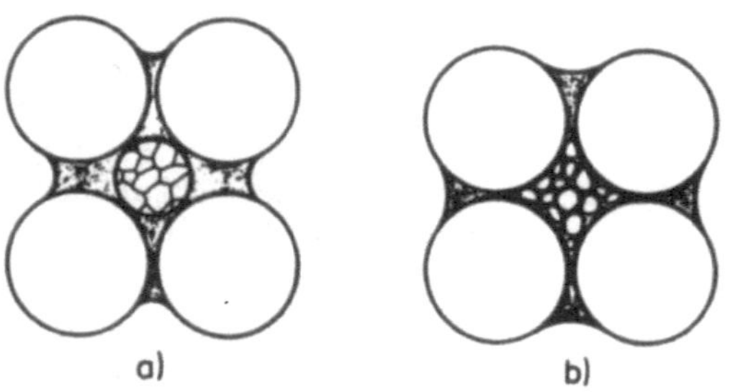

Fig. 6: Shrinkage after particle disintegration.

INFLUENCE OF "CHEMICAL" AND "STRUCTURAL" ENERGY REDUCTION ON SHRINKAGE

Up to now we have discussed some essential mechanisms on the basis of the geometry changes which occur when material is dissolved, transported and reprecipitated. Driving force for these mechanisms thus was simply the decrease in free energy of the interfaces due to the change of interface area. The free energy reduction is in the order of 1 to 10 $Jmol^{-1}$ if the initial and the dense state of a sintering body produced from particles of 10 µm diameters is compared. Chemical reactions which occur in many systems during liquid phase sintering provide a decrease in the free energy of the system of typically between 100 and 1000 $Jmol^{-1}$. This comparatively large driving force may become effective for the initiation and acceleration of solution-reprecipitation processes when differences in the chemical composition or phase composition exist between the dissolved and the reprecipitated material. The influence may be pronounced when the material transport via liquid phase is leading much faster to a chemical equilibrium of the material than other material transport processes (e.g. bulk diffusion). Similar arguments are valid for differences in the defect density (e.g. dislocations) of dissolved and reprecipitated matter. It may be noted, however, once more, that chemical or structural driving forces include no part which leads to densification but that only the decrease in interfacial energy provides the driving force necessary for the densification phenomena.

The influence of chemical driving forces during liquid phase sintering becomes especially obvious for <u>directional grain growth</u> where the interfacial energy is increased and hence solid solution or compound formation is a necessary prerequisite for the mechanism to occur.

When single crystal W spheres of uniform size are sintered in the presence of liquid Ni, growth of one sphere at the expense of its immediate neighbour occurs (25) (Fig. 7a). This phenomenon has also been found to occur in Fe-Cu and Mo-Ni systems and is therefore believed to be a rather general process connected with the compositional difference between the pure component dissolving and the precipitating solid solution. As shown in fig. 7b, the microprobe analysis demonstrates that the shrinking grains consist of 100% W while the precipitated material is a W-0.15 wt.% Ni solid solution (25). This compositional difference provides a decrease in free energy of the order of 72 $Jmol^{-1}$ and overcompensates for the increase in interfacial energy produced by the liquid solid area. Recently experimental evidence for directional grain growth in ceramic systems has been obtained in

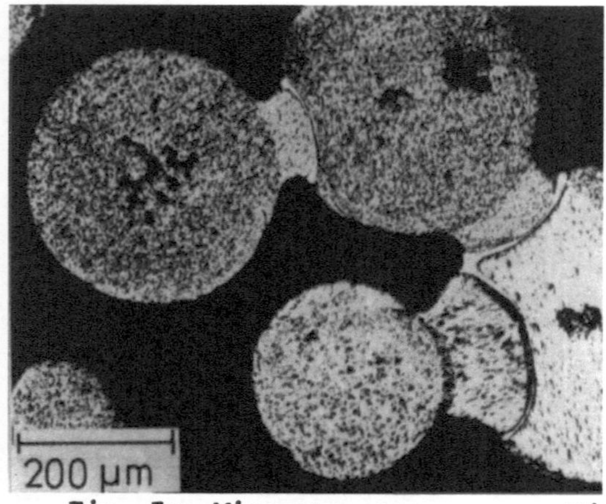

Fig. 7a: Microstructure

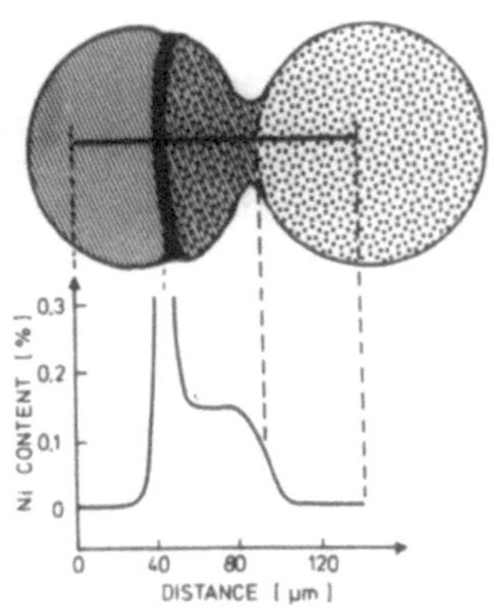

Fig. 7b: Microprobe analysis

Fig. 7a and b: Directional grain growth during liquid phase sintering of single crystal W-spheres with Ni at 1943 K (25).

$ZnO-Bi_2O_3-Li_2O$ (26). Figure 8 shows a fish-like particle shape which is similar to the shapes of directionally grown metal particles in fig. 7a. Directional grain growth itself contributes little to densification. Shrinkage may occur in connection with the filling of the concave neck regions with material originating from contact regions and from small particles in the vicinity.

A change in the chemical composition was also observed due to shape accomodated Ostwald ripening in W-Ni. Microprobe measurements of the reprecipitated layers in fig. 7b show that they

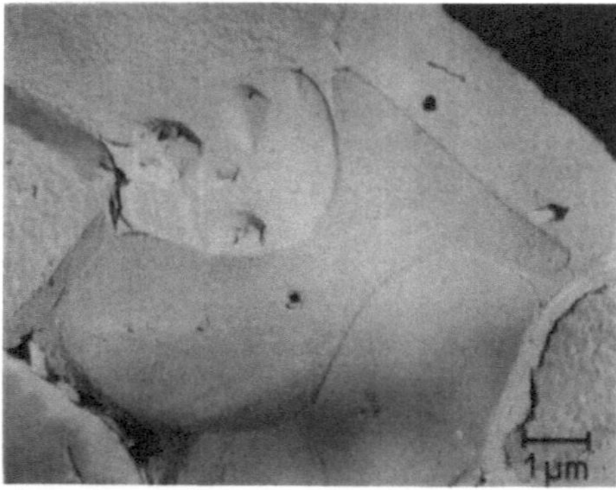

Fig. 8: Directional grain growth during liquid phase sintering of $ZnO-Bi_2O_3-Li_2O$ (26).

consist of a solid solution of Ni in W whereas the inner spheres are still pure W (8, 27, 28). It was recently demonstrated in W-Fe-Ni that the kinetics of solution-reprecipitation and thus grain growth accompanied by the formation of the solid solution are accelerated by the decrease of free energy (29).

Grain boundary penetration and particle disintegration have been observed in many systems where no measurable solubility in the solid phase exists (e.g. Cu-Bi, Al_2O_3-glass). In other systems particular in Fe-Cu it was shown that the pure component Fe was taken into solution at the tip of the advancing melt (ß), transported through the liquid phase and reprecipitated as Fe(Cu) solid solution (ß) (12) (schematically described in fig. 9). It is not yet clear whether a chemical driving force may induce or enhance the solution reprecipitation process of this mechanism.

SINTERED DENSE Si_3N_4 CERAMICS

After describing the densification phenomena during liquid phase sintering and their correlation to various mechanisms some attention shall be paid on liquid phase sintering of Si_3N_4. Most investigations so far were aimed at studying the influence of composition and the parameters time and temperature on macroscopic shrinkage. Little attention was payed to clarify the various mechanisms. We here shall concentrate on the attempt to correlate several of the mechanisms mentioned above to some experimental observations during liquid phase sintering of Si_3N_4.

DENSIFICATION BY PRIMARY REARRANGEMENT

After cold compaction of the Si_3N_4 powder mixture the particles are in contact by adhesion and other attaching features

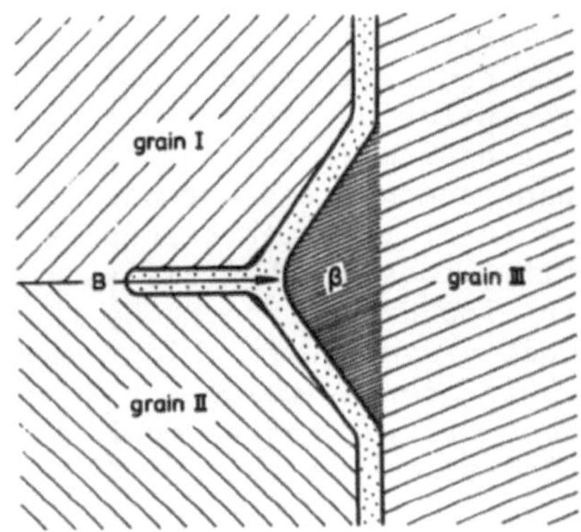

Fig. 9: Solution-reprecipitation model of grain boundary penetration.

which refer to the situation a, b, c and d in Fig. 1. During heating liquid phase occurs due to the various additives and primary rearrangement starts. The melting temperature and the amount of liquid phase depend on the sort and composition of the additives. During the first minutes the amount and composition of the melt undergoes still a continuous change by the dissolution of residual additions and of some Si_3N_4. The change in composition is accompanied by a change of the wetting, spreading and flow behaviour of the melt during this time interval. Due to the high viscosity of the silicate melts (30, 31) the spreading rate of the melt on the particles and its penetration in the particle contact areas is extremely slow compared to the metal liquids (comp. table 1). As a consequence particle movement and densification rate by rearrangement will directly depend on the viscosity of the melt as shown for sintering of Si_3N_4 with silicate melts containing various additions. The initial shrinkage rate during isothermal annealing rose rapidly for higher annealing temperaures where the viscosity of the melt became lower. The final wetting angle of the melt on the Si_3N_4 particles is known to be different for various additions and lays between 9° for MgO additions and 43° for CaO (33). Variations of the wetting angle in this range are, however, in general of minor influence for the densification by primary particle rearrangement.

The high viscosity of the melt brings about that primary particle rearrangement will occur as a simultaneous interaction of many particles without any local restriction to a melting (or for glasses to a weakening) front. For the treatment of densification by rearrangement in Si_3N_4 produced of fine powders the viscous flow model thus seems to be preferable to the liquid bridge model.

DENSIFICATION BY SECONDARY REARRANGEMENT AND CENTER TO CENTER APPROACH

When the major part of primary rearrangement is completed, secondary rearrangement and centre to centre approach, which depend on the slower dissolution-reprecipitation kinetics, control the further densification rate. As mechanisms both contact flattening of the Si_3N_4 particles or shape accomodated Ostwald ripening may be considered. As a consequence of the high viscosity of the melt which means a low diffusivity, these mechanisms should be extremely slow with respect to the low driving force by the reduction of interface area. The rapid grain growth which occurs (34, 35, 36) has to be caused by driving forces which are several orders of magnitude larger. These driving forces could be provided by the dissolution of particles consisting of α Si_3N_4

TABLE 1

Comparison of diffusion-rates in solid and liquid state for various metals and silicon based ceramics.

Literature	compound	temperature (K)	diffusing species	D (cm^2/s)	activation energy (J/Mol)
(46)	βSi_3N_4	1473	N^{3-}	$2.012\cdot10^{-21}$	776970
(31)	SiO_2 quartz	1473	O^{2-}	$2.56\cdot10^{-16}$	230120
	SiO_2 glass	1473	O^{2-}	$2.11\cdot10^{-13}$	234300
(21)	40wt%CaO,20wt% Al_2O_3,40wt%SiO_2 glass	1473	O^{2-}	$1.58\cdot10^{-7}$	227191
	"	1773	O^{2-}	$3.65\cdot10^{-6}$	227191
	"	1773	Si^{4+}	$2.35\cdot10^{-16}$	292880
	"	1773	Al^{3+}	$2.17\cdot10^{-7}$	251040
	"	1773	Ca^{2+}	$1.38\cdot10^{-16}$	334720
(38)	SiO_2-MgO-Si_3N_4	1823	(N^{3-},Si^{4+})	$1.28\cdot10^{-14}$	450000
		1823	(N^{3-},Si^{4+})	$1.22\cdot10^{-15}$	690000
(32)	Fe-Cu Fe in liquid Cu	1473	Fe	$5.17\cdot10^{-5}$	51630
(32)	Co-Cu Co in liquid Cu	1473	Co	$4.70\cdot10^{-5}$	47656

and the reprecipitation of the material as ßSi_3N_4 (30, 34, 37, 38). The growth of the ß grains at the expense of the α particles might include two mechanisms. These are directional grain growth*, as long as the ß particles are much smaller than the particles and eventually a modified Ostwald ripening if the ß particles became much larger than the adjacent α grains. The assumption that directional grain growth occurs is supported by the observation that the rate of the α-Si_3N_4 to ß-Si_3N_4 transformation is proportional to the initial volume fraction of the ß modification (34, 39). In addition it was found, that shrinkage is directly related to the amount of α-Si_3N_4, already transformed into ß-Si_3N_4 (37, 38). This behaviour may be explained in analogy to the shrinkage during shape accommodated Ostwald ripening of the large and small W particles with Ni. When the small α-Si_3N_4 grains are dissolving in the neighbourhood of the ß-particles, both ß-grains and residual α particles undergoe secondary rearrangement due to the increased mobility **. The presence of a sufficient amount of liquid phase favours the prismatic growth habit of ß Si_3N_4 grains caused by a prefered growth of the hexagonal layer structure into the < 0001> direction.

Figure 10 shows a schematic description of the possible densification of Si_3N_4 during the α-ß transformation. The model in fig. 10a demonstrates the state after primary rearrangement representing a monolayer of cylindrical (equiaxed) small α particles. The small pores between the particles are completely filled by melt (21.5 vol.-%). Several single larger pores account for a porosity of 18 vol.-%. Figure 10b shows prismatic ß-Si_3N_4 grains grown on the expence of α grains. The melt initially distributed between these α grains becomes available for redistribution. Figure 10c presents the state, when this liquid has filled the pores partially. The pore filling, which means shrinkage, necessitates a simultaneous rearrangement of small α and the large ß particles. It should be emphasized that grain growth and rearrangement take place coincidentally and continously but were here separated for the simplicity of the scheme. After rearrangement all ßSi_3N_4 grains impinge each other (fig. 10c) and

---

* If directional grain growth is defined as growth of grains at the expense of their intimate neighboured grains regardless to the increase in interface area going along with the process.

** A better use of the increased particle mobility due to the dissolution of small Si_3N_4 grains is provided by hot pressing where shrinkage during the transformation period increases with increasing pressure (37, 38).

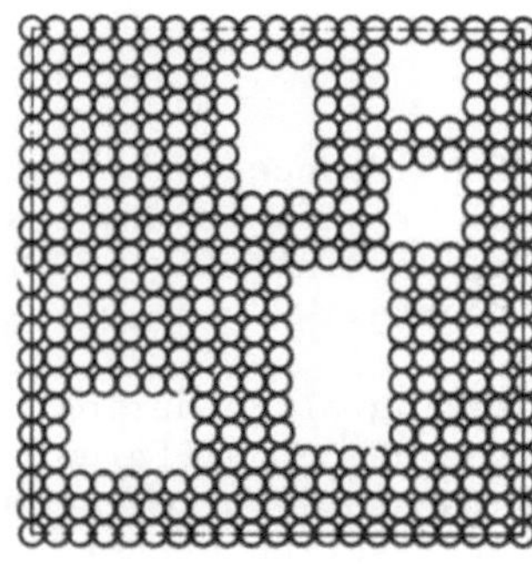

Fig. 10a: Initial stage (21.5 vol% melt, 17.2vol% porosity, $\beta/\alpha+\beta=0$)

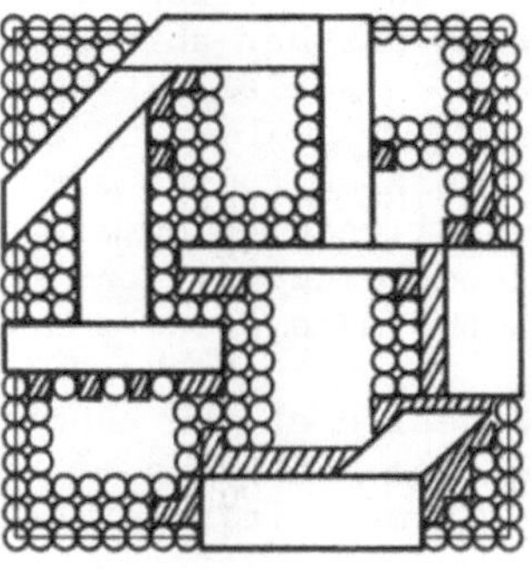

Fig. 10b: Grain growth (21.5vol% melt, 17.2vol% porosity, $\beta/\alpha+\beta=0.43$)

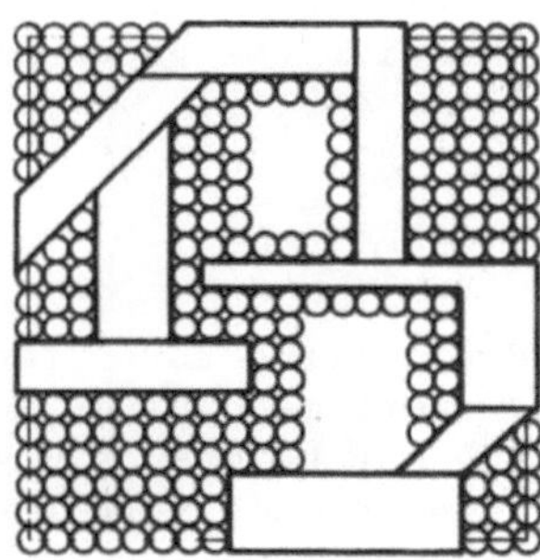

Fig. 10c: Rearrangement (21.5 vol%melt, 9.8vol% porosity, $\beta/\alpha+\beta=0.43$)

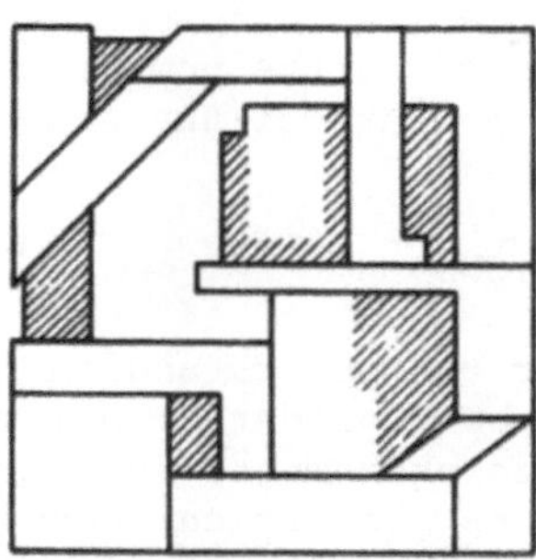

Fig. 10d: Final stage (21.5vol% melt, 9.8vol% porosity $\beta/\alpha+\beta=1.$)

Fig. 10 a to d: Schematic description of grain growth and shrinkage during the α Si_3N_4 - βSi_3N_4 transformation (○ α grain, ▭ ß grain, //// redistributed liquid).

thus have lost the possibility for further rearrangement. Therefore the transformation of the residual α grains into ß particles provides no further shrinkage (fig. 10d).

In addition to the simple considerations deduced from the experimental results on simple binary metal or ceramic systems for the complex conditions during liquid phase sintering of Si_3N_3 at least two more factors need to be considered:

1) The kinetics of the chemical reaction taking place during sintering: Sintering of Si_3N_4 enables the use of different starting materials for maintainance of the same overall composition. The differing value of heat of reaction brings about a differing influence on the sintering behavior: Important parameters are liquid composition and amount and further the melting temperatures (40 to 43).

2) The partial pressure ratio of the sintering atmosphere: In order to avoid vapourization from the condensed phases an apropriate sintering atmosphere has to be maintained during sintering. The most important vapour species thereby are N_2 and SiO (44, 45). Most commonly the partial pressures are provided by the powder bed method by embedding the green body into powders of the same composition (40 to 44).

CONCLUSIONS

For the understanding of liquid phase sintering three factors have to be considered: Driving forces, transport mechanisms and shrinkage phenomena. The shrinkage phenomena may be most simply treated as geometrical changes by primary and secondary rearrangement, as well as centre to centre approach. The later two phenomena require that dissolution-reprecipitation proceeds. Thereby various mechanisms are possible leading to different changes of particle size and shape: Particle disintegration, contact flattening and shape accomodated grain growth. The sole driving force for shrinkage is the reduction in interface energy. The kinetics of the mechanisms, however, may change remarkably if differences in the composition or structure between dissolved and reprecipitated material provide additional driving forces. If these driving forces are large even new mechanisms such as directional grain growth may occur. The later mechanism gives an important contribution for densification of Si_3N_4 in the presence liquid phase.

ACKNOWLEDGEMENT

The authors wish to thank Prof. Dr. Dr. h.c. G. Petzow for the continuous support of this work. The financial support of the Bundesministerium für Forschung und Technology (BMFT) is greatfully acknowledged.

REFERENCES

1. F.V. Lenel; in "The Physics of Powder Metallurgy", ed. W. Kingston, McGrow-Hill Book Company, New York (1951) 238.

2. H.S. Cannon and F.V. Lenel; 1. Planseeseminar, Metallwerk Plansee, Reutte (1953) 106.

3. W.A. Kaysser; Ph.D. Thesis, University Stuttgart (1978).

4. H.E. Exner; "Grundlagen von Sintervorgängen", Gebr. Bornträger, Berlin und Stuttgart (1978).

5. W.A. Kaysser, S. Takajo and G. Petzow; "Modern Development in Powder Metallurgy", MPIF/APMI, 12-14 (1980).

6. W.D. Kingery, M.D. Narasimhan; J. Appl. Phys. 20 (1959) 307.

7. W.J. Huppmann and G. Petzow; "Modern Development in Powder Metallurgy", eds. H.H. Hausner and P.W. Taubenblat, MPIF/APMI 9 (1976) 77.

8. D.N. Yoon and W.J. Huppmann; Acta Met. 27 (1979) 693.

9. W.D. Kingery; J. Appl. Phys. 20 (1959) 301.

10. A.L. Prill, H.W. Hayden and J.H. Brophy, Trans AIME 233 (1965) 960.

11. D. Berner, H.E. Exner and G. Petzow; "Modern Developments in Powder Metallurgy", eds. H.H. Hausner and H.E. Smith, MPIF/APMI, Princeton (1974) 237.

12. W.A. Kaysser, W.J. Huppmann and G. Petzow; Powder Met. 23 (1980) 86.

13. W.J. Huppmann, H. Riegger, W.A. Kaysser, K. Smolej and S. Pejovnik; Z. Metallkde. 70 (1979) 707.

14. M. Hansen and K. Anderko; "Constitution of Binary Alloys", Mc Graw-Hill, New York (1978).

15. W.A. Kaysser; "Contemporary Inorganic Materials 1978", eds. G. Petzow and W.J. Huppmann, Dr. Riederer Verlag, Stuttgart (1979) 41.

16. V. Smolej, S. Pejovnik and W.A. Kaysser; to be published in Powd. Met. Int. 14 (1982).

17. R.B. Heady and J.M. Cahn; Met. Trans. 1 (1970) 185.

18. V. Smolej and S. Pejovnik; Z. Metallkde. 67 (1976) 603.

19. W.J. Huppmann and H. Riegger; Acta Met. 23 (1975) 965.

20. W.D. Kingery; "Kinetics of High Temperature Processes", ed. W.D. Kingery, M.I.T., Cambridge (1959) 198.

21. W.D. Kingery; J. Appl. Phys. 30 (1959) 301.

22. G. Petzow and W.A. Kaysser; "Science of Ceramics 10", ed. H. Hausner, DKG 10 (1980) 269.

23. S. Pejovnik, D. Kolar, W.J. Huppmann and G. Petzow; "Science of Ceramics 9", 9 (1978) 87.

24. S. Pejovnik; Ph.D. Thesis, University of Ljubljana (1978).

25. D.N. Yoon and W.J. Huppmann; Acta Met. 27 (1979) 973.

26. W.A. Kaysser and A. Lenhart; to be published.

27. H. Riegger; Ph.D.Thesis, University Stuttgart (1977).

28. L. Kozma, W.J. Huppmann, L. Bartha and P. Mezzei, Powder Met., 24 7 (1981).

29. D.N. Yoon, private communication.

30. R.E. Lohmann and D.J. Rowcliff,; J. Am. Ceram. Soc. 69 (1980) 144.

31. G.H. Frischat, "Ionic Diffusion in Oxide Glasses", Trans. Techn. Publications 1975, Diffusion and Defect Monograph Series 3-4.

32. T. Eijima and K. Kameda, J. Jap. Inst. of Metals 33 (1969) 96.

33. G.R. Terwilliger and F.F. Lange; J. Am. Ceram. Soc. 57 (1974) 25.

34. H. Knoch and E.E. Gazza; Ceramurgia Int. 6 (1980) 51.

35. M.H. Lewis, A.R. Bhatti, R.J. Lumby and B. Norta; J.Mat.Sci. 15 (1980) 103.

36. M. Kuwabara, M. Benn and F.L. Riley; Proc. Cimtec IV, ed. P. Vincenzini, Elsevier Scientific Publ., Amsterdam (1980).

37. D.R. Messier, F.L. Riley and R.J. Brook; J. Mat. Sci. 13 (1978) 1199.

38. L.J. Bowen, R.J. Weston, T.G. Carruthers and R.J. Brook; J. Mat. Sci. 13 (1978) 341.

39. F.F. Lange; Nitrogen Ceramics, ed. F.L. Riley, Noordhoff, Leyden (1977) 491.

40. S. Boskovic, L.J. Gauckler, G. Petzow and T.Y. Tien; Powd. Met. Int. 9 (1977) 185.

41. S. Boskovic, L.J. Gauckler, G. Petzow and T.Y. Tien; Powd. Met. Int. 10 (1978) 184.

42. S. Boskovic, L.J. Gauckler, G. Petzow and T.Y. Tien; Powd. Met. Int. 11 (1979) 169.

43. G. Petzow, L.J. Gauckler, T.Y. Tien and S. Boskovic; in Factors in Densification and Sintering of Oxide and Non-Oxide Ceramics, ed. by S. Somiya and S. Saito, Assoc. for Sci. Documents Int. Tokyo (1978) 23.

44. M. Mitomo, N. Kuramoto, Y. Inomata, J. Mat. Sci. 14 (1979) 2309.

45. P. Dörner, L.J. Gauckler, H. Krieg, H.L. Lukas, G. Petzow and J. Weiss; J. Mat. Sci. 16 (1981) 935.

46. K. Kijima and S. Shirasaki; J. Chem. Phys. 65 (1976) 2668.

DISCUSSION

Mangels: How do you obtain higher shrinkages with higher green densities?
Weiss: It is assumed that shrinkage only occurs by particle rearrangement. A high green density means a more homogeneous powder packing, which is easier to rearrange.
Greskovich: Are your conclusions on pore growth in 2 dimensional arrays of particles equally applicable to 3-dimensional arrays?
Weiss: Yes - see reference (13).

SINTERING OF SILICON NITRIDE, A REVIEW

P. Popper

The British Ceramic Research Association,
Queens Road, Penkhull, Stoke-on-Trent, ST4 7LQ,
England.

ABSTRACT

Although sintering of pure Si_3N_4 has been found difficult, considerable progress has been made with liquid state sintering. The following sintering aids have been used with particular success, either singly or in combination: MgO, Y_2O_3, Al_2O_3 and $BeSiN_2$. The main problem with sintering above 1500°C is decomposition of Si_3N_4 and weight loss by evaporation of Si or SiO. The weight loss can be suppressed by sintering in a high pressure N_2-atmosphere and/or using a powder bed. Devitrification of the glassy phase improves the high temperature properties. Whilst very fine powders often result in high shrinkages, "post-sintering" of RBSN leads to very low shrinkage. In this process the sintering aid may be admixed with the Si-powder before nitridation or the RBSN body can be infiltrated cold in a soluble form or hot from the powder bed. One can now foresee the mass production of engineering components with good high temperature properties.

1. INTRODUCTION

In reviewing the progress made since the last NATO Seminar on Nitrogen Ceramics one finds the developments in the field of sintering most outstanding. Then, anyone wishing to obtain a component in silicon nitride had the choice between a complex shape made of relatively weak reaction-sintered material and a hot pressed cylinder from which a more complex shape could be machined with diamond tools at very great expense. Today, we can foresee the commercial availability of components of complex shapes formed by mass production techniques, such as injection moulding, to close dimensional tolerances and with properties similar to those of the

Riley, F.L. (ed.) Progress in Nitrogen Ceramics

hot pressed material.

In this review the three basic states of matter will be considered: solid, liquid and gas. These may be simply characterized by stating: a considerable amount of energy is required to change the shape of a solid; a given volume of liquid readily changes its shape; and a gas fills spontaneously any volume to which it is confined.

Sintering has been defined in various ways, e.g. as consolidation or densification of a powder compact. Since consolidation can take place without densification and densification without consolidation, it is suggested for the present purpose that "sintering is the process which a powder compact undergoes in order to take up a configuration of lowest surface energy."

Phenomena concerning surface energy are much more evident in the liquid state. Everyone knows from every-day experience that if two liquid droplets are brought into contact with one another they will coalesce into a larger drop. Coalescence does not happen spontaneously to solid particles because a considerable energy barrier has to be overcome to change the shape of the particle agglomerate by the movement of atoms. This energy has to be supplied to a powder compact in the form of heat (sintering) or mechanical work (forging) or possibly both, i.e. by pressure sintering or hot-pressing.

In order to widen the scope and to relate more closely to reality one must consider mixtures of powder particles of different chemical composition and one must also take account of the gaseous environment in which the sintering process is carried out, particularly if a chemical reaction between the different powders and atmosphere takes place. It is worth noting that even when no chemical reaction occurs, such as during the sintering of alumina to a high density, the elimination of the final porosity depends on whether the atmosphere is oxygen or air (1). In this connection, minor amounts of unavoidable secondary constituents are usually referred to as impurities but if deliberately added, as sintering or nitriding aids or inhibitors. If, on the other hand, there is a chemical reaction between major amounts of the powder mixture, or between the powder and the atmosphere, as in the formation of silicon nitride when heating silicon in nitrogen, then one may justifiably use the term "reaction sintering".

In the Introductory Lecture to the 1976 NATO Seminar on Nitrogen Ceramics in Canterbury (2) the preference for this term, compared with "reaction bonding", was argued, as bonding is a useful term to describe the joining of dissimilar materials, e.g. metal/ceramic bonding. However, the American and German workers in the field of engineering ceramics have recently agreed the standard abbreviations indicated below. Thus the materials produced by different sintering processes, in sequence of their historic development, are:

reaction-sintered (-bonded) silicon nitride	RBSN
pressure-sintered silicon nitride, hot pressed or isostatically hot-pressed	HPSN, HIPSN
sintered silicon nitride	SSN
post-sintered reaction-sintered (-bonded) silicon nitride	PSRBSN

2. REACTION SINTERING

This review will deal primarily with SSN and PSRBSN because other reviewers will cover RBSN (3) and HPSN (4). However, the mechanism of densification of SSN and HPSN are very similar, and much can be learned about sintering from studies concerned with HPSN. The nitridation process of forming silicon nitride from silicon is basic to all the varieties of sintered silicon nitride: in powder form it provides the raw material for SSN and HPSN and in its consolidated monolithic reaction-sintered form it is RBSN and as such the starting material for PSRBSN. In spite of the many investigations, stretching now over two decades, the nitridation of silicon is still incompletely understood (3), (5). A full explanation must explain the uniqueness of the fact that the dimensions of a powder compact of silicon do not change during nitridation; this is contrary to the usual phenomenon of solid gas reaction and is of greatest practical importance, because densification takes place without shrinkage. A fuller understanding of the process would most likely lead to a substantial decrease of the time necessary to achieve full nitridation, which is an important factor in the economics of the process. The observation by Fate & Milberg (6) that high purity Si (99.999%) powder, if pretreated in vacuum, can be fully nitrided in 1 hour, seems very significant and has been undeservedly neglected. There appears to be a scarcity of new ideas which are needed in order to make progress on this subject. An interesting and novel suggestion is the influence of the thermal conductivity of the atmosphere, proposed by Mangels (7).

3. SOLID STATE SINTERING OF "PURE" SILICON NITRIDE

Of the various mechanisms of transport of matter occurring during sintering only those of lattice diffusion from the surface and grain boundary diffusion lead to densification (8); the others may result in consolidation without shrinkage by neck formation. The low self-diffusion rate of covalently-bonded materials used to be put forward as the explanation of why they would not sinter. Today, one considers these materials not as unsinterable but as being "difficult" to sinter.

The "difficulty" can be overcome in various ways:

(a) use of very fine powders with high surface energy, which is the driving power for sintering;

(b) use of very high temperature;

(c) the application of high pressures.

The above approaches may be used singly or in combination.

(a). Very Fine Powders

These can be obtained by comminution, e.g. by milling in an attritor or by forming powders in a vapour phase reaction, e.g. silane or silicon halide plus ammonia. Two types of powder made by the latter process have become available commercially, i.e. SN402 and SN502 made by GET Sylvania. Both are 99.9% pure with respect to cation impurities and have an oxygen content of 1.5-2% as SiO_2 on the powder surface. SN402 is amorphous and the particle size is ~0.15μm. SN502 has been calcined and is typically constituted of 57% α-Si_3N_4, 3% β-Si_3N_4 and 40% amorphous. What can be achieved by using fine powders with silicon, a truly covalent material, was demonstrated by Greskovich et al (9). With a powder of 44 m^2/g surface area (or 0.06μm average particle size) relative densities of 92% and 99% were obtained by sintering at 1350°C and 1380°C, respectively.

(b). Use of High Temperatures

The limit set to the temperature in the above experiments was the melting point of Si (1410°C); in the case of silicon nitride the decomposition $Si_3N_4 \rightarrow 3Si + 2N_2$ limits the temperature to ≃1500°C or large weight losses are observed. Greskovich and Prochazka (10) have constructed a stability diagram for Si_3N_4 as a function of the partial pressures of nitrogen and silicon. At any given temperature increasing the nitrogen pressure decreases the vapour pressure of silicon and in order to obviate decomposition of Si_3N_4 at high temperatures it is necessary to prevent transport of silicon vapour out of the system. Very substantial gas pressures would be required to effectively suppress the decomposition of silicon nitride and permit sintering. Using temperatures up to 2100°C and gas pressures up to 8MPa it has not been possible to achieve high densities. The best result reported on very fine powder is 20% shrinkage but with a simultaneous 20% weight loss (9). Equipment for temperatures of the order of 2000°C and gas pressures of up to 200 MPa have only recently become available in the form of hot isostatic presses and particularly the "Minihipper". No results have as yet been published.

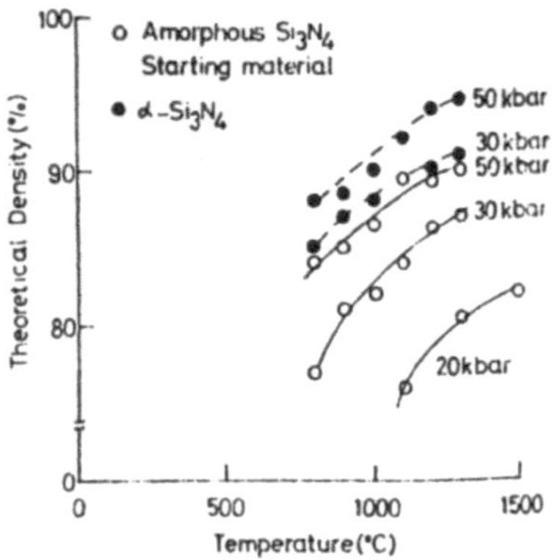

Figure 1. Effect of temperature at constant pressure on final density

(c). High Pressure Sintering

Hot pressing at very high pressures (up to 5000 MPa) at more moderate temperatures (up to 1600 °C) is experimentally more readily achieved. The results obtained (11) on two different starting powders are shown in Fig.1. The best result was 95% of theoretical density using a high α-Si_3N_4-containing powder. It is interesting to note that the amorphous powder, prepared by sputtering techniques, yielded lower densities, and that in the case of the α-Si_3N_4 powder the $\alpha \rightarrow \beta$ transformation under these conditions was insignificant. High density was only achieved when chemical purity or stoichiometry was in doubt (12).

One must conclude from the above that no commercially viable process of forming high-purity, high density components of complex shape exists, and one must invoke a further measure to make this "difficult" material sinter to a high density, i.e. the introduction of additives which act as sintering aids.

4. LIQUID STATE SINTERING BY USING SINTERING AIDS

"Pure" silicon nitride can be considered as of high purity in the cation sense only. It is frequently assumed that, like silicon, silicon nitride also is covered with a layer of silica. Therefore, the larger the surface area of a powder, i.e. the finer it is, the larger is its oxygen content. The presence of this silica layer has played an important role in explaining the action of the additives as "sintering" and "hot-pressing" aids. Little is known about the kinetics of the formation of this oxide layer; in fact direct evidence for these layers is scanty.

As with silicon, it is often assumed that all oxygen determined by chemical analysis is in the form of amorphous silica on the surface. Oxygen can dissolve in silicon nitride and it can form a silicon oxynitride. If one comminutes silicon nitride, how long does it take for a freshly cleaved silicon nitride surface to become covered with silica and what is the thickness of the layer? These are the sort of arguments which one may have to consider to explain the different behaviour of different batches of powder even if they appear identical by chemical analysis, and very often they are not even that.

The general idea of the mechanism of a metal oxide sintering aid is that it forms a liquid phase by reaction with surface silica and that the resulting silicate melt will also dissolve silicon nitride. Kingery (13) in his theoretical treatment discerns three - possibly overlapping - stages of liquid phase sintering:

Stage 1. Particle rearrangement, immediately following the formation of the liquid phase, is brought about by the remaining solid particles sliding over each other under the action of capillary forces. The rate and extent of shrinkage will depend upon the viscosity and quantity of the liquid and its wetting properties.

Stage 2. A solution-reprecipitation process will become operative if the solid particles have some solubility in the liquid phase. Since the solubility at the contact points of solid particles is greater than the solubility of other solid surfaces, material transport away from the contact points will allow the "centre to centre" distance between particles to decrease. The volume changes resulting from this can be described by

$$\frac{\Delta V}{V_o} \propto t^{1/n} \quad \ldots \quad \ldots \quad \ldots \quad \text{equation (1)}$$

where n = 3 if the rate controlling process is solution into or precipitation from the liquid, and n = 5 if diffusion through the liquid is rate controlling, assuming non-spherical particles.

Stage 3. This stage, described as coalescence or closed pore elimination, requires a solid state sintering mechanism. From what has been said before, any shrinkage due to this is unlikely in the case of silicon nitride unless very high pressures are involved.

On cooling, the liquid phase often solidifies as a glass, which shows up as a halo on the X-ray diffractogram. Any glass present will soften at high temperature and this viscous grain boundary phase is responsible for the deterioration of mechanical properties at high temperatures, most readily shown as a decrease of the creep resistance; it also has an adverse effect on the oxidation resistance. During natural cooling or by annealing at a certain temperature part of the glassy phase may be devitrified.

In many of the results using different sintering aids the degree to which complete sintering has been achieved is indicated by how far the measured density falls short of the theoretical density. However, the theoretical density cannot be calculated unless the chemical composition and crystallography of all the different phases present and their relative quantities are accurately known. Thus it would be highly desirable to quote always the theoretical density and indicate how it was derived. The presence of impurities, e.g. WC, or sintering aid, e.g. Y_2O_3, often lead to densities greater than the theoretical density of silicon nitride.

4.1. Single Component Sintering Aids

The choice of sintering aids closely follows the successful hot-pressing aids. From an empirical study of hot-pressing aids by Deely et al (14) in 1961, MgO emerged amongst a large number of oxides and non-oxides as the optimum hot-pressing aid. One of the reasons for its success is probably the great volatility of MgO in a non-oxodizing atmosphere. The uniform distribution of a small quantity of one kind of a powder in a large amount of another powder is a considerable technological problem. Volatility of a substance is a great help in achieving the uniform distribution. Other effective hot-pressing aids are: Y_2O_3, CeO_2, ZrO_2, BeO, Al_2O_3, Sc_2O_3, Mg_3N_2, La_2O_3, $BeSiN_2$, SiO_2.

Although there must have always been a considerable incentive to produce complex shapes of silicon nitride by sintering it was not until 1975 that a serious interest was taken in sintering. It is generally possible to sinter materials at a temperature a few hundred degrees above the hot-pressing temperature. One would imagine that many workers in this field tried to sinter Si_3N_4 but did not succeed because of its rapid thermal decomposition. It was only as a consequence of a hot pressing experiment that had gone wrong at the Westinghouse Labs. that Terwilliger & Lange (15) achieved a reasonably high density. In the experiment referred to, which used MgO as a hot pressing aid, a higher than normal temperature had been applied without any simultaneous pressure; a loose-fitting cylinder of silicon nitride was found in the graphite die, i.e. shrinkage had taken place. The reason for this partial success was that decomposition in the confined space of the hot-pressing die was much reduced, compared with that taking place in an open sintering furnace. The available evidence suggested that two competing phenomena took place during these experiments: (a) densification by shrinkage, and (b) thermal decomposition which limited the final density to less than 80%. The effect of sintering time and temperature is shown in Fig. 2(a) and (b) and curve "a" of Fig. 3 and 4. An obvious way of slowing down the decomposition is to raise the pressure of the sintering atmosphere; by increasing the nitrogen pressure to 1MPa Mitomo (16) was able to obtain densities greater than 90% (see Fig. 3 and 4, curve "b"). Another method of slowing down the decomposition is to use a powder bed as a buffer. By embedding the powder compacts in a powder bed of Si_3N_4 densities very similar to those of Mitomo were obtained by Giachello et al (17) (Fig. 3 and 4 curve "c"). However, the samples appeared to be inhomogeneous; they had an outside skin, about 1mm-thick, which was a lighter grey and of lower density. At first it was thought that this was caused by decomposition of silicon nitride but when the samples were examined by EPMA it was found that there was no MgO near the surface; thus the outside of the specimens was devoid of the sintering aid. The longer the sintering time and the higher the temperature the less MgO remained in a sample. Any observed weight loss previous ascribed entirely to the decomposition of the Si_3N_4 must be partly due to loss of the sintering aid. Fig.5 shows the distribution of various clements. When MgO was incorporated in the powder bed (composition: 50% Si_3N_4, 40% BN, 10% MgO; the BN was incorporated to prevent sintering of

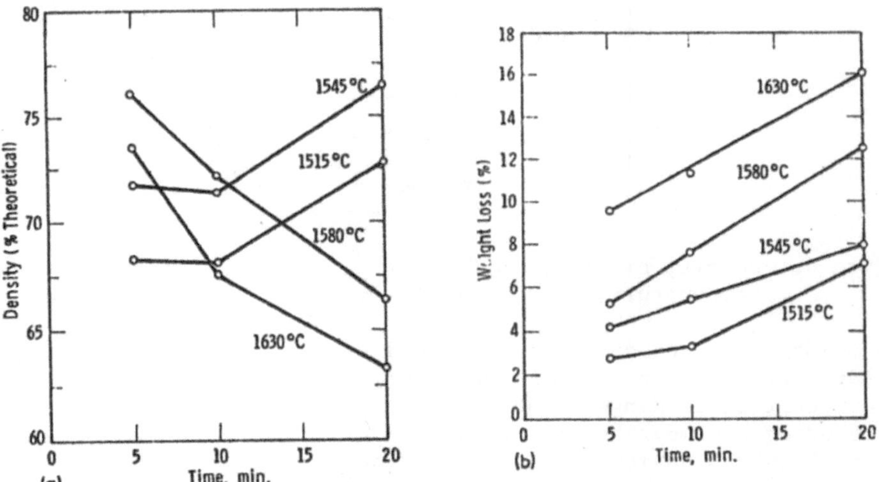

Figure 2. Densities (a) and weight losses (b) obtained for isothermal experiments at three time periods and four temperatures (15)

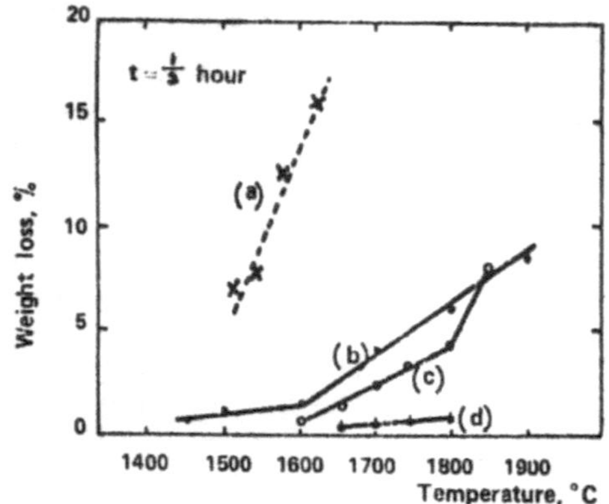

Figure 3. Weight loss during sintering versus sintering temperature: (a) obtained in nitrogen at atmospheric pressure by Terwilliger and Lange (15); (b) obtained at 1 MPa nitrogen by Mitomo (16); (c) obtained at atmospheric pressure in powder bed (17); (d) as (c) but powder bed containing MgO (17)

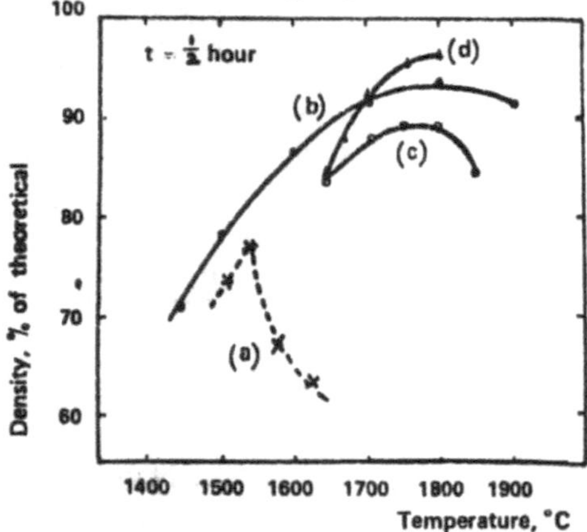

Figure 4. Density of sintered Si_3N_4 + 5% MgO versus sintering temperature; (a), (b), (c) and (d) as in Figure 3.

the powder bed) the densities obtained were higher and the weight losses lower than those of Mitomo (16), Figs. 3 and 4, curve "d". In a recent paper Mangels and Tennenhouse (18) describe experiments using a powder bed containing MgO and a N_2-sintering atmosphere of 2MPa; obviously, a combination of a high pressure atmosphere and a powder bed will give the best results, though the use of high pressure furnaces is a more costly solution. In their paper on the stability of Si_3N_4 Greskovich and Prochazka (10) point out that decomposition of Si_3N_4 in the presence of SiO_2 and

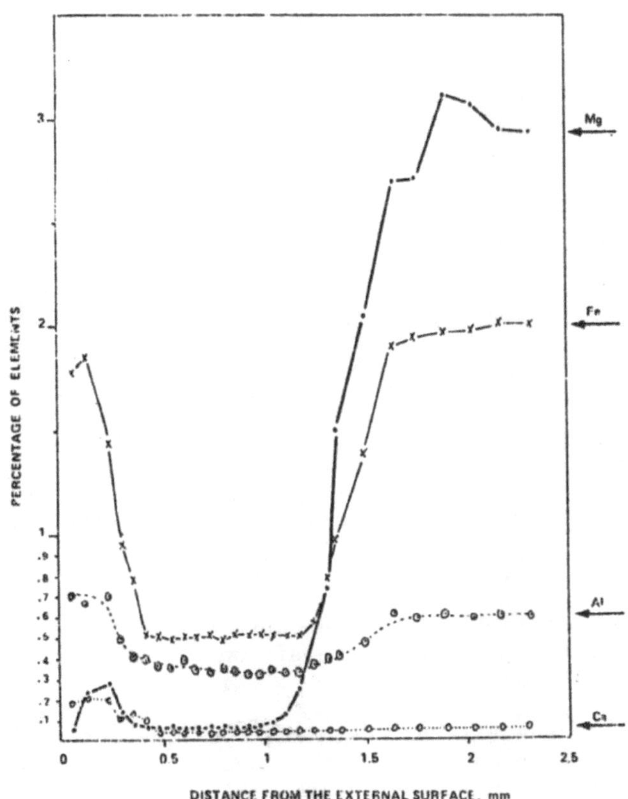

Figure 5. Concentration profile of different elements vs. distance from external surface (17)

other metal oxides may occur via reactions yielding SiO and N_2. As with Si-vapour, the partial pressure of SiO can be reduced by increasing the nitrogen pressure and, similarly, steps must be taken to avoid loss of SiO by transport out of the system, e.g. by use of a powder bed. Another implication of the powder bed technique will be mentioned later, when discussing post-sintering of RBSN. Pompe et al have recently studied the role of protective powders in the sintering of Si_3N_4-based materials (19), and show that substances like Y_2O_3 and Al_2O_3 also migrate (Figure 6). It must be appreciated that at temperatures of the order of 1800-2000°C everything becomes very mobile and volatile and one may be very misled by thinking that a substance is the same after firing as before. It has been found in the production of translucent alumina, which requires a small amount of MgO, that after a furnace has been used for some time it becomes so contaminated with MgO that there is no need to add MgO at all. The design parameters of powder beds (chemical composition, pore structure, thickness, etc.) are worthy

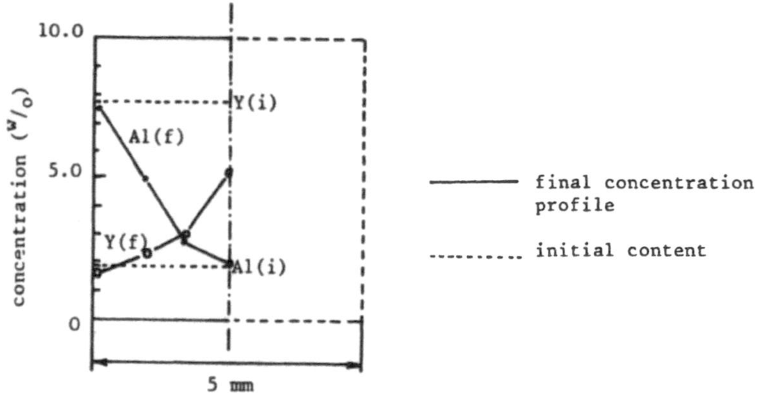

Figure 6. Y and Al concentration in doped-Si_3N_4 sintered in a protective powder of β'-sialon composition (x = 1.9) for 2.5h at 1750°C (19)

of further investigation. A good demonstration of the influence of experimental arrangements on decomposition was given by Gauckler et al (20) at the last Seminar.

Hampshire & Jack (21) have studied the kinetics of sintering using MgO and Y_2O_3 as additives. They conclude from the figures n = 3 and 5, respectively, (in equation 1) that the rate control process is different in the two cases. Before drawing conclusions on rate control one wonders whether the chemical compositions and the mass remained unaltered during these experiments.

Figures 7 and 8 show details of density, weight loss and shrinkage vs. time and temperature, and typical results obtained by Giachello and Popper (22) under optimum sintering conditions with 5% MgO as additive and sintered at atmospheric pressure in a powder bed containing MgO are shown in Table 1.

The contaminating Fe was the residue from milling in a steel mill. It is interesting that the iron is contained as an impurity phase which is a solid solution of fayalite, Fe_2SiO_4 and forsterite, Mg_2SiO_4.

Since better high temperature properties were obtained on HPSN using Y_2O_3 instead of MgO as hot pressing aids, Y_2O_3 has also been used as sintering aid (23), Fig. 9 & 10. The different behaviour of the two powders tested, i.e. SN402 amorphous vapour phase-derived and KBI nitrided silicon, is worth pointing out. The fine GTE powder sinters much more readily but because of its fibrous nature

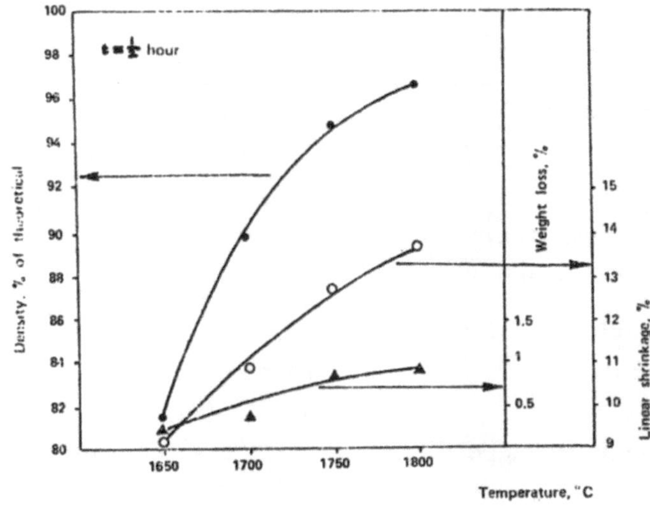

Figure 7. Density, weight loss and shrinkage vs. temperature for Si_3N_4 + 5% MgO sintered in a powder bed containing MgO (17)

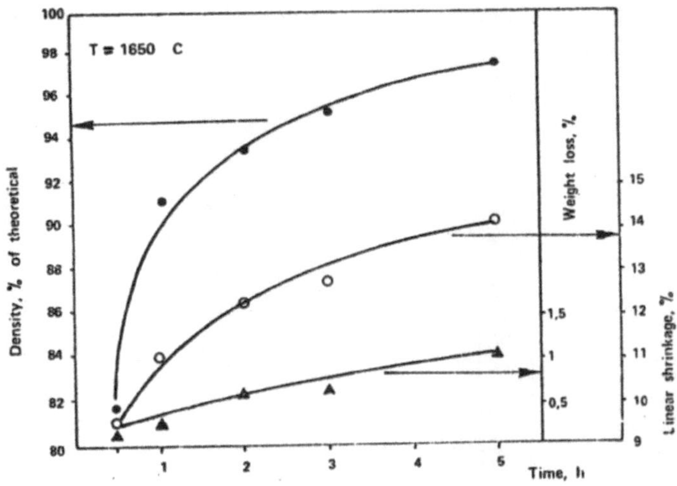

Figure 8. Density, weight loss and shrinkage vs. time for Si_3N_4 + 5% MgO sintered in a powder bed containing MgO (17)

Table 1

Pressureless Sintered Si_3N_4

Starting material	: A.M.E. silicon nitride powder
Additives	: MgO - Fe
Green density	: 2.0 Mg/m^3
Sintering conditions	: In powder bed at 1650°C for 5 hours
Linear shrinkage	: ~14%
Characteristics of the sintered material:	
- Density	: 3.10 Mg/m^3
- Phases	: β-Si_3N_4, $(Mg_{0.6}, Fe_{0.4})_2SiO_4$
- Flexural strength	: at 25°C = 500 MN/m^2 at 1250°C = 280 MN/m^2
- Weibull modulus	: 9.8
- Oxidation resistance (weight gain after 100h at 1300°C)	: 5 mg/cm^2

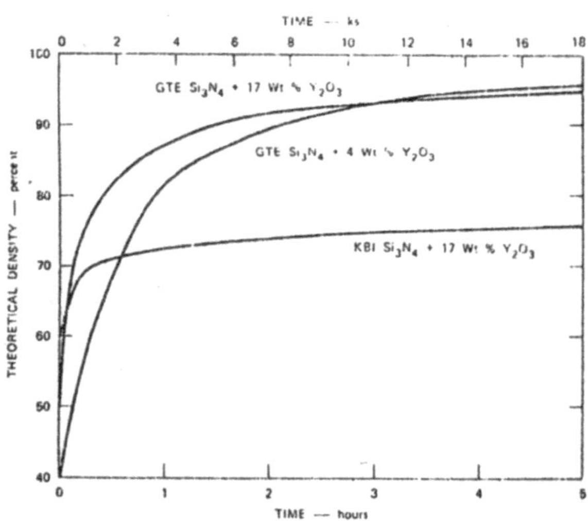

Figure 9. Densification curves for Si_3N_4-Y_2O_3 compacts sintered at 1650°C in 0.2MPa N_2 (23)

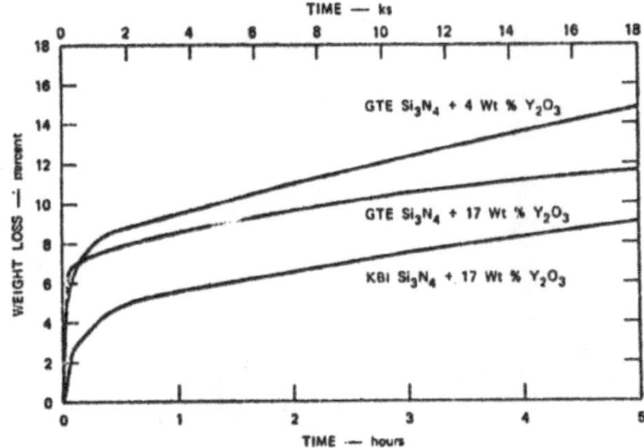

Figure 10. Weight loss during sintering of Si_3N_4-Y_2O_3 compacts at 1650°C in 0.2 MPA N_2(23)

has a low compactability (green density 40%) and consequently a very high sintering shrinkage. Because of its large surface area the weight loss is also high. In this connection the development of high surface area Si-derived powders with good packing properties is to be welcomed (24).

Because of different experimental techniques used by different investigators it is not possible to draw very valid conclusions by comparing the results using different sintering aids. Rae et al (25) have put forward reasons based on structural evidence why certain impurities, e.g. Ca, are more damaging to the high temperature properties with MgO additions than with Y_2O_3 additions. Catastrophic oxidation near 1000°C has been found in the Si_3N_4-Y_2O_3 system which has been explained by the presence of $Y_2O_3.Si_3N_4$ and other phases in the Si-Y-O-N system (26). More recently, however, some arguments have been put forward which challenge this explanation (27).

4.2. Multi-Component Sintering Aids

Impurities in the starting material are without doubt important, the least controllable being the SiO_2 content. Impurities may also be introduced during processing. Figure 11 shows how easily and how irreproducibly Al_2O_3 may be introduced during milling (28). This is undoubtedly the way Al_2O_3 was first introduced into Y_2O_3-containing mixes and in fact x% Y_2O_3 + 2% Al_2O_3 has become a favoured composition (Fig. 12) to achieve a high degree of densification (29). Similarly, the addition of $Al_2O_3$to MgO either separately or preferably as spinel results in good densification. Weight loss figures indicate a lower weight loss for the spinel (30).

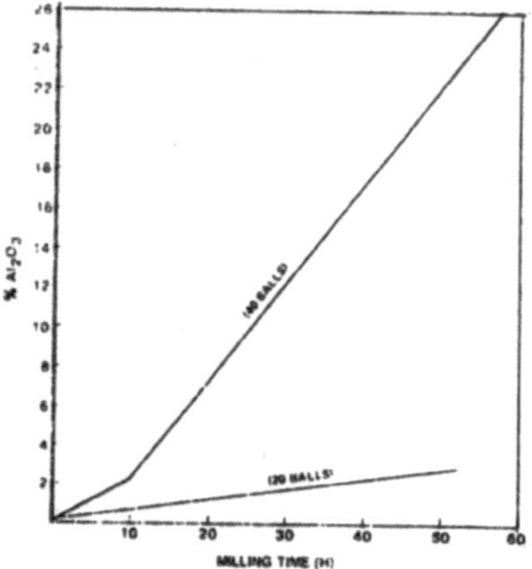

Figure 11. Al_2O_3 pickup as a function of ball-milling time (28)

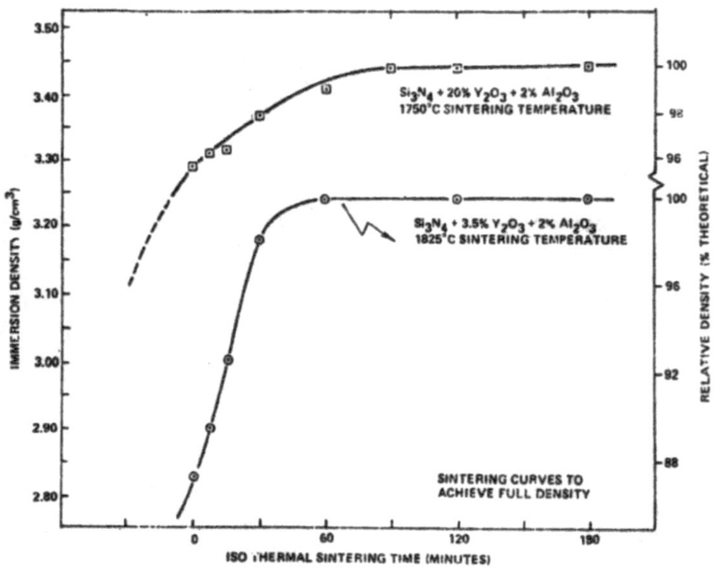

Figure 12. Isothermal Si_3N_4-Y_2O_3-Al_2O_3 sintering curves for low volume inter-granular liquid compositions (lower curve) and large volume intergranular compositions (higher curve) where full density is achieved (29)

A combination of yttria and magnesia as sintering aid, i.e. Si_3N_4 + 8% Y_2O_3 + 1% MgO, was studied by the Central Res. Labs. of FIAT (31). It was chosen for a sintering study since it represented at the time a good compromise between the two most favoured HPSN compositions. Fig.13 shows the sintering data, i.e. density,

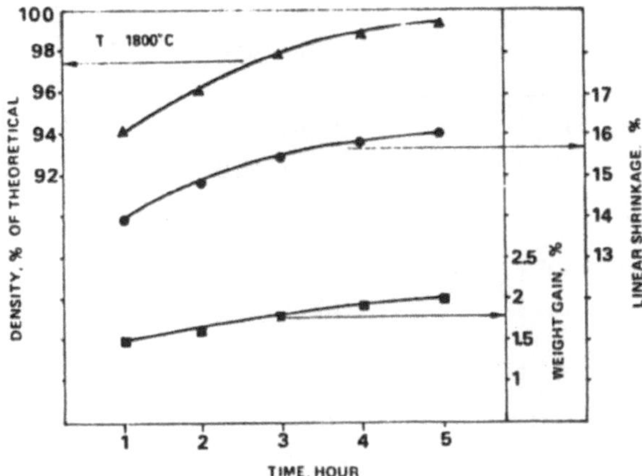

Figure 13. Density, weight gain, and shrinkage vs. sintering time for the mixture Si_3N_4 + $8Y_2O_3$ + 1% MgO (31)

shrinkage and weight changes. Here we observe an actual weight gain; this may result from the nitridation of unreacted silicon, present in the Si_3N_4 powder, but as it is possible for matter to be transported from compact to powder bed one has also to consider a possible exchange of matter in the reverse direction. Table 2 gives the general properties and Fig.14 the strength vs. temperature. It is interesting to observe at 1000°C an increase in strength with increasing temperature. Martinengo et al (32) explain this by the devitrification of the amorphous phase during testing. Fig.15 shows the X-ray diffractograms before and after devitrification. The situation is very complex and as the Si_3N_4-SiO_2-MgO-Y_2O_3 system has not been studied in detail, one is rather in the dark but the existence of the $Mg_5Y_6Si_5O_{24}$ phase is worth noting.

Complete devitrification of the glassy phase and the formation of refractory crystalline phases is a target to be achieved; this process is called "transient liquid phase sintering". The nearest to this was obtained with hot-pressed Si_3N_4 (containing approx. 3% oxygen) and an addition of 7 wt.% $BeSiN_2$. It resulted in a nearly stoichiometric solid solution of composition $Si_{3-x}Be_xN_{4-2x}O_{2x}$, which had excellent mechanical properties and high oxidation resistance. In equivalent sintering experiments at 2080°C and at 8.2MPa nitrogen pressure, 93% relative density has been obtained (33). It is, however, unlikely that a system based on the use of Be-containing compounds will ever find wide commercial use because of possible toxicity problems.

It is not intended to deal in this review with the very wide range of SIALONs (34) and the even more complex systems (35), but recently Boskovic (36) has shown that it is possible to obtain

Table 2

Pressureless Sintered Si_3N_4

Starting material	: A.M.E. silicon nitride powder
Additives	: Y_2O_3 + MgO
Green density	: 2.0 Mg/m^3
Sintering conditions	: In powder bed at 1800°C for 5 hours
Linear shrinkage	: ~16%
Characteristics of the sintered material:	
- Density	: 3.25 Mg/m^3
- Phases	β-Si_3N_4, amorphous
- Flexural strength	: at 25°C = 600 MN/m^2 at *1250°C = 380 MN/m^2
- Weibull modulus	: 9.3
- Oxidation resistance (weight gain after 100h at 1300°C)	: 2.5 mg/cm^2

*Suitable devitrification heat treatments can improve high temperature mechanical properties and oxidation resistance.

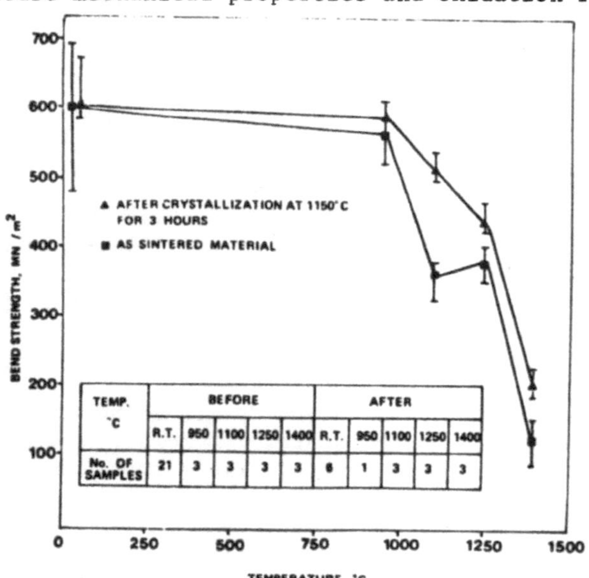

Figure 14. Bend strength vs. temperature before and after crystallization (31)

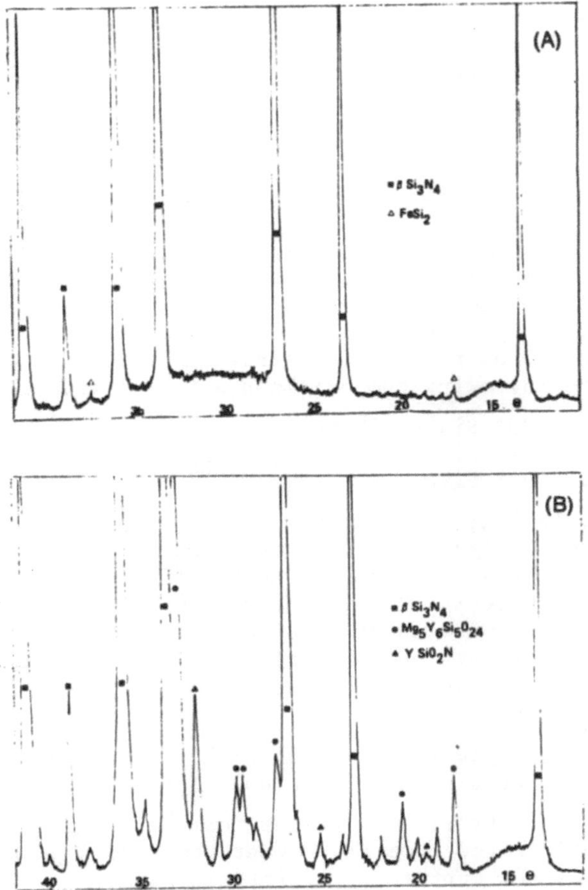

Figure 15. Diffractograms for the material Si_3N_4 + $8Y_2O_3$ + 1% MgO (A) as-sintered and (B) heated at 1400^oC for 15h (31)

sintered materials close to Si_3N_4 which consist of only two phases, i.e. the solid solution $Si_{8-z}Al_zO_zN_{6-z}$ and $Y_3Al_5O_{12}$. The latter Y(ttrium) A(luminium) G(arnet) is a refractory compound. The structural changes can be observed from the X-ray diffractograms taken at different temperatures during processing.

5. POST-SINTERING OF RBSN

Giachello & Popper (37) showed in 1979 that it is possible to densify not only Si_3N_4-powder but also RBSN. Similar work has also been carried out by Mangels and Tennenhouse (18) who also used powder beds as already mentioned and a nitrogen pressure of 2MPa. In this way it is possible to start the high temperature sintering process on a reaction sintered component of density ~2.5 g/cm^3 compared with sintering a Si_3N_4-powder compact of density ~2.0 g/cm^3. The sintering shrinkage can thus be substantially reduced, e.g. from about 16 to 6%.

The sintering aids (MgO, Y_2O_3) can be added in different ways:

(a) they can be mixed with the silicon powder before nitridation, when they are found to be present either as a magnesium-iron silicate (Fe is an impurity in the Si) or as the yttrium silicon oxynitride $10Y_2O_3.9SiO_2.Si_3N_4$ after nitridation (see Table 3). The starting compositions of the materials subjected to the post-sintering were such as to have the following compositions after nitridation:

A: 95% Si_3N_4, 5% MgO
B: 91% Si_3N_4, 8% Y_2O_3, 1% MgO
C: 92% Si_3N_4, 8% Y_2O_3

Table 4 gives details for the post-sintering treatment. Figure 16 shows the sintering behaviour of the three materials. The benefit of the 1% MgO, i.e. material B compared with C, is quite noticeable. Some properties of material C are shown in Table 5. Further data on material C and some experience of applications in the automotive area are being published (38);

(b) by multiple impregnation with a liquid containing the sintering aid in soluble form (18);

(c) by infiltration in vapour form, e.g. MgO can be infiltrated from the powder bed into a porous RSSN body not containing a sintering aid.

Figure 17 shows the weight changes, depending on the type of powder bed, which occurred during post-sintering of RBSN. The results of post-sintering treatment by infiltration of MgO from the powder bed are shown in Table 6 for RBSN's of different density. Post-sintering doubled the strength at room temperature and maintained an improvement up to 1400°C (37).

The main advantage of the post-sintering method lies in the reduction of sintering shrinkage, which stems from the fact that advantage can be taken of the reaction-sintering process which results in densification without shrinkage. Large shrinkage can cause considerable technological problems, particularly with large components when very close dimensional tolerances are required. In general, very fine powders are difficult to compact to high density, which leads during their sintering to higher than average shrinkage. The post sintering treatment allows the use of all types of forming and Mangels & Tennenhouse (18) have shown how a complex shape like a gas turbine rotor can be produced from an injection-moulded Si-body.

Table 3

Data for materials A & B before and after nitridation

	Unnitrided density Mg/m^3	Weight gain %	%ge conversion Si - Si_3N_4*	Nitrided density Mg/m^3	Crystalline phases by X.R.D.
Material A	1.60	54.9	92.4	2.45	α-Si_3N_4 β-Si_3N_4 ($\alpha/\beta \approx 75/25$) tr. FeSi $(Mg_{0.6}Fe_{0.4})_2SiO_4$
Material B	1.64	51.8	94.7	2.52	α-Si_3N_4 β-Si_3N_4 ($\alpha/\beta \approx 90/10$) $10Y_2O_3.9SiO_2.Si_3N_4$ tr. FeSi

Table 4

Data for materials A & B before and after post-sintering

	Density before sintering Mg/m^3	Heat treatment		Linear shrink. %	Sint. dens. Mg/m^3	%ge of theor* dens.	Phases present by X.R.D.
		h	^oC				
Mat. A	2.45	7	1600	7.6	3.10	96.9	β-Si_3N_4 tr. $(MgO_{0.4}Fe_{0.6})$ SiO_4, FeSi, SiC
Mat. B	2.52	5	1800	6.5	3.22	98.5	β-Si_3N_4 amorphous tr. $FeSi_2$, SiC

*based on a theoretical density of 3.20 mg/m^3 for material A and 3.27 Mg/m^3 for material B.

Table 5

Data for post-sintered RBSN containing sintering aids

Starting material	: RBSN containing Y_2O_3 and MgO as additives
Density	: 2.5 Mg/m^3
Sintering conditions	: In powder bed at 1800°C for 7 hours
Linear shrinkage	: ~6.5%
Characteristics of the sintered material:	
- Density	: 3.22 Mg/m^3
- Phases	: β-Si_3N_4, amorphous
- Flexural strength	: at 25°C = 1000 MN/m^2 at 1250°C = 500 MN/m^2
- Oxidation resistance (weight gain after 100h at 1300°C)	: 0.9 mg/cm^2

*Suitable devitrification heat treatments can improve high temperature mechanical properties and oxidation resistance.

Table 6

Data for sintered RBSN's without admixed sintering aid.

Density before sinter. Mg/m^3	Sinter. aid in powder bed	Sintering treatment		Linear shrink. %	Weight gain %	Mg content %	Final density Mg/m^3
		h	°C				
2.35	5% MgO	1	1800	6.9	0.1	0.75	3.01
2.47	5% MgO	3	1800	6.0	0.9	0.9	3.05
2.64	5% MgO	5	1800	4.7	0.9	0.67	3.05

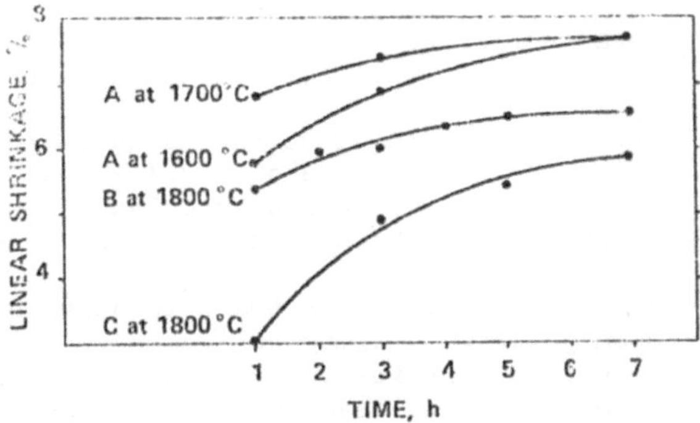

Figure 16. Linear shrinkages for materials A, B and C vs. sintering time (37)

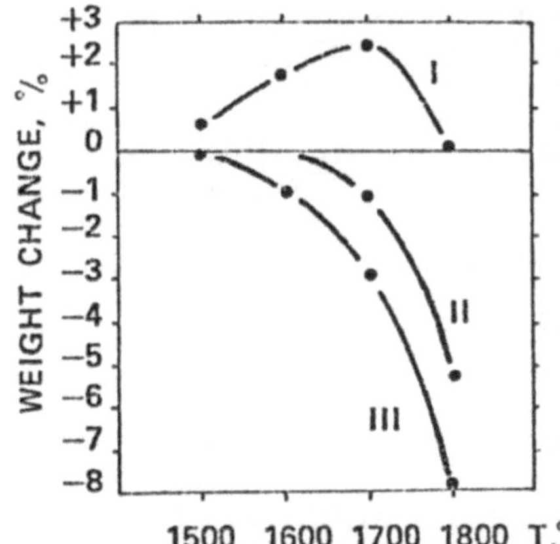

Figure 17. Influence of the powder bed on weight change for RBSN d. 2.35 Mg/m^3. Composition of powder bed for: (I) 50% Si_3N_4 - 45% BN - 5% MgO, (II) 50% Si_3N_4 - 50% BN, (III) without powder bed (37)

At first sight one might think that there is a considerable economic advantage in the post-sintering process because of the much lower cost of silicon powder compared with silicon nitride powder. However, the present price of silicon nitride powder is most probably artificially high because of the low demand. In principle, it should be faster to form silicon nitride powder by nitriding a loose powder of silicon rather than a high density compact, but one must add to this the cost of comminution and purification treatment following comminution. The post-sintering process may be difficult to apply to components of great thickness because of the problem of obtaining in-depth nitridation. However, if one allows a greater shrinkage (e.g. 10%) then a RBSN body of lower density can be used which will be more easily nitrided.

In conclusion it can be stated that the advances in sintering permit one to visualize the fabrication of complex-shaped engineering components by mass- production processes; the properties of SSN or PSRBSN will be similar to those of HPSN.

For the most demanding application the component, once it has achieved zero open porosity, can be subjected to a 'cladless' hot isostatic pressing treatment for the elimination of closed pores or any other flaws present, leading to higher strength and greater reproducibility.

REFERENCES

1. Coble, R.L. J.Amer.Ceram.Soc., 45, 123, 1962.
2. Popper, P. Nitrogen Ceramics*, p.3.
3. Riley, F.L. This volume.
4. Mocellin, A. This volume.
5. Moulson, A.J. J.Mater.Sci. 14, 1017, 1979.
6. Fate, W.A. and Milberg, M.E. J.Amer.Ceram.Soc., 61, 531, 1978.
7. Mangels, J.A. Amer.Ceram.Soc.Bull. 60, 613, 1981; also this volume.
8. Stuijts, A.L. Nitrogen Ceramics*, p.331.
9. Greskovich, C., Prochazka, S. and Rosolowski, J.H. ibid, p.351.
10. Greskovich, C. and Prochazka, S. Comm.Amer.Ceram.Soc., C96, 1981.
11. Shimada, M., Ogawa, N., Koizumi, M., Dachville, F. and Roy, R. Amer.Ceram.Soc.Bull. 58, 519, 1979.
12. Proçhazka, S. and Rocco, W.A. High Temperature-High Pressures, 10, 87, 1978.
13. Kingery, W.D. J.Appl.Phys. 30, 301, 1959.
14. Deely, G.G., Herbert, J.M. and Moore, N.C. Powder Met. No.8, 145, 1961.
15. Terwilliger, G.R. and Lange, F.F. J.Mater.Sci. 10, 1169, 1975.
16. Mitomo, M. J.Mater.Sci. 11, 1103, 1976.
17. Giachello, A., Martinengo, P.C., Tommasini, G. and Popper, P. J.Mater.Sci. 14, 2825, 1979.
18. Mangels, J.A. and Tennenhouse, G.J. Amer.Ceram.Soc.Bull. 50, 1216, 1980; also this volume.
19. Pompe, R., Eklund, L. and Hermansson, L. Special Ceramics 7. D.Taylor and P.Popper, Ed., British Ceramic Society, Stoke-on-Trent, 1981, p.97.

20. Gauckler, L.J., Boskovic, S. and Petzow, G. Nitrogen Ceramics*, p.405, Fig.2.
21. Hampshire, S. and Jack, K.H. Special Ceramics 7, as Ref.19, p.37.
22. Giachello, A. and Popper, P. Ceramics for High Performance Applications, III. (To be published).
23. Rowcliffe, D.J. and Jorgensen, P.J. Proc. of the Workshop on Ceramics for Advanced Heat Engines, Orlando, Florida, 1977, p.191.
24. Schwier, G. This volume.
25. Rae, A.W.J.M., Thompson, D.P., Pipkin N.J. and Jack, K.H. Special Ceramics 6 , Ed. P. Popper, British Ceramic Research Association, 1975, p.347.
26. Lange, F.F. Nitrogen Ceramics*, p.491.
27. Knoch, H. and Schlichting, J. Sprechsaal, 114, 99, 1981; see also this volume, Besson, J.L. et al.
28. Galasso, F.S. and Veltri, R.D. Comm.Amer.Ceram.Soc. C15, 1981.
29. Smith, J.T. and Quackenbush, C.L. Proc.Intern.Symp. in Densification & Sintering of Oxide & Non-Oxide Ceramics, Japan, 1978, p.126.
30. Masaki, H. and Kamigaito, O. J.Ceram.Soc.Jap. 84, 508, 1976.
31. Giachello, A., Martinengo, P., Tommasini, G. and Popper, P. Amer.Ceram.Soc.Bull. 59, 1212, 1980.
32. Martinengo, P.C., Giachello, A., Popper, P., Buri, H. and Branda, F. As Ref.29, p.516.
33. Prochazka, S. and Greskovich, C. As Ref.29, p.489.
34. Jack, K.H. Science of Ceramics 11 (To be published).
35. Hohnke, H. and Tien, T.Y. This volume.
36. Boskovic, S. Science of Ceramics 11. (To be published).
37. Giachello, A. and Popper, P. Ceramurgia Int. 5, 110, 1979.
38. Giachello, A., Martinengo, P.M. and Tommasini, G. Science of Ceramics 11. (To be published).

*Nitrogen Ceramics, Riley, F.L. Ed., Noordhoff, Leyden, 1977.

ACKNOWLEDGMENTS

The author would like to thank Dr. D.W.F. James, Director of Research, B.Ceram.R.A., for permission to publish this paper.

DISCUSSION

Billy: We tried to sinter AlN in a powder bed containing MgO and found loss of MgO, which we attributed to reduction of MgO by Al formed from AlN, to Mg(v).

Popper: Under low oxygen partial pressure MgO will decompose to Mg(v) + O_2 at sufficiently high temperature. Al is not specifically required.

Lange: MgO could be lost by diffusion from the sintering compact into a (purer) packing powder.

INFLUENCE OF POWDER PROPERTIES AND PROCESSING PARAMETERS ON THE SINTERING OF SILICON NITRIDE

G. Wötting[*] and H. Hausner[**]

[*]Deutsche Forschungs- u. Versuchsanstalt für Luft- u. Raumfahrt e. V., Porz

[**]Technische Universität Berlin

1. INTRODUCTION

The sintering of silicon nitride is impeded by small self-diffusion coefficients (1), and the high dissociation pressure (2) at the temperatures required for a complete densification. The investigation of various processing parameters and the selection and testing of suitable sintering additives have been pursued during the past years in order to achieve sintering densities in excess of 95% th. d.

2. SINTERACTIVE POWDERS

It is well known that the powder preparation route has a significant influence on the characteristics of the product. Powders in use are made either by the nitridation of silicon (3), by the reaction between silica, carbon and nitrogen (4) or by a vapour phase reaction between a volatile silicon compound and ammonia (5). The properties of such powders are shown in Table 1; they can be varied by changes in the processing steps of the different fabrication routes; therefore further improvements of the available powders are expected due to an active research in this field.

In general powders prepared by vapour phase reaction (VPR) have a higher purity with the exception of oxygen; sometimes they contain small amounts of chlorides, indicating the use of silicontetrachloride as raw material for their preparation. The crystalinity of these powders may cover a wide range, whereas in the powders made by nitriding silicon almost no amorphous constituents are present. Typical differences exist in the morphology of the

Riley, F.L. (ed.) Progress in Nitrogen Ceramics

powders (Fig. 1). The VPR powders have a needle-like appearance and the morphology of the Toshiba powder is an indication that it has been formed also via a gas phase, possibly SiO (4).

Specific surface areas are widely dependent on the processing parameters. The same is true for the α/β ratio, although it should be noticed that the commercially available powders have a high α-content. Very pure powders do not show an appreciable densification during sintering (6). In order to achieve high densities an oxygen content of 1 - 2% seems to be necessary (7-9) together with an optimum concentration of a sintering additive for the formation of a liquid phase. The mechanisms in liquid phase sintering, e.g. rearrangement, solution and reprecipitation are assumed to be responsible for densification. The two most important parameters concerning the sintering behaviour are the specific surface area and the oxygen content. In many cases they are correlated (Fig.2).

An increasing surface area and oxygen concentration of the powders results in higher densities due to a higher amount of liquid phase formation, but a subsequent negative influence on the high temperature mechanical properties of the sintered product can be expected.

The carbon content of a powder is of importance since oxygen may be removed as CO and the oxygen content is lowered with a subsequent decrease in the concentration of the liquid phase. Besides the specific surface area and the oxygen content of a powder its crystallinity and the α/β ratio have an influence on the sintering behaviour but play a secondary role.

Powder	HCST-LC 12	GTE SN-502	Toshiba
Preparation route	Nitridation of Si	Vapour-phase reaction	Reaction SiO_2+C+N_2
Composition (wt %)			
N	37.7	38.0	37.8
O	2.8	3.0	1.7
C	0.23	-	1.05
Metallic	0.13	0.03	0.06
Impurities ...	(Fe, Al, Ca)	(Al, Ca, Mg, Mo, Ti)	(Fe,. Al, Ca)
Crystallinity (%)	100	60	100
α/β ratio	97/3	95/5	93/7
Spec. surface area (m^2/g)...	21	4	4.8

Table 1: Properties of Si_3N_4 - Powders

HCST - LC 12 GTE SN - 502 Toshiba

Fig. 1 Scanning electron micrographs of Si_3N_4 powders

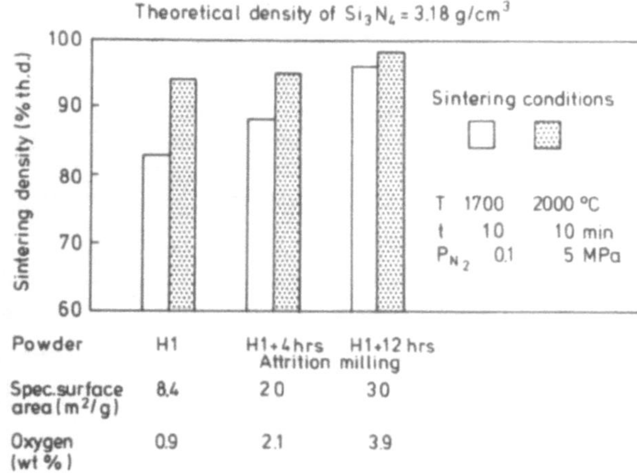

Fig. 2 Sintering density of Si_3N_4 samples (sintering additive 2 wt % Mg 0 as Mg $(NO_3)_2 \cdot 6\ H_2O$)

3. PROCESSING

Processing routes, applied in the laboratories of the authors are shown in Fig. 3. By attrition milling powders with high specific surface areas can be prepared; however, a subsequent chemical treatment with acid is necessary to remove the iron wear, resulting in an increased oxygen content. Additives and pressing aids can be mixed with the Si_3N_4 powder efficiently in a planetary mill. A very homogeneous distribution of small amounts of additives resulting in higher densities is achieved by spray-drying if the additives are used in a soluble form (Fig. 4).

The dissociation of the silicon nitride at the sintering temperature counteracts the densification process. Therefore processing steps are of importance which decrease the decomposition with the associated weight loss. An increased nitrogen pressure during sintering is here of advantage (10-13). Its positive influence is especially pronounced if powders of low oxygen content and a comparatively low surface area are sintered, (Fig. 2).

The application of "powder beds" with selected compositions during sintering has a similar effect since vapourisation is retarded and the establishment of an equilibrium pressure is facilitated. Fig. 5 shows the influence of an encapsulation of the sample on its density and weight loss in case of 2wt% MgO as the sintering additive.

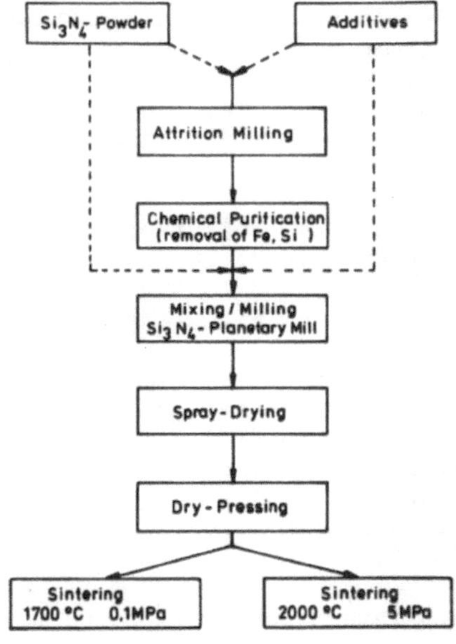

Fig. 3 Processing of sintered Si_3N_4

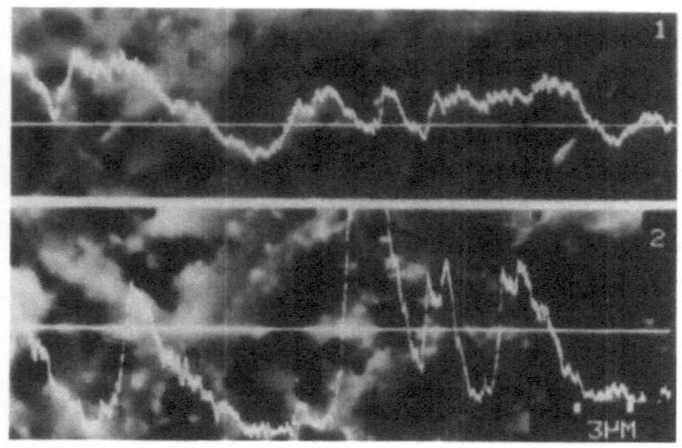

Fig. 4 Yttrium-distribution in sintered silicon nitride (SEM line-scan)
Additives: $Y\,Cl_3 \cdot 6\,H_2O$ (1); Y_2O_3 (2)

4. ADDITIVES

MgO (14), Y_2O_3 (9), Al_2O_3 (15), CeO_2 (11), ZrO_2 (5) and $BeSiN_2$ (12) have been used alone or in mixtures in various concentrations as sintering additives in order to achieve high densities. Since many of these additives form a liquid phase during the sintering process the possible influence on the mechanical properties at high temperatures has to be analysed carefully. Fig. 6 shows results obtained with various compounds added in a soluble form and processed under identical consitions.

It is evident that with MgO even in low concentrations high sintering densities can be obtained; the difference in the behaviour of various additives is probably related to the amount and the properties of the liquid phase formed at high temperatures.

For the investigation of the sintering process a high-temperature dilatometer has been developed, which can operate up to 2000°C and up to a nitrogen pressure of 5 MPa (16). In Fig. 7 the shrinkage curves are shown for samples made from the silicon nitride powder HCST-LC 12 (Table 1) with additions of 2 wt % MgO and 5 wt % Y_2O_3 resp. In the case of MgO a rapid densification can be observed at a temperature of about 1450°C.

The first maximum in the sintering rate which is expected to occur at 1500°C (based on experiments with other powders) and which is caused probably by liquid phase formation and the re-arrangement of particles is overshadowed by a second peak at 1770°C which could be an indication for the densification by a solution-precipitation mechanism. In the case of an Y_2O_3-addition an increase in the sintering rate can be observed at 1500°C. The shape of the curve and the location of the maxima is an indication that the liquid phase

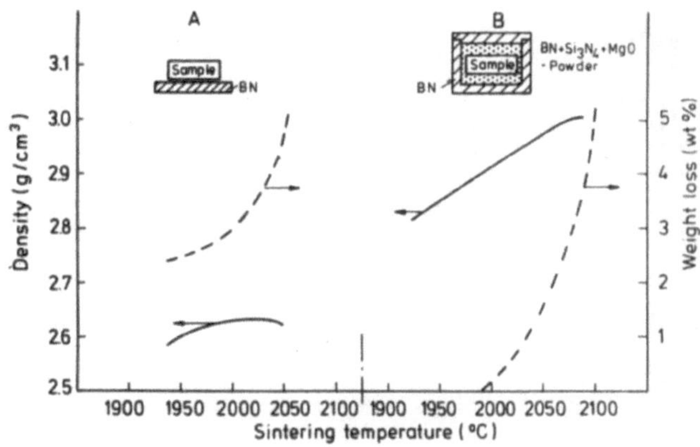

Fig. 5 Influence of sintering conditions on density and weight loss (P_{N2} = 5 MPA)

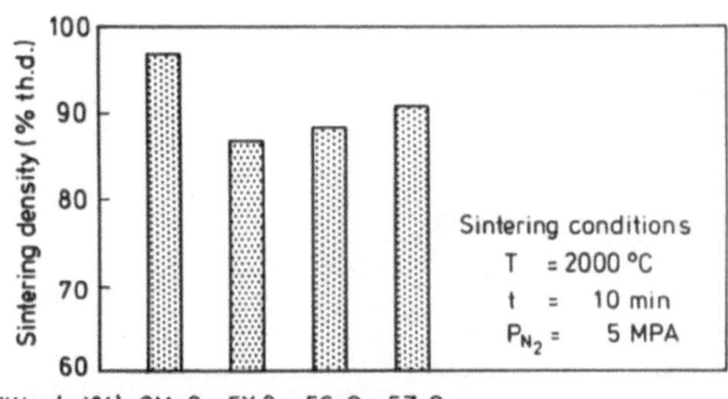

Fig. 6 Sintering density of Si_3N_4 with various additives (the sintering densities are related to the theor. density of the mixtures).

formation takes place rather slowly and that it is probably present in small amounts and with a high viscosity.

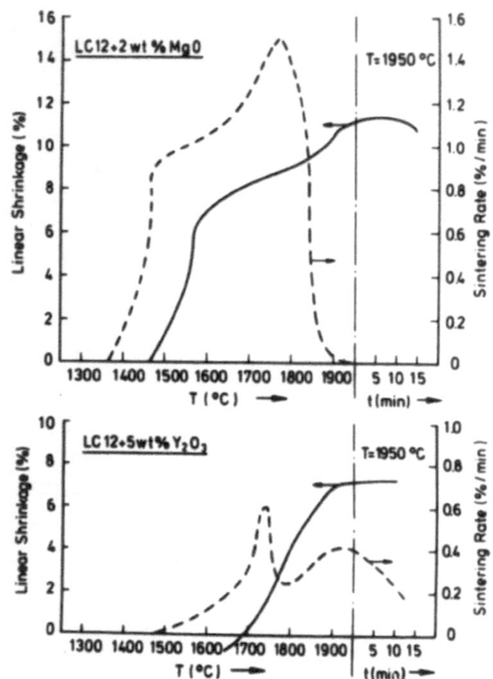

Fig. 7 Shrinkage of Si_3N_4 - samples

REFERENCES

1. K. Kijima and S. Shirasaki. Nitrogen Self-Diffusion in Silicon Nitride, J. Chem. Phys. 65 (1976) 2668-2671.
2. S.C. Singhal. Thermodynamic Analysis of the High-Temperature Stability of Silicon Nitride and Silicon Carbide, Ceramurgia Int. 2 (1976) 123-30.
3. G. Schwier. On the Preparation of Fine Silicon Nitride Powders (this volume).
4. K. Komeya and H. Inoue. Synthesis of the α-Form of Silicon Nitride from Silica, J. Mater. Sci. 10 (1975) 1244-46.
5. High Density High Strength Si_3N_4 Ceramics prepared by Pressureless Sintering of Partly Crystalline, Partly Amorphous Si_3N_4 Powder, U.S. Patent 4.073.845 (14.2.1978).
6. C. Greskovich and C. O'Clair. Effect of Impurities on Sintering Si_3N_4 containing MgO or Y_2O_3 Additives, Am. Ceram. Soc. Bull. 57 (1978) 1055-1056.
7. G.E. Gazza and R.N. Katz. Development of Advanced Sinterable Si_3N_4, AMMRC Sp 78-6 (Oct. 1978).
8. S. Prochazka and C.D. Greskovich. Development of a Sintering Process for High-Performance Silicon Nitride, AMMRC TR 78-32 (July 1978).

9. D.J. Rowcliffe and P.J. Jorgensen. Development of a Low-Cost Process for the Fabrication of Fully Dense Silicon Nitride High-Temperature Gas Turbine Components, Report NSF/Ra-770 443 (Dec. 1977).
10. M. Mitomo. Pressure Sintering of Si_3N_4, J. Mater. Sci. 11 (1976) 1103-1107.
11. H.F. Priest G.L. Priest and G.E. Gazza. Sintering of Si_3N_4 under High Nitrogen Pressure, J. Am. Ceram. Soc. 60 (1977) 80-81.
12. S. Prochazka and C.D. Greskovich. Development of a Sintering Process for High-Performance Silicon Nitride, Report AMMRC TR 78-32, July 1978; SRD-77-178.
13. G.E. Gazza and R.N. Katz. Development of Advanced Sinterable Si_3N_4, Report AMMRC SP 78-6, Oct. 1978; EC-76-A-1017.
14. G.R. Terwilliger and F.F. Lange. Pressureless Sintering of Si_3N_4, J. Mater. Sci. 10 (1975) 1169-1174.
15. K.H. Jack. Review: Sialons and Related Nitrogen Ceramics, J. Mater. Sci. 11 (1976) 1135-1158.
16. R. Peitzsch, G. Wötting and H. Hausner. Investigation of the Sintering of Silicon Nitride by Dilatometry (to be published).

DISCUSSION

Clarke: How did you measure shrinkage rates at 2000°C?

Wötting: Dimensional changes were determined with a differential transformer. The continuously recorded values were differentiated electronically. The dilatometer system was constructed of boron nitride rods, with slices of BN at contact points to prevent reaction. The system was heated in vacuum to 600°C, under 1 atm to 1700°C, when pressure was brought up to 5 MPa before raising the temperature to maximum.

Thompson: What was the form of the ZrO_2 at the end of a run?

Wötting: ZrO_2 was present as zirconia.

SINTERING OF Si_3N_4-BASED MATERIALS USING THE POWDER BED TECHNIQUE

R. Pompe and R. Carlsson

The Swedish Institute for Silicate Research,S-41258 Göteborg,SWEDEN

ABSTRACT. The work on nitrogen ceramics at the institute is chiefly aimed at developing a technologically feasible process for fabrication of stationary parts in heat engines.

The working procedure involves powder characterization and treatment, development of suitable forming methods, pressureless sintering and testing of mechanical properties.

In the sintering work isostatically pressed compacts of various Si_3N_4 powders with yttria and alumina sintering aids have been obtained with better than 99% of the theoretical density using the protective powder bed technique. Sintering dilatometry and thermogravimetry is used in optimizing the processing parameters (protective powder composition, heating rate, sintering temperature/time).

INTRODUCTION

Densification of the Si_3N_4 powder compacts has been found (1,2) to involve a mechanism of dissolution-reprecipitation of Si_3N_4 within the intergranular liquid phase formed between the nitride, the surface silica on the nitride particles and suitable sintering aids (usually metal oxides such as MgO,Y_2O_3,Al_2O_3 etc.). When using the hot pressing techniques in sintering of these materials the densification rate is almost exclusively determined by the pressure applied. In the absence of the mechanical pressure parameter as in the conventional (pressureless) sintering the driving force for densification is given by the difference in grain boundary and surface free energy. In order to obtain a uniform and sufficiently high shrinkage rate in this case it is very important to control (a) the chemical composition and (b) the physical properties of

Riley, F.L. (ed.) Progress in Nitrogen Ceramics

the starting powders (especially the particle size,form and distribution of Si_3N_4 and the sintering aids), (c) the homogeneity of the powder mixture and (d) the uniformity of density of the green body.

A complicating factor in sintering of these materials is the tendency of the nitride to decompose the pure Si_3N_4 being unstable already above 1900°C at 0.1MPa pressure of nitrogen. The main decomposition products are typically N_2 and SiO(g). The decomposition is highly undesirable for at least two reasons: (a) difficulty to keep material tolerances and (b) loss of oxygen (as SiO) which decreases the amount of the intergranular liquid phase essential to densification.
The methods frequently employed to decrease the extent of decomposition consist in increasing the N_2 pressure in the sintering atmosphere and/or putting the compacts into a protective powder bed.

In the latter method the powder bed may provide a <u>passive</u> protection in that it decreases the diffusivity of the gaseous decomposition products (SiO,Si(g)) from the specimens sintered and CO(g) (graphite furnace + oxygen impurity) towards the specimens. CO(g) can convert the nitride to carbide in a self-sustaining reaction according to the scheme below:

$$CO \curvearrowright Si_3N_4 \longrightarrow SiO \curvearrowright C \longrightarrow CO$$

Furthermore,an <u>active</u> protection can be achieved by using powders generating the same kind of gaseous species as those given off by the specimens sintered. The effect of powder bed was demonstrated in several works (3,4). Giachello and Popper (5) used the powder bed for infiltration of the sintered specimens by sintering aids.

The present work is concerned with some aspects of the preparation of the green bodies and of the sintering in powder bed when using differrent starting Si_3N_4 powders. Yttria (6 $^w/o$) and alumina (2 $^w/o$) were used as sintering aids in all runs.

EXPERIMENTAL

The nitride powders used were from two producers: SN 502 of GTE Sylvania (gas-phase prepared and calcined) and the qualities HCST 2893 and LC 10 of HC Starck,Berlin (prepared by nitridation of Si). The main characteristics of these powders are given below:

Designation	SN 502	HCST 2893	LC 10
Phase composition (%)	60α /40 am.	95α /5β	- " -
BET surf. area (m^2/g)	4.5	8.4	15.9
Oxygen content ($^w/o$)	1.6	0.84	1.6
Carbon content ($^w/o$)	-	0.32	0.14

The powder mixtures were prepared by wet-milling (mixing) in ethanol during ca 20 hours. Various types of mills and milling media

were examined including a McCrone vibro-mill, centrifuge ball mill and the media of agate, alumina and silicon nitride.
The powder compacts were made by isostatic pressing (280 MPa) with or without pressing aids.
Sintering was carried out in a graphite resistance furnace in nitrogen at 0.1MPa with the specimens suspended in a powder bed. The temperature was measured with a W5Re/W26Re thermocouple and a regulator Honeywell DCB-700 was employed for temperature control.
Initial stages of decomposition and densification, respectively, were studied in N_2 gas at 0.1 MPa (heating rate:15°/min) using a thermobalance (Mettler TA 1) and a sintering dilatometer of own construction.

RESULTS AND DISCUSSION

Preparation of specimens

The powders milled with the alumina media showed in general a large pick-up of Al_2O_3. Moreover, when running the mills witout powder the BET surface area of the Al_2O_3 released from the media was much lower (4.1m^2/g) than that of Al_2O_3 added as sintering aid (7-8m^2/g). The agate media were found most effective and thus were used in most cases. The SiO_2 pick-up normally obtained corresponded to an increase of O content in the powder by 0.5-0.9 $^w/o$ which was judged as beneficial for densification.
The BET surface area of the SN 502-based mixtures milled in the vibro-mill could be increased from 4.5 to about 8m^2/g while that of the HCST-powders remained essentially unchanged. Best milling results were obtained for all powders used in the centrifuge ball mill - 20-30m^2/g after 15-20 hours.

The pressed density seemed to be significantly affected by the morphology of the starting powders.

(a) (b)

Fig.1a,b. SEM micrographs of pressed SN 502- and HCST 2893-based powder mixtures

The SEM micrograph (Fig.1a) shows a cross-section area of a compact pressed of the SN 502-based powder (milled from 4.5 to 20.9m^2/g). The needle-shaped particles present in the starting powder can still be clearly identified. These mixtures could be pressed to only about 45% of theoretical density(TD) the presence of these particles probably having decreased the compactibility of the powder.The similarly milled 2893-based powder show a much better particle-to-particle contact (Fig.1b).The pressed density could be above 55% of TD.
By adding pressing aids (up to 18^{w}/o) to the SN 502-based powders and/or by granulation the density could be increased to about 50% of TD. However,the sintered compacts of the granulated powders could display density inhomogeneities such as shown in Fig.2a (SEM cross-section area). A close examination of the pressed compacts showed regions of increased density already at this stage,see Fig.2b.This type of macro-defects which thus seems to be derived from the microstructure of the green body could even appear in specimens which could be sintered to better than 93% of TD.

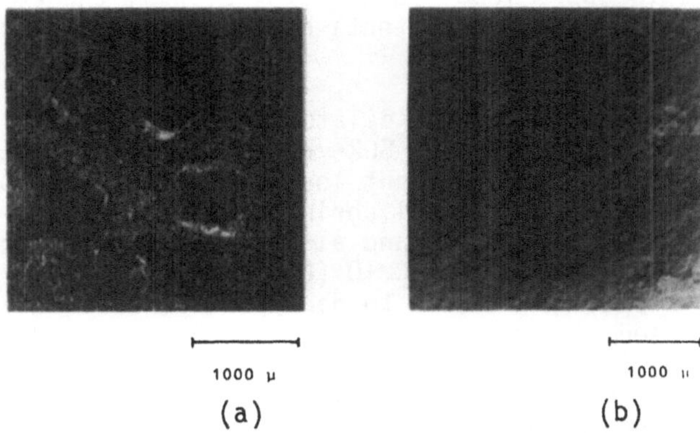

(a) (b)

Fig.2a,b. SEM micrograph of a SN 502-based compact with density inhomogeneities (a) after and (b) before sintering.

Sintering in powder bed

In order to decrease the extent of decomposition a protective powder mixture was found suitable consisting of Al_2O_3,AlN and Si_3N_4 of a β'-SiAlON composition with x=1.9 (in $Si_{6-x}Al_xO_xN_{8-x}$). This powder, even when loosely packed, densified slightly on sintering thus forming a protective capsule around the specimens.The SN 502-based compacts (even those milled to only 8m^2/g) densified to better than 99% of TD at about 1730°C (2-3 hrs) with a weight loss below 3%.The much finer HCST-powders could not be obtained with better than about 90% of TD at sintering temperatures below 1800°C.

SEM-EDS examination of certain specimens showed a considerable build-up of Y and Al gradients (Y depletion and Al enrichment) because of interaction with the β'-SiAlON protective powder as reported in more detail elsewhere (6). These gradients, when controled, may be useful for the oxidation resistance but probably cannot be accepted for all applications. In order to avoid this interaction another protective powder was selected (6), based on pre-oxidized Si_3N_4 and BN. The specimens sintered did not show any interaction in this case but this protective powder did not tend to densify and the weight loss of the specimens was 15-20%. When, however, this powder was pressed isostatically on the specimen the weight loss could be reduced to 2-3%.

To improve the sinterability of the HCST-based powders it has been attempted to increase the amount of the intergranular liquid by increasing the oxygen content of the nitride. Oxygen was introduced (as SiO_2) up to an amount of 4 w/o either by pre-oxidation by the nitride powder, by addition of fine-grained silica ($74 m^2/g$) or by prolonged milling with agate media. However, this had only a marginal effect when sintering for a maximum of 3 hours at temperatures below 1800°C. On the other hand, by increasing the sintering temperature to above 1860° (e.g. 1870°/2.5hrs) the LC 10-based mixture (with the usual O content) could be densified to better than 99% of TD.

Thermogravimetry and sintering dilatometry were used to identify the differences between the SN 502- and HCST-based powders. Fig. 3 shows schematically the weight loss curves (w/o) recorded up to 1600° and the relative initial shrinkage rates(% per hr) up to 1800°C measured with the yttria and alumina-doped Si_3N_4 compacts based either on SN 502 ($8m^2/g$) or LC 10 ($20.7m^2/g$).
As can be seen both specimens start to decompose already at about 1300°. The LC 10-based specimen appears more stable which may, of course, be due to its higher pressed density (55% of TD compared to 48% for the SN 502-based specimen). The differing densification behaviour of these specimens is illustrated by the shrinkage rate curves. The LC 10-based compact attains a maximum initial rate at a lower temperature (1720°) than that of the SN 502-based one (1780°) probably because of its finer particle size. However, the isothermal shrinkage rate at 1800° is seen to be much lower for the LC 10 specimen suggesting thus the need of using a higher sintering temperature for this type of powder.

As may also be seen in Figure 3 there is a critical temperature region of 1300-1600°C where these particular yttria,alumina-doped specimens decompose without any significant shrinkage.
The use of a protective powder bed coupled with an adequately high heating rate within this particular region seems desirable in order to avoid material degradation.

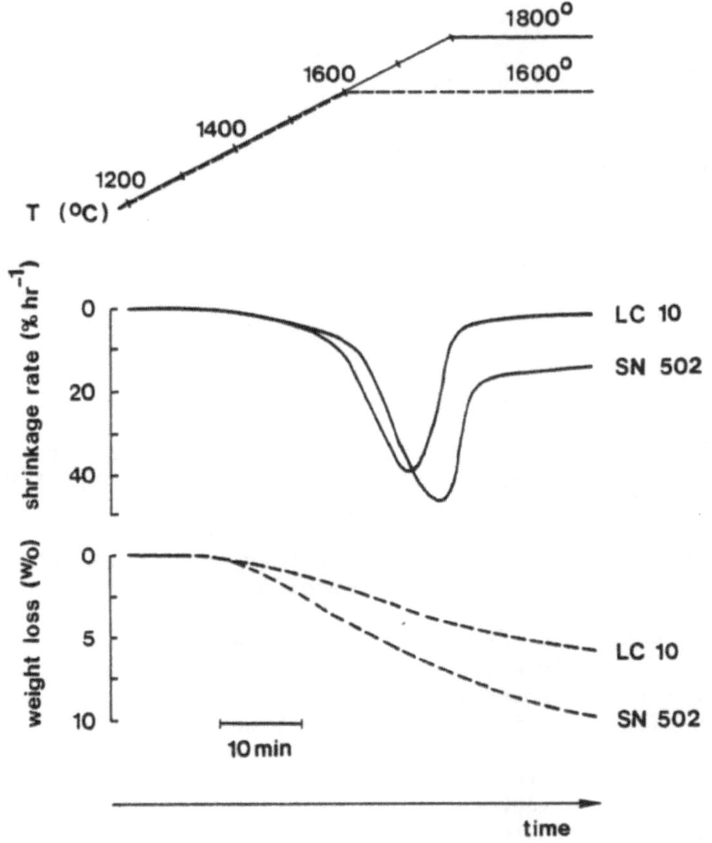

Fig.3. Weight loss curves when heating (15°/min) up to 1600° and relative shrinkage rate curves up to 1800° for the SN 502- and LC 10-based compacts

ACKNOWLEDGEMENT

This work is part of a project "High Temperature Ceramics for Heat Engines" supported financially by The Swedish Board for Technical Development.

REFERENCES

1. L.J.Bowen,T.G.Carruthers and R.J.Brook,J.Am.Ceram.Soc.61(1978)335
2. S.Wild,P.Grieveson and K.H.Jack,Spec.Ceramics 5(1972)377
3. L.J.Gauckler,S.Boskovich and G.Petzow,Nitrogen Ceramics (1977)405
4. K.Nishida and T.Miyamo,Japan. Kokai 7852,518
5. A.Giachello and P.Popper,in Energy and Ceramics, (1980)620
6. R.Pompe,L.Eklund and L.Hermansson,Special Ceramics 7,(1980)97

DENSIFICATION AND TRANSFORMATION MECHANISMS IN NITROGEN CERAMICS

S. Hampshire* and K.H. Jack

Wolfson Research Group for High-Strength Materials
Crystallography Laboratory, The University of Newcastle upon Tyne, U.K.

*(*now at College of Engineering & Science, The National Institute for Higher Education, Limerick, Eire)*

ABSTRACT. Densification of nitrogen ceramics occurs by a process of liquid-phase sintering as a result of the formation of an oxynitride liquid phase that promotes shrinkage and at the same time the phase transformation from α-Si_3N_4 to β-Si_3N_4 or β'-sialon. The kinetics of densification are interpreted by Kingery's model of liquid-phase sintering and the amount of densification during each of the three successive stages (i) particle rearrangement; (ii) solution-diffusion-precipitation and (iii) a solid-state interaction, depends on the particular additive used. The contribution to densification during the rearrangement stage (i) is greater with MgO than with Y_2O_3 because a larger volume of lower viscosity liquid is formed. Stage (ii) involves the solution of α and formation of β (or β') and this can occur without major transport of material if diffusion through the liquid is slow (Si_3N_4 with Y_2O_3) or may be accompanied by rapid transport and hence densification if the slow, rate-determining process is the formation of β (Si_3N_4 with MgO and with Li_2O; β'-sialon with MgO and Y_2O_3).

The kinetics of the transformation from α to β(or β') confirm the proposed mechanisms and similar activation energies for the different additives correspond with the Si-N bond energy suggesting that α - β is a reconstructive transformation requiring a solvent and the breaking of Si-N bonds.

Riley, F.L. (ed.) Progress in Nitrogen Ceramics

1. INTRODUCTION

Significant progress has been made in understanding the factors governing the processing of nitrogen ceramics (1-3) and it is now well established that to produce high density materials, whether by hot-pressing or pressureless sintering, an additive is required to provide conditions for liquid-phase densification (3,4).

Silicon nitride powder always contains an appreciable amount of surface silica and all oxide additives react with this silica and some of the nitride to give an oxynitride liquid that promotes shrinkage and, at the same time, the phase transformation from α to β-Si_3N_4(or β'-sialon).

The temperature of initial liquid formation observed with additions of metal oxide to silicon nitride containing 4w/o surface silica is appreciably lower than the lowest solidus temperature in the corresponding metal oxide-silica system (3,4) and confirms that nitrogen, as an additional component, lowers the eutectic temperature. For example, liquid formation occurs with MgO at 1390°C and with Y_2O_3 at 1440°C. In the Si-Aℓ-O-N system, the lowest eutectic is at 1470°C and extra additives will lower this further.

Shrinkage usually commences at the temperature of liquid formation and is subsequently accompanied by the α-β phase transformation.

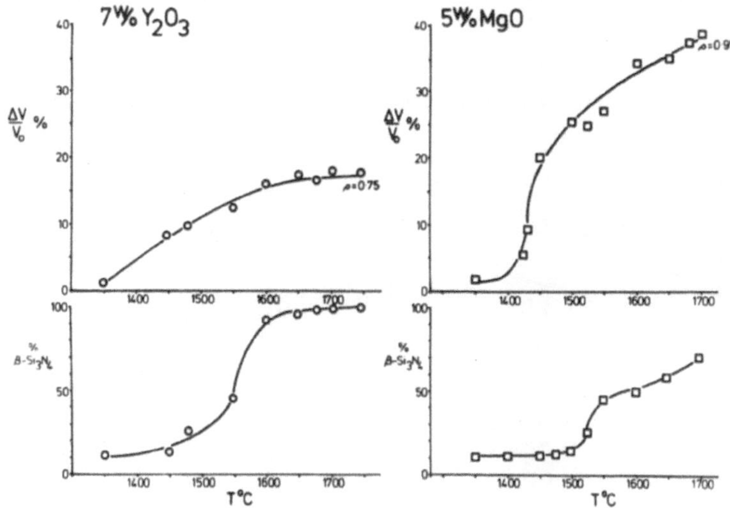

Fig. 1 Shrinkage and transformation after 30 minutes at various temperatures for pressureless sintering of Si_3N_4 with Y_2O_3 and MgO

Fig. 1 shows the shrinkage and transformation at different temperatures during pressureless sintering of silicon nitride and emphasizes the differences between magnesia and yttria as additives. With MgO, nearly complete densification is achieved with partial transformation to β whereas with Y_2O_3, complete transformation occurs with only limited densification.

2. THE KINETICS OF DENSIFICATION OF SILICON NITRIDE

The experimental details are as published previously (4). The densification kinetics are interpreted by Kingery's model of liquid phase sintering (5). Kingery describes three stages, (i) particle rearrangement, (ii) solution-precipitation and (iii) coalescence, which are conveniently plotted on log shrinkage- log time plots as shown in Fig. 2 for (a) 5w/o MgO and (b) 7w/oY_2O_3 additions. The rapid densification of the rearrangement stage is observed and with MgO accounts for 50% of the shrinkage required for full densification whereas with Y_2O_3 it is responsible for less than 25% of the total shrinkage required. The reason for the difference is the larger liquid volume and lower liquid viscosity with magnesia (Magnesium containing oxynitride liquids generally have lower viscosities than the corresponding yttrium liquids) (6). Different silicon nitride powders densify to different extents during this stage since the amount of surface silica and other impurities affects the volume of liquid.

Stage (ii) involves solution - diffusion - precipitation and

$$\Delta V/V_o \propto t^{1/n} \qquad (1)$$

where n = 3 if the rate-controlling process is solution into, or

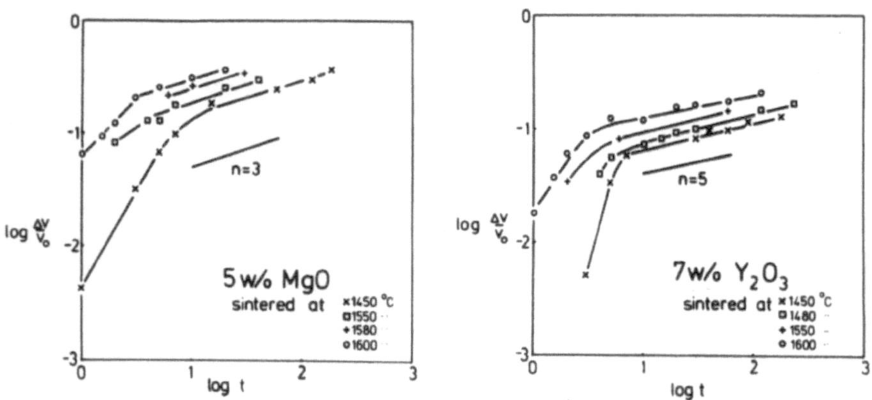

Fig. 2 Log shrinkage V. log.time (minutes) plots for sintering silicon nitride with (a) MgO and (b) Y_2O_3.

precipitation from, the liquid; n = 5 if the rate-controlling process is diffusion through the liquid, assuming that the particles are non-spherical. For MgO (and also Li_2O), n = 3 and for Y_2O_3, n = 5. Because the second stage occurs by solution of α and precipitation of β (7), the effect of the α-β transformation on densification must be considered.

3. KINETICS OF THE α - β TRANSFORMATION

Fig. 3 shows isothermal experimental α - β transformation curves during pressureless sintering of silicon nitride with (a) 5 w/o MgO and (b) 7 w/o Y_2O_3. The curves are sigmoidal and characteristic of a grain-boundary nucleated reaction with an incubation period corresponding to the particle rearrangement of stage (i) during which there is negligible transformation. The activation energies for transformation (see 4, 8) are the same for both additives and are similar to the dissociation energy of the Si-N bond i.e. 435 ± 38 $kJmol^{-1}$. The mechanism of transformation seems to be the same for both additives, each merely providing a solvent for a reconstructive transformation to occur (9) by the breaking of Si-N bonds.

4. COMPARISON OF DENSIFICATION AND TRANSFORMATION KINETICS.

To confirm the mechanism outlined, the kinetics of densification and of the α-β transformation are compared. Fig. 4 shows log shrinkage and extent of transformation plotted against log. time at 1450°C and 1600°C for Y_2O_3. Transformation starts immediately after stage (i) rearrangement and at 1600°C is complete at a relative density of 0.75. Diffusion through the highly viscous liquid is relatively slow, solution-precipitation is more rapid and the α-β transformation occurs with little material transport and hence with little densification. With MgO, the transformation

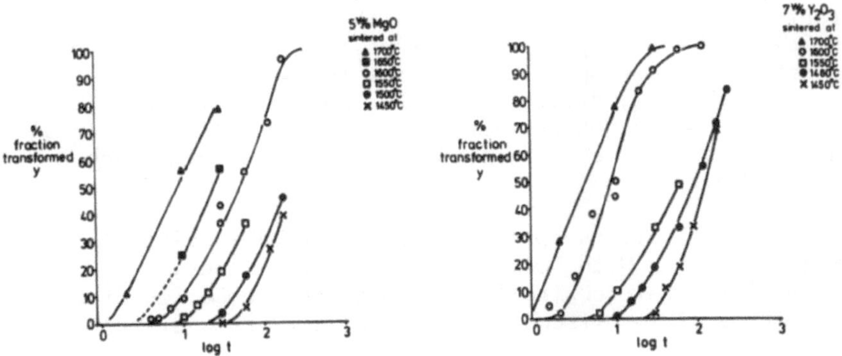

Fig. 3. Isothermal experimental α-β transformation curves during sintering with (a) MgO and (b) Y_2O_3. (t in minutes)

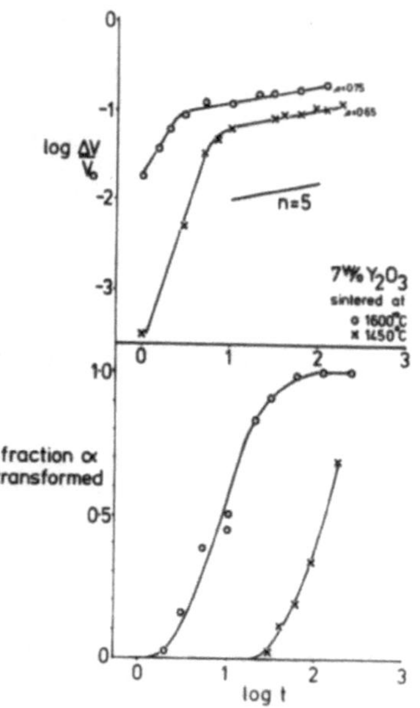

Fig. 4. Comparison of Densification and transformation with Y_2O_3

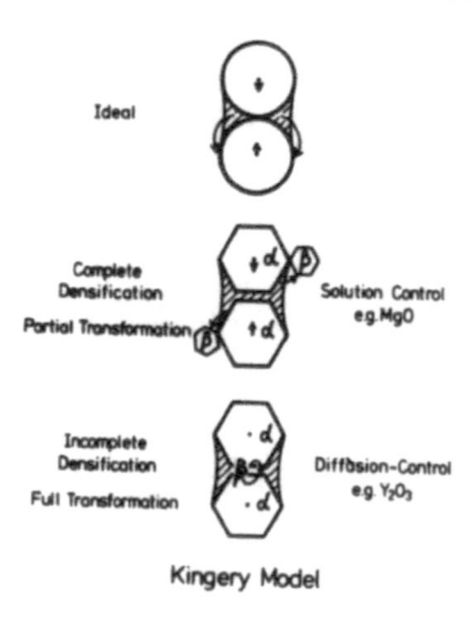

Fig. 5. Physical representation of liquid-phase densification and transformation during stage (ii).

is negligible during stage (i) as for Y_2O_3 so that about one-half of the densification is achieved without any transformation. During stage (ii) the relatively rapid transport of material through the low viscosity liquid ensures that transformation is accompanied by shrinkage. Over 90% theoretical density is attained with only partial transformation. Fig. 5 illustrates the physical processes that occur. For both additives, solution of α into the oxynitride liquid occurs preferentially at the contact areas between particles. With MgO, rapid transport of material allows precipitation of β on the free surfaces so that the distance between particle centres is reduced and densification occurs. With Y_2O_3, diffusion is slow and appreciable precipitation of β occurs in the contact areas without significant material transport. Thus, transformation takes place without much densification.

Similar considerations apply to the pressureless sintering of β'-sialons (4,8). With both MgO and Y_2O_3 additives, the reaction to form β' is rate-controlling because the formation of a large volume of low viscosity liquid allows rapid diffusion.

To summarize, the sintering of nitrogen ceramics requires the formation of an oxynitride liquid. The densification kinetics are interpreted by Kingery's liquid-phase sintering model and the contributions of the successive stages vary with the type of additive. The transformation kinetics confirm the proposed mechanisms.

REFERENCES

1. Rae, A.W.J.M., D.P. Thompson and K.H. Jack, in J.J. Burke, E.N. Lenoe and R.N. Katz (eds.) Ceramics for High Performance Applications 11 (Brook Hill Pub. Co: Chestnut Hill, Mass. 1977) p.1039.
2. Lewis, M.H., B.D. Powell, P. Drew, R.J. Lumby, B. North and A.J. Taylor. J. Mater.Sci. 12 (1977) 61.
3. Jack, K.H. in P. Vincenzini (ed.) Energy and Ceramics (Elsevier: Amsterdam, 1980) p.534.
4. Hampshire, S. and K.H. Jack, in D. Taylor and P. Popper (eds.) Special Ceramics 7 (Proc. Brit. Ceram. Soc. No. 31, June 1981) p.37.
5. Kingery, W.D. J. Appl. Phys. 30 (1959) 301.
6. Drew, R.A.L., S. Hampshire and K.H. Jack, in D. Taylor and P. Popper (eds.) Special Ceramics 7 (Proc. Brit. Ceram. Soc. No. 31, June 1981) p.119.
7. Drew, P. and M.H. Lewis. J. Mater. Sci. 9 (1974) 261.
8. Hampshire, S. Ph.D. Thesis 'The Sintering of Nitrogen Ceramics' (University of Newcastle upon Tyne, 1980).
9. Messier, D.R., F.L. Riley and R.J. Brook. J. Mater. Sci. 13 (1978) 1199.

DISCUSSION

Clarke: How were viscosities of the nitride oxide glasses determined?

Hampshire: By measuring the time required for 100 µm deformation of 4 mm diameter cylinders using high temperature deformation under load apparatus. "Spectrosil" vitreous silica was used over 1000-1300°C as standard. See R.A.L. Drew et al. in "Special Ceramics 7" (Proc. Brit. Ceram. Soc. 31 June 1981) p.119, and R.A.L. Drew "Nitrogen Glasses", Ph.D. Thesis, University of Newcastle upon Tyne, 1980.

McCauley: Due to kinetic factors we observed melting in AlON compositions of Al_2O_3+AlN ~100°C lower than the melting of bulk samples.

Hampshire: Initial liquids may be transient if the overall composition lies in a phase region with a higher eutectic. In the Si_3N_4 / 7w/o Y_2O_3 system liquid is formed at 1440°C with N-apatite, but the final equilibrium composition is β-Si_3N_4+$Y_2Si_2O_7$, and very little liquid remains.

SINTERING OF REACTION BONDED SILICON NITRIDE

J. A. Mangels

Ceramic Materials Department
Research Staff
Ford Motor Company
Dearborn, Michigan 48121

ABSTRACT

A process to produce sintered reaction bonded Si_3N_4 (SRBSN) articles has been developed. This process consists of the addition of an appropriate sintering aid to reaction bonded Si_3N_4 followed by sintering between 1780 and 2000°C, using an over pressure of nitrogen. The principal advantage of this process is the low sintering shrinkages of 5 to 10%. The properties and microstructure of two SRBSN systems sintered with MgO and Y_2O_3 additives are described and were found to be comparable to corresponding hot pressed Si_3N_4 systems. Examples of applications of both systems are illustrated, demonstrating near net shape fabrication capability of the process.

Introduction

Silicon nitride is a candidate material for a number of advanced heat engine applications. A need exists for a ceramic material which can combine high strength with a near net shape fabrication process. Reaction bonded Si_3N_4 offers this near net shape fabrication capability, however, the strength of this material is limited. Conversely, hot pressed Si_3N_4 can meet high strength requirements, but can only be produced in simple shapes. Sintered Si_3N_4 offers an optimum combination of strength and near net shape fabrication. However, there are a number of problems associated with the sintering of Si_3N_4 powders. These are primarily due

Partially funded by DOE/NASA under Contract DEN 3-167, Subcontract P1928150.

Riley, F.L. (ed.) Progress in Nitrogen Ceramics

to the fine particle size and high surface areas of the sinterable grade powders, which make component fabrication difficult. The low green density of these components results in high sintering shrinkages, often approaching 20 percent. Consequently, accurate shape fabrication is difficult.

Sintered reaction bonded Si_3N_4 (SRBSN) is a relatively new material processing technique for production of sintered Si_3N_4 (1-5). In this process the initial preform is reaction bonded Si_3N_4 (RBSN), which includes the addition of a sintering aid so that the RBSN article, which can then be pressureless sintered to form a high density sintered Si_3N_4 (SRBSN) component.

There are a number of advantages to the SRBSN approach when compared to the conventional sintering of Si_3N_4 powder:

- The starting material, primarily silicon, is readily available.
- RBSN fabrication technology, slip casting and injection molding, is relatively well developed.
- The high green densities of the RBSN (72 to 85%) result in linear sintering shrinkages of only 5 to 10%.

The advantages of this processing technique can circumvent many of the problems associated with the conventional sintering of Si_3N_4, making high strength, near net shape Si_3N_4 components a reality.

The SRBSN Process

Reaction bonded Si_3N_4 is typically produced by the reaction of a silicon powder compact with a nitrogen-containing atmosphere. Specifically, silicon powder of 98.5% purity (major impurities are 0.6 - 0.8% iron, 0.2% aluminum, and 0.5% oxygen) with an average particle size of 3 microns was mixed with either a thermoplastic organic binder for injection molding or water plus a deflocculant for slip casting. In some cases, a nitriding aid (2.5% Fe_2O_3) was also added. Following fabrication and binder removal, the parts were nitrided using a nitrogen demand nitriding cycle with a helium/hydrogen/nitrogen nitriding atmosphere (6).

Two techniques were employed to introduce the sintering additive into the Si_3N_4 component. The first technique consists of the impregnation of the RBSN component with an alcohol-salt solution of the desired sintering aid, with the salt eventually being decomposed into the oxide. Repeated impregnations were often required to attain the desired concentration of oxide sintering aid. This technique has been successfully employed to introduce up to 4 wt% MgO into RBSN articles with thin cross-sections.

A second technique involves the addition of the sintering aid to the silicon powder prior to nitriding. This technique offers the best opportunity for close composition control throughout a component and is preferred. This technique has been used to incorporate up to 14 wt % Y_2O_3 into RBSN, and is especially suitable if large cross section (> 1.5 cm) components are to be produced.

The sintering experiments were performed in a graphite resistance-heated furnace capable of operating at 10.33 MPa of nitrogen gas pressure. The samples were imbedded in packing powders having a composition identical to the composition of the component. The function of the packing powder was fourfold: 1) to support the components inside the crucible, 2) to protect the components from reacting with the carbonaceous furnace atmosphere and forming silicon carbide, 3) to help retard the thermal decomposition of the Si_3N_4, which can occur at the temperatures employed, and 4) to prevent migration of the sintering aid from the samples.

The sintering cycles were a function of the additive composition and concentration. The MgO containing materials were sintered between 1780 and 1820°C for 2 hours with a 2.07 MPa nitrogen over pressure. The Y_2O_3 containing materials were sintered between 1875 and 2000°C for up to 6 hours with nitrogen over-pressures of 0.103 to 8.26 MPa.

Material Characterization

Two SRBSN materials have been studied in detail; 3-4 wt% MgO and 8 wt% Y_2O_3. The properties of these materials are summarized in Table I. Both of these compositions have been sintered to at least 98% T.D. The room temperature strengths of both materials can be expressed using Larsen's relation (7)

$$S = 623 \exp \quad (-5.65P)$$

where S modulus of rupture and P is the volume fraction porosity, which was generated for all types of Si_3N_4.

Table I

Characterization of SRBSN Materials

	MgO	Y_2O_3
Additive Concentration	3-4 wt%	8 wt%
Phase Composition	β-Si_3N_4	β Si_3N_4 + Y SiO_2N + $Y_4Si_2O_7N_2$
Density	98% T.D (3.18 g/cc)	98% T.D (3.26 g/cc)
Characteristic Strength & Weibull Modulus		
- RT	480 MPa (11.6)	699 MPa (10.1)
- 1200°C	310 MPa (-)	565 MPa (14.9)

The high temperature strength of the MgO-SRBSN material showed a 35% loss in strength at 1200°C, compared to a 20% loss for the Y_2O_3-SRBSN. These results are similar to those of hot pressed Si_3N_4 of similar compositions, and illustrate the principal advantage of the Y_2O_3-Si_3N_4 system; it's good high temperature behavior.

The microstructure of the MgO-SRBSN material is compared to an equivalent hot-pressed Si_3N_4 composition in Fig. 1. The structure of the sintered material is composed of needle shaped grains, 1-3 microns in length and having an L/D ratio of about 5 to 1. The structure is identical to that of the hot pressed material, indicating that the overall properties of the MgO-SRBSN should be similar to those of MgO hot pressed Si_3N_4. The microstructure of the Y_2O_3-SRBSN (Figure 2A) is also composed of a mixture of fine 1-2 micron grains with some 5-10 micron needle shaped grains. This structure is similar to the Y_2O_3 doped hot pressed Si_3N_4 structure (Figure 2B).

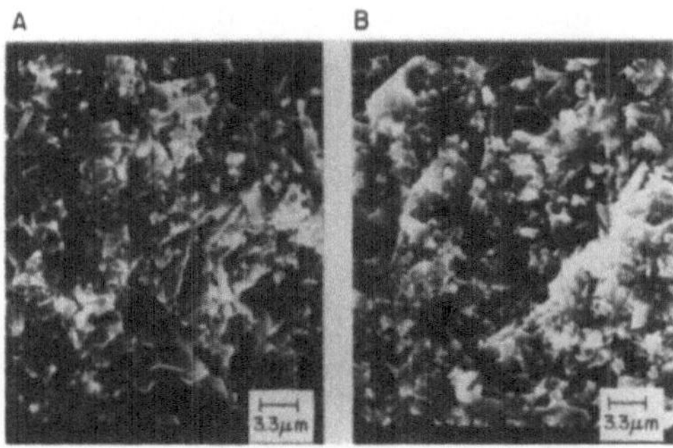

Figure 1 Microstructure of A) Hot-pressed Si_3N_4 and B) SRBSN, both produced using 3% MgO.

Figure 2 Microstructure of A) SRBSN and B) Hot-pressed Si_3N_4, both produced using 8 wt% Y_2O_3.

A major problem of Y_2O_3-Si_3N_4 materials is their low temperature (700-1000°C) oxidation resistance. Studies have shown that this problem is due to certain phases in the Si_3N_4-Y_2O_3-SiO_2 system (7) and that impurities of W, WC or carbon are major contributors to the problem (8). The oxidation resistance of this material was examined at 700° and 1000°C in a static air environment. No cracking was observed in the samples; however, a color change, from black to gray, was noted.

Component Fabrication

The SRBSN process has been applied to a variety of complex structural ceramics. The turbine stator (Figure 3A) produced with 3 wt% MgO was sintered to 98% T.D., exhibiting only 5% linear shrinkage. The turbocharger rotor (Figure 3B) and the simulated turbine rotor (Figure 3C) were both produced using 8 wt % Y_2O_3 and were sintered to 99% and 98% T.D. respectively. The respective shrinkages of 8 and 10%, reflect the differences in nitrided densities of the components required due to section size. All of these components are shown in their "As-sintered" condition, and illustrate the net-shape capability of the SRBSN process.

Figure 3 A) Gas Turbine Stator (14 cm dia.) produced using MgO-SRBSN. B) Radial Turbocharger Rotor (6.5 cm dia.) and C) Radial Turbine Rotor Hub (10.1 cm dia.) both produced using Y_2O_3-SRBSN.

References

1. J. A. Mangels, G. J. Tennenhouse, "Densification of Reaction Bonded Silicon Nitride," Bull. Amer. Ceram. Soc., Vol. 59, No. 3 (1980) p. 356.

2. J. A. Mangels, G. J. Tennenhouse, "Sintering Behavior and Microstructure Development of Yttrium-Doped, Reaction Bonded Si_3N_4," accepted for publication in the Bull. of the Amer. Ceram. Soc.

3. J. A. Mangels, "Sintered Reaction Bonded Si_3N_4 for the AGT 101 Turbine Rotor," DOE Automotive Technology Development Contractors' Coordination Meeting, Dearborn, Mi., Nov., 1980.

4. J. A. Mangels, "Sintered Reaction Bonded Si_3N_4," to be published in the proceedings of the 5th Annual Conference on Composites and Advanced Materials, sponsored by the American Ceramic Society, Ceramic-Metal Systems Division, Meeting held in Cocoa Beach, Florida, Jan, 1981.

5. A. Giachello, P.. Popper, "Post-Sintering of Reaction Bonded Silicon Nitride," Ceramurgia International, Vol. 5, No. 3, (1979) p. 110.

6. J. A. Mangels, "The Effect of Rate Controlled Nitriding and Nitriding Atmospheres on the Formation of Reaction Bonded Si_3N_4," accepted for publication in the Bull. of the Amer. Cer. Soc.

7. D. C. Larsen, "Property Screening and Evaluation of Ceramic Turbine Engine Materials," AFML-TR-79-4188, Oct, 1979.

8. F. F. Lange, S. C. Singhal, R. C. Kuznicki, "Phase Relations and Stability Studies in the Si_3N_4-SiO_2-Y_2O_3 Pseudoternary System," J. Amer. Ceram. Soc., Vol. 60, No. 5-6, (1977), p. 249.

9. H. Knoch, G. E. Gazza, "Carbon Impurity Effect on the Thermal Degradation of a Si_3N_4-Y_2O_3 Ceramic," AMMRC TR 79-27, (1979).

DISCUSSION

Quackenbush: Can RBSN be sintered without additives?

Mangels: I believe this is possible. We have found relative densities of > 3.0 in the centre of 2.7 RBSN without using a powder bed.

Popper: How many impregnations are needed to introduce the sintering aid? What are the thickness limitations in the nitriding stage?

Mangels: 3-5 are needed. Non-uniform distribution is obtained if the cross-section is > 2-3 mm. 100 mm diameter rotors can be obtained at the 2.3-2.4 relative density level. 10% linear shrinkage is obtained, which is the limit for a reasonable component.

Greskovich: What kind of grinding media did you use?

Mangels: Burundum(R) media and mills. No Al contamination was detected from dry milling, even after 144 hours. Iron contamination is a problem, however.

MICROSTRUCTURE OF DENSIFIED REACTION BONDED SILICON NITRIDE

J.R.G. Evans and A.J. Moulson

Department of Ceramics, University of Leeds, U.K.

ABSTRACT. A high temperature - high pressure apparatus to measure kinetics of densification of RBSN is described and problems relating to the thermal stability of samples and furnace components discussed. Preliminary results using a commercial RBSN are reported, together with electron micrographs of the densified material.

INTRODUCTION

The advantages of RBSN in terms of the ease of manufacture of complex shapes are well known and the two disadvantages, namely, the low strength and low oxidation resistance, are related to the 20% residual porosity. Thus, attempts have been made to fill the porosity with silicon (1) and to densify the material by sintering (2,3). Although the room temperature strength is not the most appropriate parameter for comparison of materials intended for advanced high temperature engineering applications such as RBSN, the data are more readily available. Figure 1 shows strength as a function of porosity, covering a range of different RBSN materials and, mainly unspecified, testing techniques drawn from a review (4). The figure also includes a value for fully dense HPSN and for RBSN densified by Giachello and Popper (2) and by Mangels and Tennenhouse (3).

It is well known that silicon nitride powder can be sintered by hot-pressing or by pressureless sintering. Additives are incorporated which react with the native silica layer on the nitride particles to give a liquid phase at the sintering temperature. A logical extension is to make comparable additions to RBSN and it is found (2) that RBSN with 8% Y_2O_3 + 1% MgO can be densified to 98.5% theoretical density with only 6% linear shrinkage.

Riley, F.L. (ed.) Progress in Nitrogen Ceramics

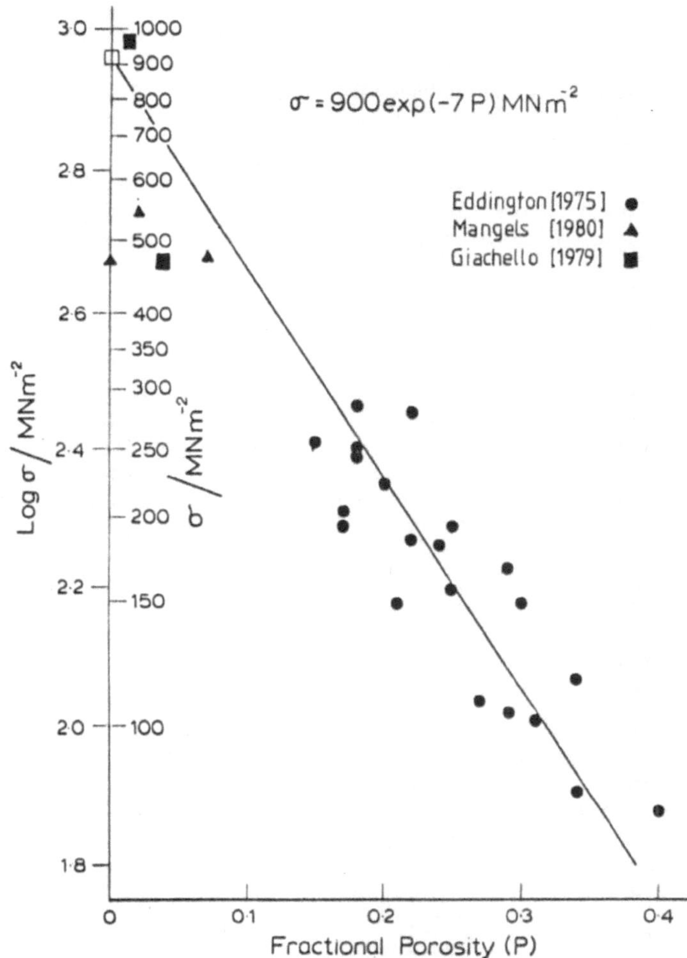

Fig. 1. Room temperature modulus of rupture as a function of porosity for RBSN. After Eddington (7), with results from Giachello (2) and Mangels (3) for densified RBSN.

There are a number of possible reasons why silicon nitride fails to sinter in the absence of a liquid phase. It is commonly held that mass transfer coefficients are low, although no convincing measurements have been made. Furthermore it is possible that material transport by surface diffusion may relieve the driving force for densification. As silicon nitride is ~ 88% covalent in character (5), it is likely that strongly directional bonding leads to high grain boundary energy and this, coupled with the relatively low surface energy for native surface silica, may yield a high value of γ_{gb}/γ_{sv} giving low dihedral angles and the possibility of stable pores. Terwilliger and Lange (6) have argued that it is a competition between decomposition and densification which controls the sinterability. If large β-Si_3N_4 grains are present or grow during heat treatment in plate or columnar form, lattice or grain boundary diffusion distances may be large and constitute a further hindrance to densification.

The aims of the present programme are to investigate the kinetics of densification of pure RBSN. This exploratory study uses a typical commercial material of density 2.5 Mg m^{-3} with an α/β phase ratio of 60:40.

EXPERIMENTAL DETAILS

Additives may be incorporated into RBSN in three ways. Firstly, the oxides may be milled with silicon powder prior to nitridation. In this case the powder must be fine compared with the silicon, and agglomeration avoided, otherwise large pores may be created. Secondly, the additive may be incorporated as a soluble salt. The concentration of salt solution to fill a 20% porous Si_3N_4 body with W% by weight of oxide is 0.127 Wz kg l^{-1}, where z is the ratio of chemical equivalence of salt to oxide. Thus to incorporate 5% MgO using $Mg(NO_3)_2.6H_2O$ requires 4 kg l^{-1} solution. In general, several impregnations of a more dilute solution are needed, each followed by a decomposition treatment. Thirdly, magnesia in a powder bed surrounding the specimen can migrate via the vapour phase into the sample during processing.

The furnace (Figure 2) comprises an induction heated unit inside a water-cooled pressure vessel. Temperature control is by a pyrometer focussed on a graphite susceptor through a glass viewing port. The top dome contains a transducer connected to an RBSN push-rod and the sample is contained in an RBSN tube. Initial difficulties with the RBSN components at high temperature and with graphite present have still to be overcome before kinetic data can be obtained. There is a tendency for the susceptor surface and the RBSN tube to convert to SiC because of, inter alia, the following reactions:

at a silicon nitride surface -

$$2Si_3N_4(c) + 3O_2(g) \longrightarrow 6SiO(g) + 4N_2(g)$$

$$Si_3N_4(c) + 3SiO_2(c) \longrightarrow 6SiO(g) + 2N_2(g)$$

$$Si_3N_4(c) + 6CO(g) \longrightarrow 3SiC(c) + 2N_2(g) + 3CO_2(g)$$

at the graphite susceptor -

$$2C(g) + O_2(g) \longrightarrow 2CO(g)$$

$$CO_2(g) + C(c) \longrightarrow 2CO(g)$$

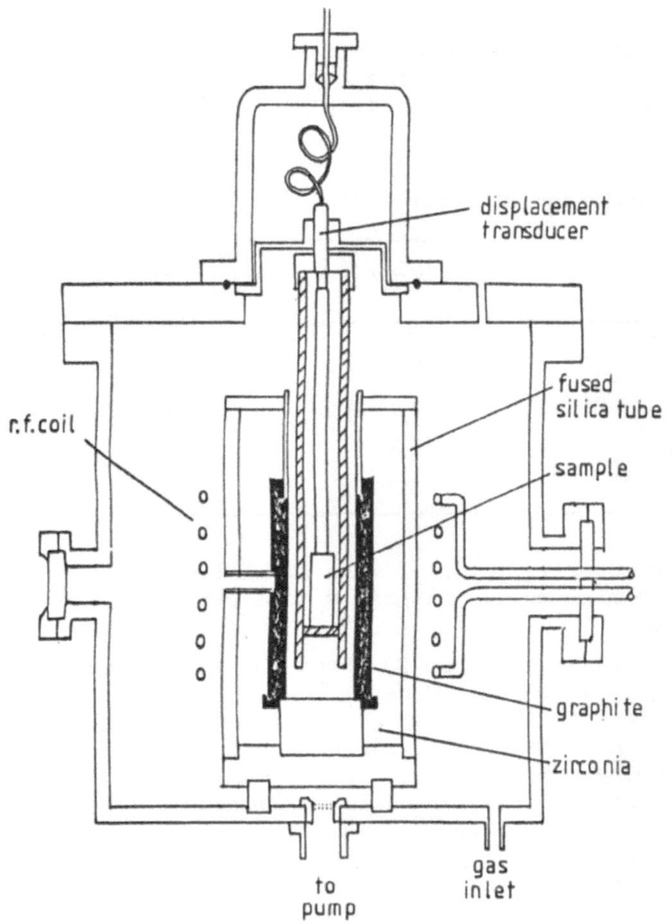

Fig. 2. High temperature, high pressure furnace for measurement of densification kinetics of RBSN.

$$SiO(g) + 2C(c) \longrightarrow SiC(c) + CO$$

Thus traces of oxygen act as a carrier for carbon and silicon. Through efforts to expel oxygen by thorough degassing and incorporation of zirconium oxygen getters, the problem has been largely overcome, as has been the allied problem of generating troublesome amounts of "smoke" in the vessel. For the present preliminary work the apparatus was run with a nitrogen over-pressure of 1 MPa and the sample incorporated in a powder bed of silicon nitride.

RESULTS AND DISCUSSION

The commercial RBSN shows some densification in the absence of additives (Table 1) in agreement with the observations of Mangels (3). There may be some glass-forming impurities in the

TABLE 1: Commercial RBSN - no sintering aids
Surround: Si_3N_4 powder and nitrogen 1 MN m^{-2}

Sintering temp./ °C	Time/ min.	- L/L_o/ %	Weight Loss /%
1700	60	0.6	1.3
1750	30	1.0	n.m.
1750	75	2.0	n.m.
1800	30	1.2	1.7

sample or in the surrounding powder. This emphasises the importance of working with high purity materials if a reliable base line is to be established.

The effect of additives (Table 2) shows that modest additions to the sample or the powder bed aid densification. Full density has not been obtained with this material for these processing times and temperatures but at intermediate stages optical microscopy shows that although the micropores are removed the larger (20 µm) pores are not so easily eliminated. This suggests that the larger pores introduced during nitridation, either by iron-rich regions melting or residual silicon melting out, must be avoided in material for post-nitriding heat treatments. This is especially the case if the larger pores are the strength-limiting defects.

TABLE 2: Commercial RBSN - with sintering aids
Surround: Si_3N_4 powder and nitrogen 1 MN m^{-2}

Sample additive	Powder additive	Temp. /°C	Time /min.	L/L_o / %	Density Mg m^3 Initial	Final	Weight loss/%
5% MgO	none	1750	30	2.0	-	-	1.1
2.5% Y_2O_3	5% MgO	1700	40	2.2	2.6	2.78	0.0
7% MgO	5% MgO	1750	60	3.4	-	-	1.8
5% MgO) 1% Y_2O_3)	5% MgO	1850	30	4.0	2.65	2.87	0.3
none	10% MgO	1750	100	3.6	2.53	2.85	1.2(gain)
4% MgO	5% MgO	1750	200	5.0	2.63	3.01	2.5

The net weight losses are made up of decomposition of the RBSN and exchange of additives with the powder bed.

For Si_3N_4 powder a BET area of typically 14 m^2g^{-1} contains 8% by weight SiO_2 so that $5^w/o$ MgO would be fully reacted:

$$2MgO + SiO_2 \longrightarrow Mg_2SiO_4$$

For RBSN with a SSA of typically 1 m^2g^{-1} (7), only 0.6% by weight SiO_2 is obtained for a 2.5 nm film thickness. Thus it may be that RBSN is deficient in SiO_2 if all the initial grain boundary area is to be wetted.

Figure 3 compares the microstructure of the starting material with the partially densified RBSN. Considerable grain growth can be seen resulting in columnar and plate-like β grains. After 30 min at 1800°C the sample was fully converted to β-Si_3N_4.

Figure 4 compares the microstructure by TEM. The starting material has a wide grain size distribution. After densification the grain size is larger and more uniform and grain boundaries and triple points show an amorphous phase suggesting that the liquid phase is capable of penetrating most of the grain boundaries in RBSN.

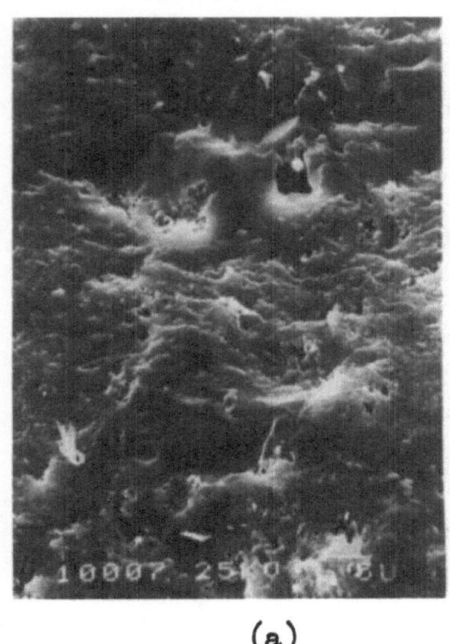

(a)

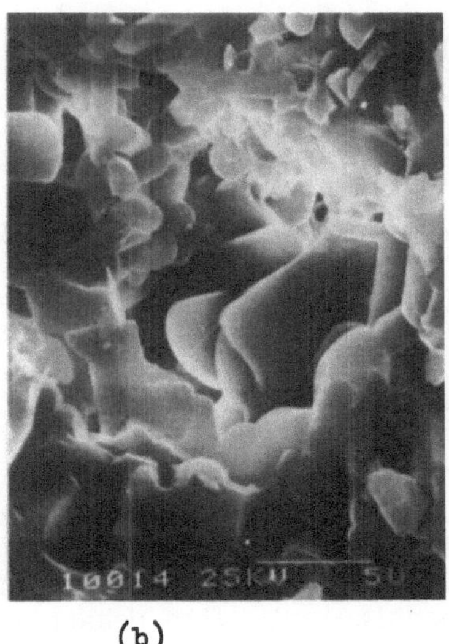

(b)

Fig. 3. Scanning electron micrographs of fracture surfaces of RBSN (a) as-received, (b) 50 min at 1800°C, 2.5% MgO 2.5% Y_2O_3.

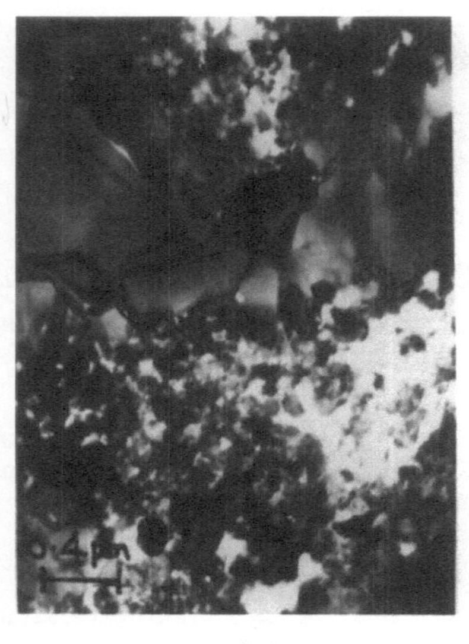

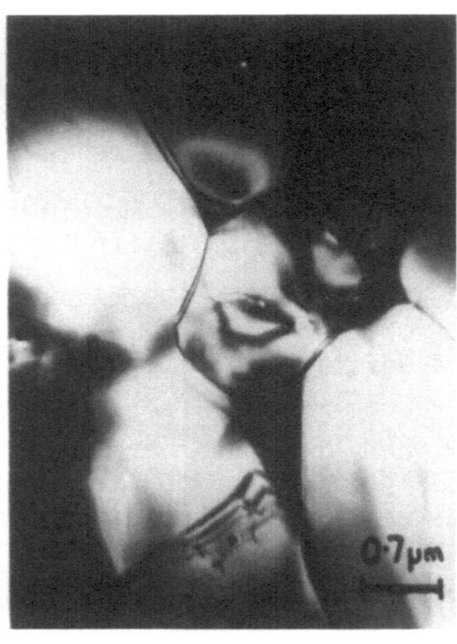

(a) (b)

Fig. 4. Transmission electron micrographs of RBSN.
(a) as received; (b) 50 min at 1800°C.

This programme is continuing with attempts to obtain densification rate data and to investigate the sintering and grain growth behaviour of RBSN using high purity starting materials.

REFERENCES

1. Y. Inomata. Yogyo Kyoshai-Shi 83 (1) 9 (1975).
2. A. Giachello and P. Popper. Ceramurgia 5 (3) 110 (1979).
3. J.A. Mangels and G.J. Tennenhouse. Amer. Ceram. Soc. Bull. 59 (12) 1216 (1980).
4. J.W. Eddington, D.J. Rowcliffe and J.L. Henshall. Powder Met. Int. 7 82 (1975).
5. P.E.D. Morgan. This volume.
6. G.R. Terwilliger and F.F. Lange. J. Mater. Sci. 10 1169 (1975).
7. J.C. Uy. Amer. Ceram. Soc. Bull. 57 (8) 735 (1978).

DISCUSSION

Lumby: How do you propose to control the microstructure of the sintered RBSN?

Evans: Uniformity of the microstructure of the starting material, particularly with regard to pore size distribution, is very important. Larger (20 μm) pores are not easily eliminated. Most commercial RBSN has an α-phase content of 60-85w/o; it would be interesting to take a high β RBSN and try to densify it, but there are two opposing arguments, the proposed contribution of the α-β transformation to the densification driving force, and the tendency to develop exaggerated β-phase grain growth.

SINTERING OF ALUMINIUM NITRIDE WITH LOW OXIDE ADDITION

K.A. Schwetz, H. Knoch and A. Lipp

Elektroschmelzwerk Kempten GmbH,
8960 Kempten, W. Germany.

ABSTRACT

By using single additions of Li_2O, CaO, MgO, MgO-SiO_2 (talc), SiO_2, B_2O_3, NiO, Cr_2O_3 and Y_2O_3 in the amount of 1 wt% to partially oxidized AlN-submicron powder (4-6 wt% O) almost complete densification could be achieved through sintering under nitrogen at comparatively low firing temperatures.

Flexural strength values of 300-500 MN/m^2 could be measured. Optical microscopy, EPMA and X-ray diffraction studies revealed a two-phase structure in the dense compounds, the second phase being an aluminium-oxynitride. It is shown that either equiaxed cubic γ-AlON (spinel) grains or plate-like rhombohedral 27R-AlN polytype grains ($Al_9O_3N_7$ or $Si_{6-x}Al_{12+x}O_xN_{20-x}$) of high aspect ratio are formed, dependent on the additive used. The different microstructures did not seem to have a marked effect on the room temperature strength of the oxynitride-bonded AlN ceramics studied.

1. INTRODUCTION

AlN exhibits a number of conspicuous properties. Among these are the excellent high temperature strength and oxidation resistance, the thermal shock resistance, the relatively low density (3.26 g/cm^3 TD), and its resistance against the attack of liquid metals. In order to take advantage of the excellent properties it is necessary to develop pressureless sintering techniques for the fabrication of dense and high strength material. This would offer a more economic way of production as compared with hot-pressing. As a consequence of its mainly covalently bonded structure, pure AlN cannot be sintered to high densities, analogous to the nonoxide ceramics Si_3N_4 (1), SiC (2) and B_4C (3). For pressureless sintering

Riley, F.L. (ed.) Progress in Nitrogen Ceramics

the addition of small or moderate amounts of sintering aids is necessary, which promote densification by increasing the diffusivity in the solid state or by forming liquid phases.

Trontelj and Kolar (4) showed that with 99 AlN-1 Ni (wt.%) powders dense compounds could be sintered, having a bend strength of 300 MN/m^2. The starting AlN powder had a specific surface area of 14 m^2/g. To activate sintering, the metallic phase must be added in a finely divided form. The high strength is discussed to be a result of the formation of elongated AlN grains in the material.

Komeya et al. (5) studied pressureless sintering of AlN containing Y_2O_3. The highest strength values (328 MN/m^2) could be obtained with a composition of 75 AlN-25 Y_2O_3 (wt.%). For hot pressing a SiO_2 addition to the AlN-Y_2O_3 mixtures proved to be advantageous (6), since a fibrous microstructure developed showing a bend strength of up to 720 MN/m^2. In his early papers Komeya discusses the formation of the fibrous microstructure to be a result of the Y_2O_3 addition; later, however, he attributes this behaviour to the SiO_2 (7). X-ray diffraction analysis of the sintered compounds revealed the lines of AlN, $Y_3Al_5O_{12}$ (YAG), and an "Al-Si-O-N" phase, which was later on identified by Jack (8) as the 27R-AlN polytype.

Ochiai et al. (9) recommend Al_2O_3 as well as Al- and Mg-silicates as sintering aids. The Soviet work of Poluboyarinov et al. (10) demonstrated that excellent powders for pressureless sintering could be obtained by planetory ball milling. The powder had a surface area of 14 m^2/g, and final densities of 98% TD could be achieved. Fe impurities were in the range of 0.5 wt.%; however no information had been given in regard to the oxygen content. In a recent study on the physical and thermomechanical properties of such dense AlN ceramics (97-98% AlN) (11), the authors report an average grain size of 1 μm and a room temperature bend strength of 250 MN/m^2. Finally Prochazka and Bobik (12) showed that the addition of 1 wt.% CaO (or SrO) is sufficient to sinter AlN powder (9.2 m^2/g; 1.8 wt.% O) into pore-free compacts. Densification is explained by the formation of a liquid phase in the system Ca-Al-O-N. During cooling $CaAl_4O_7$ precipitates and surrounds the AlN grains as a bonding phase.

In the present work the influence of different oxides (1 wt.% addition) on the sintering behaviour of finely divided AlN of technical purity (spec. surface area > 7 m^2/g) has been studied. Such powders contain as the main impurity 4-6 wt.% oxygen in the form of the hydrolysis product $Al(OH)_3$. Density and bend strength of the sintered compounds have been measured as criteria for the technological quality of the resultant AlN(O) products.

2. EXPERIMENTS

2.1 AlN powder

Some properties of the used powders are summarized in Table 1. The powder labelled FSt is basically a submicron grain fraction and was obtained by air-classifying a -325 mesh production lot of AlN powder. As shown in Figure 1, there is only a very minor fraction of grains exceeding the size of 1 µm.

The fabrication of the powder labelled RWK was done by attrition milling powder - FSt under liquid cyclohexane using tungsten carbide balls. A specific surface area of ~15 m^2/g could be achieved. As a consequence of the intensive milling process the powder was contaminated with 2.9 wt.% tungsten carbide and a final oxygen content of 5.7 wt.% was measured.

2.2 AlN with Oxide Additions (MeO)

If possible, the oxides Li_2O, CaO, MgO, talc, SiO_2, B_2O_3, NiO, Cr_2O_3 and Y_2O_3 were added as submicron powders; in some cases the grain size of the additive was < 5 µm. In order to obtain a homogeneous distribution, the AlN powder (label FSt) and the oxide were milled in the attrition mill under cyclohexane for one hour. After mixing about 1 wt.% of WC and an oxygen content of 5-6 wt.% was analysed in the powder (compare Table 1).

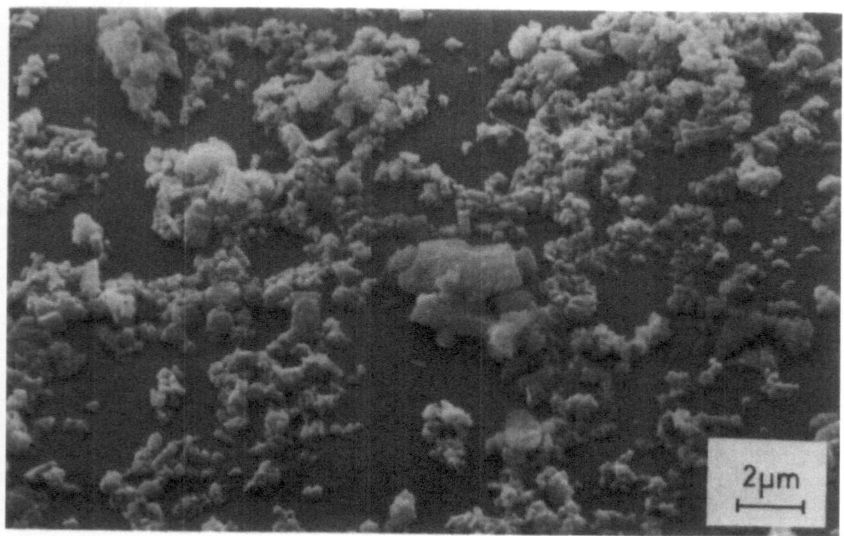

Fig. 1. SEM picture of AlN powder (7.7 m^2/g) obtained from -325 mesh ESK powder after air classification.

Label	Av.particle size FSSS (μm)	Spec.surf. area BET (m^2/g)	Powder density (g/cm^3)	N (wt.%)	O* (wt.%)	metals** (wt.%)
FSt	0.7	7.7	3.19	30.0	4.4	0.2 Fe
RWK	0.6	14.8	3.12	26.1	5.7	0.5 Fe 2.9 WC
F-St +1%MeO	0.7	10.8	3.17	28.5	5.7	0.3 Fe 1.2 WC

* vacuum extraction; ** Mg and Si < 100 ppm.

Table 1. Characterization of AlN and AlN-MeO powders used in this study.

2.3 Sintering Experiments

Green compacts of dimensions 45 x 9 x 9 mm^3 were isostatically pressed in rubber liners without pressing aids at 180 MN/m^2. After presintering the samples for 30 minutes at 800°C under flowing N_2 (0.15 m^3/h) in a SiC-tube furnace, they were exactly machined to dimensions 37 x 7 x 7 mm^3. The bars were finally sintered in a vertical resistance heated graphite-tube furnace under a stationary N_2 pressure of 1 atm. In order to minimise carbon pick-up from the graphite crucible or from CO in the atmosphere, the bars were embedded into coarse AlN powder. The time for heating up to the sintering temperatures of 1600, 1700 and 1800°C was 45 to 60 minutes, the holding time was always 60 minutes, and the furnace cooled down within 90 minutes. 5 bars were sintered for every mixture and temperature. After sintering the linear shrinkage $\Delta L/L$ was determined. Before density (Archimedes principle) and bend strength (RT, 3-point, span width 25 mm, 5 samples) were measured, the bars were ground to their final dimensions of 30 x 5 x 5 mm^3 with a 90 μm diamond grinding wheel and subsequently polished with SiC polishing media to a maximum surface roughness of 0.5 μm. For microstructural investigations strength tested bars were polished with 1 μm diamond paste on a plane perpendicular to the bar axis. Etching to reveal the grain structure was done thermally in air at 1000°C.

3. RESULTS AND DISCUSSION

The single experiments and the results are summarized in Table 2.

3.1 AlN-FSt

The densification of this powder is unsatisfactory, even at high temperatures of ≥ 1800°C (maximum density achieved 3.0 g/cm^3). Weight losses of > 6% are observed. Etched microstructures of samples sintered at ≥ 1800°C exhibit a high volume fraction of platelike grains.

AlN powder	Oxide added	Sintering* temp. (^{o}C)	Density (g/cm^3)	Linear shrinkage ΔL/L	3-pt. bend strength (MN/m^2)
FSt	-	1800	3.00	0.155	293
		1700	2.59	0.104	163
		1600	2.28	0.091	129
RWK	-	1800	3.34	0.179	291
		1700	3.39	0.176	298
		1600	3.00	0.123	180
FSt	1% CaO	1800	3.28	0.164	285
		1700	3.32	0.190	485
		1600	2.85	0.155	185
FSt	1% Y_2O_3	1800	3.30	0.192	310
		1700	3.30	0.200	300
		1600	3.20	0.176	190
FSt	1% NiO	1800	3.21	0.164	320
		1700	3.15	0.150	310
		1600	2.56	0.114	160
FSt	1% B_2O_3	1800	3.14	0.172	321
		1700	3.09	0.178	228
		1600	2.47	0.134	152
FSt	1% Li_2O	1800	3.24	0.152	272
		1700	3.05	0.129	160
FSt	1% talc	1800	3.11	0.175	355
		1700	3.07	0.153	239
FSt	1% MgO	1800	3.07	0.168	303
		1700	2.99	0.151	196
FSt	1% SiO_2	1800	3.16	0.164	351
		1700	3.06	0.144	287
FSt	1% Y_2O_3+	1800	3.37	0.187	287
	5% SiO_2	1700	3.38	0.196	461
FSt	1% Cr_2O_3	1800	3.37	0.168	369
		1700	3.21	0.158	341

*60 mins hold, nitrogen atmosphere.

Table 2. Summary of sintering experiments and results on the AlN(O) compounds.

3.2 AlN-RWK

With this powder (14.8 m^2/g) pore-free bodies could be obtained at sintering temperatures of $\geqslant$ 1700°C. The 1700°C samples show equiaxed grains under the optical microscope, whereas the 1800°C samples contain dark grey plate-like grains (2 µm thick, 5-20 µm side length), which are embedded in a light grey AlN matrix. A small volume fraction of a metal-like phase can be observed in the form of white spots ($\leqslant$ 1 µm), which disappear on etching. By electron probe microanalysis (EPMA) and X-ray diffraction the plate-like crystallites were identified as 27R-AlN polytype. Energy dispersive X-ray analysis revealed the white spots to be enrichments of iron and tungsten, i.e. they are impurity concentrations resulting from the intensive attrition milling, compare Table 1. Due to the oxygen content of 4.5 wt.% of the sintered compound and the formula $Al_9O_3N_7$ for the 27R polytype, the volume fraction of the oxynitride phase can be calculated to be about 36%.

3.3 AlN-FSt with Metal Oxide Additions

3.3.1 MgO, talc (Mg-silicate), B_2O_3, SiO_2, NiO, Cr_2O_3, Y_2O_3+SiO_2

The above additives form a two-phase structure (comp. Fig.2, C+D) consisting of AlN and the 27R polytype, as described in the preceding section. With the exception of the Cr_2O_3 and the $1Y_2O_3$+ $5SiO_2$ additions, only densities below 3.21 g/cm^3 and bend strength values below 350 MN/m^2 are obtained. The pore-free samples, sintered at 1700°C with $1Y_2O_3/5SiO_2$, showed a surprisingly high bend strength of 461 MN/m^2. In this case the interlocked plate crystals of the 27R-AlN polytype belong to the Si-Al-O-N system, and following Jack (13) their stoichiometric composition can be formulated $Si_{6-x}Al_{12+x}O_xN_{20-x}$.

3.3.2 Li_2O, CaO and Y_2O_3. Pore-free, ultrafine grained compounds are achieved at 1700°C with the addition of 1% CaO and 1% Y_2O_3. Adding 1% Li_2O 1800°C is necessary to obtain full densification. Again, a two-phase structure is observed, consisting of AlN and the oxynitride; however, in this case the aluminium oxynitride appears in the form of the cubic spinel, $5AlN.9Al_2O_3$ (14), with an equiaxed grain morphology (dark grey crystals, size 1-5 µm, Fig. 2A). In contrast to the 27R polytype containing microstructure, mainly intergranular fracture is observed. Remarkable is the high bend strength of 485 MN/m^2 for the material sintered with CaO at 1700°C. Chemical analysis of these samples resulted in an oxygen content of 5 wt.%, corresponding to 13% spinel, and a 0.5% Ca content. According to the EPMA results Ca was homogeneously distributed and could not be detected as a second phase by X-ray analysis, as reported by Prochazka (12). Only for higher CaO additions (up to 5 wt.%) we find a calcium aluminate phase on the cost of the AlN. Al_2O_3 spinel phase.

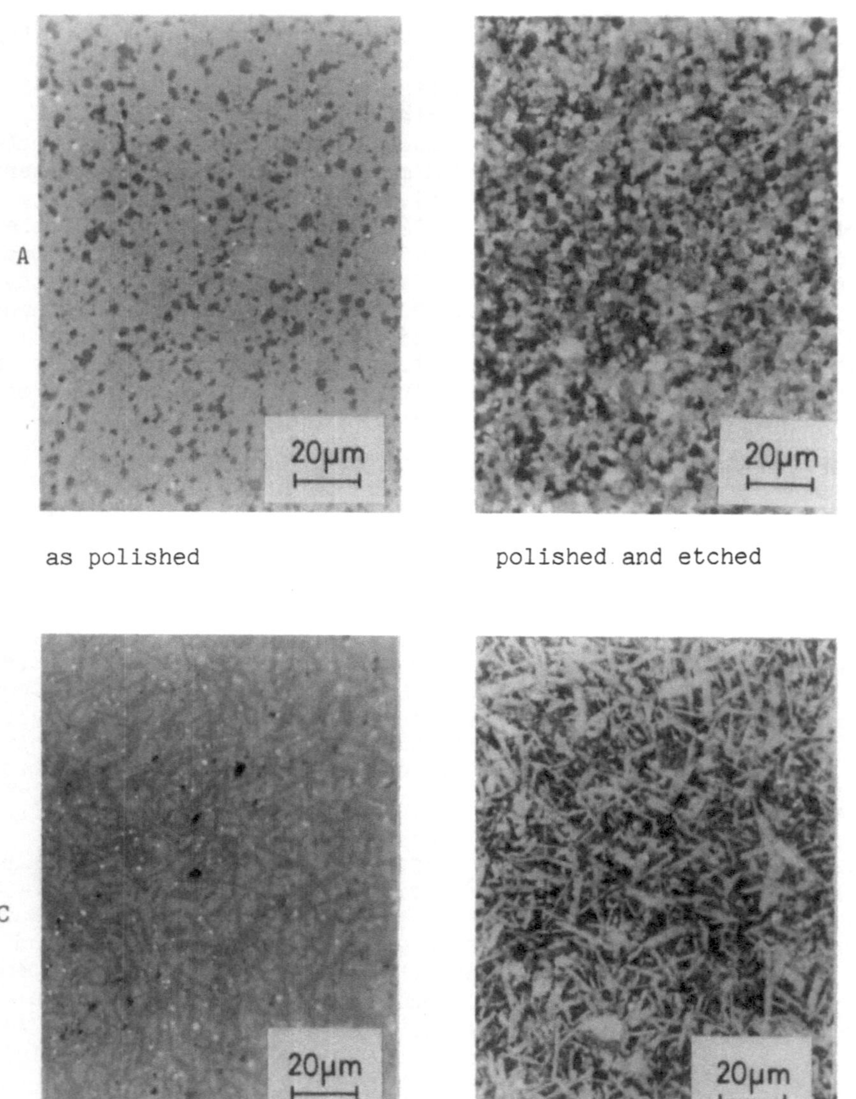

Figure 2. Microstructures of sintered oxynitride-bonded AlN.

A+B: AlN (10.8 m^2/g) + 1% CaO, 60 min 1700°C density 3.32 g/cm^3, σ_B = 485 MN/m^2.

C+D: AlN (10.8 m^2/g) + 1% Cr_2O_3, 60 min 1800°C density 3.27 g/cm^3, σ_B = 369 MN/m^2.

Sintering behaviour and the resulting grain morphology at high temperatures (rounded AlN grains next to the binder phase) suggest a liquid phase sintering mechanism, where the liquid is formed in the respective Me-Al-O-N system. A comparison between the compounds containing the 27R-AlN polytype or the AlON-spinel shows that the microstructure with the plate-like grains favours transgranular fracture modes, whereas intergranular fracture dominates, when equiaxed microstructures are observed. For both grain morphologies comparable bend strength values are obtained at room temperature.

LITERATURE

1) C. Greskovich, S. Prochazka, J.H. Rossolowski, in Nitrogen Ceramics, p.351, F.L. Riley Ed., Noordhoff, Leyden (1977).
2) S. Prochazka, in Special Ceramics 6, p.171, P. Popper Ed., Manchester (1975).
3) K.A. Schwetz and W. Grellner, "The Influence of Carbon on the Microstructure and Mechanical Properties of S-B_4C", paper presented at the 7th Symp. on Boron, Borides and Related Compounds, June 1981, Uppsala, Sweden, to be published in J. Less Common Metals (1982).
4) M. Trontelj and D. Kolar, in Special Ceramics 6, p.39, P. Popper Ed., Manchester (1975).
5) K. Komeya and H. Inoue, Trans. J. Brit. Ceram. Soc. 70 (1971), No. 3, p.107.
6) K. Komeya and F. Noda, SAE 740237 (1975).
7) K. Komeya, H. Inoue and A. Tsuge, J. Am. Ceram. Soc. 57 (1974) p.411.
8) K.H. Jack, in Nitrogen Ceramics, p.109, F.L. Riley Ed., Noordhoff, Leyden (1977).
9) T. Ochiai, M. Hirano and H. Inoue, DT-OS 2320887 (1973).
10) D.N. Poluboyarinov, M.R. Gordova, I.G. Kuznetsova and N.V. Zakharov, Neorgan. Mat. 11 (1975), p.72.
11) D.N. Poluboyarinov, M.R. Gordova, I.G. Kuznetsova, M.D. Bershadskaya and V.G. Avetikov, Neorgan. Mat. 15 (1979) p.2055.
12) S. Prochazka and C.F. Bobik, in Sintering Processes, p.321, G.C. Kuczynski Ed., Plenum, N.Y. and London (1980).
13) K.H. Jack, J. Mat. Sci. 11 (1976), p.1135.
14) J.W. McCauley and N.D. Corbin, J. Am. Ceram. Soc. 62 (1979), p.476.

STRESS ASSISTED HOT FORMATION OF CERAMICS

A. Mocellin

Laboratoire de céramique
Ecole Polytechnique Fédérale
34, chemin de Bellerive
CH - 1007 Lausanne

ABSTRACT

Brief accounts are given of current understanding and recent work in stress assisted hot-formation of ceramics. Attention is focused in turn to: i) fundamental densification mechanisms in powder compacts under load (particle and grain rearrangements, plastic deformation and creep, diffusional atomic transport); ii) conventional pressure sintering (either uniaxial or isostatic hot-pressing); iii) complex formation under stress (reactive pressure sintering and deformation processing). Concluding remarks outline a few problem areas where more work would seem desirable.

I. INTRODUCTION

In contrast with metals, relatively few ceramic materials are being industrially produced through the use of elevated temperature shaping or forming operations. There are good reasons, mostly related to high fabrication costs and technological difficulties, to explain such a situation. But there have also been indications in recent years that conventional hot-pressing and more complex stress-assisted ceramic processing will attract more and more interest in the future. This is especially so for materials of critical importance for the electronic industries or for energy conservation, with non oxides such as silicon nitride and related ceramic alloys occupying a key-position.

The principal reason why hot consolidation and shaping appear

Riley, F.L. (ed.) Progress in Nitrogen Ceramics

attractive is related to their recognized potentials for producing fully dense polycrystals of a very fine uniform grain size. Such microstructural characteristics for example are usually associated with superior mechanical strength. Also, some significant results have been obtained at various places, demonstrating the feasibility of achieving both substantial geometric changes and controlled grain orientation or texturing through so-called press-forging operations, in much the same way as is commonly practiced with metals.

Several reviews have been published (1-5) which summarize modern work in ceramic hot-forming and -pressing, up to the years ~ 1975/1976. They outline major directions of effort in theory, practice and materials applications. We shall thus limit our purpose here to briefly describing work and advances that have taken place in the last five years or so, i.e. since the first "Nitrogen Ceramics" Conference. The first part of the paper will be concerned with mechanisms and theoretical models. The basic phenomena occuring in powders submitted to heat-treatments coincident with the application of some external load will be recalled qualitatively, and a few current problem areas will be mentioned. For obvious simplicity reasons, most treatments of the subject have been restricted to either one-component or one-phase systems. This is rarely the case in practical situations because of effects associated with residual impurities of the starting raw materials or interactions with the gas phase in contact with the powder compacts. But the assumption is useful, in particular as a help for setting up some ideal reference scheme against which such effects may be evaluated.

The second part of the paper will deal with conventional pressure sintering, either uniaxial or isostatic. A short account will be given of the recently published literature, as it may appear to the "outside observer". That is to say for example that no attempt will be made at including in this review such information which may only be found in contract reports and other such documents.

In the third part of the paper, more complex processes will be dealt with: reactive hot-pressing and press-forging. In the former case, the aim is to achieve densification under load while pursuing the chemical synthesis of the ultimate ceramic composition. A typical example is provided by silicon nitride base ceramic alloys, sometimes referred to as "Sialons" . A variant of reactive hot-pressing which will also be mentioned, may be found to occur in systems undergoing a phase transformation, during which increased densification rates can be observed. Finally, some recent advances in the deformation processing of ceramics, i.e. post densification treatments carried under stress, will be brought to attention.

II. DENSIFICATION MECHANISMS

In the present practice, sintering under load frequently involves uniaxial pressing a loose or precompacted powder with high strength tooling: typically graphite dies and plungers. Loading pressures are usually around 40 MPa, and are applied at temperatures several hundred °C below normal sintering temperatures, for times ranging from a few minutes to a few hours. The operation is conducted in a chamber under either vacuum or some protective atmosphere and is thus discontinuous: an a priori handicap for industrial production. In the absence of chemical or phase transformations, the powdered system undergoes essentially geometrical changes aimed at reducing and eliminating its built-in porosity, thus leading to a dense polycrystal.

Three classes of phenomena may a priori be claimed to operate either individually or in combination, either successively or simultaneously. Depending upon mechanistic detail, they can be subdivided further and one objective of research in the field has been to identify and characterize them through theoretically derived rate equations that could model their effects.

1. Particle and grain rearrangements

Early attempts to establish theoretical expressions for hot-pressing have often considered the powder under compaction as a continuum. This may be justified to some extent from a macroscopic point of view and phenomenological rheological equations may be produced from case to case. They have usually consisted in analytical expressions depending upon two or three adjustable parameters, chosen to fit experimental densification curves. Obviously, such an approach is by itself unable to answer questions pertaining to the mechanisms involved. But it may be of value in specific cases (6) or on a relative basis.

On a microscopic (i.e. single particle) scale, the first mechanism to bring densification upon application of a load is the displacement of particles with respect to one another, by sliding along their contact areas. Depending upon peculiarities of: local stress fields, geometry and interfacial properties, the ease and extent of these particle rearrangements may vary. They do not seem to have been studied in detail yet for systems submitted to sintering treatments under load, although the room temperature conventional die-pressing of ceramic powders has been the object of much renewed interest in recent years (7). Presumably, it would be rather difficult to devise practical conditions where particle rearrangement as it takes place in cold pressing would be the only active mechanism. Due for example to the occurrence of diffusional processes, neck growth would tend to take place concurrently. As this develops, it may be expected that the response of the system will tend more and more to resemble that observed in deformation via grain boundary sliding, such as may occur in superplastic polycrystals (8). It could be interesting then, to look for ways of adapting some of the classical strain rate equations for superplasticity, to the geometry of highly porous compacts and compare their predictions with experiment. A first attempt along this line has been made (9), based on the simplified geometry of a hexagonal grain array, with pores sitting at all multiple junctions.

An alternative approach to rearrangement kinetics (10) has been based on the empirical assumption that the rearrangement rate is proportional to the amount of densification still to be achieved by rearrangement, at any instant within a given compact. Experimental evidence in the $ZrSiO_4 - Al_2O_3$ system also suggests that this assumption is not unrealistic (10).

2. Plastic deformation and creep

Dislocation related processes can also contribute to densification. The same fundamental mechanisms as have been found to operate in dense polycrystals submitted to conventional testing can take place in a powder compact. The derivation of theoretical equations to quantify their effects is, however, much more difficult because:

i) geometrical modelling of the microstructure (solid versus pore phases) is not easy and the microstructure itself is in rapid evolution: assuming steady state geometries may be subject to criticism. Modern automatic image analyzing systems and recent progress made in the theories of mathematical morphology, however, have begun to offer promise for significant progress (11 - 13).

ii) only the total load applied to the system is known. No direct information is available on the average state of stress which prevails locally within the compact. Neither is it known how this distributes over the areas of individual necks. An effective stress expression, therefore, must be postulated on more or less solid grounds in formulating strain rate or densification equations.

A classical approach has been to greatly simplify the actual geometry of the compact and derive functional relations in which estimates of the load bearing cross-section area and/or porosity are explicit parameters (9; 14). But since the analytical tractability of the problem is restricted to rather ideal cases, the need arises of some experimental means for estimating effective stresses (15) which also should account for die confinement effects.

Keeping in mind these yet incompletely solved difficulties,

theoretical densification rates can be derived from the conventional constitutive equations for plastic deformation or creep. Indeed, there have even been indications (16) that dislocation arrays may be seen nearby pores, in ceramics undergoing sintering. Plastic flow, following elastic deformation should produce finite contact areas between particles as well as their elongation within rather short times after the load has been applied. Its relative importance in the overall process, however, remains to be more systematically evaluated and experimentally documented.

Such is also the case for power law creep ($\dot{\varepsilon} = B\,\sigma^n$ where σ and $\dot{\varepsilon}$ are effective stress and strain rate respectively, whereas B and n are material constants), which may a priori be suspected to be significant at moderately high temperatures and effective stresses, rather typical of the state of affairs during most of the intermediate stage of hot-pressing. It has been shown (17) that a neck growth rate equation for power law creep, based on an extension of Hertz's contact problem to non linear deformations could be derived and yielded better agreement with some experimental observations than a previously suggested model (18).

3. Diffusional atomic transport

Individual atom or ion migration provides the basis for a third class of mechanisms in hot-pressing. This is also the ultimate scale, after the microscopic (i.e. particle rearrangements) and submicroscopic ones (i.e. cooperative atomic displacements needed in all dislocation processes), which requires consideration. Generally speaking, diffusional mechanisms can superimpose their contributions to the previously mentioned ones and they have often been considered, more or less explicitely, to play a dominant role in the overall densification kinetics; particularly so during the later stages of the process.

An important point should be emphasized here, which has been strikingly phrased in Kuczynski's definition of sintering as: "diffusional creep under the action of capillary forces" (19). In other words, no fundamental distinction needs be made between stress induced matter transport (Nabarro-Herring or Coble creep) and that resulting from the presence of curved interfaces in the densifying compact. In either case, the driving force arises from chemical potential changes with position which, for a single component system can be expressed simply as follows:

$$\mu - \mu_o = \gamma_{sv}\Omega K + \sigma\Omega \qquad \text{where}$$

μ_o is the reference chemical potential at zero curvature and stress

Ω is the atomic volume
γ_{SV} is the solid-vapor specific surface energy
$K(x; y; z)$ and $\sigma(x; y; z)$ are the local interface curvature
$(K = \frac{1}{R_1} + \frac{1}{R_2})$ and stress respectively.

In pressureless sintering, obviously $\sigma = 0$ and the above equation clearly suggests that much of the current understanding of diffusion controlled sintering is also applicable to hot-pressing. Consequently, no attempt is made in this paper to give a more detailed account of this particular aspect of the matter.

Another important point to be noted is that the fluxes of matter which are triggered by curvature differences or stress, may bring about either densification or coarsening of the microstructure, without much shrinkage of the compact. Such a competition between both these effects is due to:

i) the existence of several mechanisms for matter transport. If an atom is transferred from one position to another both located on the solid vapor interface for example, then a smoothening of that interface may result, but no shrinkage.

ii) the formation and subsequent evolution of the grain boundary network which provides high diffusivity paths and sites for vacancies or other point defects emission or absorption. In fact, the above expression for the chemical potential does not give a fully adequate picture since the energy contribution of grain boundaries is not taken into account. The interactions between grain boundary-driven and surface-driven phenomena, remain insufficiently understood.

4. Modelling and mapping

Following similar representations for deformation or sintering mechanisms which have now become popular, the concept of "pressure sintering maps" was introduced several years ago (20). It has since received some theoretical developments focusing on either the initial (14) or intermediate and final (21) stages of the process. But up to now, not many complete maps for specific ceramic systems have been published, one reason being that insufficient information about their corresponding deformation and creep behaviours has been generated.

A pressure sintering map aims at graphically representing the theoretical densification kinetics of a particular powder as they depend upon internal (i.e. material dependent) parameters such as particle size or relative density, and external variables such as temperature or applied pressure. It is based first on the set of rate equations assumed to describe the density changes which each of all conceivable elementary mechanisms would produce. As such, a map can only be as adequate to reality as are the models which it makes use of. But the major hypothesis on which all such maps have been assumed to rest is that every elementary mechanism acts independently of the others in such a way that they all add their effects and the net result is the sum of all individual contributions:

$$\dot{\rho} = \sum_i \dot{\rho}_i \qquad \text{or} \qquad \dot{x} = \sum_i \dot{x}_i \qquad \text{where:}$$

$\dot{x}$ and $\dot{\rho}$ are neck growth rate and densification rate respectively, and the summation signs refer to all a priori active mechanisms. Within an appropriate range of experimental conditions, one of the mechanisms usually is thought to dominate. To a reasonable approximation then, the contributions of all the others may be neglected. It follows that on graphical plots of density (or neck size) versus temperature or applied stress (all other parameters being fixed) the relevant mechanisms can be associated with more or less extended fields. The field boundaries are determined by solving equations of the type: $\dot{\rho}_k = \dot{\rho}_l$ with $k \neq l$, which delineate conditions under which either of the "k" and "l" mechanism contributes 50% to the overall process.

It seems useful at this stage to make a few comments about some important limitations to the present theoretical framework and understanding of pressure sintering and to suggest areas where more research efforts might be desirable. Reference has previously been made to the complex geometry of a densifying powder compact. Besides improvements in the modelling of microstructure itself, it would be necessary to adapt the available models to allow for the gradual evolution of more morphological parameters than presently accounted for. It is well realized for example that the transition between the intermediate and final stages of densification is spread over a significant range of densities (~o.85 to o.95 relative) and times. Explicit account therefore should be made for the fact that the microstructure is some variable mixture of volume elements with an interconnected cylinder-like porosity and zones with closed bubble-like pores. A first attempt to explicitely incorporate this effect in an analytical treatment has been made (21) which awaits some experimental confirmation. Also depending from a better assessment of the geometry is the evaluation of the internal distribution of stresses within the compact, it being well recognized that the identification and rationalization of conditions leading to uniform microstructures are of the utmost practical importance.

On the other hand, it has become generally appreciated in recent years that single process models may be insufficient or even inadequate for describing actual behaviours. Quite frequently sequential mechanisms are involved the effects of which cannot a priori be studied independently in the light of single process theories. Also, the hypothesis of additivity of mechanisms could prove untenable, and synergistic effects might have to be considered explicitly. Examples include:

i) the interdependence of densification and grain growth, the latter phenomenon still not being specifically accounted for in theories of the former one.
ii) such sequences as boundary diffusion - surface diffusion, the latter being necessary to redistribute matter supplied by the former.
iii) the relative contributions of the shear (favoring particle re-arrangement and sliding) and normal (favoring creep and diffusional mechanisms) components of the stress field at the local level.
iv) chemical effects which have been largely overlooked by theoreticians of sintering although their importance is well recognized, even in pressumably single phase systems. For example, the possible changes of specific surface energies due to impurity segregations concurrent with densification or the limitations arising from interface kinetics, have not been much studied.

In summary, considerable effort has been devoted to constructing models for elementary mechanisms of diffusion-controlled sintering and hot-pressing, while such processes as particle or grain rearrangements were receiving less attention than they perhaps deserved in view of their practical importance.

III. CONVENTIONAL PRESSURE SINTERING

1. Uniaxial hot-pressing

The largest fraction of recent literature on the subject is concerned with SILICON NITRIDE and is focused on two major issues:

i) the influence of starting powder characteristics (22).
ii) the role of additives of varying natures and amounts.

It had been recognized previously (23) that the Si_3N_4 $\alpha \rightarrow \beta$ structural transformation is somehow interrelated with the densification mechanism and that a high α content in the starting powder is essential for achieving high performance. It had also been suggested (24) that both processes are controlled by Si_3N_4 solution into and diffusion through a grain boundary phase in MgO-doped powders. Further evidence has since been offered which substantiates this hypothesis (25 - 29) and the concurrent development of microstructure was studied in greater detail. Grain morphology could be correlated with the $\alpha \rightarrow \beta$ transformation (27 - 29), and it has been pointed out that knowing the kinetics and morphology of the transformation, allows an optimum set of hot-pressing conditions to be identified (27 - 28) which for a given powder and MgO level will maximize the low temperature strength of the densified material.

The high temperature mechanical and oxidation behavior in turn is influenced by the composition and amount of the grain boundary phase. A number of studies, therefore, have had the purpose of increasing the refractoriness of the reaction products between the densifying aid and the unavoidable SiO_2 layers at the surface of the starting Si_3N_4 particles. Additives such as Al_2O_3 (30), Be_3N_2 and $BeSiN_2$ (32), CeO_2 (33 - 36), Li_2O (31), Sc_2O_3 (37), Y_2O_3 (30; 31; 35; 36; 38; 39), ZrO_2 (36) have been considered. Some attention has been paid to compounds and/or hot-pressing procedures such that densification could be achieved with only small amounts of added impurities (40). The use of beryllium compounds together with appropriate hot-pressing conditions (32) could result in such a reaction sequence that a transient liquid phase probably was formed and reacted with the starting α-Si_3N_4 to give a β-Si_3N_4 solid solution of approximate composition: $Si_{2.9}\,Be_{0.1}\,N_{3.8}\,O_{0.2}$.

Broadly speaking, it seems that regardless of the type and amount of additives used so far, the pressure sintering of silicon nitride is essentially controlled by the development of some boundary liquid phase which favors particle rearrangements and a solution reprecipitation mechanism, even though this boundary phase might in some cases be very thin (39) and hardly detectable. Further improvements of properties now should rely on a better knowledge of the phase limits and relationships for each multicomponent nitride - oxide system. The available information (41) certainly is far from complete yet, but in the particular case of Y_2O_3 containing compositions for example, concurrent research efforts on phase relationships (42) and material processing have led to significant progress, illustrated for example by the development of new high performance pressureless sinterable compositions (43; 44).

Besides silicon nitride, a number of OTHER MATERIALS have been pressure-sintered in a conventional way. In general, the aim was to generate experimental information on poorly sinterable systems, usually non oxides (45 - 46). For particular application - material combinations, hot-pressing has been recognized as being the only feasible fabrication route (47). Advantage has been taken also of the unique potential of hot-pressing in achieving high density and fine grain size for the preparation and property investigation of special electronic ceramics in the PZT (48) and beta-alumina (49 - 51) systems, which had been under active study for a number of years.

Various, more TECHNOLOGICALLY ORIENTED efforts should be mentioned now as indicative of some trends for future applications. Again, a high proportion have been concerned with process development or optimization studies on silicon nitride. The powder handling and batch composition for example have been systematically adjusted (52 - 54) in order to establish conditions leading to property improvements. Other work was aimed at manufacturing complex components or parts of well defined geometries (55; 56)in particular by "post hot-pressing" previously shaped and consolidated parts of reaction-bonded silicon nitride (57). More generally speaking, technical developments in hot-pressing, some of which were reviewed a few years ago (58), have tried to widen practicable ranges for some important operational parameters such as pressure. Also of concern have been the design of special tooling, the replacement of standard graphite dies and plungers by other high performance materials (59) and the modification of equipment to allow for the fabrication of larger size pieces, or for increased productivity. Results have been reported on novel techniques for high pressure densification by dynamically impacting metallic (60) or non-oxide ceramics (61). Static pressures in the GPa range on the other hand were used to produce ultrafine grain (~200 Å) TaC (62), and to investigate self-bonding in silicon nitride powders (63; 64).

2. Isostatic hot-pressing

Isostatic hot-pressing a priori offers in a single one-step technique, three important technological requirements for an industrially viable fabrication process to yield materials with superior properties. It is intrinsically capable of:

i) delivering net (or near net) shaped parts of complex geometry (e.g. turbine blades, tools, radomes, bearings, ...).

ii) ensuring that uniform densities are obtained and that a number of gross defects, associated for example with non-homogeneous particle packing are absent from the finished parts.

iii) achieving consolidation and densification as a result of the termal treatment.

It has, however, been discontinous and per se, it would obviously not be suited for a whole range of forming operations such as forging or extrusion. It has been available for some time and perhaps may now be considered a conventional pressure sintering method, in particular because the necessary equipment has become available commercially, and because solutions have been found to ease or even eliminate canning or sealing problems which have imposed major drawbacks to the development of this technique. Reviews on isostatic hot compaction have been published (65 - 66). It appears that historically, applications involving metallic (superalloys, tool steels) or mixed (cemented carbides) powders have provided most of the driving force for the development of isostatic hot-pressing. But much interest also has been generated for ceramics as is more extensively discussed elsewhere (67), but briefly examplified by the following recently reported results.

For silicon nitride, it has first been demonstrated (68) that the reaction sintered material prepared with densification aids and placed in a glass encapsulation to seal off its open porosity, could be brought up to 100% theoretical density with concurrent strength improvements. It has also been shown feasible (69) to reach high (>95%) densities, large degrees of $\alpha \rightarrow \beta$ conversion at lower temperatures and also some grain refinement on hot isostatically pressed commercial powders without additives. And more recently, it has been reported (70) that because of differences in the chemistry of the grain boundary phase, such ceramics exhibited higher high temperature strength and creep resistance than any other additive containing silicon nitride.

Oxide ceramics also are amenable to hot isostatic pressing with concurrent significant improvements in their properties and/or better control of the chemistry of the densifying systems. For example, PZT bodies have been produced, with a glass as a pressure transmitting medium (71). Because of the encapsulation, vaporization or interference with the environment were non existent. Also, a number of magnetic (72) and beta-alumina (73) ceramics were brought up to full density, starting from preforms partially sintered to a closed porosity. No encapsulation was necessary then, which brings a significant advantage in practice.

IV. COMPLEX FORMATION UNDER STRESS

It has been implicitly assumed so far that the problem of chemical effects could be reduced to an understanding of the behaviours of residual impurities and their consequences. The approximation may be sufficient for such ceramics as Si_3N_4 or "pure" oxides but more complex situations than have been considered so far may arise and need some discussion.

Attempting to draw some parallel with common metallurgical practice on the other hand, it may a priori be argued that, as has already been the case, applications or needs could appear for:

i) more or less complex ceramic alloys for the formation of which several compounds should somehow react together on a large scale.
ii) special combinations of microstructural features or of component shapes and properties, not readily accessible in a one step operation.

In the former case reactive pressure sintering, and in the latter deformation processing (or thermomechanical treatments) might a priori be considered.

1. Reactive pressure sintering

It has long been known qualitatively that the sintering or hot-pressing behaviour of systems undergoing some sort of crystalline lattice reconstruction could be substantially affected, from either a kinetic or microstructural point of view, or both. The transformation(s) of concern can be:

i) a polymorphic phase change. Included in this case might be transformations from metastable to stable crystalline modifications, such as occur in the lower temperature transitions of alumina toward the corundum structure (74 - 76). Transformational superplasticity appears likely to be involved in such cases.
ii) a thermal decomposition reaction such as the elimination of CO_2 or H_2O from carbonates or hydroxides respectively.
iii) a chemical reaction (or sequence of reactions) aimed at synthesizing a new compound from several starting raw materials. Also included in this category could be the formation of solid solutions by such processes as diffusion alloying or some cases of transient phase sintering.

Type iii) cases perhaps are of greater economic significance and, depending upon the kinds of chemical reaction(s) involved, various situations may be encountered and have been studied. For example, transparent magnesium aluminate spinel has been obtained (77) according to the $MgO + Al_2O_3 \rightarrow MgAl_2O_4$ reaction, and it was pointed out that difficulties with the overall molar volume expansion (+ 7.9%) could be reduced with operation pressures $\geqslant$ 33 MPa. But in general, an important problem arises when interpreting reactive pressure sintering experiments: the discrimination between contributions from the chemical reaction and from pore eliminating mechanisms to the overall density changes, and the identification of successive stages in the process. Although no general solution has been offered yet, a simple way of graphically representing reaction rate versus experimental densification rate has been proposed (78) and used for investigating the reactive hot pressing system (79):
$2\ ZrSiO_4 + 3\ Al_2O_3 \rightarrow 2\ ZrO_2 + 2\ SiO_2.3\ Al_2O_3$.

When more than two starting reactants are put in contact, things may become more complex as is examplified by the reaction hot-pressing in the Si - Al - O - N and similar systems. Early work on Si_3N_4 - Al_2O_3 (80; 81) showed that pore-free and β' containing materials could be prepared, but with secondary phases which would impair their properties. Subsequent studies, started from $Si_3N_4 + AlN + SiO_2$ (+ 1% MgO) (82) and from $Si_3N_4 + AlN + Al_2O_3$ (+ ~3.5% SiO_2) (83 - 85) mixtures. Broadly speaking, depending upon compositions and operating conditions, two different types of behaviours have been observed in high Si_3N_4 base compositions. At low overall $^O/_N$ ratios, densification either is very limited or is substantially delayed, while nonetheless some β'-solution is formed (84). It was argued that surface diffusion and/or vapor transport could be dominant mechanisms. For $^O/_N$ ratios exceeding those in theoretically pure β'-Si_3N_4 solution grain boundary liquids form in amounts which increase with the overall $^O/_N$ ratio, temperature or MgO additives (82). As a result, the overall reactive hot-pressing behaviour is in first approximation claimed to be qualitatively similar to that observed in hot-pressing Si_3N_4: firstly fast particle rearrangement upon initial liquid formation; secondly further densification (concurrent with β' solid solution formation) via a "Coble creep type" mechanism (83 - 84). The latter actually consists in two sequential sub-mechanisms with the slower being rate controlling, namely grain boundary sliding (at low applied loads and/or liquid contents) and the diffusion step of a solution - diffusion - precipitation process (for higher liquid contents and/or loads) (85). For the more highly substituted β'-Si_3N_4 compositions, the situation is considerably more complex as suggested by the very limited information published so far (86).

2. Deformation processing of ceramics

The use of externally applied stresses, finally offers a whole range of possibilities for the hot-forming of ceramics with markedly improved or unusually combined properties. Although not extensively investigated experimentally, such principle possibilities have long been recognized and a brief account has recently been published (5) of some significant advances in this area. In particular, for a number of magnetic and electronic oxides, there has been interest for possibly more highly textured polycrystals than are obtainable with common techniques such as field alignment of powder slurries. Several studies therefore have recently been undertaken to press-forge such ceramics, starting from predensified slugs and submitting them to a compressive creep type of test. Investigated systems include barium and strontium ferrites (87 - 88) and a number of ferroelectric bismuth compounds with layer structures (89 - 91). Generally speaking, evidence is provided that physical properties can be significantly improved as a result of more structurally and/or crystallographically textured microstructures. It is concurrently demonstrated that substantial deformations (up to -1 or more true strains) can be reached by selecting appropriate combinations of hot-forging conditions. Although the deformation mechanisms which were operating are not discussed in detail, it may be suspected that grain boundary sliding and grain rearrangement are likely to have contributed to the overall strains, particularly in cases of pre-hot-pressed fine grained materials (e.g. 90 - 91).

Large compressive strains have also been reported recently for several fine grained highly dense oxides. So far, values of about - o.8 for MgO (8), - o.7 for UO_2 (92 - 93) and - o.5 for Al_2O_3 (94) have been reached. In all these cases, grain rearrangement according to an Ashby-Verall (95) or similar model was claimed to have been the dominant deformation mode. In such a process an important role is being played by those residual pores located at multiple grain junctions, in connection with the built-up of stress incompatibilities there. It is qualitatively recognized (93) that the overall ductility depends upon the size and amount of intergranular pores when diffusional or dislocational accommodation cannot take place fast enough, given the imposed strain rates.

It now clearly appears that structural superplastic behaviour is not restricted to metallic materials. Neither is transformational superplasticity (96). It is the author's opinion that the possibilities of superplasticity as a viable shaping technique for special high performance ceramics could receive serious consideration in the future, provided difficulties associated with cavity formation and interlinkage can be overcome (97). Under such circumstances, conventional hot-pressing should remain a choice fabrication technique because of its ability to deliver materials with a fairly controllable fine grained structure, the other major requirement for practical superplastic behaviour.

V. CONCLUDING REMARKS

This review of recent work on stress assisted hot formation of ceramics has identified several areas where either rapid development has been experienced in recent years or more systematic research would appear desirable in the future. In particular:

1. Advances have been made for example in high temperature tooling, particularly for rendering hot isostatic pressing more practicable.
2. Complex processes such as reactive hot-pressing or deformation processing are either receiving renewed attention in relation to special ceramic systems, or generating interest as a result of some promising preliminary work.
3. The importance of estimating local stress states within a densifying compact and evaluating die confinement effects has been pointed out.

4. So has the need for more refined geometrical modelling of microstructures and of their gradual changes with progress of densification: assuming steady state structures with respect to all but one morphological parameters may lead to inconsistencies.

5. Chemical effects which may be of importance in determining basic parameters (e.g. changes in surface energies with impurity segregation) or which may be rate controlling, also ought to receive more attention from theoreticians, it being recognized that here as in the previous two items the task is rather formidable.

6. The major importance of sequential or coupled elementary processes is now more generally appreciated. It has been re-emphasized here in the light of several examples. It follows that:
   i) some caution should be exercized when interpreting experimental results in the light of single process theories only.
   ii) theoretical models, possibly allowing for synergistic effects, would be needed which would more systematically treat pressure sintering as a series (or sequence) of coupled elementary mechanisms.

7. Finally, the mechanism of rearrangement via sliding at interfaces under the influence of local shear forces clearly appears to be of central importance for most if not all situations of processing ceramics from powdered raw materials.

REFERENCES

1. M.H.Leipold, "Hot-pressing" in Treatise on Materials Science and Technology, Vol. 9 (F.F.Y.Wang Ed[r];Academic Press, 1976, 95-134)

2. T.Vasilos, "Densification of nitrides by hot-pressing" in Nitrogen Ceramics (F.L.Riley Ed[r]; Noordhoff, Leyden, 1977, 367-382)

3. M.R.Notis, "Advances in ceramic hot forming and pressing: theory and practice", Ceramurgia International 3 (1977) 3 - 9

4. M.R.Notis and R.M.Spriggs, "Emerging areas in hot-pressing", in Sintering, new developments (N.N.Ristić Ed[r]; Elsevier, 1979, 295-306)

5. R.C.Bradt, "Recent advances in the deformation processing of ceramics", in Adv. in Deformation Processing (J.J.Burke and V.Weiss Ed[rs]; Plenum Press, 1978, 405-423)

6. R.L.Brown, "A volumetric constitutive law for snow based on a neck growth model", J. Appl. Phys. 51 (1980) 161-165

7. A.Broese van Groenou, "Pressing of ceramic powders: a review of recent work", Powder Metall. Int. 10 (1978) 206-212

8. J.Crampon and B.Escaig, "Mechanical properties of fine-grained MgO at large compressive strains", J.Amer.Ceram.Soc. 63 (1980) 680-686

9. W.Beere, "The role of sequential processes in hot-pressing of Cu/Al_2O_3", Acta Metall. 24 (1976) 277-283

10. E.di Rupo, M.R.Anseau and R.J.Brook, "Particle rearrangement kinetics", J. Amer. Ceram. Soc. 62 (1979) 531-532

11. J.L.Chermant et al., "Morphological analysis of sintering", J. of Microscopy 121 (1981) 89-98

12. E.H.Aigeltinger and H.E.Exner, "Stereological characterization of the interaction between interfaces and its application to the sintering process", Metall. Trans. 8 A (1977) 421-424

13. A.Miro and M.R.Notis, "Quantitative image analysis of microstructure development during pressure sintering of CoO", in Sintering processes (G.C.Kuczynski Edr; Plenum Press, 1980, 457-469)

14. J.R.Matthews, "The initial sintering and hot-pressing in vibrocompacted fuels", J. Nucl. Mater. 87 (1979) 356-366

15. P.S.Gilman and G.H.Gessinger, "A method for the experimental determination of the effective stress in hot-pressing", Powder Metall. Int. 12 (1980) 38-40

16. L.Ogbuji et al., "Plastic deformation during the intermediate stages of sintering"; Ref. 1, pp 135-140

17. J.R.Matthews, "Indentation hardness and hot-pressing", Acta Metall. 28 (1980) 311-318

18. D.S.Wilkinson and M.F.Ashby, "Pressure sintering by power law creep", Acta Metall. 23 (1975) 1277-1285

19. H.H.Hausner and D.Duzevič, "Definition of the term: sintering", Science of Sintering 13 (1981) 1-6

20. D.S.Wilkinson and M.F.Ashby, "The development of pressure sintering maps", in Materials Science Research Vol. 10 (G.C.Kuczynski Edr; Plenum Press, 1976, 473-492)

21. M.R.Notis and P.Wingert, "Densification mapping for the intermediate and final stages hot pressing", Science of sintering 10 (1978) 35-44

22. W.Engel, "Starting powder requirements for hot pressing of silicon nitride", Powder Metall. Int. 10 (1978) 124-127

23. G.R.Terwilliger and F.F.Lange, "Hot-pressing behaviour of Si_3N_4", J.Amer. Ceram. Soc. 57 (1974) 25-29

24. R.J.Brook et al., "Mass transport in the hot pressing of α silicon nitride", Ref. 2, pp. 383-390

25. L.J.Bowen et al., "Hot pressing and the α - β phase transformation in silicon nitride", J.Mater. Science 13 (1978) 341-350

26. D.R.Messier et al., "The α/β silicon nitride phase transformation", ibid., pp.1119-1205

27. H.Knoch and G.E.Gazza, "On the α to β phase transformation and grain growth during hot-pressing of Si_3N_4 containing MgO", Ceram. Intern. 6 (1980) 51-56

28. H.Knoch and G.Ziegler, "Influence of MgO content and temperature on transformation kinetics, grain structure and mechanical properties of hot pressed silicon nitride", Science of ceramics 9 (1977) 494-501

29. J.L.Iskoe and F.F.Lange, "Development of microstructure and mechanical properties during hot pressing of Si_3N_4", in Ceramic Microstructures '76 (R.M.Fulrath and J.A.Pask Ed[rs]; Westview Press, 1977, 669-678)

30. A.Tsuge and K.Nishida, "High strength hot pressed Si_3N_4 with concurrent Y_2O_3 and Al_2O_3 additions", Am.Ceram.Bull. 57 (1978) 424-431

31. L.J.Bowen et al., "Hot-pressing of Si_3N_4 with Y_2O_3 and Li_2O as additives". J. Amer. Ceram. Soc. 61 (1978) 335-339

32. C.Greskovich, "Hot-pressed β-Si_3N_4 containing small amounts of Be and O in solid solution", J.Mater.Science 14 (1979) 2427-2438

33. J.P. Guha et al., "Hot-pressing and oxidation behavior of silicon nitride with ceria additive". J.Amer.Ceram.Soc. 63 (1980) 119-120

34. G.N.Babini et al., "Hot pressing of silicon nitride with ceria additions", Ceram. Intern. 6 (1980) 91-98

35. Tai-Il Mah et al., "Characterization and properties of hot-pressed Si_3N_4 with alkoxy-derived CeO_2 or Y_2O_3 as sintering aids", Amer. Ceram. Soc. Bull. 58 (1979) 840-844

36. A.W.J.M. Rae et al., "The role of additives in the densification of nitrogen ceramics", in Ceramics for high performance applications - II (J.J.Burke et al. Ed[rs]; Brook-Hill Pub.Co.(1978) 1039-1067)

37. J.Dodsworth and D.P.Thompson, "The role of scandia in the densification of nitrogen ceramics", in Special ceramics 7 (D.Taylor and P.Popper Ed[rs], 1981, 51-61)

38. G.E.Gazza et al., "Hot pressed Si_3N_4 with improved thermal stability", Amer. Ceram. Soc. Bull. 57 (1978) 1059-1060

39. D.R.Clarke and G.Thomas, "Microstructure of Y_2O_3 fluxed hot-pressed silicon nitride". J.Amer.Ceram.Soc. 61 (1978) 114-118

40. R.Becker and F.Thümmler, "Hot pressed silicon nitride with very low amounts of additives", in Energy and ceramics, P.Vincenzini Ed[r]; Elsevier, 1980, 610-619

41. F.F.Lange, "Silicon nitride polyphase systems: fabrication, microstrucutre and properties", Int.Metals Reviews No. 247 (1980)1-20

42. F.F.Lange, "Phase relations and stability studies in the Si_3N_4-SiO_2-Y_2O_3 pseudoternary system", J. Amer. Ceram. Soc. 60 (1977), 249-252

43. J.T.Smith and C.L.Quackenbush, "Phase effects in Si_3N_4 containing Y_2O_3 or CeO_2", Amer. Ceram. Soc. Bull. 59 (1980) 529-537

44. M.H.Lewis, R.J.Lumby and E.Butler, "Lucas Syalons: Composition, structure, properties and fabrications" - this conference

45. a) K.Takatori et al., "Pressure sintering of highly dense and pure non oxide materials", Ref. 40, pp. 525-533

45. b) M.Beauvy and R.Augers, "Mechanisms of hot pressing of boron carbide powders", in Science of ceramics 10, H.Hausner Ed[r], 1980, pp. 279-286; L.B.Ekbom and C.O.Amundin, ibid., pp. 303-310

46. F.F.Wang et al., "Hot pressing of silicon", Ref. 13, pp.289-294

47. J.B.Ainscough at al., "The hot pressing of cubic europia", Ref. 40, pp. 847-856

48. V.L.Balkevich and C.M.Flidlider, "Hot pressing of some piezo-electric ceramics in the PZT system", in Adv. in Ceramics Processing, P.Vincenzini Ed[r], 3rd CIMTEC Meeting, RIMINI, Italy, May 1977, 155-161

49. G.N.Babini et al., "Densification kinetics of vacuum hot-pressed sodium beta-aluminas", Ceram. Int. 3 (1977) 147-151

50. W.J.McDonough et al., "Hot pressing and physical properties of Na beta alumina", J. Mater. Science 13 (1978) 2403-2412

51. A.V.Virkar et al., "Hot-pressing of lithia-stabilized β"-alumina", Ceram. Int. 5 (1979) 66-69

52. G.Q.Weaver and J.W.Lucek, "Optimization of hot-pressed Si_3N_4-Y_2O_3 materials". Amer. Ceram. Soc. Bull. 57 (1978) 1131-1136

53. H.Knoch, "The influence of processing parameters on development of microstructure in hot-pressed Si_3N_4 and its correlation with mechanical properties", Ref. 40, pp. 737-751

54. a) R.J.Bratton et al., "Hot-pressed Si_3N_4 developments", Ref. 36, pp. 805-825

b) J.C.Uy et al., "Study of the processing and bond strength of hot pressed Si_3N_4", ibid., pp. 131-149

55. P.Walzer et al., "Development of multi-density silicon nitride turbine rotors", ibid. 503-514

56. R.R.Baker et al., "Developments in press bonding of duo density rotors", ibid. 207-230

57. E.Gugel and H.Kessel, "Post hot pressing of reaction bonded silicon nitride", ibid. 515-526

58. A.L.Stuijts and J.G.M. de Lau, "Some developments in hot pressing, Ref. 48, pp. 166-170

59. a) H.Sheinberg, "Hot pressing spinel hemishells with retract-able tooling", Amer. Ceram. Soc. Bull. 58 (1979) 719-721

b) F.Rigby, Ref. 37, pp. 249-258

60. D.Raybould et al., "A new powder metallurgy method", J. Mater. Science 14 (1979) 2523-2526

61. K.Kawada and A.Onodera, "Effect of residual strain on high pressure densification of non oxide ceramics", Amer. Ceram. Soc. Bull. 59 (1980) 1151-1152

62. W.C.Yohe and A.L.Ruoff, "Ultrafine grain tantalum carbide by high pressure hot pressing". Amer.Ceram.Soc.Bull. 57 (1978) 1123-1130

63. M.Shibata et al., "High pressure sintering of silicon nitride", Ceram. Int. 6 (1980) 146-147

64. S.Prochazka and W.A.Rocco, "High pressure hot pressing of silicon nitride powders", in Materials Science Research Vol. 11 (H.Palmour III et al. Ed[rs]; Plenum Press, 1978, 615-625)

65. H.Fischmeister, "Isostatic hot compaction - A Review", Powder Metall. Int. 10 (1978) 119-123

66. R.L.Coble, "Hot consolidation of rapidly solidified powders: sintering, hot pressing and hot isostatic pressing, in relation to the superalloys" , ibid. 128-130

67. H.T.Larker, "Hot isostatic pressing of ceramics" - this conference

68. J.C.Uy and E.R.Erman, "Isostatic densification and strengthening of reaction sintered Si_3N_4", Ref. 36, pp. 1011-1021

69. H.C.Yeh and P.F.Sikora, "Consolidation of Si_3N_4 by hot isostatic pressing", Amer.Ceram.Soc.Bull. 58 (1979) 444-447

70. R.R.Wills et al., "Preliminary observations on the hot isostatic pressing of silicon nitride", Ceram.Eng.Sci.Proc. 1 (1980) 534-539

71. M.Koizumi et al., "Fabrication of translucent ceramics by isostatic hot-pressing", Ref. 48, pp. 150-154

72. T.Berben and C.Büthker, "Isostatic hot pressing and its influence on microstructure", Science of ceramics 9 (1977) 176-182

73. G.J.May et al., "Hot isostatic pressing of beta alumina", J. Mater. Science 15 (1980) 2311-2316

74. Yu Ishitobi et al., "Reactive pressure sintering of alumina", Amer. Ceram. Soc. Bull. 59 (1980) 1208-1211

75. A.C.D.Chaklader, "Reactive hot-pressing of aluminas", J. Amer. Ceram. Soc. 61 (1978) 252-257

76. Yu Ishitobi et al., "Fabrication of translucent Al_2O_3 by high pressure sintering", Amer. Ceram. Soc. Bull. 56 (1977) 556-558

77. K.Hamano et al., "Fabrication of transparent spinel ceramics by reactive hot pressing", Yogyo Kyokai Shi 85 (1977) 225-230

78. E. Di Rupo et al., "Identification of stages in reactive hot-pressing", J. Amer. Ceram. Soc. 61 (1978) 468-469

79. E. Di Rupo et al., "Reaction hot-pressing of zircon alumina mixtures", J. Mater. Science 14 (1979) 705-711

80. H.C.Yeh et al., "Pressure sintering of Si_3N_4-Al_2O_3", Amer. Ceram. Soc. Bull. 56 (1977) 189-193

81. K.H.Jack, "The processing and properties of sialons and related nitrogen ceramics",Ref. 40, pp. 534-549

82. M.H.Lewis et al., "The formation of single-phase Si-Al-O-N ceramics", J. Mater. Science 12 (1977) 61-74

83. N.Benn et al., "Reactive hot-pressing of phases in the system $Si_{6-z} Al_z O_z N_{8-z}$", Science of Ceramics 9 (1977) 119-126

84. M.Kuwabara et al., "The reaction hot pressing of compositions in the system Al-Si-N-O corresponding to β'-sialon", J. Mater. Science 15 (1980) 1407-1416

85. M.N.Rahaman et al., "Mechanisms of densification during reaction hot pressing in the system Si-Al-O-N", J. Amer. Ceram. Soc. 63 (1980) 648-653

86. M.N.Rahaman et al., "Mechanisms of densification during reaction hot-pressing in the Al-Si-N-O system" - this conference

87. T.J.Curci et al., "Deformation processing of magnetic hexa-ferrites for Hc maximization through grain growth control", Ref. 64, pp. 359-368

88. S.K.Dey et al., "Effect of powder aspect ratio on the magnetic properties of hot-forged barium ferrite", Annual meeting of the Amerc. Ceram. Soc., May 6th 1981 - see Amer.Ceram.Soc.Bull. 60 (1981) 363

89. J.U.Knickerbocker and D.A.Payne, "Orientation of ceramic micro-structure by hot-forming methods", ibid., p. 363

90. K.Sakata et al., "Hot-forged ferroelectric ceramics of some bismuth compounds with layer structure", Ferroelectrics 22 (1978) 825-826

91. T.Takenaka and K.Sakata, "Grain orientation and electrical properties of hot forged $Bi_4Ti_3O_{12}$ceramics", Jap. J. Appl. Phys. 19 (1980) 31-39

92. T.E.Chung and T.J.Davies, "The low-stress creep of fine grain uranium dioxide", Acta Metall. 27 (1979) 627-635

93. T.E.Chung et T.J.Davies, "Pore behaviour in fine grained UO_2 during superplastic creep", in Creep and fracture of engineering materials and structures (B.Wilshire and D.R.J.Owen Ed^{rs}; Pineridge Press, Swansea, U.K., 1981, 395-407)

94. P.Carry and A.Mocellin; to be published

95. M.F.Ashby and R.A.Verrall, "Diffusion accommodated flow and superplasticity", Acta Metall. 21 (1973) 149-163

96. L.A.Winger et al., "Transformational superplasticity of Bi_2WO_6 and Bi_2MoO_6", J. Amer. Ceram. Soc. 63 (1980) 291-294

97. A.G.Evans et al., "Suppression of cavity formation in ceramics: prospects for superplasticity", J. Amer. Ceram. Soc. 63 (1980) 368-375

DISCUSSION

Morgan: There has been a great hold-up in the development of forming processes because of the conservative belief in inadequate models. Still one hears a $\dot{\varepsilon} \sim 6$ creep called "Nabarro-Herring" when grains are seen to remain equiaxed, which implies particle re-arrangement/grain boundary sliding. In "Ultra-Fine Grain Ceramics" the potential for ultra-fine grain forming by Rhodes, Morgan and others was well laid out.

Lange: People have for the last 20-30 years emphasized sintering kinetics for materials we now know how to densify in < 10 minutes. Theoretical models have been of little use. To continue this direction is only of academic interest. The real problem is achieving uniformity of the microstructure to produce uniform properties, with high Weibull modulus for example.

Greskovich: A difficulty with understanding conventional "sintering maps" for many materials lies with the competitive process of grain growth.

INFLUENCE OF COMPOSITION AND PROCESS SELECTION ON DENSIFICATION OF SILICON NITRIDE

George E. Gazza

U.S. Army Materials and Mechanics Research Center
Watertown, Massachusetts 02172

1 INTRODUCTION

Until approximately five to six years ago, dense Si_3N_4 was primarily produced by hot pressing with the use of an appropriate densifying additive. Recently, due to economic advantages and the necessity to produce complex shapes, sintering and hot isostatic gas pressing have received increasing attention. Adopted modifications of the sintering approach include the use of N_2 gas overpressure to 10 MPa to suppress thermal decomposition of the Si_3N_4, and the development of a dual N_2 gas pressure procedure where densification from the closed pore stage proceeds under a higher N_2 gas pressure than used for the initial sintering step. The sintering approach has also been facilitated by the development of improved Si_3N_4 starting powders, i.e., having higher surface area and greater uniformity. The selection of the additive composition for densification is still primarily based on prior successful hot pressing results. Initial screening and evaluation of new compositional systems is quickly performed by hot pressing. It has been demonstrated that Si_3N_4 is adaptable to compositional and/or microstructural alterations which can improve the properties and performance of the material. The primary material parameters considered for alteration are shown in Table 1. The compositional approach involves additive selection, impurity effects, and phase equilibria/behavior studies. Microstructural improvement relies on starting material characteristics and optimum selection of processing parameters.

Most studies concerned with the densification of Si_3N_4 have concentrated on the use of MgO, Al_2O_3, or Y_2O_3 as densification aids. More recently, CeO_2 and ZrO_2 have received attention. The criteria for additive selection generally emphasized resistance

Riley, F.L. (ed.) Progress in Nitrogen Ceramics

Table 1. Material Parameter

Goal	Approach: Compositional Control (Grain Boundary Eng.)	Approach: Microstructural Control
Improved Performance Reliability of Si_3N_4 by Process Control	Additives (MgO, Y_2O_3, etc.)	Grain Size
	Impurities (Metal Cations Carbon)	Grain Morphology (Equiaxed/Prismatic)
	Phase Behavior (Multicomponent Systems)	Phase Control (α/β ratio)

to high temperature creep and oxidation. Often, the selection of additive composition and quantity was problematic in that promoting ease of densification of Si_3N_4 bodies resulted in a degradation of their high temperature properties.

2 MICROSTRUCTURAL EFFECTS

Studies on the use of MgO additive have been well documented in the technical literature. Within the past few years, these studies have been increasingly concerned with microstructural development and its influence on resultant properties of Si_3N_4. Since the early observation that high alpha phase starting powders were required to produce the highest strength Si_3N_4 product, the effect of microstructure and microstructural modifications produced by changes in composition and process parameters were generally neglected until Lange (1) reported on the morphological development of beta grains in hot-pressed Si_3N_4. Knoch and Gazza (2) subsequently investigated the influence of Si_3N_4 starting powders with different alpha/beta phase content on the modulus of rupture of hot-pressed Si_3N_4-5%MgO composition using different time, temperature, and pressure parameters. In this study, the most dramatic effect of process parameter variation on the resultant microstructure and properties was produced by hot pressing Si_3N_4-5%MgO for various times from 10 minutes to three hours at 1600°C using 70 MPa uniaxial pressure. The room temperature modulus of rupture values and resulting β-fraction associated with these hot-pressing parameters are shown in Figure 1. Maximum strength is reached at nearly full conversion of α phase to β phase with interlocking elongated grain morphology. Further coarsening of the grain size by extending the hot-pressing time produces a reduction in strength.

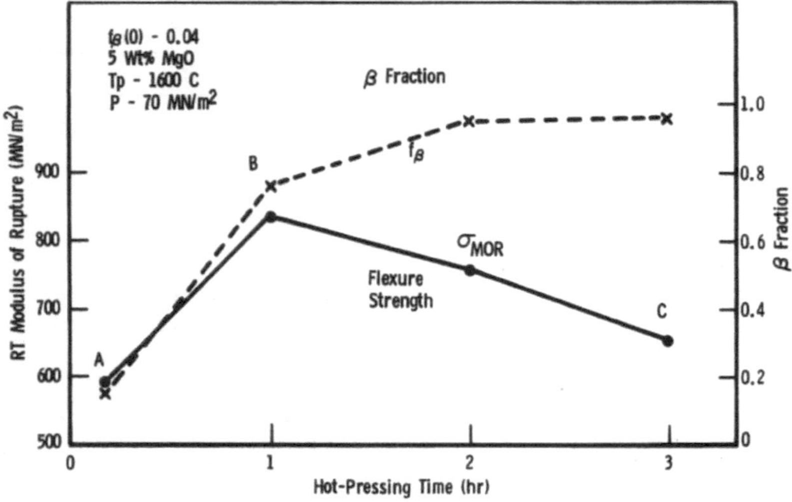

Figure 1. Effect of hot-pressing time on flexural strength and β-fraction of hot-pressed Si_3N_4-5%MgO.

Fracture toughness (K_c) of these samples, shown in Figure 2, were determined by the indentation method (3). The fracture toughness increased to a maximum of approximately 7 $MNm^{-3/2}$ with increasing α to β transformation and subsequently dropped to 4.5 $MNm^{-3/2}$ as the grain size coarsened and the aspect ratio decreased. Fracture toughness results from other studies (4,5) are also shown in Figure 2. Grain coarsening effects on K_c appear to be similar to those observed by Himsolt, et al (5).

3 PHASE BEHAVIOR AND IMPURITY EFFECTS

The development of more refractory Si_3N_4 materials using Y_2O_3, CeO_2, etc., tended to concentrate more on compositional alterations and phase relations. In particular, Y_2O_3 additions were found to form a higher viscosity glassy phase and quaternary oxynitrides. Although the properties of the Y_2O_3-doped Si_3N_4 were found to be excellent, particularly at temperatures above 1200°C, certain intermediate temperature phase instability was found at temperatures between 700°C and 1100°C in an oxidizing environment (6). The instability was attributed to large changes in the molar volume of the secondary phase and its oxidation product. However, the phase instability and resultant crack formation was not always found in oxidized Si_3N_4 bodies containing these quaternary phases. In related studies, Knoch and Gazza (7) at AMMRC and Schoun (8) at NASA found impurity influences associated with this behavior. In the former study, the presence of carbon or silicon carbide in

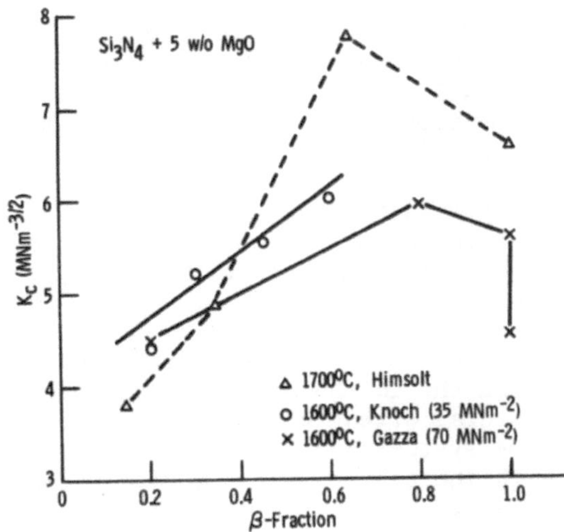

Figure 2. Variation of fracture toughness with β-fraction in hot-pressed Si_3N_4-5%MgO.

hot-pressed Si_3N_4-Y_2O_3 (melilite) specimens caused extensive cracking during oxidation at temperatures of 800°C to 1000°C while carbon-free material exhibited parabolic oxidation kinetics. When a carbon-containing Si_3N_4 powder with a carbon content of approximately 0.6% was used to form the melilite compound, degradation occurred during oxidation. Subsequently, it has been found that a carbon content as low as 0.2 to 0.3% in Si_3N_4 containing melilite phase will crack during oxidation. The extent of degradation may also depend on the amount and distribution of melilite phase in Si_3N_4. Schoun (8) demonstrated that the presence of tungsten in sintered Si_3N_4 containing the melilite phase produced severe oxidation and cracking at 750°C while similar Si_3N_4-8%Y_2O_3 material containing 2%WC or no intentional impurity showed only small weight gain.

Whether severe oxidation and cracking of Si_3N_4-Y_2O_3 compositions containing melilite phase, K-phase, or J-phase is observed or not, most studies indicate that the oxidation rates of Si_3N_4-Y_2O_3 compositions within these phase fields are higher than for compositions within the Si_3N_4-Si_2N_2O-$Y_2Si_2O_7$ triangle (6,9) or Si_3N_4-$Y_2Si_2O_7$-$Y_5Si_3O_{12}N$ triangle (10). Optimum Si_3N_4-Y_2O_3 compositions with regard to the best combination of strength, oxidation, and creep resistance should be found within the latter two compatibility triangles.

4 COMPOSITIONAL MODIFICATIONS

The successful use of Y_2O_3 and CeO_2 as densification aids to Si_3N_4 suggested that other rare earth oxides might also be effective additives. Andersson and Bratton (11) studied the rare earth oxides of Y, La, Ce, Pr, Nd, Sm, Gd, Dy, Er, and Yb with compositions within the compatibility triangle Si_3N_4-Si_2N_2O-$M_2Si_2O_7$. Additionally, they included Sc_2O_3, NiO, Cr_2O_3, and ZrO_2 in their study. The highest silicate composition ($M_2Si_2O_7$ for most additives studied) was chosen to maintain stability with respect to SiO_2 and retain good oxidation resistance. Incipient melting points were determined for Si_3N_4-M_xO_y-SiO_2 compositions as well as elevated temperature flexural strengths. Both parameters are shown plotted in Figure 3 as functions of the ionic radii of the rare earth elements used for each composition.

The ionic radii were selected because they are indicative of the bond strengths of those rare earths in the $M_2Si_2O_7$ structures. Both the elevated temperature flexural strength and eutectic melting points of the Si_3N_4-Si_2N_2O-$M_2Si_2O_7$ systems increase with decreasing rare earth ionic radii, i.e., with increasing bond strength. The high temperature property potential of Sc-containing Si_3N_4 is suggested on the plot and recent work by Morgan, Lange, et al (12) illustrates the excellent properties achieved with the Si_3N_4-Sc_2O_3-SiO_2 system. The cost of Sc_2O_3 may limit its usefulness, however, beyond the research stage. Further property improvement was suggested by alloying various rare earth pyrosilicates or by substituting smaller ions, e.g., Al^{+3}, Cr^{+3}, within solubility limits in the monoclinic structures.

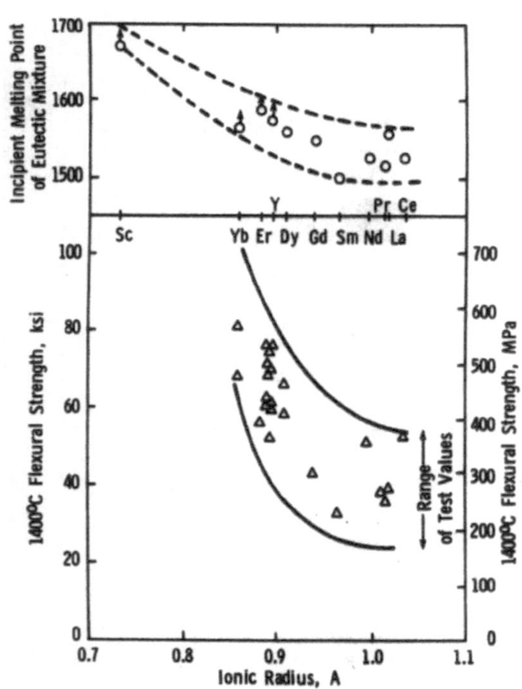

Figure 3. Effect on ionic radius of rare earth elements on the eutectic melting points and elevated temperature strengths of Si_3N_4-M_xO_y-SiO_2. (Ref. 11)

As processing emphasis of dense Si_3N_4 gradually shifted from hot pressing to sintering, accommodations in additive selection were necessary to insure a sufficient amount of liquid phase formed to promote densification. Additionally, the liquid had to form at temperatures where the system was thermally stable, and be sufficiently reactive with the Si_3N_4 to produce densification by the mechanism of solution-reprecipitation. The addition of Al_2O_3 to the Si_3N_4-Y_2O_3 system is effective in promoting sinterability over Y_2O_3 additions alone. Excellent RT strength values are obtained using the combined Y_2O_3-Al_2O_3 additions to Si_3N_4 but high temperature properties can be adversely affected by the formation of low viscosity Y-Si-Al-O-N glasses. The high temperature property studies of Smith and Quackenbush (13) have shown that for Si_3N_4 containing 4 to 13% Y_2O_3, the Al_2O_3 addition should be less than 2% to approach high temperature property requirements of gas turbine engine components. For maximum high temperature creep and oxidation resistance, the Al_2O_3 should be minimized or eliminated. The effect of small Al_2O_3 additions on the stress-rupture life of a commercially pure Si_3N_4-7.5%Y_2O_3 composition is shown in Figure 4. A stress-rupture temperature of 1300°C and a step temperature-stress rupture (STSR) cycle of 1000°C to 1300°C was used with applied stresses of 140 MPa (20 ksi) and 200 MPa (30 ksi). As shown in the figure, a reduction in the amount of added Al_2O_3 from 2% to 0.5% dramatically increases the time to failure under the given stresses. Reducing the stress level from 200 MPa (30 ksi) to 140 MPa (20 ksi) produces a $2\frac{1}{2}$ to 3 times increase in stress-rupture life. The slight difference in time to failure between the stress-rupture and STSR values may be due to a small contribution to slow crack growth at temperatures below 1300°C. In the STSR test, the specimen were held under stress at temperatures of 1000°C, 1100°C, and 1200°C for 24 hours at each temperature before reaching the final test temperature of 1300°C. As a comparative point of reference, STSR conditions (1300°C, 200 MPa) resulted in an approximately 15-hour failure life for hot-pressed NC-132 (14). The use of higher purity Si_3N_4 starting material rather than the commercially pure grade should shift the curves shown in Figure 4 toward longer time to failure.

5 GAS PRESSURE SINTERING

As processing emphasis shifts from hot pressing to sintering, modification to conventional sintering procedures must be explored to overcome densification problems associated with low diffusivity and dissociation. The use of high N_2 gas pressure for sintering was found to be effective in promoting densification while limiting decomposition of the material (15-17). The need to densify complex shapes of Si_3N_4 requires that isostatic gas pressure be used. Conventional hot isostatic pressing where pressures greater than 70 MPa are normally used has been successful in producing Si_3N_4-5%Y_2O_3

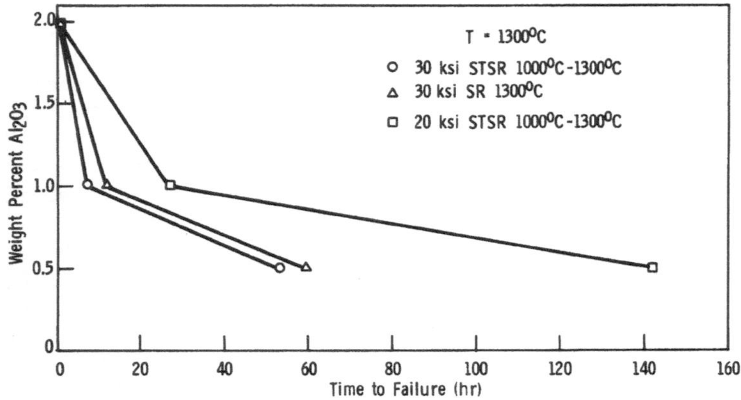

Figure 4. Effect of Al_2O_3 additions on the stress-rupture life of a Si_3N_4-7.5%Y_2O_3 base composition.

material with excellent high temperature properties (18). A cladless dual N_2 pressure technique has also shown good results in producing near fully dense ceramics (19). Greskovich and Palm (20) reported using the two-step method to produce 99+% dense $Si_3N_4/SiBeN_2$ composition. Recently, Gazza et al (21) densified $Si_3N_4/Y_2O_3/Al_2O_3$ compositions using 0.1 MPa N_2 pressure for initial sintering, then raising the pressure to 2.0 MPa to produce final densification.

The compositions used in this study are shown in Figure 5, a plot of density versus N_2 pressure. Some additive compositions were prereacted to form compounds, e.g., $Y_3Al_5O_{12}$, $Y_4Al_2O_9$, before mixing with the Si_3N_4 powders. Cold-pressed specimens were initially sintered under 0.1 MPa N_2 pressure. Some compositions sintered to the closed pore stage. In the second sintering step, the pressure was raised to 2.0 MPa while the temperature was held at 1770°C to 1780°C. Ninety-minute hold times were used for each step. As shown in Figure 5, all compositions exhibited increased densification. Those which were initially sintered to the closed pore stage were densified to near maximum density. The poor initial sintering behavior of the 117 and 113 compositions is probably related to lack of sufficient liquid formed with the available silica in the system.

Differences in resultant properties and microstructures were observed which were influenced by the starting Si_3N_4 powder. In Figure 6, microstructures obtained from replicas on polished and etched surfaces and scanning electron micrographs on fracture surfaces are shown for dual-pressure sintered Si_3N_4-10%YAG compositions. Finer grain size and higher grain aspect ratios appear to be developed in the LC-10 powder for the processing conditions used. Room temperature modulus of rupture measurements on both 112 and

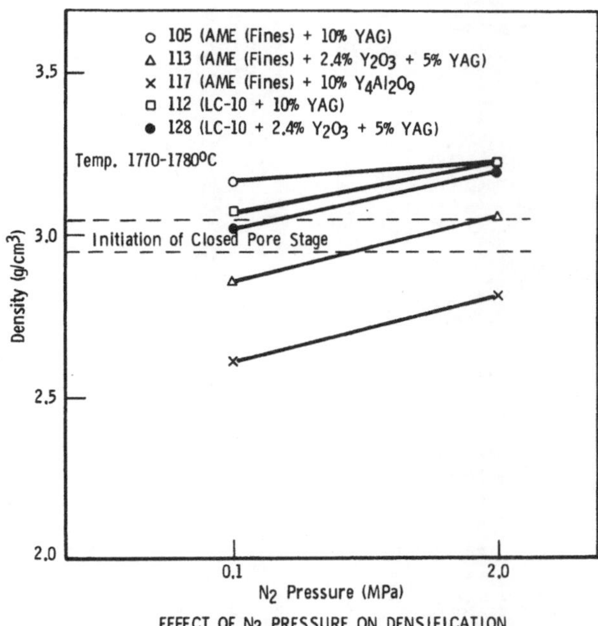

Figure 5. Effect of Dual N_2 pressure process on densification of various Si_3N_4-Y_2O_3-Al_2O_3 compositions.

105 materials reflect the microstructural differences. An average MOR of 779 MPa (113 ksi) was observed for the finer grain specimen while the coarser 105 specimen had an average MOR of 538 MPa (78 ksi).

6 SUMMARY AND CONCLUSIONS

Studies on the hot-pressing of various Si_3N_4/metal oxide systems demonstrate that both compositional and microstructural alterations can be used to produce significant property differences within a given materials system. For Si_3N_4/MgO, microstructural studies demonstrate the need for judicious selection of starting materials and a more exact determination of optimum process parameters required for densification. For the Si_3N_4/Y_2O_3 materials system, it appears that composition restricted to the Si_3N_4-Si_2N_2O-$Y_2Si_2O_7$ or Si_3N_4-$Y_2Si_2O_7$-$Y_5Si_3O_{12}N$ phase fields will have the best combination of high temperature properties. The role of impurities within this compositional system needs further elucidation.

The addition of Al_2O_3 to the Si_3N_4/Y_2O_3 system facilitates densification but appears to degrade high temperature properties

Figure 6. Microstructures of sintered Si_3N_4-10%YAG; fractographs (upper) and surface replicas (lower).

with some dependence on the amount of Al_2O_3 added. This paper did not consider grain boundary crystallization techniques which may alleviate the formation of low viscosity phases.

Emphasis on densifying high purity Si_3N_4/Y_2O_3 compositions appears to be emerging in the area of hot isostatic gas pressing. Techniques requiring cladding have been successful in producing dense, high strength bodies. Cladless hot isostatic gas pressure methods using N_2 pressures less than 10 MPa have also been effective in producing dense bodies with strength and microstructure similar to hot-pressed material.

7 REFERENCES

1. Lange, F. F. "Fracture toughness of Si_3N_4 as a function of initial α-phase content." Technical Report No. 4, ONR Contract No. N00014-77-C-0441, July 1978.
2. Knoch, H., and G. E. Gazza. Ceramurgia International 6 [2] (1980) 51.
3. Evans, A. G., and E. A. Charles. Journal of American Ceramics Society 59 (1976) 371.
4. Knoch, H., G. E. Gazza, and R. N. Katz. Proceedings of the 4th CIMTEC Meeting, Saint Vincent, Italy (1979).
5. Himsolt, G., H. Hubner, W. Kleinlein, and H. Knoch. Journal of American Ceramics Society 62 [1-2] (1979) 29.
6. Lange, F. F., S. C. Singhal, and R. C. Kuznicki. Journal of American Ceramics Society 60 [5-6] (1977) 249.
7. Knoch, H., and G. E. Gazza. Journal of American Ceramics Society 62 [11-12] (1979) 634.
8. Schuon, S. NASA TM-81528, Presentation at American Ceramics Society Meeting, Chicago, Illinois, April 28-30, 1980.
9. Rae, A. W. J. M., D. P. Thompson, and K. H. Jack. "Ceramics for high performance applications-II." Proceedings of the 5th Army Materials Technology Conference, Newport, Rhode Island, March 1977.
10. Quackenbush, C. L., and J. T. Smith. American Ceramics Society Bulletin 59 [5] (1980) 533.
11. Andersson, C. A., and R. Bratton. "Ceramic materials for high temperature turbines." Final Technical Report, U. S. Energy Res. Dev. Adm. Contract No. EY-76-C-05-5210, August 1977.
12. Morgan, P. E. D., F. F. Lange, D. R. Clarke, and B. I. Davis. Journal of American Ceramics Society 64 [4] (1981).
13. Smith, J. T., and C. L. Quackenbush. American Ceramics Society Bulletin 59 [5] (1980) 529.
14. Quinn, G. D., and J. B. Quinn. Proceedings of the International Symposium on the Fracture Mechanics of Ceramics, Penn State University, July 1981.
15. Mitomo, M., M. Tsutsumi, E. Bannai, and T. Tanaka. American Ceramics Society Bulletin 55 [3] (1976) 313.
16. Priest, H. F., G. L. Priest, and G. E. Gazza. Journal of American Ceramics Society 60 [1-2] (1977) 81.
17. Mitomo, M. Journal of Materials Science 11 [6] (1976) 1103.
18. Wills, R. R., M. C. Brockway, L. G. McCoy, and D. E. Niesz. Proceedings Ceramic Engineering and Science 1 [7-8] (1980) 534.
19. Hardtl, K. H. American Ceramics Society Bulletin 54 [2] (1975) 201.
20. Greskovich, C., and J. Palm. "Development of high performance sintered Si_3N_4." Final Technical Report, Contract No. DAAG-46-78-C-0058, Under AMMRC/DOE Interagency Agreement EC-76-A-1017-002.
21. Gazza, G. E., and R. N. Katz. Proceedings of the Department of Energy Automotive Technology Development Contractor's Coordination Meeting, Dearborn, Michigan, November 1980 (AMMRC SP 80-5).

A GAS PRESSURE SINTERING PROCESS FOR PRODUCING DENSE Si_3N_4

C. Greskovich

Ceramics Branch
Physical Chemistry Laboratory
Corporate Research and Development
General Electric Company
Schenectady, NY 12301

1 INTRODUCTION

Nitrogen pressure can reduce thermal decomposition and increase the density of sintered compacts of Si_3N_4 (1-3). The effectiveness of nitrogen pressure may be better understood from the stability diagram(4) for Si_3N_4 illustrated in Fig. 1. Based on sintering experiments for Si_3N_4 compacts containing $\approx$7 wt% SiO_2 and $\approx$7 wt% $BeSiN_2$ as densification aids, the "region of sinterability" in Fig. 1 is outlined for the conditions which permit compacts to achieve relative densities $\geq$92%. A "rule-of-thumb" is to carry-out the sintering process at an N_2 pressure at least a factor of 10 higher than the equilibrium pressure of N_2 for the reaction $Si_3N_4(s) \rightleftarrows 3Si(l) + 2N_2$. For example, in the temperature region of 1900 to 2100°C, the respective minimum N_2 pressures of $\sim$1 and 5 MPa ($\sim$10 and 50 atm) permit good sinterability and weight losses $<$2%.

The present work describes a gas pressure sintering (GPS) process for producing highly-dense ($>$99%), Si_3N_4 ceramics. A two-step, nitrogen pressurizing technique using N_2 pressures between $\sim$2 and 7 MPa and temperatures between 1900 and 2050°C was developed to facilitate the attainment of nearly full density of sintered compacts.

Riley, F.L. (ed.) Progress in Nitrogen Ceramics

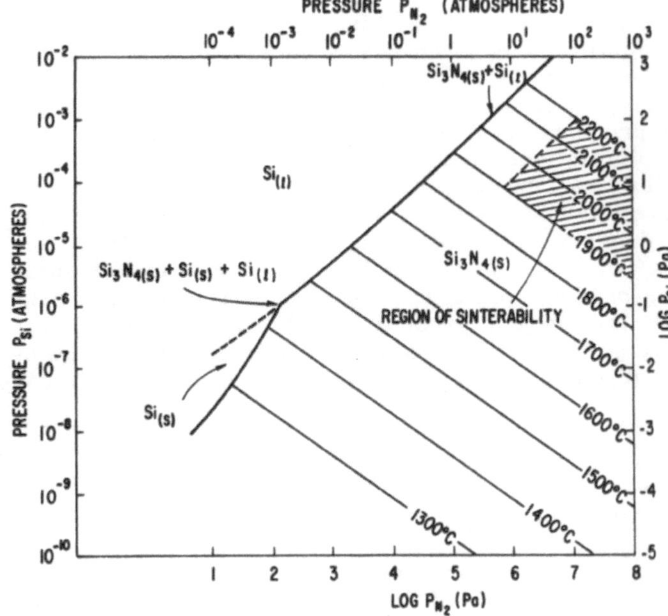

Fig. 1 Stability diagram for Si_3N_4 in equilibrium with Si and N_2.

2 EXPERIMENTAL PROCEDURES

The sintering behavior of two types of Si_3N_4 powders was examined in this study. Type I was high purity Si_3N_4 1) composed of ~60 wt% crystalline (~92% α-Si_3N_4 + 8% β-Si_3N_4) and ~40% amorphous Si_3N_4. This powder was processed with 7 wt% $BeSiN_2$ to a specific surface area of ~11 m^2g^{-1} (0.17 µm equivalent particle size) and an oxygen content of 1.8 wt%. The major impurities were (in ppm): Ca (<60), Fe (400), Al (200) and Cl (500). A standard purity grade of Si_3N_4 (Type II) 2) was also studied and was composed of ~90 wt% α-Si_3N_4, 7% β-Si_3N_4, <2% free Si and a trace of β-SiC. This powder was processed with 3.5 wt% $BeSiN_2$ to a specific surface area of ~14 m^2g^{-1} and an oxygen content of 1.5 wt%. The major impurities were Ca (600), Fe (30), Al (2100) and C (6000).

During this investigation high sintering rates were achieved when Type I Si_3N_4 powder contained greater than ~3 wt% oxygen, and when Type II Si_3N_4 powder contained >2 wt% oxygen. The oxygen (SiO_2) content of each powder composition was increased above its initial value by a controlled oxidation process (5). The sintering experiments were carried-out in a graphite resistance furnace capable of operating up to 2300°C at 10-15 MPa of N_2 gas

1) SN502 Grade, GTE Sylvania, Towanda, PA. USA.
2) Standard Grade, Hermann C. Starck Berlin, Goslar, West Germany.

pressure. A typical sintering run involved inserting a disk-shaped powder compact into a BN crucible with lid, evacuating the system up to 1000°C or back-fill with N_2 gas and pressurize while heating to 1000°C, simultaneously heating and pressurizing with N_2 gas to the desired pressure, holding at the soak temperature for the desired time, modifing the soak temperature and/or pressure if desired, and cooling and depressurizing the system. Sintered samples were characterized by weight loss, shrinkage and immersion (H_2O displacement) density. The theoretical density of Si_3N_4 was taken to be 3.18 g/cc. X-ray diffraction analysis and distribution of the phases present and solid solution effects. The size and distribution of the phases present, including porosity, was determined from polished sections using reflected light microscopy. TEM examination helped to reveal intergranular liquid phases(s).

3 RESULTS AND DISCUSSION

3.1 Sintering in a Constant N_2 Pressure

Si_3N_4 compacts of both powder compositions could be sintered to limiting relative densities ≈93% by using N_2 pressures between ∿2 and 7 MPa and temperatures between 1950 and 2050°C for a soak time of 15 min. Compacts of the Type I Si_3N_4 composition exhibited an average linear shrinkage and weight loss of ∿ 18% and 0.7-1%, respectively, whereas compacts of the Type II Si_3N_4 composition showed shrinkages of ∿ 15% and similar weight loss. The lower shrinkages observed for the lower purity Si_3N_4 compacts were directly related to their higher (58% vs. 48%) green density.

The attainment of limiting relative densities (90-93%) during sintering is illustrated in Fig. 2 for compacts of high purity

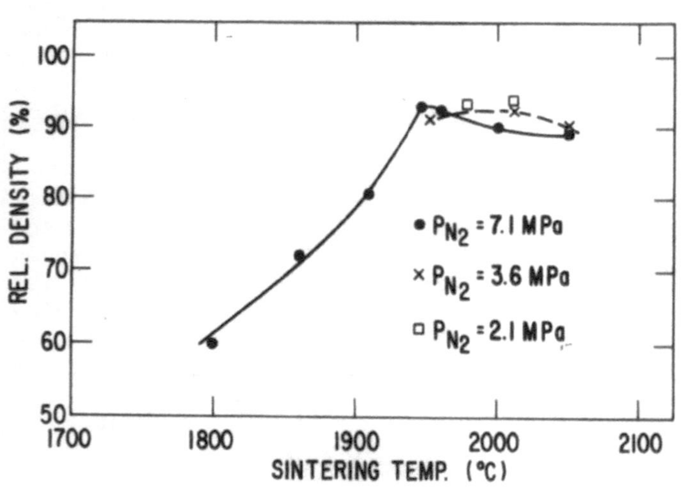

Fig. 2 Relative density vs. temperature for Type I powder compacts sintered 15 min in various N_2 pressures.

Si_3N_4 containing 7 wt% $BeSiN_2$ and 3.7 wt% O (∿7 wt% SiO_2). There was a marked increase in relative density from ≈60% at 1800°C to ≈93% at 1950°C for compacts sintered in an N_2 pressure of 7.1 MPa, followed by a relative density drop to ≈ 89% as the temperature increased to 2050°C. The relative density of compacts sintered near 2000°C increased from ≈90% to 93.5% by decreasing the N_2 pressure from 7.1 to 2.1 MPa. The highest relative densities of 92-93.5% occurred in sintered compacts that had just achieved the closed pore stage because these sintered compacts exhibited no water absorption during density measurement via the Archimedes method. This result was also confirmed by microstructural observations on polished samples (Fig. 3). The 93% dense sample shown in Fig. 3A contains a uniform distribution of irregularly-shaped, discrete pores (black phase) having an average equivalent diameter of ∿ 1.5 μm. By increasing the sintering temperature from ≈1975 to 2050°C in an N_2 pressure of 7.1 MPa, Fig. 3B shows that pores grow from ∿1.5 μm to 4 μm. The majority of the pores are distinct and still irregularly-shaped but some localized interconnectivity of the pore phase can be seen in this 89% dense sample. Close examination of the β -Si_3N_4 grains shows that the residual pores are located at grain boundary intersections and grow most likely by a pore coalescence process. The increase in average pore size may be associated with rapid grain growth, resulting in density regression of the sintered body. For this to happen the closed pores must be filled with N_2 gas having a low solubility in the matrix. Thus, the limiting densities and density regression shown in Fig. 2 for samples sintered at T > 1950°C in constant N_2 pressure appear to be caused by coalescence of gas-filled pores which have an internal gas pressure which is equal to or exceeds the sum of surface tension plus the applied N_2 pressure.

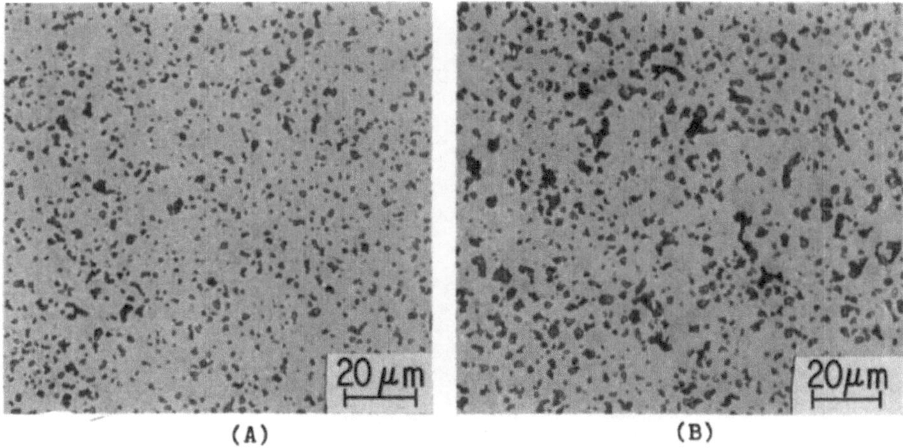

Fig. 3 Microstructures of Type I compacts sintered for 15 min in 7.1 MPa of N_2 at (A) 1975°C and (B) 2050°C.

An understanding of the sintering mechanism was provided by X-ray and optical microscopy examination. Debye-Scherrer photographs of specimens sintered at 1800 (60% relative density) and 2000°C (91% relative density) for 15 min in 7.1 MPa of N_2 showed the presence of strong β-Si_3N_4 lines and the same weak lines of other crystalline phases. The lattice parameters of the β-Si_3N_4 phase were nearly the same for both specimens and were significantly smaller than those for pure β-Si_3N_4, indicating that the β-Si_3N_4 phase contains Be and O in solid solution as previously reported (6-8). This information showed that nearly all densification occurs after the α-Si_3N_4 particles transform to a β-Si_3N_4 solid solution. Many of the weak diffraction lines observed could be indexed to trace amounts of Si_2N_2O, Be_2SiO_4 and Si. The Si_2N_2O and Be_2SiO_4 phases probably crystallize during cooling from an intergranular glassy phase (see section 3.2). Consequently, the mechanism of sintering for both Type I and II Si_3N_4 compositions involves densification of particles of β-Si_3N_4 solid solution in an oxygen-bearing liquid phase.

3.2 High Density (>99%) Si_3N_4 via a Two-Step Sintering Process

A two-step, nitrogen pressurizing technique was used to increase the driving force for densification to overcome the undesirably low limiting densities of 92-94%. First, the powder compact was sintered to the closed-pore state by heating for 15 min at 2000°C in ∿ 2 MPa of N_2 pressure. After this step was completed, the N_2 pressure was increased to a higher value, typically ∿7 MPa, for rapid densification. A similar method of increasing

Table I Sintering Results On Sintering Si_3N_4 Compacts Exposed To A Two-Step, GPS Process

Expt. No.	Powder Composition	Presintering Treatment	Sintering Treatment	Rel. Density	Linear Shrinkage (%)	Weight Loss (%)
1	Type I	Compact having an 0-content of 1.8 - 3.0 wt%.	15 min-2000°C-2.0 MPa N_2 + 18 min-2085°C-7.1 MPa N_2	91.9	19.7	1.3
2	"	Compact oxidized to 0-content ∿ 3.7 wt%	"	99.6	22.5	1.0
3	"	"	15 min-1940°C-2.1 MPa N_2 + 10 min-2000°C-7.1 MPa N_2	91.6	19.5	1.9
4	"	"	15 min-2000°C-2.1 MPa N_2 + 12 min-2000°C-7.1 MPa N_2	99.4	22.8	1.8
5	"	Compact oxidized to 0-content ∿ 3.5 wt%	15 min-2000°C-2.3 MPa N_2 + 20 min-1900°C-7.1 MPa N_2	99.7	22.8	1.6
6	Type II	None	15 min-1990°C-2.3 MPa N_2 + 12 min-2000°C-7.1 MPa N_2	92	15	+0.7
7	"	Compact oxidized to 0-content ∿ 2.6 wt%	"	98.6	17.6	2.7

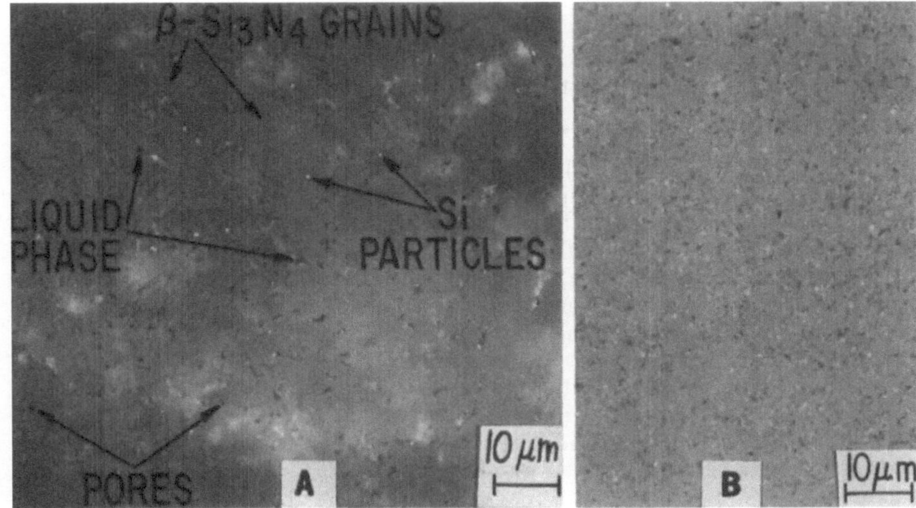

Fig. 4 Microstructures of GPS Si_3N_4 (A) Type I, 99.7% dense and (B) Type II, 98.6% dense.

grain intersections (Fig. 6). The dihedral angle between the solid and liquid phases appears to be relatively low because some liquid spreading is observed between two β-Si_3N_4 grains in Fig. 6. However, many grain boundaries did not show an "obvious" liquid phase even when viewed by higher resolution, lattice fringe techniques.

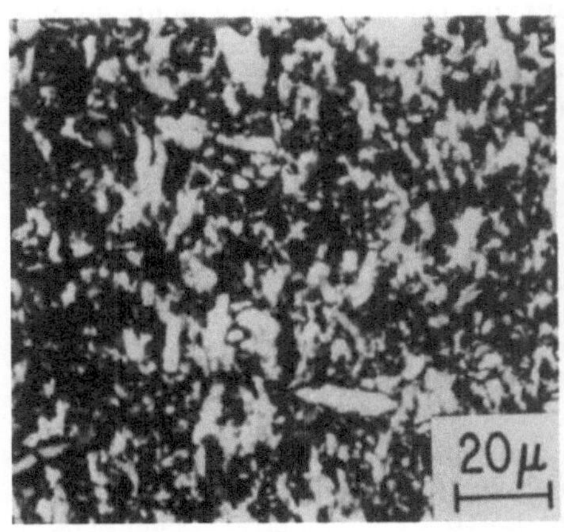

Fig. 5 Grain structure of GPS Si_3N_4 Type I viewed in thin section using transmitted polarized light.

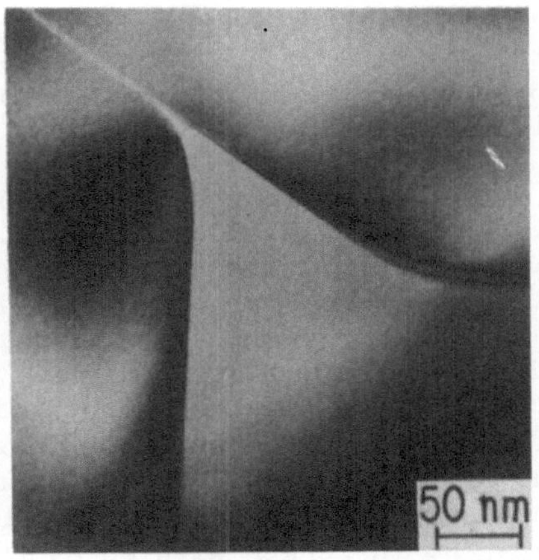

Fig. 6 Typical TEM photomicrograph of a liquid phase pocket at a 3-grain intersection.

final density via low gas pressures (< 21 MPa) was first reported on oxide ceramics(9) but the present work represents the first reported application of this technique to sintered nitrogen ceramics.

A summary of some selected sintering results for Si_3N_4 powder compositions of Types I and II are present in Table I. The sintering results of Expts. 1 and 2 show that compacts of Si_3N_4 Type I composition can be densified to 99.7% of theoretical density by the gas pressure sintering (GPS) conditions specified provided that the starting oxygen content was higher than 3 wt%. Apparently during the first sintering step the compacts used in Expt. 1 did not develop the closed pore stage because the oxygen content was too low to form sufficient oxygen-rich liquid phase to cause high final densities. Expt. 3 shows that a decrease in temperature from 2000 to 1940°C during the first step at 2.1 MPa of N_2 pressure resulted in a low relative density of 91.6% because, again, the closed pore stage did not develop before the higher N_2 pressure of 7.1 MPa was applied to the compact. This idea is supported by the results of Expt. 4 which show that a relative density ≥ 99% can be achieved when the temperature is 2000°C for both steps. Experiment 5 illustrates that the second step of the GPS process can occur at a temperature at least 100°C below the initial (2000°C) sintering temperature and still permit relative densities ≥ 99%. Under these experimental conditions the weight losses for the Type I Si_3N_4 composition were < 2%. Experiments 6 and 7 illustrate that the impure Type II Si_3N_4 compacts can be also sintered to relative densities near 99%, provided the oxygen (SiO_2) content is sufficient to develop the closed pore stage during the first step. The weight gain of 0.7% found in Expt. 6 is probably due to nitridation of the free Si in the starting Type II powder whereas the high weight loss of 2.7% observed for the sample of Expt. 7 appears to be associated with greater instability of the oxygen-rich liquid phase due to the impurities (Ca, Al and C) present.

The microstructures of dense, GPS Si_3N_4 are shown in Figs. 4-6. The polished sections in Fig. 4 are characterized by submicron-micron Si particles (bright phase), submicron β -SiC particles (light grey phase) in Fig. 4B only, β $-Si_3N_4$ grains up to 20 µm in length (grey phase), liquid phase pockets at grain intersections (dark grey phase) and submicron pores (black phase). Interference fringes and subsurface light scattering in Fig. 4A are indicative of highly dense, translucent regions. The size and morphology of the β$-Si_3N_4$ grains were revealed conveniently by a ceramographic thin section (Fig. 5). The elongated β $-Si_3N_4$ grains are up to ≈25 µm in length with an average width of ≈ 3 µm. Intergrowth of these elongated β$-Si_3N_4$ grains is apparent and typical of Si_3N_4 with high mechanical strength. High resolution TEM revealed more information about the nature of the intergranular liquid phase. A typical liquid phase pocket is generally observed at 3 -

4 CONCLUSIONS

A gas pressure sintering (GPS) process was developed to produce sintered Si_3N_4 with relative densities $\geq$ 99% and an elongated, intergrowth of β-Si_3N_4 grains typical of high strength Si_3N_4 ceramics. This process involved rapid densification of β-Si_3N_4 grains via liquid phase sintering at high temperatures ($\sim$2000°C) and low N_2 pressures ($\sim$10 MPa) whereby the N_2 pressure ($\sim$2 MPa) required to permit the development of closed porosity (at $\approx$ 92% relative density) in a sintered compact was then increased by a factor of $\sim$ 3 to 4 to reproducibly attain relative densities $\gtrsim$ 99%. The GPS process was demonstrated with standard purity and high purity grades of Si_3N_4 powders using 3.5 to 7 wt% $BeSiN_2$ and 2.5 to 3.7 wt% oxygen (in the form of SiO_2) as densification aids. The limiting relative densities of 92-94% found for compacts sintered under constant N_2 pressure appear to be caused by pore coalescence of gas-filled pores but complex chemical reactions and internal gas generation can not be presently ruled out.

ACKNOWLEDGMENTS

The author thanks the Army Materials and Mechanics Research Center for financial support of this work under an AMMRC/DOE Interagency Agreement with George Gazza as contract monitor.

REFERENCES

1. M. Mitomo, "Pressure Sintering of Si_3N_4," J. Mat. Sci. 11 (1976) 1103-1107.
2. H.F. Priest, G.L. Priest and G.E. Gazza, "Sintering Si_3N_4 Under High Nitrogen Pressures," J. Am. Ceram. Soc. 60 (1977) 81.
3. S. Prochazka and C. Greskovich, "Effect of Some Impurities on Sintering Si_3N_4 Under Nitrogen Pressure," pps 489-502 in Factors in Densification and Sintering of Oxide and Non-Oxide Ceramics, (S. Somiya and S. Saito, Gakujutsu Bunken Fukyukai, Tokyo, Japan, 1978).
4. C. Greskovich and S. Prochazka, "The Stability of Si_3N_4 and Liquid Phase(s) During Sintering," General Electric Report 81CRD048 (1981).
5. C. Greskovich and J.A. Palm, "Controlling the Oxygen Content of Si_3N_4 Powders," Am. Ceram. Soc. Bull. 59 (1980) 1155-1156.
6. K.H. Jack, "Nitrogen Ceramics," Trans. J. Brit. Ceram. Soc. 72 (1973) 376-384.
7. I.C. Huseby, H.L. Lukas and G. Petzow "Phase Equilibria in the System Si_3N_4-SiO_2-BeO-Be_3N_2," J. Am. Ceram. Soc. 58 (1975) 377-380.

8. C. Greskovich, "Hot-pressed β-Si_3N_4 Containing Small Amounts of Be and O in Solid Solution", J. Mat. Sci. 14 (1979) 2427-2438.
9. K.H. Hardtl, "Gas Isostatic Pressing Without Molds", Am. Ceram. Soc. Bull. 54 (1975) 201-207.

DISCUSSION

McCauley: How do you physically tell the difference between residual (original) porosity and 'evolved gas' pores? And have you heat-treated your close to fully dense samples to see if you can get the nitrogen to come back out of solution?

Greskovich: It is difficult to determine the origin of porosity in many polycrystalline ceramics but it seems to me that a careful study of the effect of atmosphere on the amount, size, shape and distribution of the pore phase would be required. I have annealed the gas pressure sintered Si_3N_4 at 1800 to 2000°C for 1 h in lower nitrogen pressures than used to effect densification. There was very little density drop, suggesting that dissolved nitrogen does not reform into pores and bloat the sample.

Weiss: Normally the original pores have an irregular shape, while those caused by volatilization should be spherical. This behaviour should change, however, if you only have small amounts of liquid present.

Greskovich: I have also seen non-spherical, angular, shaped pores in liquid phase sintered Si_3N_4 that has undergone substantial weight loss.

Lange: Pore coalescence can produce swelling. There are also the reactions

$$Si_3N_4 + SiO_2 \longrightarrow 6SiO + N_2$$

$$Si_3N_4 + {}^{3}/a\ M_2O_a \longrightarrow {}^{2}/a\ M_3N_a + 3SiO + N_2$$

which can produce a p_{SiO} or p_{N_2} that could exceed the external applied pressure p_{ex} at the temperatures bloating is seen. By increasing p_{ex} to $>(p_{SiO} + p_{N_2})$ this problem could be overcome. In fact reformation of Si_3N_4 and oxide should be favoured (Le Chatelier).

Greskovich: Thermal decomposition reactions which form internal gas inside closed pores can certainly cause swelling or density regression during sintering. However under isobaric conditions there was no increase in swelling with decrease in p_{N_2} between 70 and 20 atmospheres, suggesting that the decomposition reactions are not overriding or controlling. Nonetheless, complex chemical reactions inside closed pores cannot yet be excluded.

Lumby: Pumping in nitrogen should eliminate glass as Si_3N_4; thus glass and nitrogen-containing pores should not be found together.

Greskovich: With SiO and N_2 pressure increase inside closed pores equilibrium will be established between these pores and the glass-Si_3N_4 phases. At this point no further increase in gas pressure should occur inside the pore due to decomposition reactions. Hence glass and N_2-containing pores can be in equilibrium, without complete disappearance of the glass.

Katz: I think the fact that the pressurizing gas process is not inert but is "reactive" gas pressure sintering will affect μ_{N_2} entering pore pressure terms, so that the calculations shown will only be approximations. Recent Japanese work has shown that increasing p_{N_2} increases its solubility in glass.

Jack: I have assumed that you have a "balanced" composition, i.e. that you add just the amount of $BeSiN_2$ to balance the SiO_2 to give a β-structure of composition

$$(Be_xSi_{1-x})_3(O_{1.5x}N_{1-1.5x})_4$$

where x ~ 0.03. The N and O are combined in the structure and the system may well be more stable than Si_3N_4+silicate. It will not give the reaction

$$Si_3N_4 + 3SiO_2 \longrightarrow 6SiO + 2N_2$$

for example. N_2 gas in pores will not all go into the lattice as Dr. Lumby suggests, when p_{N_2} is increased. There will be dissolved N_2 and O_2 in the lattice (very small), and structural N and O, and although these will be related we probably don't know enough about the complex relationships between them. We do know that there must be a relationship between p_{O_2} and p_{N_2} at equilibrium:

$$0.1Be + 2.9Si + 0.1(O_2) + 1.9(N_2) \longrightarrow Be_{0.1}Si_{2.9}O_{0.2}N_{3.8}$$

with $K^{-1} = a_{Si}^{2.9}\ a_{Be}^{0.1}\ p_{O_2}^{0.1}\ p_{N_2}^{1.9}$.

At equilibrium p_{O_2} and p_{N_2} will be related:

$$\log p_{O_2} = -19 \log p_{N_2} + \text{Constant.}$$

The stability of the "Be" densified "Si_3N_4" depends not only on p_{N_2} but also on p_{O_2}.

Greskovich: The starting composition (Si_3N_4 + 7w/o $BeSiN_2$ + 7w/o SiO_2) is not exactly "balanced", but is slightly SiO_2 rich so that a small amount of liquid is probably in equilibrium with $Si_{3-x}Be_x$-$N_{4-2x}O_{2x}$ solid. Hence it is still possible for solid/liquid/gas reactions to occur and the extent of these reactions will depend on p_{N_2}, p_{O_2}, temperature and time.

HOT PRESSING OF ALUMINIUM NITRIDE

J.P. Lecompte, J. Jarrige and J. Mexmain

Laboratoire de Céramiques Nouvelles,
CNRS LA 320, U.E.R. des Sciences, 87060 Limoges Cedex,
France.

1. INTRODUCTION

Aluminium nitride ceramics have a good high-temperature chemical stability in contact with metals (1), and dielectric studies of hot-pressed aluminium nitride (2) have shown that the dielectric losses are particularly important at lower frequencies. It is, however, difficult to produce high purity aluminium nitride free from oxygen contamination (3) and it is now recognized that an oxygen content of about 2 wt % is necessary for obtaining full density. We discuss here investigations of the kinetics of hot pressing of commercial aluminium nitride powder, and the oxidation behaviour and mechanical properties of the dense material.

2. ALUMINIUM NITRIDE POWDER

For this study we utilised a powder fraction obtained by sieving commercial powder* in petroleum ether to prevent hydrolysis. The apparent grain size measured with a Sedigraph Micrometrics was 9 µm, while the average grain size, calculated using the equivalent sphere method from B.E.T. measurements, was 0.7 µm (BET surface area: 2.7 m^2/g). The difference between these two values was due to the fact that grains were non-spherical agglomerates, as shown by S.E.M. X-ray analysis showed only aluminium nitride, but neutron activation analysis revealed the presence of 2.7 wt % oxygen. It seemed that the oxygen existed as an oxide film of non-crystalline Al_2O_3. For the study of the grain-size influence on the rate of densification, we obtained a powder of different average grain size by the micronising method. Contamination by

*AlN, 99% pure, < 50 µm, Koch-Light Laboratories, U.K.

Riley, F.L. (ed.) Progress in Nitrogen Ceramics

alumina of the aluminium nitride powder was determined from measurements of weight loss of the grinding pellets (~3 wt % Al_2O_3, 1.44 wt % oxygen). The average grain size of the powder was 0.5 μm (B.E.T. surface area 3.3 m^2/g). The true grain size ratio of the two powders was therefore 1.4.

3. KINETIC STUDY OF DENSIFICATION BY HOT-PRESSING

A. Experimental

Hot-pressing equipment at the University of Leeds, described in detail by R.J. Weston (4), was used. Hot-pressing was carried out in inductively heated graphite dies. To avoid reaction between aluminium nitride and the graphite die (12.7 mm internal diameter) at high temperature, graphite in contact with the powder was coated with high purity boron nitride suspended in an aqueous solution of methyl cellulose. Hot-pressing was carried out at 1700°C at pressures between 5 and 30 MPa on 3 g of powder. Rates of densification were calculated for a constant relative density of 0.8 on the relative density-time plot. The density of each pressed pellet was measured with a Doulton mercury densitometer.

B. Densification rates as a function of applied pressure and grain size

The rates of densification at a relative density of 0.8 are plotted in Figure 1 for 0.7 and 0.5 μm grain size, in the range of applied pressure 5-30 MPa. We observed, for both powders, two linear curves, one covering the applied stress range of 5-24 MPa and the other from 24 MPa upwards where the densification rate was very fast. The threshold stress (σ_0) determined graphically was 6 MPa for both powders. If the kinetics of sintering are expressed by the general relationship, $d\rho/dt = k\ (\sigma - \sigma_0)^n$ n, the stress exponent in each regime can be calculated.

In Figure 2 is plotted the logarithm of the densification rate as a function of the logarithm of the difference between the applied pressure and the threshold stress. For both powders, the stress exponents were nearly 1 in the low stress range and nearly 10 in the range 24-30 MPa. The ratio of densification rates between the two powders in the lower stress region is 2.7.

C. Discussion

We use here the well-established procedure of comparing the experimental stress exponent of the densification rate with the values predicted by theoretical models (5). For pressures above 24 MPa the stress exponent is nearly 10, and results from the generation and movement of dislocations. Several mechanisms can explain the stress exponent of 1 observed in the low stress regime; for example, densification by plastic flow or densification by diffusion. The densification equations for a plastic flow

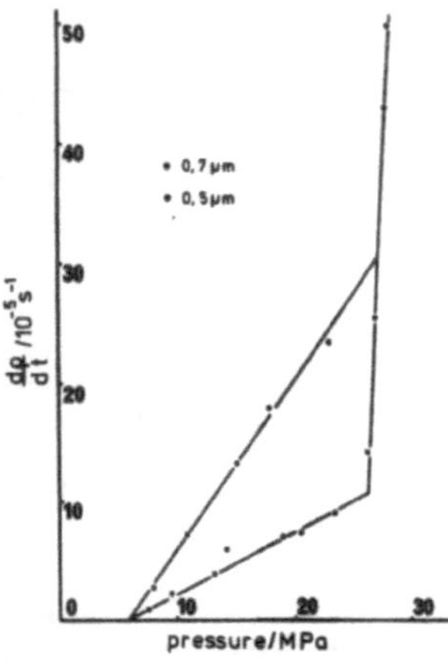

Fig. 1: Densification rates as a function of applied pressure at a relative density of 0.8.

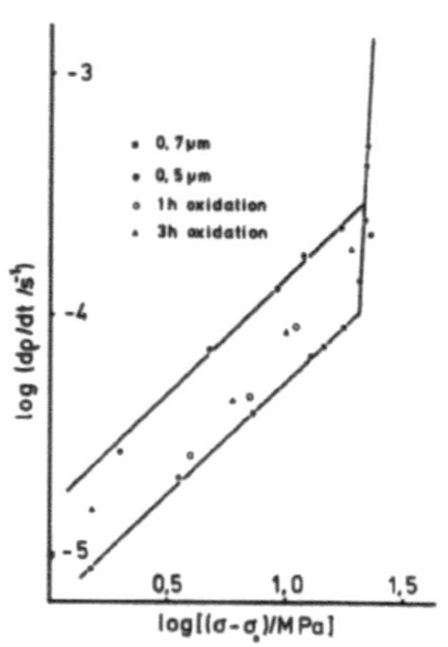

Fig. 2: Densification rates as a function of applied stress minus threshold stress (log-log scale). Data taken from Figs. 1 and 4.

mechanism of material transport during hot-pressing have been based on Mackenzie and Shuttleworth's analysis (6). The formulation proposed by Murray and adapted by Sakai and Iwata (7) did not prove helpful; the logarithmic plotting of the porosity $(1-\rho)$ above the initial porosity $(1-\rho_o)$ as a function of time did not yield that of the predicted straight line.

The equation developed for the prediction of creep rates based on diffusion by Nabarro-Herring (8) assumes a lattice diffusion mechanism, and that by Coble (9) grain boundary diffusion. Both processes are linear with applied pressure. Thus, for a grain size exponent of 2 the indicated process is that of lattice diffusion; an exponent 3 would suggest grain boundary diffusion. Figure 2 shows that the instantaneous rate of densification increases by a factor $\simeq 2.7$ when the grain size decreases by a factor of 1.4.

These results give a grain size exponent of 3 corresponding to Coble diffusion. In order to confirm this mechanism we tried to increase the thickness of the oxide film around the aluminium nitride grains by adding alumina powder* and by oxidizing the powder.

The addition of 5 wt% of alumina does not show (Figure 3) an increase in the densification rate. It can be seen in Figure 4, however, that the rates of densification increase with oxide formation. Oxide contents were analysed by neutron activation and the powders contained 6.45 (1 h oxidation at 800°C) and 14.2 wt% (3 h oxidation at 800°C) of oxygen. Increasing densification rates with both grain boundary thickness growth, and decrease of grain

*α-Al_2O_3, 99.99% pure, grain size 0.3 μm, Linde 0.3A.

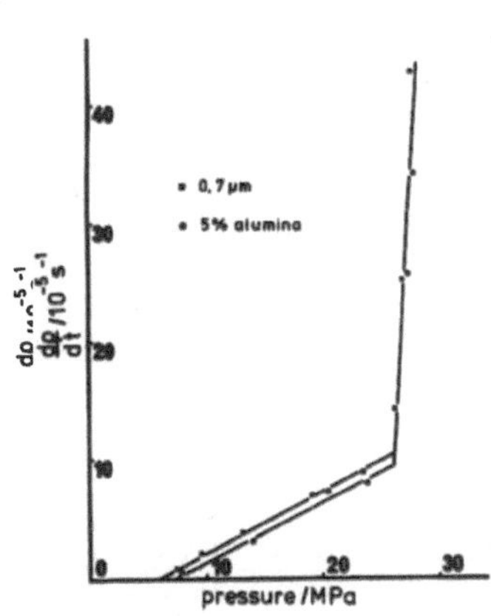

Fig. 3: Rates of densification as a function of applied pressure for the 0.7 µm powder and the powder containing 5 wt% alumina.

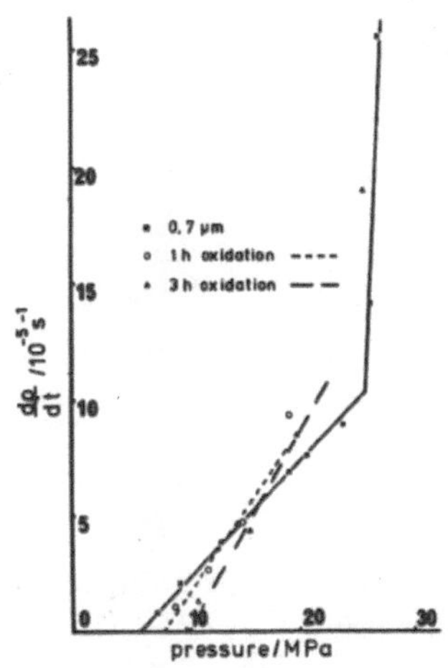

Fig. 4: Rates of densification as a function of applied pressure for powders oxidized 1 h and 3 h.

size, enable one to conclude that the grain-boundary diffusion is the rate-controlling step in the pressure range less than 24 MPa. For pressures between 24 and 30 MPa, dislocation creep is dominant. Oxidation pre-treatment appears more effective at distributing the grain boundary oxide phase than the addition of oxide powder.

4. MECHANICAL PROPERTIES

Fully densified aluminium nitride has a microhardness of 1560 Vickers, which is comparable to those of silicon oxynitride and silicon nitride. The strength of the hot-pressed aluminium nitride was measured as a function of porosity and temperature by the 3-point bending method with a cross-head speed of 0.2 mm/min. The results are given in Figure 5, in comparison with other materials. At room temperature the strength is greater than that of RBSN but decreases when the temperature increases, to reach a stable value from 800 to 1400°C. At 1400°C, the strength is of the same order as those of the other materials. We also studied at room temperature the strength as a function of the porosity. The results show that the strength remains almost constant for $P < 10\%$ and decreases afterwards. The high strength value required for R, the resistance parameter for thermal shock, is acceptable but aluminium nitride has a high coefficient of thermal expansion (5.1×10^{-6} °C^{-1}, 0-450°C) and a high Youngs modulus (315 GPa). Theoretical calculation of the R parameter gives 70°C. This R value is rather low but $R' = kR$ is important, according to the result given by Knoch (this meeting), because AlN has a good thermal conductivity. However, the thermal shock resistance (Δ Tc) determined by a non-destructive method was about 250°C. For a porosity $\sim$10%, the value observed was 310°C because the strength is constant and it has been shown (10) that

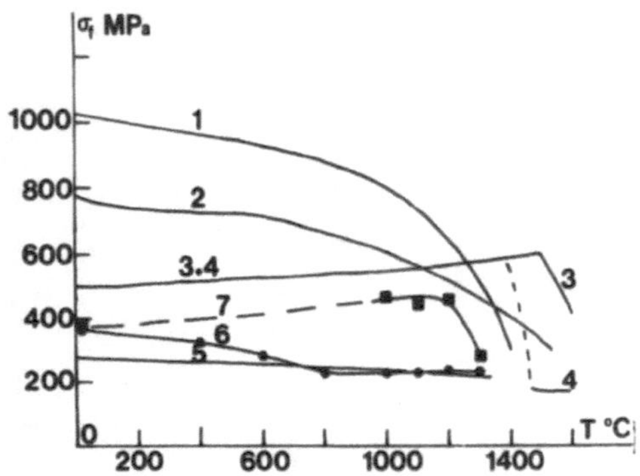

Fig. 5: Temperature dependence of 3-point bend strength. 1 - Si_3N_4 (+ MgO), 2 - SiC (+Al_2O_3), 3 - SiC (+B), 4 - SiC R sintered, 5 - RBSN, 6 - AlN, 7 - SiYON (70-25-5) mol.%.

the Young's modulus and the thermal expansion coefficient decrease as the porosity increases.

5. HOT CORROSION IN AIR

Oxidation was carried out in air between 1300 and 1700°C. Cubes (4 mm edge) of fully dense HP-AlN were suspended on the thermobalance by means of fine platinum wire. Figure 6 shows the relative increase in mass per unit surface area as a function of time. The curves obtained show that the material displays little oxidation up to 1490°C (4 mg/cm^2) after 24 hours. Up to 1600°C, oxidation becomes important. At 1300°C the oxide film has slight porosity and is adherent to the substrate. After a few hours at 1490°C, the scale becomes increasingly porous and non-protective. In the α range 0.01 - 0.2 (α = fraction of nitride reacted), the experimental results are well fitted by a parabolic law. In the range 1300 to 1700°C, the activation enthalpy determined from the Arrhenius plot was 326 $\pm$ 67 kJ mol^{-1}. In order to compare the corrosion behaviour in air of different materials, all the samples were prepared in the same geometrical conditions. The corresponding curves of α as a function of time are shown in Figure 7 for the temperature 1400°C. The oxidation resistance of AlN at this temperature is rather high compared with other silicon based ceramics presumably due to the different plasticity between the oxidation products (alumina, silica and silicates).

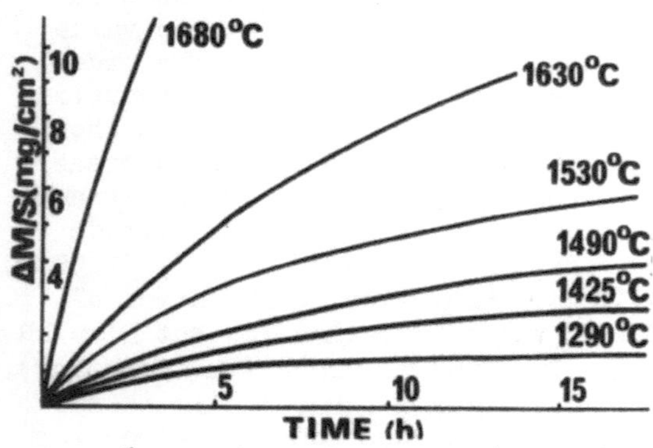

Fig. 6: Weight gain curves by unit surface area as a function of time.

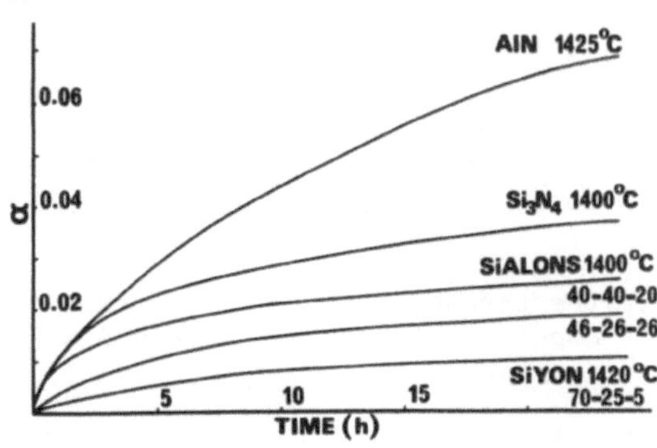

Fig. 7: α as a function of time for different materials. HPSN (11) (Tripp and Graham), Sialons (Si_3N_4, Al_2O_3, alN % in moles) from Brossard (12), Syon (Si_3N_4, SiO_2, Y_2O_3 % in moles) from Bouarroudj (13).

ACKNOWLEDGMENTS

The authors wish to thank Professor R.J. Brook and Dr. F.L. Riley of Leeds University for their helpful collaboration and discussion.

REFERENCES

1. F.S. Garibyan and G.K. Kozina. Sov. Powder Met. Metal. Ceram. 12 (5), 413 (1978).
2. A. Kumar, J.P. Lecompte and A. Moliton. Rev. Phys. Appl., 1981, to be published.
3. T. Sakai and I.M. Iwata. J. Mat. Sci. 12, 1659 (1977).
4. R.J. Weston and T.G. Carruthers. Proc. Brit. Ceram. Soc. 22, 197 (1973).
5. M.N.S. Rahaman, F.L. Riley and R.J. Brook. J. Amer. Ceram. Soc. 65, 648 (1980).
6. J.K. Mackenzie and R. Shuttleworth. Proc. Phys. Soc., 62, 833 (1949).
7. T. Sakai and I.M. Iwata. Jap. J. Appl. Phys. 15, 537 (1976).
8. C. Herring. J. Appl. Phys., 21 (5), 437 (1950).
9. R.L. Coble. J. Appl. Phys. 41 (12), 4798 (1970).
10. P. Boch, J.C. Glandus, J. Jarrige, J.-P. Lecompte and J. Mexmain. Ceramics Int. (1981), to be published.
11. W.C. Tripp and H.C. Graham. J. Am. Ceram. Soc. 59 (9), 399 (1976).
12. J.G. Desmaison, M. Brossard, M. Desmaison-Brut and P. Goursat. This Institute.
13. A. Bouarroudj. Thèse de 3e Cycle, Limoges (1981).

DISCUSSION

Morgan: I feel we know from the behaviour of pure covalent materials that bulk diffusivity is very low and that so-called "Nabarro-Herring" creep is a priori not possible and need not be considered in these systems.

Clarke: What is the origin of the threshold stress, for onset of densification, and why does it appear to increase when powders are preoxidized?

LeCompte: Ashby suggests that rate of climb of dislocations in the boundary is responsible. The dislocations would be forced to adopt local curvature for by-passing the particles. Theory requires σ_o to be proportional to the second phase volume.

Porz: Why do you see no influence of porosity on strength below 10% porosity?

LeCompte: We explain this behaviour in terms of a particular porosity distribution; we have a small percentage of open porosity, which will be responsible for fracture initiation.

Section E

VITREOUS STATE

STRUCTURE-PROPERTY RELATIONSHIPS OF THE VITREOUS STATE

Helmut A. Schaeffer

Institut für Werkstoffwissenschaften (Glas und Keramik), Universität Erlangen-Nürnberg, Erlangen, Germany

1 INTRODUCTION

The importance of the vitreous state in ceramic materials such as high-temperature oxides or nitrides, is indicated by the existence of non-crystalline (mostly silicate composed) grain boundary regions and moreover by the formation of non-crystalline silica or silicate scales during the oxidation of silicon-containing materials.
Of special interest are relationships between glass structure and glass properties, since a variety of materials properties are governed by kinetic processes within the "glassy" phase. Therefore, studies of kinetic phenomena in non-crystalline silicates are required for a deeper understanding of diffusion-controlled properties not only of glasses but especially of "crystalline/vitreous composites".

The treatment of kinetic processes will be limited here to the diffusion of those species which are responsible for the "cohesiveness" of the silicate glass network, i.e. oxygen and silicon. These network-forming species belong to the less mobile species as compared to the alkali and the alkaline earth ions (network modifiers) and are thus rate-controlling in transport processes which consist of formation or rearrangement of the glass network, e.g. reaction kinetics of glass phases at grain boundaries, occurrence of immiscibilities and crystallizations within the glass, and viscous flow. It thus becomes evident that the rate-controlling step for various transport phenomena is the diffusivity of either oxygen or silicon.

Riley, F.L. (ed.) Progress in Nitrogen Ceramics

2 NETWORK DEFECTS IN SILICATE GLASSES

The concepts of defect structure of non-crystalline silicate networks are intimately connected with the characteristic features of the vitreous state, i.e. the existence of short-range order and the absence of long-range order. In the case of vitreous SiO_2 the glass structure is characterized by the SiO_4 tetrahedron with a Si-O bond length of 0.16 nm and by a rather large distribution of the Si-O-Si bond angles between adjacent SiO_4 tetrahedra (120-180° with the maximum of the distribution at about 145°). This broad distribution is responsible for the unique flexibility of the SiO_2 glass network.

Deviations from the Si-O bond length (for instance due to the incorporation of network modifiers) or deviations from the mean bond angle (due to local ordering or phase separation) can then be envisaged as defects.
The crucial difference between a diffusion process in a crystalline (rigid) matrix and in a vitreous (flexible) network consists of the fact that the motion of a defect through a glass structure causes an irreversible displacement of the network-forming and/or network-modifying species, whereas in a crystal the original structure is restored after the passage of a defect. This implies that a fixed lattice site in a glass network cannot be defined. The structural flexibility is caused by the wide range of Si-O-Si bond angels which can be accomodated with little changes in energy (1). Vitreous SiO_2 is in this respect unique, having a range of bond angles of 60° as compared to 40° in vitreous GeO_2, for instance. In Si-N compounds the distribution of the corresponding bond angles is even narrower ($\approx 15^o$) (1).

Another typical feature of vitreous systems is that they are frozen-in undercooled melts and thus are in a thermal-structural non-equilibrium. With respect to defect formation the process of freezing-in implies that at temperatures in the vicinity of T_g (glass transformation temperature) the corresponding equilibrated structural state will be fixed and therefore also the concentration of defects, i.e. the defect concentration becomes temperature-independent below T_g: $c_{def} \propto \exp(-\Delta H_f/RT_g)$, with ΔH_f the enthalpy for defect formation (2). It follows that diffusion processes below T_g can then be characterized by a diffusion coefficient

$$D = D_o \exp(-Q/RT) \propto \exp(-\Delta H_f/RT_g) \cdot \exp(-\Delta H_m/RT) \qquad \text{Eq. (1)}$$

The experimentally observed activation energy Q can be identified with the enthalpy of migration ΔH_m, whereas the preexponential factor D_o contains the temperature-independent defect concentration. Diffusion processes below T_g are therefore linked with a smaller activation energy as compared to temperatures above T_g, where additionally the energy for defect formation has to be raised.

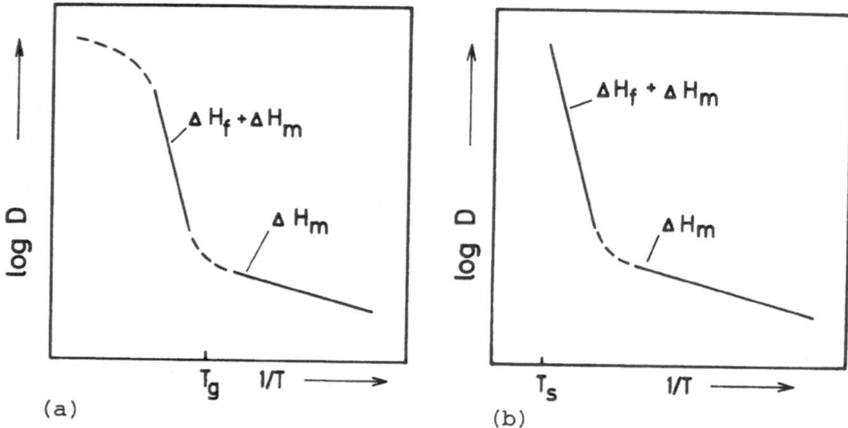

Fig. 1. Comparison between temperature dependence of diffusion coefficients in vitreous and crystalline materials
(a) glass: frozen-in defects for $T < T_g$ (≙ glass transformation temperature)
(b) ionic crystal with impurities: transition from extrinsic to intrinsic defect formation (T_s ≙ melting point)

This is phenomenologically similar to diffusion processes in ionic crystals where diffusion coefficients change from regions of low activation energy (controlled by extrinsic defects) to regions of high activation energy due to intrinsic defect formation at higher temperatures, as in Fig. 1. With increasing temperature a depolymerization of the glass network will occur so that the assumption of a "paracrystalline" solid no longer holds, i.e. an estimation of the defect concentration according to $\exp(-\Delta H_f/RT)$ is no longer valid, cp. Fig. 1.

Another aspect of defect formation in glasses results from a possible relationship between the nature of the Si-O bond and the occurrence of network defects. Traditionally ionic bond interaction was a favored premise for silicate glasses resulting in the concept of a small Si^{4+} ion (radius o.o4 nm) surrounded by large O^{2-} ions (radius o.14 nm). The other conceptual extreme leads to a covalent bond interaction now reversely consisting of a large silicon atom (radius o.117 nm) and a small oxygen atom (radius o.o66 nm). As was pointed out by Noll (3) the presence of other elements, besides oxygen and silicon, affect the Si-O bond nature (degree of covalency) and thus should give rise to different ratios of silicon and oxygen radii, which in turn are expected to cause different ratios of silicon and oxygen diffusivity (4).

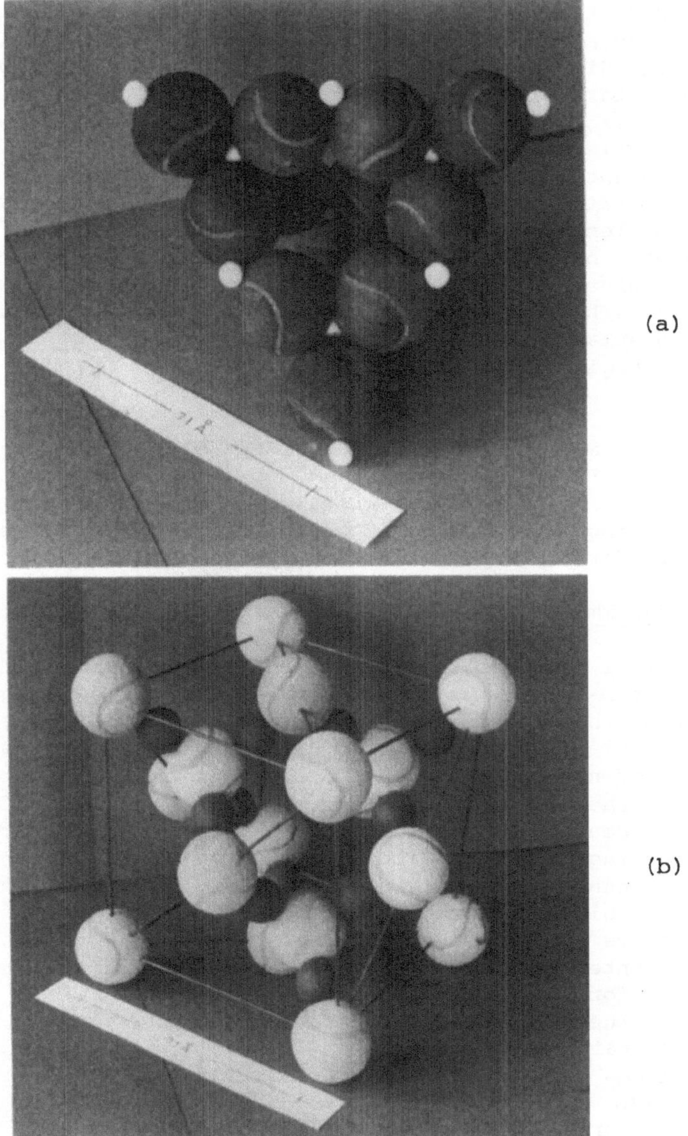

Fig. 2. Structural models of the unit cell of high-temperature cristobalite; oxygens (dark), silicons (light) (a) ionic bonding, (b) covalent bonding

Fig. 2 shows the unit cell of high-temperature cristobalite for each extreme Si-O bond case. High- temperature cristobalite is the crystalline SiO_2 structure that most closely resembles vitreous SiO_2. Qualitatively, the ionic model indicates that the diffusivity of silicon exceeds that of oxygen, whereas the reverse would be true for the covalent model.
The experimental evidence is that in SiO_2 glass and in most multi-component silicate glasses the oxygen is more mobile than silicon, so the covalent bond model seems to be more appropiate. Only in glasses with network modifiers having electronegativities larger than the value for silicon is predominant ionic bonding to be expected (4). This behavior is observed in hydrated silicates, where silicon displays a high mobility as compared to oxygen which is part of a more rigid oxygen network.

3 DIFFUSION MECHANISM OF NETWORK FORMER SPECIES IN SiO_2 GLASS

On the basis of the covalent model which appears suitable in characterizing the majority of silicate glasses possible diffusion machanisms for oxygen and silicon transport will be discussed.

3.1 Mechanism of Oxygen Diffusion

Experimentally, it has been found that the oxygen self-diffusion coefficient D_O, i.e. the diffusivity of the network oxygen, depends linearly on the oxygen partial pressure p_{O_2}. This is a strong indication for the action of dissolved oxygen molecules in assisting the migration of network oxygen. A further experimental fact to be considered is the low activation energy related to the diffusion of network oxygen, amounting to only 1oo kJ mol^{-1}. Therefore, a complete breaking of the Si-O bond during the diffusion process of network oxygen appears unlikely. Instead a diffusion mechanism was proposed (5,6) which consists of an oxygen exchange between oxygen network sites and O_2 molecules. Thus the dissolved O_2 molecules generate "interstitial defects" in the sense that they act as "carriers" for the transport of network oxygens.
Therefore, the diffusion of network oxygen becomes proportional to the concentration of O_2 molecules (c_{O_2}) dissolved in SiO_2 glass, i.e. $D_O \propto c_{O_2} \propto p_{O_2}$; the latter proportionality follows from Henry's Law.
It was further shown (6) that the proposed diffusion mechanism leads to the relationship

$$D_O \cdot c_O = D_{O_2} \cdot c_{O_2} \qquad \text{Eq.(2)}$$

with c_O standing for the concentration of network oxygen in SiO_2 glass and D_{O_2} for the diffusivity of O_2 molecules in SiO_2 glass.

Table 1. Diffusivities of oxygen and silicon in SiO_2 glass, $D = D_o \exp(-Q/RT)$

species	D_o ($cm^2\ s^{-1}$)	Q ($kJ\ mol^{-1}$)	T (^oC)	ref.
^{18}O	2.0×10^{-9}	121	850-1250	(7)
^{18}O	4.4×10^{-11}	82	1150-1430	(8)
O_2	2.8×10^{-4}	113	900-1100	(9)
^{30}Si	328	579	1100-1400	(10)

This relationship was successfully tested with available data for temperatures >1100°C (see Table 1); e.g. for 1100°C it was found that $D_O/D_{O2} = 2 \times 10^{-6}$ and $c_{O2}/c_O = 2.4 \times 10^{-6}$.

Equation (2) does not only lead to the conclusion that the diffusivity of networkoxygens occurs via exchange with O_2 molecules, but also that the diffusivity of O_2 molecules requires the exchange with network oxygen. The O_2 diffusion mechanism displays similarities with an "interstitialcy mechanism", i.e. a continuous interaction of the interstitially dissolved O_2 molecule with the oxygen network during each diffusion jump.

The oxygen diffusion mechanism is compatible with the covalent bond model, since it offers a plausible explanation for the low activation energy of oxygen transport. The directional chemical bonding and the small radius of oxygen with respect to silicon allow a more effective "penetration" of an O_2 molecule (i.e. 2 covalently bonded oxygen atoms) into the SiO_4 tetrahedron, cp. Fig. 2. It is thought that the oxygen exchange process is further facilitated energetically by the formation of an activated transitional ozone-type configuration which permits an easy oxygen exchange via rotation of this O_3 complex.

3.2 Mechanism of Silicon Diffusion

Silicon diffusion data in SiO_2 glass were obtained only recently, see Table 1. The silicon diffusivity is many orders of magnitude smaller than the diffusivity of network oxygen and exhibits an activation energy that is of the order of the Si-O single bond energy, thus suggesting the breaking of Si-O bonds during the diffusion process.

The covalent model of high-temperature cristobalite reveals that the diffusion of silicon is feasible by exploiting the adjacent unoccupied sites (unit cell consists of eight cubes whose centers are alternately occupied with silicon atoms, cp. Fig. 2). This type of diffusion mechanism takes advantage of the "openness" of the high-temperature cristobalite structure, which becomes especially evident in the case of covalent bonding. However, it is not clear if silicon migrates alone or associated with oxygen (e.g. SiO_4 unit). Contrary to the oxygen diffusion mechanism the proposed silicon diffusion enables rearrangements of the network structure and is therefore a possible rate-limiting step for viscous flow, see chapter 5.3.

4 DIFFUSION MECHANISMS OF NETWORK FORMER SPECIES IN SILICATE GLASSES

Diffusion processes of network formers in multicomponent glasses are more complicated, which can be anticipated from the variety of oxygen species: in addition to the bridging oxygens, non-bridging oxygens and even oxygen ions (in alkali- or alkaline earth-rich melts) are now present.

A different diffusion mechanism for oxygen in multicomponent glasses is immediately suggested due to the p_{O_2} independence (or at least weak dependence) of oxygen diffusivity, which would seem to imply that the O_2 molecules are no longer assisting the migration of network oxygens.

However, the diffusion data of network oxygen and O_2 gas as listed in Table 2 show similarities with respect to the activation energies, so that an interaction between network oxygen and dissolved oxygen may still be envisaged. Therefore, Eq. (2) was tested tentatively by inserting the following alterations (11): $(D_O \cdot c_O)$ was replaced by the corresponding quantities for non-bridging oxygens $(D_{O'} \cdot c_{O'})$, since these oxygens surpass the bridging oxygens in their mobility, and c_{O_2}, the "physical" solubility of oxygen, was replaced by the chemical solubility of oxygen $(c_{O2})_{chem}$, which typically exceeds the physical solubility due to the presence of impurities of polyvalent character. The chemical solubility of oxygen is given by the amount of oxygen required to achieve the transition from the lower to the higher valence state of polyvalent cations or anions and thus depends predominantly on the concentration of polyvalent species in the glass c_{poly}, therefore

$$D_{O'} \cdot c_{O'} = D_{O2} \cdot (c_{O2})_{chem} \approx D_{O2} \cdot c_{poly} \qquad \text{Eq.(3)}$$

Assuming for instance, that o.1% of the most common polyvalent impurity (iron oxide) is present in a glass that contains 1o% alkali, a ratio of $(c_{O2})_{chem}/c_{O'} \approx 1o^{-2}$ can be estimated. This ratio corresponds to the observed ratio of $D_{O'}/D_{O2}$, see Fig. 3 and Table 2.

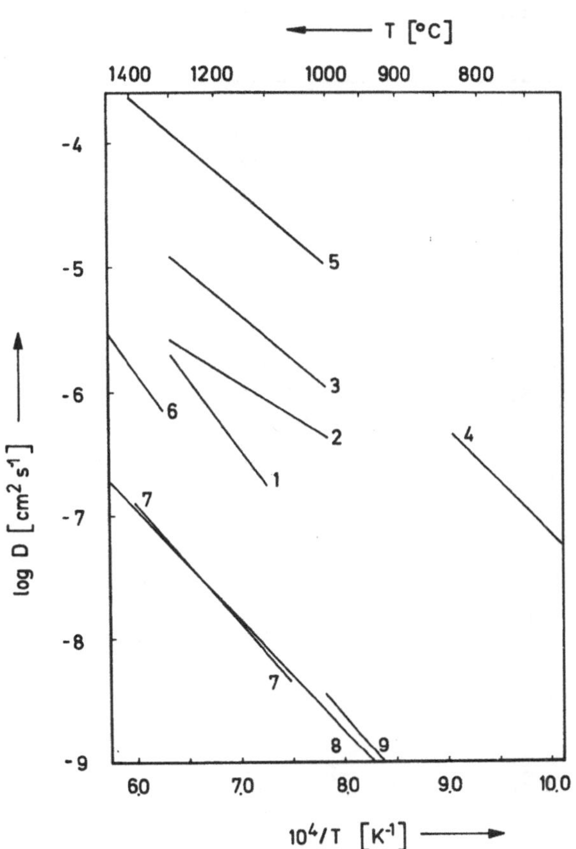

Fig. 3. Diffusion coefficients of molecular oxygen and network oxygen (tracer diffusion) in multicomponent silicate glasses, cp. Table 2.
≠≠ 1-5 O_2 diffusion; ≠≠ 6-9 network oxygen diffusion
≠≠ 1 and ≠≠ 8: Na_2O-CaO-SiO_2 glass
≠≠ 5 and ≠≠ 6: CaO-Al_2O_3-SiO_2 glass

Due to the paucity of data, especially with respect to the concentration of polyvalent impurities, a general validity for Eq.(3) cannot be established so far. Nevertheless, a dependence of network oxygen diffusion on the concentration of polyvalent species is strongly suggested, which would explain the weak p_{O_2} dependence (e.g. a change of the oxidation state from Fe^{2+} to Fe^{3+} is related with a $p_{O_2}{}^{1/4}$ dependence).

Table 2. Diffusivities of molecular oxygen and network oxygen in multicomponent silicate glasses

≠≠ in Fig. 3	composition (wt%)	D_o ($cm^2\ s^{-1}$)	Q ($kJ\ mol^{-1}$)	T (^oC)	ref.
O_2 diffusion					
1	Na_2O-CaO-SiO_2	42	222	1100-1300	(12)
2	BaO-Al_2O_3-alkali-SiO_2	$5.4x10^{-3}$	100	1000-1300	(12)
3	B_2O_3-SiO_2	0.32	134	1000-1300	(12)
4	30SrO,2Al_2O_3,29TiO_2, 39SiO_2	28.4	166	630-830	(13)
5	40CaO,20Al_2O_3,40SiO_2	4.5	138	1000-1400	(14)
network oxygen diffusion					
6	40CaO,20Al_2O_3,40SiO_2	18	227	1320-1540	(15)
7	20Na_2O,80SiO_2	$7.9x10^{-2}$	186	1060-1395	(15)
8	16Na_2O,12CaO,72SiO_2	$3.1x10^{-2}$	174	800-1470	(16)
9	36K_2O,64SiO_2	0.24	193	700-1000	(17)
	37K_2O,62SiO_2	11	250	800-900	(18)

Silicon diffusion data for multicomponent glass systems are scarcely found in the literature. The only exception is a CaO-Al_2O_3-SiO_2 glass in which oxygen and silicon diffusivities were measured, cp. Fig. 4, displaying an interesting order of diffusivities above and below T_g. It is believed that this change of order is related to increased covalent bond character at high temperatures compared to a more ionic character below T_g (4).

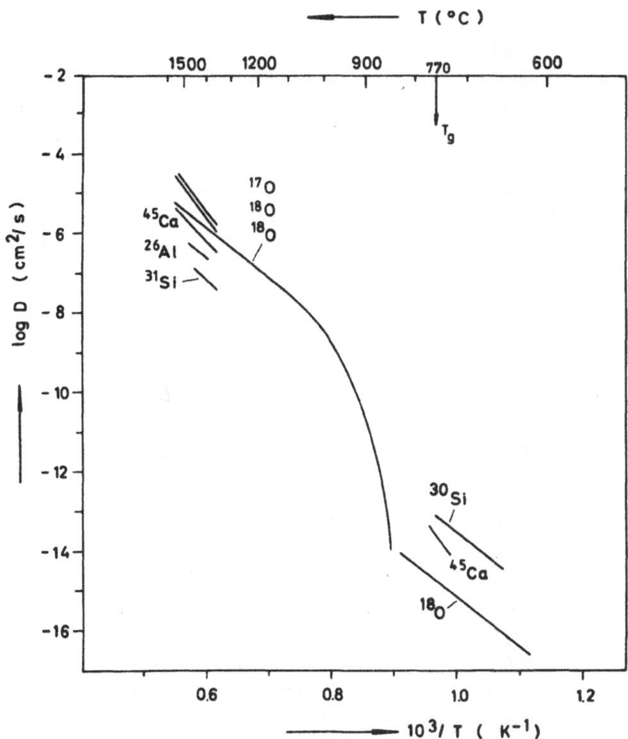

Fig. 4. Tracer diffusion coefficients in a 40 CaO-20 Al_2O_3-40 SiO_2 glass (wt%) (4).

5 GLASS STRUCTURE-PROPERTY RELATIONSHIPS

It will be shown that the network diffusion of oxygen in SiO_2 glass is rate-controlling for oxygen transport processes such as oxygen gas permeation through vitreous SiO_2 and for the oxidation of silicon-containing materials. On the other hand, silicon diffusion in SiO_2 glass seems to be the rate-controlling step of viscous flow.

In the case of multicomponent silicate glasses, diffusion-controlled properties are governed by other mechanisms. As for SiO_2 glass, oxidation processes through silicate glass layers are believed to be rate-controlled by the diffusivity of network oxygen, but here oxygen diffusion also determines the viscous flow behavior.

5.1 Oxidation Kinetics

A series of silicon-containing materials (e.g. Si, $MoSi_2$, SiC, Si_3N_4) form non-crystalline SiO_2 layers during oxidation. Apart from an initial linear time law the formation of silica is governed by a diffusion process as indicated by the parabolic growth of the SiO_2 layer (thickness x): $x^2 = k_p \cdot t$ (t ≙ oxidation time, k_p ≙ parabolic rate constant).
Due to the fact that the oxidation takes place at the substrate/silica interface and that k_p is linearly proportional to p_{O_2}, it was recognized that the diffusion of O_2 is the rate-controlling step for the oxidation process (19).
The mathematical treatment of a "tarnishing" reaction via gas permeation leads in the case of silica formation to the relationship of

$$k_p = 2\, D_{O_2} \cdot c_{O_2}/c_O \qquad \text{Eq.(4)}$$

Eq.(4) can also be written in terms of oxygen gas permeability ($P_{O_2} = D_{O_2} \cdot c_{O_2}$): $k_p = 2\, P_{O_2}/c_O$.

Combining Eq.(4) with Eq.(2) yields

$$k_p = 2\ \ D_O \qquad \text{Eq.(5)}$$

indicating a direct relationship between k_p and the diffusivity of network oxygen in SiO_2 glass. The validity of this relationship was shown to hold best for the parabolic growth of silica layers on silicon and $MoSi_2$ (6).Eq.(5) also holds with respect to the O_2 pressure dependence, since k_p was found to be linearly proportional to p_{O_2}.
For the oxidation of SiC, Eq.(5) is basically fulfilled for temperatures between 12oo and 14oo°C. However, deviations occur at lower temperatures, predominantly due to the fact that the oxide layers become so thin that the oxidation rate is governed by linear growth kinetics, which in turn is linked to a higher activation energy (comparable to the Si–C single bond energy of 318 kJ mol^{-1} (11)). A further deviation has been observed for temperatures ≥14oo°C (also in the case of $MoSi_2$) consisting of a large increase of the oxidation rate, e.g. for SiC an activation energy of about 6oo kJ mol^{-1} was observed. This activation energy is surprisingly close to the activation energy of silicon (or SiO_4) diffusion in SiO_2 glass, and the comparison of the absolute values of k_p and D_{Si} reveals that they are of the same order. Therefore, a different oxidation process must now become operative. This conclusion is further substantiated by the fact that depending on the impurity level of the SiO_2 glass the diffusivity of silicon starts to exceed the oxygen diffusivity between 17oo and 14oo°C, i.e. the silicon (or SiO_4) transport becomes faster than the oxygen transport via exchange with O_2 molecules.

In agreement with this changed oxidation mechanism is the observation that the dependence of k_p on the oxygen partial pressure decreases with increasing temperature and finally vanishes for temperatures $\geq$ 15oo°C.

Turning to the oxidation behavior of "pure" Si_3N_4 (CVD layers, powders), large deviations from Eq.(5) become evident, with the general trend being that k_p values are 1 to 2 orders of magnitude smaller than the parabolic rate constants for silicon or silicon carbide in the temperature range from 12oo to 14oo°C.
Several reasons can be responsible for the decreased oxidation rate, e.g. nitrogen permeation-controlled reaction due to the buildup of N_2 at the Si_3N_4/SiO_2 interface or the formation of a Si_2ON_2 layer affecting the transport of nitrogen and oxygen.
In this context attention is drawn to the reported formation of tridymite during the oxidation process. Tridymite is known (similar as quartz) to exhibit an oxygen diffusion coefficient which is 3 orders of magnitude smaller than in SiO_2 glass (2o). It is interesting to note that the decreased k_p values were observed to be proportional to the reported or assumed amounts of tridymite being present in the oxidation product (21-23).
Similar to the case of SiC and $MoSi_2$, higher activation energies ($\approx$ 4oo kJ mol^{-1}) were measured for temperatures >14oo°C (22), as well as a p_{O_2} independence of k_p (21,22). Again, the diffusion of silicon is suggested as the rate-determining species for the oxide growth in contrast to the interpretation given in ref.(24), where oxygen diffusion control via a vacancy mechanism in SiO_2 glass is assumed.

The oxidation process of hot-pressed Si_3N_4 and sialons in forming multicomponent silicate glass layers with various amounts of embedded crystalline silicates is debated even more (25-27). A parabolic growth rate, typically 3 to 4 orders of magnitude larger than on "pure" Si_3N_4, again indicates a diffusion-controlled process. In the literature,especially the concept of a Mg^{2+} or Ca^{2+} diffusion as the rate-limiting step has been discussed, either in the form that Mg^{2+} ions are rate-determining due to their migration through the oxide layer (25,27), or that the supply of Mg^{2+} from the grain boundary phase into the surface oxide layer represents the rate-controlling step (26).
The inward diffusion of oxygen as a conceivable rate-controlling mechanism was excluded solely on the basis that the activation energies of the parabolic rate constants appeared too high in order to be explained by the diffusion of oxygen and due to the fact that k_p is independent on p_{O_2} (25).
Unfortunately, literature data are not available for oxygen diffusion in MgO- or CaO-containing silicate glasses, however, multicomponent silicate glasses which have been investigated display activation energies for oxygen diffusion which are by a factor of

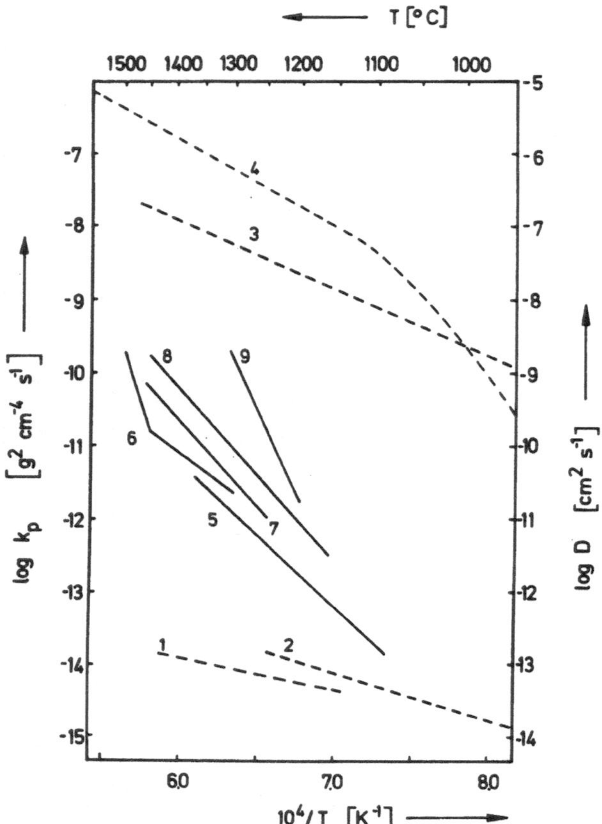

Fig. 5. Comparison between diffusion coefficients of network oxygen in SiO_2 glass (≠≠ 1(8), ≠≠ 2(7)), in multicomponent silicate glasses (≠≠ 3(16): 72 SiO_2, 16 Na_2O, 12 CaO in wt%; ≠≠ 4 (15): 40 SiO_2, 20 Al_2O_3, 40 CaO in wt%) and parabolic rate constants for the oxidation of hot-pressed Si_3N_4 (≠≠ 5-9) (25,28,26,29); activation energies (kJ mol^{-1}): ≠≠ 1: 82, ≠≠ 4: 227, ≠≠ 5: 375. It was estimated from some measurements where the oxide thickness was determined that the k_p values in the unit cm^2 s^{-1} are an order of magnitude larger than in the unit g^2 cm^{-4} s^{-1}; therefore, the right ordinate is shifted against the left one by a factor of 10.

2 to 3 larger than in SiO_2 glass and show no p_{O_2} dependence in contrast to oxygen diffusion in SiO_2 glass. Therefore, the basis for the exclusion of oxygen-controlled oxidation of for instance hot=pressed Si_3N_4 can not be maintained.

On the contrary, there is evidence that the oxidation mechanism is controlled by oxygen diffusion through the surface oxide layer, either in the form of network oxygen or O_2 molecules (this would be equivalent on the basis of Eq.(3)).
In Fig. 5 data are collected for network oxygen diffusion and parabolic rate constants. It can be seen that the k_p values of hot-pressed Si_3N_4 fall between the diffusivities of network oxygen in SiO_2 glass and those in multicomponent silicate glasses, i.e. glasses of the composition as found to be present in surface oxide layers (predominantly alkaline earth silicate glasses) are expected to possess oxygen diffusivities which are of the order of the k_p values.

Further indications for this pronounced dependence of the network oxygen diffusivity on composition are given by oxygen tracer diffusion data of a rhyolithic melt (73 SiO_2, 13 Al_2O_3, 2 (Fe_2O_3+FeO), 2 (MgO+CaO), 8 (Na_2O+K_2O) in wt%), an andesite melt (60 SiO_2, 17 Al_2O_3, 6 (Fe_2O_3+FeO), 9 (MgO+CaO), 6 (Na_2O+K_2O) in wt%), and a basaltic melt (50 SiO_2, 15 Al_2O_3, 12 (Fe_2O_3+FeO), 15 (MgO+CaO), 5 (Na_2O+K_2O) in wt%): at 1280°C the diffusion coefficients increase respectively from 4×10^{-11}, 2×10^{-9} to 3×10^{-8} $cm^2\ s^{-1}$ (30). These data are comparable with the k_p values in Fig. 5, as well as with respect to the activation energy, considering that in the basaltic melt the activation energy of oxygen diffusion amounts to 375 kJ mol^{-1}. It is interesting to note that typical crystalline silicates found in oxide layers display oxygen diffusion coefficients which also compare favorably: at 1280°C, enstatite ($Mg_2[Si_2O_6]$) 6×10^{-12} cm^2/s, diopside ($CaMg[Si_2O_6]$) 2×10^{-12} cm^2/s (30).

5.2 Viscous Flow Kinetics

Viscous flow is also a diffusion-controlled process. The relationship between viscosity η and the rate-determining "flow unit" or species is given by a Stokes-Einstein-type equation. The classical Stokes-Einstein equation neglects the atomic structure of the liquid and considers it as a continuum. The extension of this relationship on the basis of hydrodynamics (31) takes into consideration both the size of the solvent and solute molecule, i.e. when both types of species become indistinguishable the following relationship holds

$$D\eta/kT = 1/4\pi r \qquad \text{Eq.(6)}$$

with D the self-diffusion coefficient and r the radius of the rate-determining species.

It was pointed out by Glasstone et al. (32) that for the case of differently sized diffusion species or aggregates, the slower moving unit is not necessarily rate-determinig for viscous flow but instead the smaller and faster one ("ball bearing effect"). This concept was tested successfully by Oishi et al. (15) for multicomponent silicate glasses which showed that the diffusivity of oxygen is the rate-determining step for viscous flow.

In SiO_2 glass, however, the mismatch between the activation energy for viscous flow (500-700 kJ mol^{-1}) and oxygen diffusion is so large (cp. Table 1), that a species other than oxygen must be rate-controlling. The data for silicon tracer diffusion exhibit an activation energy (579 kJ mol^{-1}) which falls into the range of those for viscous flow, see Table 1. The relationship Eq.(6) can now be tested by inserting D_{Si} and η for SiO_2 glass. A radius of 0.1-0.2 nm is obtained for the "flow unit", which is of the order of the silicon radius. Therefore the proposed silicon diffusion mechanism in SiO_2 glass, see chapter 3.2, seems to be the underlying mechanism for viscous flow (6).

The kinetics of viscous flow in vitreous SiO_2 and silicate glasses can help to elucidate the deformation behavior of ceramic materials at high temperatures (creep behavior). Since practically all ceramic materials are crystalline/vitreous composites, the mechanical behavior at high temperatures will be strongly influenced by the viscous flow behavior of the glassy phase.
For a glass or a ceramic material with a high portion of glassy phase (e.g. porcelain or glass ceramic material) the steady-state creep rate $\dot{\varepsilon}$ under an applied stress σ results from Newtonian flow, i.e.

$$\dot{\varepsilon} = (1/3\eta)\sigma \; \alpha \; (r \cdot D/kT) \qquad \text{Eq.(7)}$$

Depending on the glass composition different creep rates result from different viscosities, i.e. from different diffusivities of network oxygen or silicon, the latter in the case of SiO_2 and impurity-doped SiO_2 glasses.

For a high-temperature ceramic material with small portions of glassy phase (typically present at grain boundaries only) a relationship similar to but different from Eq.(7) will hold which will take into account the ceramic microstructure, i.e. grain size d and thickness of grain boundary layer δ ($\hat{=}$ thickness of glassy layer). Such a relationship was derived by Coble to describe the deformation behavior of a dense ceramic microstructure under the assumption that the rate-determining diffusion occurs within the grain boundary phase ("Coble creep"):

$$\dot{\varepsilon} = (47\,\Omega\,\delta D/kTd^3)\sigma \qquad \text{Eq.(8)}$$

with Ω the volume of the diffusing rate-controlling species of diffusivity D.

Provided that viscous flow behavior occurs in creep experiments (i.e. $\dot{\varepsilon} \propto \sigma$), Eq.(8) should hold, i.e. it is suggested that the activation energy of the creep rate should be identical to the activation energy of viscous flow of the existing glass at the

grain boundary.
Eq.(8) can be utilized to calculate a "lower creep rate limit" by inserting the data for pure SiO_2 glass, i.e. assuming the most "refractory" glass to be present at the grain boundary.
With the following data $d = 1$ µm, $\delta = 10$ nm, $\sigma = 70$ MN/m^2, $\Omega = (4\pi/3)(0.15 \text{ nm})^3$ and $D_{Si}(1400^oC) = 2.5 \times 10^{-16}$ cm^2 s^{-1}, a creep rate of about 2×10^{-6} h^{-1} at 1400°C can be estimated. This compares favorably with the observed data of SiC (33), a material which possesses basically silica-rich glassy phase boundaries. Therefore, the estimation of a lower creep rate limit seems to be correct, and indeed the stationary creep rates of nitrogen ceramics are much higher due to less silica-rich grain boundary layers.

REFERENCES

1. Revesz, A.G. and G.V. Gibbs. "Structural and bond flexibility of vitreous SiO_2 films". Proceedings of the Conference on the Physics of MOS Insulators, Raleigh, NC, p.92 (ed. G. Lucovsky et al., New York: Pergamon Press, 198o).

2. Schaeffer, H.A. and H.J. Oel. "Oxygen-18 diffusion in lead glasses". Glastechn.Ber. 42 (1969) 493-498.

3. Noll, W. "The silicate bond from the point of view of electron theory". Angew.Chem. 75 (1963) 123-13o.

4. Schaeffer, H.A. "Silicon and oxygen diffusion in oxide glasses". Mass Transport Phenomena in Ceramics, p.311-325. (ed. A.R. Cooper and A.H. Heuer, New York: Plenum Publishing Corp., 1975).

5. Schaeffer, H.A. and K. Muehlenbachs. "Correlations between oxygen transport phenomena in non-crystalline silica". J.Mater. Sci.Lett. 13 (1978) 1146-1148.

6. Schaeffer, H.A. "Oxygen and silicon diffusion-controlled processes in vitreous silica". J.Non-Crystalline Solids 38/39 (198o) 545-55o.

7. Williams, E.L. "Diffusion of oxygen in fused silica". J.Am. Ceram.Soc. 48 (1965) 19o-194.

8. Muehlenbachs, K. and H.A. Schaeffer. "Oxygen diffusion in vitreous silica - utilization of natural isotopic abundances". Can.Mineralogist 15 (1977) 179-184.

9. Norton, F.J. "Permeation of gaseous oxygen through vitreous silica". Nature 191 (1961) 7o1.

1o. Brebec, G. et.al. "Diffusion of silicon in vitreous silica" Acta Met. 28 (198o) 327-333.

11. Schaeffer, H.A. "Oxygen and silicon diffusion in silicate glasses". Habilitationsschrift, Universität Erlangen-Nürnberg, 198o.

12. Doremus, R.H. "Diffusion of oxygen from contracting bubbles in molten glass". J.Am.Ceram.Soc. 43 (196o) 655-661.

13. Lawless, W.N. and B. Wedding. "Photometric study of the oxygen diffusivity in an alumino-silicate glass". J.Appl.Phys. 41 (197o) 1926-1929.

14. Sasabe, M. and K.S. Goto. "Permeability, diffusivity, and solubility of oxygen gas in liquid slag". Met.Trans. 5 (1974) 2225-2233.

15. Oishi, Y., Terai, R. and H. Ueda. "Oxygen diffusion in liquid silicates and relation to their viscosity". Mass Transport Phenomena in Ceramics, p.297-31o. (ed. A.R. Cooper and A.H. Heuer, New York: Plenum Publishing Corp., 1975).

16. Terai, R. and Y. Oishi. "Self diffusion of oxygen in soda-lime silicate glass". Glastechn.Ber. 5o (1977) 68-73.

17. May, H.B., Lauder, I. and R. Wollast. "Oxygen diffusion coefficients in alkali silicates". J.Am.Ceram.Soc. 57 (1974) 197-2oo.

18. de Berg, K.C. and I. Lauder. "Oxygen tracerdiffusion in a potassium silicate glass above the transformation temperature." Physics Chem.Glasses 21 (198o) 1o6-1o9.

19. Motzfeldt, K. "On the rates of oxidation of silicon and of silicon carbide in oxygen, and correlation with permeability of silica glass". Acta Chem.Scand. 18 (1964) 1596-16o6.

2o. Schachtner, R. and H.G. Sockel. "Study of oxygen diffusion in quartz by activation analysis". Proceedings 8.Intern.Symposium on the Reactivity of Solids, Gothenburg, 1976, 451-455.

21. Goursat, P. et al. "Silicon nitride and oxynitride stability in oxygen atmosphere at high temperatures". Proceedings 7.Intern. Symposium on the Reactivity of Solids, p.315-326. (ed. J.S. Anderson et al., London: Chapman and Hall, 1972).

22. Ebi, R. "High-temperature oxidation of silicon carbide and silicon nitride in technical furnace atmospheres". Dissertationsschrift, Universität Karlsruhe, 1973.

23. Horton, R.M. "Oxidation kinetics of powdered silicon nitride". J.Am.Ceram.Soc. 52 (1969) 121-124.

24. Hirai, T., Niihara, K. and T. Goto. "Oxidation of CVD Si_3N_4 at 155o° to 165o°C." J.Am.Ceram.Soc. 63 (198o) 419-424.

25. Singhal, S.C. "Thermodynamics and kinetics of oxidation of hot-pressed silicon nitride". J.Mater.Sci. 11 (1976) 5oo-5o9.

26. Cubiciotti, D. and K.H. Lau. "Kinetics of oxidation of hot-pressed silicon nitride containing magnesia". J.Am.Ceram.Soc. 61 (1978) 512-517.

27. Clarke, D.R. and F.F. Lange. "Oxidation of Si_3N_4 alloys: relation to phase equilibria in the system Si_3N_4 - SiO_2 - MgO". J.Am.Ceram.Soc. 63 (198o) 586-593.

28. Tripp, W.C. and H.C. Graham. "Oxidation of Si_3N_4 in the range 13oo° to 15oo°C." J.Am.Ceram.Soc. 59 (1976) 399-4o3.

29. Schlichting, J. and L.J. Gauckler. "Oxidation of some β-Si_3N_4 materials". Powder Metallurgy Internat. 9 (1977) 36-39.

3o. Muehlenbachs, K. and I. Kushiro. "Oxygen isotope exchange and equilibrium of silicates with CO_2 or O_2". Carnegie Institute Washington Yearbook 73 (1974) 232-236.

31. Li, J.C.M. and P. Chang. "Self-diffusion coefficient and viscosity in liquids". J.Chem.Phys. 23 (1955) 518-52o.
32. Glasstone, S., Laidler, K.J. and H. Eyring. The Theory of Rate Processes. The kinetics of chemical reactions, viscosity, diffusion and electrochemical phenomena (New York: McGraw Hill, 1941).
33. Birch, J.M. and B. Wilshire. "The compressive creep behaviour of silicon nitride ceramics". J.Mater.Sci. 13 (1978) 2627-2636.

DISCUSSION

Hampshire: To what extent would you expect cation diffusion in silicate glasses to be rate controlling?

Schaeffer: For silicate glasses there is an extensive collection of cation diffusivities in the literature, especially for cations of alkali and alkaline earth oxides (1). These cations (so-called network modifiers) are the most mobile species in the silicate network and are thus rate-controlling for transport processes which are governed by the most mobile species, e.g. ionic conductivity, electrical and mechanical relaxation phenomena, ion exchange at glass surfaces.
(1) See for instance; G.H. Frischat, "Ionic Diffusion in Oxide Glasses", Trans.Tech. Publications, Aedermannsdorf, Switzerland (1975).

Morgan: On what basis do you use the Si-O covalent bond model in the face of the belief in a 50% ionic Si-O bond?

Schaeffer: Based on the concepts by Noll I visualize the covalency of the Si-O bond as a quantity which can be increased or decreased depending on the type of the neighbouring species (degree of electronegativity). This view is substantiated by positron annihilation experiments. Previously I have attempted to correlate the ratio of silicon and oxygen diffusivities in silicate glasses with the degree of covalency of the Si-O bond. According to this increased covalency facilitates the diffusivity of oxygen (see Ref. 4). Vice versa the high diffusivity of oxygen in vitreous silica as compared to the diffusivity of silicon seems to indicate a predominant covalent bonding.

Jack: You did not mention "water" in vitreous silica. The apparent diffusivity of "water" or -OH in vitreous silica is 10^3 x that of O because ≡Si-O-Si≡ ⟶ ≡Si-OH HO-Si≡ occurs, and diffusion occurs by movement of H and OH. SiO_2 can be formed from Si_3N_4 by hydrolysis as well as by oxidation.

Schaeffer: The high temperature oxidation of silicon-containing materials via hydrolysis, i.e. "wet" oxidation, has been studied in particular for silicon. However, the increased oxidation rate as compared to "dry" oxidation is not due to an increased apparent diffusivity of "water" but caused instead by the increased "water" solubility in vitreous silica (i.e. OH solubility). The decisive quantity which determines the oxidation rate (parabolic rate constant) is the permeability which is given by the product of the solubility and diffusivity of the oxidizing agent (see Ref. 6). In the case of "wet" oxidation the diffusivity of "water" (more precisely the diffusivity of OH) is actually smaller than the diffusivity of O_2 (roughly one order of magnitude), but it is the OH solubility which accounts for the increased oxidation rate due to its three orders of magnitude increased value as compared with oxygen. Therefore, "wet" oxidation proceeds much faster, the parabolic rate constant is increased by almost the factor of 10^2 in the temperature range 1000 to 1400°C.

THE PREPARATION AND PROPERTIES OF OXYNITRIDE GLASSES

R.A.L. Drew, S. Hampshire and K.H. Jack

Wolfson Research Group for High-Strength Materials,
Crystallography Laboratory,
The University of Newcastle upon Tyne, UK.

ABSTRACT. Glasses containing more than 15a/o N in M-Si-Al-O-N systems where M is Mg, Ca, Y and Nd are obtained by fusing powder mixtures of the metal oxide with SiO_2, Al_2O_3, Si_3N_4 and AlN at 1700°C and above, and then cooling in a nitrogen atmosphere at not greater than 200°C/min. With the same cation composition, viscosity, glass transition temperature, resistance to devitrification and refractive index all increase initially with increasing nitrogen concentration. The characterization of grain-boundary vitreous phases in silicon nitride and in β'-sialons densified with MgO and Y_2O_3 is important because they often determine the high-temperature behaviour of the ceramic. Their devitrification can produce metastable phases that otherwise are not obtainable, e.g. appropriate glass compositions in the Mg-Si-Al-O-N system give (i) β"-magnesium sialons with greatly expanded phenacite structures, and (ii) nitrogen petalite, Mg_2SiAlO_4N.

1. INTRODUCTION AND EXPERIMENTAL

The accommodation of up to 10a/o N in glasses, e.g. in the Mg-Si-Al-O-N and Y-Si-Al-O-N systems was reported by Jack (1,2) at the 1976 NATO A.S.I. meeting and confirmed by Loehman (3) who found that Tg, hardness and fracture toughness all increased with increasing nitrogen content. The ease of shaping glasses, the possibility of producing glass ceramics in which the crystalline phases are refractory nitrides and oxynitrides, and the fact that oxynitride glasses occur as intergranular phases in nitrogen ceramics make it necessary to explore oxynitride glass formation and properties.

Riley, F.L. (ed.) Progress in Nitrogen Ceramics

1.1 Preparation of Glasses

Using the triangular prism representation illustrated by Fig. 1 where e/oN = 100y/(x+y), the glass-forming region on a plane of constant N:O ratio was determined, varying only the cation ratios. Planes of progressively increasing N:O ratio were then studied until the limit of nitrogen solubility was reached. SiO_2, Al_2O_3, Si_3N_4, AlN and the appropriate metal oxide powders were dry-mixed, cold-pressed into cylinders and fired in a boron nitride crucible in a tungsten-element resistance furnace in purified nitrogen for 1h at the required temperature, then cooled at about 200°C/min. Samples were examined by X-ray diffraction and scanning electron microscopy and some were analysed by e.p.m.a. for metals with an energy-dispersive analyser and for O and N with wavelength-dispersive spectrometers using "Camebax" equipment at AERE, Harwell.

1.2 Property Measurements

The glass transition temperature (Tg) and crystallisation temperature (Tc) were observed by DTA; viscosities were measured with a high-temperature deformation-under-load apparatus using the viscosity of "Spectrosil" vitreous silica at 1000°-1300°C as a standard. Devitrification was by heat-treatment in an alumina tube furnace in nitrogen, and refractive indices were measured by the "immersion technique".

2. RESULTS AND DISCUSSION

2.1 Glass Formation

Dissolved nitrogen lowers the eutectic temperature in metal oxide-silica systems (4) and also increases the viscosity of the liquid so that the tendency to glass formation might be expected to

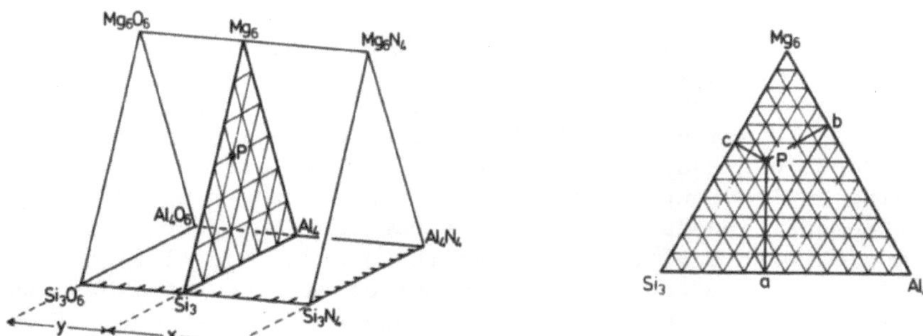

Fig. 1 Representation of a 5-component system (e.g. Mg-Si-Al-O-N) showing a plane of constant N:O ratio

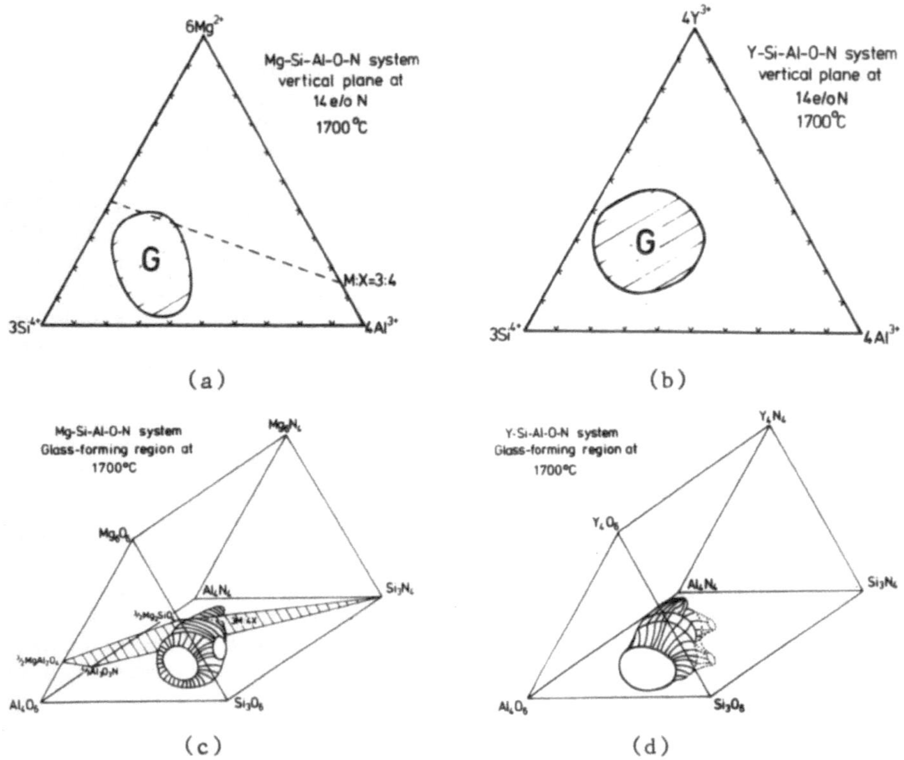

Fig. 2 Glass-forming regions in the Mg-Si-Al-O-N and Y-Si-Al-O-N systems

increase. Thus, although no glasses are formed in the $MgO-SiO_2$ system, there is a small, homogeneous glass region in Mg-Si-O-N on cooling from 1700°C. Similar limited regions are found in the Ca, Y and Ce oxynitride systems.

In M-Si-Al-O-N systems (where M is Mg, Ca, Y, Nd or other lanthanides), the vitreous regions are much more extensive; typical glass forming regions at 14e/oN for Mg and Y sialons are shown in Fig. 2 which also shows three-dimensional representations of the complete glass regions in these two systems. With increasing replacement of oxygen by nitrogen, the range of glass-forming compositions at first widens but then contracts above about 10e/oN with a simultaneous shift towards more Al-rich compositions. The coordination number of nitrogen atoms around a metal atom is, in general, smaller than that of oxygen. Thus, Mg and Ca are respectively 6-fold and 8-fold coordinated by oxygen and so are net-work modifiers in oxide systems. With nitrogen they are 4-fold coordinated and so are net-work formers. The glass forming capability therefore increases initially as nitrogen replaces oxygen, but it reaches a maximum and then decreases as the directional character of the more covalent metal-nitrogen bonding increases in the structure.

The limits of glass formation in the Mg and Ca sialon systems are at about 25e/oN (e.g. $Mg_{23}Si_{17}Al_4O_{46}N_{11}$ and $Ca_{18}Si_{18}Al_6O_{47}N_{11}$) but are greater in the Y system. Recently, a glass of approximate composition $Y_{15}Si_{15}Al_{10}O_{45}N_{15}$ has been produced, i.e. in which one of every four oxygens is replaced by nitrogen. Nd-sialon glasses with up to 25e/oN are similar in their formation to glasses in the Y-Si-Al-O-N system.

2.2 Glass Properties

Glasses with a fixed cation composition 28e/oM:56e/oSi:16e/oAl and with varying N:O ratios were prepared to allow (i) direct comparison between different M-Si-Al-O-N systems (where M = Mg, Ca, Y and Nd) and (ii) the effect of replacing oxygen by nitrogen within each system.

Figs. 3 and 4 show the variation of viscosity with temperature

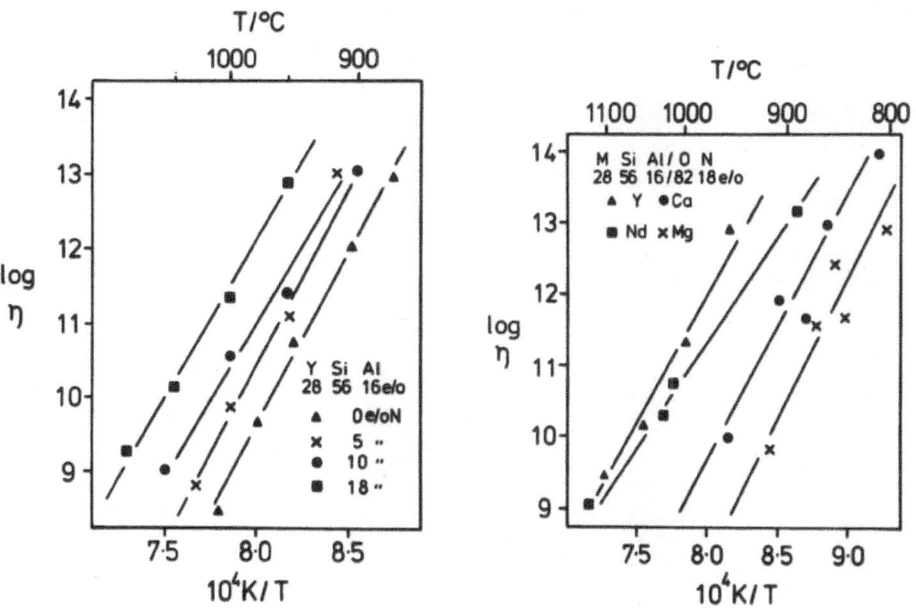

Fig. 3 Variation of viscosity with temperature for Y-Si-Al-O-N glasses of different N contents

Fig. 4 Viscosities of Y, Nd, Ca and Mg sialon glasses with 18e/oN and the same M:Si:Al cation ratios

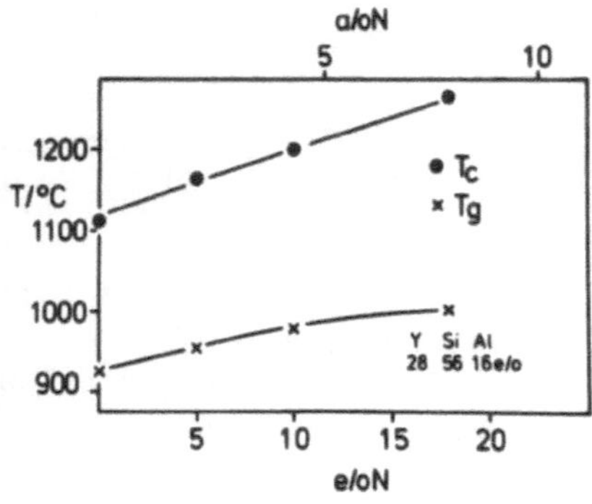

Fig. 5 Variation of Tc and Tg with nitrogen for Y-Si-Al-O-N glasses

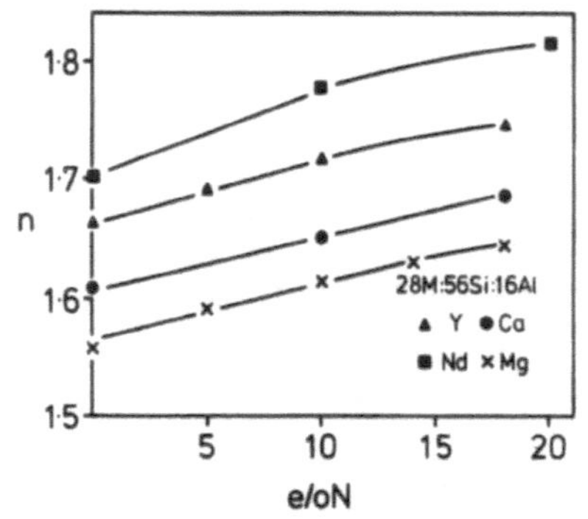

Fig. 6 Refractive indices of Mg, Ca, Y and Nd sialon glasses

for respectively (i) Y-sialon glasses of different nitrogen concentrations, and (ii) Mg, Ca, Nd and Y glasses with the same cation ratios and all with 18e/oN. Clearly, glass viscosities at any temperature increase with nitrogen content and, for a fixed nitrogen concentration, in the sequence Mg$<$Ca$<$Nd$<$Y. As might be expected, there are similar trends in the variations of glass transition temperature (Tg) and crystallisation temperature (Tc); see Fig. 5.

With increasing N:O there is a significant increase in the refractive index (n) of all glasses (Fig. 6), and by varying the cation ratio as well as N:O in Nd glasses values around 1.9 are obtainable.

Electrical properties reported (5) for Ca and Mg sialon glasses prepared at Newcastle show that the dielectric constant

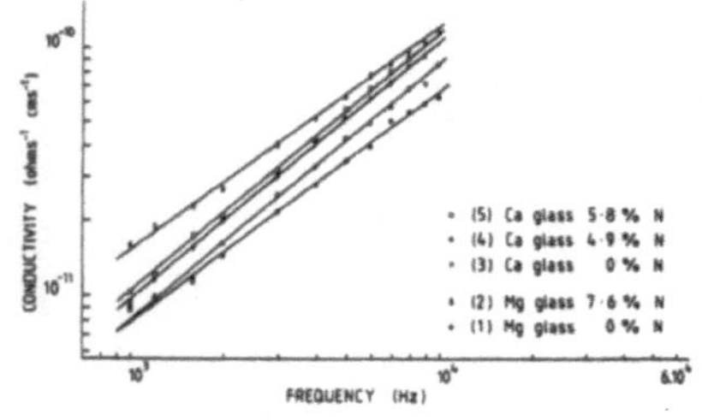

Fig. 7 Electrical conductivity (ac) of Ca and Mg sialon glasses

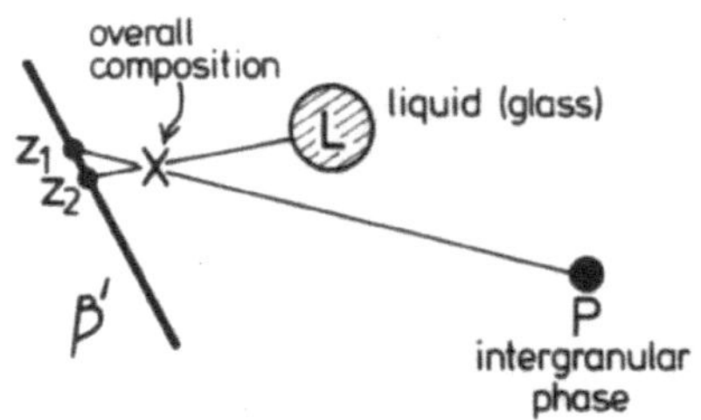

Fig. 8 Phase relationships during densification and post-preparative heat-treatment of β'-sialon

fits well with the Universal dielectric response law

$$(\varepsilon' - \varepsilon_{\infty}) \propto \omega^{(n-1)} \qquad \text{.... (1)}$$

with n = 0.99 $\pm$ 0.02; ε' increases with nitrogen concentration. The ac conductivities (Fig. 7) vary as expected for a hopping mechanism

$$\sigma_{ac} \propto \omega^{n} \qquad \text{.... (2)}$$

with n, as before, equal to unity. These properties are similar to those found for Y-sialon glasses (6).

3. DEVITRIFICATION OF VITREOUS PHASES

The liquid that is necessary for the high-temperature densification of nitrogen ceramics generally cools to give some intergranular oxynitride glass and suitable post-preparative heat-treatment to obtain refractory crystalline phases offers a method of improving properties. With β'-sialon pressureless sintered with yttria, the glass reacts with the matrix to give a slightly changed β'

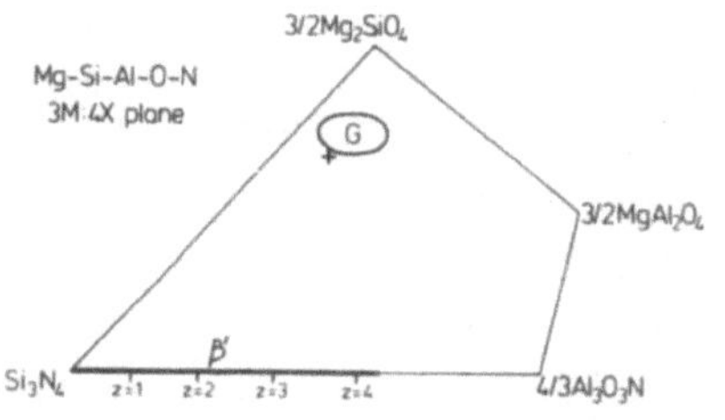

Fig. 9 The 3M:4X plane of the Mg-sialon system showing the glass-forming region

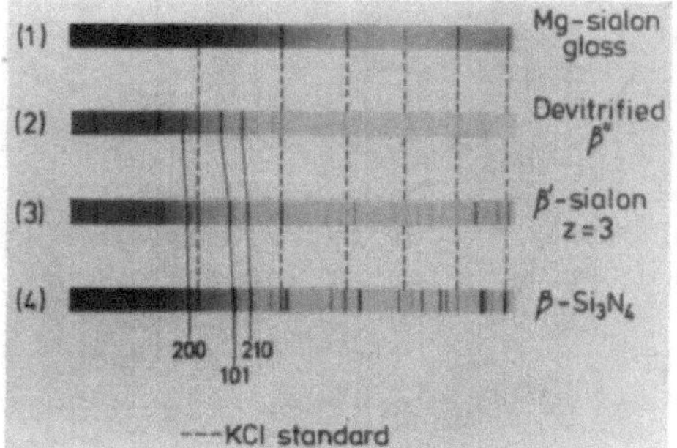

Fig. 11 X-ray powder photographs of isostructural β-Si_3N_4, β' and β'' sialons

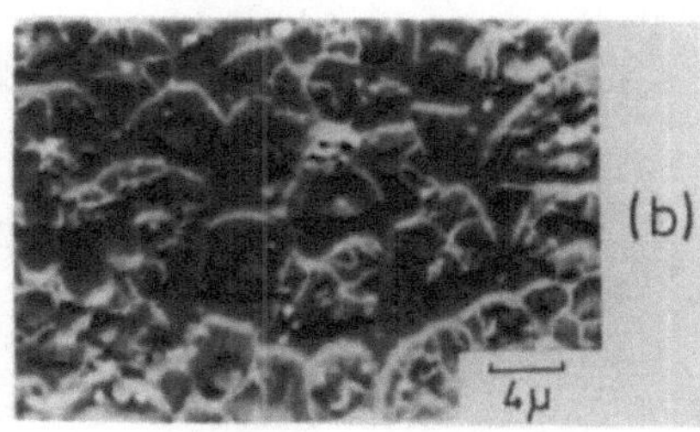

Fig. 10 Formation of β''-Mg-sialon by devitrification
(a) glass + β-Si_3N_4 nuclei by cooling liquid from 1700°C
(b) β'' grown epitaxially on β by heat-treatment at 930°C

composition and yttrium-aluminium garnet:

$$Si_5AlON_7 + Y\text{-}Si\text{-}Al\text{-}O\text{-}N \rightarrow Si_{5+x}Al_{1-x}O_{1-x}N_{7+x} + Y_3Al_5O_{12} \quad \ldots. (3)$$

β'-sialon glass β'-sialon "YAG"

The phase relationships are shown schematically by Fig. 8 and can be applied generally to other systems.

As with oxide glasses, devitrification can give phases, sometimes metastable, that are unobtainable by direct methods. The glass forming region of the Mg-Si-Al-O-N system protrudes through the 3M:4X plane (Figs. 2 and 9) and if liquid of composition just outside this region (marked + on Fig. 9) is cooled from above 1600°C the product is glass with a small amount of dispersed β-Si_3N_4. Heating at 800-950°C devitrifies the glass to give a β"-phase which grows epitaxially on the β nuclei; see Fig. 10. β" has a range of unit-cell dimensions larger than those of the highest-z β'-sialon (Fig. 11). Although it has a phenacite structure, its range of composition approaches forsterite (Mg_2SiO_4) in which only a small proportion of O is replaced by N with simultaneous replacement of Mg by Al:

$$Mg^{2+}\ O^{2-} \rightleftharpoons Al^{3+}\ N^{3-} \quad \ldots. (4)$$

The closeness in the lattice energies of the two M_3X_4 structures, forsterite and phenacite, is clearly demonstrated.

Also in the Mg-sialon system, a glass composition $2MgO.5SiO_2.3AlN$ partially devitrifies on cooling from 1750°C to give a phase Mg_2SiAlO_4N that is isostructural with lithium disilicate ($Li_2Si_2O_5$) and with petalite ($LiAlSi_4O_{10}$).

In the Y-sialon system the solid solution intermediate between N-α-wollastonite ($YSiO_2N$) and $YAlO_3$ is Y_2SiAlO_5N; it is obtained by heat-treatment at 1100°C of a glass composition $3YN.3SiO_2.Al_2O_3$ prepared by quenching from 1700°C.

4. CONCLUSIONS

Depending on the system, a limit of 17-30% of oxygen can be replaced by nitrogen in sialon glasses. Viscosity, glass transition temperature, refractive index, dielectric constant and ac conductivity increase with increasing nitrogen concentration. Since they occur as intergranular phases that affect the properties of nitrogen ceramics, further detailed study of nitrogen glasses is essential.

REFERENCES

1. Jack, K.H. J. Mat. Sci. 11 (1976) 1135.
2. Jack, K.H. in F.L. Riley ed. 'Nitrogen Ceramics' (Noordhoff: Leyden, 1977) p.257.
3. Loehman, R.E. J. Amer. Ceram. Soc. 62 (1979) 491.
4. Hampshire, S. and K.H. Jack, in D. Taylor and P. Popper eds. Special Ceramics 7 (Proc. Brit. Ceram. Soc. No. 31, June 1981) p.37.
5. Thorp, J.S. and S.V.J. Kenmuir. J. Mat. Sci. 16 (1981) 1407.
6. Leedeuke, C.J. and R.E. Loehman. J. Amer. Ceram.Soc. 63 (1980) 190.

DISCUSSION

Billy: Is there any information available on the rates of solution of Si_3N_4 or of sialons into the corresponding glasses? This will be important for their oxidation behaviour.

Jack: No information appears to exist on the rates of solution into or precipitation from these glasses.

Elias: Do you know anything about relationships between Tg values, and the glass structure?

Jack: We do not know enough about the structure of N-glasses but suppose that the increase in Tg with increasing N content is due to one N being common to 3 tetrahedra. Si-N might also be expected to be more directional than Si-O. Further the fact that Mg is a network former in N-glasses, and a network modifier in O-glasses would give fewer non-bridging non-metal atoms in the structure, and hence a higher viscosity.

Goursat: What is known about the oxidation behaviour of the N-glasses?

Jack: We have no detailed kinetic information. They oxidise quite readily at high temperature with evolution of nitrogen.

Briggs: What is the effect on the N-content of these glasses of melting under nitrogen?

Jack: A small amount of additional N will dissolve as physically dissolved molecular N_2. But again, as with Si_3N_4 etc., we should be able to stop decomposition of the N-glasses, or maintain their stability to higher temperatures, by increasing p_{N_2}.

THE CRYSTALLIZATION BEHAVIOUR OF A Mg-Si-O-N GLASS

T.M. Shaw and G. Thomas

Department of Materials Science and Mineral Engineering
University of California, Berkeley, U.S.A.

ABSTRACT

The crystallization behaviour of a glass of composition 59 wt.% SiO_2, 32 wt.% MgO and 9 wt.% Si_3N_4 has been investigated. The as-cooled glass was found to have separated into a magnesia rich and a silica rich phase on cooling. The magnesia rich phase in the glass could be readily crystallized to enstatite by heating to temperatures above 1000°C. In contrast the silica rich phase was highly resistant to crystallization at all temperatures. The microstructure developed during crystallization of the magnesia rich phase consisted of fan-like arrangements of lath shaped enstatite crystals and showed many of the features common to spherulitic crystal growth. A large volume change associated with crystallization of the magnesia rich phase resulted in the formation of porosity in crystallized glass samples. At temperatures above about 1300°C the spherulitic microstructure recrystallized to faceted grains of enstatite in regions close to the silica rich phase, the reaction apparently being initiated by softening of the silica rich phase. Also at higher temperatures silicon oxynitride crystallized out from the magnesia rich phase, appearing as small crystals situated between the lath shaped enstatite grains.

INTRODUCTION

Oxynitride glasses, present as thin intergranular films in hot-pressed or sintered silicon nitride materials, are now widely recognized as being responsible for the deterioration in properties of these materials at high temperatures. Several investi-

Riley, F.L. (ed.) Progress in Nitrogen Ceramics

gators have reported preparing bulk samples of oxynitride glasses[1-5] but so far there have been few investigations of their properties.[4] In particular there have been no systematic studies of the crystallization behaviour of oxynitride glasses. The present paper reports the results of a detailed study of crystallization in a Mg-Si-O-N glass. Emphasis is placed on identification of temperature ranges where crystallization occurs and characterization of the resulting microstructures.

EXPERIMENTAL

The glass used in the investigation was prepared by melting weighed amounts of SiO_2, MgO and Si_3N_4 powders in a molybdenum crucible in an atmosphere of nitrogen. The melting schedule consisted of heating the powder mixture to 1700°C, holding at this temperature for four hours, then cooling to room temperature by turning off the power to the graphite heating elements. Cooling was sufficiently rapid for the temperature of the melt to fall below 1000°C in about three minutes. The powder mixture was made up to give a melt with a composition of 59 wt.% SiO_2, 32 wt.% MgO and 9 wt.% Si_3N_4.

Small samples of the glass were given isothermal heat treatments in the temperature range 800°-1350°C. Polished sections of the heat treated glass samples were examined using reflected light microscopy. Thin sections of the samples were used for transmitted polarized light microscopy and also to prepare specimens for transmission electron microscopy. Final thinning of T.E.M. specimens was carried out by ion beam milling. Thinned specimens of both the as-cooled glass and heat treated glass were examined using Philips 301 and 400 microscopes operating at 100 kV. Energy dispersive and energy loss spectrometer attachments to the Philips 400 were used for composition analysis of the phases formed in the glass.

RESULTS AND DISCUSSION

Characterization of the glass.

On cooling the melt produced a light gray opaque glass. Examination of the microstructure of the glass using TEM revealed that the opaqueness was caused by separation of the glass into two distinct phases. A typical view of the microstructure, which consists of groups of droplets of a minor phase dispersed in a continuous glass matrix, is shown in fig. (1). Contrast between the two phases arises largely from preferential thinning of the minor phase during specimen preparation. Compositional analysis of the two phases, to be presented in detail elsewhere,[6] showed that the

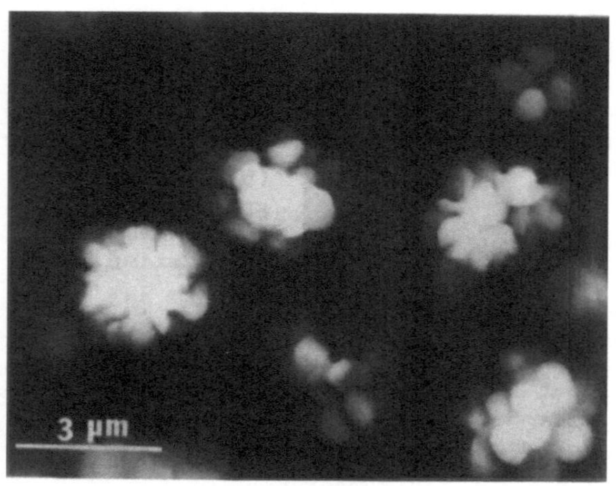

Figure 1. Bright field image of phase separation in the as-cooled glass.

minor phase contained both SiO_2 and Si_3N_4 but virtually no MgO, whereas the matrix phase contained SiO_2, MgO and Si_3N_4. From quantification of the analysis it was found that the silica-rich minor phase had a composition close to the SiO_2 - Si_3N_4 binary in the MgO-SiO_2-Si_3N_4 system and contained about 4-5 wt.% Si_3N_4. The matrix phase had a composition which lay close to the $MgSiO_3$-Si_2N_2O tie line in the MgO-SiO_2-Si_3N_4 system and contained about 8-9 wt.% Si_3N_4.

Crystallization.

Heat treatments at temperatures below 1000°C had little effect on the microstructure of the as-cooled glass. Some small crystals were observed after heating for long times at tmeperatures close to 1000°C but growth of the crystals was very slow. Once temperatures above 1000°C were reached the rate of crystallization rapidly increased and after heating for two hours at 1020°C crystallization had occured throughout the glass. The crystalline phase formed during crystallization was identified as enstatite ($MgSiO_3$) by X-ray diffraction. Fig. (2)(a) shows a transmitted polarized light micrograph of the crystallized glass. Dark cross-like features are visible in several regions suggesting a spherulitic morphology in which fiberous crystals grow radially from a central nucleation site. Another feature of the microstructure was coarse porosity that formed throughout the glass as it crystallized (see Fig. 2(b)). Comparison of the bulk density of the glass (2.68 grams/cc measured using sink float technique) with the density of enstatite (3.21 grams/cc) indicated that on the

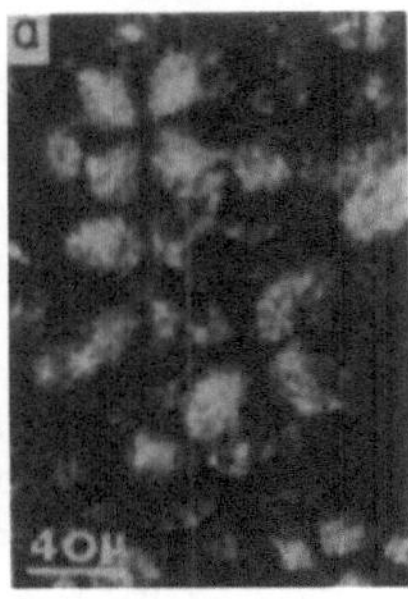

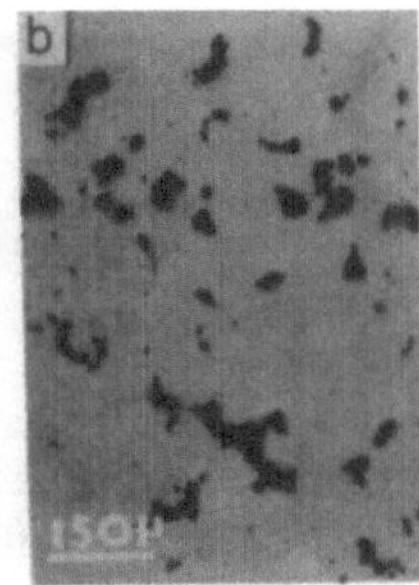

Figure 2. Optical micrographs of a glass sample crystallized at 1020°C. (a) transmitted polarized light micrograph showing "spherulitic" contrast. (b) Reflected light micrograph showing pores formed in the sample during crystallization.

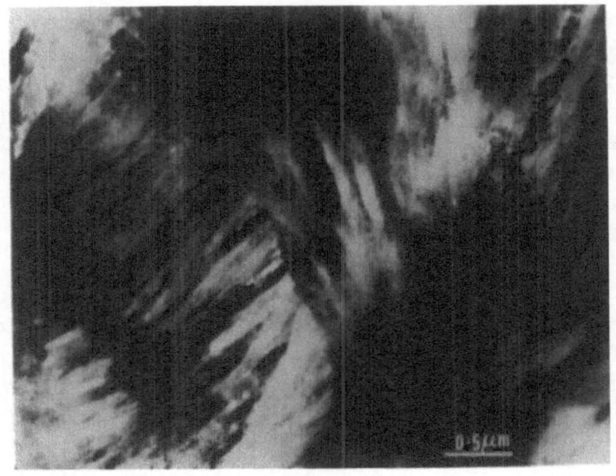

Figure 3. Bright field electron micrograph of enstatite crystals grown from the glass during a heat treatment of two hours at 1020°C.

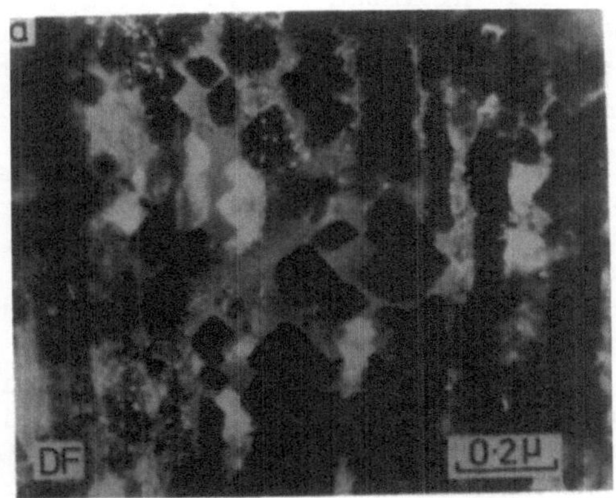

Figure 4. Dark field image showing the appearance of one of the silica droplets found in the as-cooled glass after heating for two hours at 1350°C.

order of a 15% change in volume should accompany crystallization of the glass to enstatite. About 5% of this volume change could be accounted for by shrinkage of the sample during crystallization. The remaining 10% could be adequately accounted for by the formation of porosity in the sample, indicating that porosity in the sample results from the large volume change that accompanies crystallization of the glass.

Examination of the crystallized glass using TEM revealed that only the magnesia-rich matrix phase crystallized during heat treatment; the silica-rich minor phase was unaffected and preserved its globular shape. The microstructure developed in the magnesia-rich phase (fig. (3)) consisted of lath shaped enstatite crystals several microns long and about 0.1 μm across arranged in fan-like arrays. Although the spherical clusters of crystals that are typical of a spherulitic morphology were not observed, the microstructure exhibited many of the features characteristic of spherulitic growth, i.e., diverging arrangements of fiberous crystals and noncrystallographic branching of the fibers. Similar microstructures were observed after the glass was given two hour heat treatments at 1050°C, 1100°C and 1200°C. Holding at higher temperatures produced changes in the crystallized microstructure subsequent to the crystallization of enstatite from the magnesia-rich glassy phase.

Fig. (4) shows the appearance of one of the silica-rich droplets present in the as-cooled glass after a heat treatment of two hours at 1350°C. The silica-rich phase, light gray areas in the dark field image shown in fig. (4), although uncrystallized is no longer confined to well defined droplets and has partly penetrated between the enstatite laths crystallized from the surrounding magnesia-rich glass. In addition, faceted grains of enstatite, dark regions in fig. (4), have developed in and around the glassy pocket. The appearance of the microstructure suggests that softening of the silica-rich phase has initiated recrystallization of the surrounding enstatite spherulites to a more equilibrium microstructure of faceted enstatite grains dispersed in a silica-rich glass.

A second difference observed in the microstructure after heating at temperatures above about 1300°C was the presence of a large number of small crystals in regions between individual enstatite laths that made up the spherulites. The crystals gave rise to extra reflections in electron diffraction patterns, as arrowed in fig. (5)(a), and could be clearly seen when imaged using one of these reflections, fig. (5)(b). From careful analysis of the diffraction patterns the crystals were identified as crystals of silicon oxynitride growing with a definite orientation relationship with respect to the enstatite laths. A possible explanation for this behaviour is that during crystallization of enstatite from

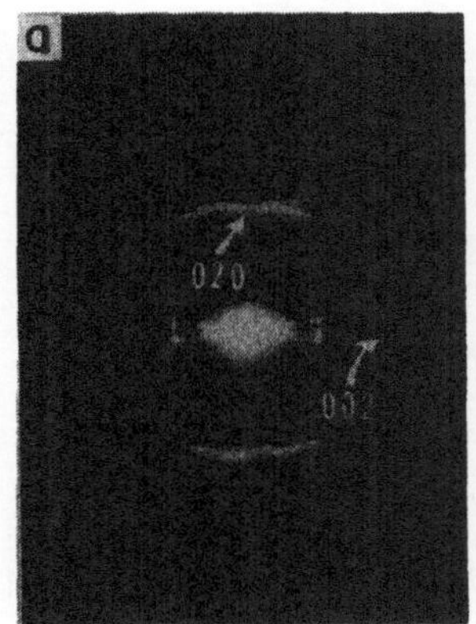

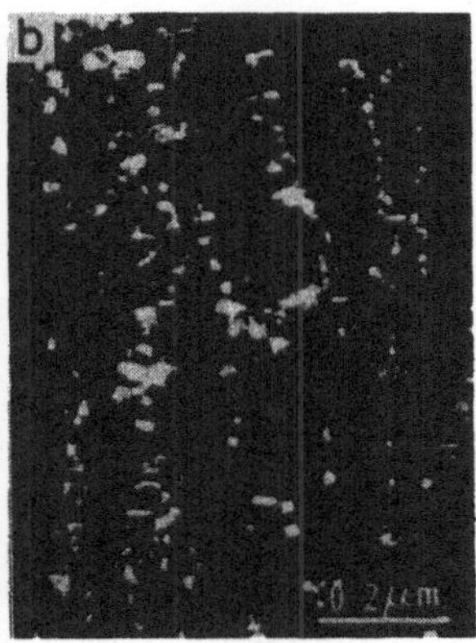

Figure 5. (a) SAD from a glass sample crystallized at 1350°C. The arrowed reflections indicate the presence of silicon oxynitride in the microstructure, (b) Dark field image of the silicon oxynitride crystals.

the magnesia-rich phase, nitrogen in the glass is rejected to regions between the enstatite laths. The nitrogen may then crystallize out as silicon oxynitride during high temperature heat treatments, the silicon oxynitride crystals nucleating heterogeneously on enstatite grains.

ACKNOWLEDGEMENTS

This work was supported by a grant from National Science Foundation. We are grateful to Dr. R.E. Loehman for supplying the material for the investigation.

REFERENCES

1. Mulfinger, H.O. J. Amer. Ceram. Soc. 49 (1966) 462-467.
2. Elmer, T.H. and M.E. Nordberg. J. Amer. Ceram. Soc. 50 (1967) 275-279.
3. Dancy, E.A. and D. Janssen. Canadian Met. Quarterly 15 (1976) 103-110.
4. Jack, K.H. in Nitrogen Ceramics, Proc. NATO Advanced Study Institute, ed. F.L. Riley 257-261. Noordhoff (1977).
5. Loehman, R.E. J. Amer. Ceram. Soc. 62 (1979) 491-494.
6. Shaw, T.M. and R.E. Loehman and G. Thomas to be published.

THERMODYNAMICS OF GRAIN-BOUNDARY GLASS CRYSTALLIZATION

R. Raj

Department of Materials Science and Engineering,
Cornell University, Bard Hall, Ithaca, N.Y. 14853
(Presented by T.M. Shaw).

ABSTRACT

Reasons why the thermodynamics of crystallization of a glass which is segregated to grain boundaries should differ from crystallization of bulk glass of the same composition are examined. The difference arises for two reasons: (a) the surface energy term in the equation for the change in free energy, in going from a glass to a crystalline state, is favourable to retaining the glass, if the dihedral angle formed between the crystal and the glass is less than $\pi/3$ (1); (b) when a crystal grows in a glass which is contained within a small crevice in a ceramic material, then the strain energy associated with crystal growth can become large which will reduce the driving force for crystallization (2). The strain energy arises because there is a volume change when glass crystallizes, and because the hydrostatic stress produced by the volume change cannot be released by fluid flow. It can in fact be shown that under most conditions the glass can be crystallized only partially.

REFERENCES

1. R. Raj. "Morphology and stability of the glass phase in glass-ceramic systems", J. Amer.Ceram. Soc. 64 245-248 (1981).
2. R. Raj and F.F. Lange. "Crystallization of small quantities of glass segregated in grain boundaries". To be published in Acta Metallurgica.

Riley, F.L. (ed.) Progress in Nitrogen Ceramics

Section F
MICROSTRUCTURE

THE MICROSTRUCTURE OF NITROGEN CERAMICS

D.R. Clarke

Structural Ceramics Group
Rockwell International Science Center
Thousand Oaks, CA 91360

INTRODUCTION

Since the nitrogen ceramics meeting[1] held in 1976, our knowledge and understanding of the role of the microstructure of this class of material has expanded and matured. At that time knowledge was scanty and restricted mainly to Si_3N_4 hot pressed with MgO; it had been established that the material consisted principally of grains of β-Si_3N_4 with impurity crystalline phases, such as WC and $WFeSi_2$, and relatively large pockets of a non-crystalline intergranular phase.[2-6] Despite a limited amount of circumstantial evidence[7,8] for the presence of a continuous intergranular phase at all grain boundaries (other than at sub-grain boundaries) its existence had not been unequivocally confirmed. By comparison our knowledge today is much more comprehensive, as this review serves to illustrate, being based on microstructural observations of a wide variety of nitrogen ceramics in both the as-fabricated condition and after use. The microstructures of hot-pressed Si_3N_4 are used for illustration since they are representative of many other nitrogen ceramics. Similar and equivalent observations of the microstructure of the β′-Sialons have been made by Lewis and his colleagues in a series of studies.[9-12]

The microstructures of nitrogen ceramics are important because they are polyphase materials rather than being single phase. Their polyphase nature not only confers a variety of possible microstructures but can also lead to a wide range of physical properties resulting from the different microstructures. Part I is devoted to a survey of the microstructural observations of the as-fabricated materials. In Part II the

Riley, F.L. (ed.) Progress in Nitrogen Ceramics

microstructural changes produced in response to potential use conditions will be described. The intergranular phase is emphasized in this review as it is at the present time (as a consequence of our present fabrication techniques) probably the most important microstructural feature. Much of our current understanding results from the development of new transmission electron microscopy techniques for the characterization of grain boundary regions in ceramics. Description of these is given in Refs. 13-16.

Part I. AS FABRICATED MICROSTRUCTURE

The microstructure of the majority of polyphase nitrogen ceramics can be considered to consist of three major elements, crystalline Si_3N_4 (or substitutionally substituted solid solution Si_3N_4 -β′-Sialon) grains constituting the principal phase, secondary crystalline phases and an intergranular non-crystalline phase. The secondary crystalline phases commonly include silicon oxynitride and quaternary metal-silicon oxynitrides, which are contingent on the particular sintering aid used and on the composition vis-a-vis the phase diagram. They also include impurity phases, such as WC, $WFeSi_2$, derived from the milling media used or phases, such as SiC, undoubtedly remaining from previous manufacturing batches made with the same processing equipment.* As will be described in Part II, both the secondary phases and the intergranular glass phase are frequently altered under potential use conditions. It is probably worth remarking that the majority, if not all, nitrogen ceramics (but not including those prepared by CVD) contain a continuous, intergranular glass phase. Early reports claiming that the intergranular phase can be crystallized, for instance in the Si_3N_4-Y_2O_3-SiO_2 system,[17] refer to the formation of the expected (from the behavior diagram) secondary metal-silicon oxynitrides from the non-crystalline phase rather than the disappearance of all the intergranular phase. The available evidence, including electron microscopy of equivalent compositions, indicates that the intergranular glass phase remains continuous throughout the microstructure, albeit of reduced volume fraction.

The morphology of the silicon nitride and secondary phases provides an indication of the processes occurring during densification. From micrographs such as Fig. 1 where the silicon nitride grains have well defined crystallographic shapes and are seemingly surrounded by the second phase ($Y_2Si_2O_7$) it can be concluded that the silicon nitride grew from a yttrium-silicon rich melt which

*These impurity phases often control the mechanical strength of silicon nitrides by acting as the flaws responsible for failure.

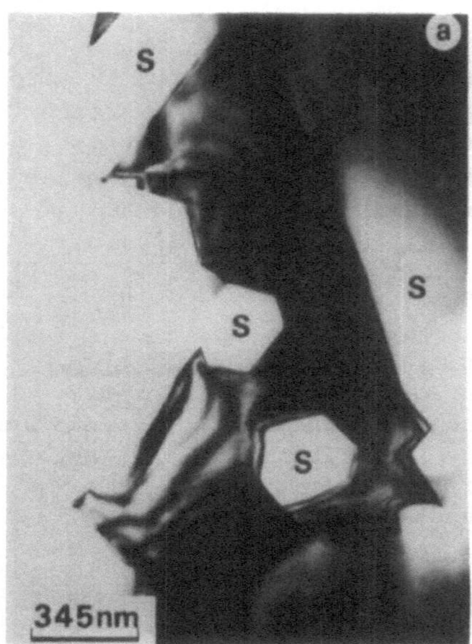

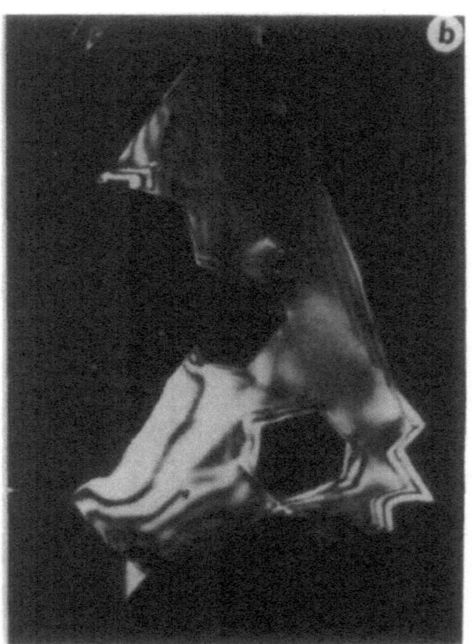

Fig. 1 Bright field (a) and dark field (b) transmission micrographs showing well-faceted grains of silicon nitride (S) surrounded by a second crystalline phase. The morphology of the Si_3N_4 grains and the dark field image (b), formed using a diffraction spot due to the second phase, suggest that the Si_3N_4 grains crystallized out of a yttrium-silicon melt that subsequently crystallized.

subsequently crystallized on cooling. Further evidence for this crystallization process comes from the observation that the secondary phase grains can each envelop several silicon nitride grains such as in Fig. 1. Such observations have been reported in both Y_2O_3 [18] and Sc_2O_3 [20] hot-pressed silicon nitrides and in the β′-Sialon system.[10] Although it is generally agreed that the silicon nitride based alloys densify in part by a solution-reprecipitation process, no morphological evidence other than that described above has so far been reported that can be attributed to the mechanism. In this context it is interesting that no instances of mixed α/β Si_3N_4 grains have been reported.

1.1 Crystal and Defect Structures

The crystal structures of both β and α silicon nitride have been the subject of extensive investigations as is surveyed by

Jack and Morgan elsewhere in these proceedings. Less attention has been directed toward the defects present in these structures. β-Si_3N_4 ($P6_3/m$) is centrosymmetric and as shown by Butler[21] and later by Evans and Sharp[2] exhibits a primary slip plane of $\{10\bar{1}0\}$ with "glide" dislocations having a Burgers vector of $\underline{b}$ = <0001>. The crystal structure of α-Si_3N_4 (P31c) is non-centrosymmetric and therefore structurally enantiomorphic domains should exist. These have recently been characterized but as yet the dislocations have not been identified although indentation studies[22] indicate that the primary slip system is $\{10\bar{1}0\}$ [0001]. The defects in silicon oxynitride remain to be characterized.

The structures of the metal-silicon oxynitrides particularly the AlN related compositional polytypes have been extensively analyzed by X-ray diffraction techniques[23,24] and some of the structural complexities clarified by electron microscopy.[25,26] These show that the wide range of solid solution exhibited by these phases is accommodated by the formation of alternating crystal structures, for instance in the Be_3N_2-Si_3N_4 system by layers of $BeSiN_2$ and β-Be_3N_2 structures. To date, study of these faults has been restricted to this system but it is a general mechanism applicable to a wide variety of other solid solution phases in nitrogen ceramics. Although not yet substantiated it appears that the mechanism may also be responsible for the apparent lack of continuous intergranular phase at the two grain junctions in hot-pressed ceramics in the Si_3N_4-$BeSiN_2$ systems,[27] implying that impurity atoms may also be accommodated in this manner.

1.2 Intergranular Non-crystalline Phase

The electron microscopy investigations have provided information about three characteristics of the intergranular phase, its microstructural location, its non-crystalline nature and its compositional range. It is now generally recognized that the intergranular phase exists in the microstructure of nitrogen ceramics as a continuously interconnected phase present at all three and two grain junctions with the exception of low angle sub-grain boundaries and special boundaries such as twin interfaces. This is illustrated by the transmission electron micrograph of Fig. 2. The intergranular phase is present in both sintered and hot-pressed materials. One characteristic feature is that the phase is very thin ~ 10Å as shown by the lattice fringe image of Fig. 3. The only nitrogen ceramic in which no intergranular phase is reported is silicon nitride prepared by chemical vapor deposition techniques.[28] Observations indicate that the majority of the volume of intergranular phase resides at the three and four grain junctions, and that as the composition is altered within the same phase compatibility field the change in volume is accommodated at the three and four grain junctions. Although the

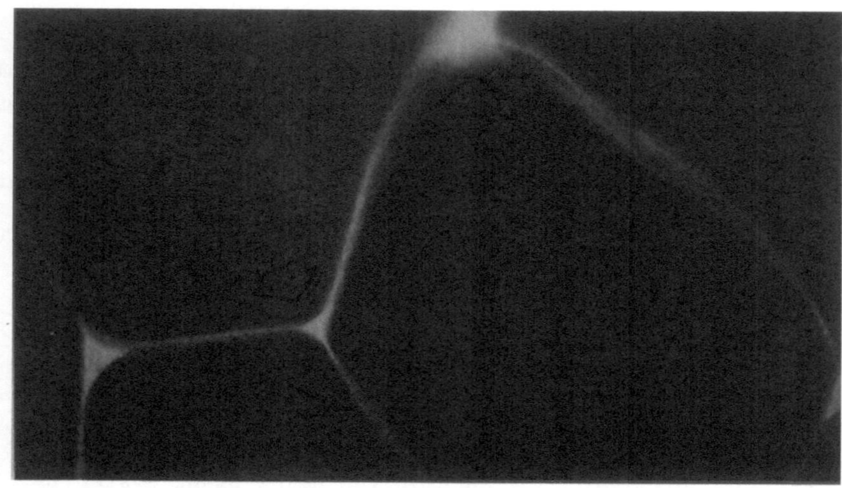

Fig. 2 Diffuse dark field electron micrograph of a MgO hot-pressed Si_3N_4 illustrating the continuous nature of the intergranular phase and its location at the triple grain junctions and between the Si_3N_4 grains.

measurements are difficult to make and the number of observations rather small, the thickness of the intergranular phase appears to be constant from one two-grain junction to another. In materials where data is available for instance in Si_3N_4 hot-pressed with MgO,[13,19] with Y_2O_3[18] and with Sc_2O_3 it is 8 - 20Å. In the 12H MgSiALON alloys it is ~ 30Å.[29] It remains to be seen whether these constant thicknesses for given materials are substantiated and whether they are affected by the nature of the solutes in the phase.

The available experimental evidence demonstrates that the intergranular phase is non-crystalline. The early work using Auger electron spectroscopy[7] indicated that the composition of the intergranular phase was one that is normally a glass. Later on the basis of transmission electron microscopy experiments in which the samples were tilted with respect to the electron beam and no contrast variation observed at the three grain junctions, the authors concluded that the phase was amorphous.[4,6] More recently convergent beam electron diffraction patterns have been recorded directly from pockets of the intergranular phase, as shown in Fig. 4, and have, in all cases, the characteristic features of diffraction from a non-crystalline solid.[30,31] Whether the intergranular phase located at the two grain junctions where

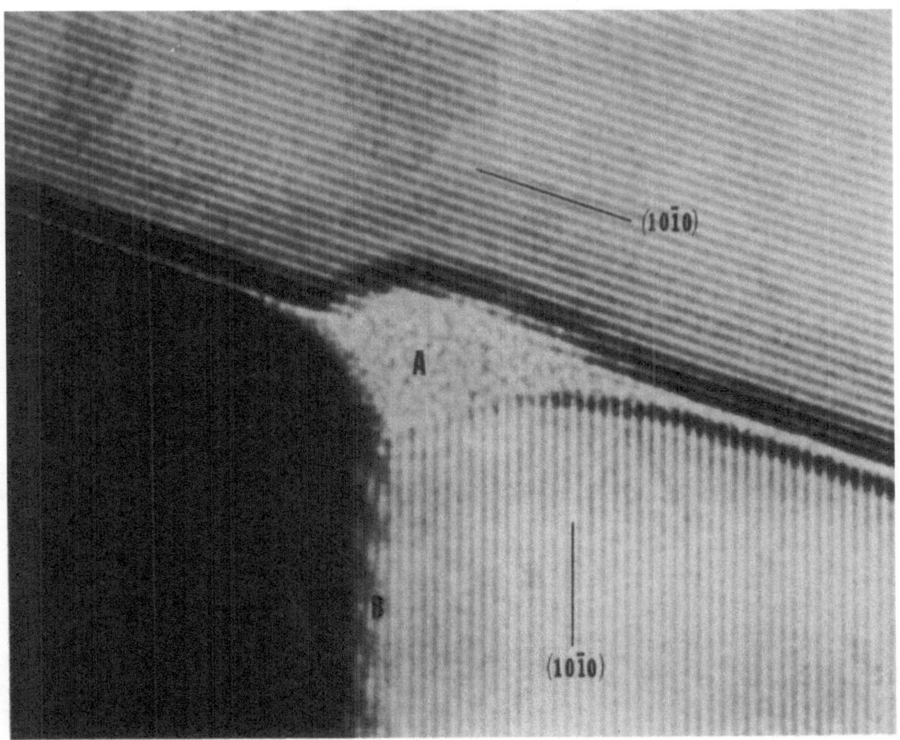

Fig. 3 Three grain junction region in a MgO hot-pressed Si_3N_4. The (10$\bar{1}$0) lattice fringes in the top and lower right grains clearly reveal the presence of the intergranular phase at the grain junction, A, and as a very thin film (~ 8Å) along the grain boundaries. The grain at the lower left is out of contrast and the Moire fringes at the boundary B indicate that the boundary is inclined to the viewing direction. (10$\bar{1}$0) spacing is 6.5Å.

its thickness is below the spatial resolution of available diffraction technqiues is also non-crystalline remains to be demonstrated. However, the indications are that it is indeed non-crystalline. In addition to the Auger results mentioned above, this material is illuminated when samples are imaged in the dark field mode of TEM using diffracted electrons having an intensity distribution corresponding to an amorphous material.[32,15] However, the nature of the dark field experiments performed to date do not rule out a form of non-randomness in the intergranular phase, such as molecular orientation adjacent to the crystalline grains. Some form of partial periodicity in the distribution of

Fig. 4 Convergent beam electron diffraction pattern recorded from the intergranular phase in a MgO hot-pressed Si_3N_4. The diffuse ring is characteristic of the intensity distribution expected from a non-crystalline solid.

atoms constituting the intergranular phase may be expected since the thickness of the phase (~ 10Å in the MgO hot-pressed Si_3N_4) is similar in magnitude to the size of the Si-O tetrahedron, the presumed structural unit of the non-crystalline phase. The issue remains unresolved.

The first results of investigations into the composition of the intergranular phase are now availabile.[30,31,33] (The early Auger electron spectroscopy results obtained from intergranular fracture surfaces lacked the necessary spatial resolution required to be sure that the recorded signal was specific to the intergranular phase alone.) By using the techniques of, X-ray microanalysis and electron energy loss spectroscopy, the elemental compositions of pockets of intergranular phase at three grain junctions have been obtained. To the author's knowledge, the analyses have so far been restricted to MgO hot-pressed silicon nitride,[30,31] a Si-Al-O-N material based on the $\beta'Si_3N_4$ phase[32] and a 12H MgSiAlON polytype alloy.[29] One general finding common to all the analyses is that the intergranular phase accommodates impurity elements. The most frequently observed impurity elements are Ca, Fe, Na and Al. The iron impurity is a consequence of it's use as a nitridation aid in the production of the starting silicon nitride powders. When powders are alternatively produced by the ammonolysis reaction of $SiCl_4$, Cl is detectable in the intergranular phase. The origin of the other impurities is unclear but they are common impurities in ceramic materials.

According to a combined X-ray microanalysis and electron energy loss spectroscopy study,[30] the composition of the intergranular phase in MgO hot-pressed silicon nitrides lying within the Si_3N_4-Si_2N_2O-Mg_2SiO_4 phase compatibility field is SiO_2 rich and lies close to the SiO_2-$MgSiO_3$ tie line. This finding is of some significance as this compositional range corresponds to that of the glass forming region in the Si_3N_4-SiO_2-MgO phase diagram.[34-37] Furthermore, the intergranular phase in a rapidly cooled silicon nitride having the composition of the eutectic in the Si_3N_4-Si_2N_2O-Mg_2SiO_4 phase field has been observed[31] to have undergone a phase separation. The phase separation and the compositions of the separated phases are consistent with the intergranular phase having a net composition lying within the glass forming regions. The intriguing observations made to date indicate that further analytical work is required. There is also a need to determine how sensitive the analytical results obtained are to irradiation effects while the analyses are being performed.

Although the majority of microstructural studies have been investigations of either hot-pressed or sintered nitrogen ceramics, dense materials produced by alternative means have been examined. The CVD material (always of the alpha form) typically consists of columnar grains and, if the manufacturing conditions are not totally correct, inclusions at three grain junctions. No discrete intergranular phase is observed. However, in the purest CVD Si_3N_4 cavities and long lengths of debonded boundaries are observed along two grain junctions[28]. It is probably too early to attribute such features to any particular mechanism but they are possibly indicative of a de-sintering phenomenon. The explosively compacted silicon nitrides examined[38] have not proved to be fully dense, and the question remains as to whether silicon nitride grain boundaries free of an intergranular phase are formed.

Part II. MICROSTRUCTURAL CHANGES PRODUCED UNDER USE CONDITIONS

A wide variety of microstructural alterations have been reported when nitrogen ceramics are exposed to exploratory use conditions, some are beneficial and others are detrimental to the overall properties of the materials. Some insight and perspective into the observed property changes can be gained by recognizing that the continuous intergranular glass phase acts as a short-circuit diffusion path through the microstructure; that it behaves in a visco-elastic manner at elevated temperatures, whereas, the principal crystalline phases (e.g., Si_3N_4, β'-Si_3N_4) remain elastic; these latter phases are also thermodynamically unstable in oxidizing atmospheres. In the following sub-sections the changes in the microstructure of nitrogen ceramics in response to a number of characteristic high temperature exposures are described. Little or no changes are produced at temperatures below ~1000°C with the important exception of oxidation induced cracking of certain yttria containing silicon nitrides in the temperature regime 700–1000°C.[39–41] Another significant generality is that the intergranular glass phase remains continuous after all the experiments described, with the exception of an argon heat treatment[42].

2.1 Sub-critical Crack Growth

Degradation in strength of all nitrogen ceramics occurs at high temperature. It is generally recognized that this is attributable to sub-critical crack growth,[43–47] a time dependent phenomenon by which pre-existing flaws or cracks slowly grow at stresses below the critical stress for catastrophic failure. Making observations of the phenomenon in order to provide evidence for the operative microstructural mechanism has proven difficult because of the fine grain size and the indistinguishability of the intergranular phase on fracture surfaces. The clearest observations have been made on room temperature fracture surfaces intercepting the fracture surfaces produced at high temperature. An example of this is Fig. 5 taken from the work of Lange[48] on slow crack growth in a MgO densified Si_3N_4. The micrograph reveals the presence of a large number of voids, some of which are linked together. These, and similar observations by Tighe,[49] are consistent with the mechanism of sub-critical crack growth in which cavities form in the intergranular phase ahead of a crack and then link together under the stress field of the crack. Further confirmation of the mechanism has been provided by identical observations of a model crystalline/glass ceramic in which the cavities in the glass can be clearly distinguished.[50] Theory suggests that as the rate of cavitation increases with both increasing volume fraction and decreasing viscosity of the intergranular phase, there should be marked differences in the size of the cavitational region around growing sub-critical cracks from one nitrogen ceramic to another. This predication has not yet been addressed in microstructural studies.

2.2 Creep Deformation

The high temperature creep deformation of nitrogen ceramics has been the subject of many studies.[51–59,9] According to the results obtained, a number of phenomena occur concurrently, with the dominate mechanism (as deduced from the stress dependence of the strain rate) depending on the material studied. This may be interpretated as meaning that the contribution of the available mechanisms to the overall deformations is microstructurally dependent. In many of the studies in the literature it is not possible to deduce how the microstructure affects or determines the creep properties of the materials as no systematic variation in the microstructure was employed. However, recent results on a

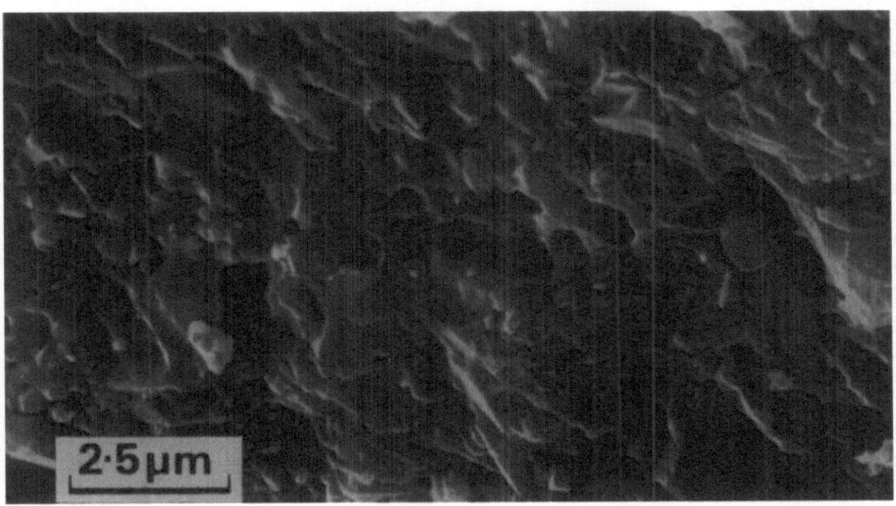

Fig. 5 Example of cavitational crack growth in a MgO hot-pressed Si_3N_4, observed beneath the 1400°C fracture surface and revealed on a room-temperature fracture surface. Scanning electron micrograph.

series of materials in the Si-Mg-O-N system in which the volume fraction of intergranular phase was varied illustrate a number of microstructural trends.[59] Four concurrent mechanisms were identified, 1) cavitational creep, 2) grain boundary sliding accommodated by elastic deformation of adjacent, constraining grains, 3) diffusional creep, and 4) hardening due to oxidation induced compositional changes. Microstructural evidence for solution-reprecipitation creep[60] was also obtained but was of a less definitive nature.

In materials containing a relatively large volume fraction of intergranular phase appreciable cavitation was observable in the microstructure. These observations were complemented by precise density measurements which indicated that the void volume increased with both the volume of the intergranular phase and the magnitude of the creep strain. The majority of the voids were formed at the three and four grain junctions. Occasional cavities along two-grain junctions, indicative of cavitation by grain separation, were also noted. Cavitation at three-grain junctions resulting from creep deformation have also been reported in commercial silicon nitride[54] and in β′ Sialon ceramics.[12] Another pertinent, but difficult to quantify observation, was the finding of increased cavity frequency in the immediate vicinity of the largest grains. Neither the author nor other workers have been able to identify nucleation sites for the cavities, but there is no observable propensity for cavities to form at inclusions, such as WSi_2 or WC, in the microstructure.

Microstructural evidence for grain boundary sliding comes from observation of creep samples cooled under the applied load thereby "freezing" in the stress field. In samples cooled under load, irrespective of the volume content of intergranular phase, strain field contrast in adjacent grains across a grain boundary and emanating from points on the boundaries throughout the microstructure were observed using the transmission microscope. The contrast was eliminated by annealing at ~ 1000°C. These observations, taken together with the strain recovery data, suggest that they arise due to grain boundary sliding accommodated by the elastic deformation of the adjacent silicon nitride grains and that the mechanism is responsible for first-stage or primary creep. The image contrast is attributed to the locking together on grain boundary steps of grains sliding in opposite directions. The observations appear to be perfectly general but require that the elastic stress state be frozen in for them to be seen.

Deformation in materials with only a small volume of intergranular phase was attributed principally to diffusional creep since they exhibited a linear stress dependence of the strain rate and no microstructural changes were noted. Diffusional creep by diffusion through the intergranular phase was presumed to occur. Likewise, no microstructural alteration was seen in support of the oxidation induced hardening phenomenon[59] although microchemical changes were measured as are described in the following section.

2.3 Environmental Reactions

A number of disparate phenomena are observed when nitrogen containing ceramics are exposed to gaseous or solid environments at elevated temperatures. For instance, oxidation treatments can result both in strength degradation or, when correctly utilized, in strength enhancement.

Of the high temperature environmental reactions the oxidation of nitrogen ceramics has been the most thoroughly documented.[61-66] Most materials exhibit passive oxidation kinetics due to formation of a protective layer on the surface during the early stage of oxidation. However, at longer exposures the oxidation kinetics are profoundly affected by the polyphase nature of the nitrogen ceramics. One of the most dramatic manifestations is the oxidation induced cracking of a number of silicon nitride based ceramics. In the worst cases the material can crumble into powder. This phenomena has been noted at 1400°C in ceramics having compositions in the Si_3N_4-MgO-Mg_2SiO_4 phase field[66] and at 700-1000°C in silicon nitride ceramics densified with Y_2O_3.[39,67] The cracking phenomenon is attributed to molar volume changes accompanying oxidation of the secondary phases in the microstructure.[39] For instance in the Si-Y-O-N system the quaternary phases undergo substantial volume changes on oxidation as exemplified by the $Y_2Si_3O_3N_4$ phase that increases in volume by 30% when oxidized to $Y_2Si_2O_7$ and SiO_2. Stresses will arise if the volume increase cannot be accommodated in some manner, such as viscous flow. This method of stress relief presumably explains why no cracking, and consequently strength degradation, occurs at 1400°C in the same Y_2O_3 densified silicon nitrides. Evidence for viscous or plastic flow during oxidation of a silicon nitride containing an unstable second phase, a cerium apatite, has been presented by Lange:[68] that the second phase extrudes from the surface during oxidation. In general, the microstructural distribution of the unstable phase is expected to influence the propensity for cracking by determining the distribution and type of stresses generated during oxidation. In the case most typical of nitrogen ceramics where the unstable phase is in isolated, unconnected grains, tensile stresses will arise within the surrounding Si_3N_4 or β′-Sialon grains to produce cracks. At the other extreme, where the phase is interconnecting and continuous, biaxial compressive stresses may develop on the surface and form the basis of surface strengthening. If the stresses generated are too high, spalling may occur.

Another closely related phenomenon is the oxidation induced cracking of partially porous nitrogen ceramics. This is particularly prevalent in reaction bonded materials but has recently also been observed in fully dense silicon nitrides sequentially heat-treated in a reducing then an oxidizing atmosphere.[42] Holes in the microstructure are filled on oxidation by SiO_2, which on cooling undergoes a phase transformation to low quartz with an accompanying volume increase.

Other than cracking, the most obvious microstructural change on oxidation is the formation of a surface scale. In the most resistant materials, the scale is a glassy patina whereas in others it can be a thick, coarse and partially crystalline crust. In all cases, X-ray microanalysis indicates that the scales are rich in both impurities and the element used as a densification aid.[61-65] These findings imply that major microstructural changes occur within the bulk of the materials. Yet until recently these changes had not been investigated or documented. When the cross section of an oxidized silicon nitride ceramic is examined visually after polishing, there appears to be a region below the scale that has been altered, but when it is studied using either the scanning electron microscope or the transmission electron microscope, no obvious microstructural changes are seen. However, when the cross section is analyzed by X-ray diffraction and X-ray microanalysis profound changes are apparent.[66] A compositional gradient, is found extending from the center of the sample to the scale. X-ray diffraction indicates that there exists a gradient in Si_2N_2O and Si_3N_4 concentration from the sample/scale interface. In addition, the microanalysis results demonstrate that there is a depletion gradient in the Mg concentration below the scale. The full details of the results are described in reference 66. The observed magnesium concentration profile suggests that the composition and volume fraction of the intergranular phase is altered by oxidation. As yet it has not proved possible to analyze the composition of the intergranular phase directly at various distances below the oxidation scale. However, there can be little doubt that the composition does change as the magnesium is known from previous studies to reside entirely in the intergranular noncrystalline phase. Also, the compressive creep properties[59] of materials containing a relatively large volume fraction of glass improve dramatically when oxidized; the strain rate decreases and the exponent in the relation $\dot{\varepsilon} = A\,\sigma^n$ changes from n ~2 to n ~1 indicating a change from cavitationally dominated creep to diffusional creep. The compositional gradient in Si_2N_2O and Si_3N_4 concentration indicates that silicon nitride is converted to silicon oxynitride in the bulk of the material below the scale.[66] How this transformation occurs and whether the silicon oxynitride grows epitaxially on the silicon nitride grain or in the intergranular phase remains to be clarified. One of the problems is that Si_3N_4 and Si_2N_2O are virtually indistinguishable in the transmission electron microscope, unless careful diffraction analysis or electron energy loss spectroscopy of each grain is performed.

The study of the microstructural response of nitrogen ceramics to oxidizing atmospheres is still at an early stage. However, on the basis of these existing observations and of the processes involved, for instance using the mechanism outlined by the author elsewhere in these proceedings,[68] further significant understanding can be expected.

CLOSING REMARKS

Studies relating microstructure, fabrication and properties clearly show the overwhelming influence of the intergranular noncrystalline phase. This is true not only of Si_3N_4 alloys, which have been subject to the most intensive study, but also of other nitrogen ceramics. It is for this reason that the intergranular phase in these materials has been emphasized. Nevertheless, as this survey has shown, our detailed knowledge of the intergranular phase is still limited. In addition to gathering more experimental data concerning the phase, a number of major microstructural challenges remain to advance our understanding. For instance, what is the atomic arrangement in the noncrystalline phase adjacent to the silicon nitride grains, what is the most appropriate description of grain boundaries containing an intergranular phase, do impurities exert any influence on the stability of the intergranular phase, does segregation within the intergranular phase occur, how are cavities nucleated in the phase and most importantly, why is the intergranular phase stable? Answering these and related questions promise not only to lead to improved nitrogen ceramics but also will aid in our understanding of othere ceramics containing intergranular phases such as silicate based refractories and glass-ceramics.

ACKNOWLEDGEMENT

The author is indebted to the U.S. Department of Energy, Division of Materials Sciences, for financial support, under contract ER-78--C-03-1885.

REFERENCES

1. "Nitrogen Ceramics" edited F. Riley (Noordhoff Press, 1976).
2. Evans, A. G. and J. V. Sharp, "Microstructural Studies on Silicon Nitride," J. Mater. Sci. 6, (1971) 1292.
3. Kossowsky, R., "The Microstructure of Hot-Pressed Silicon Nitride," ibid 8, (1973) 1603.
4. Wild, S., P. Grieveson, K. H. Jack and M. Latimer, in "Special Ceramics 5," edited P. Popper (Brit. Ceram. Res. Ass., 1972) 377.
5. Tighe, N. J., "Microstructural Aspects of Deformation and Oxidation of Magnesia Doped Silicon Nitride," in Ref. 1.
6. Drew, P. and M. H. Lewis, "The Microstructure of Silicon Nitride Ceramics During Hot Pressing Transformation," J. Mater. Sci. 9, (1974) 261.
7. Powell, B. D. and P. Drew, "Identification of a Grain Boundary Phase in Hot Pressed Silicon Nitride by Auger Electron Spectroscopy," ibid 9, (1974) 1867
8. Mosher, D. R., R. Raj and R. Kossowsky, "Measurement of Viscosity of the Grain Boundary Phase in Hot-Pressed Silicon Nitride," ibid 11, (1976) 49.
9. Lewis, M. H., B. D. Powell, P. Drew, R. J. Lumby, B. North and A.J. Taylor, "The Formation of Single Phase Si-Al-O-N Ceramics," ibid 12 (1977) 61.
10. Lewis, M. H., A. R. Bhatti, R. J. Lumby and B. North, "The Microstructure of Sintered Si-Al-O-N Ceramics," ibid 15 (1980) 103.
11. Lewis, M. H., A. R. Bhatti, R. J. Lumby and B. North, "Crystallization of Mg-containing Phases in β'-Si-Al-O-N Ceramics," ibid 15 (1980) 438.

12. Karunaratne, B.S.B. and M. H. Lewis, "High Temperature Fracture and Diffusional Deformation Mechanisms in Si-Al-O-N Ceramics," ibid 15 (1980) 449.
13. Clarke, D. R. and G. Thomas, "Grain Boundary Phases in a Hot-Pressed MgO Fluxed Silicon Nitride," J. Amer. Ceram. Soc. 60 (1977) 491.
14. Clarke, D. R., "High Resolution Techniques and Applications to Nitrogen Ceramics," ibid 62 (1979) 236.
15. Clarke, D. R., "On the Detection of Thin Intergranular Films by Electron Microscopy," Ultramicroscopy 4 (1979) 33.
16. Clarke, D. R., "Observation of Microcracks and Thin Intergranular Films in Ceramics by Transmission Electron Microscopy," J. Am. Ceram. Soc. 63 (1980) 104.
17. Tsuge, A., K. Nishida and M. Komatsu, "Effect of Crystallizing the Grain Boundary Glass Phase on the High Temperature Strength of Hot Pressed Si_3N_4 Containing Y_2O_3," ibid 58 (1975) 323.
18. Clarke, D. R. and G. Thomas, "Microstructure of Y_2O_3 Fluxed Hot-Pressed Silicon Nitride," J. Am. Ceram. Soc. 61 (1978) 114.
19. Lou, L. K. V., T. E. Mitchell, and A. H. Heuer, "Impurity Phases in Hot-Pressed Si_3N_4, ibid 61 (1978) 392.
20. Morgan, P. E. D., F. F. Lange and D. R. Clarke, "A New Si_3N_4 Material: Phase Relations in the System Si-Sc-O-N and Preliminary Studies," Comm. Am. Ceram. Soc. C77 (1981).
21. Butler, E., "Observation of Dislocations in β-Silicon Nitride," Philos. Mag. 24 (1971) 829.
22. Niihari, K., and T. Hirai, "Growth, Morphology and Slip System of α-Si_3N_4 Single Crystal," J. Mater. Sci. 14 (1979) 1952.
23. Thompson, D. P., "New Polytypes in the Be-Si-O-N System," ibid 11, (1976) 1377.
24. Huseby, I. C., H. L. Lukas and G. Petzow, "Phase Equilibria in the System Si_3N_4-SiO_2-BeO-Be_3N_2," J. Am. Ceram. Soc. 58, (1975) 377.
25. Shaw, T. M., D. R. Clarke and D. P. Thompson, "Direct Observation of the Polytype Periodicities in the Be-Si-O-N system," J. Mater. Sci. 13 (1978) 217.
26. Shaw, T. M. and G. Thomas, J. Solid State. Chem. submitted
27. C. Greskovich, private communication.
28. Clarke, D.R., in preparation.
29. Clarke, D.R., "Microstructure of a 12H Mg-Si-Al-O-N Polytype Alloy: Intergranular Phases and Compositional Variations," J. Amer. Ceram. Soc. 63, (1980) 208.
30. Clarke, D.R., N.J. Zaluzec and R.W. Carpenter, "The Intergranular Phase in Hot-Pressed Silicon Nitride Alloys: I. Elemental Composition," ibid, in press.
31. Clarke, D.R., N.J. Zaluzec and R.W. Carpenter, "The Intergranular Phase in Hot-Pressed Silicon Nitride Alloys: II. Evidence for Phase Separation and Crystallization," ibid, in press.

32. Ruhle, M.C. Springer, L.J. Gauckler and M. Wilkens, "TEM Studies of Phases in Si-Al-O-N Alloys," pp. 641 in Proc. 5th International Conference on High Voltage Electron Microscopy edited T. Mura and H. Hashimoto (Japanese Soc. Electron Microscopy, 1977).
33. Ruhle, M. to be published.
34. Jack, K.H., "Review: Sialons and Related Nitrogen Ceramics," J. Mater. Sci. 11, (1976) 1135.
35. Jack, K.H., "Sialon glasses," in Ref. 1 p. 257.
36. Tsai, R.L., PhD Thesis, Cornell University, 1981.
37. Shaw, T.M., PhD Thesis, University of California, Berkeley, 1981.
38. Hoenig, C, and C.S. Yust, "Explosive Compaction and Microstructural Analysis of Ceramcis," to be submitted J. Am. Ceram. Soc.
39. Lange, F.F., S.C. Singhal and R.C. Kuznicki, "Phase Relations and Stability Studies in the Si_3N_4-Y_2O_3 Pseudoternary System," J. Am. Ceram. Soc. 60, (1977) 249.
40. Weaver, G.Q. and J.W. Lucek, "Optimization of Hot-Pressed Si_3N_4-Y_2O_3 Materials," Bull. Am. Ceram. Soc. 57 (1978) 1131.
41. Knoch H., and G.E. Gazza, "Effects of Carbon Impurity on the Thermal Degradation of an Si_3N_4-Y_2O_3 Ceramic," J. Am. Ceram. Soc. 62 (1979) 634.
42. Clarke, D. R., "Comparison of Argon and Air Heat Treatments of Hot-Pressed Silicon Nitride," J. Am. Ceram. Soc. (submitted).
43. Lange, F. F., "High Temperature Strength Behavior of Hot-Pressed Si_3N_4: Evidence for Sub-critical Crack Growth," J. Am. Ceram. Soc. 57, (1977) 84.
44. Evans, A. G., and S. M. Wiederhorn, "Crack Propagation and Failure Prediction in Silicon Nitride at Elevated Temperatures," J. Mater. Sci. 9, (1974) 170.
45. Evans, A. G., L. R. Russell and D. W. Richerson, "Slow Crack Growth in Ceramic Materials at Elevated Temperatures," Metall. Trans. 6A, (1975) 707.
46. Lange, F. F., in "Deformation of Ceramic Materials," edited R. C. Bradt and R. E. Tressler (Plenum Press, 1976).
47. Tsai, R. L. and R. Raj., "The Role of Grain Boundary Sliding in Fracture of Hot-Pressed Si_3N_4 at High Temperature," J. Am. Ceram. Soc. 63, (1980) 513.
48. Lange, F. F., "Evidence for Cavitation Crack Growth in Si_3N_4," ibid 62, (1979) 222.
49. Tighe, N. J., "Structure of Slow Crack Interfaces in Silicon Nitride," J. Mater. Sci. 13, (1978) 1455
50. Clarke, D. R., "High Temperature Deformation of a Polycrystalline/Ceramic Containing an Intergranular Phase," Acta Metall. submitted.
51. Osborne, N. J., Proc. Brit. Ceram. Soc. 25 Mech. Prop of Ceramics 2, (1975) 263.

52. Seltzer, M. S., "High Temperature Creep of Silicon-Base Ceramics," Bull. Am. Ceram. Soc. 56, (1977) 418.
53. Kossowsky, R., D. G. Miller, and E. S. Diaz, "Tensile and Creep Strengths of Hot Pressed Si_3N_4," J. Mater. Sci. 10, (1975) 983.
54. Din, S. U. and P. S. Nicholson, "Creep of Hot Pressed Silicon Nitride," ibid 10, (1975) 1375.
55. Birch, J. M., and B. Wilshire, "The Compression Creep Behavior of Silicon Nitride Ceramics," ibid 13, (1978) 2627.
56. Iskoe, J. L., F. F. Lange, and E. S. Diaz, "Effect of Selected Impurities on the High Temperature Mechanical Properties of Hot-Pressed Silicon Nitride," 11, (1976) 908.
57. Lange, F. F., E. S. Diaz and C. A. Anderson, "Tensile Creep Testing of Improved Si_3N_4," Bull. Am. Ceram Soc. 58, (1979) 845.
58. Arons, R. M. and J. K. Tien, "Creep and Strain Recovery in Hot-Pressed Silicon Nitride," ibid 15, (1980) 2046.
59. Lange, F. F., B. I. Davis and D. R. Clarke, "Compressive Creep of Si_3N_4/MgO Alloys. Parts I, II, III," ibid 15, (1980) 601, 611, 616.
60. Raj, R, and C. K. Chyung, "Solution-Precipitation Creep in Glass Ceramics," Acta. Metall. 29, (1981) 159.
61. Kiehle, A. J., L. K. Heung, P. J. Gielisse and T. J. Rockett, "Oxidation Behavior of Hot-Pressed Si_3N_4," J. Am. Ceram. Soc. 58, (1975) 17
62. Tripp, W. C. and H. C. Graham, "Oxidation of Si_3N_4 in the Range 1300-1500°C," ibid 59, (1976) 399.
63. Singhal, S. C., "Thermodynamics and Kinetics of Oxidation of Hot Pressed Silicon Nitride," J. Mater. Sci. 11, (1976) 500.
64. Lange, F. F., "Phase Relations in the System Si_3N_4-SiO_2-MgO and Their Interrelations With Strength and Oxidation," J. Am. Ceram Soc. 61, (1978) 53.
65. Cubicciotti, D. and K. H. Lau, "Kinetics of Oxidation of Hot-Pressed Silicon Nitride Containing Magnesia," ibid 61, (1978) 512.
66. Clarke, D. R., and F. F. Lange, "Oxidation of Si_3N_4 Alloys: Relation to Phase Equilibria in the System Si_3N_4-SiO_2-MgO," ibid 63, (1980) 586.
67. Lange, F. F. and B. I. Davis, "Development of Surface Stresses During Oxidation of Several Si_3N_4/CeO_2 Materials," ibid 62, (1979) 629.
68. Clarke, D. R., "Thermodynamic Mechanism for Cation Diffusion Through An Intergranular Phase: Application to Environmental Reactions of Nitrogen Ceramics," these proceedings.

DISCUSSION

McCauley: How do you explain the observation that the composition of phases found were outside the compatibility triangle?

Clarke: I believe the composition of the glass (analyzed at room temperature) is the remnant composition and corresponds to the composition that cannot be crystallized. On cooling of the liquid from high temperature the expected phases crystallize out rejecting those species which cannot be incorporated into the crystal structures. It should also be emphasized that the phase behaviour diagrams currently used are based on the crystalline phases observed after cooling to room temperature and do not necessarily correspond to the phases actually co-existing at elevated temperature.

Jack: Our proposed phase diagram for the Mg-Si-N-O system would predict the observations you make of glass devitrifying to enstatite and Si_2N_2O. The glass composition does not correspond with the eutectic composition found by Dr. Lange. Distinction must be made between the range of composition that forms glass, and that forming a liquid - the latter is very much wider - and the glass that is obtained will depend upon the rate of cooling.

Clarke: A distinction must be made between the composition of any liquid phases present at sintering temperatures and the composition of any remnant glass found after cooling to room temperature. It is the latter composition that we measure. Just as the eutectic composition in a simple metal binary system decomposes on cooling to give the two terminal phases, one would also expect the liquid present in Si-N materials at high temperature to also decompose at least in part on cooling, and one would expect the last material to solidify to correspond to the lowest melting composition in the phase diagram: hence the finding of a composition similar to your nitrogen glass, and that investigated by Loehmann and Shaw.

Weiss: Can you say something about the area from which your chemical microanalysis is made, and its correlation to the grain size, or grain boundary layer width?

Clarke: The rule of thumb is: the probe diameter should be at least 3x smaller than the size of the region being analyzed, and the thickness of the region should be no more than ~3 the probe diameter. Normally I use a probe 5-40 nm in diameter. All the analyses to date have been performed on triple point pockets, as the two grain junction widths are too small to analyze.

Greskovich: Should we call a 1 nm thick region between two Si_3N_4 grains an intergranular "phase" or a region of "disorder"?

Clarke: I prefer the former description as it has a different composition to the grains of the major crystalline phase.

EVALUATION OF MICROSTRUCTURE IN ß'SiAlON MATERIALS BY TEM METHODS AND ITS CORRELATION TO SOME PROPERTIES

P. Greil and J. Weiss

Max-Planck-Institut für Metallforschung,
Institut für Werkstoffwissenschaften,
Pulvermetallurgisches Laboratorium
Heisenbergstraße 5
7000 Stuttgart 80, West-Germany

INTRODUCTION

The ß-SiAlON materials are well characterized in respect to phase equilibria (1,2) and densification behaviour (3-8). Several TEM investigations report the presence of an amorphous grain boundary phase (9-13). Other investigations report some properties of such materials (14-19). All these investigations, however, are carried out on materials of different compositions and processing conditions. It is the aim of this work to characterize a hotpressed ß solid-solution (ß ss) material of 11 eq.% Al^{3+} with varying oxygen content with regards to microstructure, high temperature strength, creep and oxidation resistance. These investigations were carried out to define a close correlation between material composition, densification behaviour and the resulting properties. In order to make these property measurements comparable to other Si_3N_4 type materials of different composition and processing conditions the characterization of the microcstructure is required. This paper is thought to present the latest results of our work, with a more thorough discussion left open for the future.

Riley, F.L. (ed.) Progress in Nitrogen Ceramics

EXPERIMENTAL

1) specimen preparation

The ß-SiAlON compositions, shown in figure 1, were prepared by mixing appropriate quantities of α-Si_3N_4', Al_2O_3'' and AlN'''. The milling was carried out in an attritor mill for 12 h, filled with n-hexane and Al_2O_3-balls. The wear of the balls and the container was taken into account for the final compositions.

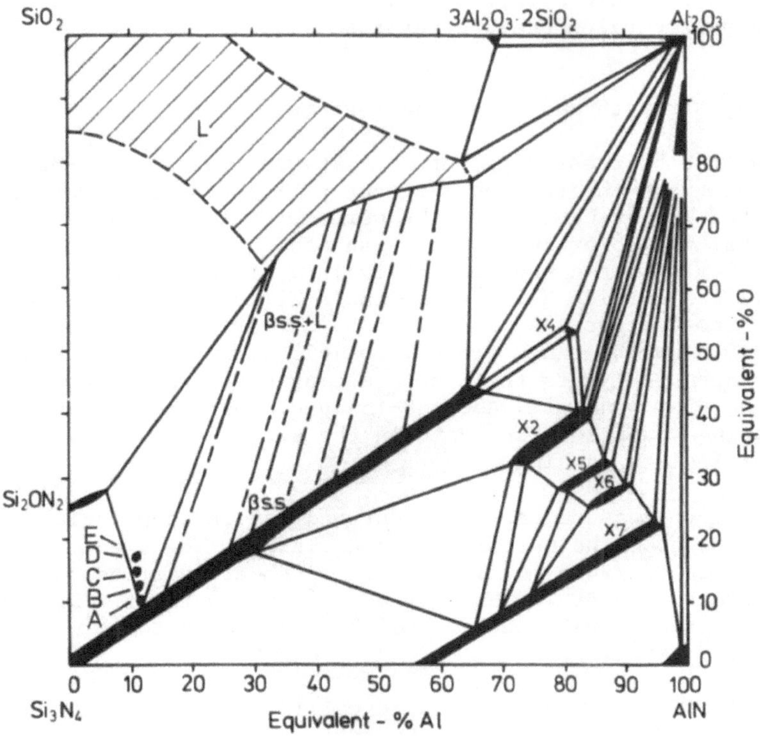

Fig. 1: The SiAlON system at 2023 K after Naik et al. (2) with the compositions of the specimens prepared.

' α-Si_3N_4 Starck, Berlin HCST 3510
'' Al_2O_3 ALCOA A16
''' AlN Starck, Berlin HCST 530

The dried and sieved powders were hotpressed at 2023 K at a pressure of 35 MPa for 45 min in BN coated graphite dies. After firing, the pellets with a diameter of 35 mm were ground off and the density was measured. An X-ray diffractometer[+] (Ni filtered CuK_{α} radiation) was used to determine the phase composition.

For transmission electron microscope (TEM) examination discs with a diameter of 3 mm were drilled out ultrasonically, ground to 30-50 µm and finally ionthinned with an Ar-beam. After coating with carbon, the specimens were examined in an 200 kV[++] transmission electron microscope (TEM). For a more detailed description of the experimental procedure refer to ref. (20).

2) Bending strength and oxidation resistence

From the hotpressed pellets rectangular bars of 2x3x30 mm were cut out and the tension surfaces polished with 2-3 µm diamond paste. The fracture stresses were determined in four point bending using a lower knife edge span of 20 mm and an upper span of 7 mm. Tests were conducted in air in a silicon carbide furnace using silicon carbide knife edges and a silicon nitride stamp for loading. Temperature was controlled by a Pt/Pt-Rh thermocouple within $\pm$ 5 K in the range from RT to 1673 K. The specimens were loaded perpendicular to their hotpressing direction with a crosshead speed of 0,1 mm/min. About 8-15 specimens were determined for each testing condition. In addition the stress strain curves were drawn parallel to loading.

The creep measurements were carried out in cooperation with G. Grathwohl from the university of Karlsruhe. The specimen dimensions were 45x3x4.5 mm. The experiments were carried out in air at a constant load of 100 MN over different length of time. From the creep curves always the minimal creep rate in the time interval was determined.

The oxidation experiments were carried out in air at 1473 and 1673 K. The weight gain of the specimens was measured continuously at the constant temperature using a thermobalance[+++] heated by SiC-heating elements. The specimens were put on a Al_2O_3-base and the temperature was measured by a Pt/Pt-Rh thermo-couple to within $\pm$ 3 K. The oxidized specimens were examined for surface and matrice by X-ray and transmission electron microscopy (TEM).

+ Philips
++ JEM 200 A
+++ Netsch

PROBLEMS IN THE EVALUATION OF TEM MICROGRAPHS

This section outlines some of the methodical problems involved in the characterization of microstructure of ß-Si_3N_4 materials. The ultimate small grain size only allows TEM methods to be used. Obtaining statistically reliable results is a difficult problem but is essential for microstructure characterization. Furthermore TEM does not in general allow to use the laws of plain stereology, but one has to take into account, that a projected image of a threedimensional signal has to be evaluated. In a good approximation, these laws can be applied here for cases of high magnification, i.e. small depth of focus and if it could be assumed that a limited section is in a position perpendicular to the electron beam such as grain boundaries with the amorphous phase in the dark field images. The lack of sharp contrast mecessitates the use of microdensitometry (20,21) and hence imposes another uncertainty factor. The grain size was measured only by the minimal prism diameter that is unaffected by variation of the prism axis relative to the image plain. Finally the preferential grain shape was characterized by using the shape parameter F_1 (22,23)

$$F_1 = \frac{4 \pi A}{U^2}$$

A = cross sectional area of one grain; U = circum ference of this grain.

and its standard deviation. The introduction of such auxiliary parameters is generally necessary, if parameters can not be expressed directly by a mathematical formula of geometrical quantities.

RESULTS

ß-ss was found by X-ray to be the predominant crystalline phase, X_1, and free Si were present in quantities at the limit of detectability. The amorphous phase was detected by TEM methods only. The amount and distribution of this phase, in grain boundaries and in triple points is given in figure 2. Figure 3 shows the minimal prism diameter with a mean value of $\sim$0.5 µm for samples of composition B and E, that is constant upon the increasing amount of amorphous phase. The shape parameter that indicates the preferred shape is given in figure 4 and indicates a preferential prismatic growth habit with increasing amount of oxygen; rough maximal values for the length of these prismatic grains are 8-10µm, another estimate can be given for the width of the amorphous grain boundary layer, measured by microdensitometry (20,21), that lays between 1.5 nm (composition B) and 2.2 nm (composition E). The

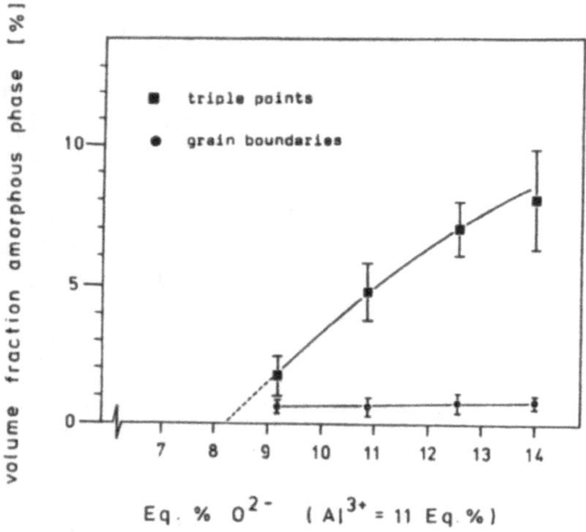

Fig. 2: Volume fraction of the amorphous phase located in the triple points and the grain boundaries.

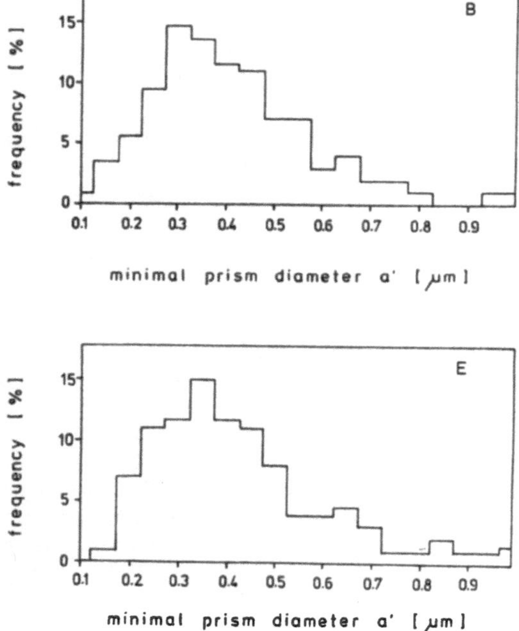

Fig. 3: Minimal prism diameter a' for materials of composition B and E.

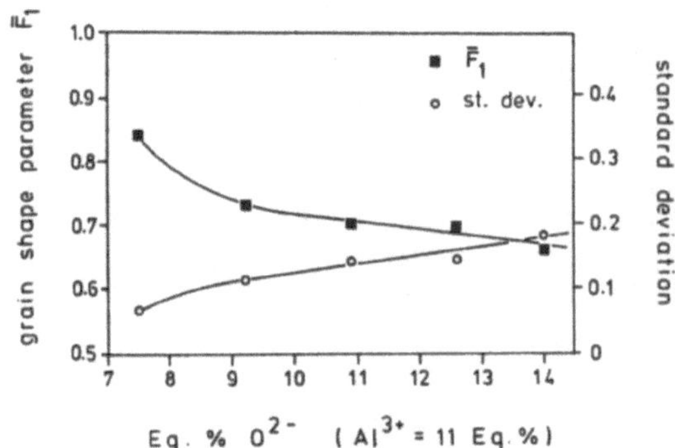

Fig. 4: Grain shape parameter $F_1 = \frac{4 \pi A}{U^2}$ and the standard deviation of F_1.

surplus of amorphous phase is contained in the triple points, that show an increase in the mean volume for about one order of magnitude.

The high temperature strength for materials of all four compositions is shown in figure 5 correlated to the temperature. Material E with the highest volume fraction of amorphous phase shows the best strength and the lowest degradation with increasing temperature. This behaviour is supported by the recorded stress strain curves, shown in figure 6 for two temperatures, where materials of composition E fracture almost brittle while those of composition B show a district plastic deformation. The results of the creep measurements are listed in table 1 and do not show such a distinct influence of the composition.

Oxidation measurements were carried out at 1473 and 1673 K. The results are shown in figures 7a and 7b. They reveal that in the time from 0-10 h - shown enlarged in figure 7c for 1673 K - materials with composition B and C show a lower oxidation rate as those of composition D or E. At longer exposure times this reverses. All materials exhibited a low fraction of cristobalite in the oxide layer, besides this no crystalline phase was detected within this layer. Figures 8a to 8d show this surface layer with an increasing bubble density from composition B to E.

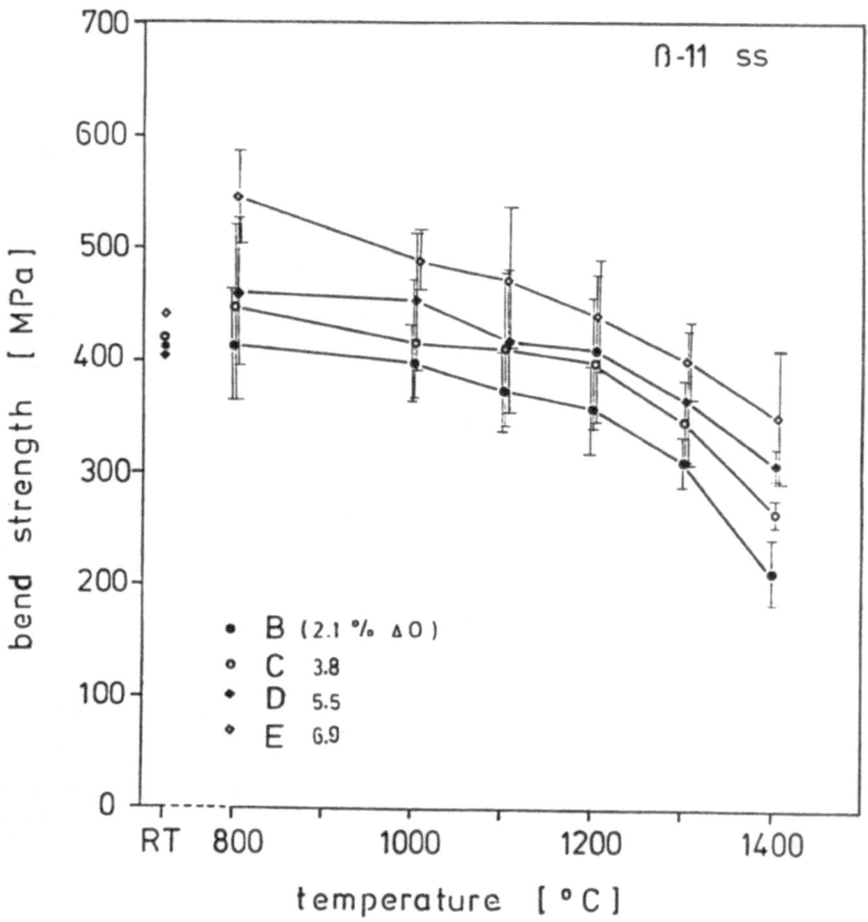

Fig. 5: Influence of composition (amount of glass) on high temperature strength.

DISCUSSION

The results of the microstructure evaluation reveal, that the prismatic growth habit is correlated to the amount of liquid present during densification. Most grain boundaries were found to include an amorphous phase of nearly constant thickness. Materials of composition A with low amount of excess oxygen show 14 % porosity and bend grain boundaries without detectable amorphous layer (20). Some of the grains adjacent to such boundaries show disloca-

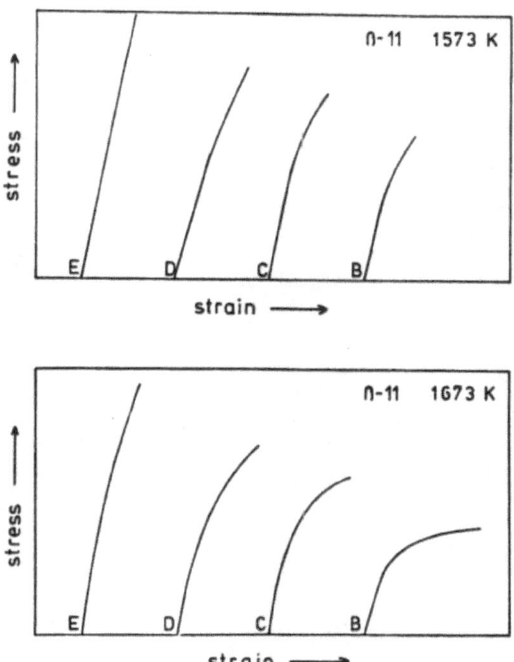

Fig. 6: Stress-strain curves for two temperatures.

TABLE 1 Creep resistance

specimen designation	composition (eq.%)				creep conditions in air, load=100MN		(h^{-1})
	Si	Al	O	N	T(K)	t(h)	
B	89.4	10.6	9.2	90.8	1573	24	$5.4 \cdot 10^{-5}$
					1573	50	$3.3 \cdot 10^{-5}$
					1473	24	$3.3 \cdot 10^{-5}$
					1473	50	$2.0 \cdot 10^{-5}$
C	89.4	10.6	10.9	89.1	1573	24	$6.7 \cdot 10^{-5}$
D	89.4	10.6	12.6	87.4	1573	24	$3.5 \cdot 10^{-5}$
E	89.4	10.6	14.0	86.0	1573	24	$5.6 \cdot 10^{-5}$
					1598	24	$1.5 \cdot 10^{-4}$
					1623	24	$3.4 \cdot 10^{-4}$
					1473	24	$1.2 \cdot 10^{-5}$

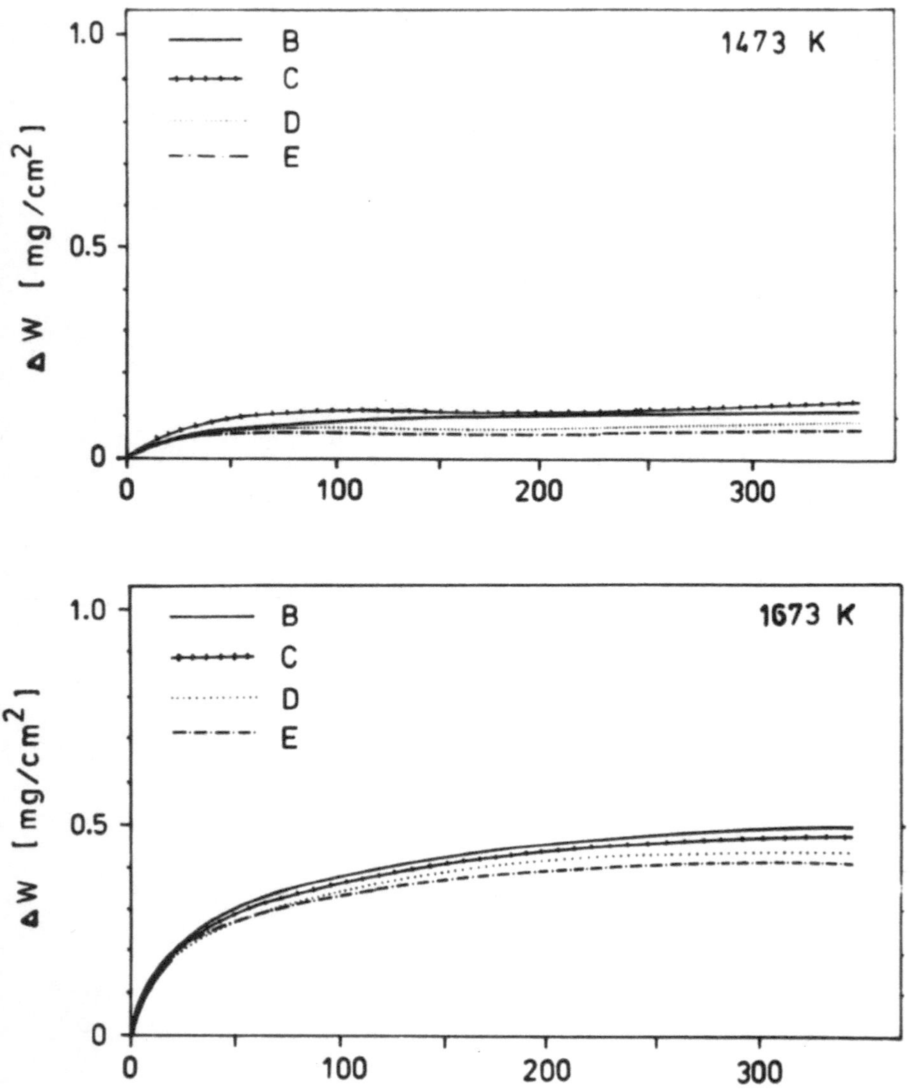

Fig. 7a: Oxidation resistance at 1473 K.
Fig. 7b: Oxidation resistance at 1673 K.

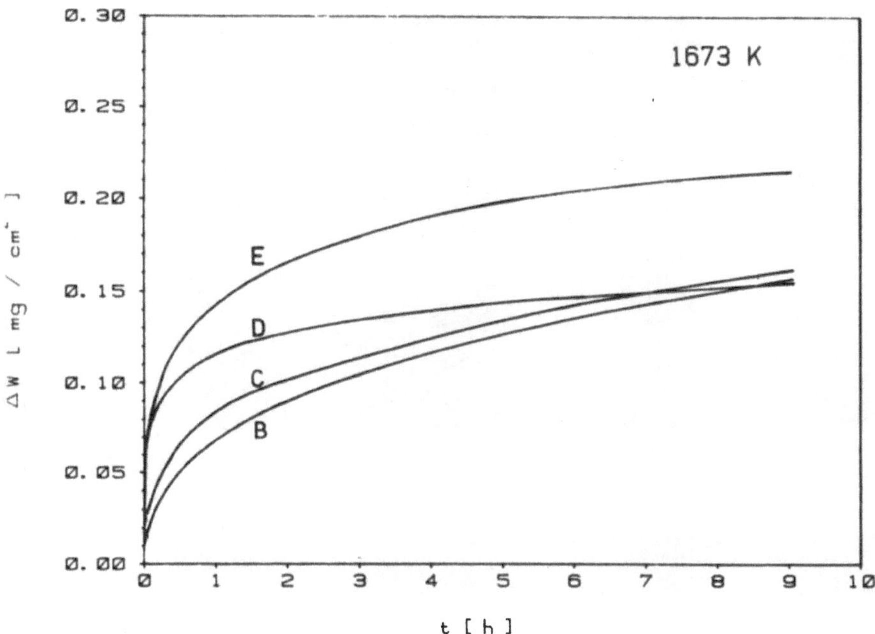

Fig. 7c: Oxidation resistance at 1673 K in the first stage.

tions along the boundary that may result from the densification process, although plastic deformation is not thought to contribute to the densification process in a significant manner (24). Estimation of the amount and the distribution of the amorphous phase was carried out from these micrographs. The values for the total amount of glassy phase were within the range obtained by density and phase diagram considerations. It should be emphasized, however, that the problems noted above only allow determination of values with a rather limited accuracy.

The high temperature strength increases parallel to the increasing volume fraction of glass. This behaviour is in contradiction to what one expects on first sight, which is that due to enhanced viscous flow the strength degradates. An explanation for the opposite type of behaviour can be given by:

1) According to the phase equilibria (25) Mg^{2+} does not dissolve in the ß ss structure. If therefore Mg^{2+} and Ca^{2+} are present

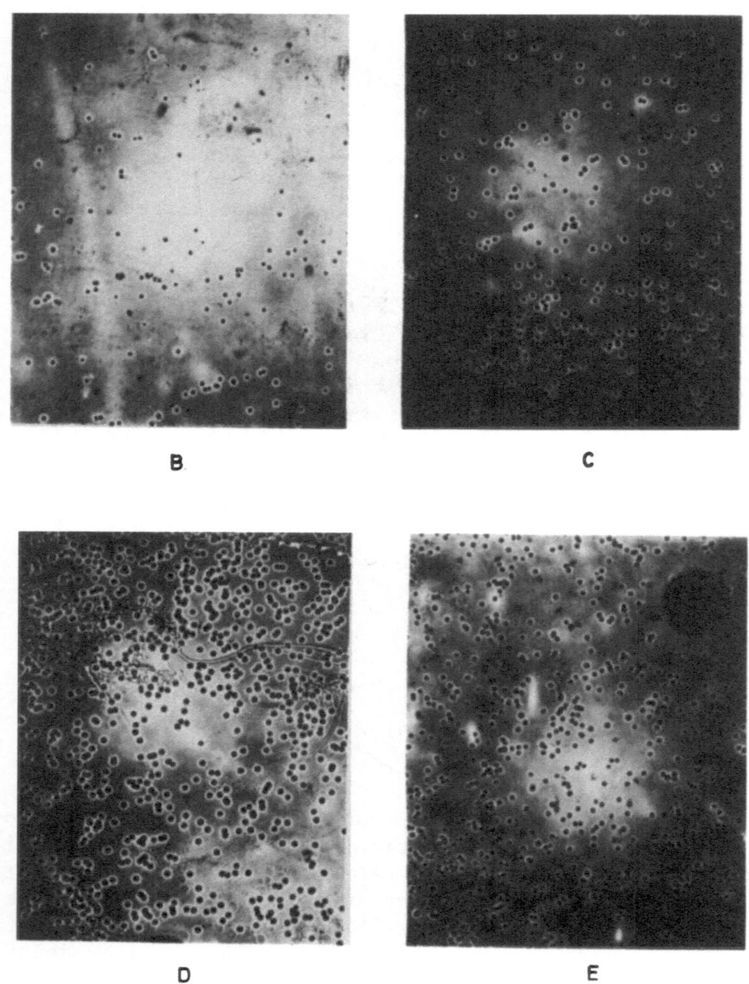

Fig. 8 a-d: Bubble density in the surface layer upon oxidation for 100 h at 1673 K in air.

as impurities, they may concentrate in the amorphous phase. In several investigations enrichment of these two impurity elements in the amorphous phase was detected by microchemical analysis (26-29). If these findings are applied on the present work, a significant variation of the impurity concentration in the amorphous phase results from the varied amount of this phase. This changes the viscosity of this phase remarkably. Table 2 lists the calculated impurity concentrations in the amorphous phase. Thus it was assumed that all the impurity content of the starting materials (see ref. 20) is contained within the volume fraction of amorphous phase noted above.

2) Different thermal expansion of crystalline and amorphous phases could yield different thermal stress concentrations due to the varied ratios of volume fraction of these two phases. The stress-strain curves in figure 6 show that for all materials at least some plastic deformation occurs, this is thought not to be an important factor because at these temperatures the stresses would have been relived already (30).

3) The enhanced prismatic morphology may require a higher work of fracture. Observations of fracture surfaces showed a predominant intercrystalline fracture for specimen of all compositions. Thus any difference in work of fracture has to result from grain size, shape and orientation. An influence of the latter microstructural factors can not be ruled out. Furthermore, the different mean size of triple points may effect the creep behaviour. The results of the creep measurements do not yet allow a conclusive statement.

TABLE 2

specimen designation	theoretical impurity concentration in the amorphous phase wt.%				
	Mg^{2+}	MgO	Ca^{2+}	CaO	Fe^{2+},Fe^{3+}
B	0.7	1.2	1.5	2.2	0.5
C	0.3	0.5	0.6	0.8	0.2
D	0.2	0.4	0.4	0.5	0.1
E	0.2	0.3	0.3	0.4	0.1

At the present time the impurities seem to be the predominant factor controlling the high temperature strength. Direct proof for this statement however will only be acquired by microchemical analysis of the amorphous phase, which is presently being carried out.

The oxidation resistance shows a behaviour similar to the high temperature strength, i.e. materials of composition B shows the highest oxidation rate indicating the formation of a lower viscous surface layer. This is supported by the observation of bubble formation in figure 8 a to d. Composition B shows a low density of bubbles while that of composition E has the highest density. This phenomenon may be explained as follows: The low viscous surface layer allows fast diffusion of N_2 outwards and of O_2 inwards, in the higher viscous surface layer of material E, however the N_2 diffusion is too slow, thus the vapour pressure inside the layer causes bubbles to form. Another significant phenomenon is, that in the first period of oxidation (0-10 h) material B shows the lowest rate of oxidation: The volume fraction of glass in material B is low. Thus beneath the initial SiO_2 layer a high amount of crystalline area is present. In the first stage this crystalline material oxidizes increasing the thickness of the initial SiO_2 layer. Gradually the impurities diffuse out of the grain boundary phase into the SiO_2 layer and increases the diffusivity hence the oxidation rate. In material E with high amount of glassy grain boundary phase the impurities migrate faster (due to higher area fraction) into the SiO_2 layer, therefore, in the beginning the oxidation rate is higher.

CONCLUSIONS

The microstructure of hotpressed ß solid solution materials with varied O^{2-}/N^{3-} ratio at 11 eq.-% Al^{3+} was investigated by TEM methods in conjunction with high temperature strength, creep and oxidation resistance. The results reveal that with increasing amount of oxygen an enhanced prismatic grain morphology combined with increasing amount of an amorphous phase is obtained. The amount and composition of the amorphous phase are thought to be the controlling factors for the properties. This statement has yet to be proven by microchemical analysis. Future work will include a thorough discussion of the results as well as a comparison to the properties of other Si_3N_4 materials.

ACKNOWLEDGEMENT

Financial support by the Bundesministerium für Forschung und Technologie (BMFT) is gratefully acknowledged.

REFERENCES

1. L.J. Gauckler, H.L. Lukas and G. Petzow, J. Am. Ceram. Soc. 58 (1975) 346.

2. I.K. Naik, L.J. Gauckler, T.Y. Tien, J. Am. Ceram. Soc. 61 (1978) 332.

3. S. Boskovic, L.J. Gauckler, G. Petzow and T.Y. Tien, Powd. Met. Int. 9 (1977) 185.

4. S. Boskovic, L.J. Gauckler, G. Petzow and T.Y. Tien, Powd. Met. Int. 10 (1978) 184.

5. S. Boskovic, L.J. Gauckler, G. Petzow and T.Y. Tien, Powd. Met. Int. 11 (1979) 169.

6. M. Kuwabara, M. Benn, F.L. Riley, J. Mat. Sci. 15 (1980) 1407.

7. M.N. Rahaman, F.L. Riley, R.J. Brook, J. Mat. Sci. 16 (1981) 660.

8. M.H. Lewis, B.D. Powell, P. Drew, R.J. Lumby, B. North, A.J. Taylor, J. Mat. Sci. 12 (1977) 61.

9. D.R. Clarke, Ultramicroscopy 4 (1979) 33.

10. D.R. Clarke, J. Am. Ceram. Soc. 63 (1980) 208.

11. M. Rühle, C. Springer, L.J. Gauckler and M. Wilkens, Proc. Fifth Int. Conf. High Voltage Electron Microscopy, Kyoto (1977) 641.

12. A.G. Evans and J.V. Sharp, J. Mat. Sci. 6 (1971) 1292.

13. P. Drew and M.H. Lewis, J. Mat. Sci. 9 (1974) 261.

14. L.J. Gauckler, S. Prietzel, G. Bodemer and G. Petzow, Nitrogen Ceramics, ed. F.L. Riley, Noordhoff, Leyden (1977) pp. 529.

15. M.S. Seltzer, Bull. Am. Ceram. Soc. 56 (1977) 418.

16. J.M. Birch, B. Wilshire, J. Mat. Sci. 13 (1978) 2627.

17. B.S.B. Karunaratne, M.H. Lewis, J. Mat. Sci. 15 (1980) 449 and 1781.

18. J. Schlichting and L.J. Gauckler, Powd. Met. Int. 9 (1977) 36.

19. M.H. Lewis, P. Barnard, J. Mat. Sci. 15 (1980) 443.

20. P. Greil and J. Weiss, to be published in J. Mat. Sci.

21. O.L. Krivanek, T.M. Shaw and G. Thomas, J. Am. Ceram. Soc. 62 (1979) 585.

22. E.E. Underwood, 'The Stereology of Projected Images', J. of Microscopy 95 (1972) 25.

23. E.E. Underwood, 'Quantitative Stereology', Addison-Wesley, Massachusets (1970).

24. L.J. Bowen, R.J. Weston, T.G. Carruthers and R.J. Brook, J. Mat. Sci. 13 (1978) 341.

25. L.J. Gauckler, J. Weiss T.Y. Tien and G. Petzow, J. Am. Ceram. Soc. 61 (1978) 397.

26. R. Kossowsky, J. Mat. Sci. 8 (1973) 1603.

27. B.D. Powell, P. Drew, J. Mat. Sci. 9 (1974) 1867.

28. S. Hofmann and L.J. Gauckler, Powd. Met. Int. 6 (1974) 90.

29. M. Kirn, Ph.D. thesis, university of Stuttgart (1979).

30. A.G. Evans, J.R. Rice, J.P. Hirth, J. Am. Ceram. Soc. 63 (1980) 368.

DISCUSSION

Katz: An explanation for the fact that the material with the greater volume of amorphous phase has better high temperature properties may be that there is stress-enhanced crystallization, which would be easier if there were more, rather than less, glass?

Weiss: I would rule out any effects of crystallization from the glass phase because our materials were heated to test temperature and fractured within 15 minutes, and we observed only minimal crystallization of X_1-phase after 400 h.

Lange: Could you have fabricated your material in two different compatibility triangles, one giving a lower viscosity glass than the other?

Weiss: We believe that our specimens lay within the same compatibility triangle at 2033 K.

Morgan: The stress/strain data indicate clearly that the effect of ledges in the boundary is rate controlling. The chemistry of boundary films is all-important; the excess liquid at the triple points is not very important.

Desmaison: Are your oxidation kinetics parabolic?

Weiss: The curves were roughly parabolic. However, different mechanisms will play a role with the different compositions; which ones is still an open question.

SIALON X-PHASE

D.P. Thompson and P. Korgul

Wolfson Research Group for High-Strength Materials,
Crystallography Laboratory,
The University of Newcastle upon Tyne, UK.

ABSTRACT. Sialon X-phase has a triclinic unit cell with *a*, 9.69; *b*, 8.56; *c*, 11.21Å; α, 90.0; β, 124.4; γ, 98.5^{o} and essentially a point composition close to $Si_3Al_6O_{12}N_2$. When liquid of this composition is cooled, well-crystalline "low"-X is formed with the above unit-cell dimensions. Its crystal structure is similar to mullite and consists of alternate chains of octahedra and tetrahedra linked to form sheets in the (100) plane; these sheets are joined together by a complex network of tetrahedra, some units of which resemble the Si_6N_8 units in the unit cell of β-Si_3N_4. Whereas the repeat distance along the chains is 2.85Å, the tetrahedral network repeats every three tetrahedra (8.56Å).

At more silica-rich starting compositions, X-phase crystallises as very fine needles ("high"-X) which give diffuse and often very weak X-ray reflections for $k \neq 3n$. This is due to frequent faulting of the structure by shifts of *b*/3 in the (100) plane which results in occupation of all available tetrahedral sites in the network and gives an apparent *b* repeat distance of 2.85Å.

1. INTRODUCTION

Sialon X-phase was discovered independently by Oyama & Kamigaito (1) and by Jack & Wilson (2) as a second phase when silicon nitride and alumina were hot-pressed together at 1700°C. Its composition is close to $Si_3Al_6O_{12}N_2$ with a very small range of homogeneity. The crystal structure is complex and for this reason many of the early attempts to characterise X-phase (see, for example, Drew & Lewis (3); Wild (4); Gugel et al. (5); Jack (6)) were unsuccessful. In the present work the crystal structure has been determined

Riley, F.L. (ed.) Progress in Nitrogen Ceramics

Table 1. Unit cell dimensions for X-phase

Form	a	b	c	α	β	γ	
Low	9.69	8.56	9.85	81	70	81	Zangvil (7)
Low	15.93	8.54	11.17	90.0	91.0	100.3	Okamura & Inoue(8)
Low	9.68	8.54	11.19	90.0	124.3	99.2	Zangvil et al.(9)
Low	9.68	8.56	11.21	90.0	124.4	98.5	Present work
High	9.66	2.84	11.17	90.0	124.4	98.5	Present work

by single crystal X-ray methods and this has been used to interpret results obtained by optical and electron microscopy.

2. UNIT CELL

Jack (6) reported that the X-ray pattern of X-phase was not always the same and two different forms, high-X and low-X, could be distinguished. Low-X occurs at the composition given above but high-X is formed at more silica-rich compositions. Electron probe microanalysis shows that high-X is a mixture of low-X plus glass but structurally low-X is a superlattice of high-X with the b edge-length of high-X one third of that of low-X. The unit cells of both forms are triclinic with the dimensions given in Table 1. The different ways of describing the low-X cell arise because the present authors and Zangvil et al. (9) have used the Delauney cell (shortest edge-lengths with all angles $\geqslant 90^\circ$) whereas Zangvil (7) used the cell with shortest edge-lengths and Okamura & Inoue (8) used a C face-centred cell with the angles nearest 90°.

3. STRUCTURE DETERMINATION AND REFINEMENT

A crystal of low-X, prepared by slow cooling from the melt, was set up on a Syntex $P2_1$ 4-circle diffractometer and 6897 X-ray intensities were collected using $MoK\alpha$ radiation. An Lp correction was applied but absorption effects were neglected. A model for the structure proposed by Okamura & Inoue (8) was used to initiate refinement. The final R-index of 9% was obtained after very little refinement of atomic parameters but the occupation scheme for tetrahedral sites required considerable adjustment. This was due to twinning and stacking faults in the crystal which resulted in occupation of certain empty sites. For the purpose of discussion it is convenient to first consider the idealised structure and then show how this is modified by defects.

4. DISCUSSION

Figure 1 shows the idealised structure projected down the b axis.

All the atoms occur in layers perpendicular to b at intervals of one-sixth. Numbers in the diagram express the heights of atoms in numbers of sixths above the plane of the paper. In this orientation octahedra project as parallelograms and tetrahedra as isosceles triangles. Because of the similarity in atomic number between silicon and aluminium and between oxygen and nitrogen, metal and non-metal ordering schemes cannot be determined from X-ray data alone. Aluminium is probably coordinated by oxygen in octahedral sites.

The structure consists of chains of octahedra parallel to b linked together in the c direction by chains of tetrahedra and in the a* direction by two distinct types of tetrahedral network. Type I networks, illustrated in the bottom left and top right quadrants of Figure 1, consist of two groups of three tetrahedra joined together with the non-metal atoms in the centre of each group in three-planar coordination. This arrangement is similar to that of the six tetrahedra in the unit cell of β-Si_3N_4 but in X-phase is only repeated once in every three layers. Type II networks are shown in the top left and bottom right quadrants of

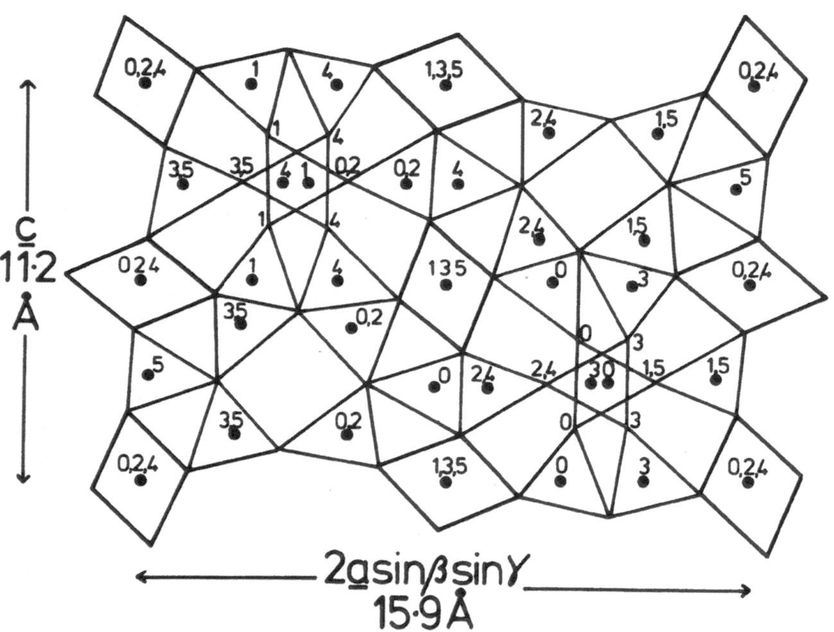

Fig. 1 The idealised atomic arrangement in low-X projected down the b axis. The C-face-centred cell has been used.

Figure 1 and consist of tetrahedra pointing inwards towards a central region which is occupied by two tetrahedra separated by intervals of b/2. Whereas the structure is similar to mullite in the linking of chains of octahedra and tetrahedra in the c direction, the increased proportion of tetrahedra in X-phase result in a larger repeat distance in the a* direction and hence a more complex cell of lower symmetry. The unit cell contents of $M_{52}X_{84}$ correspond quite well to six $Si_3Al_6O_{12}N_2$ formula units, giving a calculated density of 3010kg m^{-3} in good agreement with measured values (2850-3000 kg m^{-3}, present work; 3000 kg m^{-3}, Zangvil et al. (9)).

The actual structure differs from that described above in the occupation of additional tetrahedral sites. Thus, for example, in the Type I unit in the bottom left-hand quadrant of Figure 1, there are vacant sites at height 3 immediately below the left-hand tetrahedron and at height 2 immediately above the right-hand tetrahedron. Occupation of these sites instead of the normal sites results in the adjacent Type II units being displaced up or down by intervals of b/3. Whether the result is a twin or a fault depends on the continued occupation scheme for these vacant sites in adjacent unit cells. Both types of defect are observed on high resolution lattice images (see Figure 2) but whereas twinning occurs in almost every specimen and is difficult to eliminate, stacking faults can be removed by careful annealing. Thus low-X, prepared by slow-cooling of the $Si_3Al_6O_{12}N_2$ composition from the melt, has a large-grained microstructure (Figure 3(a)) which is submicroscopically twinned but is almost unfaulted. High-X prepared at more silica-rich compositions,

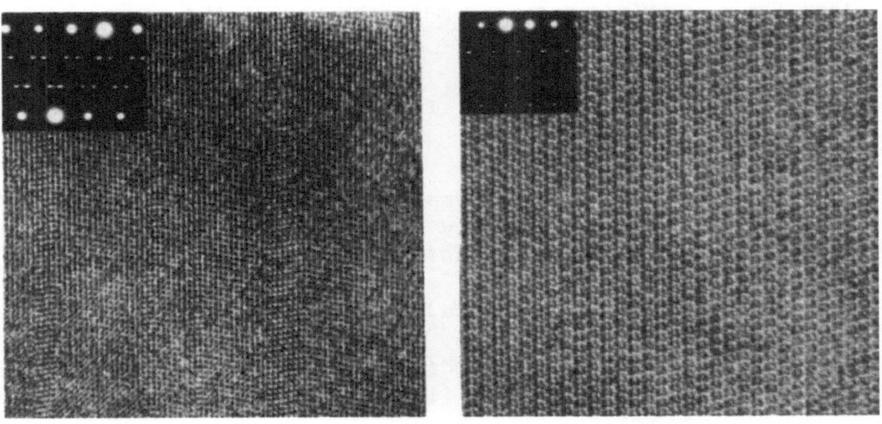

(a) (b)

Fig. 2 High-resolution lattice images of X-phase showing (a) twinning, and (b) stacking faults. The separation of adjacent vertical (100) planes in both micrographs is 7.8Å.

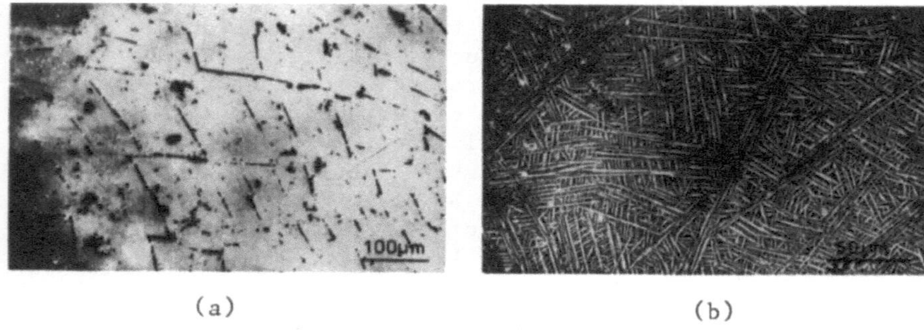

(a) (b)

Fig. 3 Optical micrographs of (a) low-X, and (b) high-X.

crystallizes rapidly at lower temperatures from a highly viscous supercooled sialon liquid and is therefore fine-grained (Figure 3(b)) and heavily faulted. This smears out the metal atoms in tetrahedral sites to give an apparent $\underline{b}$ repeat distance of one-third of that of low-X. The difference between the two forms is therefore very satisfactorily explained in terms of the frequency of stacking faults in the structure.

ACKNOWLEDGEMENTS

We wish to thank Professor N-G. Vannerberg of the Department of Inorganic Chemistry, Chalmers Technical University, Gothenburg for permission to use the 4-circle diffractometer and also Dr. Jägner, Dr. Lindquist, Mr. Anderssen and Dr. Ljungström for invaluable assistance in the use of the equipment. We also thank Professor K.H. Jack for advice and encouragement throughout the course of the work.

REFERENCES

1. Oyama, Y. and K. Kamigaito, *Jap. J. Appl. Phys. 10* (1971) 1637.

2. Jack, K.H. and W.I. Wilson, *Nature 238* (1972) 28.

3. Drew, P. and M.H. Lewis, *J. Mat. Sci. 9* (1974) 1833.

4. Wild, S. in P. Popper ed. *Special Ceramics 6* (Proc. Brit. Ceram. Res. Ass., July 1974) (1975) 309.

5. Gugel, E., I. Petzenhauser and A. Fickel, *Powder Met. Int. 7* (1975) 66.

6. Jack, K.H. *J. Mat Sci. 11* (1976) 1135.

7. Zangvil, A. *J. Mat. Sci.* *13* (1978) 1370.

8. Okamura, F.P. and Z. Inoue, (to be published) (1981).

9. Zangvil, A., L.J. Gauckler and M. Rühle, *J. Mat. Sci.* *15* (1980) 788.

DISCUSSION

Jack: Since X-phase is the only congruently melting solid in the Al-Si-N-O system, has it any technological advantages?

Thompson: Preparing X-phase is not easy - weight losses are difficult to avoid and because the composition is close to the glass-forming region. Most samples contain residual glass. Probably for these reasons very little property data is available for X-phase, but the thermal expansion coefficient is $\sim 6\ MK^{-1}$

Kizilyalli: Would I.R. spectroscopy help to clarify the nature of the high and low forms of X-phase?

Thompson: I would expect the spectra to be complex and too closely similar to be useful.

MICROSTRUCTURE DEVELOPMENT IN SILICON NITRIDE

H. Knoch, K.A. Schwetz and A. Lipp

Elektroschmelzwerk Kempten GmbH
8960 Kempten, W.-Germany

Introduction

The intensive research on silicon nitride materials in the past decade has clearly demonstrated that there are two crucial problems which have to be resolved in order to fabricate dense components, which perform reliably in low and high temperature applications. These problems may be addressed as microstructure and phase composition control during fabrication.

Since the early work of Deeley et al. (1) it is known that densification of Si_3N_4 can be accomplished by hot-pressing if the powders contain a densification aid. Many additive systems have been reported in the literature for hot-pressing (2-7) and more recently for pressureless sintering (8-11). The densification of Si_3N_4 powder is attributed to a liquid phase sintering process, when the additive, Si_3N_4, and the always present SiO_2 form a liquid at high temperatures, which promotes densification via a solution-reprecipitation mechanism. Depending on the additive used, solid solutions (12, 13) of Si_3N_4, crystalline (14) and glassy grain boundary phases (15, 16) may be formed. Although this paper will not concentrate on the various additive systems in the processing of dense Si_3N_4, it must be stated, that the secondary phases formed during fabrication influence microstructure and properties of the material. Studies of the phase relationships in the Si_3N_4-additive systems (a review has been given by Lange (17)) have been an important step on the way to high performance Si_3N_4 of controlled phase composition.

Besides the secondary phases and their distribution in the materials it is the grain structure which dominates the mechanical

Riley, F.L. (ed.) Progress in Nitrogen Ceramics

properties. Lange (18) was the first to show that an equiaxed grain morphology was obtained with high β starting powders, whereas a fibrous grain morphology developed, when high α starting powders were used. The fibrous grain morphology resulted in better mechanical properties. In several papers different investigators demonstrated, that the α to β phase transformation and the growth of elongated fibrous β grains are closely correlated (19-25). Since the α to β phase transformation necessarily implies the growth of β crystals, phase transformation and grain size must be correlated which was shown by Knoch and Gazza (26).

In the following sections some models in regard to the development of microstructure in Si_3N_4 will be discussed, with the main emphasis on the development of grain structure. As the model system Si_3N_4-MgO was chosen, where the most data are available. And it shall be demonstrated, that the processing parameters influence the resulting microstructure to a great extent, which has to be kept in mind when high strength material has to be fabricated.

2. CORRELATIONS BETWEEN MICROSTRUCTURE DEVELOPMENT AND PHASE TRANSFORMATION

Following (26), figure 1 schematically shows some parameters and mechanisms which are known to influence the microstructure of hot-pressed Si_3N_4.

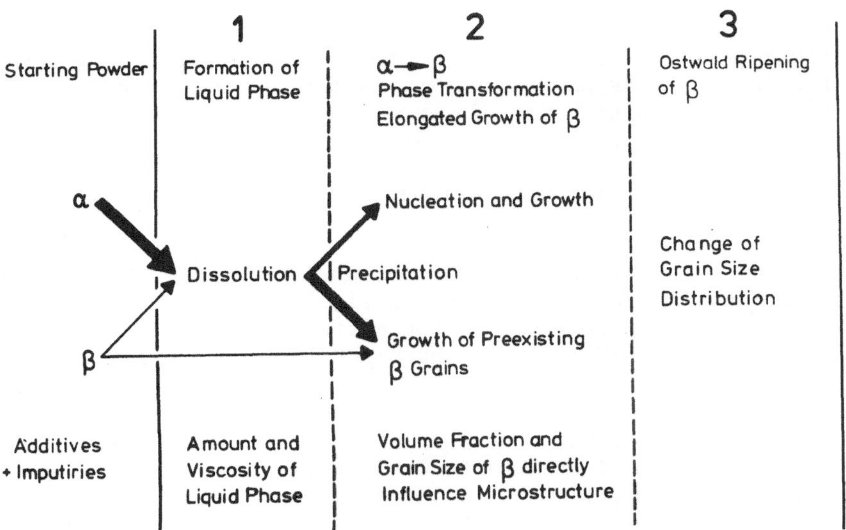

Figure 1: Parameters and mechanisms influencing microstructure during hot-pressing (26)

The hot-pressing regime is divided into 3 regions: the formation of liquid phase, the α/β phase transformation, where the principal thermodynamic driving force for the formation and growth of β is assumed to be the difference in the free energy of the α and β structure at temperature (19), and the Ostwald ripening of β . In this region all grains are β , the driving force for grain growth is assumed to be generated by the reduction of free surface energy. The three indicated regions will overlap in reality. With the formation of the liquid phase, whose amount and viscosity is determined by the additive and impurities of the system, phase transformation and densification will start. Using MgO, Bowen and coworkers (27) obtained the same activation energies for the phase transformation as for densification, suggesting the same mass-transport mechanisms for both phenomena. For the early stages of densification they assumed the diffusion step to be rate controlling in the solution-reprecipitation mechanism. Full densification can be achieved well before the α/β phase transformation has been completed, so that the microstructure development will continue in the dense state. The key processes for the development of microstructure and the correlated phase transformation are indicated by the reaction paths shown in figure 1. It has been demonstrated that the growth of preexisting β grains in the starting powder directly influences the microstructure of the hot-pressed compound (19, 20). Another important influence on the microstructure (and on transformation kinetics) originates from nucleation of β out of the liquid and growth of these grains. This reaction seems to be strongly dependent on the applied pressure during hot-pressing. In (26) a smaller grain size was observed for the same degree of transformation at high applied pressure (70 MN/m^2) as compared with lower pressure (35 MN/m^2) for otherwise constant conditions. This finding suggests an enhancement of the nucleation rate of β out of the liquid with increasing applied pressure, so that the transformed volume fraction of β simply is formed by more grains, resulting in a finer grain size. On the other hand an apparent increase of transformation rate is observed with increasing pressure (26, 29, 30).

Figure 2a and 2b (26) show, that an increase of the β volume fraction in the starting powder gives rise to an increase of the transformation rate also. At all temperatures shown, the rate of transformation is faster for the high β starting material. After 2 h at 1500 °C, the f_β = 0.04 powder only transformed to f_β = 0.18. Two hours at 1700 °C were required before complete conversion occured. In contrast, the f_β = 0.29 starting powder transformed after 2 h at 1500 °C from 0.29 to 0.63. Full transformation was observed after 2 h at 1600 °C, a 100 °C lower than that noted for the low β starting material.

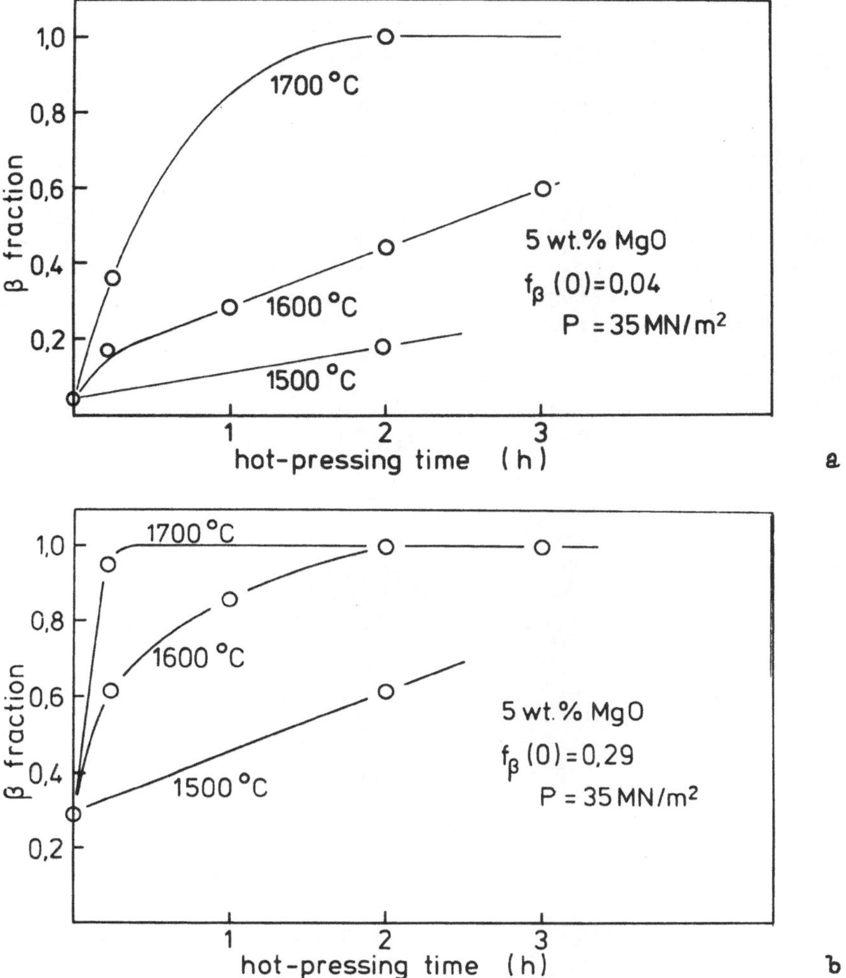

Figure 2: β fraction as a function of hot-pressing time and temperature for Si_3N_4 with 5 wt.-% MgO. The β fractions in the starting powder are 0.04 (a) and 0.29 (b) (25).

The microstructure development during phase transformation, i.e. the growth of elongated fibrous grains has been studied intensively in the literature (17-28). Summarizing the experimental findings it can be stated, that microstructures with high aspect ratio β grains are only obtained, if starting powders with small β volume fractions are used. With increasing β fraction in the starting powder the microstructure tends to become less fibrous until for β starting powder equiaxed grains are obtained.

It is significant that the transformation rate depends on the initial amount of β phase in the starting powder. The influence of preexisting β phase particles on the transformation rate, as well as the influence of enhanced nucleation of β , strongly suggests that the total α/β interface area in the material controls the phase transformation. In (26) a probable growth mechanism was suggested, which is shown schematically in figure 3. It is assumed that in the early stages of transformation of the dense body, isolated β grains, surrounded by a liquid phase, are embedded in an α matrix. During grain growth of β (and phase transformation) in a dense body three rate limiting mechanisms can be discussed, which are also shown in figure 3: the dissolution of α into the liquid, the diffusion through the liquid, and the reprecipitation at the β surface.

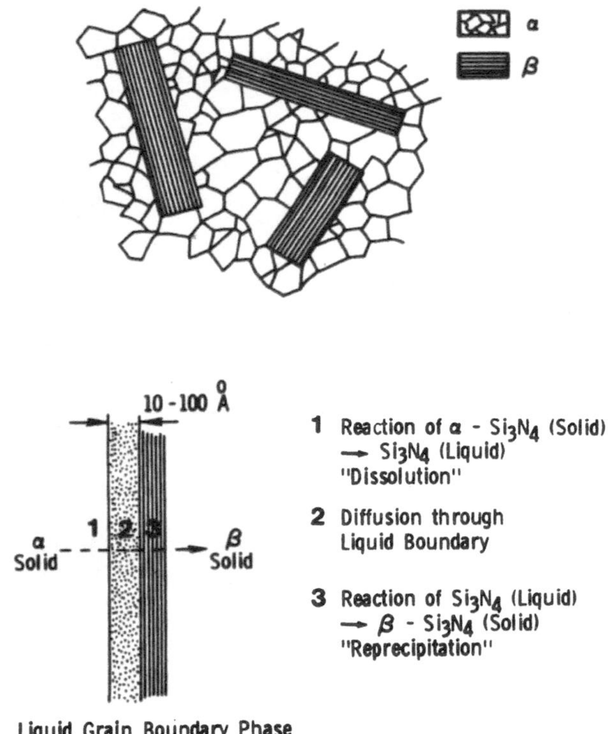

Figure 3: Schematic view of processes occuring during the α to β phase transformation of hot-pressed Si_3N_4 (26).

For the densification step, both the solution reaction (31) and the diffusion through the liquid (22, 32) have been discussed as rate limiting processes. In dense material, the diffusion distances for the solution-precipitation process during the α to β phase transformation can be assumed to be in the order of 10-100 A, i.e. in the range of the thickness of the liquid grain boundary phase (15, 16). In this case the diffusion step may be regarded as a fast one. Thus it is more likely that a reaction step is rate controlling, and it can be concluded that for the growth of β perpendicular to the c-axis (the slower growing direction) this must be the precipitation step.

The suggested model for grain growth and correlated phase transformation is only valid for unhindered growth, which can be assumed to be realistic in the early stages of phase transformation. With further phase transformation growing β grains will impinge, leading to a depletion of α in certain areas and a decrease of α/β interface. Additionally, growing β grains push the liquid phase at the growth front into the matrix. The thickness and the impurity concentration of the liquid grain boundary phase will change. The influence of impurities on the growth kinetics of β is essentially unknown. It becomes obvious that the measured transformation rate during hot-pressing is a result of rather complex microscopic mechanisms.

During the phase transformation and the development of a fibrous microstructure, one should be aware of the situation that β grains do not only grow into their elongated direction parallel to the c-axis. There is also a not negligible increase of the thickness of the growing β prisms during transformation. Taking advantage of the fact that the diameter of an elongated rod does not change in a projection to a plane, the change of the shortest diameter of all grains on a polished section will give some information how the diameter of elongated β grains changes. Figure 4 (26) shows a histogram of the shortest diameter distribution of a hot-pressed Si_3N_4, and it clearly indicates that with increasing transformation the thickness of the β prisms increases. For unhindered growth, the ratio of the growth rates parallel and perpendicular to the c-axis seems to be about 10 (26), and will determine the aspect ratio of these grains. Nevertheless, the expression for the aspect ratio $R = 1 + f_\alpha / f_\beta$ (19), (f_α, f_β are initial volume fractions), which is obtained by neglecting the growth perpendicular to the prism axis, is a good approach in most cases. In order to demonstrate realistic changes of grain size and grain morphology due to transformation induced grain growth, figure 5 (25) gives a good example. Although the hot-pressing temperatures are different, it shows the drastic differences in grain size and morphology, which can be obtained with the same starting powder in Si_3N_4 materials.

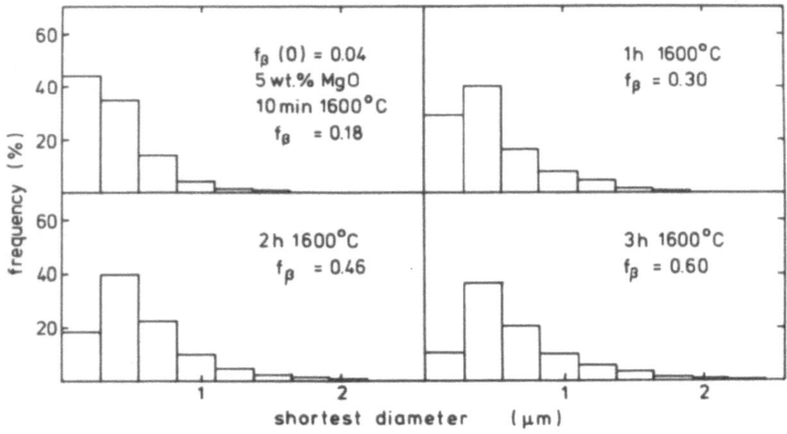

Figure 4: Distribution of the shortest grain diameters in Si_3N_4 after different hot-pressing times (26).

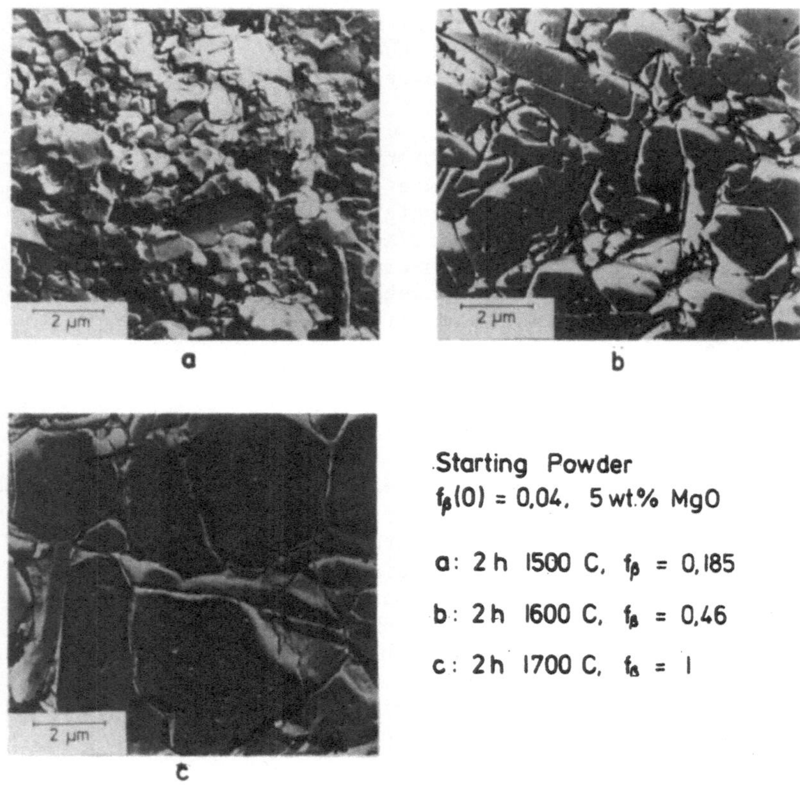

Figure 5: Typical replicated microstructures (TEM) of Si_3N_4 after hot-pressing at different temperatures (25).

3. INFLUENCES OF MICROSTRUCTURE DEVELOPMENT ON MECHANICAL PROPERTIES

As has been discussed in the preceeding sections, the microstructure development in hot-pressed silicon nitride is closely linked to the occurence of grain boundary phases in the material. Since failure of this material under stress occurs predominantly by intergranular fracture (18, 22, 24), it becomes clear that the second phase at the grain boundaries is an essential feature of this material. Dense Si_3N_4 containing sintering aids may be described as a material consisting of hard grains, bonded together by a grain boundary phase. From this point of view there are principally two approaches to understand and to control the mechanical properties.

The first is to control the quality of the grain boundary phase, or to control the bond strength between Si_3N_4 grains. This concept has been described as "grain-boundary-engineering" (33). The main object here is to improve the high temperature strength by developing highly refractory grain boundary phases (3, 34-37) and crystalline ones (14). This concept needs the understanding of the phase relationships in the various systems.

In the second approach the microstructural appearance of the material has to be examined. For a given composition of the grain boundary phase - or for a given bond strength between Si_3N_4 grains - the mechanical properties can be influenced by the grain size and the grain morphology. As first shown by Lange (18) the development of elongated β grains in hot-pressed Si_3N_4 is of major significance to obtain high strength material. The irregular growth of elongated β prisms during the α/β phase transformation leads to a mechanically interlocked microstructure. Another important factor influencing the strength of brittle materials is the grain size. In general, coarse grained materials have lower strength than fine grained microstructures. In HPSN the α/β phase transformation and grain growth of β are closely correlated, i.e. the elongated growth of preexisting and nucleated β grains determines the resulting microstructure (19, 25, 26). The transformed volume fraction of β can be used as a measure for the volume fraction of elongated grains. Figure 6 (22) shows a typical plot of the fracture stress and the transformed volume fraction of β as a function of hot-pressing time. The strength increases with increasing β and shows a maximum when the phase transformation has been completed. Although in this region grain growth of β must occur (which has principally a negative influence on strength) an increase in strength is observed due to the elongated and interlocking growth of β. The decreasing strength after completed transformation is discussed to be a result of grain growth without further enhanced interlocking growth (24-26), which is observed during the α/β phase transformation.

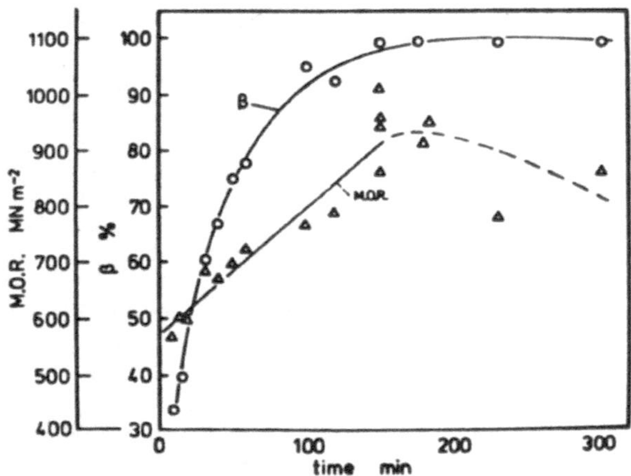

Figure 6: Transformed volume fraction of β and bend strength of Si_3N_4 with 5 wt.-% MgO as a function of hot-pressing time (22).

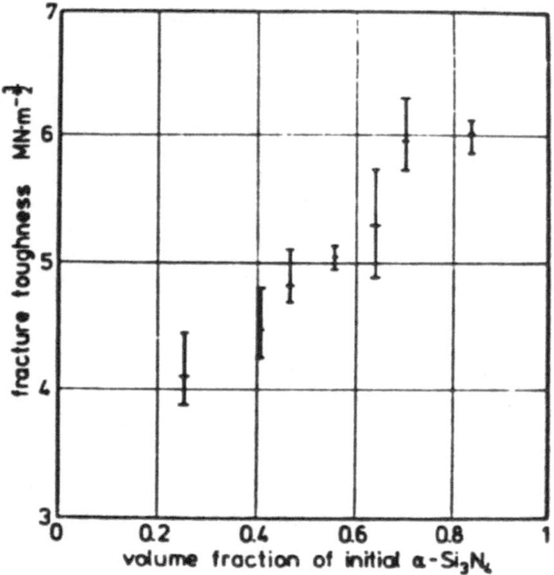

Figure 7: Fracture toughness (indentation technique) of hot-pressed Si_3N_4 containing 5 wt.-% MgO as a function of the initial α phase content in the starting powder (38).

Fracture energy (18, 24) and fracture toughness are influenced most effectively by the microstructure. Figure 7 (38) shows fracture toughness values as a function of the initial volume fraction of α Si_3N_4 in the starting powder (= volume fraction which is transformed into elongated β grains during hot-pressing). Similar microstructures are obtained when the material is hot-pressed for shorter times than needed for complete phase transformation, leaving fractions of equiaxed α in the dense compound.

For the fabrication of high strength Si_3N_4, the material has to be dense, the phase transformation has to be completed to have a high volume fraction of elongated grains, and the grain size has to be possibly small. If the microstructure development mainly is the consequence of the growth of preexisting β (compare figure 1), then the starting powder shouldcontain the β particles in a finely divided form. Total number and size distribution of growth sites will determine the microstructure. Another possibility to influence the final grain size is to enhance the nucleation of β out of the liquid during the phase transformation. Increased applied pressure could be a way to do so (26, 29, 30). Another idea is to enhance nucleation, following the principles of nucleation and growth of crystals from undercooled supersaturated liquids (39). In a study it was shown (40) that the bend strength of hot-pressed Si_3N_4 with large amounts of liquid phase (10 wt.-% MgO) could be increased to values $>$ 850 MN/m^2, when hot-pressing was performed in a way that the liquid phase was allowed to form during heating up and a 5 min. hold at 1780 °C. Then the temperature was dropped to 1650 °C and held until the α/β phase transformation had been completed. When hot-pressing was performed at 1780 °C, strength values of about 680 MN/m^2 could be obtained. The influence of controlled α/β transformation and grain size on the strength of hot-pressed Si_3N_4 are summarized in figure 8, which shows the influence of process development on strength, which in this case is basically the influence of microstructure development.

4. CONCLUSIONS

The literature data show that in general the best mechanical properties are achieved with small grain sizes and a high volume fraction of elongated β grains, i.e., when the α/β conversion is complete and no further grain coarsening after completed transformation has occured.
Processing parameters like β fraction in the starting powder, applied pressure and temperature history are important factors influencing phase transformation and correlated development of microstructure. Optimum microstructures can be achieved by process control of these parameters.
In addition to grain size and grain morphology, the mechanical properties of dense Si_3N_4 are influenced by the chemistry of the

grain boundary phase and other second phases. But as long as the grain boundary phase is the weak link, only shifts of the general microstructural influence on mechanical properties should be observed.

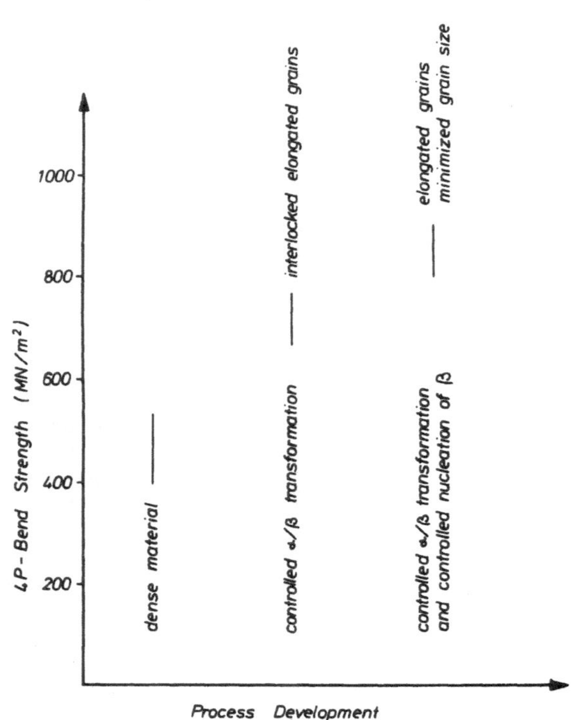

Figure 8: Strength of hot-pressed Si_3N_4 with 10 wt.-% MgO as a function of the process development stage.

REFERENCES

1. Deeley, G.G., Herbert, J.M. and N.C. Moore, Powder Metall 8 (1961) 145
2. Jack, K.A. and W.J. Wilson, Nature Phys. Sci 28 (1972) 238
3. Gazza, G.E., J. Am. Ceram. Soc. 56 (1973) 662
4. Mazdiyasni, K.S. and G.M. Cooke, J. Am. Ceram. Soc. 57 (1974) 536
5. Huseby, J.C. and G. Petzow, Powder Met. Int. 6 (1974) 17
6. Rice, R. and W.J. McDonough, J. Am. Ceram. Soc. 58 (1975) 264
7. Greskovich, C., Palm, J.A. and S. Prochazka, Bull. Am. Ceram. Soc. 57 (1978) 830

8. Terwilliger G.R. and F.F. Lange, J. Mat. Sci 10 (1975) 1169
9. Mitomo, M., Tsutsumi, M., Bannai, E. and T. Tanaka, Bull. Am. Ceram. Soc. 55 (1976) 313
10. Priest, H.F., Priest G.L. and G.E. Gazza, J. Am. Ceram. Soc. 60 (1977) 81
11. Prochazka, S. and C.D. Greskovich, General Electric, Report SRD-77-178 (AMMRC TR 78-32), July 1978
12. Gauckler. L.J., Lukas, H.L. and G. Petzow, J. Am. Ceram. Soc. 58 (1975) 346
13. Huseby, J.G., Lukas, H.L. and G. Petzow, J. Am. Ceram. Soc. 58 (1975) 377
14. Tsuge, A. and K. Nishida, Bull. Am. Ceram. Soc. 57 (1978) 424
15. Clarke, D.R. and G. Thomas, J. Am. Ceram. Soc. 60 (1977) 491
16. Lou, L.K.V., Mitchel, T.E. and A.H. Heuer, J. Am. Ceram. Soc. 61 (1978) 392
17. Lange F.F., International Metals Reviews 1(1980) 1
18. Lange F.F., J. Am. Ceram. Soc. 56 (1973) 518
19. Iskoe, J.L. anf F.F. Lange, Ceramic Microstructures 76, p. 669 (Fullrath, R.M. and J.A. Pask eds., Boulder Colorado: Westview Press 1977)
20. Knoch, H. and G. Ziegler, Science of Ceramics 9 (1977) 494
21. Knoch, H. and G. Ziegler, Ber. Deut. Keram. Ges. 55 (1978) 248
22. Bowen, L.J. and T.G. Carruthers, J. Mat. Sci. 13 (1978) 684
23. Messier, D.R., Riley, F.L and R.J. Brook, J. Mat. Sci. 13 (1978) 1199
24. Himsolt, G., Knoch H., Hübner, H. and F.W. Kleinlein, J. Am. Ceram. Soc. 62 (1979) 29
25. Knoch, H., Gazza, G.E. and R.N. Katz, Energy and Ceramics, p.737 (P. Vinvencini ed., Amsterdam: Elsevier Sci. Publ. Comp. 1980)
26. Knoch, H. and G.E. Gazza, Ceramurgia International 6 (1980) 51
27. Bowen, L.J., Weston, R.J., Carruthers, T.G. and R.J. Brook, J. Mat. Sci 13 (1978) 341
28. Lange, F.F., Nitrogen Ceramics, p. 491 (Riley, F.L. ed., Leyden: Noordhoff 1977)
29. Prochazka, S. and W.A. Rocco, High Temperatures-High Pressures 10 (1978) 87
30. Yeh, H.C., private communication and Yeh, H.C. and P.F. Sikora Bull. Am. Ceram. Soc. 58 (1979) 444
31. Amato, J., Martorana, D. and B. Silengo, Mat. Sci and Eng. 28 (1977) 215
32. Bowen, L.J., Carruthers, T.G. and R.J. Brook, J. Am. Ceram. Soc. 61 (1978) 335
33. Katz, R.N. and G.E. Gazza, Nitrogen Ceramics, p. 417 (comp.ref.28)
34. Gazza, G.E., Bull. Am. Ceram. Soc. 54 (1975) 778
35. Lange F.F., J. Am. Ceram. Soc. 61 (1978) 53
36. Lange F.F., Singhal, S.C. and R.C. Kuznicke, J. Am. Ceram. Soc. 60 (1977) 249
37. Weaver, G.Q. and J.W. Lucek, Bull Am. Ceram. Soc. 57 (1978) 1131
38. Lange, F.F., J. Am. Ceram. Soc. 62 (1979) 428
39. Salmang, H. and H. Scholze, Die Keramik, (Springer-Verlag) Berlin Heidelberg New York 1968)
40. Knoch, H., in preparation

CONTRIBUTION OF THE STRENGTH-POROSITY RELATIONSHIP OF REACTION BONDED SILICON NITRIDE

Jürgen Heinrich

Rosenthal Aktiengesellschaft, Institut für Werkstofftechnik, Wittelsbacherstrasse 49, 8672 Selb, FRG.

1. INTRODUCTION

The development of reaction bonded silicon nitride (RBSN) led to a wide use of this material in structural applications at high temperatures. Because of the porosity of about 20 vol % and the resulting low mechanical strength, however, these applications are often restricted. To improve strength, the microstructure must be optimised and/or the total porosity must be reduced.

The purpose of this investigation is to study the influence of pore size and total porosity on strength. The pore size was varied by introducing artificial pores and by changing the grain size of the initial silicon powder. The total porosity was reduced by pressureless sintering and by hot isostatic pressing of RBSN with yttria as sintering aid.

2. EXPERIMENTAL DETAILS

2.1 Sample Preparation

Reaction bonded silicon nitride specimens were produced by nitriding injection moulded and isostatically pressed silicon samples. The characteristics of the initial silicon powders are given in Table 1. In order to vary the pore size at constant total porosity and other microstructural parameters, powder D was mixed with 4 vol % of organic wax spheres with a definite grain size range. Samples with a green density of 1.50 gcm^{-3} were prepared by isostatic pressing. Three different wax fractions (0-36 µm, 63-90 µm, 125-180 µm) were used. Before the nitridation the wax was burned out. In two other sets of samples the pore

Riley, F.L. (ed.) Progress in Nitrogen Ceramics

structure was varied by using different grain sizes of the starting powders A, B and C (Table 1). These powders were injection moulded to different green densities.

All silicon samples were nitrided under static conditions, at a maximum temperature of 1400°C for 100 hrs in a mixture of 90 vol % N_2 and 10 vol % H_2 and a gas pressure of 950 mbar. In order to reduce the total porosity by post densification different amounts of yttria were added to the silicon powder before injection moulding and nitridation.

Pressureless sintering occured in a graphite furnace under nitrogen. To prevent evaporation at the sintering temperatures of 1700-1800°C the test bars were put into a boron nitride cubicle which was filled with a powder mixture of boron nitride, silicon nitride and yttria. The silicon nitride - yttria ratio was the same as that of the specimens.

For hot isostatic pressing (HIP) the samples were encapsulated in tubes of fused silica. After evacuating and sealing the specimens were hiped in an ASEA-QIH 32 hot isostatic press at 1750°C and 2000 bars.

More details of processing conditions are presented elsewhere (1,2).

	A	B	C	D
Grain size⁺ (µm)	<10	10-37	37-63	<60
d at 50% weight fraction (µm)	7	26	51	14
B.E.T. surface area (m^2g^{-1})	2.09	0.46	0.24	1.35
Impurity element content (wt.%)				
Fe	0.58	1.01	0.8	0.65
Al	0.17	0.22	0.16	0.23
Ca	0.02	0.014	0.02	0.04
C	0.04	0.04	0.05	n.d.
O	0.8	0.4	0.4	n.d.

\+ Supplier values

n.d. = not determined

Table 1: Characteristics of the starting silicon powders.

2.2 Microstructure and Mechanical Properties

The density was measured in mercury by Archimedes' principle, the α/β-ratio by x-ray diffraction. The macropore size determination occured in the light microscope. Details of the quantitative microstructural analysis are described in (3). Fracture strength was determined in four-point bending with specimens of 3.5 by 45 mm. In order to determine the fracture toughness chevron notched specimens were used as described in (4).

3. RESULTS AND DISCUSSION

The fracture strength σ_f of brittle materials in a bend test is dependent on the maximum flaw size at the surface and the crack growth resistance.

Applying linear elastic fracture mechanics the crack growth resistance can be characterised by fracture toughness K_{IC}, which can be effected by the α and β grain size, the α/β-ratio, the pore size distribution, the total porosity and the unreacted silicon. The maximum flaw size depends on the pore size distribution.

To investigate the influence of pore size, all other microstructural parameters being kept constant, artificial pores were introduced as described in section 2. With increasing pore size fracture strength decreases, whereas the crack growth resistance measured by fracture toughness K_{IC} is nearly independent of the pore size (Table II). The fracture mechanics predictions are shown in Fig. 1. The data points tell the measured strength for each specimen plotted against the radius of the largest pore at the surface the fracture originated from. The curve shows the strength calculated for a semicircular surface crack of size a (5).

Wax particle size (µm)	Density ($g\ cm^{-3}$)	$\frac{\alpha}{(\alpha+\beta)}$	$\bar{d}^*$(µm)	σ_F (MNm^{-2})	K_{IC} ($MNm^{-3/2}$)
0-36	2.39±0.04	0.71	48	140±12	1.45
63-90	2.39±0.01	0.73	66	119±18	1.51
125-180	2.41±0.02	0.64	100	101±14	1.50

*Arithmetic mean value of the artificial pores.

Table II: Characterization of microstructure of samples with artificial pores.

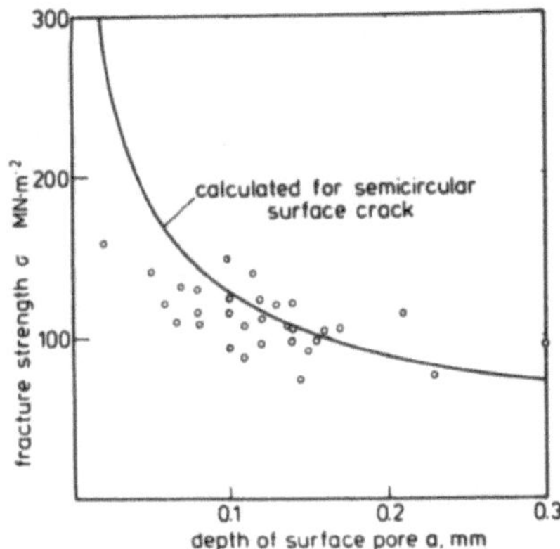

Fig. 1. Predicted and measured strength versus radius of surface pores.

As can be seen, the critical defect may be quite well characterised by the largest pore at the tensile surface. Because of the fact that the total porosity of about 20 vol % is given by the reaction bonding process and that the influence of the α/β- ratio on strength is low (6) the most effective way of increasing the strength of RBSN is to reduce pore size, which can be achieved by modifying the processing conditions.

As an example Fig. 2 shows the pore structure of three samples where the pore size distribution was changed by varying the silicon grain size of the starting powder, while the total porosity of these samples remains constant. Strength increases at constant density by about 50% with decreasing initial silicon grain size, i.e. with decreasing pore size (Fig. 3).

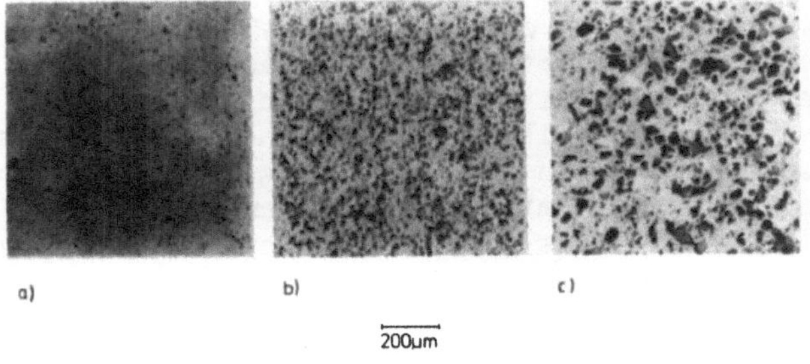

Fig. 2. Polished sections of samples with different pore structure, produced from different initial silicon powders (a) 10 µm, (b) 10-37 µm, (c) 37-63 µm.

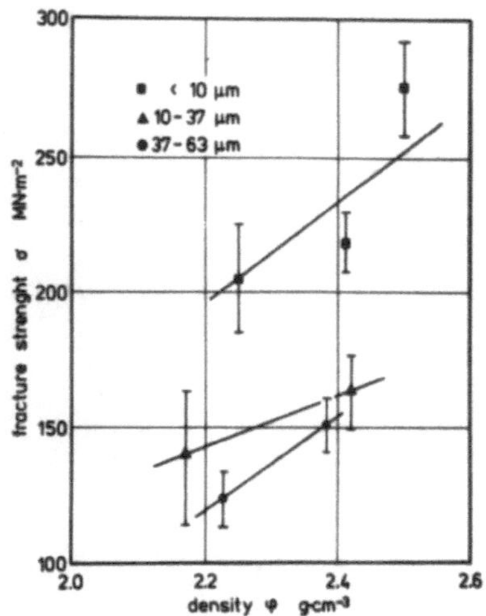

Fig. 3. Strength as a function of density and initial silicon grain size.

The highest strength level of RBSN reached is, however, around 350 MNm^{-2} which, for many applications, is still too low. A further increase in strength can only be achieved by increasing the density of the material. For that reason different amounts of Y_2O_3 were added as sintering aid to the initial silicon powder. After nitriding these samples were pressureless sintered (SRBSN) and hot isostatically pressed (HIPSN) as described in section 2. Starting with a density of about 2.3 g cm^{-3}, density grows with increasing sintering time and temperature (Figs. 4 and 5). Under the applied sintering conditions the sintering temperature could not be increased because the evaporation rate (> 10 wt %) at temperatures above 1800°C was too high.

For 4.2 and 7.0 wt % yttria the total porosity could be reduced at 1800°C/3 hrs by about 14% to a density of 2.75 g cm^{-3}. A significant difference in density between the 4.2 and 7.0 wt % materials could not be observed. With yttria amounts under 3 wt % nearly no increase in density could be observed. The highest densities achieved were 3.04 g cm^{-3} with 11.5 wt % Y_2O_3, but the starting powders were not homogeneously mixed and the samples showed some large inhomogeneities and pores so that they could not be used for the strength considerations.

Reducing the total porosity from about 28% to 14% by pressureless sintering led to an increase in strength from about 220 MNm^{-2} to 460 MNm^{-2} (Fig. 6).

Without any porosity visible in the light microscope dense samples could be produced by hot isostatic pressing. These specimens show the highest strength values of about 800 MNm^{-2} (Fig.6).

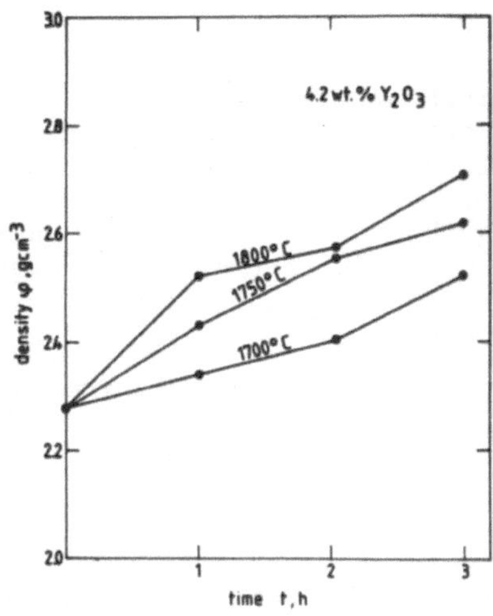

Fig. 4. Density vs sintering time and -temperature.

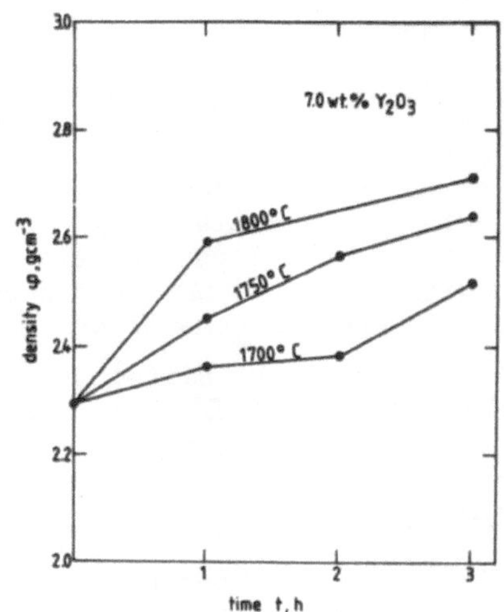

Fig. 5. Density vs sintering time and -temperature.

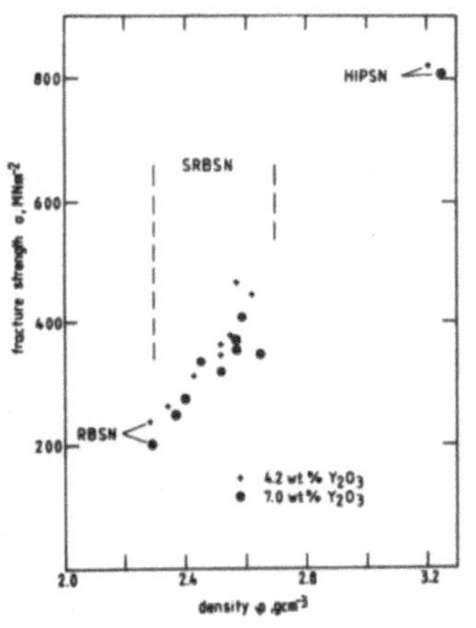

Fig. 6. Strength as a function of density for different silicon nitride qualities.

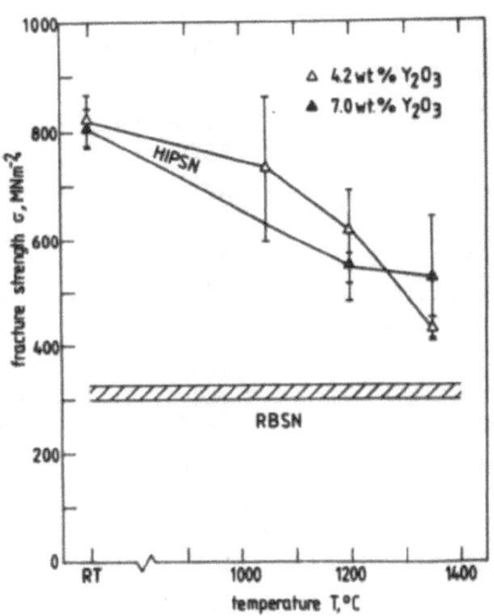

Fig. 7. Temperature dependence of strength of RBSN and HIPSN.

Because of the formation of a liquid phase by the sintering additive the high temperature strength decreases, whereas the strength of RBSN remains nearly constant up to 1400°C (Fig. 7). The hiped samples show, however, better high temperature strength up to 1350°C than our best RBSN quality.

4. CONCLUSION

Without sintering additives the mechanical strength of RBSN can be improved by optimizing the microstructure. The highest strength level that can, however, be reached is about 50% of dense commercial hot pressed silicon nitride. A further increase in strength can be achieved by decreasing the volume fraction of porosity by pressureless sintering and hot isostatic pressing. Compared with hot pressing these procedures show the advantage that complex shaped components can be produced.

Because of the different room- and high-temperature properties of RBSN, SRBSN and HIPSN one can be sure that all those three materials will find their applications as structural materials.

Literature

1. Heinrich, J. "Der Einfluβ von Herstellungsbedingungen auf das Gefüge und die mechanischen Eigenschaften von reaktionsgesintertem Siliciumnitrid". DFVLR-FB 79-32, 1979, Thesis.
2. Heinrich, J. and Böhmer, M. "Microstructure and mechanical properties of hot isostatic pressed silicon nitride". Science of Ceramics 11, submitted for publication.
3. Heinrich, J. and Streb, G. "Quantitative microstructural analysis of reaction bonded silicon nitride". J. Mat. Sci. 14 (1979) 2083-90.
4. Munz, D., Bubsey, R.T. and Shannon, Jr. J.L. "Fracture toughness determination of Al_2O_3 using four-point-bend specimens with straight-through and chevron notches", J. Am. Ceram. Soc., in press.
5. Heinrich, J. and Munz, D. "Strength of reaction bonded silicon nitride with artificial pores", Am. Ceram. Soc. Bull. 12 (1980) 1221-1222.
6. Heinrich, J. and Hausner, H. "Microstructure and mechanical properties of reaction bonded silicon nitride". In Energy and Ceramics, Proc. 4th Int. Meeting on Modern Ceramics Technologies, P. Vincenzini (ed.), Amsterdam - Oxford - New York; Elsevier Scientific Publishing Co., 1980, 780-782.

DISCUSSION

Thümmler: We carried out strength-pore size experiments on natural RBSN but could find no clear correlation. The definition of a critical flaw in this case seems to be more difficult than in yours. One should be careful in translating results from your experiments to real material.

Heinrich: The problem is that it is difficult to determine the pore that the crack originated from, when the pore size is $< \sim 50$ µm. Our artificial pores were 50-150 µm; 'natural' pore size is usually smaller. On the other hand we considered nearly spherical pores; RBSN normally has pores of complex shape, and shape also has to be taken into account.

Section G

CHEMICAL INTERACTIONS

THE KINETICS OF GAS-SOLID REACTIONS AND ENVIRONMENTAL DEGRADATION OF NITROGEN CERAMICS

M. Billy

Laboratoire de Ceramiques Nouvelles, CNRS LA 320,
UER des Sciences, 87060 Limoges, France.

ABSTRACT. Nitrides are thermodynamically less stable than oxides and are thus expected to oxidize in the usual oxygen-containing atmospheres such as air, steam, or carbon dioxide. Predicting the environmental degradation of nitrogen ceramics at high temperature is therefore a kinetic problem that falls within the scope of general kinetics of gas-solid reactions.

1. MAIN FEATURES OF A GAS-SOLID REACTION

This type of reaction has been fully developed for half a century from numerous studies on the oxidation of metals (1). Models have been proposed (2) for explaining the general kinetics, all of them being based on elementary steps which successively take place through any gas-solid reactions:

$$M + G \longrightarrow MG$$

1.1 The basic stages

The initial step is gas adsorption. The adsorbed species will dissolve into the solid, here considered to be a metal, and form on the surface two-dimensional structures called "nuclei". They grow to cover the whole surface during the stage of nucleation and growth which is nearly instantaneous in most cases, under normal conditions of pressure and elevated temperatures.

The second stage is concerned with the growth of the film separating the reacting species as soon as a continuous compact scale is formed, an oxide for example. Now the reaction will proceed if only one of both reactants can diffuse through the oxide. Take the simple case of an n-type semiconductor, where diffusional defects

Riley, F.L. (ed.) Progress in Nitrogen Ceramics

are predominantly interstitial cations M^+. These defects have been generated at the inner phase boundary:

$$M \rightleftharpoons M^+ + e^-$$

They diffuse outwards through the scale and reach the outer phase-boundary where another interfacial reaction takes place; it corresponds to an electron transfer onto chemisorbed species with formation of anions and simultaneous cation vacancies which are immediately filled in.

$$G_{ads} + e^- \leftrightarrows (G^-)^{\circ}_{-} + \square^{-}_{+}$$

$$\square^{-}_{+} + M^+ \leftrightarrows (M^+)^{\circ}_{-}$$

All these processes, i.e. gas adsorption, inner and outer interface reactions and diffusion, are in progress simultaneously, so that the slowest process determines the overall kinetics. The rate limiting process can thus be detected from the observed kinetic law, and its pressure and temperature dependence, by reference to models previously established (2) in each case.

If a non-protective scale is formed, the reacting gas can easily reach the substrate through pores or cracks, as the gaseous diffusion proceeds rapidly. The formation of the new compound MG is then located at the inner phase boundary. It is obvious that porous scales have deleterious effects on the general kinetic behaviour of a material.

Therefore, it is important to know whether a scale is protective or not. Predictions depend in fact on the creep properties of the coating which normally grows under compressive stresses due to a volumic expansion coefficient greater than unity. These stresses will be accommodated during scale growth, either by plastic deformation or by cracking of the scale, when critical growth stresses are reached. As plasticity increases with temperature, it can be expected that some oxides provide a better protective character at higher temperatures. Another consequence is that metals forming high melting point oxides with a poorer plasticity at usual oxidizing temperatures will therefore be expected to have a poor oxidation resistance. Compare for example the oxidation behaviour of transition metals such as titanium, zirconium or hafnium with that of copper or nickel.

Finally it is worth noting that induced growth stresses are very sensitive to the geometry of the oxidized sample. In particular, if conformational strains, and those due to oxidation at edges, corners or curvatures, are not accommodated by creep, departure from protective oxidation kinetics may be observed.

1.2 Particular trends in the oxidation of pure nitrides

There are many broad differences between the oxidation of metals and the oxidation of nitrides according to a reaction of the type:

$$MG' + G \longrightarrow MG + G',$$

where G' denotes nitrogen and G oxygen.

a) The chemical process at the inner interface is now a reaction involving gas formation. How can nitrogen be released outwards through a compact scale? Anion diffusion being excluded due to the size of N^{3-}, it is suggested that nitrogen atoms normally diffuse through low resistance paths such as ingrown dislocations and boundaries in the polycrystalline reaction layers and materials (3). Nitrogen may also be trapped near the inner interface which will be of great importance on kinetics and on the scale integrity.

b) Nitrides are known to be unstable at high temperatures, so that oxidation will result in additional nitrogen release still located at the inner phase boundary.

c) Another point that much be outlined is the incompatibility of the oxide scale with the substrate for silicon based materials. It is well known (4) indeed that the reaction between silica and silicon nitride:

$$Si_3N_4\ (s) + 3SiO_2\ (s) \longrightarrow 6SiO\ (g) + 2N_2\ (g)$$

is favoured at high temperatures and low oxygen pressures, such as those encountered close to the Si_3N_4/SiO_2 interface, when a compact scale is formed (not to mention the "active oxidation" of Si_3N_4 at very low oxygen partial pressures).

d) Another matter of complexity arises from the early formation of two or more oxides, such as through the oxidation of SiMON compounds where silicon has been partly substituted in the β-Si_3N_4 structure. Silica then reacts with the other oxide to produce a silicate phase at an appropriate temperature. As a result, the reaction mechanism will change with temperature and the oxidation scale will no longer be considered to be passive.

e) Contrary to most metals, nitrides are brittle materials which can hardly accommodate the strains due to the compressive stresses developed in the coating during oxidation. So, more intensive cracks are to be expected, not only through the oxidation scale but also through the substrate, the degradation of which may be enhanced.

All these parameters are discussed in the present paper. We will first consider the oxidation behaviour of pure nitrides or related compounds, before discussing the special kinetic problems raised in oxidation of sintered ceramics.

2. OXIDATION KINETICS OF PURE NITRIDES AND RELATED COMPOUNDS

2.1 Basic Kinetic Models

2.1.1 Nucleation and growth. This step is normally fast since the whole surface is almost instantaneously covered by an oxide scale under usual pressures. Thus, nucleation and growth phenomena are only known to be rate determining in some very rare cases. Among nitrides, we can just mention the nitridation of germanium (5) or GeO_2 (6) with ammonia.

Without going into details given elsewhere (7), it can be said that the mechanisms involved, whatever they may be (step or chain nucleations), can always be interpreted by the well-known Prout and Tompkins' law for a branched chain nucleation model:

$$\text{Log} \frac{\alpha}{1 - \frac{\alpha}{2\alpha_i}} = kt + \text{Cte},$$

where α corresponds to the fraction of substrate reacted and α_i to the value at the inflection point of the curve. However, it must be recalled that any autocatalytic process can be represented by a rate expression of the form

$$\frac{d\alpha}{dt} = k\,\alpha(1 - \alpha)$$

which matches a sigmoid curve but not necessarily a nucleation and growth mechanism.

2.1.2 Protective coating. Provided that a non-porous crystalline compound has been formed at the early stages of the reaction and that both interfacial processes are much faster than diffusion (8), the oxidation is expected to follow a parabolic relationship of the form

$$\alpha^2 = Kt \qquad \text{or} \qquad \left(\frac{\Delta W}{S}\right)^2 = kt$$

where ΔW represents the weight gain. This parabolic law, however, is strictly valid for plates when the surface (S) remains constant. In the case of a spherical symmetry, such as powders consisting of spherical particles, changes in reactive surface area must be taken into account. This leads to a more complex formula first proposed by Valensi (8):

$$F(\alpha) = \frac{\Delta}{\Delta - 1} - (1-\alpha)^{2/3} - \frac{1}{\Delta - 1}\left[1 + (\Delta - 1)\alpha\right]^{2/3} \frac{2V_o}{R_o^2}\,kt$$

Δ is the volumic expansion coefficient, i.e. the ratio of the molar volume of the product (V) to that of the reacting solid (V_o);

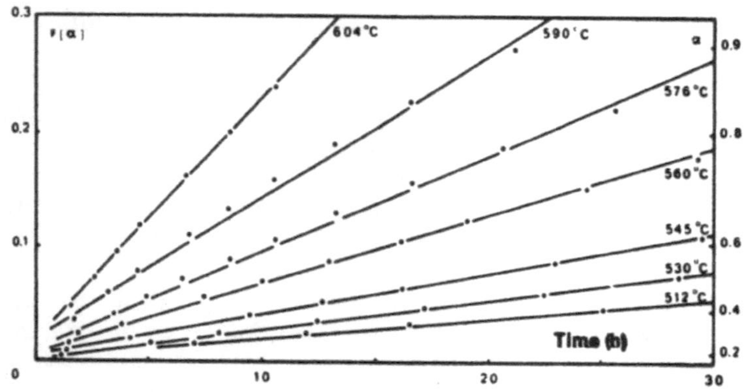

Fig. 1. Kinetic behaviour of vanadium nitride powders in oxygen (pO_2 = 400 torr). After (14).

R_o denotes the initial mean radius of each spherical grain. From this expression, it can be seen that the apparent reactivity of a solid at a given temperature (k = cte) will increase as the powder becomes finer.

The parabolic law has been reported to describe oxidation of many nitrides by oxygen. Fig. 1 is concerned with the oxidation kinetics of powdered vanadium nitride (14). The reaction is controlled by a diffusion process, as checked by plotting $F(\alpha)$ as a function of time. The rate constants are very similar at each temperature to those obtained from VN plates and from oxidation of the metal itself (15), which leads to the formation of the same oxide (V_2O_5) with the same activation enthalpy. Thus, it is clear that oxygen diffusion through the oxide scale is rate-determining and that nitrogen has no practical effect on the oxidation kinetics of the nitride.

These results can be extended to oxygen corrosion of both CVD (16) and powdered forms (17-19) of pure Si_3N_4. Here again, there is a fairly good agreement between parabolic constants obtained for plates and powders. Experimental results give an activation enthalpy of 142-146 kJ mol^{-1} and a mean parabolic rate of 10-14 g^2 cm^{-4} s^{-1} in the temperature range 1100-1300°C. These values are very close to those obtained for the oxidation of SiC (19,20), Si_2N_2O (18), Si (21) and for the oxygen diffusion through SiO_2 (22). As a consequence, the oxidation of pure Si_3N_4 only depends on the oxygen diffusion through the surface layers. In such conditions, the nature of the silica must be of importance on kinetics. Indeed discontinuities in rate and activation enthalpy have been observed at approximately 1100°C, when crystallisation of amorphous silica takes place by increasing temperature. Above 1400-1450°C, another rate discontinuity appears which seems in turn to be related to a loss in Si_3N_4 stability above the silicon melting point; the oxidation mechanism is then distinctly modified.

The oxidation behaviour of aluminium nitride is governed by a diffusion process described from experiments on plates (23) or powders (24) in the temperature range 1000-1400°C. The oxygen diffusion through the oxide scale is still rate-determining and, here again, the activation enthalpy varies (from 86-174 kJ mol^{-1}) when crystallization of alumina (α-Al_2O_3) occurs above 1100°C.

The phenomena are more complex as far as sialon phases are concerned, because their reactivity increases with aluminium content (25), and a duplex oxidation scale is formed. The oxidation diffusion through this scale is only rate-determining so long as the oxide phases remain amorphous below 1000°C for 15R sialon or 1200°C for β' compositions. At higher temperatures, crystallization of silica or alumina, or the formation of mullite and its crystallization, disturb the diffusion regime until a stabilized oxide scale is formed. In this region, it has been suggested (26) that the oxygen penetration through SiO_2 glass layers, rather than through mullite crystals, might be the limiting process since the activation enthalpy (146 kJ mol^{-1}) is very similar to that observed for oxidation of both Si_3N_4 and SiC in the temperature range 1200-1400°C.

<u>2.1.3 Non-protective scales</u>. Assuming that the gaseous diffusion through the porous coating is not the limiting process, the scale formation in a direction perpendicular to the initial surface obeys a linear law of time described by either of the expressions:

$$\Delta W/S = Kt \quad \text{or} \quad \alpha = kt$$

for planar specimens, and a kinetic law of the form:

$$G(\alpha) = 1 - (1-\alpha)^{\frac{1}{3}} = \frac{k}{R_o} t$$

in the case of powder specimens consisting of spherical grains with a mean initial radius R_o. Pressure dependence on the rate constant is then normally linear.

This type of kinetics describes the oxidation of TiN by water vapour (27), the oxidation of germanium oxynitride (28) and that of Υ-aluminium oxynitride phase (30) up to 1100°C. Examples are to be found (29) in the system Si_3N_4-SiO_2-Y_2O_3 where previous Lange and Singhal's results (31) have been confirmed for the J phase ($Si_2N_2O.2Y_2O_3$) within the temperature range 1300-1500°C and the K phase ($SiN_2O.Y_2O_3$) between 1100 and 1300°C.

The "active oxidation" of silicon-based compounds at very low oxygen partial pressures must be also included in this section. It is really a limiting case, as there is no coating at all, but simply the evaporation of a volatile species, such as SiO or N_2,

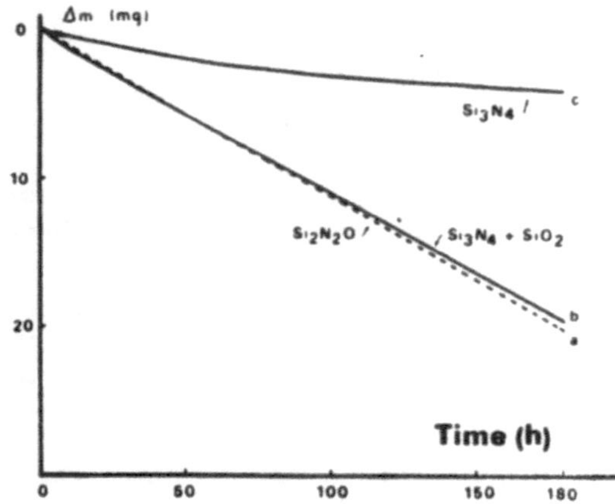

Fig. 2. Decomposition of (a) Si_2N_2O, (b) equimolar SiO_2 + Si_3N_4 and (c) Si_3N_4 at 1510°C in argon. After (32).

giving linear kinetics. Nearly the same is the kinetic behaviour of mixtures of SiO_2 + Si_3N_4 at elevated temperatures under very low oxygen pressures. Here again, weight losses still obey (32) a linear law, as shown in Figure 2. Moreover, kinetics are strictly similar to those of Si_2N_2O decomposition, through identical conditions. This implies the same mechanism and, therefore, an early fast splitting of the oxynitride which leads to the formation of SiO_2 and Si_3N_4 in equal proportions. The same is found (32) for Ge_2N_2O decomposition which results in the formation of volatile GeO.

2.2 Mixed Kinetics Regimes

2.2.1 Pseudo-linear model. A linear rate law may be related to a rate determining diffusion process as in the case of the oxygen corrosion of titanium nitride (33), between 850 and 1000°C. Here, the coating is non-protective as checked by the formation of soft rutile layers above titanium nitride. It was proved, however, that each layer is growing parabolically but separates periodically from the nitride, due to compressive growth stresses when a critical thickness e_c is reached. This thickness corresponds to a critical time t_c such as:

$$e_c^2 = K_p t_c$$

As the critical values e_c and t_c are very small, the overall kinetics appear to be linear on thermograms, so that:

$$e_c = K_L T_c = K_L \, e_c^2 / K_p$$

and

$$K_p = K_L \, e_c$$

By measuring K_L from kinetic curves and e_c from micrographic examinations, the parabolic constant was calculated and shown to be similar to that found for the direct oxidation of titanium (38), with the same activation enthalpy of 184 kJ mol^{-1}. So the linear step is related to a predominantly short-circuit diffusion of oxygen through the first rutile lamella near the inner boundary.

<u>2.2.2 Linear-parabolic model</u>. A porous coating may become protective during the course of the reaction, due to crystal growth phenomenon or a pore closure mechanism.

Examples of this type have been found for oxidation of aluminium (24) and silicon (9,16,18) nitrides, silicon oxynitride (18) and sialon phases (25). Here, the oxidation initially proceeds through a reaction controlled stage characterized by an activation enthalpy which corresponds roughly to the Si-N or Al-N bond energy. In most studies, however, this initial linear region is not observed because it lasts a rather short time (about half an hour or less) and presumably also because the specimens already have a surface oxide layer which then starts growing parabolically.

It can be noticed that the pore closure mechanism does not always lead to a parabolic law. If both reactants are unable to diffuse through the compact scale, in particular when temperatures are not high enough, the overall kinetics may become logarithmic or asymptotic. Such a case is currently observed among oxidation of sintered silicon or aluminium nitride-based ceramics at moderate temperatures. Here, the silicate or aluminate-containing coatings are highly protective, so that the oxidation is to a large extent nearly stopped.

<u>2.2.3 Paralinear model</u>. The initial compact coating becomes porous with time as a result of crystallization or fracture, so that a parabolic and then a linear law will operate successively in a so-called "paralinear" kinetic model.

A typical example is observed for nitridation of calcium (34) where calcium nitride is growing parabolically up to a critical size, above which the surface scale crystallizes and becomes permeable. The limiting step is now a diffusion process through the compact underlying scale of constant thickness e_c, with a constant rate $de/dt = kS/e_c$. In this linear region, kinetics are thus closely related to a pseudolinear mechanism.

In most cases, however, the linear stage becomes operative just as cracks appear over the whole scale, presumably due to nitrogen release, at higher temperatures in particular, or for long time exposures. Such a case has been found for the oxidation of CVD-Si_3N_4 at 1600°C (9).

2.3 Mechanical properties of the coating and kinetics behaviour

As already shown, oxidation resistance is highly dependent on the protective character of the oxide scale and must be related to mechanical properties of both oxide and nitride substrate at operating temperature. Since oxides are usually more voluminous than the starting nitrides ($\Delta > 1$), compressive growth stresses will normally develop through the substrate (10). Two alternative ways of relaxation of these stresses have to be considered:

1 - If the scale is badly adherent, stresses are usually accommodated by detachment of the coating. Here, the oxidation of TiN in oxygen (33) is a typical example, as already mentioned.

2 - If the scale is well adherent, the relaxation of stresses may proceed:

a) by plastic deformation of the oxide, such as silica during oxidation of silicon nitride and oxynitrides. Here, kinetics are normally parabolic.

b) by cracking of the coating when its plasticity and mechanical strength are poor and incompatible with the critical level of stresses in the scale. Kinetics are usually linear as for oxidation of both J and K phases of the SiYON system. In these cases, failure takes place on corners or edges especially, resulting in a Maltese cross symmetry when cubic specimens are used (29).

c) by a cooperative rupture of the scale and the substrate. A progressive cracking then spreads to the whole specimen which crumbles. Examples of this catastrophic oxidation occur for interstitial nitrides such as ZrN or HfN (37) and NbN or TaN (38). A chain fragmentation takes place after a thin adherent oxide film

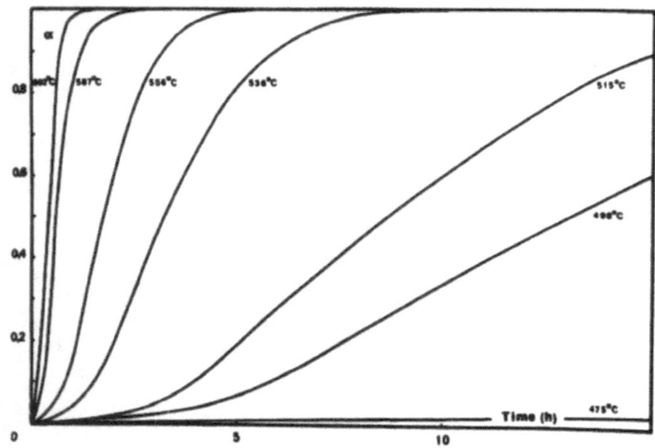

Fig. 3. Oxygen corrosion of δ NbN$_{.87}$ plates.

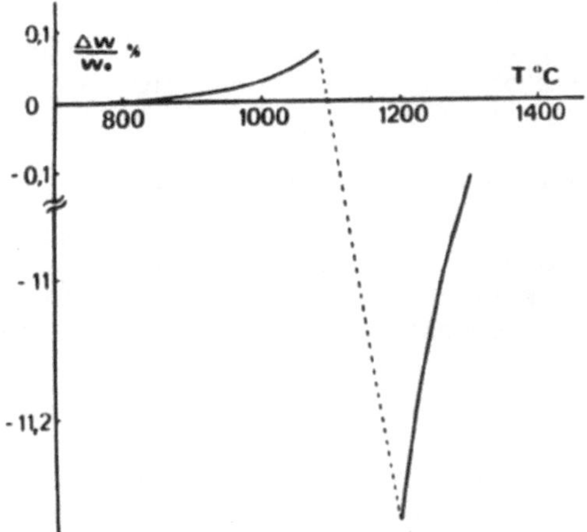

Fig. 4. Oxidation behaviour of the J-SiYON phase in air, through gradual increasing temperatures (300° /h). After (29).

has been formed. It proceeds inwards according to an autocatalytic type phenomenon and explains the sigmoidal shape of the kinetics (Fig. 3). However, this intergranular crack process, which is only due to the scale growth, must not be confused with a branched chain germination of the oxide.

From all these cases, it is obvious that kinetic behaviour is tightly connected with the ability of the scale to accommodate the increasing growth stresses developed during the reaction. This essentially depends on the plasticity of the scale and, as plasticity normally increases with temperature, it is tempting to think that the kinetics behaviour could change and that the oxidation resistance could be better with increasing temperatures.

Study of the oxidation of the J-phase in the SiYON system (29) is particularly rewarding. As already mentioned, kinetics are linear in the 1300-1500°C range, but at lower temperatures, plasticity of the oxide scale decreases to such a point that the relaxation of stresses leads to a bursting of the sample, as seen in sudden weight losses on thermograms (Fig. 4) around 1100-1200°C. With the K phase, such an anomaly does not occur because the deformation of the oxide product is easier, due to a lower melting point (1780°C) of the so-formed silicate $Y_2Si_2O_7$ than that (1950°C) of Y_2SiO_5 produced by oxidation of the J phase.

Finally, it must be kept in mind that for predicting oxidation kinetics, chemical data are of minor importance compared with the mechanical properties of the scale, strength and plastic behaviour in particular. Oxidation resistance of nitrides will thus be highly dependent on two important parameters:
- the volumic expansion coefficient Δ that gives an indication of the stress levels in both the scale and the substrate during

scale growth;
- the melting point of oxidation product that allows one to appreciate the likely plastic behaviour of the scale at working temperatures.

3. PROBLEMS RAISED IN THE OXIDATION OF SINTERED CERAMICS

As far as nitrogen ceramics are concerned, we are dealing with composite materials containing second phases at grain boundaries when additives are used. So, the kinetics will be generally more complex, and the influence of the microstructure (porosity in particular) may become dominating.

3.1 Influence of Porosity

This influence is clearly suggested by the wide scatter of oxidation data obtained for materials of different densities, such as reaction bonded Si_3N_4 (35,39,40). Since reaction rate is directly correlated with the reactive surface area, it is obvious that the higher the open porosity, the higher the rate of oxidation. A linear regression between reactivity and porosity can be observed in some favourable cases (42). However, it is difficult to predict a general influence of porosity owing to different overlapping parameters, i.e. closed or open porosity pore size, and pore size distribution.

Let us consider a typical oxidation of a sintered material possessing a large amount of porosity, as in the case of RBSN illustrated in Fig. 5 after Davidge et al. (40). At first sight, it is surprising that the high the temperature, the smaller is the rate of oxidation. The overall kinetics can be explained in fact by considering two competitive stages, the internal oxidation due to the pores and the external oxidation depending on the sample geometry, the importance of each being dependent on temperature. At lower temperatures near 1000°C, when the reaction proceeds smoothly, all the pores are progressively closed by silica. At high temperatures, surface silica layers are growing rapidly, so that superficial pores may be closed before the internal oxidation is complete.

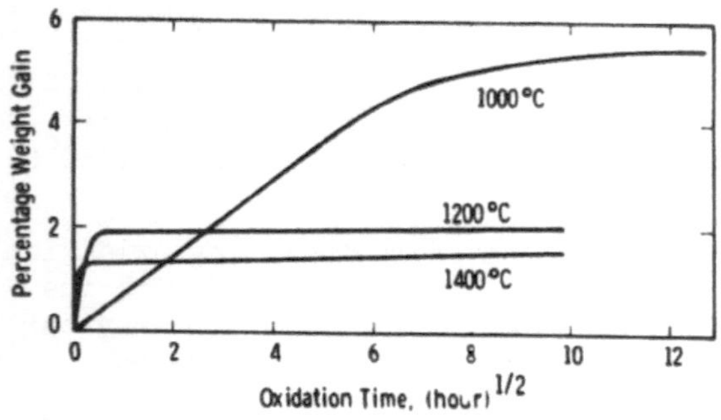

Fig. 5. Oxidation of reaction-bonded Si_3N_4. After (40).

And as this external pore closure mechanism takes place sooner with higher temperatures, it results in an oxidation resistance much better at 1400°C than at either 1200 or 1000°C.

Pore size influence can be easily explained in the same way since smaller pores, nearby the surface especially, allow smaller amounts of internal oxidation (43). On the other hand, since the presence of oxidation products in the bulk material has a deleterious effect on creep properties, a promising solution for preventing internal oxidation would consist of providing porous materials with a dense coating obtained by chemical vapour deposition or ion-plating.

3.2 Influence of the Additives

Because of the presence of secondary phases, oxidation behaviour of hot-pressed materials is generally different from that of additive free compounds, and a large increase in oxidation rate is normally observed. Though the rates of oxidation depend on chemical purity and nature and concentration of the additives, the same general influence of these parameters has been established from a number of studies on the oxidation of nitrogen ceramics. A substantial body of experimental data (10-14, 35-36, 44-45) now exists on the oxidation of MgO-containing hot-pressed Si_3N_4 materials, which gives a fairly good idea of the mechanism involved. In the present section, we will discuss in detail the case of this pioneer material, the results of which can presumably be extended to other systems.

Singhal (35) summarized the oxygen corrosion results for commercial hot-pressed Si_3N_4 containing 1 wt % MgO and impurities such as Fe (0.6%), Al (0.1%) and Ca (0.07%). This material still exhibits a parabolic behaviour similar to that of pure Si_3N_4, but with an activation enthalpy (377 kJ mol^{-1}) and parabolic constants (about 10^{-12} g^2 cm^{-4} s^{-1} in the 1100-1300°C range) much larger than the values reported for oxidation of the pure nitride (11,35). On the other hand, the additive or other impurities present in the ceramic concentrate in the surface oxide scale consisting of enstatite, silica and small quantities of glassy phases. Thus, it was first concluded that the rate controlling step was the magnesium and impurity (Ca^{2+}) diffusion through the oxide layer. This mechanism seemed to be supported by the fact that greater amounts of MgO or impurities, particularly calcium, cause a greater rate of oxidation attack (13).

This limiting diffusion process, however, was in contradiction with the morphology of the scale which contained cracks and pores, particularly for higher temperatures (1325-1475°C). The thickness was greater than that expected on the basis of the weight gain (12), which supposed amounts of dissolved nitrogen in the scale, and the unreacted nitride surface was pitted.

Such observations clearly suggested that the corrosion scale

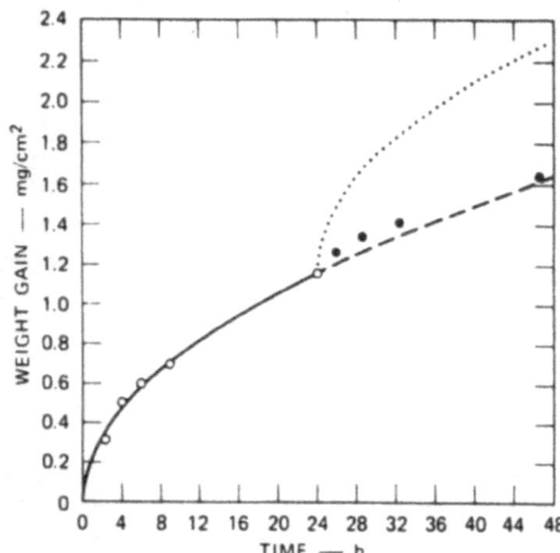

Fig. 6. Oxidation and reoxidation curves for NC132 hot-pressed Si_3N_4 at 1370°C. After Cubicciotti et al. (44).

was not protective, even though the oxidation obeys a parabolic law. By oxidizing a sample for 24 hours at 1370°C and reoxidizing it after removal of the surface scale, Cubicciotti and Lau (36) found that the curve was an extrapolation of the original parabola (Fig. 6) and not a new one (dotted curve) that would have been expected if the corrosion scale had been protective. On the basis of the magnesium concentration profile through the oxidized samples (Fig. 7), Cubicciotti and Lau (44) finally concluded that the diffusion of Mg^{2+} out of the material and into the corrosion scale was the rate controlling step.

Considering now the large concentration of magnesium at the outer surface and its decrease to small values in the bulk of the corrosion scale, it must be thought that the magnesium silicate precipitates at the outer surface, the driving force thus being the oxidation of a migrating phase, supposed to be a N-containing glassy phase due to its instability in contact with air. This is consistent with a liquid phase present in the scale during oxidation, as shown by microscopic examination (11,41,44). The formation of this phase necessarily takes place at the scale/subscale interface by dissolution of Si_3N_4 in Mg-containing silicate phase, as suggested by the pitting of the nitride surface and by the solubility of Si_3N_4 in magnesium silicates (46,47). On the basis of the glass composition reported by Jack (47) in the SiO_2-Si_2N_2O-$MgSiO_3$ compatibility triangle of the phase diagram, it can be thought that the liquid phase will allow rapid diffusion from the inner to the outer surface of the scale, where devitrification occurs by precipitation of enstatite and oxidation of the so-formed oxynitride. This oxidation, which involves nitrogen release, explains the porous morphology of the coating and also the nitrogen bubbles present in some places in the scale where oxygen pressure

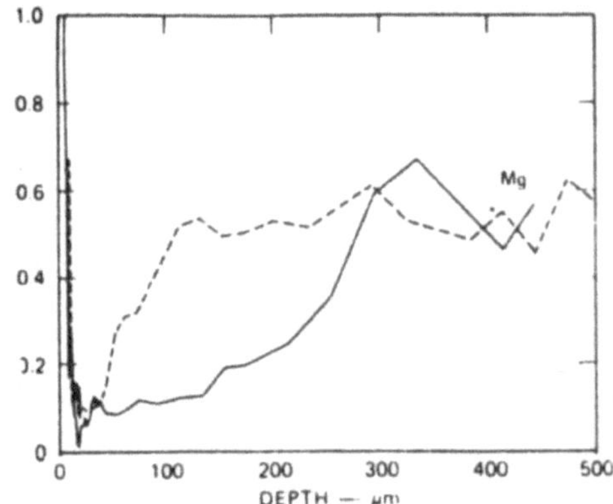

Fig. 7. Electron microprobe scans of Mg (wt %) through oxidized HPSN at (solid lines) 1370°C for 40 h and (dashed lines) 1400°C for 8 h. After (44).

is high enough to allow partial oxidation of the glassy phase.

As far as the substrate is concerned, the experiments conducted by Lange these last few years (13,45) have been decisive. It was shown that magnesium diffusion proceeds through a grain boundary glassy phase and that the presence of a liquid phase promotes diffusivity. As the lowest eutectic composition in the compatibility triangle Si_3N_4-Si_2N_2O-Mg_2SiO_4 lies close to the Si_3N_4-Mg_2SiO_4 tie line, larger Mg/Si ratios favour larger liquid contents, resulting therefore in a greater decrease of both strength and oxidation resistance of the material.

The magnesium diffusion through the subscale is strongly related to internal oxidation with formation of Si_2N_2O, as shown by Clark and Lange (45) and recently by Morgan et al. (48) in other Si_3N_4-based ceramics. Oxygen enrichment and magnesium depletion with depth have a similar profile illustrated by Fig. 8. Moreover, it depends on the initial Mg/Si ratio of the material within the compatibility triangle Si_3N_4-Si_2N_2O-Mg_2SiO_4. Both phenomena have thus the same origin in relation to the oxidation process, as proved in particular by the lack of Si_2N_2O in the material when annealing is performed in argon (48). As a consequence, Si_2N_2O formation cannot be interpreted in terms of the instability of the glassy phase in contact with Si_3N_4.

This Si oxynitride formation obviously results from an inward oxygen diffusion favoured by the presence of the glassy phase, which might be rate determining. In support of that suggestion, it may be noticed that the concentration gradient of the magnesium does not remain constant through the subscale (Fig. 7), which implies a departure from the parabolic law. Thus, it can be thought that the partial oxidation of the glassy phase, as oxygen diffusion

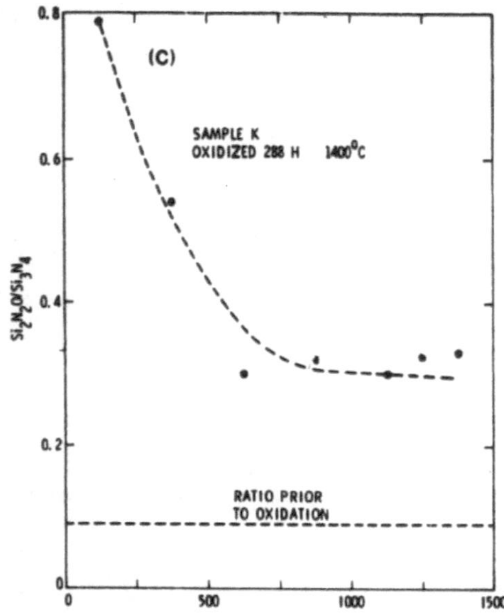

Fig. 8. Si_2N_2O/Si_3N_4 ratio as a function of depth (μm) below the oxidation scale. After Clarke and Lange (45).

proceeds, leads to Si_2N_2O precipitation and, subsequently, to an increase of the Mg/Si ratio in the glassy phase. The volume content of this phase then increases and promotes its outward diffusion, so that the magnesium depletion profile simply reflects the ability of magnesium to diffuse as a result of the parabolic progression of the internal oxidation.

It can be concluded that the large increase in oxidation rates compared with those of pure nitrides is due to the presence of a glassy grain boundary phase which normally provides high diffusivity paths for the reacting species. This glassy phase, however, allows accommodation of the internal stresses developed during oxidation, and may have favourable effects in preventing failures, when oxygen corrosion takes place at lower temperatures in particular. This is the significance of the increasing oxidation rates observed (13,45) as the composition of $MgO-Si_3N_4$ ceramics shifts away from the eutectic composition, beyond the $Si_3N_4-Mg_2SiO_4$ tie line ($MgO/SiO_2 > 2$).

4. CONCLUSIONS

In the light of the present review on the oxidation of nitrogen ceramics, it appears that the kinetic behaviour of pure nitride phases is strongly dependent on the response of the oxide scale/substrate couple to mechanical stresses developed during oxidation. Even though the qualitative influence of oxidation-induced phase changes has been amply demonstrated, further work is now required to quantify these mechanical stresses. As failure prediction for ceramics in service is closely related to developments in this field, progress can be expected in the near future.

As exemplified by MgO-containing Si_3N_4-based ceramics, the actual oxidation mechanism contrasts with the conventional view of gas-solid reactions. Two sets of processes have been recently highlighted, both of which depend on ability to form ternary composition compounds with low eutectic temperatures. In the oxide scale, the nitride compatibility with the secondary phases and oxidation products is very important. In the subscale, diffusivity of oxygen or impurities, enhanced by the presence of a glassy phase, may become predominant at appropriate temperatures. The competition between these two sets of processes determines the overall kinetics. For this reason, any progress will be conditioned by a better understanding of the phase diagrams and, especially, by a better characterization of glassy phases and eutectic compositions in each compatibility triangle.

REFERENCES

1. P. Kofstad. High temperature oxidation of metals, John Wiley and Son, New York, 1966.
2. P. Barret. Cinétique hétérogène, Gauthier-Villars, Paris, 1973.
3. W.W. Smeltzer and J. Desmaison. Nitrogen Ceramics, ed. F.L.Riley, Noordhoff (Leyden), 1977, p.219.
4. S.C. Singhal. Ceramurgia International, 2, 123 (1976).
5. J.C. Labbe, F. Duchez and M. Billy. C.R. Ac. Sciences Paris, 273 C, 1750 (1971).
6. J.C. Labbe and M. Billy. C.R. Ac. Sciences Paris, 277C, 1137 (1973).
7. M. Billy. Nitrogen Ceramics, ed. F.L. Riley, Noordhoff (Leyden), 1977, p.203.
8. G. Valensi, in Bénard, L'oxydation des métaux, Vol. 1, Gauthier-Villars, Paris, 1962, p.171.
9. T. Hirai, K. Niihara and T. Goto. J. Am. Ceram. Soc. 63, 419 (1980).
10. F.F. Lange. J. Am. Ceram. Soc. 63, 38 (1980).
11. A.J. Kiehle, L.K. Heung, P.J. Gelisse and T.J. Rockett. J. Am. Ceram. Soc. 58, 17 (1975).
12. W.C. Tripp and H.C. Graham. J. Am. Ceram. Soc. 59, 399 (1976).
13. F.F. Lange. J. Am. Ceram. Soc. 61, 53 (1978).
14. P. Lefort, J. Desmaison and M. Billy. C.R. Ac. Sciences Paris, 289, 271 (1979); Annales Chimie 5, 692 (1980).
15. P. Lefort, J. Desmaison and M. Billy. J. Less-Com. Metals, 77, 1 (1981).
16. I. Franz and W. Langeheinrich. Reactivity of Solids, Chapman and Hall, London, 1972, p.310.
17. R.M. Horton. J. Am. Ceram. Soc. 52, 121 (1969).
18. P. Goursat, P. Lortholary, D. Tétard and M. Billy. Reactivity of Solids, Chapman and Hall, London, 1972, p.315.

19. E. Fitzer and R. Ebi. Dechema, Frankfurt, 1972.
20. K. Motzfeldt. Acta Chem. Scand. 18, 1596 (1964).
21. B.E. Deal and A.S. Grove. J. Appl. Phys. 36, 3770 (1965).
22. R.M. Barrer. J. Chem. Soc. 89, 378 (1934).
23. J-P. Lecompte. Thesis, Limoges, 1981.
24. D. Tétard and M. Billy. Reaction kinetics in heterogeneous chemical systems, Elsevier, 1975, p.512.
25. M. Brossard, D. Brachet, P. Goursat and M. Billy. Annales de Chimie, 4, 7 (1979).
26. J. Schlichting Nitrogen Ceramics, ed. F.L. Riley, Noordhoff (Leyden), 1977, p.627.
27. F. Nardou. Thesis, Limoges, 1980.
28. J.C. Labbe and M. Billy. Materials Chem. 4, 159 (1979).
29. P. Goursat, C. Dumazeau and M. Billy. Environmental degradation of high temp. materials, Inst. Metallurgists Series 3, No. 13, Vol. 2, Chameleon Press, London, 1980.
30. P. Goursat, P. Goeuriot and M. Billy. Materials Chemistry, 1, 131 (1976).
31. S.C. Singhal, F.F. Lange and R.C. Kuznicki. J. Am. Ceram. Soc. 60, 249 (1977).
32. M. Billy, J.C. Labbe and P. Lortholary. Mat. Chemistry, 4, 189 (1979).
33. J. Desmaison, P. Lefort and M. Billy. Oxidation of Metals, 13, 505 (1979).
34. R. Streiff. Thesis, Nancy, 1967.
35. S.C. Singhal. J. Mat. Sci. 11, 500 (1976); Ref. 7, p.607.
36. D. Cubicciotti, K.H. Lau and R.J. Jones. J. Electrochem. Soc. 124, 1955 (1977).
37. J. Desmaison. Thesis, University of Limoges, 1978.
38. P. Lefort. Thesis, University of Limoges, 1981.
39. I. Guzman, Y. Litvin and G. Turchina. Ogneupory, 2, 47 (1974).
40. R.W. Davidge, A.G. Evans, D. Gilling and P.R. Wilyman. Special Ceramics Vol. 5, ed. P. Popper, Brit. Ceram. Res. Assoc., Stoke-on-Trent, 1972.
41. N.J. Tighe. Ref. 7, p.441.
42. M. Billy. Ref. 7, p.635.
43. G. Grathwohl and F. Thümmler. Ceramics for High Performance Applications, Vol. 2, ATMC Series, Brook Hill, 1978, p.573.
44. D. Cubicciotti and K.H. Lau. J. Am. Ceram. Soc. 61, 512 (1978).
45. D.R. Clarke and F.F. Lange. J. Am. Ceram. Soc. 63, 586 (1980).
46. E.A. Dancy and D. Jannsen. Can. Metall. Q. 15, 103 (1976).
47. K.H. Jack. Ref. 7, p.259.
48. W.J. McDonough, C.C. Wu and P.E.D. Morgan. J. Am. Ceram. Soc., 64, C45 (1981).

THERMODYNAMIC MECHANISM FOR CATION DIFFUSION THROUGH AN INTERGRANULAR PHASE: APPLICATION TO ENVIRONMENTAL REACTIONS WITH NITROGEN CERAMICS

D.R. Clarke
Structural Ceramics Group
Rockwell International Science Center
Thousand Oaks, CA 91360

ABSTRACT

A mechanistic model for the migration of cations from a continuous intergranular phase in response to environmental reaction at the surface of polyphase silicon nitrides is described. It predicts that the composition of the intergranular phase may be modified and controlled, with consequential beneficial or detrimental effects, by the deliberate formation of surface scales rich or poor in impurity cations. Two examples are illustrated, "purification" of the intergranular phase using an SiO_2 surface layer leading to high temperature strength enhancement and strength degradation and corrosion when a CaO rich surface layer is employed.

INTRODUCTION

Most, if not all, nitrogen ceramics contain a continuous, intergranular phase at all two and three grain junctions.[1] Because the phase is non-crystalline it can, above its softening point, act as a short circuit diffusion path throughout the microstructure much as grain boundaries in metals and single phase ceramics behave at temperatures below which volume diffusion is significant. An indication of the temperature at which diffusion through the intergranular phase can become appreciable is provided by internal friction measurements. Thus, in MgO hot-pressed silicon nitride (Norton NC132), where an internal friction peak[2] is measured at ~1000°C, diffusion effects should be detectable above this temperature. Oxidation studies of polyphase silicon nitride above 1000°C clearly demonstrate the Mg and Ca cations, together

Riley, F.L. (ed.) Progress in Nitrogen Ceramics

with other impurity cations such as Na, Al, Fe, K, concentrate at the surface in an oxide scale.[3-7] Since these elements are found only in the intergranular phase, migration to the oxide scale must occur through this phase illustrating its role as a short circuit diffusion path.

The purpose of this short contribution is to describe a mechanism for the migration of the impurity cations through the intergranular phase to the surface scale. Although the mechanism was originally proposed to provide an explanation for the compositional changes created by oxidation,[7] it is of wider relevance being applicable to a variety of environmental reactions as illustrated with the examples from the corrosion of a silicon nitride alloy.

MECHANISM OF CATION DIFFUSION

The basis for the mechanism proposed here and shown schematically in Fig. 1 is that on oxidation a reaction or diffusion couple is created between the silica produced on the surface (by the passive oxidation of the silicon nitride grains) and the intergranular phase in the bulk of the ceramic. A gradient in chemical potential exists because, in the interior of the ceramic, the intergranular phase is in, or close to, equilibrium

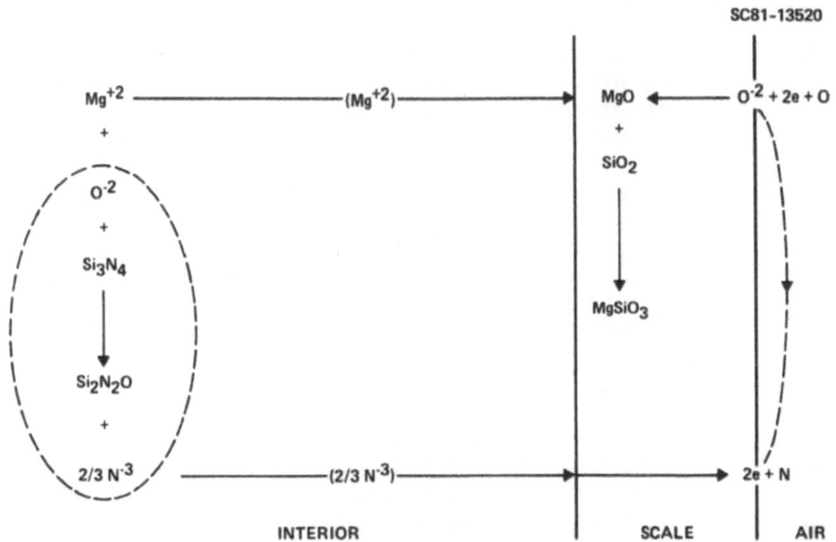

Fig. 1 Schematic reaction mechanism for magnesium diffusion in response to a SiO_2 scale formed on the surface of a silicon nitride alloy containing an intergranular phase. As shown here, the SiO_2 forms on oxidation on the silicon nitride. The mechanism is also appropriate to the migration of other cations in the intergranular phase.

with the silicon nitride and any additional secondary phases; whereas, at the surface it is not in equilibrium with the SiO_2 oxidation product. It is assumed that at the scale/sample interface the intergranular phase reacts with the silica and that at the temperatures involved the phase and the silica have relatively low viscosities. Two driving forces can be identified. Since the intergranular phase is a silica rich glass and the oxidation product is pure silica the cations in the intergranular glass will migrate to minimize the free energy of mixing (primarily by maximizing the entropy of mixing) until the cation concentration is constant in both. This situation can not be achieved as SiO_2 is continually being formed by the surface oxidation of the silicon nitride grains. The second driving force is the reduction in total free energy by the reaction of Mg^{2+} with the SiO_2 and O^{2-} to form crystalline $MgSiO_3$ at the surface. Both cause a local depletion in the Mg^{2+} concentration in the intergranular phase and thereby establish a concentration gradient below the scale. The loss of Mg from the intergranular phase frees oxygen in the phase to react with the Si_3N_4 grains forming Si_2N_2O. In turn this releases 2/3 N^{3-} which can diffuse through the intergranular phase with the Mg^{2+} to maintain charge neutrality. At the surface the nitrogen ion forms atomic nitrogen and leaves the system. The net effect is that a compositional gradient is formed beneath the scale. Similar arguments can be given for the outward diffusion of other impurity cations.

The mechanism outlined above is essentially an internal oxidation process but driven by a surface reaction. It has been assumed above that electron transfer occurs at the air/scale interface and that mass transport occurs by the diffusion of ions. Whether this is the transport mechanism or whether it is by neutral species is not known. The rationale of the cations diffusing to the scale rather than the SiO_2 into the intergranular phase is the finding that in MgO/SiO_2[8], CaO/SiO_2[9] and PbO/SiO_2[9] diffusion couples the Si ion is the much slower moving species. In the context of oxidation, the phenomenon is concurrent with the reaction produced by the inward diffusion of oxygen.

TEST OF MODEL

According to the mechanism, cation diffusion occurs in response to the formation of a reaction couple between the intergranular phase and a surface composition. Thus the most direct test is to heat a silicon nitride sample, encapsulated in SiO_2, in the absence of air and demonstrate that a compositional gradient and Mg depletion gradient are established. Such an experiment has been performed[10] and the Mg depletion gradient measured is compared with that produced in the same material under equivalent oxidation conditions (288 hrs at 1400°C) in Fig. 2. When the experiment was repeated using a single phase CVD silicon nitride no compositional change was produced.

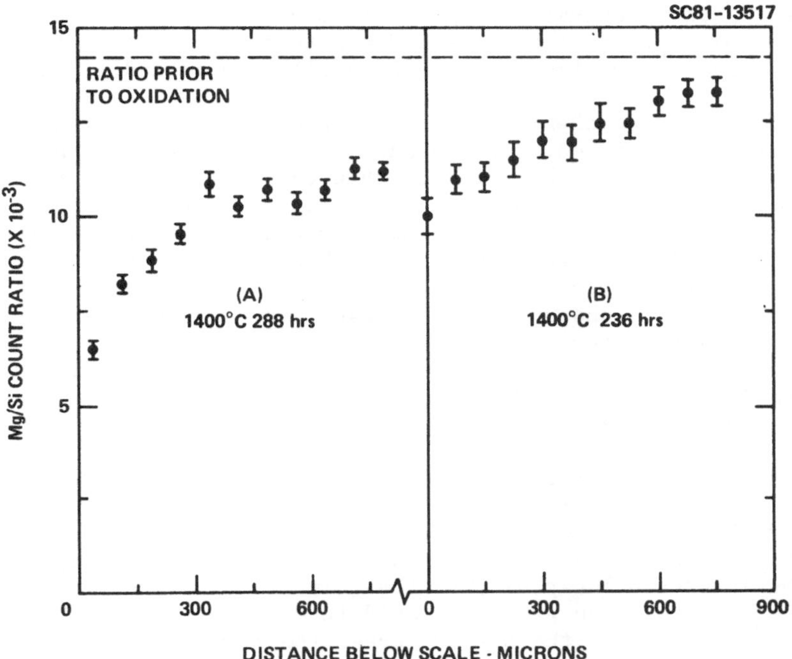

Fig. 2 Comparison of the magnesium depletion gradient created (a) on oxidation and (b) by heating in a SiO_2 powder pack in the absence of air.

IMPLICATIONS AND APPLICATIONS

Cation diffusion according to the mechanism described alters the composition and volume fraction of the intergranular phase. It therefore affords a means by which the composition of the intergranular phase might be controlled. This can be envisaged to have beneficial or detrimental consequences.

The formation of a pure SiO_2 surface layer, either by oxidation or by surrounding a sample with SiO_2 powder, causes the additive and impurity cations to migrate to the surface. One consequence is that such diffusion will act to "purify" the intergranular phase. As it has previously been demonstrated that the high temperature strength of silicon nitride alloys decreases with increasing impurity concentrations,[11,12] it is expected that "purifying" the intergranular phase will result in improved high temperature properties. This prediction has recently been borne out. By using a post-fabrication oxidation step the creep resistance[13] and flexural strength[14] of a series of MgO hot-pressed silicon nitrides and the intermediate temperature strength[15] of a Y_2O_3 sintered silicon nitride have all been markedly increased.

If a reverse chemical potential gradient for a particular cation is established, for instance by a silicate environment having a higher concentration of that cation than that of the intergranular phase, inward migration of the cation is predicted. This may lead to material degradation of two related types. The least severe is the enrichment of the intergranular phase with the cation resulting in a lowering of the viscosity of the phase and a consequential increase, for instance, in the rate of cavitational creep and sub-critical crack growth. The more severe degradation mechanism would be a compositional alteration of the intergranular phase that would result in solution or corrosion of the Si_3N_4 or β'-Si_3N_4 grains.

An example of these detrimental effects is the degradation of a MgO hot-pressed silicon nitride (Norton NC132) caused by heating for 200 hrs at 1200°C surrounded by a CaO/MgO/SiO_2 silicate. The MgO/SiO_2 ratio of the silicate was chosen to be the same as that of the intergranular phase, so that, in principal, only a Ca cation concentration gradient would be created. After the treatment and removal of the surface scale the four-point flexural strength of the material at 1400°C was measured to be 31.2 ± 1.9 as compared with 37.0 ± 1.1 ksi for the untreated ceramic. X-ray micro-analysis across the section of the exposed material reveals a Ca concentration gradient extending into the material, indicating that the inward migration of Ca was responsible, at least in part, for the observed strength degradation. No additional phases below the scale were detected by X-ray diffraction. When the experiment was repeated with the same commercial silicon nitrides but at 1400°C corrosion products were detected below the surface scale by X-ray diffraction. This observation, together with X-ray micro-analysis findings using the transmission electron microscope that the microstructure consisted of silicon nitride grains seemingly surrounded by a crystalline Si, Mg, Ca compound with pockets of glass rich in Mg and Ca, suggests that corrosion was taking place in the bulk of the material.

As these examples demonstrate the subsurface composition of polyphase silicon nitrides can be modified, via the intergranula-phase, according to the nature of the surface scale produced by environmental reaction with the silicon nitride. The model proposed can be used as a basis for understanding and predicting these compositional changes.

REFERENCES

1. Clarke, D.R., "The Microstructure of Nitrogen Ceramics," these proceedings.
2. Tsai, R.L. and R. Raj, "The Role of Grain Boundary Sliding in Fracture of Hot-Pressed Si_3N_4 at High Temperatures," J. Am. Ceram. Soc. 63 (1980) 513.
3. Kiehle, A.J., L.K. Heung, P.J. Gielisse and T.J. Rockett, "Oxidation Behavior of Hot-Pressed Si_3N_4," ibid 58 (1975) 17.
4. Tripp, W.C. and H.C. Graham, "Oxidation of Si_3N_4 in the Range 1300-1500°C," ibid 59 (1976) 399.
5. Cubicciotti, D. and K.H. Lau, "Kinetics of Oxidation of Hot-Pressed Silicon Nitride Containing Magnesia," ibid 61 (1978) 512.
6. Clarke, D.R., and F.F. Lange, "Oxidation of Silicon Nitride Alloys: Relationship to Phase Equilibria in the Si_3N_4-SiO_2-MgO System," ibid 63 (1980) 586.
7. Singhal, S.C. "Oxidation of Silicon Nitride and Related Materials," in "Nitrogen Ceramics" edited F.L. Riley (Nordhoff Press, Leyden, 1977) p. 607.
8. Brindley, G.W. and R. Hayani, "Kinetics and Mechanism of Formation of Forsterite by Solid State Reaction of MgO and SiO_2," Philos. Mag. 12 (1965) 505.
9. Lindner, R., "Silicate Formation in the Solid State," Z. Phys. Chem. 6 (1956) 129.
10. Clarke, D.R. and F.F. Lange, "Oxidation of Silicon Nitride Alloys: An Experimental Test of Cation Diffusion," J. Am. Ceram. Soc. (submitted).
11. Richerson, D.W., "Effect of Impurities On the High Temperature Properties of Hot-Pressed Silicon Nitride," Bull. Am. Ceram. Soc. 52 (1973) 560.
12. Iskoe, J.L., F.F. Lange and E.S. Diaz, "Effect of Selected Impurities on the High Temperature Mechanical Properties of Hot Pressed Silicon Nitride," J. Mater. Sci. 11 1976 p. 908.
13. Lange, F.F., B.I. Davis and D.R. Clarke, "Compressive Creep of Si_3N_4/MgO Alloys. III. Effects of Oxidation Induced Compositional Change," J. Mater. Sci. 15 (1980) 616.
14. Lange, F.F. and B.I. Davis, "Strengthening of Polyphase Si_3N_4 Materials Through Oxidation," to be submitted.
15. Clarke, D.R., F.F. Lange and G. Schnittgrund," Strengthening of A Sintered Silicon Nitride by a Post-Fabrication Heat Treatment," J. Am. Ceram. Soc. submitted.
16. Clarke, D.R., to be published.

OXIDATION KINETICS OF HOT-PRESSED SILICON NITRIDE

G.N. Babini and P. Vincenzini

C.N.R., Research Institute for Ceramics Technology, Faenza, Italy.

ABSTRACT. A simplified model for evaluation of the oxidation behaviour of hot-pressed silicon nitride (HPSN) has been derived which accounts for the amount and type of grain boundary phase and for the composition at the Si_3N_4 oxide reaction interface. The model proved suitable for explaining the oxidation behaviour of a wide range of materials including those with MgO, CeO_2, (CeO_2 + SiO_2), (Y_2O_3 + SiO_2), (MgO + Y_2O_3) additives. Some deductions are also possible on how to improve the oxidation resistance of HPSN.

1. INTRODUCTION

A substantial amount of work has been done on the oxidation resistance of HPSN. All investigators, with the exception of Ref.1, agree that additive and impurity diffusion through the grain boundary phase or through the oxide scale is the rate governing step for oxidation (2-8). No attempts have yet been made to relate thermodynamic and kinetics data to the specific properties which affect diffusion in silicate melts, in order to achieve a general understanding of the oxidation phenomenon.

Based on oxidation results on a wide range of additive systems, the present study was a search for common factors which might provide a satisfactory general picture of the oxidation behaviour of HPSN. Although a number of simplifying assumptions are required, the model seems to work satisfactorily whatever the specific system.

2. EXPERIMENTAL

Cylindrical billets were hot pressed to full density from the same silicon nitride powder with the following sintering aids: (2.5 – 10.0) wt% MgO, (5 – 20) wt% CeO_2, (6.6 wt% CeO_2 + 6.9 wt%

Riley, F.L. (ed.) Progress in Nitrogen Ceramics

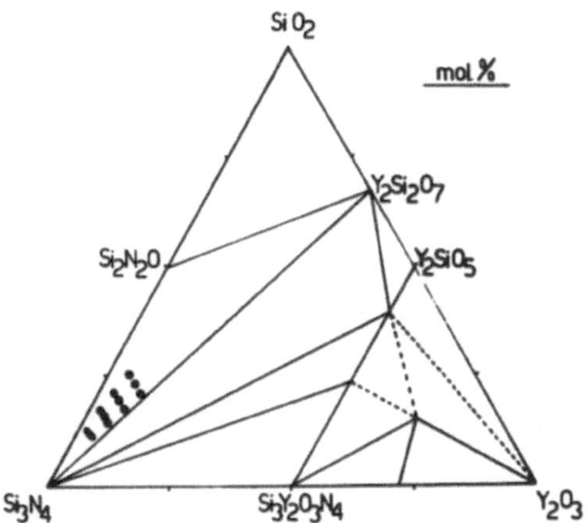

Fig. 1. Location of compositions belonging to the Y_2O_3-SiO_2-Si_3N_4 system investigated in this study.

SiO_2), (1 wt% MgO + 8 wt% Y_2O_3), (3 - 15) wt% Y_2O_3 + (3 - 10) wt% SiO_2. The silicon nitride (AME-Refractory grade) had a BET area of 3720 m^2kg^{-1}. It contained 1.9^w/o amorphous silica and 1.4^w/o silicon. Chemical analysis gave: Si (total) 58.41%, N 37.73%, Al 0.42%, Fe 0.46%, Ca 0.13%, Mg 220 ppm, Na 120 ppm, K 50 ppm. The α/β ratio was 3.8, and the TEM mean grain size was 150 μm.

All the 12 compositions of the Y_2O_3-SiO_2-Si_3N_4 system belong to the compatibility triangle Si_3N_4-Si_2N_2O-$Y_2Si_2O_7$ and can be grouped in three sets having approximately the same SiO_2/Y_2O_3 ratio (2, 3 and 5 respectively, Fig. 1). Oxidation tests were performed in air in a thermogravimetric apparatus.

3. RESULTS

Parabolic weight gains were observed for all systems whatever the amount of additive and the oxidation temperature. A comparison at T = 1573 K for various materials is shown in Fig. 2.

The following can be pointed out from a general evaluation of the experimental observations:

i) diffusion of additive and impurity cations from the bulk to the oxide scale is a common feature for all materials. Concentration of the same elements also occurs at the outer oxide skin, although with some differences among specific elements and systems;

ii) oxidation rates increase with the amount of MgO (Fig. 3), and remain approximately constant for CeO_2. No simple correlation is to be expected for the (Y_2O_3 + SiO_2)-doped materials, since all their compositions are different;

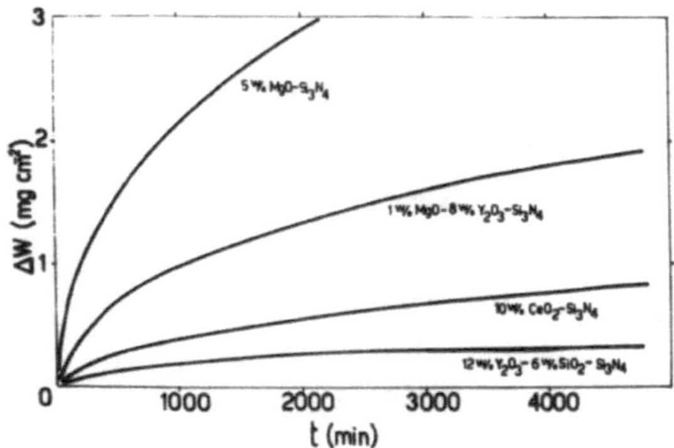

Fig. 2. Isothermal weight gain curves for HPSN obtained with various sintering aids (T = 1573 K).

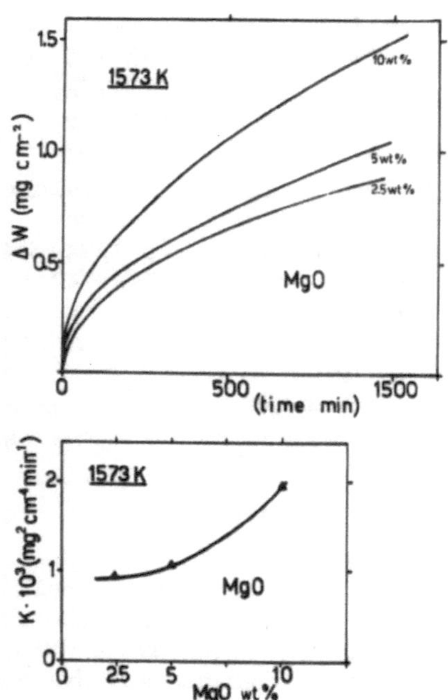

Fig. 3. Dependence of the oxidation rate constant as a function of MgO content (T = 1573 K).

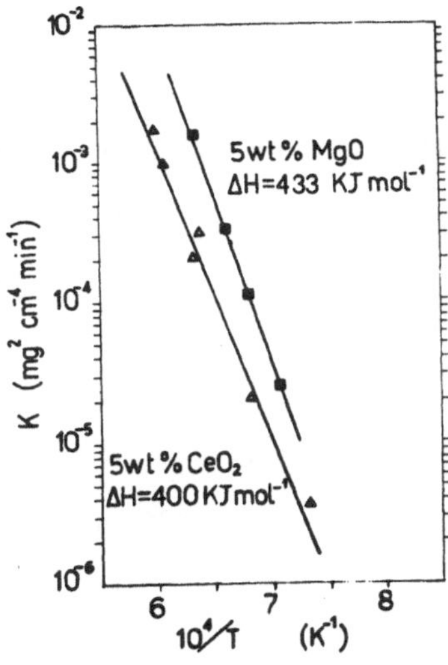

Fig. 4. Arrhenius plots of the oxidation rate constants for HPSN containing 5 wt% MgO and 5 wt% CeO_2.

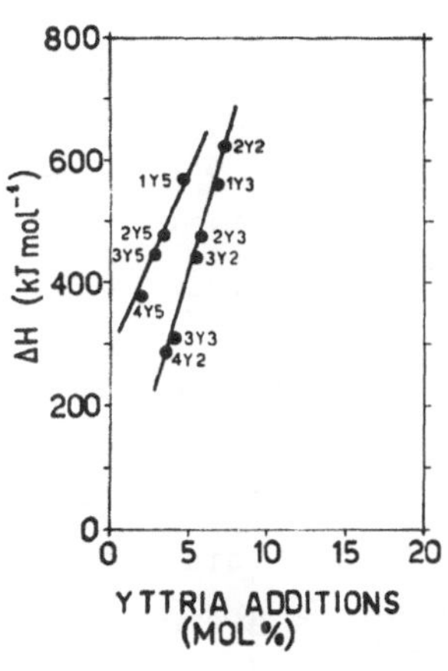

Fig. 5. Variation of the apparent activation enthalpy for oxidation for materials hot pressed from the Y_2O_3-SiO_2-Si_3N_4 system.

iii) the "apparent" activation enthalpy for oxidation, ΔH, is independent of the amount of additive for both MgO and CeO_2. Arrhenius plots of oxidation rates for the 5 wt% compositions are reported in Fig. 4. The more complex behaviour of the Y_2O_3-SiO_2-Si_3N_4 system is shown in Fig. 5. For this system ΔH values appear related to type and amount of the grain boundary phase. Thus the composition of the grain boundary phase should be approximately constant with the amount of additive for MgO and CeO_2, but not for Y_2O_3-SiO_2.

4. DISCUSSION

4.1 The simplified diffusion model

Experiments show that two types of concentration gradient are established during oxidation for diffusing additive and impurity cations: the first between the bulk and the Si_3N_4/oxide reaction interface, and the second between the reaction interface and the outer surface of the oxide scale. Our view is that the former is most directly involved in oxidation.

If the grain boundary phase is represented by an assembly of parallel cylindrical channels with longitudinal axis parallel to flow direction, and if:

i) oxidation is controlled by migration of additive and impurities from the bulk to the Si_3N_4/oxide reaction interface,

ii) the weight gain ΔW is proportional to the flux J of the diffusing species, i.e.

$$\alpha \Delta W^* = S\,J$$

where $\Delta W^* = \Delta W/t = \sqrt{K}$ at $t = 1$

α is a factor linked to the kind of interaction between the specific additive and Si_3N_4

S is the total diffusion cross section = Na

a is the average cross sectional area of the diffusion channels

N is the number of diffusion channels (which can be assumed constant for each system);

iii) the concentration profile along the diffusion path is described by a linear relation:

$$-\frac{\partial c}{\partial x} = \frac{(c_{gb} - c_r)}{\Delta x}$$

where c_{gb} is the concentration of additive in the grain boundary phase

c_r is the concentration of the additive at the Si_3N_4/oxide reaction interface

Δx is the distance between the reaction interface and the unreacted material.

This assumption is not binding for the validity of the model. Furthermore it is in line with the profile of the additive concentration in the grain boundary phase for MgO-doped materials after oxidation at 1673 K for 8h (3) and for Y_2O_3-doped materials at 1768 K for 16h (7). By applying the first Fick's diffusion law*:

$$J = -D\frac{\partial c}{\partial x}$$

where D is the diffusion coefficient of the species involved

c is concentration

x is the diffusion path length,

---

*All parameters are to be intended as representative of average values related to the different species of diffusing cations.

one obtains the following relationships:

$$\Delta W = \frac{D}{\alpha}\frac{N}{x} \cdot (c_{gb} - c_r)a = Aa \tag{1}$$

$$\frac{\Delta W}{\delta a} = \frac{D}{\alpha}\frac{N}{\Delta x} \cdot (c_{gb} - c_r) = B\,(c_{gb} - c_r) \tag{2}$$

A linear relationship between oxidation rate and amount of grain boundary phase is to be expected only when the factor A in eqn.(1) is constant, which implies constancy of D, Δ x and $(c_{gb} - c_r)$. Experimental data seem to indicate that at a given oxidation temperature the crucial variable among these parameters is the concentration gradient $(c_{gb} - c_r)$, at least for the Y_2O_3-SiO_2-Si_3N_4 system. Therefore the amount of the grain boundary phase and the concentration gradient appear as the most important parameters for oxidation when data at a given oxidation temperature are to be compared for materials belonging to the same system. It is worth pointing out that the one to two orders of magnitude differences of oxidation rate constants at equivalent temperatures between different systems are most probably linked to the general properties of the grain boundary phase, i.e. to viscosity, as suggested in a recent report (9), which strongly affects diffusion coefficients. Here we are confined to the less relevant changes deriving from variations in composition for each individual system. A further indication of the model is that materials with good oxidation resistance can be obtained also in the presence of very large amounts of grain boundary phase, provided the concentration gradient $(c_{gb} - c_r)$ is low enough.

4.2 Application to real systems

In real systems more than the oxidation mechanism is most probably acting during oxidation. Also if the direct contribution to oxidation of the direct reaction of Si_3N_4 with oxygen should be disregarded, it is to be stressed that a number of cationic species are currently involved in the process, each of which is characterized by specific diffusion parameters. Therefore all parameters appearing in Eqns. (1) and (2) represent average values. Furthermore, the variation of D as a function of the concentration profile in each material is not accounted for by the model. Finally the geometry assumed is somewhat oversimplified.

For an adequate evaluation of the grain boundary phase, reference has been made to known equilibrium phase diagrams and to results of X-ray diffraction on the hot pressed samples. An excellent agreement was obtained for all system by comparing the calculated densities with those obtained experimentally for fully dense materials.

4.2.1 Y_2O_3-SiO_2-Si_3N_4

The appropriate description of the grain boundary phase is provided by the equilibrium phase diagram of the binary Y_2O_3-SiO_2, and the X-ray diffraction data on hot pressed materials. Excess SiO_2 was assumed to react with Si_3N_4 to yield Si_2N_2O.

The grain boundary phase for the materials having SiO_2/Y_2O_3 ratio of ~2 and ~3 consists of the two yttrium silicates $Y_2O_3.2SiO_2$ and $2Y_2O_3.3SiO_2$ in addition to the silicates of the impurities and Si_2N_2O; only one of the above yttrium silicates, i.e. $Y_2O_3.2SiO_2$, is present in the samples with SiO_2/Y_2O_3 ratio ~5, the other components being the same. This difference in composition may presumably give rise to the behaviour in the ΔH values of Fig. 5.

A general interpolation of all oxidation data (Fig. 6) at T = 1573 suggests that coefficient A of Eqn. (1) is composition-dependent. More interesting indications are obtained if the amount of the grain boundary phase and the concentration gradient are both taken into account. Fig. 7a and b show the plots $\Delta W^*/a$ vs c_{gb} at T = 1373 and 1573 K respectively. The intercept of the straight lines on the x axis determines the concentration $c_r = c_{gb}$ at the interface reaction corresponding to minimum oxidation, which is a function of temperature. This accounts for the experimental behaviour of the ΔH values shown in Fig. 5. The model thus allows an experimental derivation of the equilibrium interface concentration for each temperature and consequently suggests the most adequate composition for the grain boundary phase in order to minimize oxidation by cation diffusion.

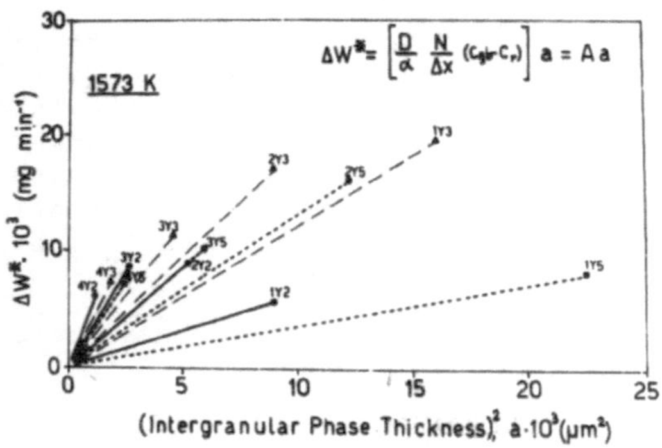

Fig. 6. ΔW^* vs s^2 relationship for materials hot pressed from the Y_2O_3-SiO_2-Si_3N_4 system (T = 1573 K).

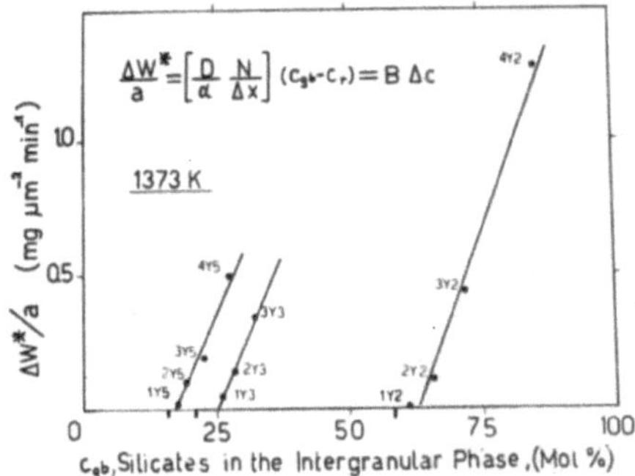

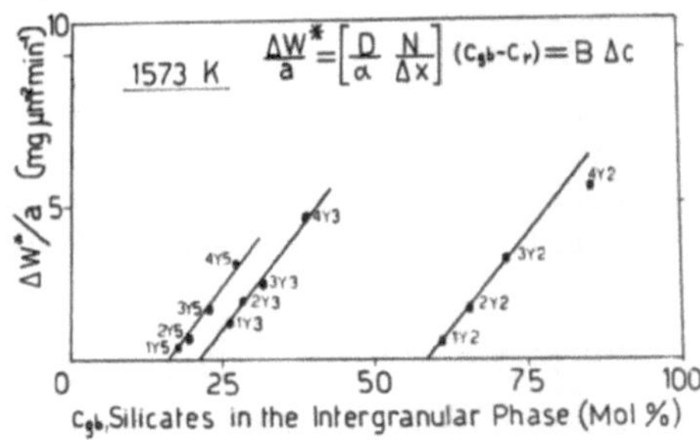

Fig. 7. ΔW^* vs c_{gb} for materials hot pressed from the Y_2O_3-SiO_2-Si_3N_4 system. (a) T = 1373 K; (b) T = 1573 K.

4.2.2 MgO-Si_3N_4

All SiO_2 in the starting nitride powder is assumed to react with magnesia to give forsterite, and the excess MgO to react with Si_3N_4 to give a magnesium silicon nitride ($MgSiN_2$).

Because of the constancy of ΔH on the amount of the grain boundary phase and of the composition of the oxide scale being approximately constant with the MgO added (10), the term A in Eqn. (1) should be approximately constant and the oxidation rate directly related to a (Fig. 8). The fact that the interpolating line does not intersect the origin is not relevant because of the drastic change in the composition of the grain boundary phase when very little amounts of additive are involved, the excess silica exerting a diluting effect which greatly affects the concentration gradient within the material.

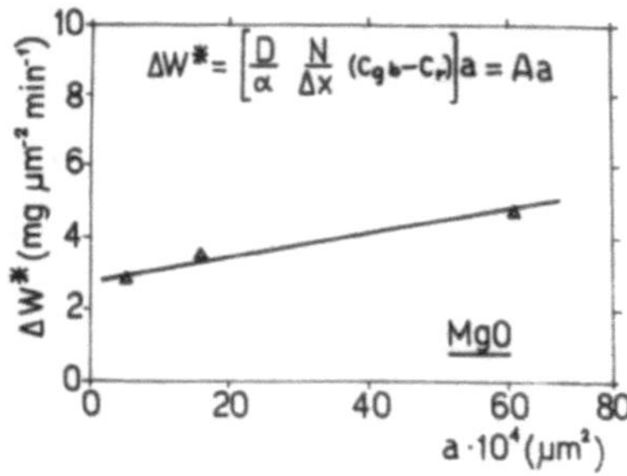

Fig. 8. ΔW^* vs s^2 relationship for materials hot pressed from the $MgO-Si_3N_4$ system (T = 1573 K).

4.2.3 $CeO_2-Si_3N_4$ and $CeO_2-SiO_2-Si_3N_4$

Of the four compositions tested in this system, three differed only in the amount of ceria, and the fourth in the addition of excess silica. Both Δ H and K values appear unaffected by the amount of CeO_2. The constancy in Δ H may be associated with the reaction:

$$12CeO_2 + Si_3N_4 \longrightarrow 6Ce_2O_3 + 3SiO_2 + 2N_2$$

which assures the invariance of composition of the grain boundary phase during hot pressing (11). On the other hand the concentration at the interface reaction in this system is maintained constant through a mechanism described elsewhere (7) which accounts for the poor solubility of Ce^{+4} in silicate melts. Therefore the concentration gradient $(c_{gb} - c_r)$ does not vary with CeO_2 additions. Furthermore, if also considerable amounts of CeO_2 are added, the volume of the intergranular phase undergoes only a small increase as a consequence of the formation of cerium orthosilicate $(Ce_{4.67}(SiO_4)_3O)$ which takes place with a relevant volume contraction. Thus the invariance of the oxidation rate is completely justified.

When excess silica is present, both volume and composition of the grain boundary phase and consequently the concentration gradient can vary substantially; this justifies the different apparent Δ H value and the lower oxidation rate constant for the (CeO_2+SiO_2)-doped material as shown in Fig. 9.

4.2.4 $MgO-Y_2O_3-Si_3N_4$

Among the materials investigated yttria-doped HPSN exhibits by far the better oxidation resistance. It has been shown that this may be related to the refractoriness (high viscosity) of the

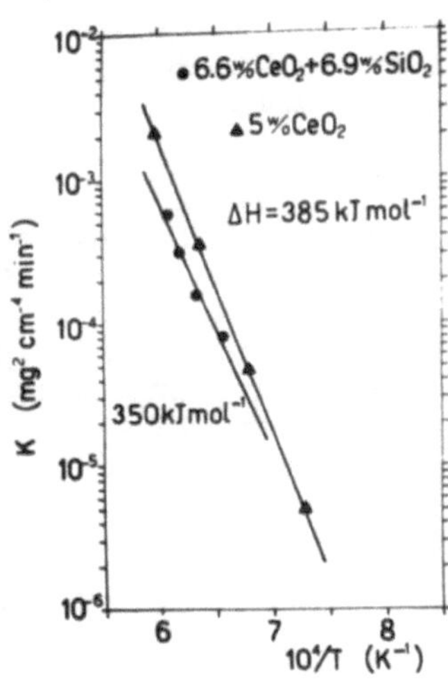

Fig. 9. Arrhenius plots of the oxidation rate constants for HPSN containing 5 wt% CeO_2 and 6.6 wt% CeO_2 + 6.9 wt% SiO_2.

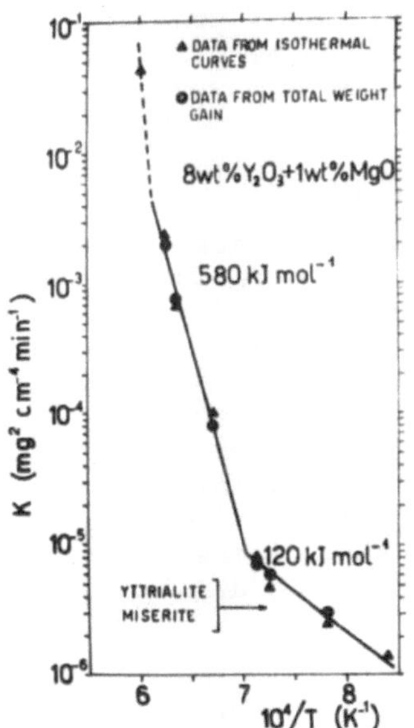

Fig. 10. Arrhenius plot of the oxidation rate constants for HPSN containing 8 wt% Y_2O_3 + 1 wt% MgO.

grain boundary phase (9) with which a poor sinterability is also associated.

One possibility for having at the same time in a material both good oxidation resistance and good sinterability, is to make use of a combined additive system in which one kind of additive lowers the sintering temperature and the other forms one or more crystalline compounds in the intergranular phase, which are capable of hindering diffusion of additive and impurity cations. This approach was used to hot-press and sinter (MgO + Y_2O_3)-Si_3N_4 compositions. Very low oxidation rate constants, similar to that of the best (Y_2O_3 + SiO_2)-doped material, are found up to approximately 1423 K for a (1 wt% MgO + 8 wt% Y_2O_3). At temperatures higher than 1423 K the oxidation behaviour is comparable to normal MgO-doped materials (Fig. 10).

The presence of yttrialite (H,Na,F) $(Y,Ln)_5Si_6O_{21}$ and miserite $(Ca,K,Na,Al,Y)SiO_3$ formed during hot pressing and stable up to 1373-1473 K, is the origin of the low oxidation rates at T 1423 K, a fraction of additive and impurity cations being held in solid solution, the diffusion of which is thus impeded. Incidentally it can be noted that ΔH = 120 kJ mol^{-1} in the stability field of yttrialite and miserite; this value approaches the activation enthalpy for oxygen diffusion in silicate glasses (~80 kJ-~140 kJ), which might perhaps support the direct oxidation of Si_3N_4 by oxygen as playing a major role in this specific case.

5. CONCLUSIONS

The model provided here is oversimplified, but it has proved useful in accounting for general aspects of the oxidation behaviour of materials containing a broad range of additives, and over a range of temperatures, including the possibility of producing high oxidation resistant materials in the presence of very large amounts of grain boundary phase. It provides also a guide to the selection of individual grain boundary phase compositions in order to achieve minimum oxidation rates. It has been shown how a proper combination of additives can result in materials of high oxidation resistance also in the presence of a sintering aid such as MgO which is generally deleterious.

REFERENCES

1. W.C. Tripp and H.C. Graham. Oxidation of Si_3N_4 in the range 1300-1500°C. J. Amer. Ceram. Soc. 59, 399 (1976).
2. S.C. Singhal. Thermodynamics and kinetics of oxidation of hot pressed silicon nitride. J. Mater. Sci. 11, 500 (1976).
3. D. Cubicciotti and K.H. Lau. Kinetics of oxidation of hot pressed silicon nitride containing magnesia. J. Amer. Ceram. Soc. 61, 512 (1978).
4. D.R. Clarke and F.F. Lange. Oxidation of Si_3N_4 alloys: relation to phase equilibria in the system Si_3N_4-SiO_2-MgO. J. Amer. Ceram. Soc. 63, 586 (1980).
5. C.L. Quackenbush and J.T. Smith. Phase effects in Si_3N_4 containing Y_2O_3 and CeO_2: II Oxidation. Am. Ceram. Soc. Bull. 59, 533 (1980).
6. G.N. Babini, A. Bellosi and P. Vincenzini. Oxidation of silicon nitride hot pressed with ceria additions. Special Ceramics 7, ed. D. Taylor and P. Popper, Brit. Ceram. Soc., Stoke-on-Trent, England, 1981, p.169.
7. D. Cubicciotti and K.H. Lau. Kinetics of oxidation of yttria hot-pressed silicon nitride. J. Electr. Soc., Solid St. Sci. and Techn., 126, 1723 (1979).
8. G.N. Babini, A. Bellosi and P. Vincenzini. Oxidation kinetics of silicon nitride hot-pressed with yttria and silica additions. Paper presented at 83rd Annual Meeting of the Am. Ceram. Soc., Washington, May 1981.
9. G.N. Babini, A. Bellosi and P. Vincenzini. Oxidation behaviour of silicon nitride hot pressed with various sintering aids. Paper presented at Science of Ceramics 11, Sweden, June 1981.
10. G.N. Babini, A. Bellosi and P. Vincenzini. La Ceramica 34 (3), 11 (1981).
11. G.N. Babini, A. Bellosi and P. Vincenzini. Ceramurgia Int., 6, 91 (1980).

DISCUSSION

Morgan: Your data for ΔH are similar to those Raj and myself would think would be in the region of ΔH (solution) of Si_3N_4 in the melt. The further extension of the ideas that the $\Delta H_{(observed)}$ is likely to be a sum of several terms also explains the fact that the ΔH is larger for the MgO system where the effect is more dependent on the amount of liquid, than that for the Y_2O_3 system.

Billy: You observed changes in activation enthalpy for Si_3N_4 oxidation, the first around 1050°C. The reaction mechanism may change at this temperature (which is not high enough to take into account the mobility of grain-boundary glassy phases) due to the formation of silica and other oxides below 1050°C, and then to the formation of silicate phases in the oxidation scale.

Babini: In the SiO_2-Si_3N_4-Y_2O_3 system the apparent ΔH change depends on the SiO_2-Y_2O_3 content and is probably due to (i) a viscosity change, (ii) concentration gradient changes, (iii) the variation of the concentration at the reaction interface with temperature. In the MgO-Y_2O_3-Si_3N_4 system, ΔH changes around 1150°C. Below this temperature the stability of yttrialite and miserite hindered additive diffusion and probably a mechanism based on the diffusion of oxygen through the oxide scale has to be taken into account.

Clarke: How did you determine the composition of the intergranular phase?

Babini: It was calculated, approximately, using the Y_2O_3-SiO_2 phase diagram, assuming that SiO_2 excess reacts with Si_3N_4 to give Si_2N_2O, and in accordance with XRD analysis, to give good agreement between the calculated theoretical density and experimental density.

OXIDATION BEHAVIOUR OF β'-SIALONS IN OXYGEN AND CARBON DIOXIDE

J.Desmaison*, M.Brossard, M.Desmaison-Brut* and P.Goursat

Laboratoire de Céramiques Nouvelles (L.A. C.N.R.S. No.320) Université de Limoges, 87060 Limoges Cédex, France.

1. INTRODUCTION

Since the last NATO Advanced Study Institute on Nitrogen Ceramics, the oxidation behaviour of the β'-sialon phase has been investigated by several researchers (1-8). Testing has generally been carried out in air at temperatures up to 1450°C on powders (4,6) and hot-pressed (1,2,5,7,8) or pressureless sintered sialons (2,3) with variable amounts of sintering aids (MgO or Y_2O_3). Simple parabolic kinetics and improved oxidation resistance compared with conventional MgO doped hot-pressed silicon nitride are usually observed. But, the oxidation behaviour seems to be very dependent upon the nature and the quantity of sintering aid. For example, it has been reported that increasing Y_2O_3 additions increase the strength of sialons at the expense of their oxidation resistance (2,3). The rate controlling process is supposed to be either the inward diffusion of oxygen through an alumino-silicate glass film (5) or the outward diffusion of metallic cations such as Mg^{2+} or Y^{3+} from the grain boundaries of the underlying material (2). However a recent study on materials formed without densification aids has demonstrated that it is unsafe to make the broad assumption that diffusion controlled parabolic kinetics are normal (1). Therefore the main objective of the present investigation was to gain a better understanding of the oxidation mechanism in β'-sialons densified without the use of liquid-generating additives.

*Two of us, J.D. and M.D.-B., wish to thank Prof. R.J. Brook and Dr. F.L. Riley from the Department of Ceramics of the University of Leeds (U.K.) for the provision of laboratory facilities. J.D. is indebted to the Royal Society, the British Council and the C.N.R.S. for financial support.

Riley, F.L. (ed.) Progress in Nitrogen Ceramics

2. MATERIALS AND EXPERIMENTAL PROCEDURE

Two sialons A and B were prepared by reaction hot-pressing of silicon nitride, aluminium oxide, and aluminium nitride powder mixtures.

Sialon A: $Si_{3.20}Al_{2.66}O_{3.20}N_{4.80}$ (ρ = 3.08 g/cm^3), obtained at Limoges (1770°C, 20 MPa, 50 mm) from the following starting materials: Si_3N_4 (Alfa-Ventron 99.5%), Al_2O_3 (Alcoa, > 99.9%) and AlN (Alfa-Ventron, 99.5%), is slightly oxygen rich with respect to the β'-sialon phase line.

Sialon B: $Si_{3.5}Al_{2.5}O_{2.5}N_{5.5}$ (ρ = 3.06 g/cm^3) hot-pressed at Leeds* (1700°C, 20 MPa, 90 mm), from a "balanced" powder mixture of Si_3N_4** (Lucas Industries Ltd., 99.5%), has a composition located on the β'-sialon phase line (z = 2.5).

Oxidation was studied by continuous thermogravimetry using an automatic electrobalance. Cubes of side 5 mm were cut from the hot-pressed discs and hand polished to < 800 mesh silicon carbide. Samples were suspended by a platinum wire basket and lowered rapidly to the pre-equilibrated hot zone. The experiments were performed either in static O_2 or in flowing CO_2 (10 l/h) under 1 atm pressure. The kinetic curves were obtained by plotting either the fractional weight gain $\alpha = \Delta w/\Delta w_\infty$ or the weight gain per unit area versus time. After oxidation the quenched oxide films were examined by optical and scanning electron microscopy and characterised by X-ray diffraction and EDAX analysis.

3. RESULTS

3.1 Oxidation in oxygen

Generally the kinetics have a parabolic shape but in some cases they exhibit a paralinear tendency (Fig. 1a). Above ~1550°C it is possible to observe a noticeable increase in the reactivity of sialon A (full line curves). The behaviour of sialon B (dashed line curves) is very close but suggests a slightly better oxidation resistance.

Below 1550°C, the oxide layer consists of a mixture of crystallized mullite plus a vitreous phase. Above this temperature, when the melting point of mullite is reached (T ~ 1590°C) only a glassy phase is observed. In both cases, but particularly in the high temperature range, the scale is not protective. Indeed, cross-section (Fig. 2a) or surface examination (Fig. 2b) of the scale shows the presence of significant open porosity.

3.2 Oxidation in carbon dioxide

Here again, an important increase of the reaction rate above 1550°C (dashed line curves) and a slightly better oxidation resis-

**The major impurites in this Si_3N_4 powder are (in wt%): Fe(0.5), Ca (0.05) and Mg (0.04). The impurity level of the Alfa-Ventron Si_3N_4 is very close.

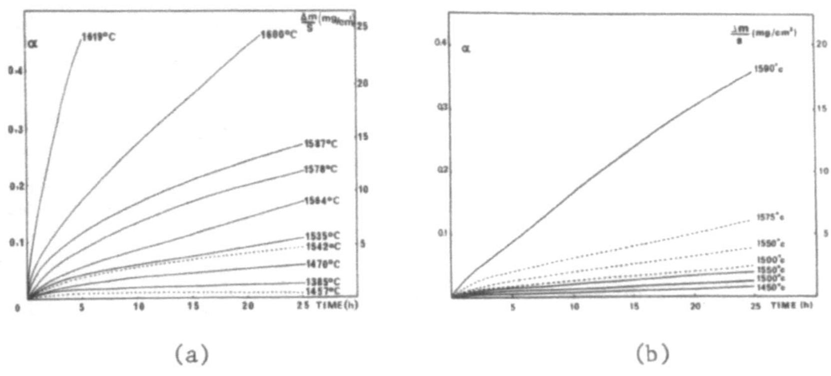

(a) (b)

. 1. - Oxidation kinetics in oxygen (a) and in carbon dioxide(b).

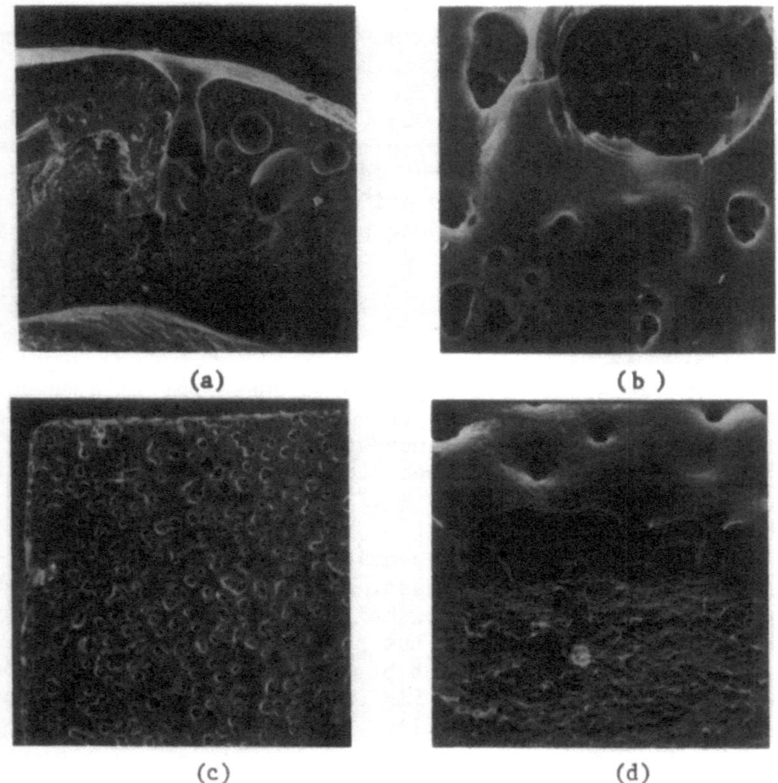

(a) (b)

(c) (d)

2. - SEM pictures of samples A oxidized in oxygen (a) T = 1619°C, X 40 ; (b) T = 1385°C, X 300 ; or of samples B oxidized in carbon dioxide (C) T = 1500°C, X 200 ; (d) T = 1450°C, X 1000.

tance of sialon B (full line curves) are observed. But this time paralinear behaviour is the general rule and the number of channels through which rapid gas access to the sialon/silicate interface can occur is much more important (Fig. 2c and 2d). In addition the formation of gas bubbles at this interface is particularly visible (Fig. 2d). Finally we notice that within the temperature range explored, the alumino-silicate scales are non-crystalline. Therefore it seems that the atmosphere plays not a negligible role in influencing the crystallinity and morphology of the oxidation products. The slight difference in impurity level and composition of the materials appears, on the contrary, to be of secondary importance.

4. DISCUSSION

In Fig. 3a and 3b the data are presented in log-log form. In both cases the plots are irregular in the lower temperature range ($T < 1550^\circ C$) but become almost linear in the upper range. These results clearly indicate that the shape of the kinetics is a function of temperature, time and also atmosphere. Furthermore such plots show that it is impossible to use the classical parabolic law:

$$\alpha^2 = k_p t \qquad (1)$$

because the mean values of their slopes are fluctuating between $\frac{1}{2}$ and 1. According to the following equations

$$\alpha^n = kt \qquad (2)$$

$$\text{Log } \alpha = \frac{1}{n} \log t + \log k \qquad (3)$$

this means that the shape of the (α, t) curves is intermediate between parabolic and linear. With increasing temperature and time the exponent n tends to approach 1, particularly so when carbon dioxide is used.

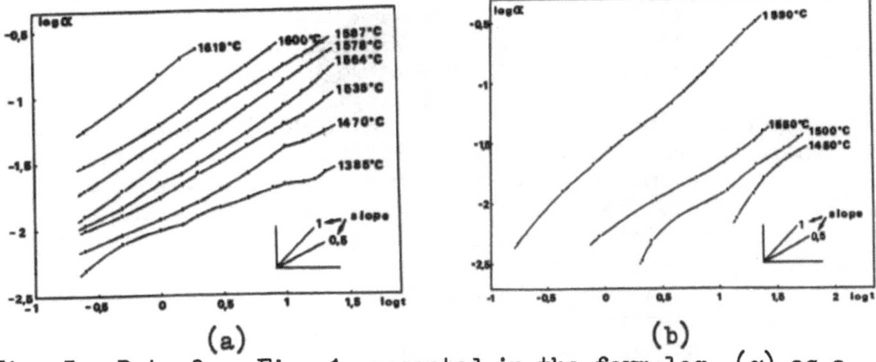

Fig. 3. Data from Fig. 1 presented in the form $\log_{10}(\alpha)$ as a function of $\log_{10}(\text{time})$.

Therefore we are faced with the impossibility of analyzing quantitatively the experimental data. Nevertheless we will give a qualitative interpretation using an oxidation model mainly based on the morphological observations.

The overall oxidation reaction of the β' sialon phase can be written:

$$Si_{6-z}Al_zO_zN_{8-z} + (6 - \frac{3}{4}z)\,O_2 \rightarrow (6-z)SiO_2 + \frac{z}{2}\,Al_2O_3 + \frac{8-z}{2}\,N_2\uparrow \quad (4)$$

There is formation of an alumino-silicate layer and evolution of nitrogen. The location of the nitrogen gas bubbles at the sialon silicate interface implies the necessity for inward oxygen diffusion through the silicate layer in order to feed the oxidation reaction (4) located at this inner interface. The flow rate of the diffusing species is given by the basic equation (5) derived from the well-known Fick's first law:

$$\frac{dQ}{dt} = k' \frac{S}{X} \quad (5)$$

where Q, S and X are respectively the oxygen uptake, the area and the thickness of the oxide film (9). If S is supposed constant and X is proportional to α and Q, the reaction rate takes the form:

$$\frac{d\alpha}{dt} = k'' \frac{S}{\alpha} \quad (6)$$

which leads to the classical parabolic law (1). However a diffusion-limiting step does not necessarily imply a parabolic law. For example, if for any reason the outer part of a corrosion scale becomes porous causing the thickness of the protective layer to remain constant, so does the flow rate according to equation (5) and by extension the observed kinetics are linear. Consequently, the present case may be explained by considering an intermediate variation of X with time. Initially protective, perhaps because of the formation of a nitrogen containing glass, the corrosion scale becomes porous when this glass is saturated and releases in molecular form its excess nitrogen. At this point of the reaction the oxygen transport may occur in five different ways schematically described in Fig. 4. In some places (I) ionic (or atomic) oxygen has to migrate through the layer. In other places (II, III, IV), due to the existence of channels or pores, both ionic and gaseous diffusion occur. Finally where the channels are running without discontinuity from the surface to the interior almost only gaseous diffusion takes place (V). Therefore the protective properties of the scale are changing from one point to another at a given time. But they are also changing with time at a given point because of the nitrogen bubbling. In fact, to take account of all the phenomena it is necessary to consider the existence of an effective diffusion barrier of thickness X_{eff} defined by the statistical mean length of all the ionic (or atomic) diffusion paths (assuming gaseous diffusion quicker). At the beginning of the reaction X_{eff}

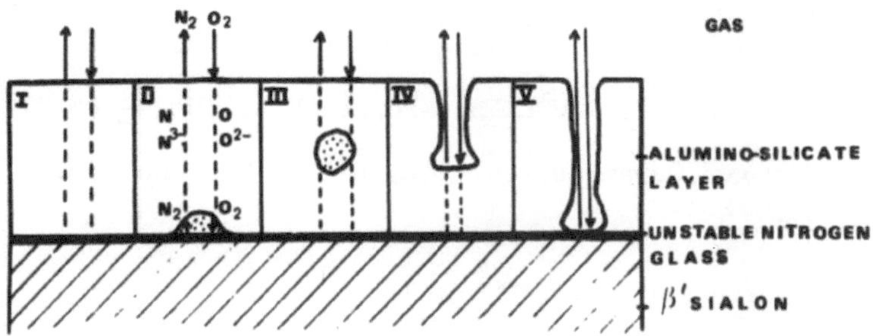

Fig. 4. Schematic representation of the oxidation model.

corresponds to the observed scale thickness X but as the oxidation proceeds X_{eff} becomes smaller than X and tends to be almost constant.

In this way assuming S constant in equation (5) it is possible to explain both the initial decrease of the oxidation rate and its stabilization for longer times. With this model it is also possible to take into account the effect of temperature and atmosphere. When T is increasing the viscosity of the silicate diminishes, hence the nitrogen bubbles are more easily released inducing augmentation of the open porosity and of the linear tendency. At a given temperature, the viscosity of the oxide seems lower in carbon dioxide than in oxygen. Consequently it is normal to observe more linear kinetics. Finally because of the evolving nature of the kinetics it is impossible to give any significance to the experimental activation energies and to use them to define the limiting step, as is often done (4).

5. CONCLUSION

This study confirms that for β'-sialon materials formed without densification aids the oxidation kinetics are not parabolic. Indeed above 1400°C, the form of the curves, which depends on the temperature and the atmosphere, is approximately paralinear. The oxidation mechanism is complex due to the existence of high diffusivity oxygen paths of variable length associated with the open porosity created by nitrogen gas evolution.

REFERENCES

1. DESMAISON, J.G. and F.L. RILEY. J. Mat. Sci. 16 (1981) 2625.
2. LEWIS, M.H. and P. BARNARD. J. Mat. Sci. 15 (1980) 443.
3. ARIAS, A. J. Mat. Sci. 14 (1979) 1353.
4. BROSSARD, M., D. BRACHET, P. GOURSAT and M. BILLY. Ann. Chim. Fr. 4 (1979) 7.
5. SCHLICHTING, J. Rev.Int.Hautes Temp. Refract. Fr. 16 (1979) 67.

6. BROSSARD, M., D. BRACHET, P. GOURSAT and M. BILLY. C.R. Acad. Sc. Paris, Serie C, 286 (1978) 345.
7. SINGHAL, S.C. and F.F. LANGE. J. Amer. Ceram. Soc. 60 (1977) 190.
8. SCHLICHTING, J. and L.J. GAUCKLER. Powder Met. Int. 9 (1977) 36.
9. BILLY, M. Nitrogen Ceramics, ed. F.L. Riley, Noordhoff, Leyden, 1977, p. 203.

DISCUSSION

Tien: Can you comment on the non-wetting of the sialon by the liquid film?

Desmaison: The only thing we can say is that the non-wetting behaviour of the silicate layer decreases as z increases.

Lange: Could oxygen in the CO_2 have been a factor?

Desmaison: We do not think that the non-wetting of the sialon is linked to p_{O_2} because we obtained almost the same behaviour in oxygen and in carbon dioxide, for a given temperature and z value.

Gugel: Did you not observe mullite crystallization in your glass - because a high Al_2O_3 glass is not very stable?

Desmaison: For z values higher than 2.5 noticeable crystallization of mullite was observed at the scale surface.

Porz: Was there no attack of the Pt specimen holder?

Desmaison: Platinum was wetted by the low viscosity oxide film, but there was no chemical attack.

Billy: Comments concerning the Clarke oxidation model for hot pressed silicon nitride. This model neglects the inwards diffusion of oxygen which does exist as proved for example by the reverse oxygen diffusion out of the material when annealing is performed in argon. Such a result shows, on the other hand, that internal formation of Si_2N_2O through reaction between Si_3N_4 and the grain boundary glassy phase must be excluded. I cannot agree with the outwards diffusion of N^{3-} ions; these are too unstable to exist in our materials.

Clarke: I have stressed in my presentation and in our publications that the model describes a situation additional to the inward diffusion of oxygen when oxygen is present. Thus for a complete picture it is necessary to take into account the effect of oxygen diffusion as well as the effects of cation diffusion.

Regarding the stability of N^{3-}; I cannot comment here, but the model is not dependent on whether nitrogen diffuses as N^{3-} ions or as molecules.

METALLIC INFILTRATION OF REACTION BONDED SILICON NITRIDE

W.G. Schmidt

Hutschenreuther AG, Central Laboratory,
D-8672 Selb, W. Germany

1. INTRODUCTION

Recent developments in reaction bonded silicon nitride (RBSN) science and technology have sustained interest in its gas-turbine applications, and as a general engineering material in room and high-temperature assemblies. Minimisation of porosity is an important feature of much research effort concerning RBSN. The technology of infiltration of the porosity, unusual for RBSN, has been successfully achieved, and open porosity in RBSN has been filled with molten metals. RBSN is wetted at temperatures above 900°C by a range of liquid metals, some having a low specific gravity. Low density (1800 kg m^{-3}) RBSN can be fully infiltrated by several metals under "vacuum" conditions. A pressure of 200 Pa is often sufficient. Medium density (2000-2200 kg m^{-3}) RBSN can be infiltrated at higher temperatures or by the use of extended heating times. High density (2400-2600 kg m^{-3}) RBSN containing submicrometer pores can be satisfactorily infilled, minimising or reducing internal oxidation in oxidising environments. Liquid metals normally active with respect to infiltration can be activated by alloying with metals such as calcium, and new compounds formed by reactions involving RBSN have been detected. Materials formed by this infiltration technique have characteristic of cermet-type materials. It is possible to obtain a continuous matrix at an infiltration level in the region of ~ 30 v/o. The electrical resistivity of RBSN infiltrated with a silicon alloy for example, can be as low as 0.06 Ωm, similar to that of commercial p- or n- type single crystal silicon.

The advantages of certain types of cermet (ceramic-metal, multi-phase) materials are well recognised. The metal - metal-

Riley, F.L. (ed.) Progress in Nitrogen Ceramics

carbide[1] system has been exploited for many years in the metal working and cutting area, and the use of fibrous, or whiskery reinforcement of metals for a range of advanced applications has been examined in some detail[2]. Materials based on silicon infiltrated silicon carbide[3-4] have reached a refined state of development and are finding many commercial applications.

The requirement for satisfactory infilling of a ceramic is that the porosity be open. RBSN inevitably contains porosity in the range 15-55%, according to the packing density of the silicon powder used in its production[5], and much of this porosity is accessible. This porosity makes the material vulnerable to "low temperature" oxidation ($<1000^{\circ}C$) and attempts have been made to coat RBSN with mixed oxides with a view to blocking access to internal porosity and improving oxidation resistance[6]. A difficulty is found when attempts are made to provide barrier films of a metallic nature, because of the general unwetability of RBSN by low melting metals, such as aluminium and other alloys[1,7-23]. This feature is indeed exploited in the successful applications of RBSN components in the aluminium industry[9,12,20-21]. For effective coating by a metal, complete wetting would be required, and a two-phase RBSN material similar to self-bonded carbide[3-4], consisting solely of silicon and silicon nitride has so far not been produced.

2. RBSN WETTING BY METALLIC ALLOYS

Until the early 1970's little information was available about liquid metal-silicon nitride wetting angles. Two U.S. Patents[24-25] claimed that silicon nitride could be wetted by alloying nickel with some titanium (and silicon). Ni(Ti) or Ni(Ti-Si) powder compacts were heated in nitrogen for short times at $1550^{\circ}C$ on a silicon nitride block. Sessile drops formed with contact angles in the region of 45°. A short notice[26] reported the infilling of ceramics, using high vacuum conditions (10^{-4} Pa) and temperatures corresponding to 1.5 Tm (the melting point of the metal in K). The possibility of infilling silicon nitride powder compacts and low density RBSN was subsequently explored in detail[27]. More recently further reports[28] have appeared on the use of titanium to enhance the wetting of silicon nitride by silicon, through the formation of a titanium nitride and silicide layer on the RBSN surface. Pre-coating with a titanium powder slurry caused complete spreading of liquid silicon on a RBSN surface at $1550^{\circ}C$ in vacuum, whereas under similar conditions liquid silicon wetted, but did not flow, on a clean RBSN surface, or one vacuum-coated with a titanium film. A degree of uncertainty about the wetability of silicon nitride by liquid Si(Ti) alloys still exists therefore.

3. EXPERIMENTAL

Experiments were carried out with commercial[+] and laboratory produced[++] RBSN, with densities up to 2600 kg m^{-3}. Porosities varied between 55% and 20%. A wide range of metals of electrolytic or similar purity (99.5%), commercial silicon alloys (28-64^{w}/o Si; 1-57^{w}/o Al; 9-33^{w}/o Ca; 5^{w}/o Fe; 0.5^{w}/o Mn; 0.5^{w}/o C; 0.1^{w}/o Ti) and commercial aluminium alloys containing Mg, Si and Mn were investigated (Table 1). A standard sessile drop arrangement was used[24-25,28], with melting being carried out in a cold-wall metal vacuum furnace with a carbon heating element. Temperatures were measured optically. The normal vacuum level used was 20-200 Pa (with an upper temperature limit of 1620-1750°C), although in some experiments on medium or high density RBSN a pressure of 10^{-3} or 10^{-1} Pa was used with an upper temperature limit of 1200 or 1300°C. At higher temperatures under vacuum thermal decomposition on the surface of the silicon nitride occurred, and argon at a pressure of 10^{4}-10^{5} Pa was substituted for the vacuum environment. The silicon nitride specimen rested in a carbon (and sometimes a porcelain) crucible. Heating rates were in the range of 10-500°C min^{-1}, with a hold time of 1-100 min. The maximum temperatures used with different systems varied from 840 to 1880°C. Contact angles between the sessile drops and the RBSN, ot the infiltration depth, were measured visually, or from concentration profiles within the RBSN after removal from the furnace. Other techniques of examination used on selected specimens included x-ray diffraction (CuK_{α} radiation), optical microscopy, electrical resistivity (indium coated), Vickers macro- (Diatestor) and micro-hardness (Reichert MeF_2) measurements, and bend strength measurements. Some oxidation tests up to 1250°C in laboratory air were also made.

4. RESULTS

The pure metals Mg, Ti, Zr, V, Cr, Mn, Fe, Co, Ni, Cu and Al showed moderate to high wetting angles. In some cases strong pitting corrosion was observed (by the metals of the iron group at 1670°C and by Zr, V, Cr at 1880°C in argon for example). There was a general tendency for chemical reaction to occur with formation of the metal silicide, and in some cases the metal nitride. The titanium surface for example reacted completely to give the golden TiN at temperatures well below its melting point (1670°C). Examples of contact angles under a range of conditions are shown in Table 2.

\+ Annawerk, Ceranox-Division, D-8633 Rödental, W. Germany: "Annasinid 98".

\++ Inst. f. Chemical Technology of Inorganic Materials, University of Technology, A-1060 Vienna.

Symbol:	Si(Ca-Al)	Si(Ca-Fe-Al)	Al(Si-Ca)	Al(Si-Ca-Fe)
Si	64	63.9	32	28
Al	2	1.4	52	57
Ca	33	30.5	16	9.7
Fe		3.4		4.9
Mn	0.5			
C	0.5	0.1_8		
Ti		0.0_7		
S		0.0_4		
M.Pt(°C) est.	1150	1150	900	900

Table 1: Chemical composition of Si and Al alloys++

Metal	Temperature(°C)	Atmosphere	Contact Angle(°)	Phase Detected
Mg	850	vacuum	90-120	(Mg-Si)
	950	argon	90-120	(Mg-Si)
	990	vacuum	-	Mg evaporation
Al	990	vacuum	120	-
Al(Si-Ca)	990	vacuum	0 +	new phase, AlN
Al(Si-Ca)	1200	vacuum	0 +	new phase, AlN
Si(Ca-Al)	1200	vacuum	0 +	new phase
Mn	1280	vacuum	10 +	(Mn-Si)
Cu	1280	argon	120	(Cu-Si)
Fe	1670	argon	90-120	Fe-Si
Co	1670	argon	135	Co-Si
Ni	1670	argon	135	Ni-Si
Si	1670	argon	10 +	-
Mn	1720	vacuum	-	Mn evaporation
Cu	1720	vacuum	60-90	Cu-Si
U	1720	vacuum	0	-
			+	infilling observed

Table 2: Examples of contact angles

++ according to supplier (w/o):

Si(Ca-Al), Al(Si-Ca): Ferrolegierungs GmbH, D-8223 Trostberg, W. Germany. ("CaSi", "Ca-Si-Al": Analyt. No. 90224/1971).

Si(Ca-Fe-Al), Al(Si-Ca-Fe)[30]: Schöller-Bleckmann, A-2630 Ternitz, Austria.

Fig. 1(a) (Magnif. 100 x):

Capillary flow of Si(Ca-Al) melt into porous RBSN+)

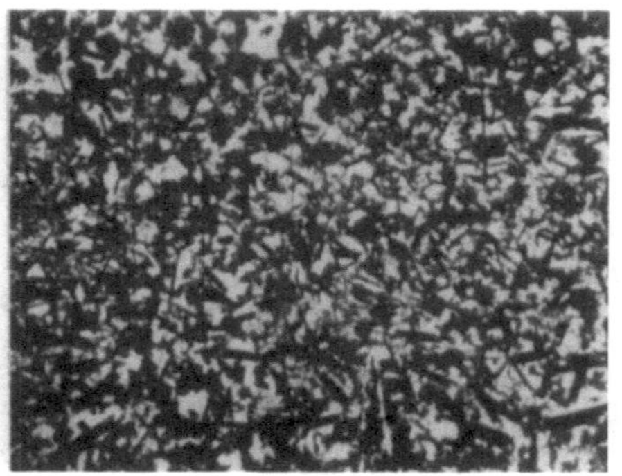

Fig. 1(b) (Magnif. 100 x):

Needle-like new phase formed in completely infiltrated RBSN+) at 1370°C and 20 Pa

In Fig. 1(a) the nature of capillary flow of the liquid Si(Ca-Al) alloy can be seen at the interface between the original part of the RBSN+) and the partially infiltrated RBSN. The micro-pores are first infiltrated under the action of capillary forces. Larger pores remain unfilled, to be filled later when sufficient liquid has been drawn into the RBSN. Fig. 1(b) shows the commercial+) RBSN (3% porosity) after complete infiltration with the same Si(Ca-Al) alloy at 1370°C and 20 Pa. The formation of a new needle-like phase can be seen. At 1200°C the Si(Ca-Al) alloy completely infiltrated medium density RBSN++) and an Al(Si-Ca) alloy infilled the surface. At 1550°C a similar Al(Si-Ca-Fe) alloy infiltrated high density RBSN+) test bars, filling 10% of the open porosity. 75% of the porosity of the high density RBSN+) bars was infiltrated at 1400°C with a Si(Ca-Fe-Al) alloy.

X-ray diffraction analyses showed the presence of α-Si_3N_4, β-Si_3N_4, Si Al, AlN, $CaSi_2$ and a new unidentifiable phase, which is identical with a phase previously described[30]. Micro and macro-indentation hardness measurements suggested the existence of at least six different phases. The residual silicon nitride showed considerable hardness ($HV_{0.05}$ = 25 GPa), and bend strengths on the material could exceed 300 MPa. Oxidation at 500° and 1000°C gave no significant weight gain. Although it had been infiltrated at 1200°C and 20 Pa, at 1250°C oxidation kinetics, initially parabolic, gave weight gains of 0.8% in 25 h. A green-grey adherent film was seen to be developed by 28 h. Resistivity measurements on RBSN infiltrated at 1370°C and 20 Pa by the Si(Ca-Fe-Al) alloy gave a value of 0.06_5 m, showing the formation of an interconnected metallic matrix. The low resistivity is similar to that of a silicon resistor (~ 0.1 Ωm) usually containing 36.5 w/o Si with 1.5 w/o Fe embedded in a porcelain-like body[29]. Silicon wetted and infiltrated RBSN at 1670°C in argon.

5. SUMMARY

Reinforcement of RBSN of a range of densities by infiltration with liquid metals is possible. Satisfactory results have been obtained with Si and Al (alloys containing some Ca). Alloying Si with Ca enhances the infilling potential by a factor of 10, compared with Al.

6. REFERENCES

1. R. Kieffer, G. Jangg, P. Ettmayer. "Sondermetalle, Metallurgie/Herstellung/Anwendung" Springer-Verlag, Wien - New York (1971).
2. G. Banik, W. Wruss, A. Vendl. Sprechsaal 113 (1980) p.261-263.
3. A. Schmidt. Ph.D. Thesis, University of Technology, A-1040 Vienna (1968).
4. C.W. Forrest, P. Kennedy, J.V. Shennan. TRG Report 2053 (s), U.D.C. No. 661.665, U.K. Atomic Energy Authority, Reactor Fuel Element Lab., Springfields (1970).
5. F.L. Riley. In Proc. 4th Internat. Mtg. "Modern Ceramics Technologies", Saint-Vincent, Italy, May 1979, "Energy and Ceramics", P. Vincenzini (Ed.), Elsevier, Amsterdam-Oxford-New York (1980) p.550-568.
6. W. Wruss. Priv. communication, University of Technology, A-1060 Vienna (1970).
7. J. Collins, R. Gerby. Metals (1955) p.612-615.
8. T. Frangos. Mat. in Design. Eng. (1958) p.115-117.
9. P. Popper, S. Ruddlesden. Trans. Brit. Ceram. Soc. 60 (1961) p.603-626.
10. K. Müller. "Wissensch. Grundl. D. Modernen Technik", Reihe A, Bd. 7, Akademieverlag Berlin (1963) p.78-84.
11. G. Yasinskaya. Ogneupory (1965) No.2, p.20-23.
12. K. Müller, H. Rebsch. Silikattechnik 17 (1966) p.279-282.
13. L. Parr, W. May. Proc. Brit. Ceram. Soc. 7 (1967) p.81-98.
14. H. Feld. Mitt. d. Annawerke Nr. 97 (1968/69) p.1-5, D-8633 Oeslau.
15. H. Feld, E. Gugel, H. Nitzsche. see (14) but (1969) folder 7.
16. Anonymous. Degussa catalogue W 110, A 110 and A 510 "Degussit Sn 34" (1969).
17. K. Hübner, F. Saure. Chemiker Zeitung 95 (1971) p.931-934.
18. R. Kieffer, G. Jangg, P. Ettmayer. see (1) p.437, 447.
19. H. Feld. Ph.D. Thesis, University of Technology, A-1040 Vienna (1971).
20. R. Felten. Sprechsaal 107 (1974) p.92-110.

21. F. Fickel. Chemie Technik 4 (1975) p.317-320.
22. Anonymous. Annawerk leaflet Ceranox NR (1979).
23. Anonymous. Annawerk leaflet Ceranox NH (1979).
24. M. Ginsberg, R. Krock. US-Patent No.3,399.076 filed April 1965, patented Aug. 1968.
25. M. Ginsberg. US-Patent No.3,428.450, filed May 1965, patented Feb. 1969.
26. Anonymous. Ceramic Industry Magazine (1973) p.7.
27. G. Leimer, E. Gugel. Zeitschrift für Metallkunde 66 (1975) p.570-576.
28. D.R. Messier. in Proc. 4th Annual Conf. on "Composites and Advanced Materials", Jan. 1980, Florida, J.D. Buckley (Conf. Dir.), Ceramic Engineering and Science Proceedings 1 (1980) p.624-633.
29. H. Heuschkel, K. Muche. "ABC-Keramik" VEB D. Verlag f. Grundstoffindustrie, Leipzig (1974) p.182.
30. G. Weissmann. Ph.D. Thesis, University of Technology, A-1040 Vienna (1971).

DISCUSSION

Greskovich: What was the thickest specimen you have infiltrated? Is there a thickness limit?

Schmidt: About 20 mm for 30% porosity RBSN. If the open porosity is high enough there should be no practical thickness limit. It is basically a very simple process.

Mangels: Did you see any reaction between the pure metals or alloys and the Si_3N_4?

Schmidt: Yes. Fe Co Ni all react at high temperature. The Ca-activated Si and Al-alloys also reacted to form new phases of different hardness and colour.

BRAZING OF SILICON NITRIDE

Johann E. Siebels

Volkswagenwerk AG Wolfsburg
Research and Development

1. INTRODUCTION

The application of ceramic high performance materials like silicon nitride requires an economic possibility of joining with surrounding parts and components.

In the case of high temperature gas turbines those joints have to survive high thermal loads and, in particular in rotating components, high dynamic loads.

This paper will concentrate on joining tasks for turbine rotors. As methods some liquid phase brazings will be described. Brazing, if successful, would be an economic process even for more complicated joining areas and could allow moderate tolerance demands.

Figure 1 schematicly shows possible respectively necessary joinings for gas turbine rotors. On one hand there are connections between ceramics with each other and on the other hand of ceramics with metallic parts.

Riley, F.L. (ed.) Progress in Nitrogen Ceramics

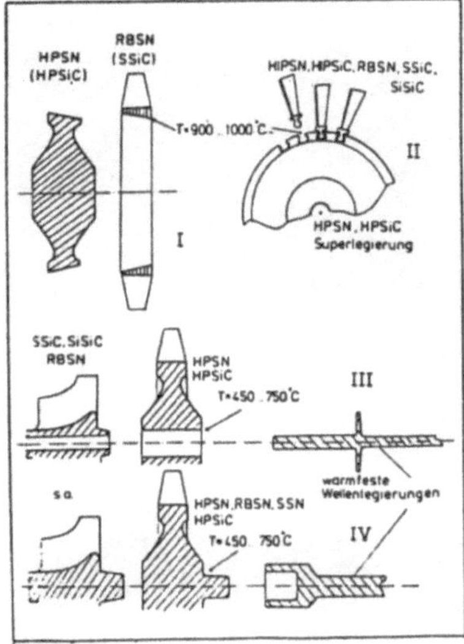

Figure 1: Joining tasks for turbine rotors:
I. all ceramic rotor
II. Hub-blade connections
III./IV. metallic shaft-ceramic rotor attachments

2. CERAMIC - CERAMIC BRAZINGS

Suitable brazing materials for silicon nitrides are eutectic alloys of elements with high affinity to oxygen (i.e. Ti, Cu, Be, Zr, Al, a.o.), so called "active brazing mixtures". Those elements allow numerous combinations of multi-component-systems for brazings of reaction bonded and hot pressed or other high density silicon nitrides. Additionally also the effects of the basic elements of the ceramics and impurities have to be taken into account for the binding reactions.

During the brazing process only low pressures between the parts to be connected are necessary. Depending on the wetting behaviour, the geometry of contact areas and the desired brazing layer thickness this pressure is in the order of $p = 1 \dots 5$ N/cm².

Exceptionally accuratly the optimized temperature schedule for Si_3N_4-brazings has to be found out. A standard procedure cannot be specified.

Figures 2 ... 4 show examples for active brazings of HPSN to RBSN with a TiCuBe-alloy (49 % Ti, 49 % Cu, 2 % Be). This alloy was the starting composition for a series of other brazing alloys. Dependent on the choise of preheating steps different types of brazing layers can be generated.

Figure 2 shows three examples for homogeneous brazing zones. In figure 3 TiCuBe-brazings with intermediate layers of different phase composition are presented.

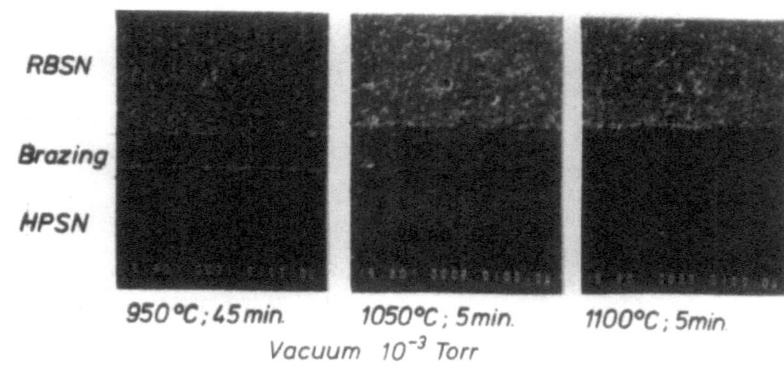

Figure 2: TiCuBe active brazing forming homogeneous brazing layers

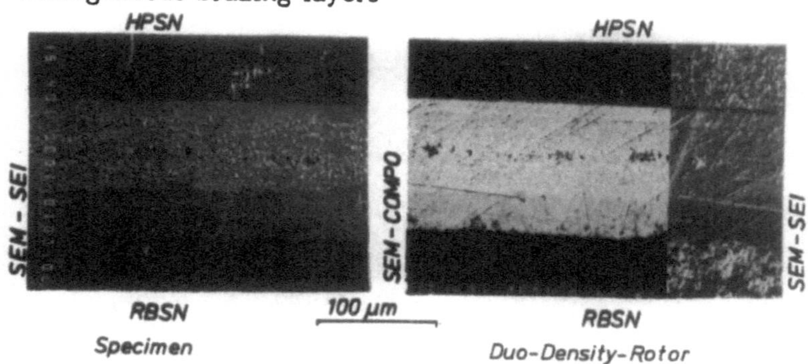

Figure 3: TiCuBe active brazing forming intermediate layers

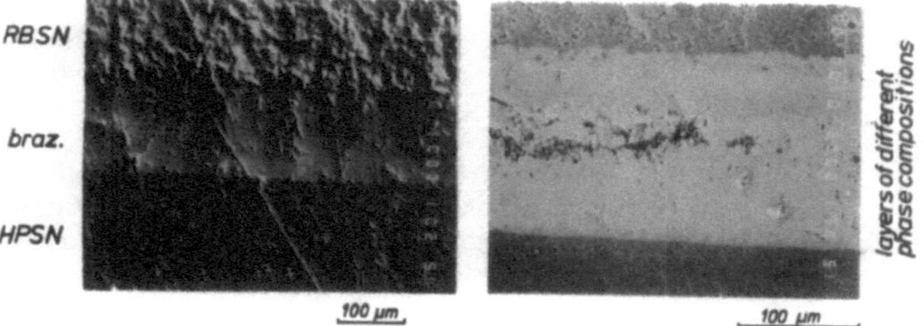

Figure 4: HPSN-TiCuBe+Zr-RBSN brazings

Figure 4 shows two examples of TiCuBe-brazings with Zr-additives. In dependence of the purity of the zirconium also different brazing layers can form under identical brazing conditions.

An interesting example of brazing HPSN to RBSN with pure aluminium shows figure 5. The final connecting layer is silicon.

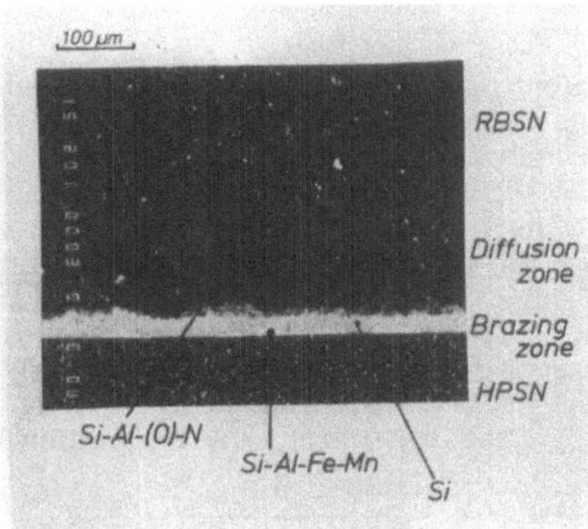

Figure 5: Vacuum brazing of HPSN/RBSN by Al (final processing: 13oo° C, 1o min.)

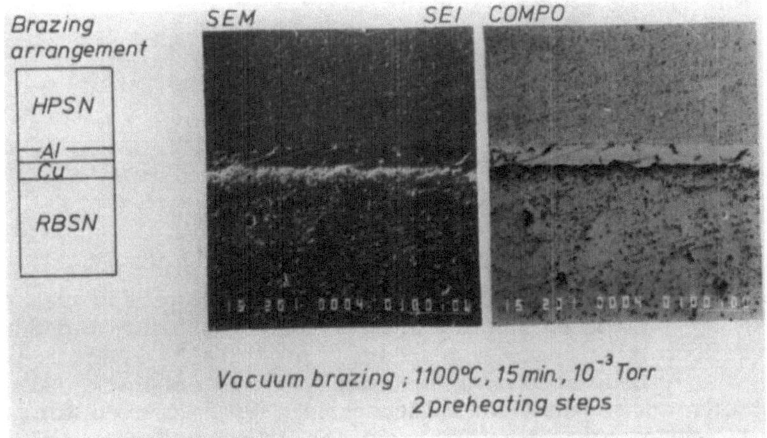

Figure 6: HPSN-Al-Cu-RBSN vacuum brazing

A variant of this brazing avoiding diffusion of Al into the RBSN is given in figure 6.

All brazings shown were performed in vacuum ($p \leq 10^{-3}$ Torr). In some cases similar results can be obtained by processing in nitrogen or argon atmospheres.

2.1 Remarks on the brazing of Si_3N_4

Besides the mentioned problem of finding the appropriate temperature schedule for proper brazings there are some factors leading to uncompleted or poor connections when not being taken into account.

It is clear, that the processing parameters have to be chosen respecting the vapor pressures and the evaporation rates of the used brazing elements. But additionally some of those elements form nitrides with dissociation temperatures far below the final brazing temperature (see table 1). This can lead to vapor films at the boundary of the brazing zone, in the case of HPSN - RBSN preferably at the HPSN side, and prevent a connection. In any case this effect causes porous layers of poor strength.

Element	Nitride	N-cont. %	Dissociation temperature
Al	AlN	34.1	1870 °C (1050 °C)
Ti	TiN	22.6	1500 °C (melt. 2950)
Zr	ZrN	13.3	melting 2930 °C
W	W_2N	3.6	800 °C
Cr	CrN	21.2	900 °C
Fe	Fe_2N	11.2	560 °C
	Fe_4N	5.9	650 °C
Si	Si_3N_4	39.9	1900 °C

Table 1: Nitrides of some brazing elements and their dissociation temperatures

Also impurities of the ceramics (W, Fe) can take part in such reactions when the process parameters allow excessive diffusion into the brazing zone (figure 7).

In static component tests some brazings have survived thermal loads (long time exposure to combustion gas, thermal shocks) exceeding the required. Rotor spin tests led to failures at maximum speeds of around 50000 1/min. The reason of those relatively early failures is the difficulty to transfer the good brazing results of specimens to the large joining surfaces of rotor hubs and blade rings. Voids of about 3 ... 5 mm² already cause early fracture in spin tests.

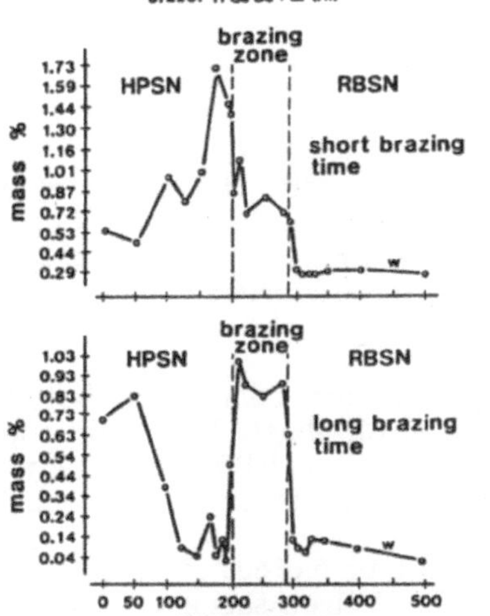

Figure 7: Diffusion of tungsten impurity

Another factor that has to be taken into account in the design are the residual stresses caused by the different thermo-mechanical properties of the joined materials. Figure 8 for example shows the residual tensile stresses calculated for a standard HPSN-RBSN specimen (bonding area 1o x 15 mm², thicknesses: RBSN 2 mm, HPSN 6 mm) with Ti-brazing.

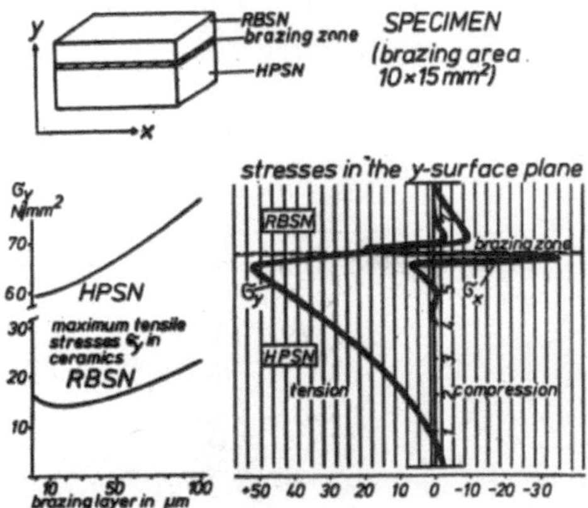

Figure 8: Residual stresses in a RBSN-Ti-HPSN-brazing

3. METAL - CERAMIC BRAZINGS

Plane brazings of metals (heat resistant steels, superalloys) to silicon nitrides need transition layers compensating the different thermal expansions of the materials. This method proved to be very difficult and risky. Other methods using the different expansions of ceramics and metals seemed to be more promising.

Figure 9 shows the principle of the connection between a ceramic rotor with a shaft pin and a metal shaft. This solution, however, was applicable for HPSN rotors only. For RBSN even with an optimized design of the joining area the residual stresses were too high.

Also the attempt to minimize the dangerous stresses by spring elements (figure 1o) did not result in the desired success. The residual stresses

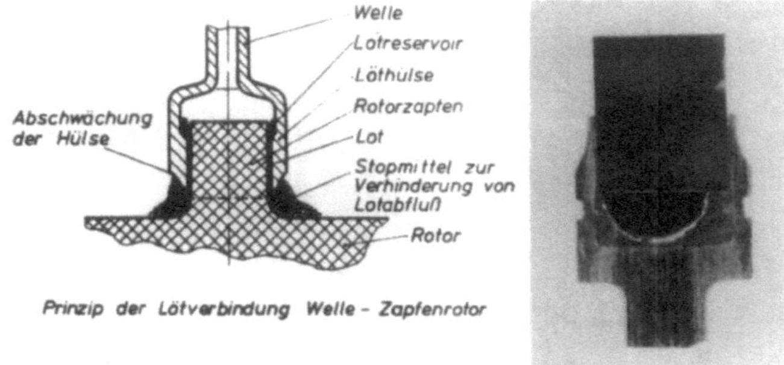

Figure 9: Brazing of turbine rotors with shaft pins

Figure 1o: Brazing with spring elements

were still too high for RBSN to allow additional operational stresses.

More advantageous are tie bolt brazings for hubs with bores. Figure 11 shows the principle of such brazings for axial rotor hubs. This design allows to lead residual tensile stresses into hub areas not being highly stressed by operational stresses.

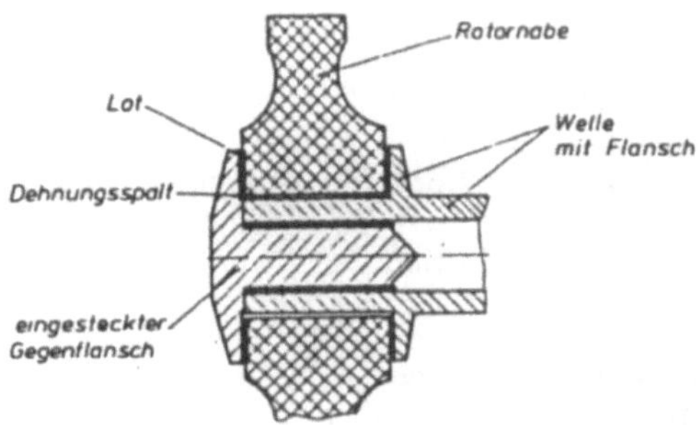

Figure 11: Shaft - hub brazing of axial turbine rotors with hub bores

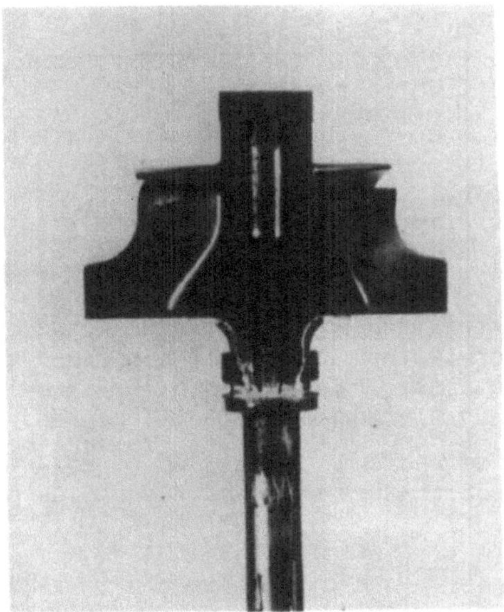

Figure 12: Tie bolt brazing of a ceramic radial turbine rotor (Turbocharger)

Figure 12 shows an optimized design for radial rotors where brazing of metal and ceramic is no more necessary.

For the shaft pin brazing of HPSN and the tie bolt brazings it could be demonstrated that these connections were able to transmit the desired torques at operation temperatures.

Engine tests are planned for the near future.

4. CONCLUSIONS

The development of brazing methods led to some promising possibilities to join silicon nitride materials with each other. A problem still is the transferability of results obtained at small specimens to larger components.

The application of brazing to connect metallic and ceramic components also is possible. Some work, however, has to be done to improve the design of the joining areas.

A critical point for the application of high temperature brazings is the thermal instability (strength degradation after annealing) of silicon nitride materials. Thus further progress in joining techniques by brazing will depend very much on the improvement of - in particular - reaction bonded silicon nitride properties.

DISCUSSION

Mangels: How did you minimize the stresses present in the bore area of your radial turbine rotor?

Siebels: The tensile operational stresses in the bore are partially compensated by the compressive stresses from the die bolt. For axial rotors the operational tensile stresses are compensated by up to 30-50% at operational speed and temperature, and induced tensile stress can be led into hub areas being stress free in operation.

Gugel: Is the prestressing by your special kind of mounting of the radial turbine rotor (with a hole) to the shaft due to the high application temperature? The difference in thermal expansion of the ceramic rotor and the metal shaft will decrease the pre-stress?

Siebels: The desired residual stress can be adjusted by the design of the free length of the tie bolt and the choice of material combinations.

REACTIONS OF BORON NITRIDE, AS A POWDER BED AND CRUCIBLE MATERIAL FOR SINTERING AND HOT PRESSING OPERATIONS

DISCUSSION:

Tien: The use of boron nitride in contact with hot silicon nitride specimens is open to the objection that the boron nitride may react to affect the properties of the silicon nitride. However, the effect has not been demonstrated to be severe. At present BN seems to be the best material available. B_2O_3 could be removed from the BN if this is essential.

Clarke: At the high temperatures involved (>1800°C) and the long residence times, I would predict that B will diffuse from a BN powder bed into the ceramic because of the chemical potential gradient of B between the bed and the compact. This may have profound effects on the high temperature properties of the compact. The evidence that B can enter the intergranular phase comes from current work on the reaction between Si_3N_4, SiO_2 and BN; at 1850°C densification and a sluggish reaction occurs in which the $\alpha \rightarrow \beta$ Si_3N_4 transformation partially occurs, and the microstructure consists of partially reacted BN in a boro-oxynitride glass. It is worth noting that (i) commercial BN can contain appreciable B_2O_3 and (ii) that the B_2O_3-SiO_2 eutectic occurs at very low temperatures (451°C).

Knoch: Commercial BN can contain up to 10% B_2O_3; but we suppose that the small contact area will limit the effect of B_2O_3 on the specimens being sintered. It is also possible that B_2O_3 is lost by volatilisation before the appearance of liquid.

Hampshire: We have prepared oxynitride glasses in BN crucibles; these liquids do not wet the crucibles, and at 1700°C no BN is dissolved in the glass.

Jack: We have tried to introduce B into oxynitride glasses and have not been able to get much in even at 1700°C. There is no doubt that N goes into borosilicate glasses but the reaction of BN with Si_3N_4 and oxynitride glasses is another matter. It is very slow and in the times required for densification the amount of B going in is probably very small.

Weiss: We dissolve Si_3N_4 in lithium and sodium borate glasses at 1100°C (proportion 1:20) in the preparation of X.R.F. analysis samples.

Jack: BN would not be as reactive as a borate glass.

Clarke: At very high temperatures BN will react with SiO_2, and in many hours of heating at 1850°C the kinetics will not be zero. This is a problem to be watched.

Section H

MECHANICAL AND PHYSICAL PROPERTIES

HIGH TEMPERATURE DEFORMATION AND FRACTURE PHENOMENA OF POLYPHASE Si_3N_4 MATERIALS

F. F. Lange

Structural Ceramics Group
Rockwell International Science Center
Thousand Oaks, CA 91360

1 INTRODUCTION

Three factors are currently hindering progress in adapting structural ceramics to high performance engineering practice: (1) variable strength, (2) lack of in-depth design practice, and (3) degradation of high temperature mechanical properties. The problem of strength variability is directly linked to fabrication and machining, both of which introduce the wide distribution of flaws and cracks responsible for failure. Directions to overcome this most important problem have been identified and within the next five years, one might expect significant progress on the commercial scale (pending market applications). In-depth design practice requires experience which can only be gained, as it has in the last ten years, mainly by directed trial and error, i.e., programs do need failures from which one can learn. Degradation of material properties at high temperatures, the subject of this review, is not a serious problem for certain structural ceramics, e.g., some forms of SiC. But for dense Si_3N_4 ceramics, which can better minimize thermal stresses due to lower thermal expansion and elastic modulus, high temperature degradation has limited the material's appeal to those who were led to expect too much without further materials understanding and development.

During the last 5 years, the progress in understanding relations between fabrication, composition, microstructure, and properties has led to significant material improvements at high temperatures. Today, one knows how to make a bad material good, how to formulate the chemistry to make improved materials, and how to pressureless sinter some of these compositions to bring them into the realm of engineering economics. The objective here is to

Riley, F.L. (ed.) Progress in Nitrogen Ceramics

convey this current understanding and to demonstrate material improvement.

2 POLYPHASE NATURE OF DENSE Si_3N_4

Attempts to densify pure Si_3N_4 powder, which decomposes in 1 atmosphere of N_2 at ~ 1850 C, have not been successful. Densification requires additives. The role of the additive is to react with Si_3N_4 and impurities to produce a liquid at high temperatures which allows mass transport through solution-reprecipitation to consolidate the solid Si_3N_4 particles in equilibrium with the liquid (see Ref. 1 for a general review of this subject). Oxygen, in the form of either SiO_2 or Si_2N_2O, is by far the largest impurity in "purer" Si_3N_4 powders. Thus, the general reaction may be expressed as

$$Si_3N_4 + SiO_2 + \text{additive (metal oxide)} \xrightarrow{\text{(Temp)}} Si_3N_4\ \text{(solid)} + \text{liquid} \xrightarrow{\text{cooling}} \text{Dense } Si_3N_4 + \text{Second Phases} \quad .$$

On cooling, the liquid solidifies; Si_3N_4 and other crystalline phases may (and do in some systems) partition from the liquid. But since the liquid is largely siliceous, solidification invariably results in a residual, continuous glassy phase. Although the glassy phase is chiefly responsible for high temperature mechanical property degradation, some secondary crystalline phases also cause severe problems at moderate temperatures.

Neglecting the glassy phase for a moment, examination of phase equilibria can quickly illustrate the polyphase nature of dense Si_3N_4. Figure 1 illustrates two subsolidus phase diagrams in which MgO and Y_2O_3 were used as densification additives, respectively. For compositions prepared in either of these two systems, the resulting material would either contain two phase, i.e., compositions reaching equilibrium on a tie line, or three phases, i.e., composition reaching equilibrium in one of the compatibility triangles containing Si_3N_4 as an end-member. When impurities are included, phase equilibrium must be expanded to compositional space, and, in theory, dense Si_3N_4 could contain more than three equilibrium phases. Thus, the number and type of equilibrium secondary phases depends on the amount of each of the starting constituents (Si_3N_4, SiO_2, additive(s), impurities) and their phase relations. When one fabricates material close to the Si_3N_4 corner of any of these systems, small changes in any of the other constituents can dramatically change the secondary phase. For example, small changes in the oxygen content (or SiO_2 content) can

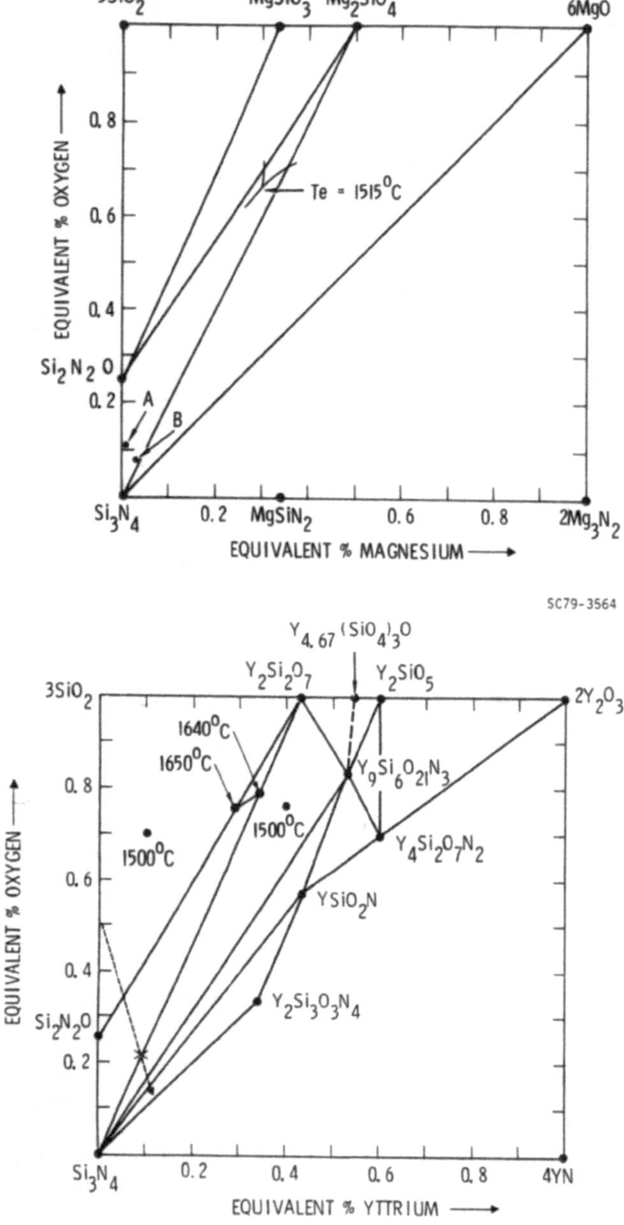

Fig. 1 (a) The Si-Mg-O-N system (2) and (b) the Si-Y-O-N system (3-6).

shift the end-point composition from one compatibility triangle to another. Thus, control of the oxygen content in the starting powder and during fabrication (control of volatilization reactions involving SiO) is crucial for controlling the type and amount of the secondary phases.

Theoretically, single phase Si_3N_4 can only be achieved with a densification aid in two systems, viz. the Si-Al-O-N (7) and Si-Be-O-N (8) systems, in which Si_3N_4 forms an extended solid-solutions of the types $Si_{3-x}Al_xN_{4-x}$ ($x < 2.1$) and $Si_{3-x}Be_xO_{2x}N_{4-2x}$ ($x < 0.5$), respectively.

Although equilibrium may be obtained during the high temperature densification process (indicative of the complete $\alpha \rightarrow \beta$ transformation, i.e., all material taking part in the reaction), equilibrium is not achieved during cooling, as exemplified by the residual glassy phase observed in all polyphase Si_3N_4 materials to date. Despite the non-equilibrium nature of these materials, the equilibrium phase diagram can be used to estimate the composition and volume content of the glassy phase. These estimates require knowledge of the system's eutectics (1).

Let us use the Si_3N_4-Mg_2SiO_4-Si_2N_2O compatibility triangle of the Si-Mg-O-N system to illustrate this estimation. Compositions fabricated in this compatibility element close to the Si_3N_4 end-member (mole % Si_3N_4 > 0.85) are observed to contain two crystalline phases, β-Si_3N_4 and Si_2N_2O; the Si_2N_2O/Si_3N_4 ratio increases from ~ 0 for composition along the Si_3N_4-Mg_2SiO_4 tie line to a maximum value as one approaches the Si_3N_4-Si_2N_2O tie line. Electron microscopy reveals a continuous glass phase (9,10) but fails to reveal a crystalline phase containing Mg. Auger spectroscopy (11,12), high resolution, non-dispersive x-ray energy analysis (13) and electron energy loss spectroscopy (13) revealed that the glass contains Mg, Si, (impurities such as Ca, Na, etc.), O, and some nitrogen. That is, one estimate of the glass composition (excluding impurities) is to suggest it is the Mg_2SiO_4 which has failed to crystallize. The volume fraction could be determined by the lever rule, developed to analyze phase diagrams. A second approach would be to assume that the glass is equivalent to the eutectic composition, i.e., the last bit of liquid to solidify. Again knowing the eutectic, one can estimate the volume content of the glassy phase. The volume fraction of the glassy phase increases as the bulk composition is chosen closer to the eutectic composition. If we use this approach, we would expect that on fabricating a series of materials, each containing the same mole % of Si_3N_4, but different MgO/SiO_2 ratios, the material in this series containing the largest amount of glass would be expected to lie on the Si_3N_4-eutectic join. For the compatibility triangle under consideration, this join occurs for the MgO/SiO_2 molar ratio of ~ 1.5 (14).

Since the viscosity of glass is proportional to its "equilibrium melting temperature", this approach suggests that the temperature where degradation begins can be increased by choosing systems with higher eutectic temperatures and by choosing compositions within this system that are far from the eutectic composition. These two criteria are in direct competition with fabrication requirements. That is, it appears that the densification kinetics can be increased by lowering the eutectic temperature and/or by increasing the volume fraction of the liquid.

Considerations concerning the dihedral angle between the liquid (or viscous glass) and the crystalline phases and its volume fraction strongly suggest that although some glass will be located at two grain junctions (grain boundaries) the majority will be found at 3 and 4 grain junctions. Theoretical arguments (15) concerning the crystallization of these 3-grain junction "pipes" and 4-grain junction pockets of glass indicate that if the liquid is constrained from moving in and out of these pipes and pockets, the strain energy arising from the molar volume changes associated with crystallization can prevent crystallization. These analyses show that the larger the pipe or pocket, the easier the crystallization. Thus, some crystallization is expected as we move closer to the eutectic composition. That is, crystallization of Mg_2SiO_4 does occur when the bulk composition is shifted toward the eutectic, which of course allows us to experimentally determine the sub-solidus tie lines shown in Fig. 1.

Before moving on to examine the effect of the glassy phase on mechanical properties, let us first examine the effect of the unstable, secondary crystalline phase on moderate temperature degradation.

3 EFFECT OF UNSTABLE CRYSTALLINE PHASES

As first discovered by Lange et al (3), some quaternary nitrogen compounds readily oxidize at moderately low temperatures. Large molar volume changes accompany these oxidation reactions. The quaternary yttrium oxy-nitride compounds illustrated in Fig. 1 are four such phases unstable relative to moderate temperature oxidizing environments. Table 1 lists these and other unstable phases known to this author, their oxidation products, and the molar volume change accompanying the oxidation reaction.

Table 1. Molar Volume Change of Unstable Secondary Phases Compatible with Si_3N_4

Secondary Phase	Oxidation Product*	Volume Change, %
$Y_2Si_2O_3N_4$	$Y_2Si_2O_7 + SiO_3$	+30
$YSiO_2N$	$0.5Y_2Si_2O_7$	+12
$Y_9(SiO_{3.5}N_{0.5})_6$**	$1.5Y_{4.67}(SiO_4)_3O + 0.5Y_4Si_3O_{12}$	+5
$Y_5(SiO_4)_3N$**	$0.5Y_{4.67}(SiO_4)_3O + 0.75Y_2SiO_5$	+4
$Ce_5(SiO_4)_3N$	$5CeO_2 + 3SiO_2$	+8
$CeSiO_2N$	$CeO_2 + SiO_2$	+14
$Ce_2Si_2O_7$***	$2CeO_2 + 2SiO_2$	+7
$ZrO_{2-x}N_{4x}$ $(x = 0.2)$	ZrO_2 (monoclinic)	+5

*Oxidation product can be determined directly from equivalence diagrams (see section on "Oxidation resistance").
**Both formulas (indicated in the section on "Rare-earth systems") for yttrium N-apatite structures are shown.
***All Ce silicates are unstable in air.

Since oxidation of dense bodies is a surface reaction, the molar volume increase accompanying these surface reactions gives rise to surface compressive stresses. These surface compressive stresses can result in initial strengthening and, depending on factors discussed below, longer term degradation.

If the oxidation product of the unstable phase exhibits a gradient from the surface to the interior caused by diffusion, one can analyze the stress state in a similar manner to the way one analyzes thermal stresses caused by a thermal gradient. For the case of an infinite cylinder of radius b, the stress components in cylindrical coordinates are (16)

$$\left.\begin{aligned}
\sigma_r &= \left(\frac{\Delta V}{V}\right)\left(\frac{E}{1-\nu}\right)\left(\frac{1}{b^2}\int_0^b V_f r\,dr - \frac{1}{r^2}\int_0^r V_f r\,dr\right)\\
\sigma_\theta &= \left(\frac{\Delta V}{V}\right)\left(\frac{E}{1-\nu}\right)\left(\frac{1}{b^2}\int_0^b V_f r\,dr - \frac{1}{r^2}\int_0^r V_f r\,dr - V_f\right)\\
\sigma_z &= \left(\frac{\Delta V}{V}\right)\left(\frac{E}{1-\nu}\right)\left(\frac{2}{b^2}\int_0^b V_f r\,dr - V_f\right)
\end{aligned}\right\} \quad (1)$$

V_f is the volume fraction of the reaction products. Its value as a function of the radial coordinate r defines the concentration gradient of the reaction products developed by the diffusion of oxygen. This gradient is a function of the oxidation kinetics. $\Delta V/V$ is the molar volume change, and E, ν are the elastic properties of the polyphase material.

The moment oxidation proceeds, an infinitestimal surface layer will completely oxidize. The concentration of the reaction product in this layer is equal to and cannot exceed the volume fraction V_o of the unstable phase. Thus, the surface concentration of the reaction product is fixed by the volume fraction of the unstable phase.

At the surface (r = b), the stresses determined from Eq. (1) are

$$\sigma_r = 0$$

and

$$\sigma_\theta = \sigma_z = \left(\frac{\Delta V}{V}\right)\left(\frac{E}{1-\nu}\right)\left(\frac{2}{b^2}\int_0^b V_f r\,dr - V_o\right) \qquad (2)$$

The maximum compressive surface stress occurs the instant oxidation proceeds. After this instant, the magnitude of the stresses decreases, but at the same time the compressive stress extends to a greater depth. From a viewpoint that these biaxial compressive stresses could help prevent surface cracks from extending under a subsequently applied tensile stress, there would appear to be an optimum oxidation period that would place the entire length of the surface crack in compression. On the other hand, it should be noted that radial, subsurface tensile stresses also arise during oxidation. These tensile stresses could lead to surface spalling. The tensile stresses arising in the cylinder's center can be neglected. It can be seen that the magnitude of the surface stresses depends on the volume fraction of the unstable phase, the molar volume change, the oxidation kinetics, and the dimension of the body.

The development of these surface stresses and the phenomena associated with them have been characterized for materials in two different Si-Mg-O-N systems (16,17). Figure 2 illustrates the data from one material which was part of a series fabricated in the Si-Zr-O-N system (16) (Si_3N_4 + 25 volume % ZrO_2) containing the unstable Zr oxynitride which oxidizes to monoclinic ZrO_2 with an associated molar volume increase of ~ 5%. Figure 2a illustrates the amount of the Zr-oxynitride reacted to the monoclinic ZrO_2 on the surface (XRD results) as a function of the oxidation (air)

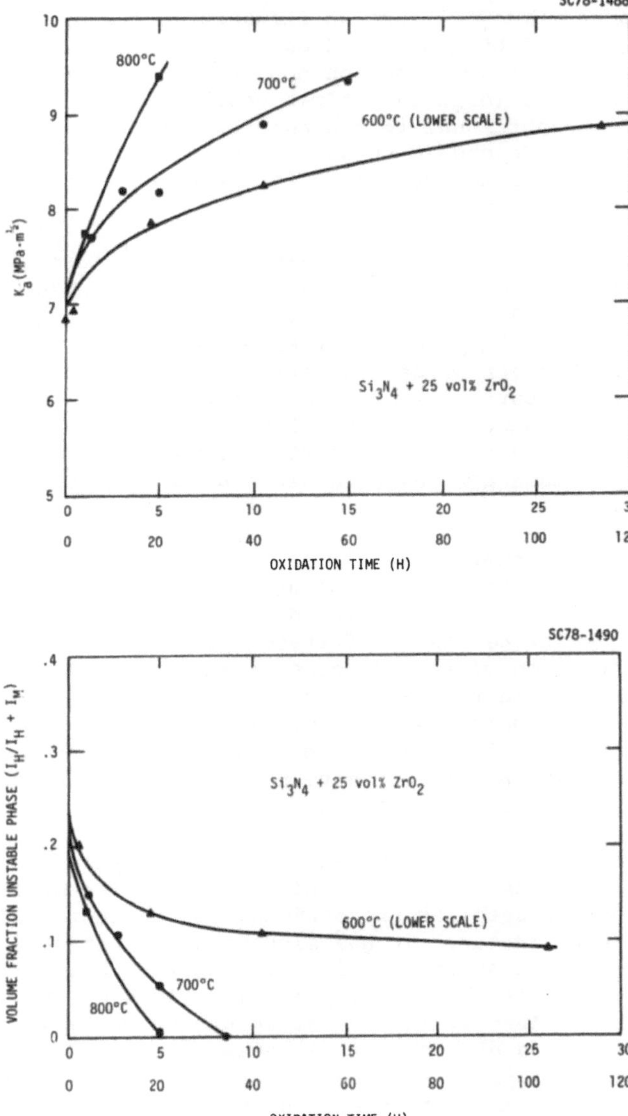

Fig. 2 (a) Volume fraction of Zr-oxynitride (10) oxidized on surface to monoclinic ZrO2, (b) K_a at surface vs oxidation treatment (16).

temperature and time. Figure 2b illustrates the apparent critical stress intensity factor measured on the surface by the indentation technique for the same oxidation conditions. The effective compressive stress (magnitude times depth) arising during oxidation is proportional to the apparent critical stress intensity factor. It should be noted that Si_3N_4, the major phase in this polyphase material does not, for practical purposes, oxidize in this temperature range.

As illustrated in Fig. 2, the effective compressive stress increases with the same kinetics as the unstable phase reacts to form monoclinic ZrO_2. After relatively short periods at 700°C and 800°C, surface spalling was observed which prevented further measurements. Longer oxidation periods at these temperatures caused the whole surface to detach itself from the specimen. Thus for these conditions, viz. volume fraction of the unstable phase and oxidation temperature, the surface compressive stresses were sufficient to cause general mechanical disintergration and break-away oxidation.

Data for the other materials in this series of Si_3N_4/Zr-oxynitrides showed that the magnitude of the effective compressive surface stress was directly related to the volume fraction of the initial Zr-oxynitride phase. Surface spalling did not occur when the volume fraction of the unstable phase was $< 10\%$. Prolonged oxidation causes the reaction product gradient to diminish, diminishing the compressive stress. Thus, depending on the magnitude of the molar volume change for the reaction under consideration, there exists a volume fraction below which the unstable phase will not cause degradation.

It is obvious that the surface compressive stresses can be useful in increasing strength. This has been demonstated (16,17); the increased strength was proportional to the increased apparent stress-intensity factor. It has also been demonstrated that these surface compressive stresses will increase the threshold velocity to induce surface cracking during particle impact (18). Despite these useful attributes, the reader should be reminded that when the volume fraction of the unstable phase exceeds a critical value, prolonged exposure to moderate temperatures in the reacting environment will cause catastropic degradation.

The degradation phenomena associated with an unstable phase disappears at higher temperatures, i.e., a critical temperature exists, above which degradation is not observed despite the oxidation reaction. The reason for this was not made clear until SEM observation with Si_3N_4 containing unstable Ce-apatite showed that the reaction products (viz. CeO_2 + SiO_2) were extruded to the surface to relieve some of the surface compressive stresses, as shown in Fig. 3 (17). For these materials, extrusion was observed for temperatures > 900°C. Above ~ 900°C, degradation was not observed, but below ~ 900°C, the oxidation reaction resulted in severe spalling.

Fig. 3 Extrusion of reaction products of unstable phase during oxidation (17).

This critical temperature can also be altered by impurities and additional additives which appear to govern the deformation characteristics of the reaction products. For example, material fabricated by Toshiba by hot-pressing Si_3N_4 with the combined additives of Y_2O_3 and Al_2O_3 was observed by this author (19) to contain a second crystalline phase, viz. $YSiO_2N$, one of the unstable phases in the Si-Y-O-N system. Experience of the author had shown that small amounts of $YSiO_2N_2$ (similar to that found in the Toshiba material) would cause degradation ≤ 1100°C. But degradation was not observed for the Toshiba material in oxidizing environments < 1100°C. Careful oxidation experiments were conducted and coupled with SEM observations. XRD analysis confirmed the oxidation of the $YSiO_2N$, but SEM observations showed that the extrusion process occurred to temperatures as low as 600°C. Below 600°C, the oxidation kinetics were negligible. Thus, the addition of Al_2O_3 appeared to lower the critical temperature to a range where one could neglect the degradation phenomena due to negligible oxidation kinetics.

4 DEFORMATION PHENOMENA

It is well recognized that the high temperature strength degradation of polyphase Si_3N_4 is caused by cavitational creep.

Therefore before we discuss strength, let us examine the deformation phenomena associated with these materials.

The bending, tensile, and compressive creep behavior of commercial and various experimental Si_3N_4 materials have been studied by numerous investigators (see Ref. 1). These results show that the creep resistance is ~ 7 times greater in compression than in tension, creep resistance depends on impurities and the Si-Mg-O-N system in which the material was fabricated, steady-state stress experiments range between 1.5 and 2, and cavitation appears to be a dominant microstructural feature produced by creep. The definitive experiments directly relating creep resistance to composition have only recently been published. These results have directed materials improvement. Two groups have been responsible for these new results: Warwick University (Lewis and students) and Rockwell Science Center (Lange and co-workers), which will be reviewed here.

The work at Rockwell (20) was primarily directed toward understanding the relations between creep resistance and composition. Compressive creep experiments were performed at 1400°C in air on four materials fabricated in the Si_3N_4-Si_2N_2O-Mg_2SiO_4 compatibility triangle of the Si-Mg-O-N system. These four materials were chosen because of their extreme compositional differences relative to one another and the eutectic composition. Using the hypothesis developed in Section 2, the volume fraction of the glass was estimated. Compositions A, being furthest from the eutectic, was expected to contain the least amount of glass, whereas composition B, being much closer to the eutectic was expected to contain much more glass. Experiments included precise (sink-float) density measurements before and after creep, and dilation measurements at different stress levels which were analyzed with the empirical steady-state relation, $\dot{\varepsilon} = A\sigma^n$.

Figure 4 illustrates that composition B is less creep resistant and has a stress exponent of ~ 2, whereas composition A has a stress exponent of ~ 1, suggesting a mechanism controlled by diffusion. Figure 5 shows that composition B exhibits extensive cavitation, whereas composition A does not cavitate. Figures 4 and 5 show that materials fabricated in the same system can not only have different creep resistances, but can also have different dominant creep mechanisms. These results suggest that two concurrent creep mechanisms(viz. cavitation and diffusion) can control the sustained deformation of polyphase Si_3N_4 and that the dominance of one mechanism over the other depends on the volume fraction of the glassy (or liquid) phase.

Electron microscopy reveals that most cavities are observed at 3 grain junctions, although as shown in Fig. 6, grains are

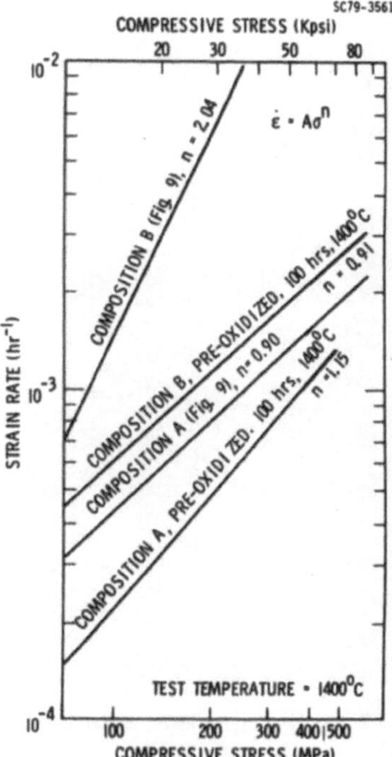

Fig. 4 Steady-state compressive creep vs stress for Si-Mg-O-N materials (20).

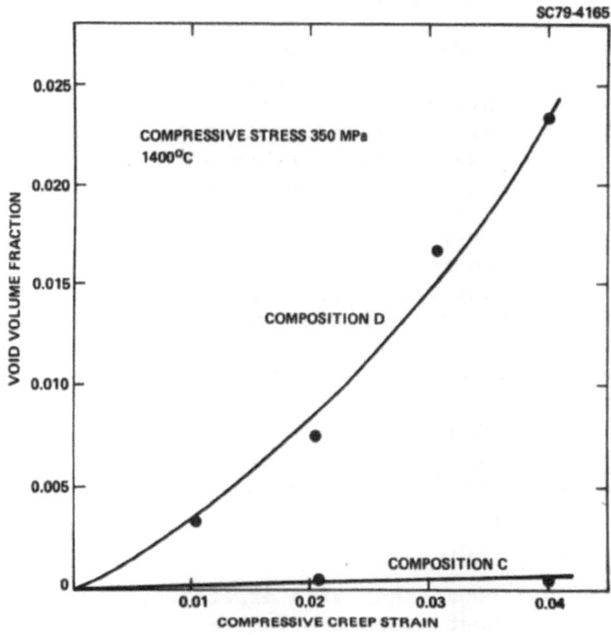

Fig. 5 Fractional void volumes vs creep strain.

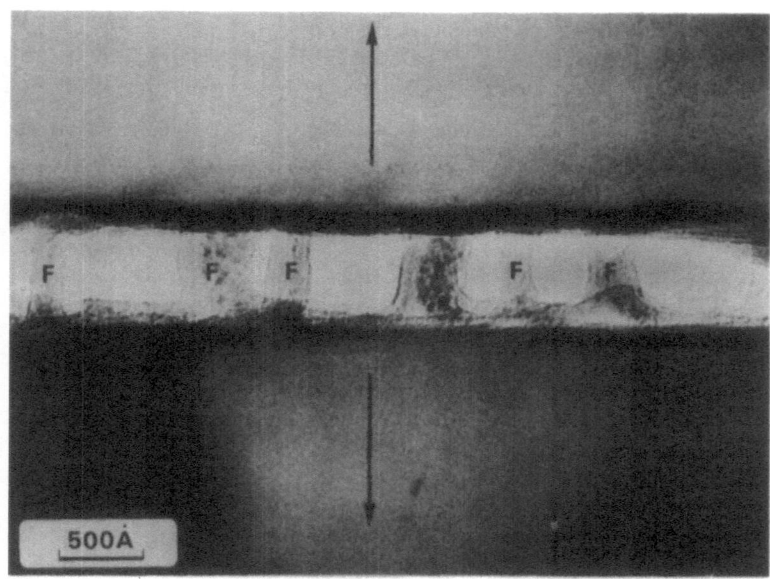

Fig. 6 Separation of Si_3N_4 grains showing frozen glassy fibrils (20).

observed to separate normal to their common boundary, leaving frozen glassy fibrils joining the two grains.

The Warwick group (21) has obtained similar results from two materials based on β'-Si_3N_4 ($Si_{3-x}Al_xO_xN_{4-x}$, $\chi \simeq 0.5$). The difference between the two materials was the additive used to enhance densification. In one, ~ 1 wt % MgO was used to produce a liquid for densification which subsequently disappeared via solution into the β'-Si_3N_4 structure. This material contained no detectable glass phase.* The other materials, a mixed MgO/Mn_3O_4 additive was used which resulted in a residual glass phase located at three grain junctions. Creep experiments showed that the β'-Si_3N_4 material without the glassy phase was much more creep resistant and exhibited a stress-exponent of 1. Material with the residual glass had a stress exponent > 1.5, suggestive of creep dominated by cavitation. The Warwick group also reported that the apparent activation energy for cavitational creep is ~ 500 KJ/mole, consistent with values obtained by others for a variety of Si_3N_4 exhibiting cavitational creep. The apparent activation energy for diffusional creep was reported as ~ 830 KJ/mole. More

---

*This claim is inconsistent with experimental creep data indicating that heat treatment was required to obtain improved creep resistance.

recently (22), the Rockwell group determined an apparent activation energy of ~ 660 KJ/mole for a material (composition B) known to cavitate and ~ 1080 KJ/mole for diffusional dominated creep (composition A).

One model of cavitational creep (23) relating creep strain (ε_{cav}), volume fraction of liquid (V_i), applied stress (σ), liquid viscosity (η), and time (t) results in the following expression:

$$\varepsilon_{cav} = \left(\left(1 - \frac{V_i^2 \sigma t}{\eta}\right)^{-1/4} - 1\right)\frac{V_i}{3} \qquad (3)$$

This expression shows that cavitational strain cannot be simply related to stress by a power law expression and that a steady-state creep rate should not be observed for cavitational creep. Thus, for materials that are dominated by both difusional and cavitational creep, power law analysis of the apparent steady-state condition as a function of stress would result in an apparent stress exponent > 1, consistent with experimental values ranging between 1.5 and ~ 3.*

One can also combine Eq. (3) with relations for diffusional creep models (20) (where the glass phase is the diffusional path) to show that cavitation creep will dominate at high volume fractions of glass, large stresses, lower viscosities, and long test periods. For example, a material can be dominated by diffusional creep at low stresses and by cavitational creep at high stresses. Figure 7 show results on a material fabricated on the Si_3N_4- $Y_2Si_2O_7$ tie line in the Si-Y-O-N system. At lower stresses, creep appears to be dominated by diffusion (n ~ 1), whereas at higher stresses, cavitation dominates (n ≈ 2.4). A similar behavior was found for material fabricated with Sc_2O_3 (24).

These observations just confirmed the suspicions of most investigators, i.e., one should get rid of the glassy phase. But, if the glass cannot be completely eliminated, due to fabrication requirements, these results do indicate that improvements could be obtained by compositional changes which (1) lower the volume fraction of the glassy phase (choosing compositions furthest from the eutectic), (2) increase the viscosity of the glass (chosing

---

*Higher values might be precluded from experiments unless the investigator attempts to define tertiary creep as a steady-state condition.

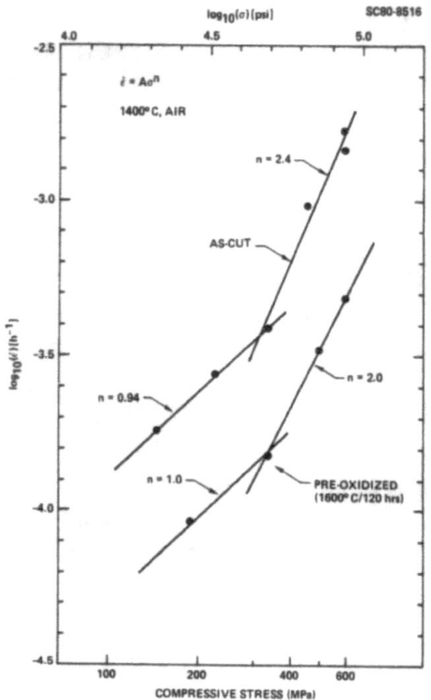

Fig. 7 Steady-state compressive creep vs stress for a Si-Y-O-N material (22).

systems with high eutetic temperatures), and (3) attempt to form β'-Si_3N_4 single phase solid solutions, as pioneered by Lumby (25).

The glass phase is also responsible for viscoelastic effects that account for primary creep and strain recovery after unloading (20,26).

The most significant observation, reported by the Rockwell (20) and Warwick (21) groups, was that oxidation significantly improved creep resistance. It has been demonstrated by both groups that a preoxidation treatment can not only lower the creep resistance, but can also change the dominant creep mechanism from cavitation to diffusion. This is illustrated in Fig. 4 for materials fabricated in the Si-Mg-O-N system and in Fig. 7 for a material in the Si-Y-O-N system.

As detailed elsewhere (27), the oxidation of polyphase Si_3N_4 results in compositional changes. The viscous glass phase is a continuous path for fast diffusion, which is one of its roles in the oxidation process. The second role is its part in the oxidation reaction itself. Since the glass phase is not compatible with SiO_2, it forms a diffusion couple with the SiO_2 produced on the surface through the oxidation of Si_3N_4. That is, the internal glass and the surface SiO_2 would like to equilibriate their composition. Cations in the glass (excluding Si) thus diffuse to the surface in an attempt to equilibriate composition. Oxygen and SiO_2 left behind by the outward moving cations react with Si_3N_4 to form Si_2N_2O. External oxygen diffusing inwards can also react with Si_3N_4 to form Si_2N_2O. Thus, both the outward diffusing cations and the inward diffusing oxygen drives the composition toward the Si_3N_4-Si_2N_2O tie line. Namely, the glass phase is a fugitive of the oxidation process. Since this process rarely achieves equilibrium during reasonable experimental time periods, a compositional gradient is produced from the surface to the center of the specimen. The steepness of this compositional gradient depends on the oxidation kinetics. Figure 8 illustrates the extent of this gradient on a cross section of a specimen fabricated in the Si-Mg-O-N system (close to composition B, Fig. 4) which was oxidized at 1400 C/300 hrs, sectioned and oxidized again for ~ 1/2 hr. Since the oxidation kinetics (or diffusional flux) are also proportional to the volume fraction of glass, material closer to the surface which is depleted of glass relative to the center will form a thinner oxide scale relative to the thicker (and whiter) scale formed on the inner portion of the cross-section. From the viewpoint of creep resistance, the outside will be stiffer than the inside and, for prolonged oxidation treatments, the inside can be made much stiffer than its initial state.

Fig. 8 Cross-sectional area of oxidation specimen (Si-Mg-O-N system) showing compositional gradient.

Thus, the post-fabrication oxidation treatment decreases the volume fraction of the glassy phase, which, in turn, increases the creep resistance and shifts the dominant mechanism from cavitational creep to diffusional creep. As will be seen in the next section, this post-fabrication oxidation treatment can lead to a significant strengthening of an initially unsatisfactory material.

5 FRACTURE PHENOMENA

Strength degradation of polyphase Si_3N_4 is attributed to the onset of subcritical crack growth, i.e., the slow growth of preexisting cracks at stress levels which do not cause catastrophic fracture. In the region of sub-critical crack growth, a material's strength depends on its stress history, e.g., stressing rate. Strength is lower in this regime because a low stress will extend a small crack to a larger crack which causes catastropic fractures. The susceptibility of a material to subcritical crack growth can be determined directly through crack velocity vs stress intensity measurements and indirectly through strength/stressing rate and/or creep-rupture measurements. The relation between crack velocity (V) and the stress intensity factor (K) can be expressed by the empirical power function

$$V = BK^n \quad , \qquad (4)$$

where the exponent n is a measure of the material's susceptibility to subcritical crack growth. For commercial polyphase Si_3N_4, $n < 10$ at temperatures > 1250 C, indicative of extensive subcritical crack growth (see Ref. 1).

Cavitational creep of material within the high stress field of a crack is responsible for subcritical crack growth in polyphase Si_3N_4 (28). Cavities are not only observed ahead of an extended crack stopped prior to catastrophic fracture, but also to a considerable distance into the material from the fracture surfaces (28), i.e., a cavitational zone surrounds the propagating crack. Subcritical crack extension occurs by the linking of cavities. Work done by the loading system to produce the cavitational zone is responsible for the increased critical stress intensity factor for these materials when they enter this regime of subcritical crack growth.

The relation between cavitational creep and strength degradation is thus obvious.

The flexural strength of a series of materials fabricated in the Si-Mg-O-N system containing 0.91 mole fraction of Si_3N_4 and different MgO/SiO_2 molar ratios is shown in Fig. 9 (29). Strengths at 1400 C are lower due to the presence of a continuous glass phase which results in slow crack growth due to cavitational

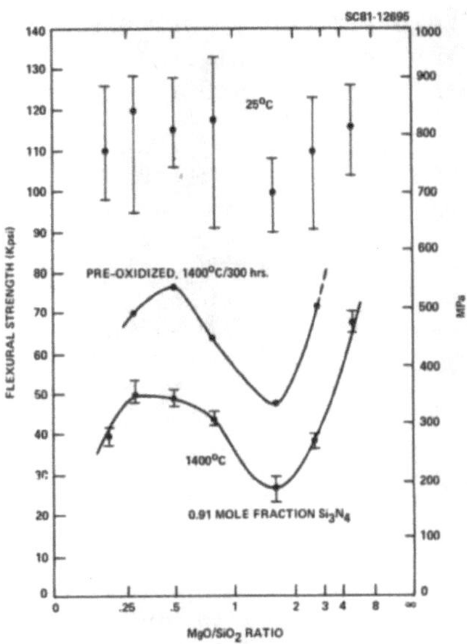

Fig. 9 Flexural strength vs MgO/SiO_2 molar ratio (29,32).

creep in the high stress field of the crack causing failure. Note the lowest strength corresponds to a composition close to the eutectic composition, i.e., containing the largest amount of glass. Within the relatively low stress regime used to examine creep behavior, this composition was observed to exhibit extensive cavitation; that is, compositions containing more glass are more susceptible to cavitational induced subcritical crack growth, and thus exhibit greater strength degradation.

Before discussing the strengthening of these materials by an oxidation treatment, it should be noted that oxidation produces detrimental surface pits on materials fabricated in the Si-Mg-O-N system. As first characterized by Singhal (30), these pits lower the strength at all temperatures. Studies strongly suggest that the pits are produced as a result of a localized oxidation reaction involving iron-rich inclusions (31). Materials fabricated in other systems (e.g., the Si-Y-O-N system) which do not react with iron during oxidation do not exhibit surface pitting (31).

Specimens from each of the materials shown in Fig. 9 were oxidized at 1400°C/300 hrs, surface ground to remove pits produced during oxidation, and then fractured at 1400°C (32). As discussed in Section 4, oxidation produces compositional changes and reduces the volume fraction of the continuous glassy phase. Despite the fact that most of the more improved material was ground off during pit removal, XRD results showed that the composition shifted toward the Si_3N_4-Si_2N_2O tie line, indicative of a reduction in the glass content. Results shown in Fig. 9 (plotted with their initial MgO/SiO_2 molar ratio) indicate significant strengthening.

Table 2 shows the strengthening that can be achieved by subjecting commercial NC132* to a post-fabrication oxidization treatment. After grinding to remove surface pits, strength at 1400°C can be more than doubled by a treatment at 1500°C/300 hrs. Also, the oxidation treatment does not significantly degrade the room temperature strength. Figure 10 shows that compositional change (i.e., Si_2N_2O/Si_3N_4 ratio) observed on the ground surfaces as a function of the oxidation treatment.

Table 2. Fluxural Strength Results for Commercial Si_3N_4

Oxidation Treatment		Test Temp	Number	Wt. Gain/ Area	Depth Removed	Average Strength	Weibull Parameters	
Temp (°C)	Time (hr)	(°C)		(mg/cm^2)	(cm)	(MPs)	m	σ_0 (MPa)
None		25	7	-	-	878	8	924
None		1400	8	-	-	282	(±4%)**	
1400	50	1400	3	1.61	0.015	254	(±4%)	
1400	100	1400	3	2.13	0.020	256	(±7%)	
1400	256	1400	4	-	-	288	(±8%)	
1400	332	25	8	2.39	0.020	982	15	1014
1400	332	1400	8	2.39	0.020	280	24	286
1400	548	1400	3	-*	0.020	376	(±6%)	
1500	206	25	7	-*	0.050	805	11	834
1500	206	1400	7	-*	0.050	482	17	496
1500	306	1400	5	-*	0.050	571	(±3%)	

*Scale flaked off during oxidation.
**Numbers in brackets denote percent difference of either maximum or minimum strength values from average.

Several comments should be made concerning these data. First, it should be recognized that flexural strength determinations can greatly over estimate strength when the material does not have a linear stress-strain response. This is the case for the specimens that were not subjected to an oxidation treatment, i.e., their true average strength is less than their apparent strength of 282 MPa reported in Table 2. Second, relatively short periods (50 hrs) of oxidation at 1400°C produced a nearly linear stress-strain response but a slightly lower apparent strength (254 MPa). Third, further oxidation at 1400°C did produce a linear stress-strain behavior and true strength values. Thus, despite the

*Norton Company

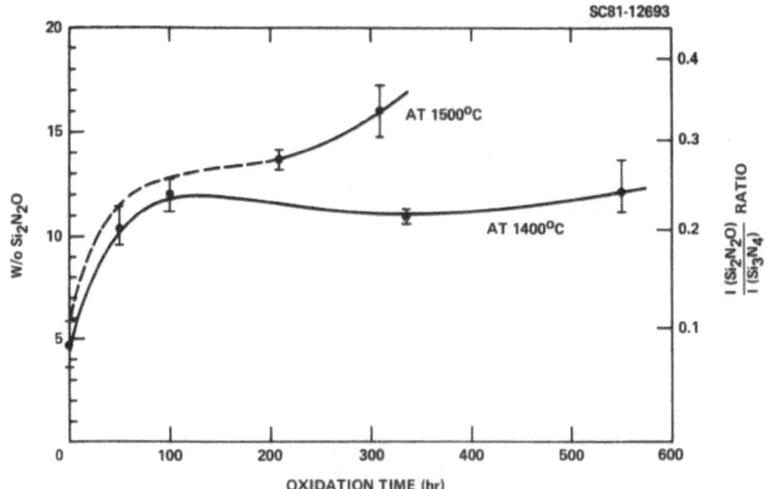

Fig. 10 Ground surface compositional change vs oxidation treatment (32).

values reported in Table 2, an oxidation treatment at 1400 C does increase strength. Fourth, the oxidation kinetics at 1400 C were not sufficient to produce the large strength increases sought; oxidation kinetics were thus increased by raising the temperature to 1500 C.

It should be noted that the room temperature strength was increased after the 1400 C/332 hr oxidation treatment. As discussed in Section 3, molar volume increases associated with oxidation reactions can place the surface in compression. Figure 11 reports the apparent critical stress intensity factor of the surface as a function of the oxidation period at 1400 C. As shown, the large molar volume change associated with the oxidation of Si_3N_4 to Si_2N_2O can lead to modest compressive surface stresses despite the apparent relief of most strain by deformation. Uneven oxidation should result in shape change.

The Warwick group (33) have taken a different approach to relate strength to the volume fraction of glass phase by studying crack growth kinetics at high temperatures. Their crack growth kinetics experiments were preformed on the same two β′-Si_3N_4 materials that they had studied under flexural creep, i.e., one containing residual glass (labled β′-B) which exhibited cavitational creep, and one where "no glass was detected" (labled β′-C) which exhibited diffusional creep. As expected, their susceptibility to subcritical crack growth was quite different; as shown in Fig. 12, the stress intensity exponent (see Eq. (4)), n = 7 for material β′-B and n = 13 for material β′-C; that is, the material less susceptible to subcritical crack growth contains less glass.

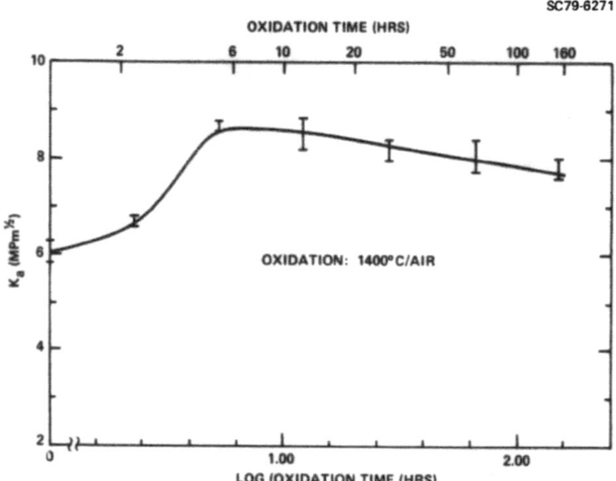

Fig. 11 K_a on surface vs oxidation period (32).

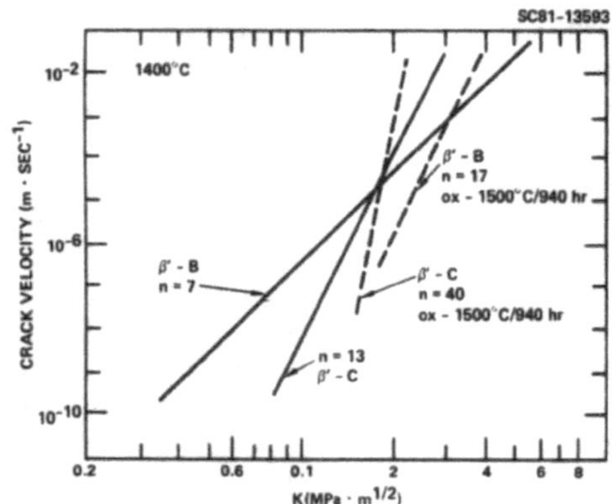

Fig. 12 K vs crack velocity results for two β′-Si_3N_4 materials (33).

The Warwick group, (33) also recognizing the effect of oxidation on creep resistance (see Section 4), pre-oxidized their fracture mechanics specimens at 1500°C for 940 hrs. As shown in Fig. 12, the susceptibility of each material to subcritical crack growth decreased, viz. n for β′-B increased from 7 to 17, and, for β′-C, increased from 13 to 40. A value of n = 40 indicates that subcritical crack growth is difficult to detect, suggesting that if other strength controlling factors do not change upon oxidation (viz. K_c and crack size distribution), one would expect that the oxidized β′-C material would not exhibit any strength degradation from room temperature to 1400°C.

6 CONCLUDING REMARKS

Qualitative relations between fabrication, phase equilibria, microstructure, and properties have been used to direct materials improvement. These directions have resulted in significant material improvements. Unstable crystalline phase should be avoided by selecting compositions in phase areas that avoid the unwanted phase. It is obvious that residual glass should be avoided. Lacking such control, composition selection in phase systems with high eutectic temperatures will increase the temperature were degradation begins. This has been demonstrated for compositions for the Si_3N_4-Si_2N_2O- $Y_2Si_2O_7$ compatibility triangle of the Si-Y-O-N system and compositions in the Si_3N_4-Si_2N_2O-$Sc_2Si_2O_7$ area of the Si-Sc-O-N system. Selection of compositions furthest from the eutectic composition will minimize the glass phase content. Impurities that lower the eutectic temperatures and thus increase the volume fraction of glass should be eliminated. Fabrication of a single phase β'-Si_3N_4 solid solution will of course eliminate both unstable phase and residual glass.

In practice, following the rules listed above leads to a material that is difficult to fabricate and impossible to pressureless-sinter with today's tehnology. That is, residual glass phases are pactically unavoidable. Here, the post-fabrication oxidation treatment can be used to significantly improve a poor material. It has been demonstrated that the glass phase can be a fugitive of the oxidation process. This treatment not only increases the materials resistance to creep and sub-critical crack growth (and thus, strength degradation), but also increases its resistance to oxidation. Thus, compositions that can easily be sintered can be improved by a post-fabrication oxidation treatment. Besides decreasing the glass content, this treatment also results in predictable phase changes and can thus be used to eliminate the unstable phases, viz. through oxidation treatments above the critical temperature where surface compressive stresses are relieved by deformation.

These improvements should be reduced to commercial practice.

ACKNOWLEDGEMENTS

This work was supported by the Air Force Office of Scientific Research Contract No. F4962-77-C-0072.

REFERENCES

1. F.F. Lange, Int. Met. Rev. No. 1, Rev. 247, pp. 1-20 (1980).
2. F.F. Lange, J. Am. Ceram. Soc. 61, 53 (1978).
3. F.F. Lange, S.C. Singhal and R.C. Kuznicki, ibid 60, 149 (1977).
4. R.R. Wills, S. Holmquist, J.M. Wimmer and J.A. Cunningham, J. Math. Sci. 11, 1305 (1976).
5. K.H. Jack, "Processing of Crystalling Ceramics," (eds. H. Palomar, R.F. Davis and T.M. Hare) 561, Plenum (1978).
6. L.J. Gauckler, H. Hohnke and T.Y. Tien, Bull. Am. Ceram. Soc. 57, 828 (1978).
7. L.J. Gauckler, H.L. Lukas and G. Petzow, J. Am. Ceram. Soc. 58, 346 (1975).
8. I.G. Huseby, H.L. Lukas and G. Petzow, ibid, 377.
9. D.R. Clarke and G. Thomas, ibid, 60, 491 (1977).
10. L.K.V. Lou, T.E. Mitchell and A.H. Hueur, ibid, 61, 392 (1978).
11. R. Kossowsky, J. Mat. Sci. 8, 1603 (1973).
12. B.D. Powell and P. Drew, ibid, 9, 1867 (1974).
13. D.R. Clarke, this volume.
14. F.F. Lange, J. Am. Ceram. Soc. 62, 617 (1979).
15. R. Raj and F.F. Lange, Acta Met (in press); also in this volume.
16. F.F. Lange, J. Am. Ceram. Soc. 63, 38 (1980).
17. F.F. Lange and B.I. Davis, ibid, 62, 629 (1979).
18. D.A. Shockey, private communication.
19. F.F. Lange, B.I. Davis (unpublished).
20. F.F. Lange, B.I. Davis and D.R. Clarke, (Parts 1, 2, 3) J. Mat. Sci. 15, 601, 611, 616 (1980).
21. B.S.B. Karanaratne and M.H. Lewis, ibid, 449.
22. F.F. Lange and B.I. Davis (unpublished).
23. F.F. Lange, "Deformation of Ceramics," (ed. R.C. Bradt and R.E. Tressler) pp. 361-81, Plenum (1976).
24. P.E.D. Morgan, F.F. Lange, D.R. Clarke and B.I. Davis, J. Am. Ceram. Soc. 14, C-77 (1981).
25. M.H. Lewis, B.D. Powell, R. Rew, R.J. Lumby, B. North and A.J. Taylor, J. Mat. Sci. 12 16 (1977).
26. R.M. Arons and J.K. Tien, ibid 15, 2046 (1980).
27. D.R. Clarke and F.F. Lange J. Am. Ceram. Soc. 63, 586 (1980).
28. F.F. Lange, ibid, 62, 222 (1979).
29. F.F. Lange, ibid, 61, 53 (1978).
30. S.C. Singhal in Brittle Material Design, "High Temperature Gas Turbine," Interim Tech. Rept. AMMRC-CTR-75-28, Sept. 1975.
31. F.F. Lange, J. Am. Ceram. Soc. 61, 270 (1978).

32. F.F. Lange, B.I. Davis and M.G. Metcalf, Final Rept. to AFOSR, April 1981, Science Center Rpt. No. SC5099.4FR.

33. B.S.B. Karunaratne and M.H. Lewis, J. Mat. Sci. 15, 1781 (1980).

DISCUSSION

Weiss: Is the volume fraction of glass diminished by outwards cation diffusion only, or by parallel crystallization?

Lange: As I see it cation diffusion to the surface leaves behind oxygen and SiO_2, which react with Si_3N_4 to form Si_2N_2O. This is a reaction and not a pure crystallization in the sense of your comment, although one can think of it in either terms.

Jack: Regarding the glass composition, I accept that the intergranular glass ought to be stabilized in the way Dr. Raj suggests, but there is no reason why this should be the eutectic composition. The liquid region is much more extensive than the glass and a glass of the eutectic composition would be obtained only by very slow cooling and the amount of it would be small. Also in systems with impurities, the real ones, the eutectic composition is not the one you have taken. It seems that the way you have applied the lever rule cannot be valid.

Lange: For the composition examined choice of the eutectic composition for the glass was most consistent with X-ray analysis of materials, strength degradation and oxidation results. I have neglected impurities in choosing the eutectic (ca. < 200 ppm). Any other choice for the glass composition for $1 < MgO/SiO_2 < 2$ would also have been consistent with data (Si_3N_4-eutectic join: $MgO/SiO_2 = 1.6$). Of course this may not strictly apply for other M-Si-N-O systems, but why not!

Desmaison: In addition to the indentation fracture technique what other methods can be used to characterize the growth stresses present in corrosion scales?

Lange: High angle X-ray diffraction techniques are state-of-the-art methods for surface stress determination. One needs to calibrate the material to reduce error since the elastic modulus perpendicular to the measured plane spacing is usually not known. This is done by selecting the hkl to be examined and stressing the material to determine the calibration constant.

TIME-TEMPERATURE EFFECTS IN NITRIDE AND CARBIDE CERAMICS

R. NATHAN KATZ AND GEORGE D. QUINN

Army Materials and Mechanics Research Center
Watertown, MA 02172 (USA)

INTRODUCTION

High-performance ceramics based on the nitrides and carbides of silicon are being actively evaluated in a large number of heat engine and industrial heat exchanger demonstration programs (1-4). In most of these applications the ceramic components will be exposed to high temperatures, in oxidizing environments, for times ranging from a few thousand to several tens of thousands of hours. For successful application, it is essential that the designer have a full understanding of the time, temperature, and stress dependence of the strength (and/or retained strength after environmental exposure) for these materials. Unfortunately, in the 1975-1977 time period when the preliminary design studies for many of today's engine demonstration programs were initiated, the available data on silicon nitride and carbide ceramics were very sparse. Recognizing the need for a more extensive data base in this important area, the U. S. Department of Energy funded our laboratory and others to address the problem. This paper presents the methodology used and some of the principal results obtained by our laboratory during the past four years. More detailed results are presented in references 5 through 8.

One of the major reasons for the scarcity of such data was (and remains) the extremely rapid rate of materials development and improvement for these nitrides and carbides during the past decade. Since major improvements in materials and materials processing were occurring approximately every 6 to 12 months, there were problems with the economic costs of carrying out extensive testing on each materials variant, and also problems in obtaining sufficient samples for tests due to the developmental

Riley, F.L. (ed.) Progress in Nitrogen Ceramics

nature of many of the materials. Therefore, a screening test was required to survey the time, temperature, and stress behavior of a material, which would use relatively few specimens. This lead to the development of the stepped-temperature stress-rupture (STSR) test described below. For materials which showed promise upon screening and for which sufficient material was available, conventional stress-rupture (static fatigue) testing was carried out. Such tests can be used to determine the time of survival of a material at a given temperature, stress, and stressed volume. Alternatively, assuming slow crack growth, key parameters for life prediction can be extracted by the appropriate mathematical treatment of the stress rupture data (9). A third area of concern was (and is) to determine if, or to what degree, combined high temperature oxidative exposure and thermal cycling will reduce the retained strength of the candidate materials. A discussion of these three areas of study follows.

THE STEPPED-TEMPERATURE STRESS-RUPTURE TEST

The STSR test is an extension of the common flexural stress-rupture test to include a range of temperatures. One typical temperature cycle chosen for our requirements (and which conveniently fits into the usual laboratory work week) is illustrated in Figure 1. A specimen is loaded into a furnace equipped with a four-point bend fixture (spans 3.8 x 1.9 cm; see Ref. 8 for details) and the furnace is heated to 1000 C in air with no load applied to the sample. At 1000 C a deadweight load is applied. Should the sample survive 24 hours at that temperature, the furnace is then heated (in ≈ ½ h) to 1100 C and again allowed to soak for 24 hours. This cycle is repeated for 1200 C, 1300 C and

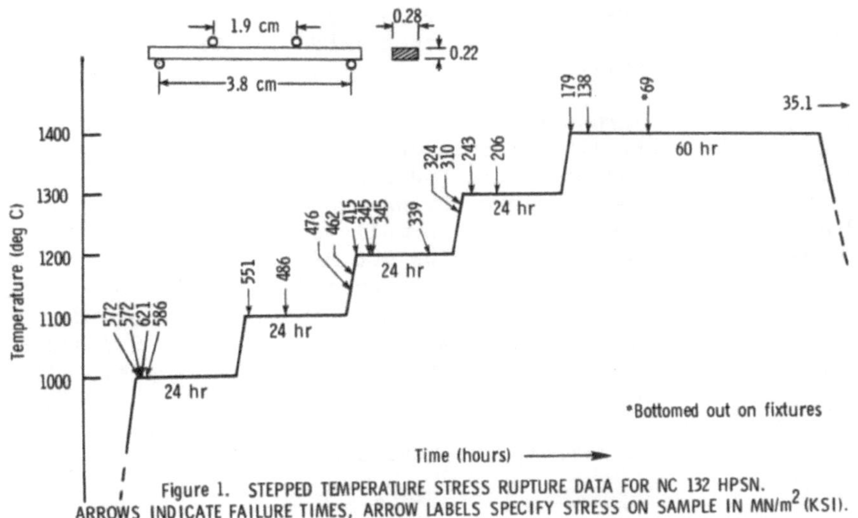

Figure 1. STEPPED TEMPERATURE STRESS RUPTURE DATA FOR NC 132 HPSN. ARROWS INDICATE FAILURE TIMES, ARROW LABELS SPECIFY STRESS ON SAMPLE IN MN/m^2 (KSI).

1400 C, but in the last case, the soak is maintained for 60 hours. Throughout the test, the sample is subjected to a constant dead-weight load. If a sample breaks, the furnace is cooled and unloaded and the time of failure is denoted by an arrow on the STSR plot. The arrow is labeled with the stress that was applied to the sample. A series of tests was executed with differing loads corresponding to stress levels calculated from the elastic beam formula.

For example, as illustrated in Figure 1, a sample loaded to 486 MPa survived the 1000 C soak, but failed at 10 hours into the 1100 C soak. Another sample, loaded to 324 MPa survived 1000 C and 1200 C soaks, but failed during heatup to 1300 C. As a last example, one sample survived the entire heat cycle, while sustaining a stress of 35.1 MPa.

Although the temperature history of a STSR specimen is more complex than that in a conventional stress-rupture test, the key intention is that any unusual temperature sensitivity will be identified quickly with a minimum number of specimens. Figure 1 illustrates a "normal", well-behaved material, i.e., time-dependent failures occur at lower stresses as one increases temperature and the transitions occur rather gradually as one goes from one temperature to the next. Such behavior is not always observed. One striking example is hot-pressed Si_3N_4 with 13% Y_2O_3. The Y_2O_3 additive is effective in promoting sintering yet minimizing the high temperature creep that is characteristic of the MgO additive grades of Si_3N_4 (10). Initial test results were promising; however, a catastrophic instability at 1000 C was identified for some compositions (11-12). The 13% Y_2O_3 material was susceptible to this instability (11), and, as a result, the manufacturer withdrew it from the market. Nevertheless, it was chosen for the present study to verify the ability of the STSR test to isolate such areas of instability.

Conventional stress-rupture tests of 13% Y_2O_3 material at 1200 C revealed very little time-dependent behavior: samples loaded to 500 MPa tended to break during loading or survive hundreds of hours without failure (at which point the specimen was unloaded intact). Thus one would normally expect that this material would be outstanding at loads below 500 MPa and temperatures below 1200 C. The STSR results shown in Figure 2 illustrate otherwise. Of eight samples tested, seven failed in the 1000 C range at stresses below 350 MPa. The 69 MPa sample failed at 19.7 hours and had gross secondary cracks through the entire sample. All of these samples would likely have survived the 1200 C stress-rupture trials for hundreds of hours. The STSR trials succeeded in identifying the unstable temperature regime in this case.

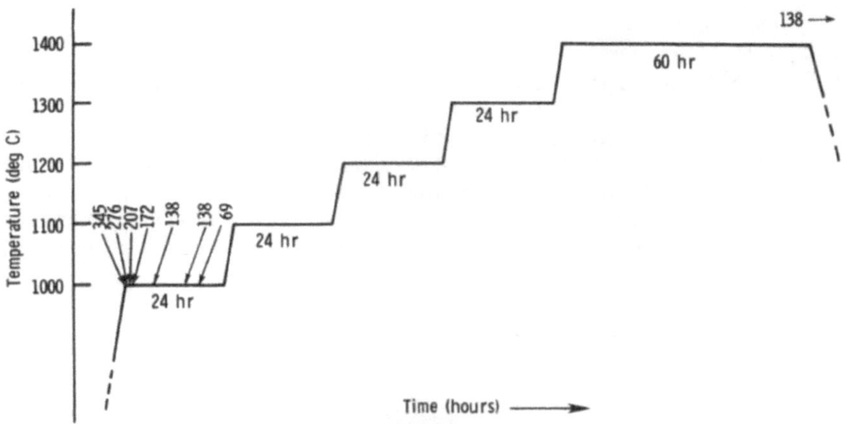

Figure 2. STEPPED TEMPERATURE STRESS RUPTURE DATA FOR HOT-PRESSED SILICON NITRIDE WITH YTTRIA ADDITIVE.

Evaluation of three grades of RBSN (ρ2.53 to 2.77 g/cc) indicated that the most critical temperatures from a time-dependent strength standpoint are the 1000 to 1100 C and 1300 to 1400 C ranges (8). Time dependence in the former range has been attributed to oxidation (13) and in the latter range to creep (14). Again the STSR test quickly isolated the critical temperature and stress ranges where time-dependent behavior occurred even though different mechanisms were involved.

STRESS-RUPTURE TESTS

As previously stated, S-R testing enables one, in many cases, to evaluate materials at temperatures, stresses, and stressed volumes approximating those to be encountered in service. Our laboratory is presently engaged in such a materials evaluation of four candidate silicon-based ceramics for application as erosion resistant trailing edges in a small radial gas turbine nozzle. S-R testing was carried out to times in excess of 1000 hours. As a result of these tests, three materials were judged acceptable from the standpoint of time-dependent strength retention. Final materials selection will be made on the basis of erosion and cost data. Such systems specific applications of S-R testing are relatively unusual for Si_3N_4or SiC ceramics at this point in time.

A more general use of S-R testing is to obtain data for life prediction calculations (15). The reciprocal slope of the S-R curve (in log-log form) is the exponent n in the slow crack growth power law:

$$V = AK^n \quad (1)$$

where: V is the crack grow rate, A is a constant, and K is the stress intensity factor. S-R data at 1200 C obtained on NC 132, hot-pressed Si_3N_4 by various investigators (8, 16, 17) with different lots of material and differing test geometries is shown in Figure 3. The data and the value of n obtained by these investigators are significantly more consistent than results obtained in double torsion tests or variable strain rate strength testing.

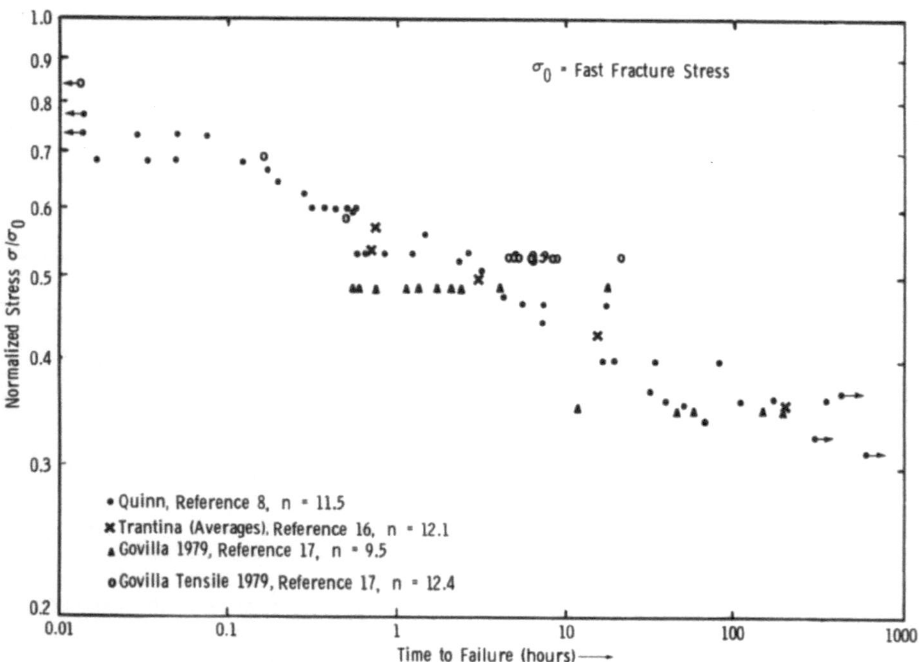

Because of the above considerations, we have concentrated on the use of S-R testing to generate our life prediction data base. Figures 4 through 6, summarize some of the S-R data obtained on Si_3N_4 s and SiC in our laboratory during the past four years.

COMBINED-EXPOSURE/THERMAL CYCLING EFFECTS

As shown above, most silicon nitrides and carbides exhibit a decrease in RT strength after exposure to oxidation in the 1000 C to 1400 C range for several hundred hours (18). If static oxidation presents a problem, what will occur under cyclic oxidation? Work is currently underway at AMMRC (8), AiResearch (19), Volkswagen

(20), and elsewhere to study effects on retained strength after cyclic exposure of engine ceramics at high temperatures, in oxidizing environments, and with large members of thermal cycles. Benn and Carruthers at AiResearch (19) utilize a combustor rig with an oxidizing combustion gas at temperatures to 1375 C combined with cyclic air quenching. Quinn at AMMRC (8) has used a stepped temperature oxidation cycle in a furnace, in air, coupled with multiple thermal shocks accumulated in a MAPP gas/O_2 flame. Siebels at VW (20) utilizes stepped temperature cycling in a furnace, in air, between RT and 900 C to 1260 C. Table 1 shows the range of materials response to such thermal cycling studies. Of particular

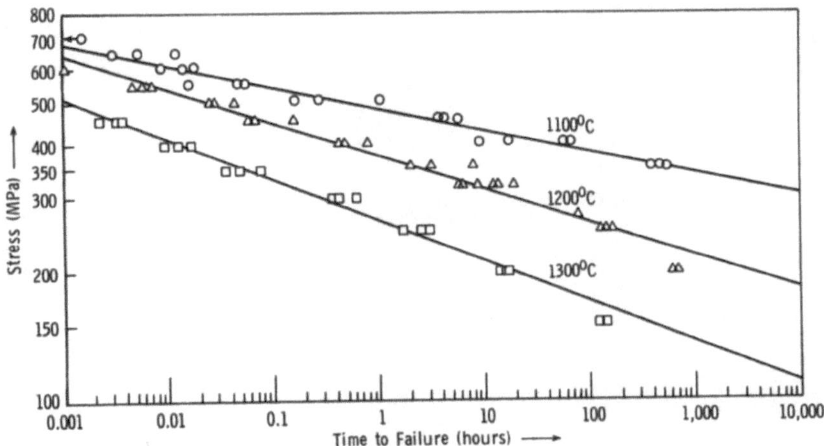

Figure 4. STRESS-RUPTURE OF NC 132 (HPSN) AS A FUNCTION OF TEMPERATURE.

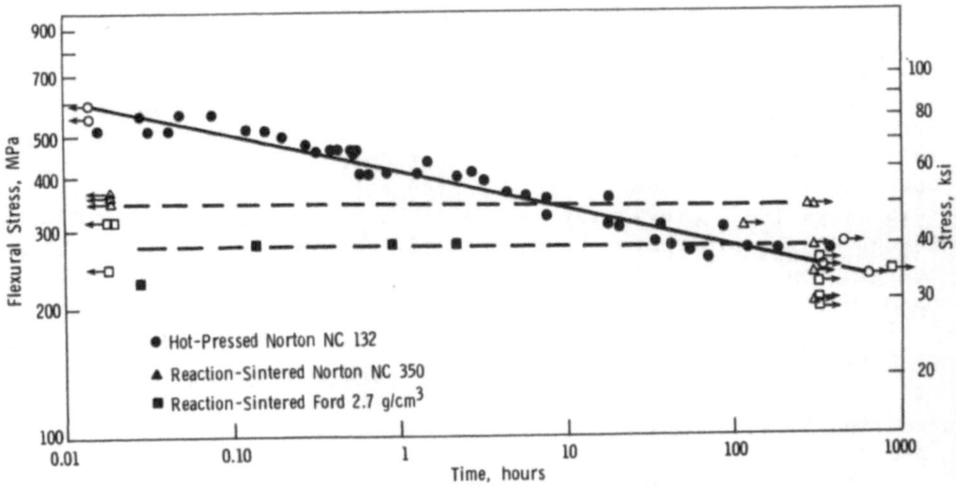

Figure 5. FLEXURAL STRESS RUPTURE AT 1200°C FOR THREE SILICON NITRIDES.

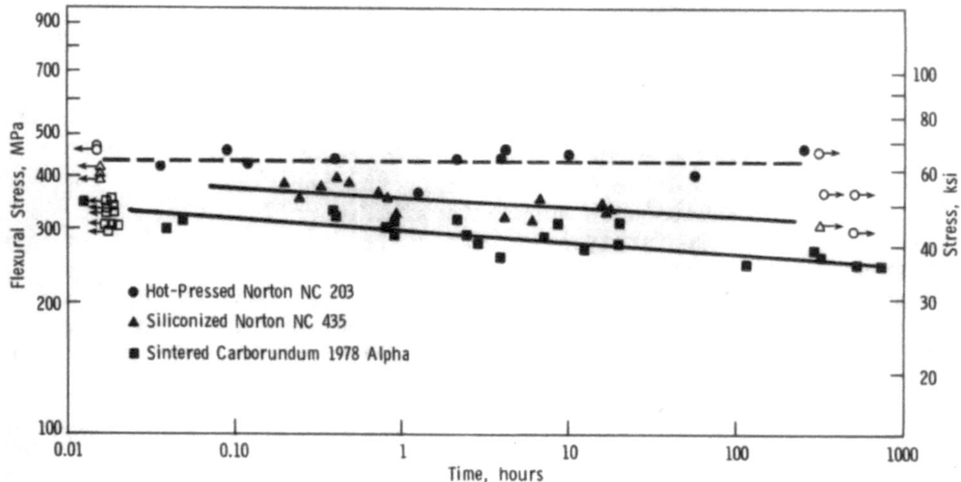

Figure 6. FLEXURAL STRESS RUPTURE AT 1200°C FOR THREE SILICON CARBIDES.

TABLE 1. Combined Thermal Exposure - Thermal Cycling Tests of Engine Ceramics to 2500°F

Material	Virgin MOR (ksi)	Exposed MOR (ksi)	% Change in MOR	Laboratory and Test Condition
NC-132 HPSN	104	50.5	-51	AMMRC - 360 hour/500 cycles in air and flame [30]
NC-203 HP SiC	99	102	+3	
NC-350 RBSN	43	35	-19	
KBI-RBSN	30	24	-20	
Silcomp - Si/SiC	47	32	-32	
Ford - RBSN	42	36	-15	
ACC RBSN-101	37.4	29.4	-21	AiResearch - 350 hour/1700 cycles in combustor gas [31]
Sintered α-SiC	45.8	45.9	0	

interest is the comparison of the RBSN data. In spite of different manufacturing procedures, differing densities, and differing test procedures, the percent strength degradation for similar times falls in the 15% to 20% range. Recent data of Siebels (20) indicates that newer grades of RBSN tested at VW now also fall into this range (RBSN previously tested at VW showed much greater degradation). It appears that a mutually consistent data base may be emerging on cyclic oxidation of RBSN. At present, multiple data bases do not exist for the response of other engine ceramics to cyclic oxidation. Also specimen data on the effect of cyclic oxidation under load are nonexistent.

SUMMARY AND CONCLUSIONS

This paper has reviewed some of the current work in our laboratory and elsewhere on the time and temperature dependence of strength in Si_3N_4 and SiC based ceramics. While most of these materials exhibit some degree of time-dependent behavior, we should put this in perspective by contrasting the behavior of these high-performance ceramics with superalloys. Figure 7 shows such a comparison. One immediately sees that designers of metal systems have learned to live with time-dependent behavior significantly worse than that exhibited by the Si_3N_4's and SiC's. Nevertheless, the ceramic materials developers can improve performance, and in order to gain acceptability for these alternate, brittle materials with the design community, we must do so.

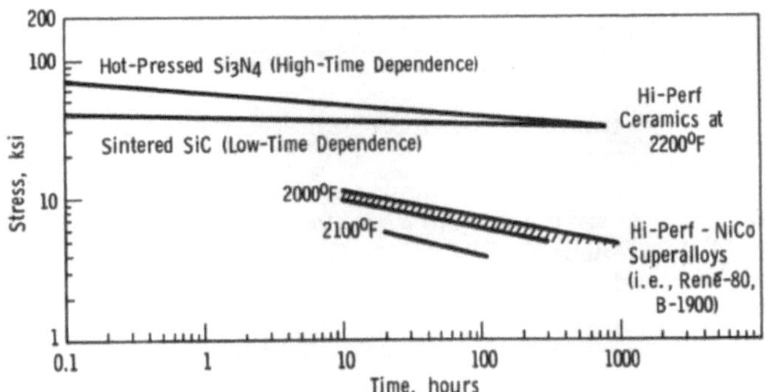

Figure 7. STRESS-RUPTURE BEHAVIOR OF HI-PERFORMANCE CERAMICS COMPARED TO HI-PERFORMANCE SUPERALLOYS.

The major conclusions are as follows:

1) The STSR test has proven its value as a method of screening for time-dependent behavior in a short time with few specimens.

2) The S-R test is highly reproducible from laboratory to laboratory and test configuration to test configuration. This reproducibility manifests itself in the consistency of the slow crack growth exponent n.

3) The limited data obtained so far shows the importance of combined oxidative exposure/thermal cycling tests. It is important that such testing be extended to include tests with an applied load, in contrast to the tests described above, which were carried out without a load.

ACKNOWLEDGMENT

The authors would like to acknowledge the U. S. Department of Energy, which provided partial support for this work under DOE/AMMRC Interagency Agreement E (49-28)-1017.

REFERENCES

1. Ceramics for Turbines and Other High Temperature Engineering Applications, Ed. D. J. Godfrey, Proceedings of the British Ceramic Society, 22 (1973).
2. Ceramics for High Performance Applications, Ed. J. J. Burke, A. E. Gorum, and R. N. Katz, Brook Hill Publishing Co., Chestnut Hill, MA (1974).
3. Ceramics for High Performance Applications II, Ed. J. J. Burke, E. M. Lenoe, and R. N. Katz, Brook Hill Publishing Co., Chestnut Hill, MA (1978).
4. Keramische Komponenten für Fahrzeug-Gasturbinen, Ed. W. Bunk and M. Bohmer, Springer-Verlag, Berlin (1978).
5. Quinn, G. D., Katz, R. N., and Lenoe, E. M., Proceedings of the DARPA/NAVSEA Ceramic Gas Turbine Demonstration Engine Program Review MCIC 78-36, August 1977, 715-737.
6. Quinn, G. D. and Katz, R. N., Am. Cer. Soc. Bull., 51 (1978) 1057-8.
7. Quinn, G. D. and Katz, R. N., J. Am. Cer. Soc., 63 (1980), 117-119.
8. Quinn, G. D. "Characterization of Turbine Ceramics After Long-Term Environmental Exposure", AMMRC TR 80-15, April 1980.
9. Jakus, K. and Ritter, J. E., J. Am. Cer. Soc., 61 (1978) 274-5.
10. Gazza, G. E., J. Am. Cer. Soc., 56 (1973) 662.
11. McLean, A. F., et al, "Brittle Materials Design, High Temperature Gas Turbine", AMMRC CTR 75-28 (1975) 138-39.
12. Lange, F. F. Singhal, S. C., and Kuznicki, R. C., J. Am. Cer. Soc., 60 (1977) 249-52.
13. McLean, A. F. et al. "Brittle Materials Design, High Temperature Gas Turbine", AMMRC CTR 75-8, (1975) 81-84.
14. Din, U. and Nickolson, P. S., J. Am. Cer. Soc., 58 (1975) 500-502.
15. Davidge, R., McLaren, J., and Tappin, G., J. Mat. Sci, 8 (1973) 1699-1705.
16. Trantina, G. G., J. Am. Cer. Soc. 62 (1979) 377-380.
17. Govilla, R., "Ceramic Life Prediction Parameters", AMMRC TR 80-18, May 1980.
18. Miller, D. G., et al. "Brittle Materials Design, High Temperature Gas Turbine - Materials Technology", AMMRC CTR 76-32, v. 4, December 1976.

19. Benn, K. W. and Carruthers, W. D. "3500 Hour Durability Testing of Commercial Ceramic Materials", 5th Interim Report on Contract DEN 3-27, June 15, 1979.
20. Siebels, J. E. "Oxidation and Strength of Silicon Nitride and Silicon Carbide", Presented at the 6th Army Technology Conference: Ceramics for High Performance Applications-III, Orcas Island, WA. July 1979, (to be published by Plenum Press).

DISCUSSION

Clarke: One of your graphs indicated that the stress-rupture curves of HPSN crossed and fell below those of RBSN at ~1000 h. Is this a real effect?

Katz: It is, but we do not have sufficient data for times beyond 1000 h to extrapolate with any confidence.

Fields: At NBS we are doing long-term tests and find considerable creep in some of these materials. Rather than breaking they deform excessively over a period of 1000 h. How do you handle this in your stress rupture diagrams?

Katz: We try to use the STSR test to screen out the exceptionally deformable materials. Where we have to test materials showing significant creep, we accept the inaccuracy that is being introduced.

Popper: You talked about 'degradation' but Lange discussed 'upgrading' through oxidation. This may lead to early failures in some samples, followed by long life after upgrading occurs?

Katz: Lange treated oxidation under load; we are testing under load. This is not an inconsiderable difference. We find that survivors of S-R tests often do exhibit enhanced R.T. strengths. However, running a part under load at high temperatures in air for hundreds of hours would not appear to be a cost effective method of strength enhancement at R.T. One other important point is that because critical flaw populations are different at low and high temperatures there is no reason to suppose that one will get strength enhancement at high temperatures.

HIGH TEMPERATURE FATIGUE FAILURE IN PRESSURELESS SINTERD SILICON NITRIDE

I. Oda, M. Matsui and T. Soma

Research and Development Laboratory
NGK INSULATORS, LTD., Nagoya, Japan

ABSTRACT. Fatigue failure mechanism of pressureless sintered Si_3N_4 was studied. The specimen used was a Si_3N_4 added with SrO, MgO and CeO_2. The fatigue failure time under 4-point bending stress, was measured in ambient atmosphere up to 1300°C. The temperature effects on strength, toughness, elastic moduli and oxidation resistance were also measured. The behavior in fatigue failure and other mechanical properties were found to change above a transition temperature of about 1000°C. The causes of the fatigue failure below and above the transition temperature were attributed to slow crack growth and creep deformation, respectively.

1. INTRODUCTION

Si_3N_4 is a leading candidate for use in high temperature structural component such as rotors in advanced gas turbine engines, because it has high strength, low thermal expansion coefficient, low specific gravity and high heat resistance. Various kinds of Si_3N_4 ceramics have been developed for this purpose. Among them, pressureless sintered Si_3N_4 (PS-SN) is a most promising one because the components with complicated shapes can be easily fabricated using this material. In this paper, the fatigue failure mechanism was investigated on a PS-SN added with sintering aids of SrO, MgO and CeO_2. The fatigue failure behaviors under static bending stress were observed. The temperature effects on strength, toughness, elastic moduli and oxidation resistance were measured. The microstructural changes were observed on fatigue failured specimens. Using above results, the fatigue failure mechanism was discussed in terms of slow

Riley, F.L. (ed.) Progress in Nitrogen Ceramics

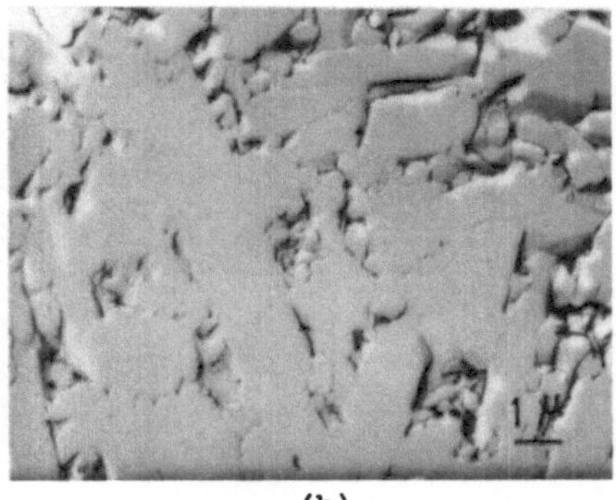

(a) (b)

Fig. 1. Microstructures of a pressureless sintered Si_3N_4 added with SrO, MgO and CeO_2. (a) fracture surface, (b) polished and etched surface.

crack growth and creep.

2. EXPERIMENTAL

The specimen was prepared by the following procedure. The submicron Si_3N_4 powder was mixed with 1 wt% SrO, 4 wt% MgO and 5 wt% CeO_2 powders in ball mill. The mixed powder was molded and isostatically pressed at 2500 kg/cm^2. The molded body was fired at 1700°C in N_2 atomosphere and machined to each specimen shape with diamond tools. The x-ray diffraction pattern showed α and β modifications of Si_3N_4. The microstructure consists of both blocky and column-like crystals, as shown in Fig. 1. Static fatigue and creep were measured by 4-point bending device [1] of 30 mm outer span and 10 mm inner span. Specimens were bars with the dimensions 2 mm thick, 4 mm wide and 40 mm long. The surface finishing was made with 800 grit diamond wheel. The creep strain rate in Si_3N_4 generally decreases with time [2]. In this experiment, creep strain rate was determind from the tangential line at 0.3 % strain in the strain-time curve. Strength was measured by 4-point bending device of 30 mm outer span and 10 mm inner span. The specimen was a bar with dimensions 3 mm thick, 4 mm wide and 40 mm long. The crosshead speed of the testing machine was 0.5 mm/min. K_{IC} was measured using single edge notched beam specimen [3]. The specimen was a bar with the dimension 4 mm thick, 3 mm wide and 40 mm long. A notch (o.1 mm wide and 1 mm deep) was made at the center of the specimen. Velocities of longitudinal and shear waves were measured by using pulse-echo-overlap method [4] and Young's modulus, shear modulus and Poisson's ratio were calculated. Oxidation resistance was estimated by measuring the weigth gain of strength specimens at various temperature.

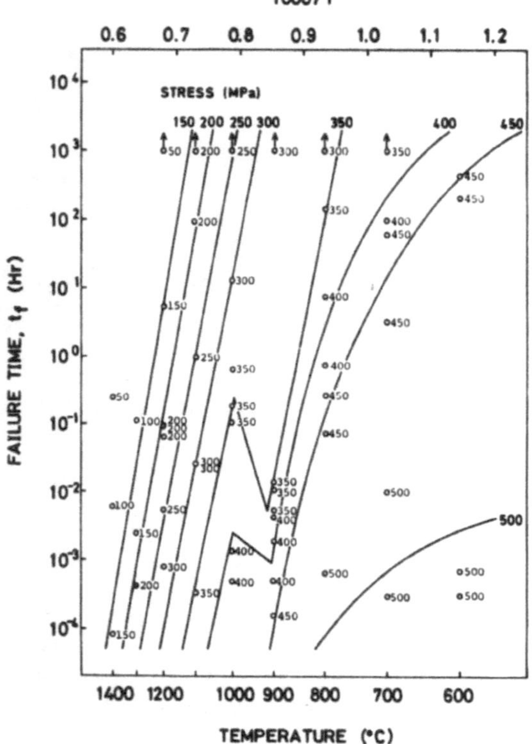

Fig. 2. Arrhenius plot of fatigue failure time under static 4-point bending stress. Arrows indicate that the specimen survived for 1000 hrs.

3. RESULTS

The Arrhenius plots of fatigue failure time are shown in Fig. 2, where an anomalous change is seen at 1000°C. The activation energy for fatigue failure increased with temperature up to 1000°C. Above 1000°C, the activation energy for fatigue failure was between 900 and 1200 kJ/mole, and the fatigue failure time (t_f) can be fitted to the empirical relation

$$t_f = A \exp\left(\frac{a-b\cdot\sigma}{RT}\right) \qquad (1)$$

where T is the absolute temperature, R the gas constant, A = -85 (hr), a = 1200 (kJ/mole) and b = 0.9 × 10^{-3} (m^3/mole). The solid lines above 1000°C in Fig. 2 shows the relationship given by equation (1). The strains at fatigue failure were less than 0.05 % below 1000°C, above which those were between 0.3 and 1 % and had a tendency to be large when stress was small and temperature was high. The temperature effect on strength is shown in Fig. 3. The strength at room temperature was 540 MPa and the Weibull modulus was 14. The strength was 400 MPa at 1000°C, above which

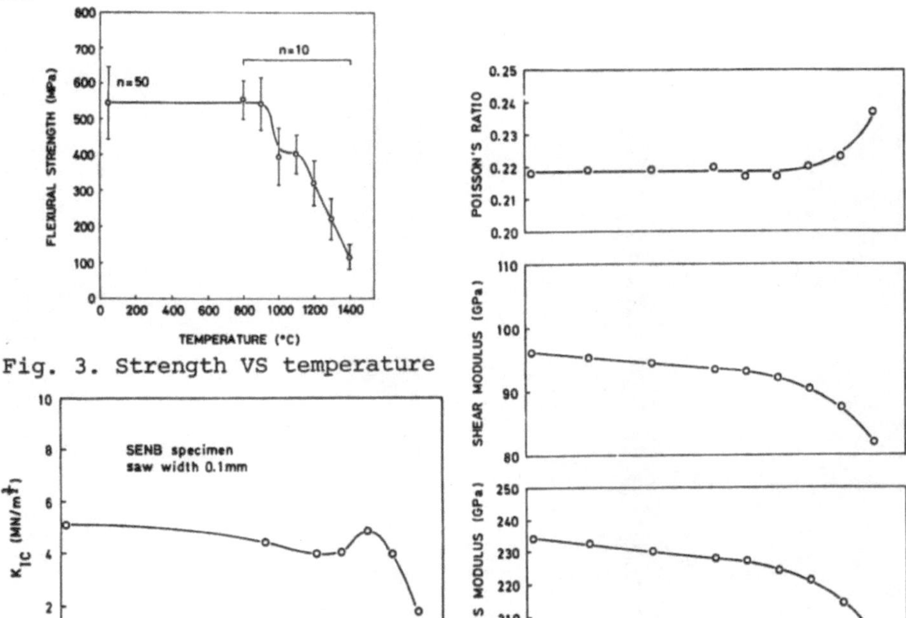

Fig. 3. Strength VS temperature

Fig. 4. Temperature effect on fracture toughness

Fig. 5. Temperature effect on elastic moduli

the strength decreased. The activation energy for creep was 1100 kJ/mole. The temperature effect on K_{IC} is shown in Fig. 4. The value is 5.2 MN/m$^{3/2}$ at RT and decreases gradually with temperature up to 1000°C, above which that increases again and shows maximum at 1200°C. Temperature effects on Young's modulus, shear modulus and Poisson's ratio are shown in Fig. 5. Young's modulus and shear modulus decrease gradually with temperature up to about 1000°C, above which the decreasing rate increase. Poission's ratio is almost constant between RT and about 1000°C, above which that increases abruptly and become 0.24 at 1100°C. The elastic moduli could not measured above 1100°C because of the attenuation of ultrasonic pulse. The relationship between weight gain (Δw) by oxidation and time (t) was fitted to the parabolic relation

$$(\Delta W)^2 = B \cdot t \cdot \exp\,(-\Delta H/RT) \qquad (2)$$

where ΔH is the activation energy for oxidation and B the constant. ΔH was estimated at 310 kJ/mole, from the experimental results.

The fracture surface and the microstructure of fatigue failured specimens are shown in Fig. 6. A fracture mirror is

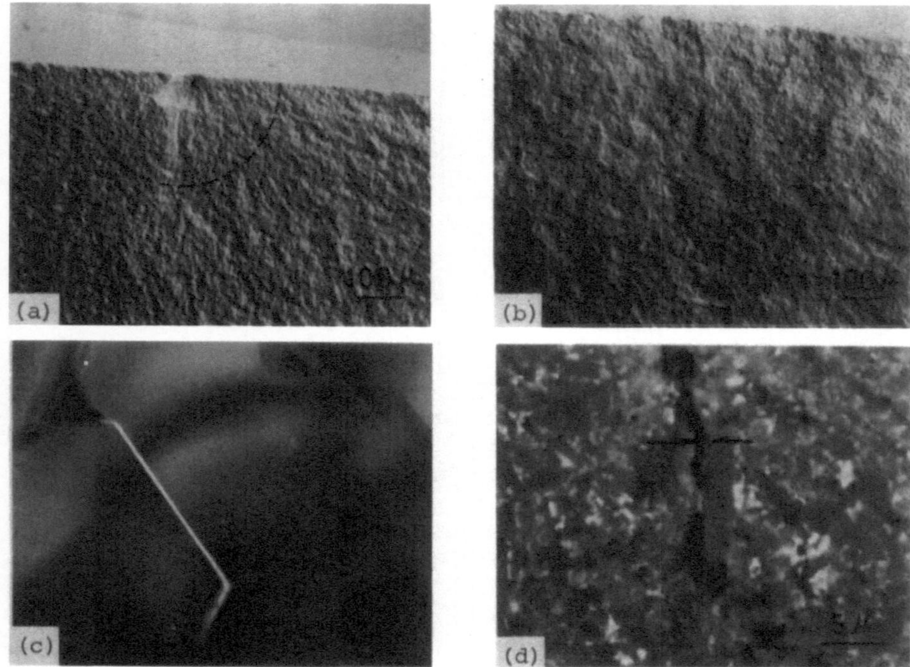

Fig. 6. Fracture surface and microstructure of the fatigue failured specimen. (a) fracture surface failured at 700°C. (b) fracture surface failured at 1200°C. (c) cavities at the grain boundary in the specimen (b). (d) macrocracks in the specimen (b).

clearly observed in the specimen fractured below 1000°C, while in that above 1000°C it is not observed and the fracture surface is very rough. In the specimen fractured above 1000°C, cavities at grain boundaries and macrocraks were observed. Starting from the surface or pores formed during sintering, the macrocrack propagates perpendicular to the tensile surface and reaches up to the extent of hundreds microns. The branching of the macrocrack was also observed.

4. DISCUSSION

From the experimental results, mechanical properties were found to change above the transition temperature (T_t) of about 1000°C. The observed behaviors of elastic moduli around T_t indicate that softening of the grain boundary is the principal cause of this transition phenomena. The increase of K_{IC} above T_t may indicate that stress relaxation at crack tip occurs by creep deformation of grain boundaries or crack branching in this temperature range. By this effect, the growth of a single large crack is considered to be arrested and many microcracks may be generated in whole of the stressed region.

The cause of fatigue failure below T_t is considered to be slow crack growth of pre-existing flaw because the fracture mirror clearly appeared and oxidation was negligibly small in this temperature range. On the other hand, that above T_t is considered to be creep phenomenon because the strain at fatigue failure reaches as large as 0.3 to 1 % and cavities at grain boundaries and microcracks are seen in crept specimens. The activiation energy of fatigue failure above T_t was 900 to 1200 kJ/mole and almost the same as creep. On the other hand, that of oxidation was about 310 kJ/mole and much smaller than that of fatigue failure. Therefore, the cause of fatigue failure above T_t was confirmed to be attributable not to oxidation but creep. From the microstructure of crept specimen, the fatigue failure above T_t is considered to occur as follows. Grain boundary sliding occurs by the softening of grain boundary and cavities are formed. Macrocracks initiate from near surface or pores and grow combining cavities and pores with the increase of creep strain of the specimen. When the creep strain or the length of macrocrack reaches to the limit, the fracture occurs instantaneously.

5. CONCLUSION

The behavior of fatigue failure in a PS-SN was found to change above a transition temperature. This phenomenon was related to the changes of elastic moduli, strength and fracture toughness. Below the transition temperature, the cause of the fatigue failure was attributed to the slow crack growth of pre-existing flaw. Above the transition temperature, it was attributed to the creep phenomenon accompanied by the formation of cavities and macrocracks at grain boundaries.

REFERENCES

1. G. W. Hollenberg, G. R. Terwilliger and R. S. Gordon, J. Amer. Ceram. Soc., 54 [4] 196-99 (1971).
2. R. M. Arons, J. K. Tien, J. Mater. Sci. 15, 2046-58 (1980).
3. L. A. Simpson, J. Amer. Ceram. Soc., 57 [4] 151-54 (1974).
4. E. P. Paradakis, J. Acoust. Soc. Am. 42 [5] 1045-50 (1967).

EFFECT OF DEFORMATION ON THE FRACTURE OF Si_3N_4 AND SIALON

R.J.Fields, T.J.Chuang, E.R.Fuller,Jr., and N.J.Tighe

National Bureau of Standards, Washington, D.C.

INTRODUCTION

At elevated temperatures, all materials creep. This is true in nitrogen ceramics (1-3) and is particularly important for them because of their intended applications at high temperatures. In general, the total strain, ε_T, in a body can be separated into an elastic component, ε_e, which is independent of time at constant load and a creep component, ε_c, which increases with time:

$$\varepsilon_T = \varepsilon_e + \varepsilon_c$$

Since the creep rate can be characterized by

$$\dot{\varepsilon} = A\sigma^n$$

where σ is the stress and A and n are constants, we see that creep occurs throughout a stressed body. In cracked bodies, the deformation due to creep is concentrated at the crack tip because of the high stresses there. One can define a creep zone (analogous to the plastic zone in metals) by determining that region within which the creep strain exceeds the elastic strain (Fig. 1). This zone grows with time because the creep strain at constant stress increases with time, while the elastic strain is independent of time. This creep zone has the same effect on the measured fracture toughness as the plastic zone in metals: the toughness is increased by the deformation (4,5). Since creep deformation increases with time, one would expect to find a strong dependence of toughness on loading rate. In what follows, such a dependence is found in a creeping silicon nitride. In a creep resistant sialon, no such dependence is observed. Additional

Riley, F.L. (ed.) Progress in Nitrogen Ceramics

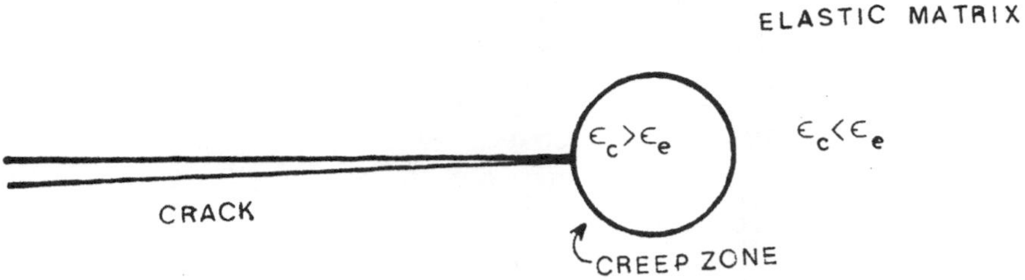

Fig. 1. The creep zone around a crack.

experimental evidence is presented to show that the dependence found in the silicon nitride is due to deformation and not to oxidation or microstructural changes.

EXPERIMENTAL PROCEDURE:

The materials tested here were a commercial silicon nitride (NCX34) which is hot pressed with 8% yttria and a sialon, nominally of composition Si_5AlON_7. The sialon was prepared by reacting SiO_2, Si, and Al in nitrogen at 1400°C. to obtain a porous body. This was then densified by hot pressing at 1800°C. until a density between 3.10 and 3.14 was obtained. The blocks of material were cut into bars 5x10x50 mm. in size for testing. Each bar was notched with a thin (.5 mm.) cutting wheel to a depth of 2 to 5 mm. in the 10 mm. direction. Compression specimens 5x5x10 mm. were also fabricated.

The bars were tested in flexure at cross-head speeds of 1.27 x 10^{-2} to 1.27 mm./min (0.0005 to 0.05 in./min.) on a universal testing machine. Testing was performed in air inside a high temperature furnace using a silicon carbide 4-point flexure rig (major span = 40 mm., minor span = 10 mm.).The silicon nitride was tested at 1400°C. and the sialon was tested at 1500°C..

In general, the bars were slowly loaded at a constant displacement rate until they fractured. To investigate the effect of time at temperature, some bars were slowly loaded at 1.27 x

Fig.2. Fracture surfaces of NCX34 (left) and sialon, showing the notch (top), slow crack growth, and rapid fracture surface.

10^{-2} mm. (0.0005 inch/min.) until crack growth was observed as indicated by distinct non-linearity in the load-displacement curve. At this point the specimens were rapidly unloaded. After holding at temperature for some time ,Δt, the specimens were reloaded at 0.05 inch/min. to fracture.

Finally, compression tests were performed on these materials at cross-head displacements rates between .002 and .00002 inches per minute. From the resulting load-displacement curves and the specimen dimensions the stress at which $\varepsilon_c=\varepsilon_e$ was determined. Machine compliance was accounted for in the above procedure.

RESULTS

Typical fracture surfaces of these materials are shown in Fig. 2. In general, slow crack growth preceded fracture so that the rapid crack growth started from a naturally "sharp" crack. The critical stress intensity factor for rapid fracture, K_{Ic} , was determined from the crack length at the onset of rapid fracture (as deduced from the fracture surface) and the load at which rapid fracture took place (as deduced from the load-displacement curve) (4). This critical stress intensity factor is plotted in Fig. 3 as a function of temperature and displacement rate. In the case of the silicon nitride, the critical stress intensity factor increases with increasing temperature and decreasing displacement rate. The sialon, in contrast, shows no such dependence. At high

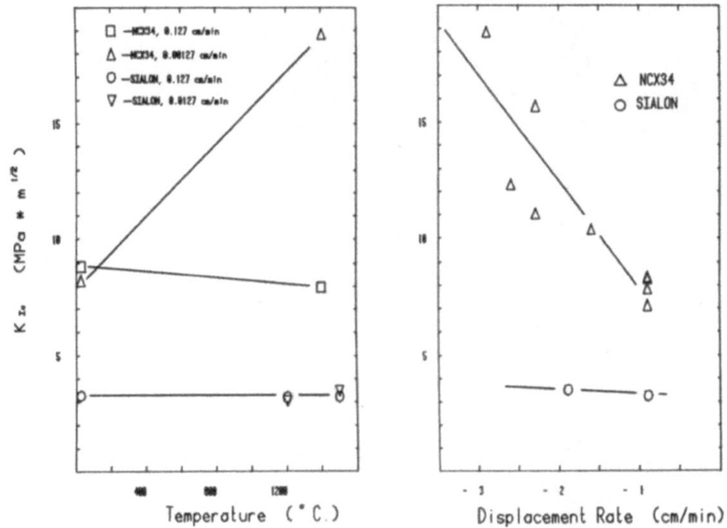

Fig. 3. K_{Ic} as a function of temperature and displacement-rate.

loading rate where no creep is expected, the silicon nitride appears inherently tougher than the sialon. Micrographs of the fast fracture surfaces of these two materials are shown in Fig. 4.. The silicon nitride exhibits an extremely rough microstructure while the fracture surface of the sialon is extremely flat. One would expect that the tortuous path followed by the crack in silicon nitride would raise the toughness above that for the sialon even in the absence of creep. In Fig. 6, the stress for which $\varepsilon_c = \varepsilon_e$ is plotted as a function of strain rate. In general, the sialon was so creep resistant that the SiC compression platens failed before any noticeable creep occurred. Therefore, the sialon is considerably more creep resistant than the silicon nitride even though the sialon was tested at a higher temperature.

DISCUSSION

From the results presented above, it can be seen that the toughness of NCX34 is loading rate dependent for the range of rates used here. The sialon material showed no such dependence at these rates. We have attributed this increasing toughness with decreasing rates to time dependent deformation at the crack tip. Such deformation would blunt and/or shield the crack just like a plastic zone, and raise the applied stress intensity factor required for fracture. The stress required to obtain a given

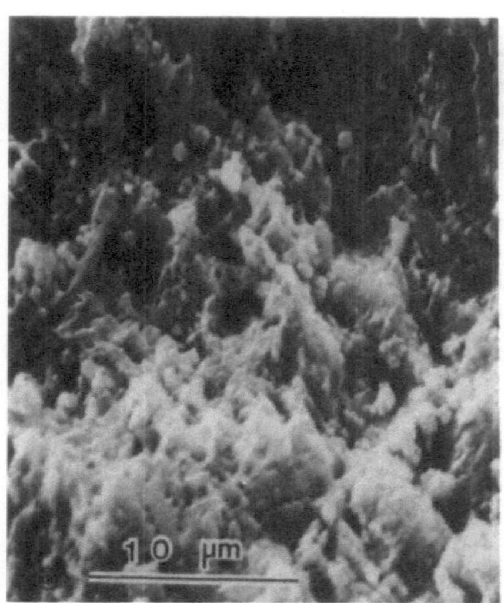

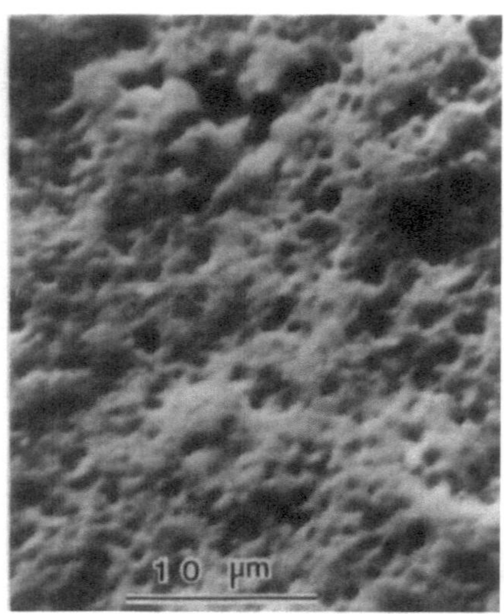

Fig.4. Fast fracture surfaces of NCX34 (left) and sialon.

creep strain is higher in the sialon than in the silicon nitride . This is consistent with the idea that more creep strain is occurring at the crack tip in the silicon nitride and that this strain raises K_{Ic}.

There is also the possibility that oxidation or some time dependent microstructural change could be occurring in the silicon nitride material (1,6,7). However, on the time scale of the present experiments (2 - 3 hours), we do not think that these influences are as significant as the creep effects. The tests conducted with the delay time before reloading support our contention. If it were the time at temperature that effected the raised K_{Ic}, then the longer the hold time, the higher the K_{Ic} should be. This in not found to be the case. Fig.5 indicates that the reverse is true. The results in Fig.5 can be explained in terms of another manifestation of time dependent deformation: stress relaxation. When the rapid unloading occurs at the beginning of the holding period, the non-uniformly strained enclave around the crack tip is placed in compression by the elastic region in which it is embedded. The creep strain which developed during the tensile loading of the crack is now reversed and will in time return to near zero, thus relaxing the residual crack tip compressive stresses. The K_{Ic} is seen to decrease with increasing hold time from the high values associated with the low loading rates to a low value consistent with the applied rapid rate.

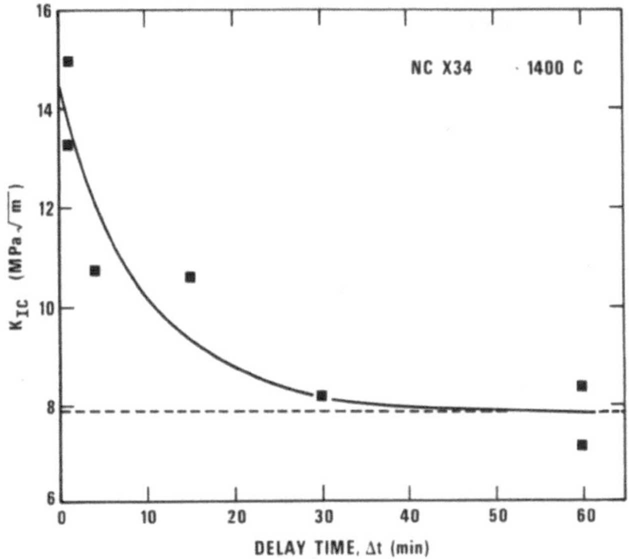

Fig.5.Effect of holding at temperature before rapidly fracturing.

It is instructive to speculate on what would happen at loading rates considerably below those applied here. In the case of the sialon, one might expect it to begin to show some of the effects we presently see in the NCX34. What about the NCX34; would its K_{Ic} keep rising? As the loading rate decreases, the creep zone increases in size. As it does so, the stress intensification at the crack tip is reduced by blunting and the load is shed from regions near the crack to regions further from the crack. When this happens , time-dependent degeneration processes, like viscous or diffusional cavitiation, cause a weakening of the remaining ligament. Fracture will then occur at a relatively low applied load, and this fracture load will decrease with decreasing loading rates. However, it is important to note that, when the creep zone becomes a substancial proportion of relevant specimen dimensions, the stress intensity factor does not adequately describe the stress distribution ahead of the crack and is no longer a valid parameter. In this regime a more appropriate field parameter describing the state of time-dependent crack-tip stress is C* (8).

CONCLUSION

The critical stress intensity factor for fracture is raised by deformation at the crack tip. The K_{Ic} for NCX34 is displacement-rate sensitive in the range .05 to .0005 inches/min.

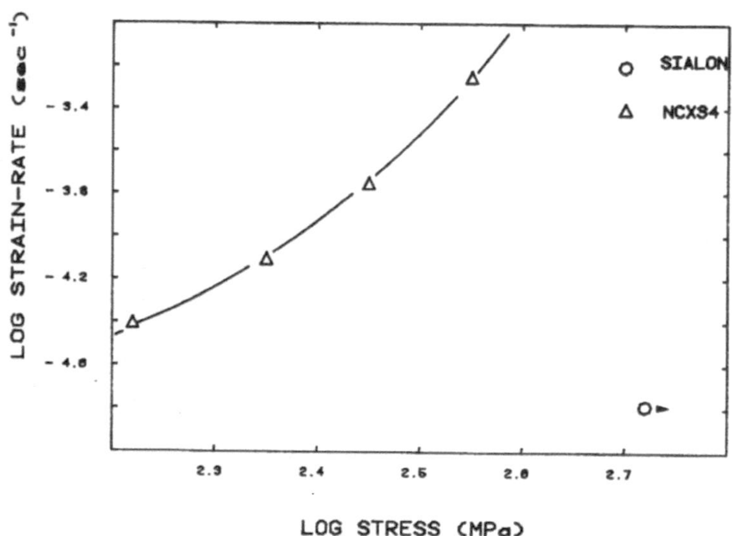

Fig.6. Creep behavior of NCX34 and sialon.

while that of Si_5AlON_7 is not. At similar stresses the silicon nitride is definitely creeping more than the sialon. Holding at temperature for various lengths of time before fracture gives results which support the conclusion that the effect is due to loading rate and not due to oxidation or other microstructural changes.

ACKNOWLEDGEMENTS

We would like to thank Dr.K.Kobayashi for preparing the sialon material and D.Harne for testing it. This work was supported by the U.S. Department of Energy under contract number DE-A105-800R20679. The authors gratefully acknowledge this support.

REFERENCES

1. F.F.Lange,B.I.Davis, and D.R.Clark."Compressive Creep of Si_3N_4/MgO Alloys,Parts 1,2,and 3."J.Mater.Sci. 15, 601 (1980).
2. M.S.Seltzer."High Temperature Creep of Silicon-Base Compounds." Bul.Am.Ceram.Soc. 56, 418 (1977).
3. J.M.Birch and B.Wilshire."Compressive Creep Behavior of Si_3N_4 Ceramics."J.Mater.Sci.13, 2627 (1978).

4. S.T.Rolfe and J.M.Barsom. Fracture and Fatigue Control in Structures; Applications of Fracture Mechanics (Prentice Hall, 1977).
5. G.R.Irwin."Linear Fracture Mechanics, Fracture Transition, and Fracture Control." Eng.Fract.Mechs.1, No.2, Aug(1968).
6. N.J.Tighe and S.M.Wiederhorn. Air Force Materials Lab. Tech. Rept. AFML-TR-78-83 (1978).
7. F.F.Lange,S.C.Singhal, and R.C.Kuznicki. "Phase Relations and Stability Studies in the Si_3N_4-SiO_2-Y_2O_3 Pseudoternary System." J.Amer.Ceram.Soc. 60, 249 (1977).
8. J.D.Landes and J.A.Begley."A Fracture Mechanics Approach to Creep Crack Growth." Westinghouse Corp. Scientific Paper 74-1E7-FESGT-P1(1974).

CYCLIC FATIGUE BEHAVIOR OF CERAMICS

Y.Matsuo, Y.Hattori, Y.Katayama and I.Fukuura

NTK Technical Ceramic Division
NGK Spark Plug Co., Ltd., Komaki, Aichiken, Japan

1. INTRODUCTION

In recent years, ceramics have been investigated because of their application as structural components at high temperature. A great deal of this research has dealt with various mechanical properties. Although, considerable effort has been made in the study of fracture strength and creep properties, comparatively a little is known about the fatigue behavior under cyclic stress. Fatigue studies of alumina are reported by S.Pearson(1), F.Guiu (2), etc., and silicon nitride by R.Kossowsky(3), T.M.Yonushonis and K.E.Fofer(4), and life prediction of ceramics under cyclic heat stress is reported by N.Kamiya and O.Kamigaito(5).

The present paper is concerned with "Cyclic Fatigue" that represents the behavior of characteristic degradation with increasing the number of cycles. The object of the present investigation is to study the cyclic fatigue behavior of silicon nitride. At the same time, life prediction which is necessary for engineering and designing of the structural components will be made based on Weibull statistics.

2. EXPERIMENTAL PROCEDURE

2.1 Test Materials

IN this study, three kinds of silicon nitride: hot pressed (H.P.SN), sintered(S.SN) and reaction sintered silicon nitride (R.S.SN), were used for test materials. H.P.SN was fabricated at 1800 °C, 20 MPa for 30 min, with Al_2O_3 and Y_2O_3 for additives. S.SN was fabricated at 1700°C for 60 min, with Al_2O_3 and MgO for

Riley, F.L. (ed.) Progress in Nitrogen Ceramics

Property \ Material	R.S. Si_3N_4	S. Si_3N_4	H.P. Si_3N_4
Bulk Density (g/cm^3)	2.70	3.15	3.35
Flexual Strength r.t. (MN/m^2) 1200C	350 350	670 250	1000 600
Weibull Modulus, m	14	11	20
Fracture Toughness ($MN/m^{3/2}$) r.t.	2.4	5.9	6.9
Young's Modulus ($x10^5$ MN/m^2) r.t.	2.5	2.7	3.3

Table 1. Mechanical properties of test materials.

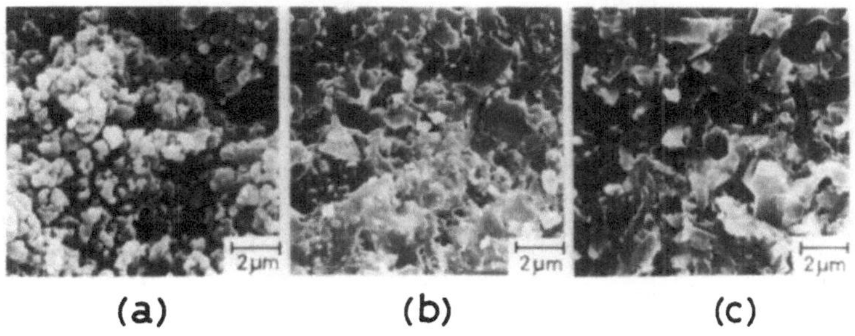

Fig. 1. Scanning electron micrographs of silicon nitride: (a) reaction sintered (b) sintered (c) hot pressed.

additives. R.S.SN was fabricated at 1400°C for 20 hrs with no additive. All these sintering were done in nitrogen atmosphere. Mechanical properties of these silicon nitride are listed in Table 1 and microstructures are shown in Fig. 1.

2.2 Dimension of Specimen

As is well known, it is difficult to make complex shape specimen with ceramics because of their poor machinability. Therefore, rectangular specimen was selected to improve the reliability of the measurement in this study. The dimension of specimen was 5 mm in width, 3 mm in thickness and 30 mm in length. Surface of specimen was ground along the length with 200 grit diamond wheels and chamfered 0.3 mm in radius.

2.3 Measurement

Relationships of cyclic stress and number of cycles to failure were evaluated by the cyclic fatigue test machine as shown schematically in Fig. 2. The specimen was placed over a 20 mm outer span on silicon nitride pins of 4 mm diameter in the cyclic fatigue machine. A sinusoidal mode of load was applied by eccentric motor rotating at 3300 rpm. Applied load was controlled by eccentric angle, detected by load cell and recorded on a electro-

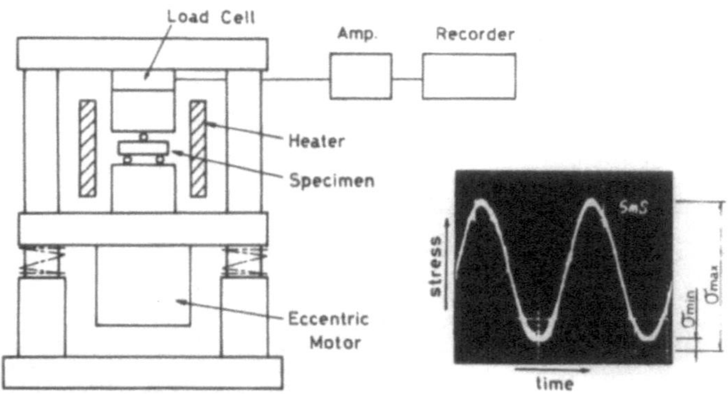

Fig. 2. Schematic of test apparatus.

magnetic oscillograph. The range of applied load ($R=\sigma_{min}/\sigma_{max}$) was 0.1 ~ 0.15. Number of cycles was given by measuring the time to failure.

The furnace provided stable temperature of up to 1000°C for this study, three levels of temperature were selected, i.e. room temperature, 800°C and 1000°C. Fixtures in the furnace were made of high density reaction sintered silicon nitride. All tests were done in ambient air.

3. RESULTS AND DISCUSSION

3.1 Relation between Cyclic Stress and Number of Cycles to Failure

Fig. 3 (a) shows the cyclic fatigue S-N curves for three kinds of silicon nitride at room temperature and Fig. 3 (b) shows the same for 800 °C and 1000 °C respectively. The arrowed test points indicate interrupted tests after 10^7 cycles.

A straight line can be drawn through the test points on the logarithmic graph by the least squares method. Then, the cyclic fatigue behavior gives the following relation between cyclic stress (σ) and number of cycles to failure (N); $\sigma^n \times N = \text{const.}$ An exponent value (n) is an important fracture mechanics parameter since it indicates cyclic fatigue resistance of the materials, i.e. higher the value of n gives higher fatigue resistance. Table 2 shows the result of the cyclic fatigue test from the present study. The value in parenthesis shows the correlation coefficient.

It was concluded from the experiment discribed above; at room temperature, the value of fatigue resistance (n) are 132, 63 and 55 against R.S.SN, H.P.SN and S.SN respectively. Also, these values seem to indicate the stress intensity exponent in the subcritical crack velocity vs stress intensity relation,

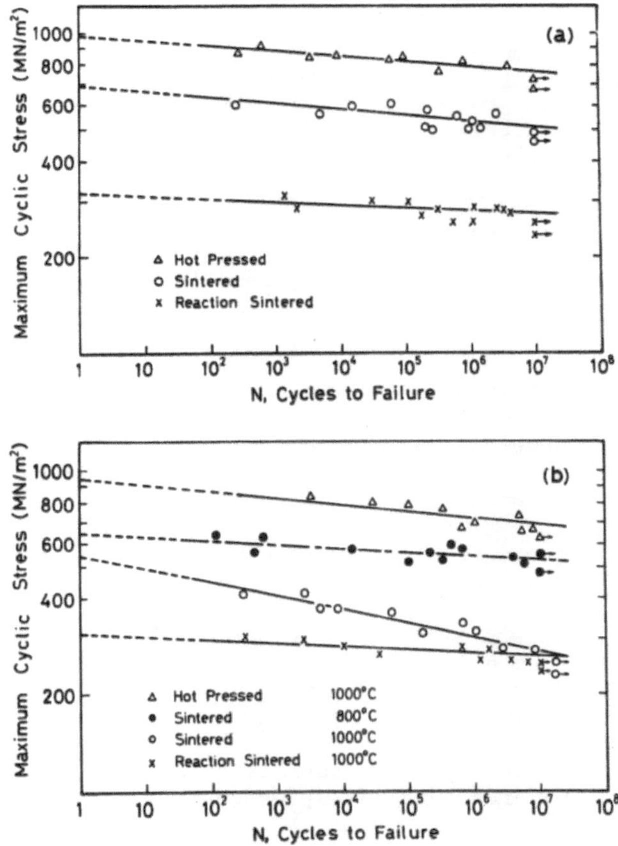

Fig. 3. Cyclic fatigue S-N curves for R.S., S. and H.P.Si_3N_4, (a) at room temperature (b) at 800 °C and 1000°C.

$V=A\cdot K_I^n$. Because these values shown in Table 2 demonstrated near agreement with the value of n which was given by the two most generally used techniques, i.e. double torsion test and double cantilever beam test.

At high temperature, in the case of R.S.SN and H.P.SN, the value of n at 1000°C were 106 and 50, 20% and 21% decrease compared with room temperature. In the case of S.SN, the value of n were 75 and 23 against at 800 °C and 1000 °C respectively, 36% increase and 58% decrease compared with room temperature. Oxidation layer was observed clearly on surface of the specimen which failed more than 10^5 cycles at 1000 °C, where as for R.S.SN and H.P.SN this was scarcely observable.

Therefore, the fatigue resistance of these silicon nitride depends on the total amount of grain boundary glassy phase both at room temperature and high temperature, i.e. larger amount of

Material \ Test Temp.		R.T.	800°C	1000°C
Si_3N_4	H.P.	$\sigma^{63} \times N = 985^{63}$ (-0.95)	——	$\sigma^{50} \times N = 950^{50}$ (-0.85)
	S.	$\sigma^{55} \times N = 682^{55}$ (-0.67)	$\sigma^{75} \times N = 658^{75}$ (-0.72)	$\sigma^{23} \times N = 562^{23}$ (-0.58)
	R.S.	$\sigma^{132} \times N = 330^{132}$ (-0.58)	——	$\sigma^{106} \times N = 324^{106}$ (-0.61)

Table 2. Relation between applied stress (σ) and number of cycles to failure (N).

glassy phase gives lower fatigue resistance. This tendency become to be more sensitive at 1000 °C like the behavior of high temperature strength shown in Table 1. While, the improvement of fatigue resistance observed on S.SN at 800°C is considered to depend on the relief mechanism of localized tesile stress.

3.2 Relation between Fracture Surface Mode and Theoretical Critical Crack Length

Typical fracture surface of S.SN after cyclic fatigue test at room temperature and 1000 °C are shown in Fig. 4. The character of fracture surface was found that the size of fracture mirror increase with decreasing cyclic stress and increasing number of cycles to failure.

Generally, for brittle materials such as silicon nitride, the criterion for fracture is expressed by Griffith's equation as follows;

$$\sigma = \sqrt{2 E \gamma_s / \pi a}$$

where, σ , E, γ_s and a are fracture stress, Young's modulus, effective surface energy and critical crack length.

Here, substituting the value of E, γ_s and σ under tested condition shown in Fig. 4, into this Griffith's equation gives the critical crack length. Table 3 shows the calculated critical

room temperature		1000°C	
applied stress σ (MN/m^2)	crack length a_{cal} (μm)	σ (MN/m^2)	a_{cal} (μm)
630	28	460	72
590	32	350	124
490	46	280	194

Table 3. Calculated critical crack length (a_{cal}) against applied stress shown in Fig. 4, where at room temperature E=2.7 x10^5MN/m^2, γ_s=64 J/m^2 and at 1000°C, E=2.6x10^5MN/m^2, γ_s= 92 J/m^2.

Fig. 4. Scanning electron micrographs of fracture surface after cyclic fatigue test for sintered Si_3N_4 at room temperature and 1000 °C.

crack length (a_{cal}) against various applied stress. These values shown in Table 3 are larger than the size of fracture origin ($a \simeq 10 \sim 20$ μm). This indicates that subcritical crack grew up a_{cal} during cyclic testing and these values are directly proportional to fracture mirror size (F.M.) examined in Fig. 4; $a_{cal} \fallingdotseq 0.3$ F.M. Therefore, it seems that the fatigue life of ceramics under cyclic stress is controlled by the subcritical crack growth.

4. PREDICTION OF FATIGUE LIFE

4.1 Theory

The subcritical crack growth of ceramics can be described by the following equation (1).

$$V = da/dt = A \cdot K_I^n = A \cdot Y^n \cdot \sigma^n \cdot a^{n/2} \quad (1)$$

where a, Y, σ and a are material constant, geometrical factor, applied stress and crack length. In the case of cyclic stress, applied stress (σ) is expressed as follows,

$$\sigma = \sigma_{max} \cdot g\ (t) \quad (2)$$

where g(t) is periodic function. Equation (1) can be integrated, gives equation (3)

$$N \cdot \sigma_{max}^n = (\ a_i^{1-n/2} - a_c^{1-n/2})\ /\ G \quad (3)$$

$$(G = (n-2)/2 \cdot A \cdot Y^n \int g(t)^n\, dt)$$

where N,G and a_i are number of cycles to fatilure, constant and initial crack length. In equation (3), $a_c \gg a_i$ and $n \gg 1$ the follwing approximation can be made.

$$N \cdot \sigma_{max}^{n} = a_i^{1-n/2} / G = Const. \tag{4}$$

Then, as is well known, the fracture strength of ceramics follows the Weibull statistics.

$$F = 1-\exp\left[-V \cdot (\sigma_f / \sigma_0)^m\right] \tag{5}$$

where F, m, V, σ_0 and σ_f are failure probability, Weibull modulus, stressed volume, normalization constant and fracture stress. The fracture stress (σ_f) is related to a through the following equation (6).

$$\sigma_f = K_{Ic} \cdot Y^{-1} \cdot a_i^{-1/2} \tag{6}$$

Substitution equation (6) into equation (5) gives equation (7)

$$F = 1-\exp\left[-V \cdot (a_i / a_0)^{-m/2}\right] \tag{7}$$

where,

$$a_0 = (K_{Ic} \cdot Y^{-1} \cdot \sigma_0^{-1})^2 \tag{8}$$

Then, substitution equation (4) into equation (7) gives the following equation (9).

$$\ln\ln(1/1-F) = m/n \cdot \ln N + m \cdot \ln\sigma_{max} + C \tag{9}$$

where C is constant. Equation (9) is indicated the relation between applied stress, number of cycles and failure probability.

4.2 Experimental Verification

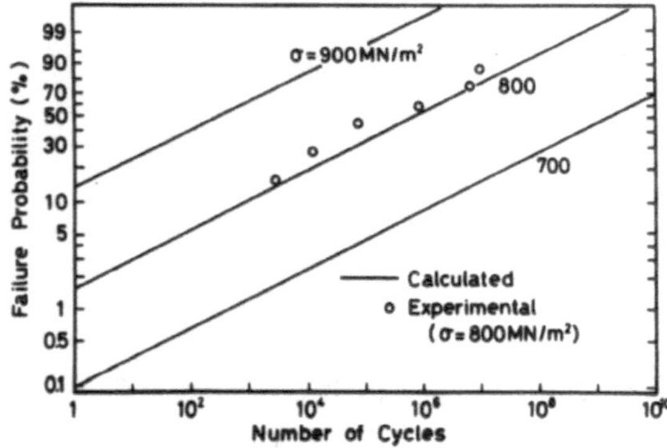

Fig. 5. S-P-N diagram for H.P.Si_3N_4 under cyclic stress at r.t.

Fig. 5 shows the S-P-N diagram for H.P.SN at room temperature. Straight lines are the theoretical lines with the calculation of equation (9). The plotted circles indicate the experimental values under 800 MN/m^2 cyclic stress. These values are in good agreement with the theoretical line. Then, this results suggest that it is possible to predict the fatigue life under cyclic stress by this method. In the future, further tests are necessary to verify this method at various stress levels.

5. CONCLUSIONS

1) Cyclic fatigue behavior of Si_3N_4 can be represented by the formula;

$$\sigma^n \times N = \text{const.}$$

2) Si_3N_4 used in present study are susceptible to mechanical fatigue damage even at room temperature and R.S.Si_3N_4 has the highest fatigue resistance in this tested temperature range.

3) Cyclic fatigue resistance of Si_3N_4 at high temperature should be altered due to the relief of localized tensile stress and the plasticity of glassy phase.

4) Cyclic fatigue life of Si_3N_4 is considered to be predicted by some formula. Experiments on H.P.Si_3N_4 showed the validity of the formula.

REFERENCES

1) S.Pearson, "Delayed Fracture of Sintered Alumina," Proc. Phys. Soc., London, Sect. B, 69 (444) 1293-96 (1956).
2) F.Guiu, "Cyclic Fatigue of Polycrystalline Alumina in Direct Push-Pull," J.Mater. Sci., 13 1357-61 (1978).
3) R.Kossowsky, "Cyclic Fatigue of Hot-pressed Si_3N_4," J. Amer. Ceram. Soc., 56 (10) 531-35 (1973).
4) T.M. Yonushonis and K.E. Hofer, "Cyclic Fatigue of Silicon Nitride Materials," in Proceedings of the 1977 DARPA/NAVSEA Ceramic Gas Turbine Demonstration Engine Program Review, Report No. MCIC-78-36 235-45 (1978).
5) N.Kamiya and O.Kamigaito," Prediction of Thermal Fatigue Life of Ceramics," J.Mater. Sci., 14 573-82 (1979).

THE FRACTURE BEHAVIOUR OF HOT-PRESSED SILICON NITRIDE BETWEEN ROOM TEMPERATURE AND 1400°C

Karel Kříž

Institut für Werkstoffwissenschaften I
Universität Erlangen-Nürnberg, 8520 Erlangen, Germany

ABSTRACT. The high temperature fracture behaviour and other mechanical properties of hot pressed silicon nitride are correlated to chemical composition and microstructure. It will be shown that the mechanical behaviour is controlled by the refractory properties of the present intergranular phase and by the degree of linking of the grain structure.

1. INTRODUCTION

Hot pressed silicon nitride has emerged as a possible material for high temperature technologies. However, such applications require comprehensive knowledge and control of its properties. Intergranular phases, formed during processing - as a consequence of the reaction between sintering aids and everpresent SiO_2 - layers on Si_3N_4 - powder particles - influence the creep strength [1] , the oxidation resistance [2] and the fracture behaviour [3] at elevated temperatures. It can be assumed that thermally activated processes in these grain boundary phases enable slow crack growth [4] .

The purpose of the present investigation is to correlate the fracture behaviour - characterized by the fracture toughness K_{Imax} and by measurements of slow crack growth - of two grades of hot pressed silicon nitride between 20 and 1400°C to the chemical composition and to the microstructure.

2. MATERIAL

The investigations were performed with two grades of commercially available hot pressed silicon nitride manufactured by Annawerk/Rö-

Riley, F.L. (ed.) Progress in Nitrogen Ceramics

dental with two different sintering aids: Grade 1 with MgO and grade 2 with Y_2O_3. The microstructure of both materials mainly consists of elongated β-Si_3N_4 prisms. However, the grade 2 appears to show a smaller grain size and a better linking of the structure than grade 1.

3. ELASTIC CONSTANTS AND INTERNAL FRICTION

High temperature elastic constants as well as internal friction were examined by a resonance frequency technique in order to determine the behaviour of the bulk material and for separating the effect of stress concentrations due to notches and cracks.

The shear moduli of both grades fall to low values above 1000 - 1100°C. In grade 1, this decrease continues up to 1400°C. In contrast, grade 2 shows a transition from the unrelaxed to the relaxed modulus [5] between 1100 and 1200°C. These relaxation processes, clearly indicated by internal friction peaks at 900°C in grade 1 and at 1100°C in grade 2, are probably a consequence of grain boundary sliding, which can be correlated to the viscous shear resistance of the intergranular phase [6].

4. HIGH TEMPERATURE FRACTURE TOUGHNESS

The fracture toughness measurements were conducted between 20 and 1400°C in air using a high temperature four point bending apparatus described elsewhere [7]. The specimens ($2,5 \times 3,8 \times 28$ mm^3) with straight through notches (width: 60 - 70 µm, depth: ~ 1 mm) were tested at a crosshead speed of 0,05 mm/min.

The observed load-deflection-records are shown in fig. 1. No measurable slow crack growth could be detected in grade 2 specimens in the temperature range investigated. Grade 1 specimens showed only

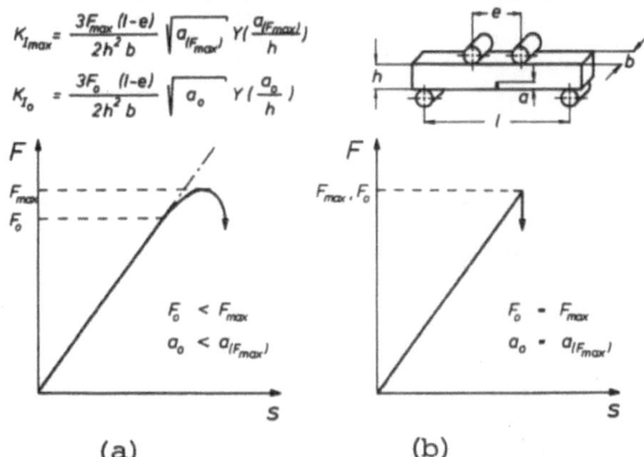

Fig. 1. Load-deflection-records with (a) and without (b) slow crack growth.

such behaviour (Fig. 1b) below 900°C. In this case, the obtained fracture toughness K_{Imax} mainly depends on the crack velocity and can be compared to the usual "K_{Ic}-values"[8].

Above 900°C an increasing slow crack extension was observed in grade 1 before reaching the maximum load F_{max} (Fig. 1a). After the applied load exceeds F_o (according to the stress intensity K_{Io}), the crack starts to grow. The K_{Imax}-value, calculated from the maximum load F_{max} and the actual crack length a F_{max} (controlled by optical fractography; e.g. Fig. 3b) depends on the crack velocity and on the amount of crack extension Δa between F_o and F_{max} [8]. If the detectable crack growth starts at F_{max}, K_{Io} and K_{Imax} will become identical.

The K_{Io}- and K_{Imax}-results obtained are shown for both grades in fig. 2. Only a slight decrease of K_{Io} and K_{Imax} can be observed in grade 1 between 20 and 900°C. Above 900°C, the softening of the intergranular phases enables slow crack growth at lower and lower loads, resulting in a strong decrease of K_{Io}. In contrast, K_{Imax} increases above 900°C, forming a maximum at 1200°C and falling again until 1300°C is reached. The K_{Imax} results in that temperature range are interpreted as follows:

Grain boundary sliding processes, which start at 900°C, lead to the formation and growth of microcracks/voids in the stress field near the crack tip (Fig. 3a) and to macroscopical crack branching[9]. These processes provide an additional contribution to the energy dissipation and therefore increase K_{Imax}. The morphology of fracture surfaces changes appreciable due to these processes. The example of a precrack, introduced at 1100°C (Fig. 3b - range A) shows, that the high temperature fracture surface exhibits a higher roughness than the room temperature fracture (range B). Moreover, above 1000°C the fracture surfaces are covered with a layer of softened intergranular phase, or oxidation products.

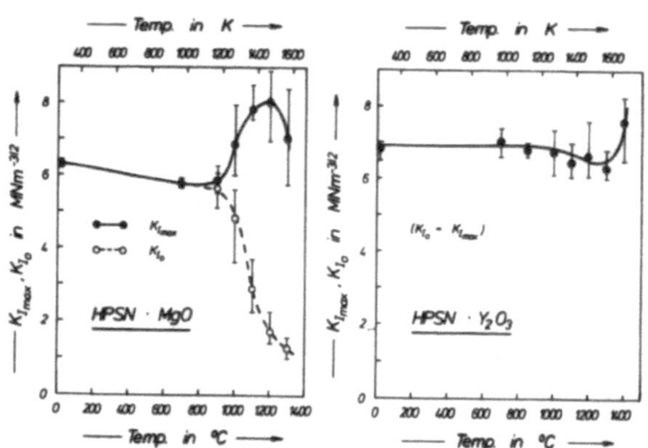

(a) (b)

Fig. 2. Temperature dependence of K_{Io} and K_{Imax} of grade 1 (a) and grade 2 (b).

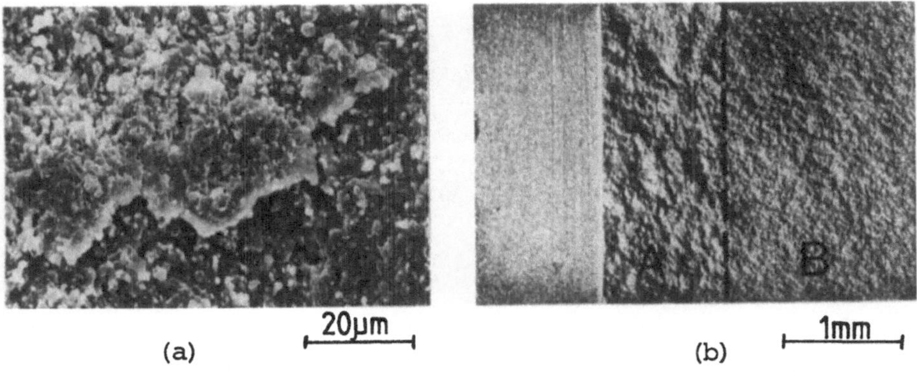

(a) (b)

Fig. 3. Fracture surfaces of grade 1 specimen: (a) secondary microcrack, (b) high temperature precrack (range A) and room temperature fracture (range B).

Above 1200°C the energy dissipative contribution of these additional fracture events seems to be overcompensated by the softening of the intergranular phase itself. The cohesion of the structure decreases more and more; K_{Imax} decreases with rising temperature.

In the Y_2O_3 containing grade 2, the K_{Io}- and K_{Imax}-values are identical up to 1400°C. With regard to the experimental scatter they remain constant below 1000°C. Their slight decrease up to 1300°C is probably due to the decrease of Young's modulus in the same temperature range. The increase above 1300°C can be understood as the beginning of microplasticity and microcracking near the notch root, which does not affect the shape of the F-s-curves (Fig. 1 b).

Between room temperature and 1000°C no change in fracture surfaces could be observed. The contribution of transcrystalline fracture appears to be higher in grade 2 material than in grade 1 [9]. While no slow crack growth could be detected up to 1400°C, the fracture structures were covered by the softened grain boundary phase or oxidation products above 1100°C [9] . Inhomogeneities in the chemical composition and in the distribution of this phase are probably responsible for the relatively large scatter of data [9].

5. SLOW CRACK GROWTH

Further important information concerning the fracture behaviour may be obtained from measurements of the temperature effect on slow crack growth. Such investigations were performed only on the MgO containing grade 1, using the double torsion technique at constant load [10] . Crack velocities, calculated from compliance change of DT-specimens (2 x 22 x 70 mm^3), were examined at K_I = const., while the temperature was increased in 100°C - steps up to 1400°C.

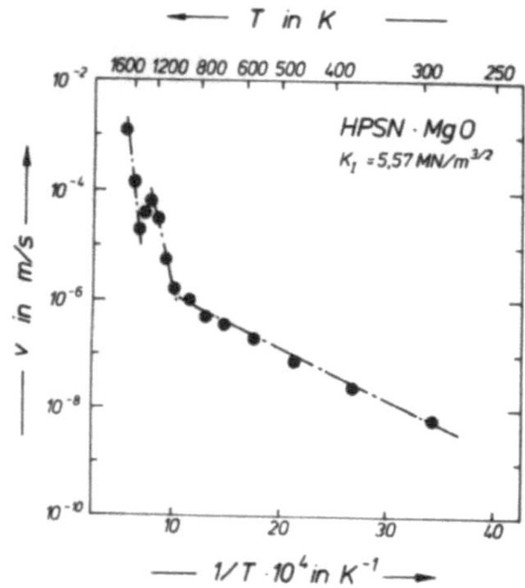

Fig. 4. Temperature dependence of the crack velocity in grade 1

The temperature dependence of the crack velocity is shown in fig 4; one is lead to conclude from the curve shape obtained that several mechanisms are responsible for slow crack growth in the temperature range investigated. Between room temperature and about 700°C, an activation energy of 16 kJ/mol could be determined. Such low values indicate that water vapour affects the slow crack growth - frequently occuring within the grain boundary phase- in this temperature range [11] . Activation energies of 140 kJ/mol were found between 700 and 1000°C. Similar activation energies were obtained from creep tests on grade 1 [12] between 900°C and 1200°C (90 to 140 kJ/mol). These values are in good accordance with data reported for several hot pressed silicon nitrides [1] , and for diffusion of O^{2-} as well as of alkali metals in glasses.

The decrease in crack velocity between 1000 and 1200°C seems to be caused by the same energy dissipative secondary cracking near the crack tip, used for the interpretation of the $K_{Imax}(T)$-curve (Fig. 2a).

Above 1200°C these processes will be superposed by enhanced viscoplastic deformation. Therefore the activation energy comes close to the values for viscous flow of silicate glasses.

6. CONCLUSION

Our fracture toughness results have shown that the high temperature fracture behaviour of HPSN can be significantly improved by using Y_2O_3 as sintering aid (instead of MgO).

The high temperature slow crack growth in MgO containing grade 1 seems to be governed by the same microprocesses, which are responsible for creep deformation; i.e. by the decreasing viscosity of the intergranular glassy phase. At lower temperatures, local, stress controlled adsorption and corrosion processes probably play a crucial part.

ACKNOWLEDGMENT

The work has been performed with financial support from the Federal Ministry for Research and Technology, Federal Republic of Germany, which is gratefully acknowledged.

REFERENCES

1. R. Becker, cfi/Ber. DKG 58 (1981), 93-101
2. S.G. Singhal, J. Mater. Sci. 11 (1976), 1177-1186
3. F.F. Lange, J. Am. Ceram. Soc. 57 (1974), 84-87
4. A.G. Evans, L.R. Russell, D.W. Richerson, Met. Trans. 6A (1975) 707-716
5. C. Zener, Phys. Rew. 60 (1941), 906-910
6. D.R. Mosher, R. Raj, R. Kossowsky, J. Mater. Sci. 11 (1976), 49-53
7. W. Grellner, Ph. D. Thesis, Univ. of Erlangen (1980)
8. H. Brettfeld, F.W. Kleinlein, D. Munz, R.F. Pabst, H. Richter, Z. Werkstofftechnik 12 (1981), 167-174
9. K. Kříž, B. Ilschner, Proc. of the 10th Plansee-Seminar, ed. H.M. Ortner, Reutte (1981), Vol. 1, 331-347
10. D.P. Williams, A.G. Evans, J. Test. Eval. 1 (1973), 264
11. T. Devezas, Ph. D. Thesis, Univ. of Erlangen (1981)
12. K. Kříž, Ph. D. Thesis, Univ. of Erlangen (1982)

PARAMETER STUDIES ON THE OXIDATION AND THE STRENGTH BEHAVIOUR OF SILICON NITRIDE

Johann E. Siebels

Volkswagenwerk AG, Wolfsburg
Research and Development.

1. INTRODUCTION

Silicon nitride materials have potential use in various engineering contexts. One possible application is for components of high temperature gas turbines. Reaction bonded silicon nitride (RBSN) is a very promising material because of its formability to complicated shape by mass production methods (injection molding). The development objective is a material having reproducible high strength, good thermal shock resistance, good resistance against oxidation and corrosive attack, as well as a low high-temperature creep rate. Those characteristics, being essential for the application, also have to be stable after long exposure under operating conditions. One of the material characteristics being intensively investigated at Volkswagenwerk AG is the stability of non-oxide ceramics after long oxidation times, and especially of RBSN.

2. TEST CONDITIONS

The oxidation test conditions were chosen in respect of the temperature schedule of an all-ceramic turbine rotor. Figure 1 shows the different temperature cycles applied to bend test bars. The 1260°C level represents the average maximum temperature in the middle section of the blade of an axial turbine rotor. 900°C is the average operational temperature of the rim. The oxidation resistance is expressed as the weight gain per geometric surface area, by cycle II.

Evaluation of the long-time-stability of the ceramic materials is made by measurement of 4-point bend strength. Oxidation experiments are performed in electric furnaces, in air.

Riley, F.L. (ed.) Progress in Nitrogen Ceramics

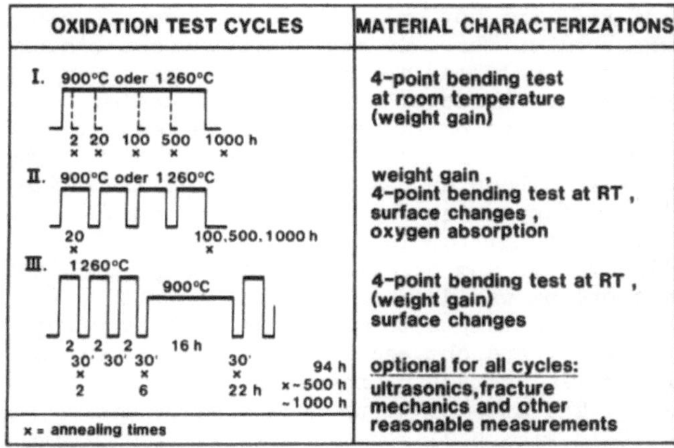

Fig. 1. Test cycles and material characterization.

3. ROOM TEMPERATURE STRENGTH AFTER ANNEALING AND OXIDATION RESISTANCE

Figure 2 shows the strength behaviour after annealing of three injection molded RBSN materials. RBSN-A shows strength degradation at 900°C and is stable at 1260°C. Material B is stable at 900°C and has a drop of strength to very low values after annealing at 1260°C. RBSN-C changes strength only slightly after annealing. The lowest values after 1000h are in the order of 230-250 MPa, which is promising for further development.

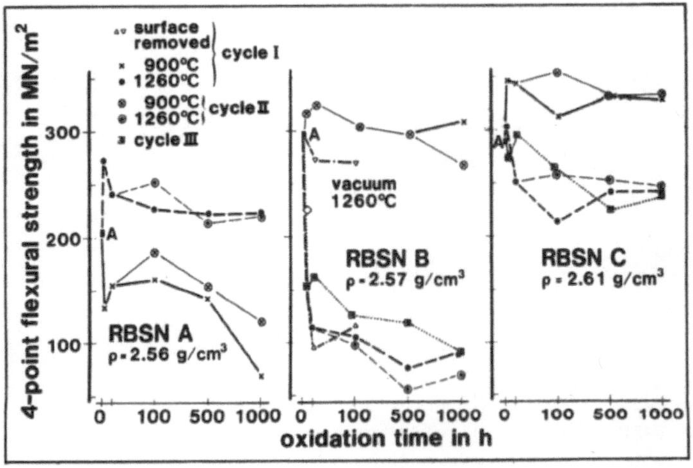

Fig. 2. Strength behaviour of injection molded RBSN after annealing.

That strength changes of RBSN are not only due to surface effects is shown for RBSN-B. Removal of the oxidized surface areas (thickness of layers 100 µm) does not lead to significant change in the strength of the test bars. Moreover annealing in vacuum (< 0.01 Pa), or in a powder bed as oxygen getter, also leads to strength degradation. But it is not valid to draw conclusions from the lower amount of changes. These annealing procedures in addition cause changes in the surface area of the specimen. RBSN-C proved to be a good starting point for further development of a high strength, thermally stable, and oxidation resistant, injection molded RBSN. (This development was done in cooperation with DEGUSSA.) Figure 3 shows two attempts to meet this goal by using a new silicon starting powder, and changed fabrication parameters. The advantage of a high initial strength, however, was lost after annealing (material 020). The attempt to produce a structure closer to 018 (material C in Figure 2) gave a certain improvement, but was not completely satisfactory. Also the oxidation resistance was influenced adversely, Figure 4.

The attempt to reproduce a material with the structure of RBSN 018 with the new silicon powder was a failure, due to its lack of thermal stability, Figure 5. RBSN 022 (see also Figure 5) with a very fine microstructure (see later) finally had the desired stability, at least in a short-time test. With this success a continuous series of RBSN development materials was available to study various parameters to find the reasons for the strength behaviour (and preferably the degradation effects after annealing).

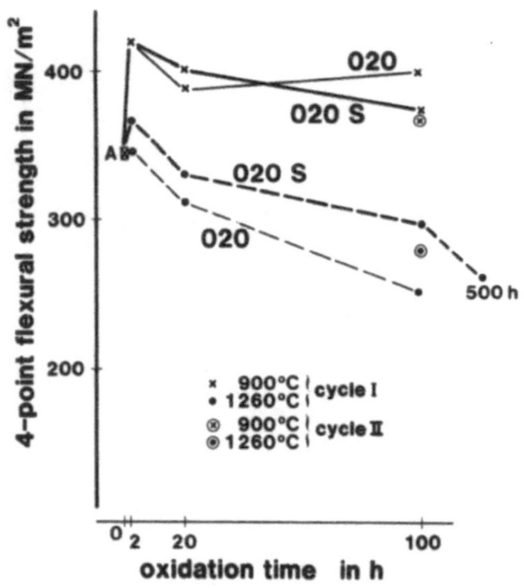

Fig. 3. Strength of RBSN 020 after oxidation.

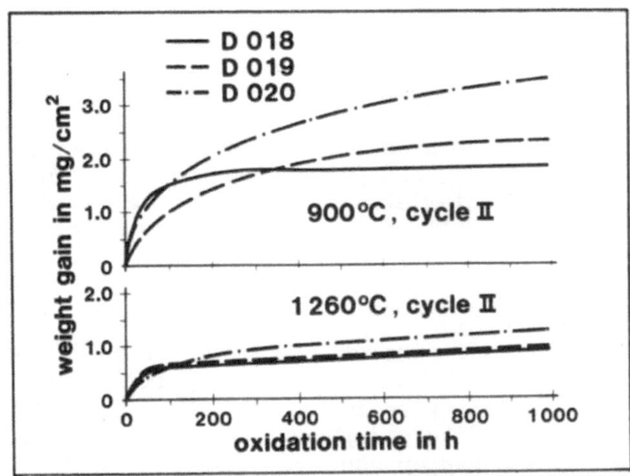

Fig. 4. Weight gain curves of oxidized RBSN.

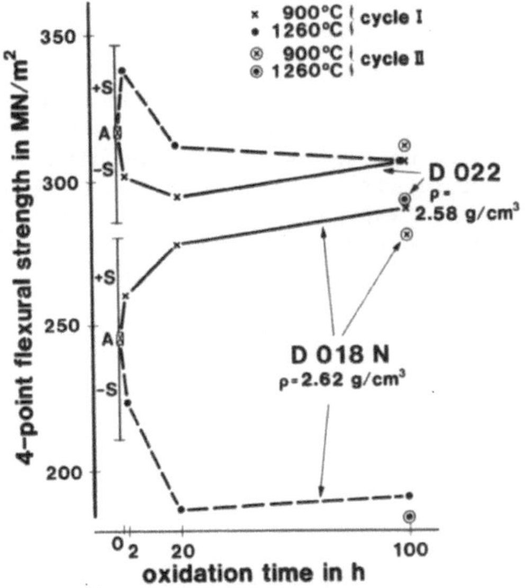

Fig. 5. Reproduction attempt for RBSN 018 and optimized RBSN 022.

4. INVESTIGATIONS ON MATERIAL PARAMETERS

Many theories identify surface effects as responsible for the strength degradation of RBSN after annealing. Some experimental results, however, show that this cannot be valid in all cases. Some authors also presume an influence of iron and calcium impurities. Investigations were therefore made to isolate the effective materials parameters.

4.1 Microprobe analysis

In order to have a rapid method for the determination of impurity elements, an existing microprobe procedure for the analysis of light elements was improved. Figure 6 compares the chemical compositions of some injection molded RBSN of a development series. The materials 018 N - 022 are made from the same starting powder and a significant difference to 017 and 018 cannot be seen. The reaction on oxidation of those materials in some cases is quite different. Thus the probability of a decisive influence on the strength behaviour is low.

4.2 Microstructure, oxygen content

Although structural changes in the cross-section of some RBSN test bars could be observed with the naked eye, X-ray diffraction analysis showed that the materials were identical before and after annealing. Also an influence of the α-β phase ratio can be ruled out. The same is valid for the oxygen content. Material with quite different strength properties had the same low contents of oxygen (0.5 - 1.5%).

4.3 Density, Porosity

Figure 7 compares the densities, open porosities and average pore sizes of the RBSN of the development series. One result of this development is that it is not necessary to achieve high densities (= 2.7 g/cm^3) to attain high strength materials. A density of 2.5 - 2.6 g/cm^3 is sufficient. Temporarily the open porosity value seemed to be an influencing factor on the strength behaviour after annealing, but the behaviour of RBSN 022 again made this assumption questionable.

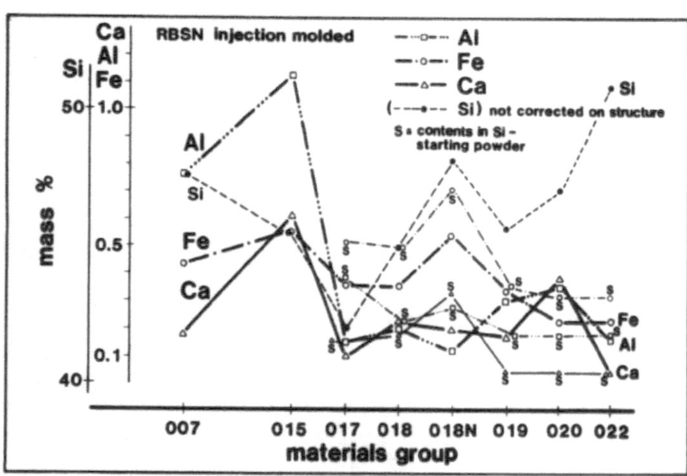

Fig. 6. Microprobe analysis of impurities in RBSN.

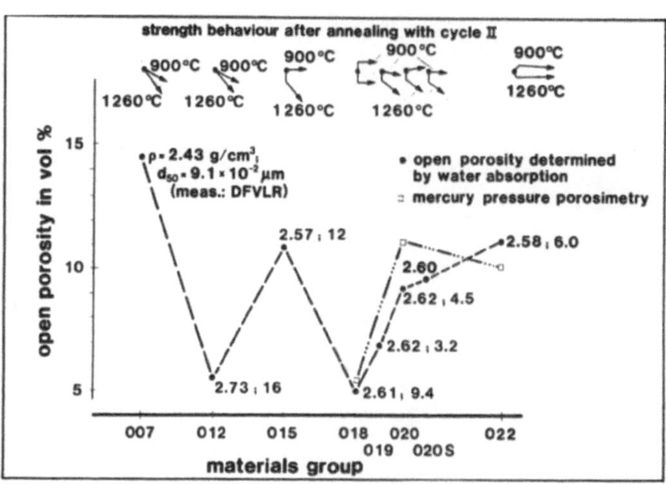

Fig. 7. Density, porosity and average pore size of the investigated RBSN.

4.4 Microstructure

Figure 8 shows the microstructures of the RBSN 015, 018 and 022. A material with the structure of 018 would be expected to have moderate properties only, and contradicts the actual properties. Only the creep resistance is poor due to the amount and distribution of the free silicon. It can be expected that the very fine microstructure of 022 also will meet the creep requirements.

4.5 Surface

As surprising as the comparison of the microstructures is the comparison of the oxidized surfaces of bend test bars (Figure 9). The surface of 018 has the worst conditions observed of all RBSN qualities tested up to now. Also it could be shown that impurity elements of the bulk material are present in the oxide layers. This is another indication that the strength after annealing is affected by factors other than the surface conditions.

4.6 Fracture mechanics

To find out whether surface effects or bulk material effects are responsible for the strength behaviour of the RBSN material, investigation methods have to be applied enabling conclusions to be drawn regarding volume effects. One way could be the application of fracture mechanics techniques, e.g. the determination of crack propagation velocity as a function of the stress intensity $(v = f(K_1))$.

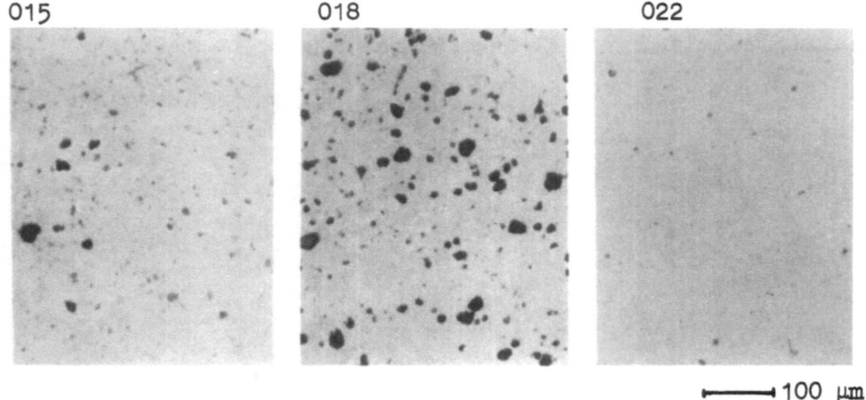

Fig. 8. Structures of the RBSN 015, 018, 022.

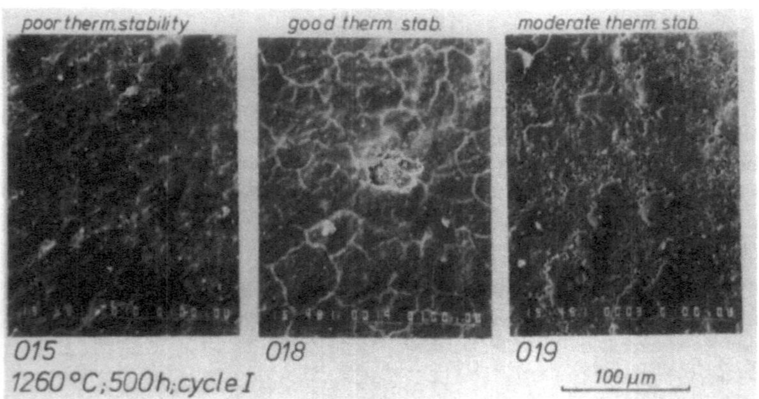

Fig. 9. Oxidized RBSN surfaces.

Figure 10 shows an example of such measurements in a double torsion test. The differences between the curves for undamaged and damaged test bars are not significant, so that this does not prove a volume effect. But for those experiments isostatically pressed RBSN was used and this material does not react in the same way on oxidation as injection molded RBSN. These measurements are being continued with other materials.

4.7 Sound velocity measurements

Another method allowing conclusions to be drawn regarding changes in the bulk material is the determination of the velocity

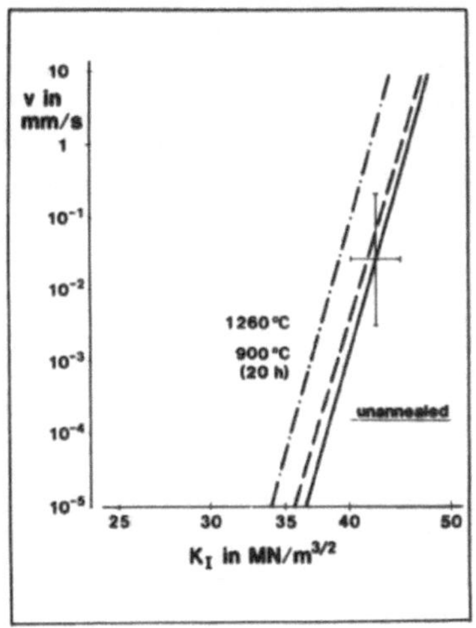

Fig. 10. v-K_1 curves from double torsion tests of isostatically pressed RBSN.

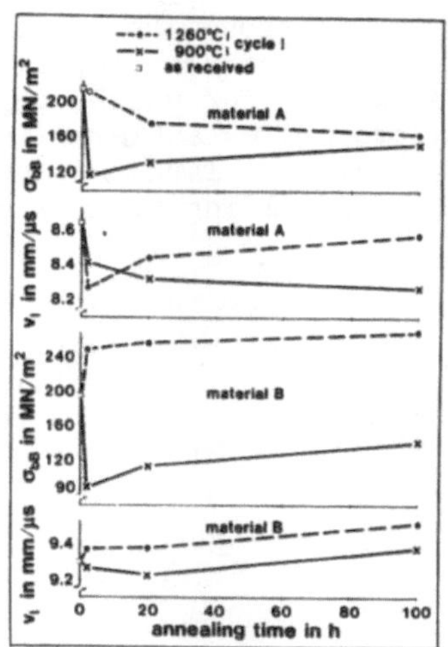

Fig. 11. Changes of strength and sound velocity by annealing.

of sound. Figure 11 shows examples for two different injection molded RBSN's. This result is very promising, because different strength changes result in different velocities of sound. This technique is therefore being extended to all RBSN materials under investigation. This programme is not yet completed.

5. HIGH TEMPERATURE STRENGTH

Figure 12 shows the high temperature strengths of two RBSN materials compared with room temperature strengths after 2 hours annealing (measurements of 3 different laboratories). In this case room temperature and high temperature strength show contrary behaviour. Nevertheless RBSN 018 also proved to be the most stable in the high temperature bend test. For the last development step, 022, this test has not yet been made. A measurement in another laboratory, however, shows that this material again has a drop in high temperature strength at about 800°C.

6. CONCLUSIONS

The development of injection molded RBSN has led to high strength, oxidation resistant, materials. However, a satisfactory explanation for the mechanisms responsible for thermal stability or instability has not yet been found. Possibly the investigation of the morphology of the α-phase, planned for the near future, will provide the required explanation. There is a satisfactory level of reproducibility in the more recent materials, which augers well for the further development of components.

Figure 13 shows the present state of development of RBSN and the required materials properties for application in radial and axial turbines. For small components the transferability of material properties achieved in test bars already has been demonstrated. But some larger components need more development work, especially in molding technology.

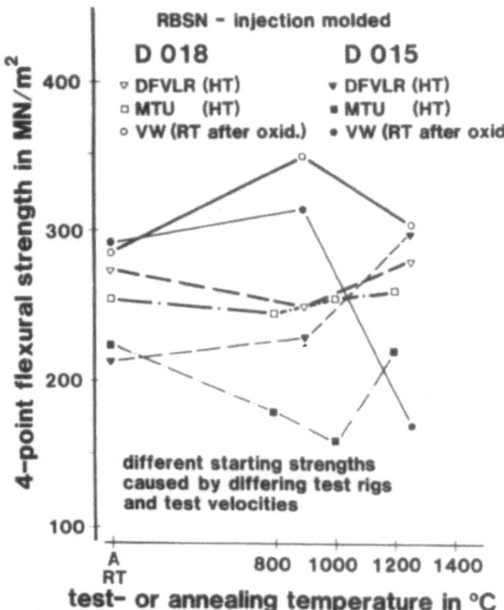

Fig. 12. High temperature strength and strength after short time (2h) annealing of injection molded RBSN.

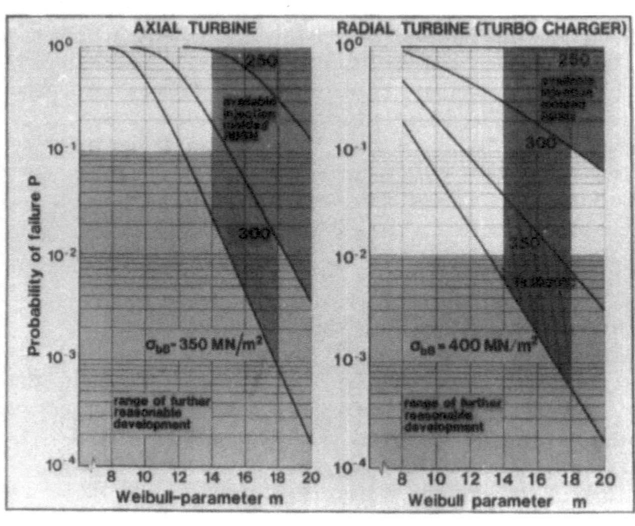

Fig. 13. State of RBSN development and strength requirements for turbine rotors.

DISCUSSION

Thümmler: Your results indicate that with RBSN the correlation between porosity, pore size distribution and oxidation behaviour, on the one hand, and strength development with time on the other hand, seems to be much more complicated than with respect to creep. The fact that you could not achieve clear cut correlations with all your materials is obviously due to the difficulties in adequate characterization of RBSN because of its complicated microstructure. I think more work in this area is called for.

Siebels: I agree: to isolate the decisive parameters governing post-oxidation strength behaviour is a difficult task. We hope to get there by investigations of the morphology of the α-phase in the groups of materials shown.

Mangels: Did you employ a 'flash oxidation' pre-treatment prior to testing?

Siebels: We tried 1 h at 1400°C with some materials but the advantages get less after short testing times according to our cycles - so we no longer apply 'flash-oxidation' prior to our tests.

CHARACTERIZATION, OXIDATION AND MECHANICAL BEHAVIOUR OF REACTION BONDED SILICON NITRIDE

F. Porz

University of Karlsruhe, D-7500 Karlsruhe 1,
Federal Republic of Germany.

ABSTRACT

A characterization of different RBSN qualities is presented and some data on oxidation behaviour and strength after cyclic oxidation are given. The results are discussed relating the observed phenomena with the microstructure of the materials.

INTRODUCTION

Reaction bonded silicon nitride (RBSN) is of special interest for high temperature applications, because of the ability to fabricate nearly every shape without additions of sintering aids, which may degrade the high temperature properties. The properties of RBSN are strongly dependent on the microstructure of the porous material. For qualities with small channel-radii the internal oxidation is limited only to an outer section. These materials show an excellent creep behaviour (1,2). The strength degradation of RBSN is also dependent on the amount of oxide phase present.

CHARACTERIZATION

The investigations were conducted on several commercially-available RBSN-materials. Table 1 summarizes their properties. The green silicon compacts for materials 1,2 and 6 were fabricated by isostatic pressing and for materials 3 to 5 by injection moulding.

The ceramographic sections for the photomicrographs (Fig. 1) were prepared according to the method indicated in (3). They show the very different appearance of the macro-porosity: the homogeneous microstructure with only small pores of material 5 contrasts with the relatively large pores of materials 2 and 3; the pictures of materials 1 and 6 indicate a large amount of interconnected

Riley, F.L. (ed.) Progress in Nitrogen Ceramics

material	density [g/cm³]	open porosity [% total vol.]	closed porosity [% total vol.]	fraction of trapped Hg [%]	$r_{50,e}$ [nm]	specific surface area [m²/g]
1	2.48	20	2	46	59	3.9
2	2.67	13	3	56	37	2.0
3	2.62	6	12	46	38	0.3
4	2.59	11	8	61	28	1.6
5	2.57	9	10	67	20	1.2
6	2.47	21	1	69	108	1.1

Table 1. Characteristics of the RBSN materials investigated.

porosity. In addition some unreacted silicon or silicides can be seen in materials 2, 3 and 5.

The distribution of the micropores, which can not be observed under the microscope, was determined by Hg-porosimetry. It is assumed that these pores are principally channels connecting the large pores with each other. With some limitations, mercury porosimetry permits the characterization of the pore-channels by radius of opening, volume, surface and length (4).

Conducting a porosimetry-test two typical branches of the volume-distribution-curve are observed (Fig. 2): during pressure increase the mercury enters via small channel-openings and fills the large pores; during extrusion the mercury does not completely drain because of voids which are not enlarged towards the outside of the specimen (4). The intrusion-curve gives the distribution of

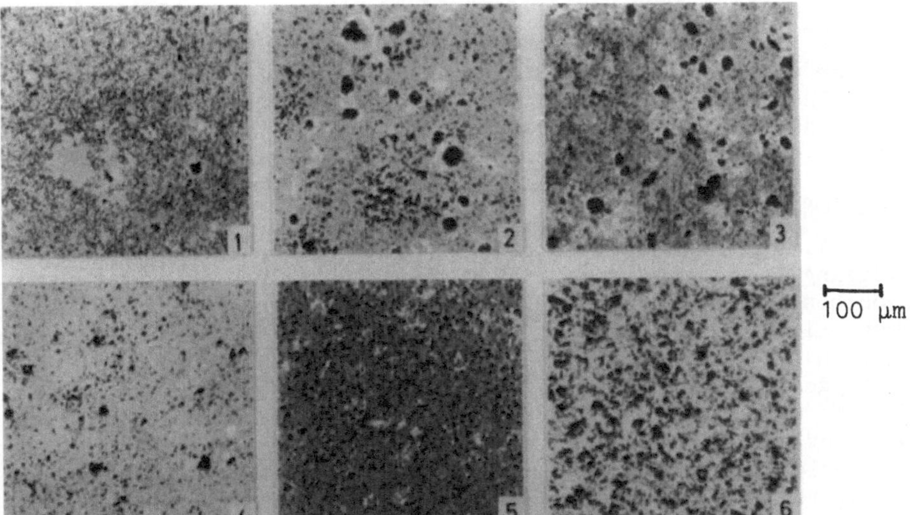

Fig. 1. Optical micrographs of the RBSN materials.

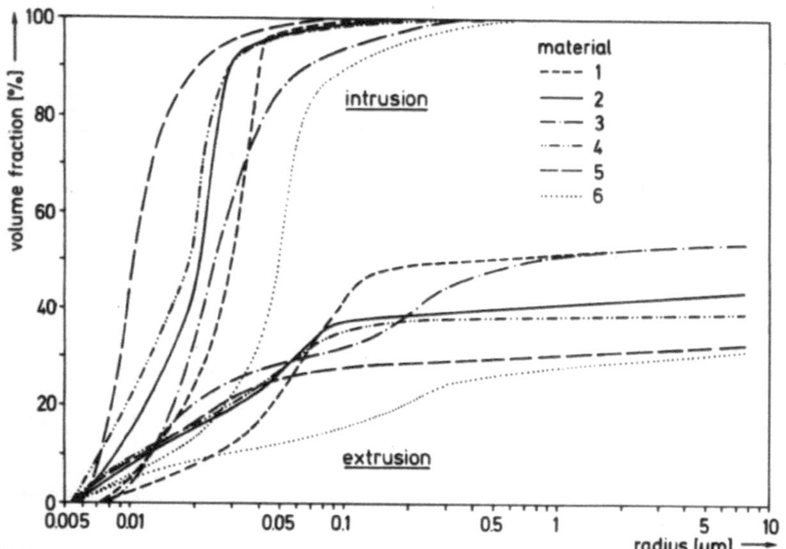

Fig. 2. Pore size distribution curves by Hg-porosimetry.

the channels opening into a certain pore-volume; from the extrusion curve the distribution of the channel-volume is obtained.

The radii of these pore channels, which allow oxygen to penetrate the material during oxidation, is given in Table 1 together with the amount of open porosity determined from the total mercury-volume intruded. As a characteristic channel radius the radius $r_{50,e}$ at 50% of the total extruded mercury-volume was chosen. The specific surface area was determined by nitrogen adsorption by the BET technique.

For materials, 1, 2 and 6 porosity consists mainly of open pores. In materials 3 to 5, which have been nitrided from injection moulded silicon compacts, only less than about 55% of the porosity is open.

OXIDATION BEHAVIOUR

Silicon nitride is unstable with respect to SiO_2. Under normal oxidizing conditions, with high oxygen partial pressure, the oxidation obeys a parabolic rate law arising from the formation of a dense oxide layer the oxygen has to diffuse through to the reaction site. In the porous RBSN two different oxidation steps occur: first the oxygen attacks the geometrical surface and the surface of the pores as well (external and internal oxidation). This stage is completed by sealing the pore-channels near the surface. In the second stage oxidation is taking place only at the geometrical surface, leading to a strongly reduced overall oxidation rate. This behaviour can clearly be seen in Fig. 3 showing the isothermal oxidation of material 6 in air. To avoid oxidation during heating, the

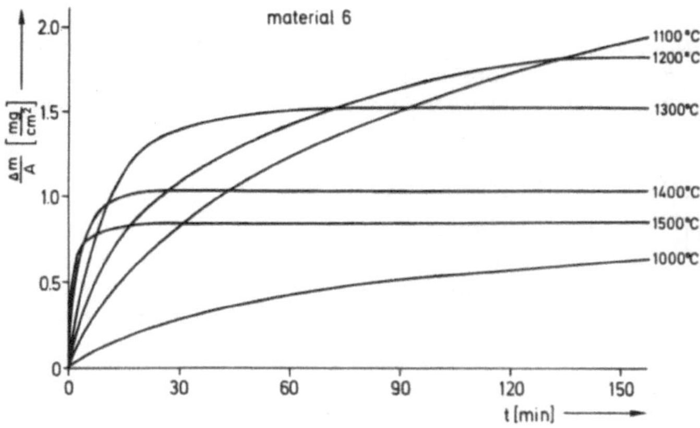

Fig. 3. Isothermal oxidation of RBSN in air.

specimens were heated in nitrogen to test temperature, then 10 l/h of dried air was passed across the specimens and the mass increase recorded by a thermal balance. The mass increase m is related to the geometrical surface A of the specimens (dimensions 12 x 1.2 x 3.5 mm). At temperatures of above 1200°C the pores are sealed in less than 3 hours. With increasing temperature the oxidation rate increases leading to a rapid blocking of the pores and a reduced overall oxidation attack. The subsequent blocking of pore-channels leads to a decreased surface exposed to the oxygen. For this type of reaction an asymptotic rate-law holds (5):

$$\frac{\Delta m}{A} = C\ (1-e^{-kt}) \qquad (1)$$

C is the final mass increase per geometrical surface, t the isothermal oxidation time and k the asymptotic rate constant. The oxidation isotherms of Fig. 3 have been fitted to this rate-law using a computer program and the rate constant was calculated. From the Arrhenius plot of the rate constant (Fig. 4) an activation energy of Q = 247 kJ/mol is obtained. The oxidation product was determined to be mostly cristobalite, with no other crystalline SiO_2-phase or oxynitride present. If any, only small amounts of amorphous silica are formed during the oxidation of RBSN (2).

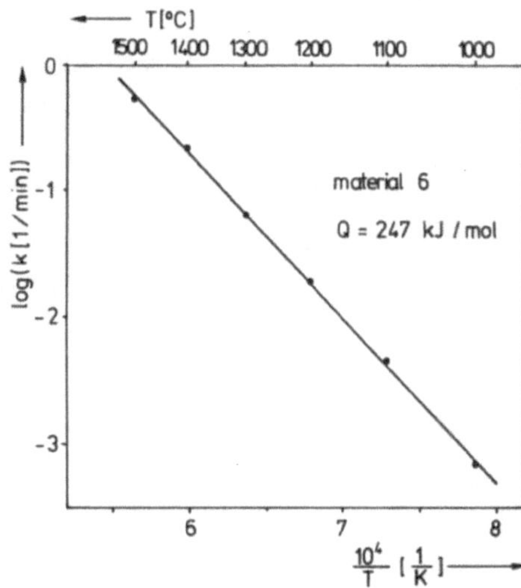

Fig. 4. Arrhenius plot of the asymptotic rate constant.

X-ray measurements of cristobalite profiles across the specimen cross section and REM examination showed that internal oxidation for material 6 was limited to a subsurface region of less than 0.6 mm at 1100°C to 1300°C and to only 0.2 mm at 1400°C oxidation temperature.

ROOM TEMPERATURE STRENGTH AFTER CYCLIC OXIDATION

Flexure strength of oxidized RBSN (dimensions 3.5 x 4.5 x 45 mm, with inner and outer spans of 20 and 40 mm respectively, five specimens for each data point) was measured at room temperature. As-fired RBSN specimens were oxidized in laboratory-air using a hydraulically operated specimen carrier, which inserted the specimens into the hot furnace, held them for 15 mins. at the test temperature and removed the specimens from the furnace to cool down for 15 mins. in ambient air.

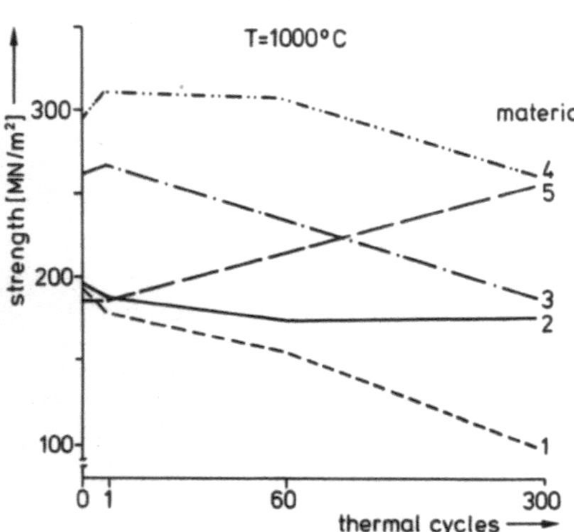

Fig. 5. Room-temperature strength after cyclic oxidation at 1000°C.

Figs. 5-8 present the room-temperature strength and the mass increase after cyclic oxidation at 1000°C and 1260°C respectively. For most of the materials flash oxidation of 15 mins. (i.e. one cycle) increases the strength. Further oxidation decreases the strength at both oxidation temperatures for all materials but one (material 5) which showed a marked tendency of crack healing especially at 1000°C. The mass gain after cyclic oxidation is low for materials 3 to 5. Materials 1 and 2 undergo severe oxidation.

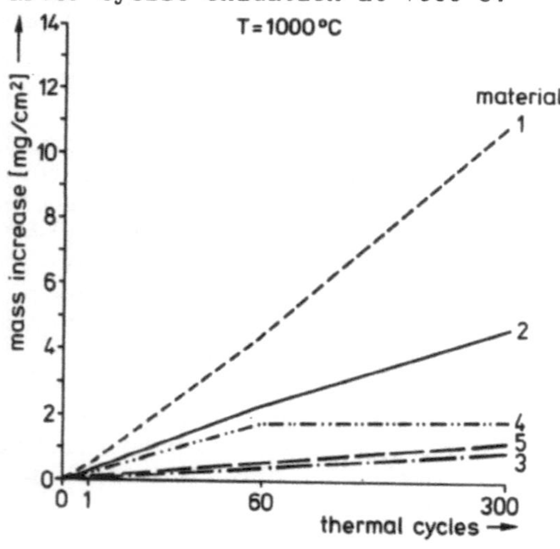

Fig. 6. Mass increase after cyclic oxidation at 1000°C.

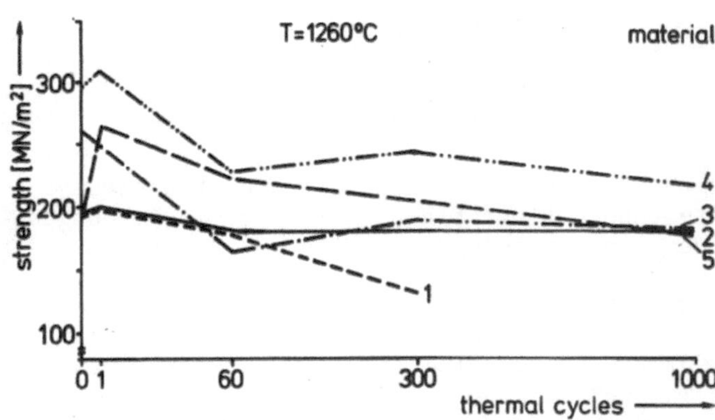

Fig. 7. Room temperature strength after cyclic oxidation at 1260°C.

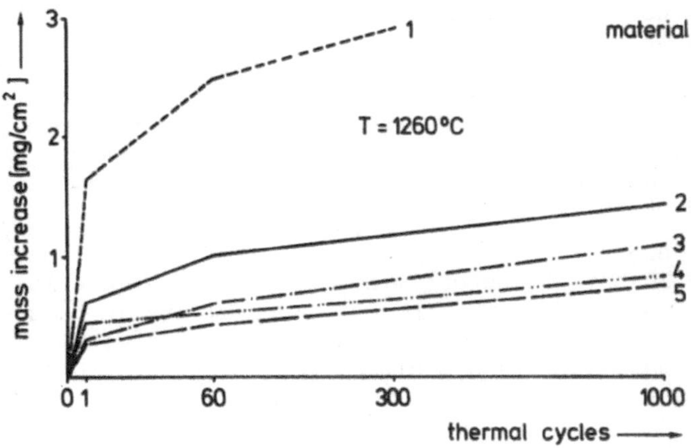

Fig. 8. Mass increase after cyclic oxidation at 1260°C.

DISCUSSION

The microstructure of RBSN controls the oxidation and the mechanical behaviour of RBSN; pores allow the oxygen to penetrate the specimen and they act as flaw limiting the strength. The internal oxidation of RBSN is limited by the sealing of the pore channels thus severely restricting subsequent attack. This is due to the 82% volume increase during the conversion of silicon nitride to silica. The blocking of pores near the surface of the material can be explained by the oxygen gradient in the pore channels. In small channels the oxidation reaction consumes more oxygen than can be carried to the reaction site via Knudsen-diffusion through the channel. The oxidation of RBSN can be described by an asymptotic rate law leading to asymptotic rate constants. The activation energy of 247 kJ/mol over the temperature range from 1000 to 1500°C is consistent with published oxidation data on Si_3N_4 powder (6,7). This value also corresponds to the activation energy for the oxidation of silicon, when crystalline SiO_2 scales are formed (8).

Since the strength of ceramics is controlled by the stress necessary to propagate the critical flaw, it is necessary, besides increasing the bulk density, to reduce the size of the critical defects. These can be single pores or pore-clusters, inclusions of second phases or large inhomogeneities and processing flaws, all at or near the surface of the bend specimen. Oxidation can influence the flaw geometry by rounding the flaw-tips or surface grooves. The volume increase during oxidation, the thermal expansion mismatch between SiO_2 and Si_3N_4 and the transformation contraction of the cristobalite during cooling have a weakening effect on the room temperature strength of RBSN.

The strength of material 5 with a relatively uniform microstructure and small pores is relatively low in the as-delivered state because of surface defects. After oxidation these defects heal and the strength values are improved. The high amount of interconnected porosity together with large pore channels leads to severe oxidation of material 1 and to an extreme strength degradation. Material 4 with narrow pore channels and a relatively homogeneous microstructure exhibits the highest strength values in the as-delivered state and after oxidation tests.

As illustrated above the microstructural features governing the properties of RBSN have to be optimized to produce a material with a good engineering capability. Recent improvements in production technology and reproducibility leading to materials with high density, very small pores and a homogeneous microstructure show that this is possible.

ACKNOWLEDGMENT

This work was supported by the Bundesministerium für Forschung und Technologie, Bundesrepublik Deutschland.

REFERENCES

1. Grathwohl, G. and F. Thümmler. J. Mat. Sci. 13 (1978) 1177.
2. Grathwohl, G., F. Porz and F. Thümmler. Proc. Brit. Ceram. Soc. 26 (1978) 129.
3. Porz, F. and E. Martin. Practical Metallography, 18 (1981) 66.
4. Cohrt, H., F. Porz and F. Thümmler. Powder Met. Int. 13 (1981) 121.
5. Evans, U.R. The Corrosion and Oxidation of Metals, London: Edward Arnold, 1960.
6. Mitomo, M. and J.H. Sharp. Yogyo Kyokai Shi, 84 (1976) 33.

7. Horton, R.M. J. Am. Ceram. Soc. 52 (1969) 121.
8. Tripp, W.C., J.W. Hinze, M.G. Mendiratta, R.H. Duff, A.F. Hampton, J.E. Stroud and E.T. Rodine. "Internal Structure and Physical Properties of Ceramics at High Temperatures"; Final Report, No. ARL-TR-75-0130, June 1975.

CREEP AND INTERNAL OXIDATION OF REACTION BONDED SILICON NITRIDE

F. Thümmler and G. Grathwohl

Nuclear Research Center and University of Karlsruhe
D - 7500 Karlsruhe
Federal Republic of Germany

ABSTRACT

It has been shown in the past, that the high-temperature deformation behaviour of Si_3N_4 is determined generally by the presence of oxygen containing phases in the microstructure and that of RBSN by internal oxidation during high temperature exposure. The oxidation and creep response of RBSN including the creep mechanism, with the stress exponent (n= 1.7-1.8) and the activation energies (360-390 kJ/mole), are outlined. The marked influence of pretreatment parameters on internal oxidation and creep can be understood by the decisive influence of pore size distribution parameters. The redistribution of porosity during creep deformation give raise for a special aspect to explain the creep mechanism. Impacts of internal oxidation on other high temperature properties are mentioned.

INTRODUCTION

In the following paragraphs some experimental results and theoretical considerations are summarized in order to characterize internal oxidation and the high-temperature-long-term deformation behaviour of reaction bonded silicon nitride (RBSN) under the latest point of view. The fracture phenomena will not be considered in this paper. Some preceding publications on RBSN from the authors laboratory (1-4) give as well additional results as more detailed information to some special aspects.

EXPERIMENTAL FACTS

It is well known that RBSN from different sources, i.e. with different densities and microstructures may exhibit very different

Riley, F.L. (ed.) Progress in Nitrogen Ceramics

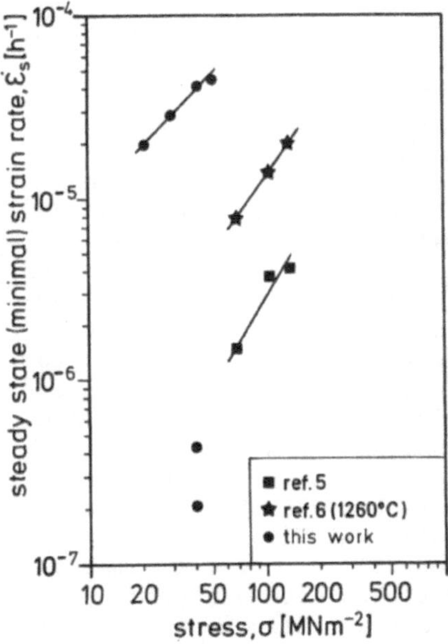

Fig.1 Steady state (minimal) creep rate of various RBSN materials at 1300°C (1260°C)

primary as well as secondary (steady state) creep behaviour under similar creep conditions. An example is shown in Fig.1 where steady state creep data from 4-point-flexure tests are shown. This behaviour has sometimes been roughly correlated with the different densities of the material, however, internal oxidation processes in the sample, influenced by microstructural parameters are more relevant to understand the creep response. A former "key experiment" to compare the creep of a low density RBSN (σ = 2,2 g/cm^3) in ambient air and in vacuum (Fig.2) demonstrates the surprising resistance of this early material, when the environment is (nearly) non-oxidative. Thus, oxidation processes via open porosity (internal oxidation) must be active to influence high-temperatue deformation. This is clearly demonstrated by the investigation of four different materials, as shown in Fig.3. The upper curves show the distribution of the micropores, obtained by penetration of mercury under increasing pressure into the sample (Hg-porosimetry). This method is described many times in literature and has been improved and explained in more detail recently in this laboratory (7), described also by F.Porz in this volume (8). The lower part of Fig.3 represents the creep curves at 1300°C, 40 MN/m^2 in ambient air, showing also the densities of the samples. Obviously, the materials with the lower micropore-size distribution and the higher densities are more creep resistant than the large-pore-size

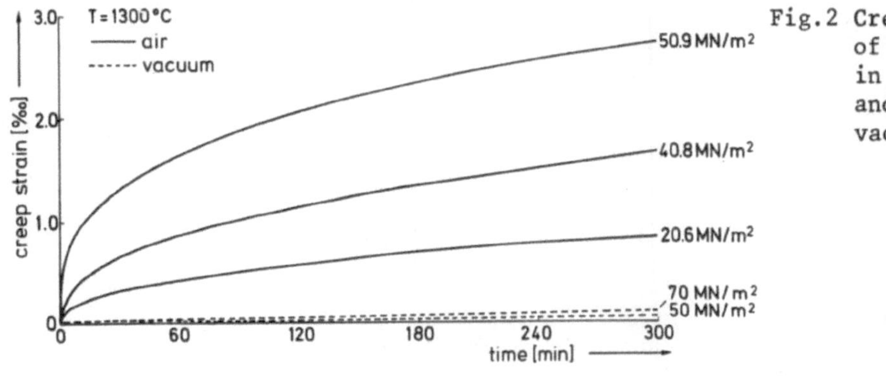

Fig.2 Creep of RBSN in air and vacuum

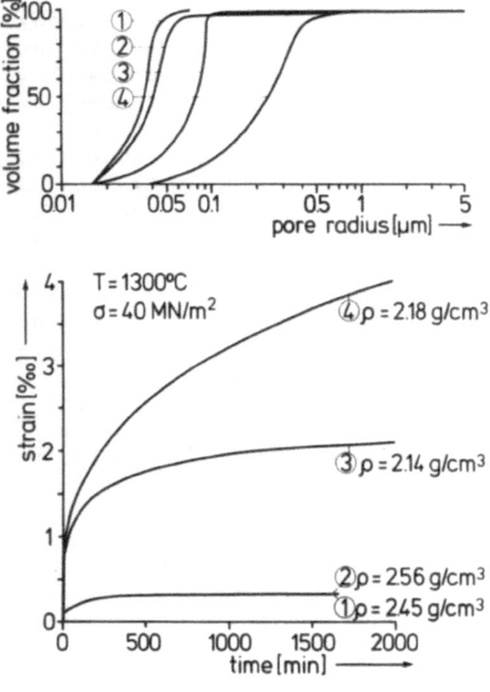

Fig.3 Creep and pore-volume distribution of RBSN

and low-density materials. A quantitative analysis of the internal oxidation of two materials (one low and one high density RBSN) after a creep experiment is demonstrated in Fig.4. The cristobalite profiles have been measured by X-ray-diffraction (RDA) beginning from the surface and going stepwise to a depth of about 2 mm; also the total SiO_2 content was determined by electron microprobe analysis (EMA)(and by Rutherford α-particle back sattering, not shown in this figure (9-10)). The SiO_2 content in the central part of the samples has also been analysed by chemical analysis (CA). One recognizes the severe internal oxidation of the large-pore-size and the moderate (up to a depth of 0.5 mm) oxidation of the small-pore-size sample. At least in the case of severe internal oxidation nearly all SiO_2 is present in form of cristobalite. When the oxidation was moderate, cristobalite could not be detected below a depth of 0.5 mm, the small amount of entire SiO_2 in the interior is, at least partly the oxygen content present before high temperature exposure.

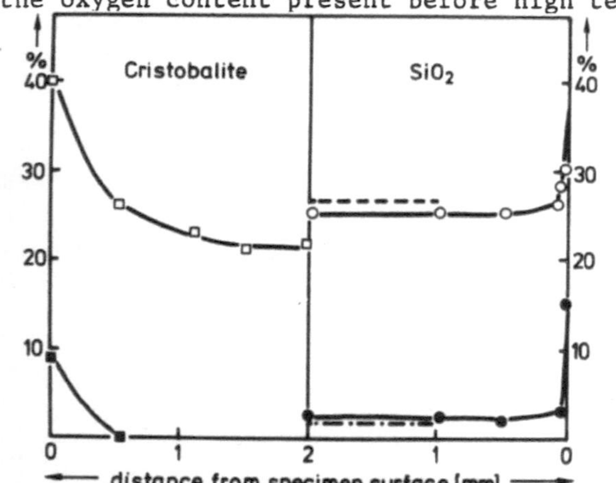

	RDA	EMA	CA
1	□	○	- - - - -
2	■	●	-·-·-·-

1) material I: heating rate: 2°/min; 1300°C, 70 MN/m², 105 h

2) material II: heating rate: 2°/min; 1400°C, 100 MN/m², 105 h

Fig.4 Cristobalite- and SiO_2-profiles in RBSN after creep tests (mat I: ρ= 2.14 g/cm³, mat II: ρ= 2.56 g/cm³)

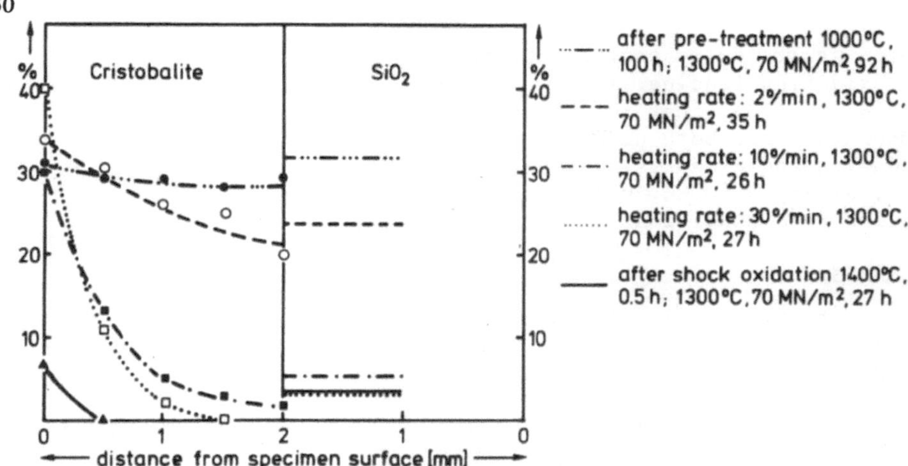

Fig.5 Cristobalite and SiO_2-profiles in RBSN (ρ= 2.14 g/cm^3) after different pretreatments and creep tests at 1300°C, 70 MN/m^2

Fig.5 shows, that even a material of a low density can be influenced drastically in its internal oxidation behaviour by a pretreatment. The heaviest oxidation takes place during annealing at an intermediate temperature, namely 1000°C. The lowest internal oxidation was observed after a shock annealing at 1400°C, both experiments were followed by a creep test. When different heating rates (2°/min - 30 °/min), without any pretreatment are used in creep testing, considerably different cristobalite profiles could be measured after creep test. Especially these results indicate the main occurrence of the internal oxidation during the heating-up period and not during the constant high temperature period itself. In Fig.6 the creep curves are collected for the samples analysed according to Fig.5. These

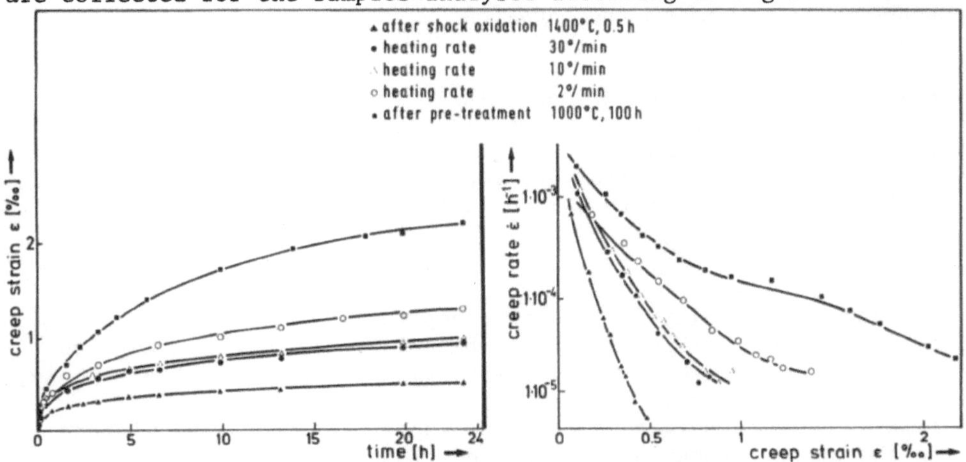

Fig.6 Creep of RBSN of low density (2.14 g/cm^3) at 1300°C, 70 MN/m^2; after different pre-treatments.

creep results fully confirm the internal oxidation finding. It has further been shown (10), that the starting temperature for both processes, i.e. oxidation and creep is virtually the same, namely ∿ 800°C, under the same heating rate conditions.

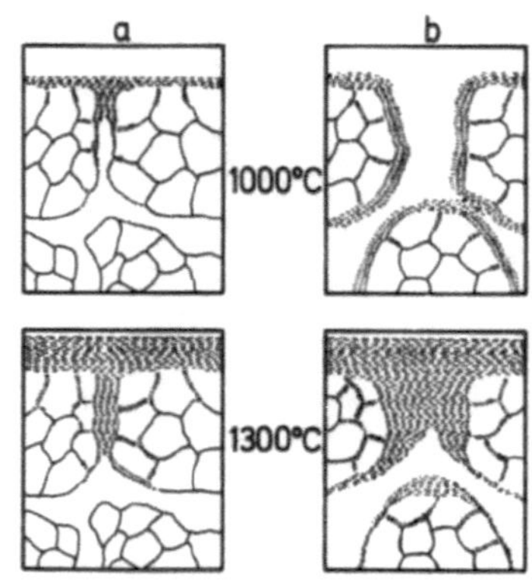

Fig.7 Model of RBSN Oxidation Material with (a) small and (b) large pore size

MECHANISMS

A principal model of internal oxidation is shown in Fig.7. It demonstrates the competition of the protective scale formation and oxidation via pore channels and grain boundaries in 2 RBSN microstructures with different pore channel sizes at 2 oxidation temperatures. It should be emphasized, that pore channels and grain boundaries must be included into the model, otherwise the creep properties could not be affected in the way mentioned before. This model cannot consider impurity effects, which are well known for the high-temperature-long-term behaviour of HPSN, but only very little investigated of RBSN. There is some evidence, that e.g. the influence of Ca cannot be ruled out for RBSN (10).

It is generally agreed that grain boundary effects including grain boundary separation and crack opening are dominating in high-temperature deformation of RBSN. The Si_3N_4 itself seems to be completely undeformed. In the steady state creep range, which can be approximated in RBSN only after a relatively long time, the stress exponent n according to the equation

$$\dot{\varepsilon}_s = A \cdot \sigma^n \cdot \exp(-Q/RT) \qquad (1)$$

was 1.7 to 1.8 in our experiment. This is compatible with the grain boundary effects mentioned before. The apparent activation energy for creep, Q, was found to be between 360-390 kJ/mol. A most interesting feature is the often very long extended transient creep range in RBSN, up to 100 h or even more. According to the equation

$$\dot{\varepsilon}_t = a \cdot t^{-C} \qquad (2)$$

Fig.8 shows a steadily decreasing creep rate to very low values between 10^{-5} and 10^{-6}/h. A similar behaviour can be observed also for some types of HPSN. During deformation an apparent strengthening effect occurs, the following explanation can be proposed for:

- Impurity elements (e.g.Ca,Mg) from the grain boundaries migrate to the surface under oxidizing conditions. This leads to a reduction of the concentration of these elements in the intergranular phases combined with an increase of their viscosity. As the creep

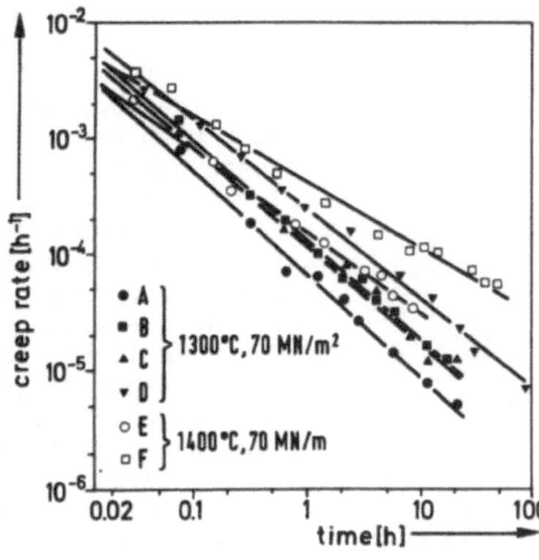

Fig.8 Transient creep of RBSN (ρ= 2,14 g/cm^3)

phenomena are controlled by viscous processes, this atomic transport is combined with a decreasing creep rate. It has also been found that a higher fraction of transgranular fracture is found in a microstructure after a heat treatment process; an increasing creep strength is caused by an increase in creep activation energy in a material with a higher degree of covalent grain boundary bonding.
The internal oxidation, described before, leads to considerable amounts of silica (or silicates) at the interfaces in the material. These amorphous phases, with their weakening effects at high temperatures, are mainly formed in the early stages of high temperature exposure. Crystallization of amorphous intergranular phases, i.e. amorphous silica to cristobalite, during the high-temperature creep test can increase the creep strength and induces a transient component of the creep rate.

Creep strain at fracture generally is very low in RBSN: In early materials exhibiting high creep rates, values up to 1 % were measured. Recent high density, low pore size materials with minimum creep rates of about 10^{-6}/h at 1300°C exhibit only about 0.2 % strain at fracture. These data are functions of temperature, stress and pretreatment and are not determined systematically.

REDISTRIBUTION OF POROSITY AND CREEP MECHANISM

It has been measured at a low density RBSN, that the creep deformation is accompanied by a considerable change in the pore size distribution, illustrated again by mercury intrusion curves (Fig.9). The upper part of this figure indicates that the open porosity is diminished by the common creep and internal oxidation processes. As a main result, the (open) pore size distribution is changed

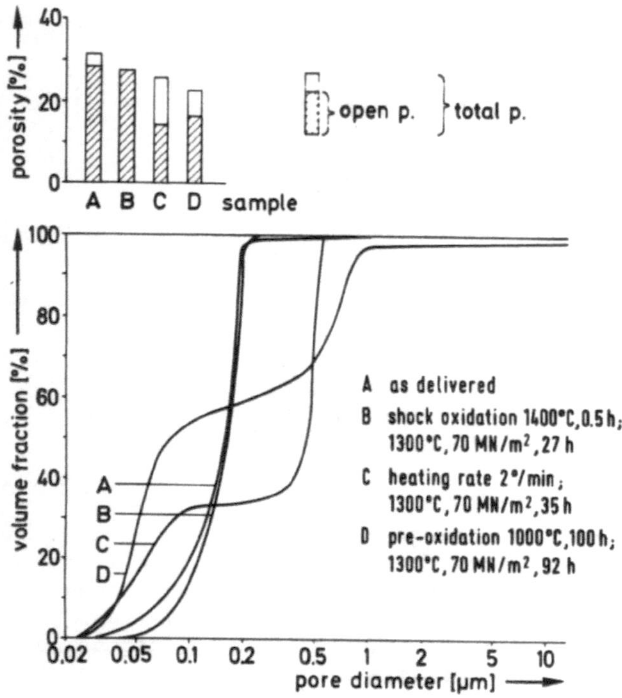

Fig.9 Porosity and pore size distribution after creep test (ρ = 2,14 g/cm^3)

very little (curve A and B), when the creep test is performed after shock oxidation, by which a high creep strength has been achieved (see Fig.6). After creep test with very low heating rate or when a 1000°C pre-treatment annealing has been applied before creep, a strong deviation of the porosity curves (C and D) from the original curve(A) has been observed. The curves split up into a part with very narrow and another part with wide pore channels.

Especially this phenomenon can be correlated to the creep mechanism. The viscous flow of the glassy phase in a dense microstructure leads to grain boundary sliding and accomodation processes like grain boundary separation or cavitation occur. In the RBSN, however, the extremely fine distributed porosity may bring an additional event to the creep phenomenon: The possibility that two adjacent grains may be separated by a pure sliding process. These processes would lead to a redistribution of the porosity and not to its increase e.g. by cavitation. Obviously, this process can be able to partly replace the accomodation, which is necessary in a dense microstructure beeing deformed by grain boundary sliding.

RELEVANCE OF INTERNAL OXIDATION TO OTHER MECHANICAL PROPERTIES

A clear-cut correlation has been worked out so far only for the influence of internal oxidation on creep. There is strong evidence, however, that the impact of this oxidation phenomenon on high temperature properties is more general. Time dependent failure (delayed fracture) as well as material degradation after long isothermal or cyclic treatment should be influenced in this way. Work on strength and oxidation after thermal cycling is presented by F. Porz in this volume (8), showing that only materials with narrow pore-channels leading to a limited oxidation attack are able to retain the room-temperature strength even after long term cyclic exposure.

CONCLUSIONS

According to the treatment in the foregoing paragraphs the following conclusions can be drawn:

- The role of dislocation motion under practical creep conditions (say 1300°C and 100 MN/m^2) is negligible, it becomes more important only at higher temperatures and high stresses.
- The pure and monophase Si_3N_4 does nearly not creep at all under practical conditions.
- The high-temperature non-elastic deformation is initiated by oxygen containing phases in the microstructure, which can be formed either after material preparation, during the exposure at high temperature (RBSN) or during materials preparation due to the sintering aids used (HPSN). These oxide phases are necessary to obtain a considerable high temperature deformation. In this context, the state of grain boundaries is extremely important as well as the stability of the microstructure in an oxidative environment during high temperature exposure.
- Creep of RBSN is, perhaps apart from the very beginning, not preferably a real deformation process, but is accompanied by initiation of cracks and crack growth, even in the primary and secondary stage. A redistribution of the microporosity has been observed during creep which leads to a special explanation of the creep mechanism.
- While the primary (transient) creep can be very extended, a tertiary creep range is normally not observed in RBSN.
- The creep strength of the recent high quality RBSN materials is no longer a problem for nearly all perspective applications. The main problem is that of strength and of strength degradation during high temperature exposure and after thermal cycling, being also influenced by internal oxidation.

ACKNOWLEDGEMENT

This work was partly supported by the Bundesministerium für Forschung und Technologie, Bundesrepublik Deutschland.

REFERENCES

1. Grathwohl, G. and Thümmler, F., J. Mater.Sci. 13 (1978) 1177.
2. Grathwohl, G., Porz. F. and Thümmler, F., Proc.Brit. Ceram. Soc. 26, (1978) 129.
3. Grathwohl, G. and Thümmler, F., Ceramurgia International 6 (1980) 43.
4. Cohrt, H., Grathwohl, G. and Thümmler, F., Res Mechanica Letters 1 (1981) 159.
5. Din, S.U. and Nicholson, P.S., J.Am. Ceram. Soc. 58 (1975) 500.

6. Mangels, J.A., "Ceramics for high-performance applications" ed. by J.J. Burke, A.E. Gorum and R.N. Katz, Brook Hill, Chestnut Hill, 1974, p. 195.

7. Cohrt, H., Porz, F. and Thümmler, F., Powder Met. Int. 13 (1981) 121.

8. Porz, F., this volume

9. Lombaard, J.M. and Meyer, D., Nucl. Instrum. Methods 145 (1977) 347.

10. Grathwohl, G., Kriechen von reaktionsgesintertem Siliziumnitrid,KfK Report 2675, 1978.

11. Porz, F., Grathwohl, G. and Thümmler, F., Proc. Brit. Ceram. Soc. 31 (1981) 157.

INFLUENCE OF CORROSION AND MICROSTRUCTURE ON MECHANICAL PROPERTIES OF SiYON CERAMICS

P. Goursat*, A. Bouarroudj* and J.L. Besson**

*Université de Limoges, Centre de Recherches et d'Etudes Céramiques, L.A. CNRS 320, Limoges.
**Ecole Nationale Supérieure de Céramiques Industrielles, L.A. CNRS 320, Limoges.

ABSTRACT. Considerable work has been undertaken to develop a new class of high performance ceramics (1), mainly for power engines. Nitrogen ceramics offer a better thermal shock resistance than oxides (2) and they hold great promise for high temperature uses. For such applications, the chemical and mechanical properties are essential. The studies carried out on hot-pressed Si_3N_4 with magnesia (3) prove that the oxidation resistance and the creep behaviour depend essentially on the properties of the intergranular phases. To hinder high temperature boundary sliding, leading to cavitation and rupture (4), the use of yttria as a densification aid is recommended by many researchers (5).

The aim of the present study is to investigate the influence of microstructure and corrosion effects on mechanical properties, especially toughness and creep resistance. Particular attention is paid to stresses induced by corrosion in the oxide scale and in the substrate.

1. EXPERIMENTAL

The powders used were 99.5% pure for Si_3N_4, 99.9% for SiO_2 and 99.99% for Y_2O_3. The median particle size was about 5 µm for all the compounds. The samples were uniaxially hot pressed at 1750-1850°C and 20 MPa in a graphite die, coated with boron nitride. Sintered materials tested had more than 95% of the theoretical density. Phases were detected by the X.R.D. technique and the microstructure investigated by S.E.M. The distribution of impurities was determined by X-ray microanalysis (E.D.A.X.). Two compositions were studied (Fig. 1):

Riley, F.L. (ed.) Progress in Nitrogen Ceramics

ceramic A = Si_3N_4-SiO_2-Y_2O_3
(90-5-5 mol %)

ceramic B = Si_3N_4-SiO_2-Y_2O_3
(70-25-5 mol %)

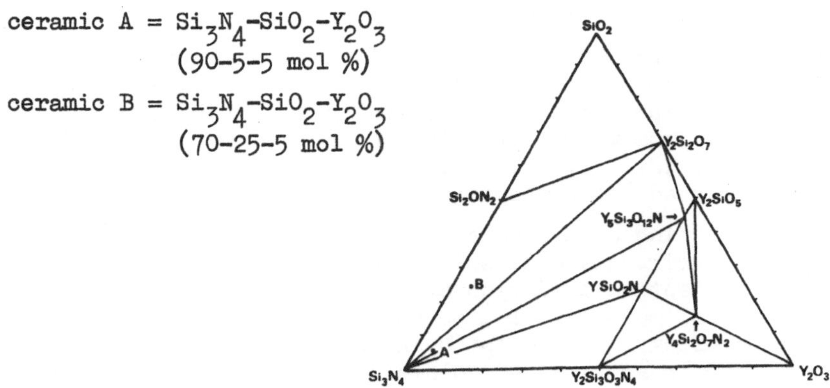

Fig. 1. The Si_3N_4-SiO_2-Y_2O_3 system.

Specimens for three point bending tests were bars (4 x 4 x 25mm) cut from discs (Ø = 35 mm, e = 6 mm). The tensile face was polished with diamond paste, and the edges chamfered. Strength and creep experiments were performed in air.

2. INFLUENCE OF CORROSION ON STRENGTH

Ceramic A.

Two batches of samples, with the same initial composition, were prepared according to the following conditions:

- batch A_1: sintered 2 hours at 1800°C; porosity 1-2%
- batch A_2: sintered 2 hours at 1850°C; porosity 4-5%.

X.R.D. analyses of the specimens showed very important changes in composition (see Table 1) for set A_2; this may be explained either by thermal decomposition of silicon nitride at 1850°C, or by preferential reactions.

Table 1

Materials	Reaction products approximate composition mol %			
	βSi_3N_4	$Y_2Si_3O_3N_4$	$Y_5Si_3O_{12}N$	$\alpha Y_2Si_2O_7$
A_1	92	traces	traces	traces
A_2	75	7	12	6

Bars, placed on a disc of the same material, were oxidised in air for 48 hours at various temperatures and then tested in 3 point bend at room temperature. The moduli of rupture and the weight gains after oxidation are summarised in Figure 2.

A_1 samples had good oxidation resistance at temperatures up to 1250°C. S.E.M. observations (Fig. 3a) indicated the presence of an oxide film which coated the residual porosity and surface defects resulting from machining. Improvement in the surface smoothness, the outward diffusion of impurities, and grain boundary depletion induced strengthening of the material.

A_2 samples, which contained a high proportion of the oxynitride phases, were strongly oxidised in the 650-1300°C temperature range. The preferential oxidation of these phases, enhanced by a noticeable residual porosity, led to an intergranular reaction and the penetration of oxygen into the bulk of the material. The formation of oxides with an expanded volume creates cracks and fresh surface area (Fig. 3c) which allow the reaction to continue. Moreover at these temperatures, silicon nitride does not react sufficiently to produce silica which would stop the reaction, and the oxide is not yet viscous enough to provide an overlapping scale. Hence, these phenomena induce a decrease in the modulus of rupture.

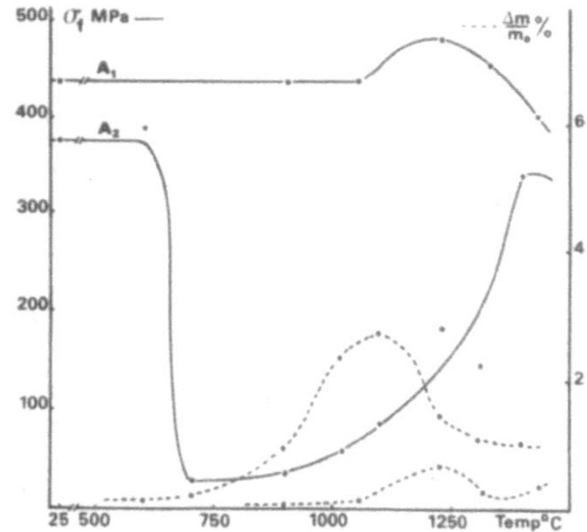

Fig. 2. Room temperature bend strength, and weight gain, after oxidation at various temperatures.

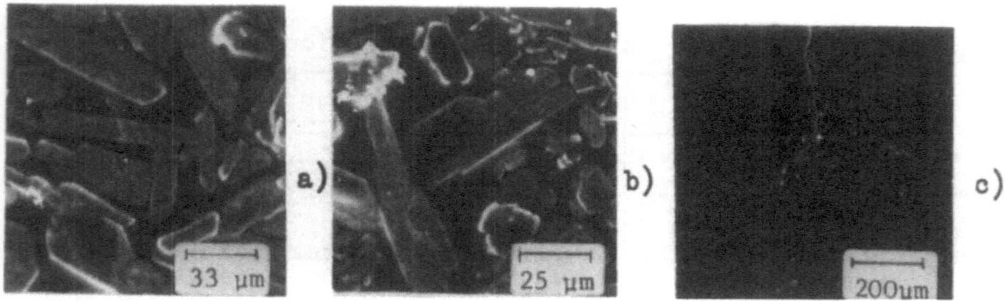

Fig. 3. Oxidation layers (a): A_1 sample at 1400°C, (b) A_2 sample at 1400°C, (c) A_2 sample at 1100°C.

But at higher temperatures, the reaction rate of silicon nitride with oxygen increases, and the silica formed gives some plasticity to the oxide scale (Fig. 3b), the strength being enhanced. However at temperatures higher than 1380°C oxidation restarts, due to the high reactivity of the intergranular phases with oxygen and to the fact that the viscosity of the overlapping product diminishes, leading to a marked decrease in the strength of the material.

Ceramic B.

Samples were hot pressed for 2 hours at 1800°C. The residual porosity was less than 1%. The X.R.D. patterns indicated the presence of Si_3N_4, Si_2N_2O and partially crystallized $Y_2Si_2O_7$. The experimental processes were the same as for ceramic A. The results are summarized in Figure 4.

The modulus of rupture, and the weight gain, change in the same way up to 1380°C. The oxidation remains fairly limited up to 1200°C, increases slightly and reaches a plateau between 1280 – 1380°C; then it increases rapidly above 1400°C. The bend strength remains the same up to 1100°C, and is enhanced when oxidation begins. This increase is due to the smoothing of surface defects such as residual porosity and machining damage. As long as oxidation is blocked by a thin protective oxide layer, the strength stays the same. At higher temperatures, the viscosity of the oxidized scale is lower, which facilitates corrosion, and the rapid increase in reaction rate leads to a drastic fall in strength. As shown elsewhere (6), two mechanisms take place successively as the temperature is raised: an intergranular diffusion process when oxides are crystallised, a second process when they are vitreous.

S.E.M. studies of the surface and the bulk permit better understanding of this behaviour. At temperatures lower than 1400°C,

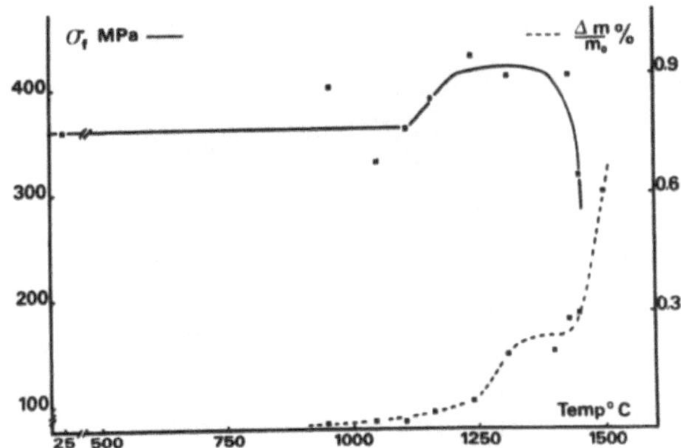

Fig. 4. Room temperature bend strength and weight gain after oxidation at various temperatures.

growing crystals are observed (Figure 5a), the size and number of which increases until the surface is entirely covered. Cross-section micrographs show that the phenomena are in fact more complex. An oxide film is observed, but also a change in the colour of the substrate above 1100°C (Figure 5b) which may be related to the oxidation of an intergranular glassy phase containing nitrogen. This penetration continues as temperature is raised, and the strength is improved. For temperatures higher than 1450°C, the oxide scale grows and the internal interface becomes irregular (Figure 5c), with a broadening of grain boundaries. Numerous bubbles which have been entrapped and which have burst at the surface can be observed.

X.R. microanalysis (E.D.A.X.) (Figure 5d) of strongly oxidized samples reveals that the scale is enriched with yttrium which is concentrated in a region near the internal interface. It is difficult to decide whether this should be considered as a heaped distribution or a layer disrupted by bubbles.

The mechanical properties of this rough, porous, scale are poor. All these defects build up a network of low resistance paths and precracks. Moreover, at the internal interface, the grain boundary broadening leads to a decohesion of the microstructure which produces a marked decrease in the modulus of rupture.

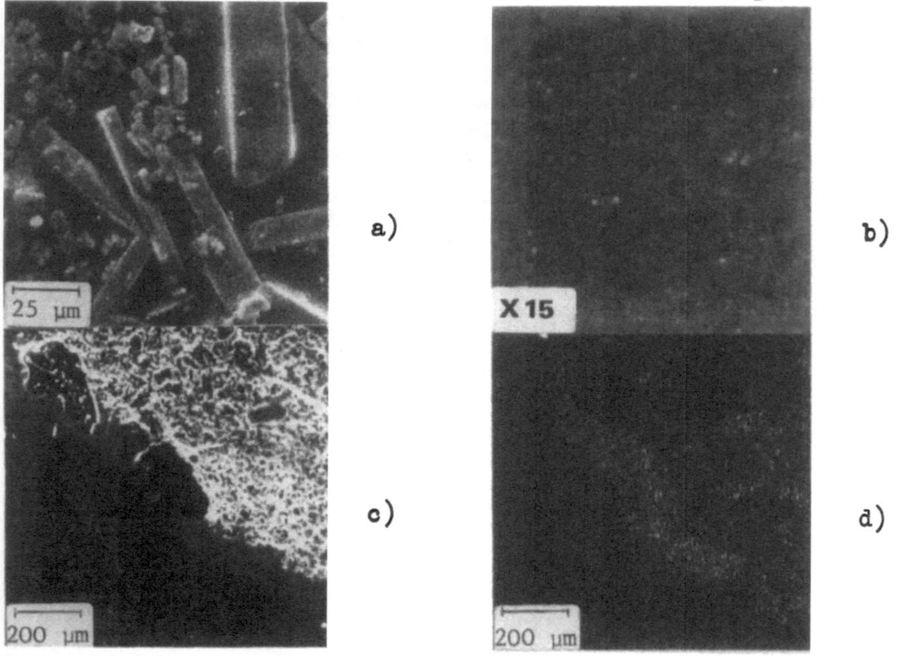

Fig. 5: (a) S.E.M. of the surface layer and (b) optic cross-section of a sample oxidized at 1400°C; (c) S.E.M. and (d) X.R. microanalysis of yttrium at the internal interface.

3. CREEP

Composition B only was studied. Three-point bend creep experiments were performed in air. Bending strength increased slightly from room temperature up to about 1200°C and then decreased rapidly (265 MPa at 1300°C). Typical creep curves are shown in Fig. 6. A "steady-state", if any, is reached only after a very long time, which suggests development of the microstructure during creep. The room temperature bend strength, determined after 100 hours creep at 1150°C, was 600 MPa, instead of 400 MPa for as-sintered material. Fracture occurs in 12 hours at 1250°C and 240 MPa, without tertiary creep being observed. Tertiary creep appears at high temperature (1300°C) and high stress (85% of the strength), but in this case, rupture occurs within a very short time.

The apparent activation energy, measured at constant creep time between 1150 and 1250°C, was ~460 kJ/mole.

The stress exponent was determined from incremental stress changes (Fig. 7). It changes from n < 1 for increasing steps to n > 1 for decreasing steps. A similar result was obtained (4) for β'-SIALON, hot pressed with 1 wt % MgO. This is interpreted in terms of a change in grain boundary microstructure with time.

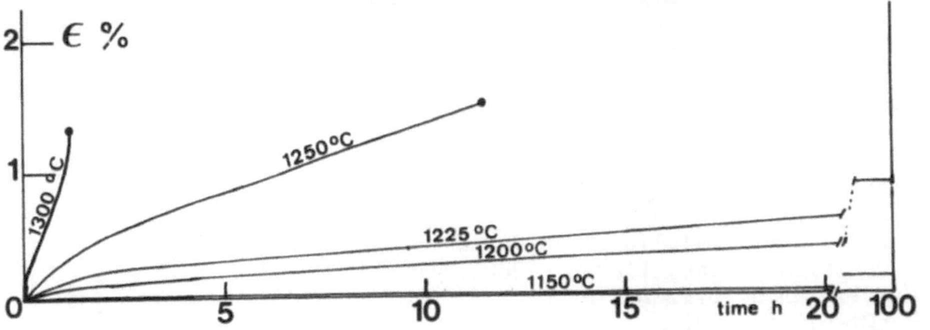

Fig. 6. Creep curves at various temperatures at a stress of 240 MPa.

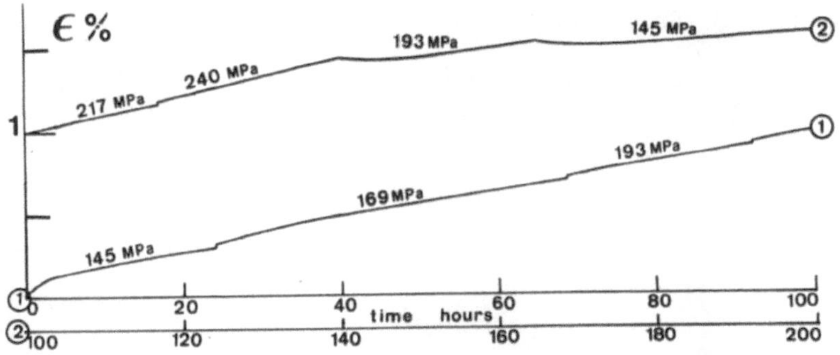

Fig. 7. Creep curve at 1200°C with incremental stress changes.

In the case of decreased applied stress "negative" creep could be observed over several hours. This behaviour suggests that the deformation includes a viscoelastic component. Similar observations are reported for RBSN (7) and for HPSN (8).

After heat treatment (100 hours, 1400°C, N_2), the grain boundary glassy phase is almost entirely crystallised. The room temperature bend strength is not affected but the material is no longer creep resistant. Fracture occurs in less than 30 minutes.

4. CONCLUSION

Results show that for SiYON ceramics the intergranular phases react with oxygen at rather low temperatures, whatever the composition may be. The difference in creep behaviour between heat-treated (N_2 atm) and as-sintered samples points to the deleterious effect of the recrystallised intergranular silicate, and the influence of oxidation on the development of vitreous phases during creep in air.

ACKNOWLEDGMENTS

Dr. P. Lortholary is gratefully acknowledged for assistance with electron microscopy (Service de microscopie, Université de Limoges).

REFERENCES

1. J.J. Burke, E.N. Lenoe and R.N. Katz, Ceramics for High Performance Applications Vol. 2 (1978).
2. Environmental Degradation of High Temperature Materials, Proc. Inst. Metallurgists Series 3, No. 13, Vol. 1, p.3/1, Chameleon Press, London (1980).
3. F.F. Lange, B.I. Davis and D.R. Clarke. J. Mater. Sci. 15, 601-610 (1980).
4. B.S.B. Karunaratne and M.H. Lewis. J. Mater. Sci. 15, 449-462 (1980).
5. G.E. Gazza. Amer. Ceram. Soc. Bull. 54, 778-781 (1975).
6. J. Desmaison, M. Brossard, M. Desmaison and P. Goursat. This Institute.
7. G. Grathwohl and F. Thümmler. Ceramurgia Int. 6, 43 (1980).
8. R.M. Arons and J.K. Tien. J. Mater. Sci. 15, 2046 (1980).

DISCUSSION

Lange: Your reaction zone in material B is likely to be a compositional gradient in Si_2N_2O, which would result in compressive stresses due to the molar volume change (and some relaxation due to deformation). Thus strengthening is expected. Your negative creep rates upon unloading are due primarily to the viscoelastic nature of the materials; that is the applied tensile stress stores an equivalent compressive stress, which when the tensile stress is reduced manifests itself by causing shrinkage. These stores stresses decay with time.

Porz: During bend tests you have a shift of the neutral axis leading to a stress redistribution after unloading and thus to negative creep rates (H. Cohrt, G. Grathwohl, F. Thümmler; Res. Mechanica Letters 1 (1981) 159).

Mangels: Can the strength reductions of "A" be related to the presence of N-Melilite? What is the distribution of this phase?

Goursat: The strength decrease is due to the presence of the $Y_2Si_3O_3N_4$ and $Y_5Si_3O_{12}N$ phases and it appears for a critical proportion of these phases. The distribution of N-Melilite is uniform throughout the bar, but stresses induced by oxidation are not randomly distributed with this symmetry.

THERMAL PROPERTIES AND THERMAL SHOCK RESISTANCE OF SILICON NITRIDE

G. Ziegler

DFVLR, German Aerospace Res. Est., Materials Lab., Cologne

1 INTRODUCTION

Heat conduction and thermal shock resistance are important properties for special high-temperature structural applications of Si_3N_4, such as the all-ceramic turbine. For example, high engine efficiency frequently requires low thermal conductivity in order to reduce heat losses. On the other hand, good thermal shock resistance calls for high values of thermal conductivity and thermal diffusivity in order to minimize thermal stresses. It is therefore essential that the values of thermal shock resistance and thermal conductivity/diffusivity of Si_3N_4, and the variables controlling them, are well understood.

Fig.1 shows the thermal diffusivity as an example of one thermal property for various forms of Si_3N_4 [1-15]: reaction-bonded Si_3N_4 (RBSN), hot-pressed Si_3N_4 (HPSN), sialon, and sintered Si_3N_4 (SSN). For example, the thermal diffusivity varies by about a factor of 10 for RBSN. There is, however, a wide range of data for the other forms of Si_3N_4. Fig.2 gives the thermal shock resistance data of Si_3N_4, characterized by the critical temperature difference ΔT_c after water and oil quench [7,9,16-18]. Again, there is a large scatter in thermal shock data of Si_3N_4. The following questions arise on the basis of these experimental data: What is the reason for the large scatter in thermal properties and thermal shock resistance of Si_3N_4? Which parameters influence the thermal properties and the thermal shock resistance of Si_3N_4? If these questions can be answered, it is possible to control these properties and to achieve the optimum values for each structural application.

The discussion of these two questions is the main objective of

Riley, F.L. (ed.) Progress in Nitrogen Ceramics

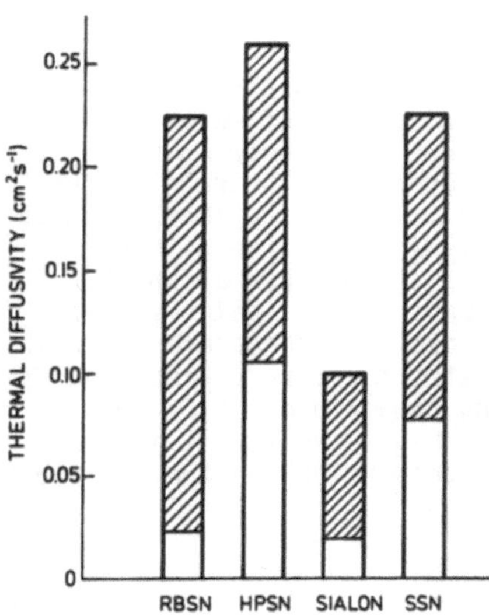

Fig.1 Room temperature thermal diffusivity of Si_3N_4; RBSN, HPSN, sialon and SSN [1-15].

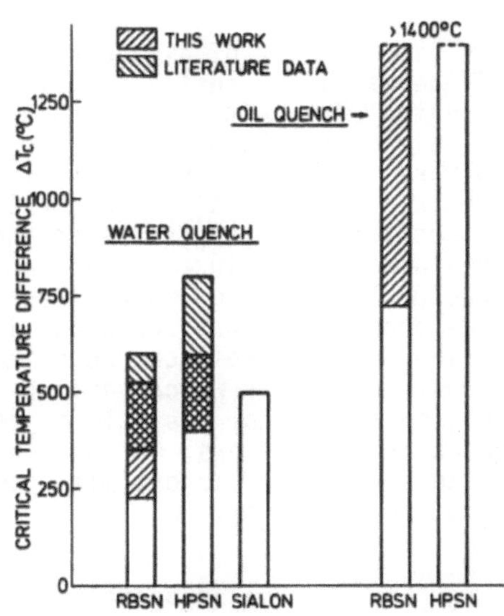

Fig.2 Critical temperature difference ΔT_c of Si_3N_4 after water and oil quench; RBSN, HPSN and sialon [7,9,16-18].

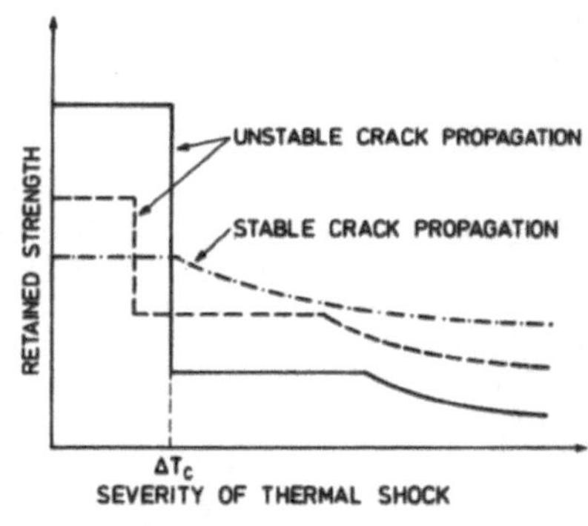

High strength ceramics ———

Low strength ceramics – – –

RESISTANCE TO FRACTURE INITIATION

$$\Delta T_c = \frac{A \cdot \sigma_F (1-\nu)}{\alpha \cdot E} \left(1 + \frac{B}{\beta}\right)$$

BIOT NUMBER $\beta = \frac{D \cdot H}{2\lambda}$

SEVERE HEAT TRANSFER ($\beta \gg 1$):

$$\Delta T_c \longrightarrow R = \frac{\sigma_F (1-\nu)}{\alpha \cdot E}$$

MILD HEAT TRANSFER ($\beta < 1$):

$$\Delta T_c \longrightarrow R' = \frac{\sigma_F (1-\nu)}{\alpha \cdot E} \cdot \lambda$$

ΔT_c	critical temperature difference
σ_F	tensile fracture stress
E	Young's modulus of elasticity
ν	Poisson's ratio
α	coefficient of thermal expansion
λ	thermal conductivity
A,B	constants, depending on the geometry
β	Biot number
D	diameter of the specimens
H	heat transfer coefficient

Fig.3 Some basic principles of thermal shock resistance.

this paper. First of all, data regarding the thermal properties and thermal shock resistance of a variety of Si_3N_4 grades are presented. The main part of this work, however, deals with the analysis of the large variations in thermal properties and thermal shock resistance on the basis of the correlation between microstructure and thermal and mechanical properties. The role of the various material properties in the prediction of the thermal shock resistance to fracture initiation of Si_3N_4 is discussed. Microstructural effects on thermal properties and on thermal shock resistance are emphasized here. These basic investigations were carried out with well-characterized laboratory grades of controlled microstructure, in which various microstructural variables were isolated as much as possible by varying powder properties and processing parameters. On the basis of these results, it is possible to control and to optimize thermal properties and thermal shock resistance of Si_3N_4.

2 SOME FUNDAMENTAL ASPECTS OF THERMAL SHOCK RESISTANCE AND THERMAL CONDUCTIVITY/DIFFUSIVITY

2.1 Thermal Shock Resistance

Fracture of ceramic materials subjected to thermal stress can occur as a result of unstable or stable crack propagation depending on the size and number of cracks participating in the fracture process (see Fig.3) [19-22]. The maximum or critical temperature difference ΔT_c, to which ceramic materials can be subjected without cracking, is dependent on the mechanical and thermal properties of the material (σ_F, E, ν, α, λ), on the specimen size and shape, and on the heat transfer coefficient of the quenching medium [19,20,22,23]. For very high Biot numbers (severe heat transfer), ΔT_c can be characterized by the thermal stress resistance parameter R; for very low Biot numbers (mild heat transfer) by the thermal stress resistance parameter R'. In most practical cases, ΔT_c can be described by a value between R and R'. High thermal stress resistance to fracture initiation calls for a high ratio σ_F/E, for low values of the coefficient of thermal expansion α and, depending on the quenching conditions, for a high thermal conductivity λ.

Quenching experiments clearly show unstable crack propagation for all investigated grades of dense Si_3N_4, characterized by relatively high initial strength, and for all investigated RBSN grades (Figs. 4 and 5) [9,16,17]. It should, however, be noted here that, for RBSN materials with rather low initial strength and with significant data scatter in the as-nitrided state, there may be a range of ΔT_c in which excessive coefficients of variation (Fig.5 - material B after oil quenching) or a pronounced bi-modal distribution of all strength data (Fig.5 - material C) is observed [24]. Never-

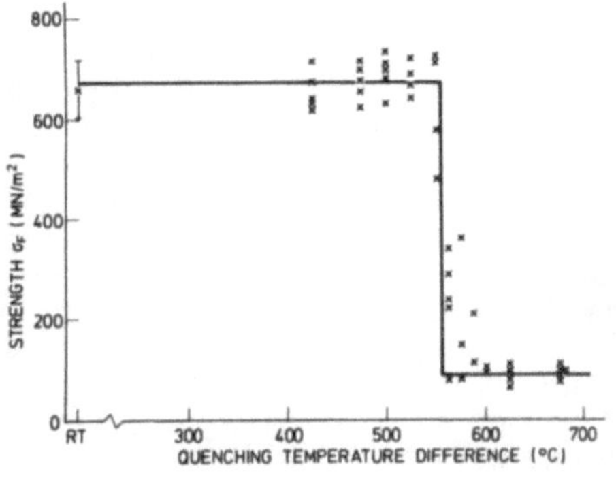

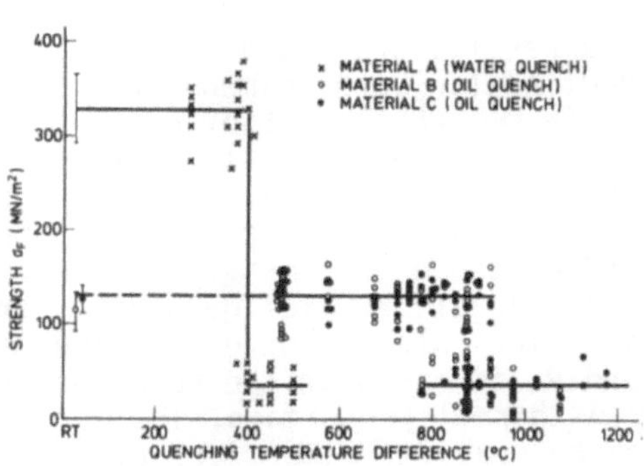

Fig.4 Strength as a function of quenching temperature difference after water quench for HPSN with 3 % MgO.

Fig.5 Strength as a function of quenching temperature difference after oil and water quench for various RBSN grades.

Material A: coefficient of variation 10.9 %
Material B: coefficient of variation 18.0 %
Material C: coefficient of variation 12.1 %

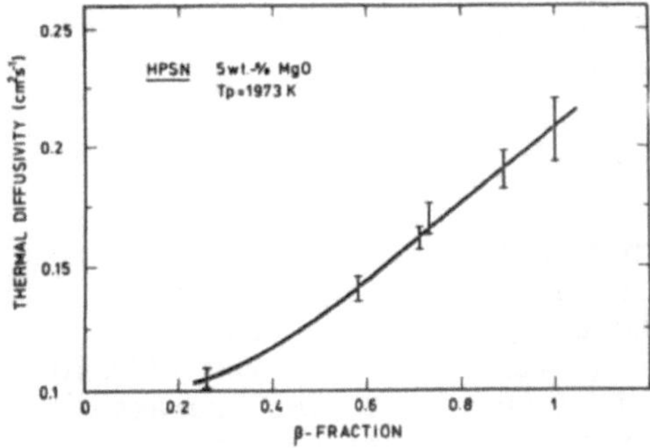

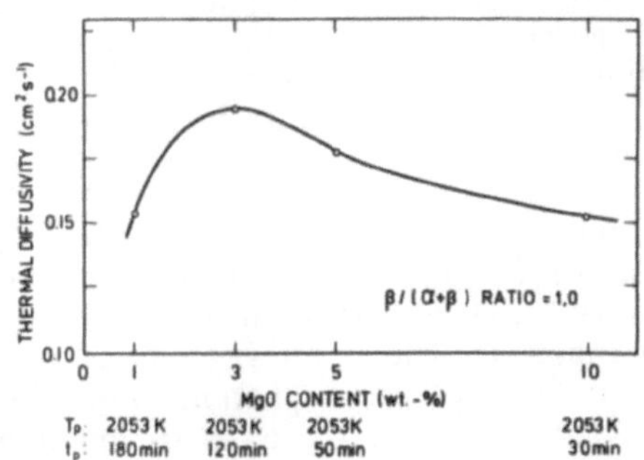

Fig.6 Thermal diffusivity of HPSN as a function of β-fraction.

Fig.7 Thermal diffusivity of HPSN as a function of MgO content.

theless, unstable crack propagation was proved in all cases.

The results presented in this investigation were obtained using rectangular bars, dimensioned 5 by 5 by 50 mm (HPSN) or 4.5 by 3.5 by 45 mm (RBSN) or 6 by 6 by 60 mm (RBSN - only materials B and C in Fig.5). The thermal shock resistance was characterized by the thermal stress resistance parameter R for severe heat transfer; by the thermal stress resistance parameter R' for mild heat transfer; and by the experimentally determined critical temperature difference ΔT_c after water quenching, only in some cases after oil quenching. The calculation of the thermal stress resistance parameters R and R' is based on the room temperature values of the mechanical and thermal properties. The calculation of ΔT_c for Si_3N_4 and for the given specimen dimensions after oil and water quench (Fig.3), as well as the experimental results show that the thermal shock resistance of Si_3N_4 can be described by a value between R and R' for the given experimental conditions. Two consequences can be drawn from these results: The thermal conductivity and thermal diffusivity of Si_3N_4 play an important role in the evaluation of thermal shock resistance to fracture initiation. Microstructural effects on thermal shock resistance should be discussed on the basis of the changes in mechanical properties, in particular the ratio σ_F/E, and also of those in thermal properties, caused by microstructural effects.

2.2 Thermal Conductivity and Thermal Diffusivity

In general, heat transfer can be effected by electron transport, by phonon transport, and by internal radiation at high temperatures [25]. Heat conduction in structural ceramics takes place at temperatures which are higher than room temperature, indeed up to about 1500 °C for dense, and about 500 °C for porous ceramics, primarily as a result of phonon transport. Thermal conductivity by phonon transport is given by this equation:

$$\lambda = \frac{1}{3} c_p \cdot v \cdot l$$

where c_p is the specific heat, v is the wave velocity and l the mean-free-path of phonons. In the temperature range mentioned above, differences in thermal conductivity are mainly due to changes in the mean-free-path of the phonons. The values of the mean-free-path of phonons reported in literature vary from 200 to 4 Å. The mean-free-path of phonons is decreased by two processes: By phonon-phonon interactions, which are important over a wide temperature range. This process describes the temperature dependence of thermal conductivity and diffusivity. For temperatures higher than the Debeye temperature, λ is proportional to the inverse of the absolute temperature. Secondly, by the scattering of phonons at various lattice imperfections. This process describes the microstructural effects on thermal conductivity and thermal

Table 1 Microstructural effects on phonon conductivity.

Single-phase crystalline ceramics:
• structure and composition of pure materials
• grain boundary effects
• impurities and solid solutions
Single-phase non-crystalline ceramics:
• phonon mean-free-path is limited to the order of interatomic distances by the random structure
• influence of composition → small effects
• temperature dependence → small effects
Multiphase ceramics:
• conductivity of each phase
• amount and arrangement of phases
• porosity, grain size, texture, impurity content, degree of crystallinity, microcracking
• example porosity: influence of total porosity, size, shape and orientation of pores

diffusivity. At temperatures higher than room temperature, the imperfection scattering is independent of temperature. As a consequence, microstructural effects on thermal conductivity become important at low temperatures, where the thermal scattering mean-free-path is large. Based on these considerations, thermal conductivity and thermal diffusivity are expected to depend strongly on microstructural and compositional effects. Some microstructural effects in single-phase crystalline and non-crystalline ceramics, and also in multiphase ceramics, are listed in Table 1 [25-32].

Which parameters might affect the thermal conductivity and thermal diffusivity of Si_3N_4? Table 2 summarizes possible microstructural effects on thermal conductivity in dense and reaction-bonded Si_3N_4. For dense Si_3N_4, solid solution lowers thermal conductivity [2,4-6]. As the phonon mean-free-path in Si_3N_4 is much smaller than the grain size at temperatures above room temperature, any effect of the size and shape of the grains in dense Si_3N_4 seems unlikely [25]. Glassy phases decrease thermal conductivity. As regards RBSN, literature data show that total porosity [7], oxidation products [8], and microcracks [8,17] caused by oxidation lead to a decrease in thermal conductivity. Metallic free silicon content is expected to increase thermal conductivity [7].
These microstructural effects are known from literature. However,

Table 2 Influence of microstructural parameters on thermal conductivity and thermal diffusivity of Si_3N_4.

Dense Si_3N_4		RBSN	
• solid solution	↓	• total porosity	↓
• residual porosity	?	• pore size	?
• phase composition α/β	?	• pore structure	?
• grain size	~	• silicon content	↑
• glassy phase	↓	• phase composition α/β	?
• orientation	?	• grain structure	?
		• SiO_2 (oxidation)	↓
		• microcracking	↓

what is the influence of the other microstructural variables? The answer is given in the following on the basis of the correlations of various microstructural characteristics and thermal diffusivity data, measured at room temperature. Thermal diffusivity, a, is defined by the expression $a = \lambda/\rho \cdot c_p$ in cm^2s^{-1}, where ρ is the density of the material in gcm^{-3}, and c_p the specific heat in $Jkg^{-1}K^{-1}$ (λ in $Wm^{-1}K^{-1}$). Thermal diffusivity describes the heat conduction under unsteady-state conditions, when the temperature at any point varies with time. On the basis of this definition, thermal conductivity can be computed from diffusivity data if the density of the material and the specific heat are known. Thermal diffusivity can be measured by the so-called laser flash technique [3,28,33]. Using this technique, a high-intensity, short-duration laser flash is absorbed on the front surface of a thin specimen. The resulting rear face temperature rise is measured by an infrared detector or by a thermocouple. More details of thermal diffusivity measurements are given elsewhere [10].

3 MICROSTRUCTURAL EFFECTS ON THERMAL DIFFUSIVITY AND THERMAL SHOCK RESISTANCE OF Si_3N_4

3.1 Hot-Pressed Si_3N_4

3.1.1 Microstructural Changes

The influence of the following microstructural parameters is discussed in this paper:

- The α to β transformation, which is closely connected to the change in grain morphology;

- the influence of grain growth after complete phase transformation;
- the influence of the amount of glassy phase (variation of the amount of the sintering aid MgO for 100 % β-phase) and
- the influence of orientation.

These microstructural parameters were changed by systematic variation of time and temperature during the hot-pressing process and by variation of the composition of the starting powder using MgO as sintering aid [13,34,35]. Microstructure was characterized by measuring the density, the β/(α+β) ratio (x-ray analysis), the mean grain intercept and the aspect ratio of the grains by scanning electron microscopy of polished and etched surfaces [13,36]). More experimental details are given in [10,11,13].

3.1.2 Thermal Properties

Fig.6 shows the changes in thermal diffusivity as a function of the β-fraction, that is also the changes in thermal conductivity, because the specific heat and the density are approximately constant for all these specimens. Thermal diffusivity strongly increases with increasing β-amount. Some explanations can be given for these experimental results: For low β-amounts (short hot-pressing time) residual pores might lower the thermal diffusivity. In the case of higher β-amounts (longer hot-pressing times), differences in the crystal structure of the α- and β-modifications may affect thermal diffusivity. The more strained crystal structure of the α-phase [37] may enhance phonon scattering and lower thermal diffusivity. The most likely reason is that an enrichment of the impurities in the grain boundaries takes place during the α- to β-transformation caused by a liquid phase sintering mechanism. It can therefore be concluded that the purity of the β-phase is higher than that of the α-phase. As discussed above, impurities are effective phonon scatterers. The higher purity of the β-phase is thought to increase thermal diffusivity with increasing β-amount.

Fig.7 gives the data for the thermal diffusivity as a function of MgO content after complete α- to β-transformation. With increasing MgO content, the thermal diffusivity first increases to a maximum at about 3 wt.% MgO, followed by a decrease at high MgO content. The increase in thermal diffusivity at low MgO content is thought do be due to the reduction of residual pores. The decrease in thermal diffusivity with higher MgO content can be attributed to the increasing amount of the glassy grain boundary phase. As pointed out above, glassy materials have much lower values of thermal diffusivity and thermal conductivity than crystalline solids.

There is one more variable which affects the thermal diffusivity and thermal conductivity of HPSN, and this is the orientation effect. In Fig.8, the thermal diffusivity is plotted as a function

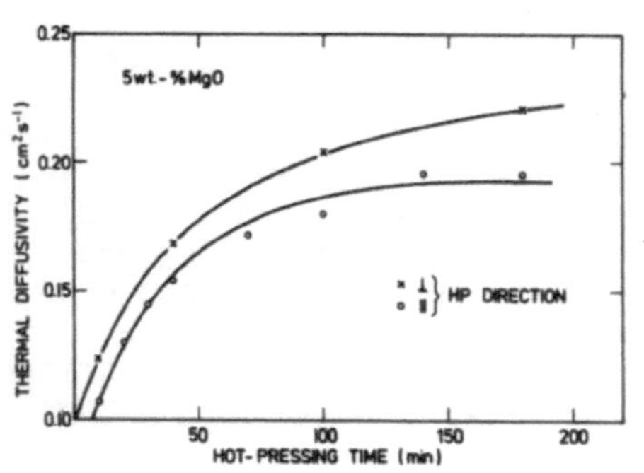

Fig.8 Effect of hot-pressing direction on thermal diffusivity of HPSN with 5 wt.% MgO.

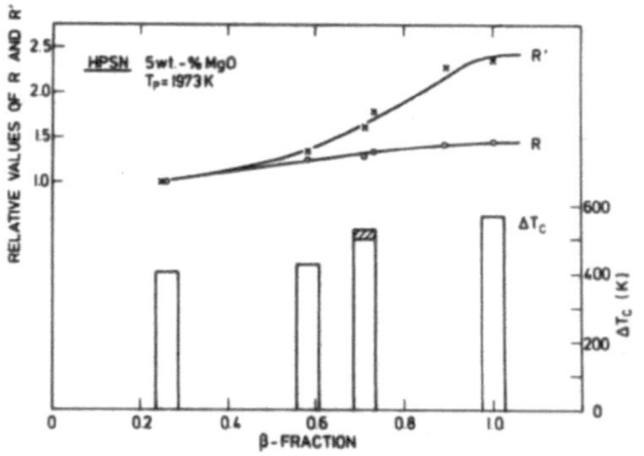

Fig.10 Relative values of R and R', and data of ΔT_c after water quench, as a function of β-fraction in HPSN.

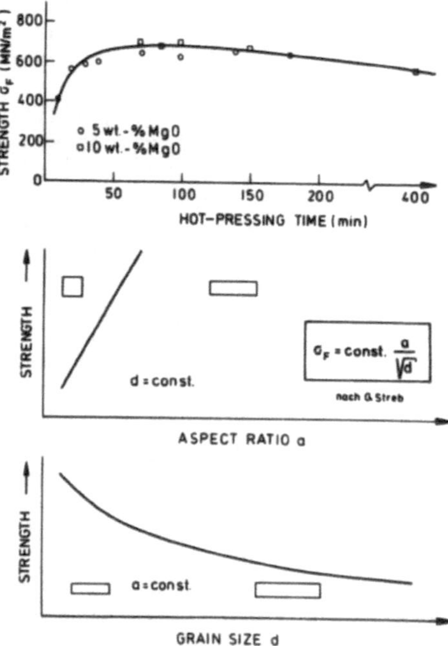

Fig. 9 Fracture strength as a function of hot-pressing time for two sets of specimens with 5 and 10 wt.% MgO, and schematic plot of the two microstructural effects aspect ratio and grain size.

of hot-pressing time, that is, as a function of β-amount, measured parallel and perpendicular to the hot-pressing direction. The thermal diffusivity values perpendicular to the hot-pressing direction are higher (about 15 %) than the corresponding values parallel to the hot-pressing direction. The reason for this observation is that the elongated β-grains exhibit a preferred orientation perpendicular to the hot-pressing direction. Similar orientation effects on mechanical properties were observed by F.F. Lange [38].

3.1.3 Thermal Shock Resistance

Important criteria for the thermal shock resistance to fracture initiation are the ratio σ_F/E and the thermal conductivity λ. As, in the case of HPSN, the variation in Young's modulus of elasticity is small (less than 5 % for the investigated materials), microstructural effects on thermal shock should be discussed both on the basis of the changes in fracture strength and in thermal conductivity. Fracture strength of HPSN is mainly controlled by two microstructural parameters (Fig.9: experimental strength data as a function of hot-pressing time for 2 sets of specimens) [13,34, 35,38-41]: Firstly, by the elongated β-grains formed during the α- to β-transformation, which lead to an improvement of the mechanical properties (increase in σ_F). Secondly, by the overall grain size; grain coarsening leads to a decrease in strength. Both effects on strength are schematically shown in Fig.9.

What are the microstructural effects on the thermal shock behaviour of HPSN? Based on the correlation of the essential microstructural variables and the mechanical and thermal properties it is possible to predict variations in thermal shock, caused by microstructural effects. The role of the various material properties in the prediction of the thermal shock resistance to fracture initiation, and microstructural effects on these properties are summarized in Table 3 (the increase of the microstructural parameters leads to the marked changes: ↑ increase; ↓ decrease): The α- to β-transformation leads to an increase in strength, to a strong increase in thermal conductivity, and to only a slight decrease in Young's modulus of elasticity. As a result of these variations, an increase in R and an essential increase in R' is expected with increasing β-amount. Grain coarsening causes a decrease in strength, in particular for long hot-pressing times. The decrease in strength at nearly constant Young's modulus and thermal conductivity should result in a slight decrease in R and R'. The change in thermal conductivity as a function of MgO content - the increase to a maximum at 3 wt.% MgO followed by a decrease at higher MgO content - should lead to a variation of R'. R is mainly controlled by the grain structure and only reveals small changes for all MgO contents because of the nearly constant ratio σ_F/E.

Table 3 Influence of microstructural characteristics on thermal shock resistance of HPSN

characterized by

$$R = \frac{\sigma(1-\nu)}{\alpha \cdot E} \text{ AND } R' = \frac{\sigma(1-\nu)}{\alpha \cdot E} \cdot \lambda \text{ WITH } \lambda = a \cdot \rho \cdot c_P$$

• α TO β TRANSFORMATION	σ↑ E↓ λ↑	R↑ R′↑
• GRAIN SIZE	σ↓ E≈ λ≈	R↓ R′↓
• MgO CONTENT	σ≈ E↓ λ↑↓	R ≈ R′↑↓

As a result of the observed variations in mechanical and thermal properties, the microstructural characteristic in dense Si_3N_4, which has the strongest influence on thermal shock resistance, is the α- to β-transformation. Fig.10 gives the relative changes of R and R', and, in addition, the experimental data of ΔT_c after water quench, as a function of the β-fraction. With increasing β-amount, R increases by a factor of about 1.4, R' by a factor of 2.5. This improvement of thermal shock resistance is confirmed by quenching experiments in water: Increase in ΔT_c with increasing β-fraction. Compared to the influence of the α- to β-conversion, the effect of grain size and the effect of the amount of glassy phase on thermal shock resistance is small.

3.2 Reaction-Bonded Si_3N_4

3.2.1 Microstructural Changes

The following effects on thermal properties and thermal shock resistance are discussed for this material group: The influence of pore fraction, pore size, phase composition, grain structure, oxidation on thermal diffusivity and thermal shock resistance to fracture initiation.

Investigations were carried out in part with commercial grades, and, in particular, with laboratory grades in which the influences of various microstructural parameters were isolated as much as possible. Microstructure was changed by introducing artificial pores, by varying the size of the silicon starting powder, the green density and by post-heat-treatment of the as-nitrided samples [9,14,42,43]. Microstructural characterization included density measurements, mercury porosimetry (characterization of micropores), quantitative x-ray analysis, and light and scanning electron microscopy of unetched and etched surfaces [44,45]. Experimental details are given in [9,10,12,14,17,43].

3.2.2 Thermal Properties

Influence of density: In Fig.11, thermal diffusivity is plotted against the relative density (as a % of the theoretical density). A linear correlation between thermal diffusivity and density was found by Larsen and Bortz [7]. In the present work, two points can be deduced from the measurements: Regarding the low density material, there is a general tendency towards increasing thermal diffusivity with an increase in density. For the density range between 68 and 83 %, however, no clear density dependence was found. For example, at constant bulk density, a variation in thermal diffusivity from 0.04 to 0.11 cm^2s^{-1} was observed. As a consequence, it may be concluded that thermal diffusivity and thermal conductivity of RBSN in this density range should mainly be controlled by other microstructural parameters.
What other microstructural parameters have a significant influence on the thermal properties of RBSN? As regards commercial grades, it was not possible to find a clear correlation between thermal properties and any microstructural parameter. The reason for this is that in most cases in commercial grades a superposition of various microstructural effects on thermal diffusivity takes place and, therefore, it is very difficult to decide which parameter is the decisive one controlling the thermal properties. For this reason, laboratory grades were investigated.

Influence of the size of macropores: Table 4 presents the influence of the size of spherical artificial macropores between 48 and 100 µm on thermal diffusivity. Thermal diffusivity data clearly indicate that, at constant bulk density, the size of the spherical artificial macropores has no effect on thermal diffusivity. This result is in agreement with theory.

Table 4 Influence of the size of macropores on thermal diffusivity of RBSN

		Density	Pore size (µm)		$\alpha/(\alpha+\beta)$	Thermal
		(gcm^{-3})	macro	micro	ratio	diffusivity ($cm^2\ s^{-1}$)
Set I	D1	2.39	48	0.137	0.71	0.099
	D2	2.39	66	0.129	0.73	0.101
	D3	2.41	100	0.097	0.64	0.098

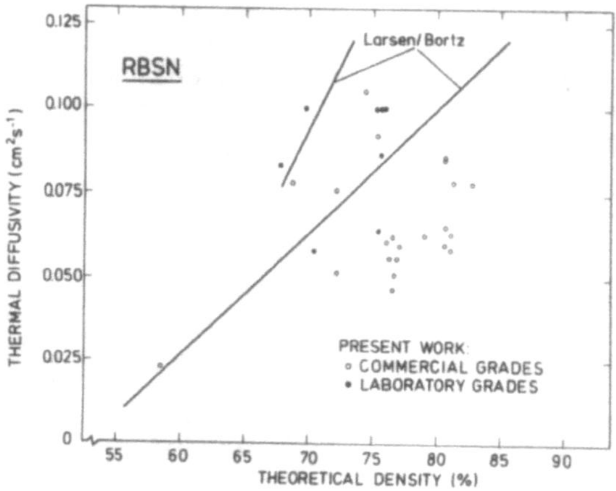

Fig.11 Thermal diffusivity of RBSN as a function of density.

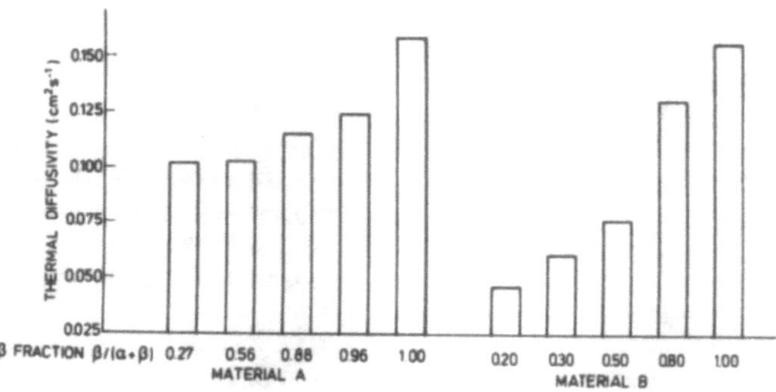

Fig.12 Influence of microstructural changes during post-annealing on thermal diffusivity of RBSN. Material A: coarse-grained; Material B: fine-grained.

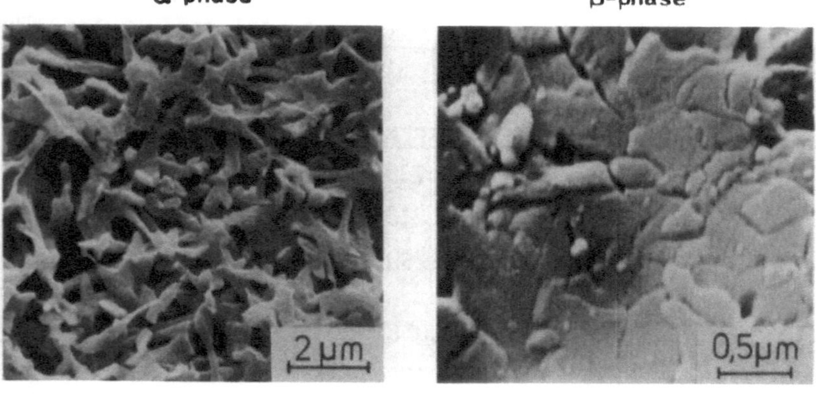

Fig.13 Morphology of the phases α and β in RBSN.

Influence of the amount and morphology of the α-phase: In two other sets, the amount of the phases α and β was changed by varying the particle size of the starting silicon powder. The total porosity in each set was nearly constant. Microstructural changes and thermal diffusivity data are summarized in Table 5. Using this preparation method, the variation of phase composition is closely connected to the variation in grain morphology (indicated in Table 5 by the size of the micropores). It can be concluded from these results that there is a correlation between the α-amount, the size of the micropores and the thermal diffusivity data: Room temperature thermal diffusivity of RBSN is strongly influenced by the amount and size of the α-needles. As indicated in both sets, increasing α-amount results in a significant decrease in thermal diffusivity. For the processing technique used here, high α-amounts are closely associated with small micropores and with fine α-needles. These results were confirmed by post-heat-treatments of as-nitrided materials in the temperature range of 1480 to 1780 °C. The main characteristic of the post-annealing is the α- to β-transformation connected with changes in grain morphology, the coarsening of the needle-like grain structure of the α-phase, and the increase in the size of micropores [14]. Thermal diffusivity data after various annealing treatments are shown in Fig.12 for two different starting materials with fine and coarse grain structure. In both cases thermal diffusivity increases with increasing β-amount. The microstructural effect on thermal diffusivity, caused by post-annealing, is much higher for the fine-grained starting material.

Some reasons may be given for these microstructural effects: As mentioned before, the more strained crystal structure of the

Table 5 Influence of the amount and morphology of the α-phase on thermal diffusivity of RBSN.

		Density (gcm^{-3})	Pore size (μm) macro	micro	α/(α+β) ratio	Thermal diffusivity ($cm^2\ s^{-1}$)
Set II	A	2.25	8	0.130	0.78	0.058
	B	2.17	22	0.240	0.75	0.083
	C	2.23	34	0.250	0.66	0.100
Set III	A	2.41	7	0.058	0.82	0.060
	B	2.42	19	0.108	0.70	0.084
	C	2.38	28	0.152	0.60	0.109

α-phase, the fibre-like morphology of the α-phase and, possibly, the size and the amount of micropores. It seems that, in particular, the morphology of the phases α and β has a strong influence. It is widely agreed that the α-phase exists in a kind of fine needle-like morphology (Fig.13) [44-46]. The complex-shaped micropores of different sizes are partly thought to correspond to the spacings between the α-needles [44,47]. In contrast, under usual nitridation conditions, the β-phase is equiaxed and more coarse-grained [44,46].
It is known from literature that fibrous microstructures exhibit lower thermal conductivity than a matrix containing isolated equidimensional pores [48]. For this reason, the thermal conductivity/diffusivity of RBSN could be decreased with an increasing amount of the needle-like α-phase. For a given α-content, the thermal properties are also dependent on the morphology of the α-phase. As demonstrated in Fig.14, fine microstructures reveal lower thermal properties, mainly due to the increase in phonon scattering at the fibre surface.

Influence of oxidation: Oxidation is a further microstructural effect on the thermal diffusivity of RBSN. A marked minimum in thermal diffusivity over the oxidation temperature range of 1000 to 1250 ^{o}C was observed by Hasselman et al. [8]. The strong variation in thermal diffusivity may be caused by different effects: The formation of the oxidation products (crystalline and amorphous SiO_2), the penetration depth of the oxidation products (the formation of a protective surface layer at high oxidation temperatures obviously has no effect on the overall thermal diffusivity), and, possibly, microcracks formed during cooling from oxidation temperature [17].

3.2.3 Thermal Shock Resistance

What correlation is to be expected between thermal shock resistance to fracture initiation and microstructural variables (Table 6)?

Influence of density: In general, higher densities lead to higher strength values. The porosity dependence of strength can be described by an exponential form ($\sigma_F = 400 \exp(-3.9\, f_p)$) [43]. Young's modulus of elasticity and thermal conductivity increase with increasing density. The porosity dependence of Young's modulus of elasticity can be expressed by a linear relationship [49,50] or also by an exponential form [51]. Nevertheless, the experimental data for σ_F/E in this work show a general tendency to increase with higher densities. As a consequence, changes in the thermal stress resistance parameter R are determined by the increasing ratio σ_F/E. Here it should, however, be kept in mind that density changes are usually superimposed by variations in other microstructural parameters. In particular, the influence of the macro-

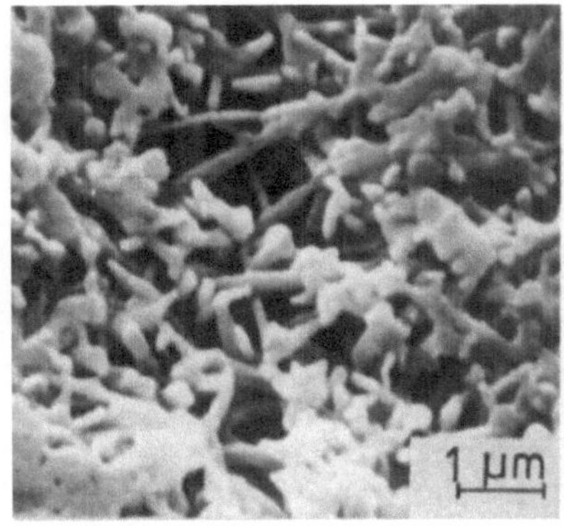

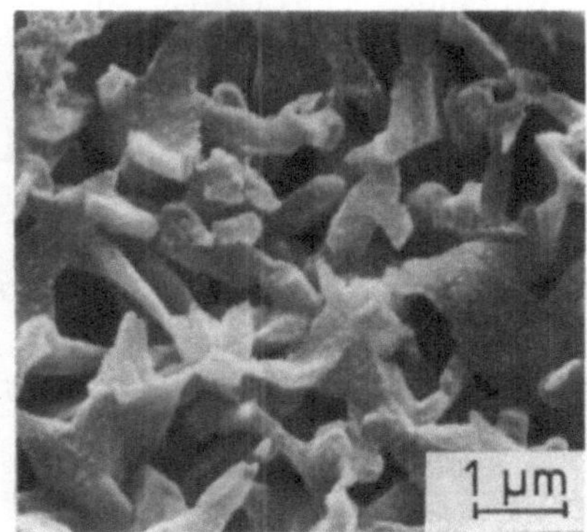

Fig.14 Microstructure of 2 RBSN grades with different grain structure, indicating large differences in thermal diffusivity.
Material A: a = 0.06 cm^2s^{-1} Material B: a = 0.11 cm^2s^{-1}

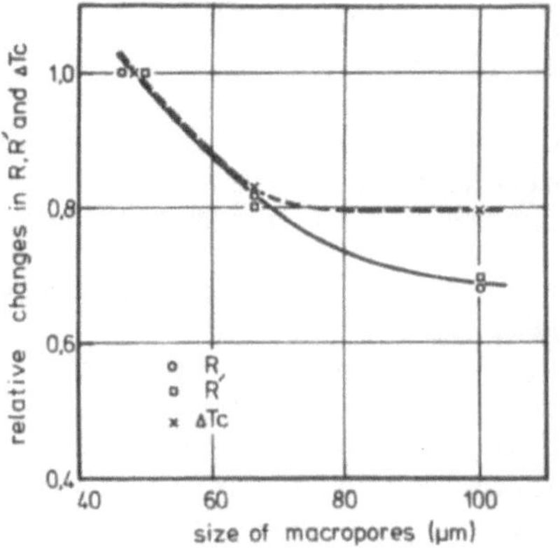

Fig.15 Relative changes of R and R', and of ΔT_c after water quench, as a function of the size of macropores in RBSN (ρ = 2.4 gcm^{-3}).

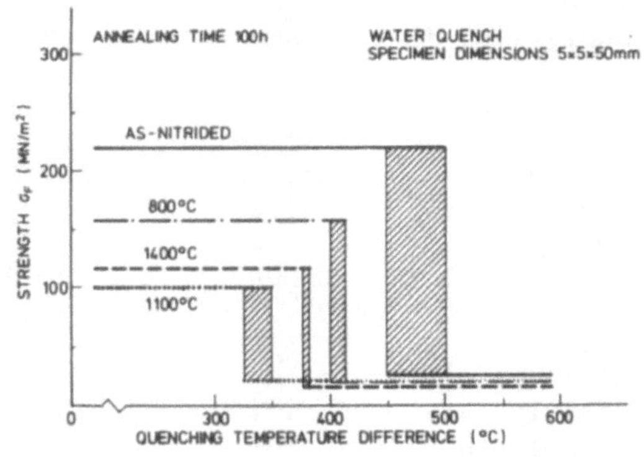

Fig.16 Influence of the extent of oxidation on the thermal shock resistance of RBSN: strength as a function of quenching temperature difference (ρ = 2.45 gcm^{-3}).

Table 6 Influence of microstructural characteristics on thermal shock resistance of RBSN

characterized by $R = \dfrac{\sigma(1-\nu)}{\alpha \cdot E}$ AND $R' = \dfrac{\sigma(1-\nu)}{\alpha \cdot E} \cdot \lambda$ WITH $\lambda = a \cdot \rho \cdot c_p$		
• DENSITY	$\sigma \Uparrow\downarrow$ $E \Uparrow$ $\lambda \Uparrow$	$R \Uparrow\downarrow$ $R' \Uparrow$
• PORE SIZE	$\sigma \Downarrow$ $E \approx$ $\lambda \approx$	$R \Downarrow$ $R' \Downarrow$
• α - PHASE	$\sigma \uparrow$ $E \approx$ $\lambda \Downarrow$	$R \uparrow$ $R' \uparrow\Downarrow$
• OXIDATION	$\sigma \Downarrow$ $E \downarrow$ $\lambda \Downarrow$ $\alpha \Uparrow\downarrow$	$R \downarrow$ $R' \Downarrow$

pore size can cause high data scatter. A strong increase in the thermal shock resistance parameter R' is expected for higher densities as a result of the increase in thermal conductivity.

Influence of the size of macropores: At constant total porosity the fracture strength decreases with increasing pore size [43,52]. There is no change in Young's modulus and thermal conductivity for nearly equiaxed pores and for constant total porosity [53]. The decrease in strength leads to a decrease in the ratio σ_F/E and therefore in R and, because of the constant values of thermal conductivity, to the same relative changes in R'.

Influence of the amount and morphology of the α-phase: In recent investigations, a slight increase in strength was observed with increasing α-amount, whereby other microstructural parameters were constant [54]. As already discussed in detail, the amount and morphology of the α-phase have a strong effect on thermal diffusivity and thermal conductivity. The higher the α-content and the finer the α-needles, the lower is the thermal diffusivity. The slight increase in strength can lead to a small increase in R. R' is mainly expected to decrease with increasing α-content. In reality, a superposition with other microstructural variables, for example changes in the macropore size, can occur depending on the processing procedure [9]. In this case, the effect of each microstructural characteristic on σ_F, λ, R and R' is to be analysed.

Influence of oxidation: The influence of oxidation on thermal shock resistance is a very complex problem [17]. As a consequence of the variations observed in microstructure, in thermal and mechanical properties during exposure to various temperatures, it is to be expected that oxidation will strongly influence thermal shock resistance. As the maximum thermal stresses are built up in the near-surface region during cooling, the thermal shock resistance of oxidized RBSN is mainly controlled by the type and amount of the oxidation products, and by the size of the flaws in the surface layer. Fracture strength can change in both directions; strength degradation has been observed in most cases [17,55]. Young's modulus of elasticity is only slightly reduced by the oxidation products. Thermal diffusivity/conductivity can be strongly decreased by the oxidation treatments. Moreover, oxidation products cause changes in the coefficient of thermal expansion α and, as a consequence, lead to variations in thermal shock resistance. Higher thermal stresses are expected because of the relatively high α-value of cristobalite, compared to Si_3N_4. However, it should be mentioned that the formation of a certain amount of amorphous silica with a low α-value may complicate the estimate. Considering the variations in mechanical and thermal properties, a significant decrease in thermal shock resistance is expected in most cases for long exposure times.

Figs. 15 and 16 give two examples proving these considerations. Firstly, experimental data illustrating the influence of the size of artificial macropores: The relative changes of R, R' and ΔT_c with pore size are plotted in Fig.15. Quenching experiments in water reveal that ΔT_c decreases with increasing pore size, as expected from the variations in R and R'. Secondly, experimental data of ΔT_c illustrating the influence of oxidation: After oxidation at various temperatures, ΔT_c is shifted very clearly to lower values, for this material from 500 to 325 °C. The strong decrease in ΔT_c after oxidation and, in most cases, the lower ΔT_c-values at higher oxidation temperature generally prove the discussed qualitative estimate. Here it should, however, be noted that the influence of oxidation on mechanical and thermal properties and on thermal shock to fracture initiation greatly depends on the microstructure of the materials under investigation and on the oxidation conditions.

4 CONCLUSIONS

The microstructural effects on thermal diffusivity and thermal shock resistance of HPSN and RBSN are summarized in Figs. 17 to 20.

4.1 Thermal Diffusivity

Thermal diffusivity/conductivity of HPSN (Fig.17) is strongly influenced by the phase composition, the α/β-ratio and the amount of glassy phase, as well as by orientation effects. High β-amounts lead to an increase in thermal diffusivity. Maximum values were observed between 3 and 5 wt.% MgO. The orientation effect, caused

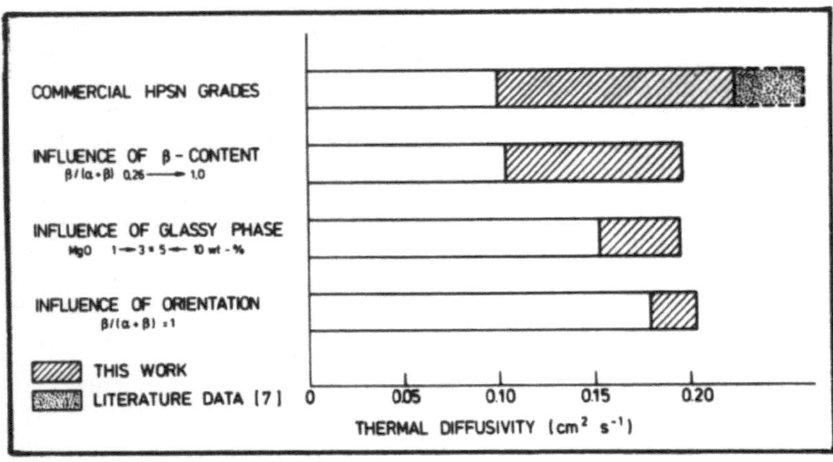

Fig.17 Microstructural effects on thermal diffusivity of HPSN.

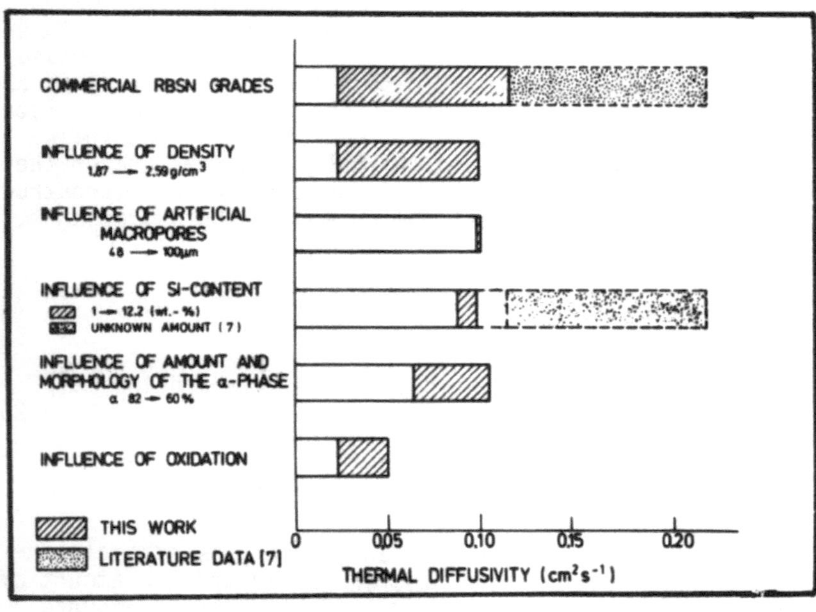

Fig.18 Microstructural effects on thermal diffusivity of RBSN.

by hot-pressing, results in variations of about 15 % in thermal diffusivity under the applied experimental conditions.

The experimental data for RBSN lead to the following correlations and conclusions (Fig.18):

- Thermal diffusivity data of commercial RBSN grades cover a wide range. It was not possible to find a clear and unequivocal correlation of the experimental values and single microstructural variables. Nevertheless, the wide variation in thermal diffusivity can be explained by investigating laboratory grades with well-prepared and well-characterized microstructures.
- There is a general tendency to increasing thermal diffusivity with an increase in density. In many cases, however, density changes are superimposed by variations in other microstructural variables. In particular, this is the case for the investigated materials in the density range between 68 and 83 % t.d.
- As demonstrated by the laboratory grades, the size of spherical artificial macropores has no effect on thermal properties at constant bulk density, and when other microstructural variables are also constant.
- Up to now, the influence of unreacted metallic silicon content is not quite clear. Very high thermal diffusivity data in literature were interpreted on the basis of high silicon amounts [7]. Thermal diffusivity measurements on laboratory grades with silicon contents between 1 and 12 wt.%, whereby other microstructural characteristics were constant, revealed only small changes [56].
- The results clearly indicate that the amount and morphology of the α-phase have a strong influence on thermal diffusivity and thermal conductivity of RBSN. An increase in the α-amount leads to a decrease in thermal diffusivity and conductivity. Microstructures with small α-needles exhibit low thermal properties.
- Literature data show that there is a strong decrease in thermal diffusivity for certain oxidation treatments.

4.2 Thermal Shock Resistance

Figs. 19 and 20 summarize the thermal shock results of HPSN and RBSN. These graphs clearly indicate the influence of various microstructural parameters on thermal shock resistance to fracture initiation, here characterized by a quality factor indicating positive or negative changes. Positive effects indicate an improvement of thermal shock resistance with increasing microstructural parameters; negative effects indicate a detrimental influence on thermal shock with an increase in a single microstructural variable. Moreover, these graphs demonstrate whether the variations in thermal shock resistance are caused by changes in fracture strength or in the thermal conductivity.

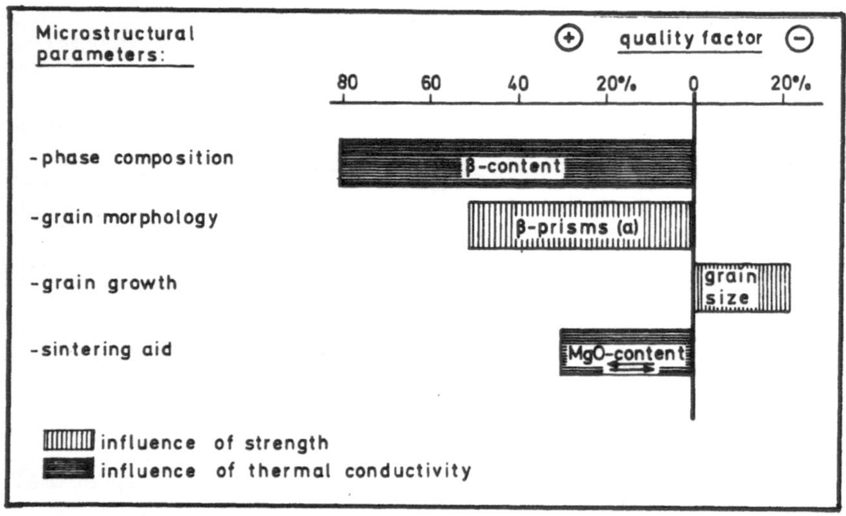

Fig.19 Microstructural effects on thermal shock resistance to fracture initiation of HPSN.

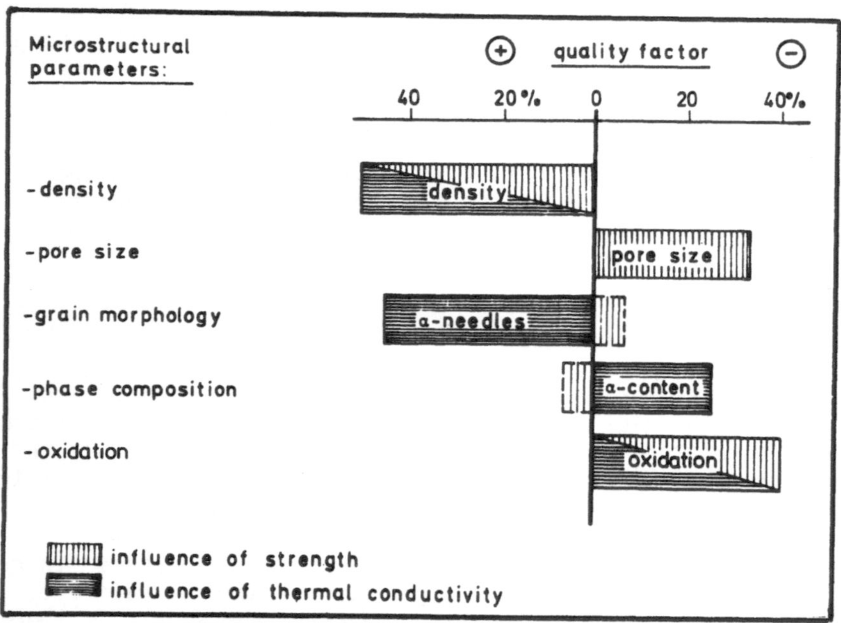

Fig.20 Microstructural effects on thermal shock resistance to fracture initiation of RBSN.

Thermal shock resistance to fracture initiation of HPSN is controlled by the changes in mechanical and thermal properties which are strongly influenced by grain morphology, by overall grain size and by phase composition, that is, the β-content and the amount of the glassy phase. Fracture strength up to about 1000 ^{o}C is mainly determined by the elongated β-grains (positive effect), and by the grain growth (negative effect). Thermal conductivity is strongly influenced by the β-content and the amount of glassy phase. The results show that the strongest microstructural effect on thermal shock resistance of HPSN is the α- to β-transformation. Maximum thermal shock resistance of HPSN has been achieved after complete α- to β-transformation, and with small grain sizes. Further improvement can be obtained by optimizing the amount of sintering aid.

Thermal shock resistance of RBSN (Fig.20) is strongly affected by the microstructure, and by the corresponding changes in mechanical and thermal properties. High density values generally improve the thermal shock resistance. When the total porosity and other microstructural parameters are constant - as in the case of the incorporation of artificial pores - the influence of pore size on thermal shock resistance to fracture initiation is determined only by the increase in flaw size with increasing pore size. As a consequence, an increase in pore size leads to a decrease in ΔT_C. High α-amounts and very fine grain structures of the α-phase cause a decrease in the thermal shock resistance. The influence of oxidation is strongly dependent on the microstructure of the material and on the oxidation conditions. In most cases, thermal shock resistance is decreased by oxidation effects. In many cases, a superposition of various microstructural effects (density, macropore size, amount and morphology of the α-phase) takes place, depending on the processing conditions. Here, it is necessary to analyse the influence of each microstructural characteristic on fracture strength, thermal conductivity and thermal shock resistance. In this connection, it should be mentioned that, for a more quantitative calculation of the thermal shock resistance to fracture initiation, the temperature dependence of the properties, in particular of the thermal conductivity, is of importance [57]. It can be concluded from these results that for RBSN good thermal shock resistance to fracture initiation calls for highly dense material, for microstructures with small macropore size, not too high α-amount and coarse-grained α-phase.

ACKNOWLEDGEMENTS

This study was partly supported by the Deutsche Forschungsgemeinschaft (D-5300 Bonn-Bad Godesberg) and partly by the DFVLR, German Aerospace Research Establishment, Cologne.

REFERENCES

1. W. George, Proc. Brit. Ceram. Soc. 22 (1973) 147.
2. W.J. Arrol, in "Ceramics for High Performance Applications", Eds. J.J. Burke, A.E. Gorum and R.N. Katz, Brook Hill Pub., Chestnut Hill, MA (1974), 729.
3. F.F. Lange et al., J. Amer. Ceram. Soc. 59 (1976) 454.
4. T. Hirai et al., Amer. Ceram. Soc. Bull. 57 (1978) 1126.
5. M. Kuriyama et al., ibid. 57 (1978) 1119.
6. R.J. Lumby, B. North and A.J. Taylor, in "Ceramics for High Performance Applications - II", Eds. J.J. Burke, E.N. Lenoe and R.N. Katz, Brook Hill Pub., Chestnut Hill, MA (1978), 893-906.
7. D.C. Larsen and S.A. Bortz, Tech.Rep. AFML-TR-79-4188 (Oct 1979).
8. W. Zdaniewski, D.P.H. Hasselman, H. Knoch and J. Heinrich, Amer. Ceram. Soc. Bull. 58 (1979) 539.
9. G. Ziegler and J. Heinrich, Ceramurgia Internat. 6 (1980) 25.
10. G. Ziegler and D.P.H. Hasselman, J. Mater. Sci. 16 (1981) 495.
11. G. Ziegler et al., J. Amer. Ceram. 64 (2), (1981) C-35.
12. G. Ziegler and R. Ziegler, Special Ceramics 7, Ed. D. Taylor and P. Popper, B. Ceram. R.A. (1981) 133.
13. G. Ziegler and H. Knoch, ibid. 7 (1981) 145.
14. G. Ziegler and J. Heinrich, Science of Ceramics 11 (in press).
15. G. Ziegler, to be published.
16. G. Ziegler and R. Leucht, Ber.Dt.Keram.Ges. 55 (1978) 105.
17. G. Ziegler, Science of Ceramics 11 (in press).
18. N. Claussen, Int. Report Max-Planck-Institut Stuttgart (1980).
19. D.P.H. Hasselman, J. Amer. Ceram. Soc. 52 (1969) 600.
20. D.P.H. Hasselman, Ceramurgia Internat. 4 (1978) 147.
21. N. Claussen and D.P.H. Hasselman, in "Thermal Stresses in Severe Environments", Eds. D.P.H. Hasselman and R.A. Heller, Plenum Press, (1980) p. 381-395.
22. G. Ziegler and D.P.H. Hasselman, to be published.
23. W.D. Kingery, J. Amer. Ceram. Soc. 38 (1955) 3.
24. G. Ziegler and D.P.H. Hasselman, Ceramurgia Internat. 5 (1979) 126.
25. W.D. Kingery, "The Thermal Conductivity of Ceramic Dielectrics" in Progress in Ceramic Science, Ed. J.E. Burke, Pergamon Press (1962) p. 182-235.
26. J.E. Matta and D.P.H. Hasselman, J.Amer.Ceram.Soc. 58 (1975) 458.
27. W.D. Kingery, H.K. Bowen and D.R. Uhlmann, "Introduction to Ceramics", 2nd edn. John Wiley, New York (1975).
28. D.P.H. Hasselman et al., J. Amer. Ceram. Soc. 58 (1976) 241.
29. R. Berman, "Thermal Conductivity in Solids", Clarendon Press, Oxford (1976).
30. H.J. Siebeneck et al., J. Amer. Ceram. Soc. 59 (1976) 84.
31. D.P.H. Hasselman et al., J. Amer. Ceram. Soc. 61 (1978) 590.
32. D.P.H. Hasselman, J. Comp. Mater. 12 (1978) 403.
33. W.J. Parker et al., J. Appl. Phys. 32 (1961) 1697.
34. H. Knoch and G. Ziegler, Science of Ceramics 9 (1977) 494.
35. H. Knoch and G. Ziegler, Ber.Dt.Kerm.Ges. 55 (1978) 242.

36. H. Knoch, R. Leucht and G. Ziegler, Sonderbände der Praktischen Metallographie 9 (1978) 255.
37. C.M.B. Henderson and D. Taylor, Trans. J. Brit. Ceram. Soc. 2 (1975) 49.
38. F.F. Lange, J. Amer. Ceram. Soc. 56 (1973) 518.
39. H. Knoch and G.F. Gazza, Ceramurgia Internat. 2 (1980) 51.
40. G. Himsolt, H. Knoch, H. Hübner and F.W. Kleinlein, J. Amer. Ceram. Soc. 62 (1979) 29.
41. H. Knoch and J. Heinrich, Z. f. Werkstofftechnik 10 (1980) 361.
42. J. Heinrich, Ber. Dt. Keram. Ges. 55 (1978) 239.
43. J. Heinrich, D. Munz and G. Ziegler, Powd. Met. Internat. (in press).
44. J. Heinrich and G. Streb, J. Mater. Sci. 14 (1978) 2083.
45. G. Ziegler and R. Ziegler, Microstructural Science 9 (1981) 91, Eds. Petzow, Paris, Albrecht, McCall, Elsevier North Holland, Inc.
46. S.C. Danforth and M.H. Richman, Metallography 9 (1976) 321.
47. S.C. Danforth and M.H. Richman, J. Mater. Sci. 14 (1979) 240.
48. H.J. Siebeneck, R.A. Plenty, D.P.H. Hasselman and G.E. Youngblood, Amer. Ceram. Soc. Bull. 56 (1977) 572.
49. A.G. Evans and R.W. Davidge, J. Mater. Sci. 5 (1970) 314.
50. J.W. Edington, D.J. Rowcliffe and J.L. Henshall, Powd. Met. Internat. 7 (1975) 82 and 136.
51. A.J. Moulson, J. Mater. Sci. 14 (1979) 1017.
52. J. Heinrich and D. Munz, Amer. Ceram. Bull. 59 (1980) 1221.
53. Z. Hashin, J. Appl. Mech. 29 (1962) 143.
54. J. Heinrich and H. Hausner, in "Energy and Ceramics", Ed. P.Vincencini, Elsevier Scient. Pull. Comp., Amsterdam/Oxford/New York, (1980) p. 780-792.
55. J.E. Siebels, presented at "Ceramics for High Performance Applications III, Reliability", Orcas Island, July 1979.
56. G. Ziegler and J. Heinrich, to be published.
57. G. Ziegler, to be published.

THERMAL SHOCK RESISTANCE OF TWO NITROGEN CERAMICS

P. BOCH and J.C. GLANDUS

LA 320 ENSCI LIMOGES (France)

ABSTRACT Thermal shock resistance of silicon oxynitride and aluminium nitride has been studied in comparison with alumina. Experimental data are discussed in relation to thermal shock resistance parameters.

1 - THERMAL SHOCK RESISTANCE PARAMETERS

Nitrogen ceramics are competitors for being used in thermal engines because of their interesting properties : a toughness K_{Ic} as high as 8 $MPa\sqrt{m}$ and a strength which can exceed 1000 MPa, a low creep velocity at high temperatures, a low thermal expansion (α_{20}^{1000} < 3.5 10^{-6} K^{-1}), rather low elastic moduli (Young's modulus $E \sim 3.10^5$ MPa) and a rather high thermal conductivity ($k \sim 50 Wm^{-1}K^{-1}$) all data leading to a good thermal shock resistance. This paper deals with thermal shock resistance of silicon oxynitride Si_2N_2O and aluminium nitride AlN, in comparison with alumina.

Many studies have described the features of thermal shock damages and specified that two analysis exist :

- The thermoelastic analysis(1) concerns the initiation of the cracks. The derivation relates the thermal gradients to internal stresses and states that cracking occurs when stresses equal a certain fracture strength σ_f. The main parameters are R and R' :

$$R = \sigma_f \frac{1-\nu}{E\ \alpha},\ R' = k\ R\ (\nu : \text{Poisson's ratio})$$

- The energetic analysis (2) concerns the stability of pre-existing flaws. According to Griffith's energy balance, micro-cracks propagate as soon as the relaxation of elastic energy exceeds the fracture surface energy. The main parameters are :

Riley, F.L. (ed.) Progress in Nitrogen Ceramics

$$R'''' = \gamma_f \frac{E}{\sigma_f^2(1-\nu)} , \quad R_{ST} = (\gamma_f/E\alpha^2)^{1/2} (\gamma_f : \text{fracture energy})$$

The first theory calls for a high value of σ_f, to reduce the crack apparition The second one calls for a high value of γ_f and accepts a diminution of σ_f to reduce elastic energy. Both theories are complementary, as the comparison between R and R_{ST} demonstrates (3).

2 - RESISTANCE TO THERMAL CRACK INITIATION

The conceptual simplicity of the thermoelastic analysis dissimulates several drawbacks :

- In the case of an infinitely fast thermal shock, the critical temperature difference ΔT_c is equal to $\sigma_f \frac{f(\nu)}{\alpha E}$, $f(\nu)$ being a function of Poisson's ratio and shape of samples : $f(\nu) = (1 - \nu)$ for cylinders, $(1-2\nu)/(1-\nu)$ for discs etc... For the usual values of ν in ceramics (0.2 to 0.3) $f(\nu)$ is generally approximated by $(1-\nu)$, whatever the geometry. This leads to the R parameter. But a question arises (4) about σ_f. Is it the tensile strength ? The flexural strength (3 points, 4 points, biaxial) ? For a batch of samples showing a scattering of data (Weibull) is σ_f equal to the mean value $\overline{\sigma_f}$ or to the minimum value σ_{mini} ?

- In the case of mild thermal shocks, the critical value ΔT_c is $R/A(\beta)$ where $A(\beta)$ is a function of Biot's number $\beta = ah/k$ (a characterizes the mean size of the sample, h is the heat exchange coefficient : β increases as the thermal shock severity increases). Two questions arise : What is the expression of $A(\beta)$ for a given geometry ? What is the value of β for a given shock ?

To measure ΔT_c we have chosen a quench into water at room temperature. Instead of using the classical method of determination of ΔT_c by measurements of the strength of quenched samples having suffered increasing thermal shocks, we have used a non-destructive method : the measurement of the variations of damping capacity and/or of elastic constants of vibrating samples (5). One sample only is sufficient to draw the thermal damage curve with good accuracy.

- For the reference material (99.5 % pure dense alumina Degussit Al_{23}) R is 70°C for beams and 60°C for discs, by taking $\nu = 0.25$, $E = 35\ 10^4$ MPa, $\alpha = 8\ 10^{-6}\ K^{-1}$, $\sigma_f = 260$ MPa (mean value in a three point bend test).

- Silicon oxynitride Si_2N_2O (5) has been hot pressed (1650°C, 33 MPa) with 5 wt% MgO additions, the density being about 99 %. R is 250°C for cylinders and 240°C for discs, by taking $\nu = 0.2$, $E = 22.2\ 10^4$ MPa, $\alpha = 3\ 10^{-6}\ K^{-1}$, $\sigma_f = 210$ MPa (3 point test).

- Aluminium nitride AlN has generally been less studied than silicon nitride but offers interesting properties (6). The samples have been hot pressed (1900°C, 33 MPa) and were nearly 100% dense.

For discs, R is about 160°C by taking $\nu = 0.245$, $E = 31.5\ 10^4$ MPa, $\alpha = 4.9\ 10^{-6}\ K^{-1}$, $\sigma_f = 360$ MPa (3 point test).

The experimental results for various sample shapes are given in Table 1.

	ΔT_c (°C)		
Sample size (mm)	Al_2O_3 +	Si_2N_2O ++	AlN ●
Prismatic beam 4 x 4 x 25	180		
Long cylinder 6(diam) x 80	180		
Short cylinder 20(diam) x 20+ or 20 (diam) x 16++	105	230	
Thin disc 30(diam)x2.5+,++ or 30(diam) x 3●	180	380	250
Thick disc 30 (diam) x 4.5	165		

Table 1 Critical temperature difference for water quench

A size effect appears : the massive specimens have a ΔT_c value lower than that of smaller specimens, due to the increase of β with specimen size ; for non massive alumina samples ΔT_c is about 180°C, which concurs with literature data (7). AlN has an intermediate thermal shock resistance. Si_2N_2O shows rather high values of ΔT_c, which may be compared with RBSN(8). For the geometry of thin discs, the hierarchy $\Delta T_{c\,Al_2O_3} < \Delta T_{c}\ AlN < \Delta T_{c\,Si_2N_2O}$ corresponds to the hierarchy of R, but we face difficulty when it comes to massive cylinders of Si_2N_2O for wich $\Delta T_c < R$ (which is impossible because $A(\beta) < 1$)

To discuss this point we must quantitatively compare the experimental ΔT_c values to the calculated ones : $\Delta T_c = R/A(\beta)$. For that one must know the $A(\beta)$ functions and the β values.

The $A(\beta)$ functions may be approximated by the case of a plate (1) for discs and that of a long cylinder (9) for beams and cylinders. With regard to β : the mean size "a" is half in diameter or half in thickness ; the conductivity k is about 5 $W\ m^{-1}\ K^{-1}$ for Al_2O_3 and seems to be superior for Si_2N_2O, SiAlONs (10) and AlN(11). The main difficulty is about h value : the scattering of literature data is very wide and it must be recalled that h quickly varies with ΔT, due to water boiling (7, 12, 13).

In the case of massive cylinders of Si_2N_2O, T_c is 250°C which corresponds to a high value of h (9) : about 5 $10^4\ W\ m^{-2}\ K^{-1}$. The corresponding values of β and $A(\beta)$ would be about 20 and 0.7 respectively : thus the R value might be 230 x 0.7 ∿ 160°C (instead of 250°C). In the case of discs T_c is 400°C which corresponds to a value of h in the order of $10^4\ Wm^{-2}\ K^{-1}$. This leads to $\beta \sim 2$; $A(\beta) \sim 0.3$: the R value might be 380 x 0.3 ∿ 115°C.

These derivations are very crude, but show that the R values of Si_2N_2O are overestimated. No important error may arise from ν, E, α, : consequently the drawback comes from σ_f.

We had chosen $\sigma_f = \sigma_{b.3} = 210$ MPa (measured on 4 x 4 x 25 mm beam) : the ΔT_c values prove that σ_f must be chosen inferior to $\sigma_{b.3}$, which corresponds to Davidge's view (14). Such conclusions are not generalizable : in the case of alumina, it seems that σ_f is close to $\sigma_{b.3}$. This disparity must be related to Weibull's modulus (the higher this modulus is, the smaller the difference between σ_f and $\sigma_{b.3}$ is). Furthermore, about the shock effect itself : the stresses increase very rapidly which implies that the strength value must be the dynamical one. The above discussion points to the fact that thermal shock resistance of ceramics can under no circumstances be evaluated without experimental data about ΔT_c. Reservations must be made about the estimation of R (what is the value of σ_f ?) and about the comparison between ΔT_c and R (what is the value of A(β) ?

Porosity effects have been studied in silicon oxynitride and aluminium nitride.

For Si_2N_2O, the comparison of P = 0,P = 8 %, P = 16 % shows that ΔT_c is not greatly modified by porosity (Fig.1). This is due to the fact that σf and E decrease nearly at the same rate when P increases. However, their behaviour in respect to thermal damages is very different : the crack propagation is restrained by pores, which leads to a less dramatic loss of resistance in porous samples (15).

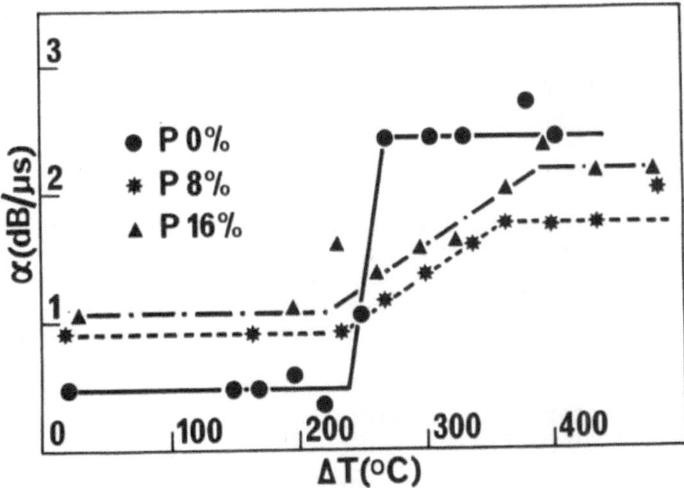

Fig. 1 Acoustic attenuation of porous Si_2N_2O versus thermal shock severity.

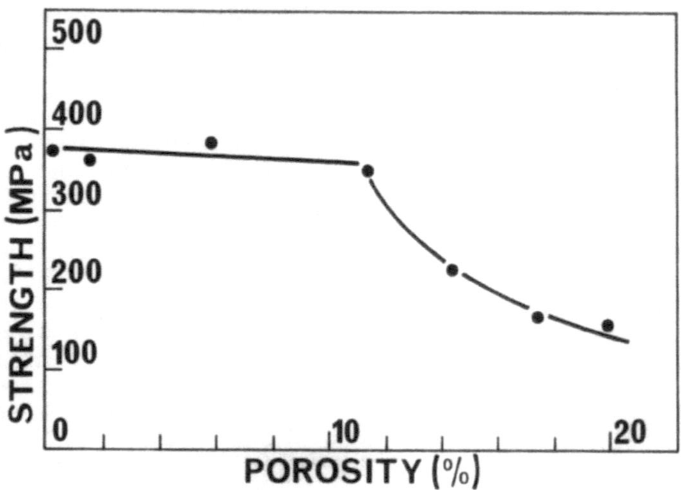

Fig. 2 Strength versus porosity for AlN

- For AlN the results are different : ΔT_c increases when porosity increases (ΔT_c = 310°C for P = 12 %, instead of 250°C for P = 0 %). This fact may be explained by the peculiar behaviour of AlN : when P increases E linearly decreases but $\sigma_{b.3}$ remains nearly a constant, up to P = 12 % (Fig.2). This is related to the fact that the volume of pore phase increases but that the size of the largest pores (critical flaws) (8) is nearly a constant (6).

3 - RESISTANCE TO THERMAL CRACK PROPAGATION

The energy concept (2) leads to various parameters, among which the most important is the thermal shock damage resistance parameter R'''' :

$$R'''' = E\,\gamma f/\sigma_f^2\,(1 - \nu)$$

R'''' applies to severe thermal shocks in which fracture cannot be avoided : it characterizes the minimum in the extent of crack propagation. There are not many quantitative applications of R'''' but this parameter gives qualitative informations about the final crack length l_f (for a given crack density) and consequently to the strength loss of ceramics for $\Delta T > \Delta T_c$: the fractionnal restrained strength is proportionnal to $(R'''')^{3/4}$. Moreover, R'''' correlates satisfactorily with multiple thermal shock resistance (thermal fatigue) (16, 17).

The main variable is the fracture energy γf, which is related to toughness (critical stress intensity factor K_{1c}).

$$R'''' = K_{1c}^2\,(1 + \nu)/2\sigma_f^2 \quad \text{(in plane strain)}$$

By taking $\sigma_f = \sigma_{b.3}$, K_{1c} = 3.7 MPa $\sqrt{m}$ (Al_2O_3), 4.3 MPa $\sqrt{m}$ (Si_2N_2O), 2 MPa $\sqrt{m}$ (AlN) the results obtained for R'''' (in μm) are respectively : 125, 250, 20.

In conclusion, Si_2N_2O appears to be better than Al_2O_3 for crack initiation and for crack propagation ; AlN offers the inconvenience of a low resistance to thermal damages.

4 - ACKNOWLEDGMENTS

The authors are grateful to Dr Dumazeau and Dr Lecompte for providing the samples.

5 - REFERENCES

(1) W.D. KINGERY J. Amer. Ceram. Soc. 38,1 (1955) 3-15
(2) D.P.H. HASSELMAN Amer. Ceram. Soc.Bull.49,12(1970)1033-1037
(3) D.P.H. HASSELMAN J. Amer. Ceram. Soc. 54,4 (1971) 219
(4) J.C. GLANDUS, P. BOCH Int.J.of Thermophysics 2,1(1981)89-101
(5) J.C. GLANDUS, P. BOCH Energy and Ceramics, P. VINCENZINI Ed., Elsevier, Amsterdam (1980) 661 - 670
(6) P. BOCH, J.C. GLANDUS et al. Ceramics International(1981)In press
(7) P.F. BECHER J. Amer. Ceram. Soc. 64,1 (1981) C 17-18
(8) G. ZIEGLER, J. HEINRICH Ceramurgia International 6,1(1980)25-30
(9) E. GLENNY, G. ROYSTON Trans.Brit.Ceram.Soc.57,10(1958)645-677
(10) R.R. WILLS, R.W. STEWART, J.M. WIMMER Amer. Ceram. Soc. Bull. 55, 11 (1977) 194 - 200
(11) K. KOMEYA, F. NODA Toshiba Rev. 6, (1974) 13 - 18
(12) J.P. SINGH, J.R. THOMAS, D.P.H. HASSELMAN J. Amer. Ceram. Soc. 63,3 (1980) 140 - 144
(13) P.F. BECHER, D. LEWIS et al. Amer.Ceram.Soc.Bull 59 (1980) 542 - 548
(14) R.W. DAVIDGE, G. TAPPIN Trans. Brit. Ceram. Soc. 66,8 (1967) 405 - 422
(15) R.D. SMITH, H.V. ANDERSON, R.E. MOORE Am.Ceram.Soc.Bull 55, 11 (1976) 979 - 982
(16) T. DARROUDI, J. HOMENY, R.C. BRADT Industrial Heating 5, (1980), 22 - 27
(17) N. KAMIYA, O. KAMIGAITO J. Mater. Sci. 14(1979) 573 - 582

NON-DESTRUCTIVE FAILURE PREDICTION IN CERAMICS

A. G. Evans

Department of Materials Science and Mineral Engineering
University of California, Berkeley, CA 94720

ABSTRACT

The role of non-destructive evaluation procedures in the reliable mechanical performance of ceramics is discussed. Non-destructive measurements capable of providing information about the dimensions of fracture initiating defects are described and fracture mechanisms associated with these defects are discussed. The application of NDE measurement procedures and fracture models to the development of accept/reject decision procedures is then presented. The decision process is illustrated by examples.

1. INTRODUCTION

The mechanical failure of ceramics occurs from defects introduced during either fabrication or surface preparation (pre-existent flaws) or during exposure to aggressive environments (e.g. oxidation, projectile impact). Failure studies reveal an appreciable sensitivity of the failure condition to the defect type [1] (fig. 1) and to the surrounding microstructure [2]. The important fracture inducing defects can be characterized by pertinent non-destructive measurements. Non-destructive techniques thus contribute vitally to the assurance of structural reliability, based upon their ability to reject defective components. Processing also exerts an important influence on structural reliability, as expressed through relations between the processing procedure, the defect population and microstructure of the material [3]. The ultimate reliability of ceramic components must be based on the appropriate combination of non-destructive procedures and processing controls suggested by a comprehensive fundamental understanding of fracture, the scattering of waves by fracture initiating defects

Riley, F.L. (ed.) Progress in Nitrogen Ceramics

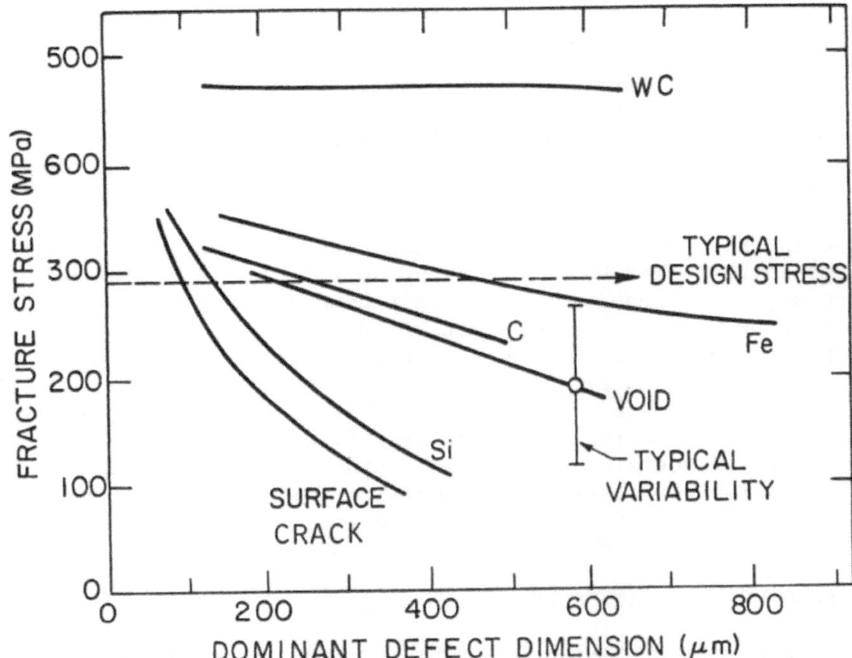

Fig. 1 The dependence of the fracture strength on defect size for fracture initiating defects in Si_3N_4.

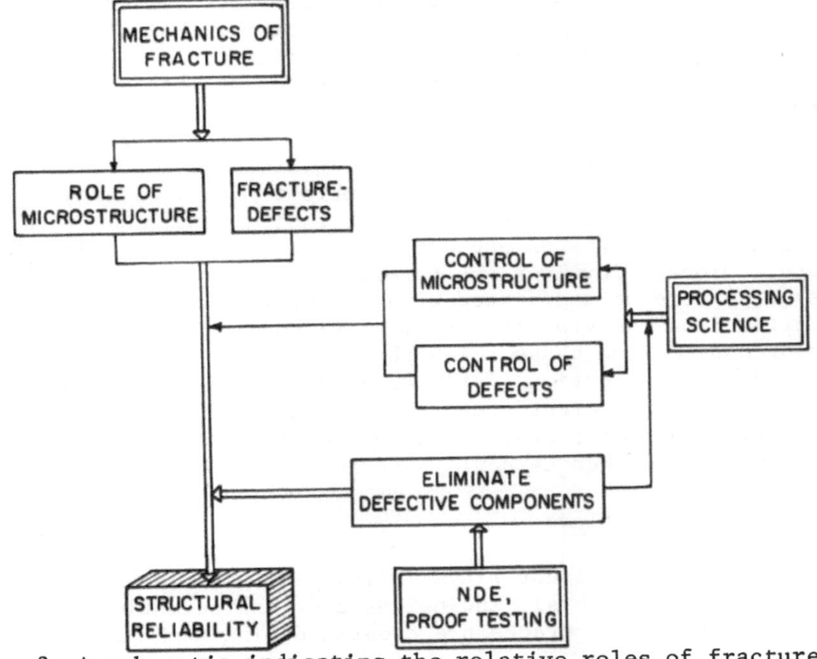

Fig. 2 A schematic indicating the relative roles of fracture studies, processing and NDE in structural reliability.

and processing (fig. 2).

The ultimate survival of a brittle structural component at an acceptable probability level involves the detection and analysis of waves scattered or absorbed by defects. The most versatile mode of flaw characterization entails the use of acoustic waves: either bulk waves or surface waves [4]. The utility of acoustic waves for providing the requisite survival information is emphasized in this paper. Other non-destructive techniques are discussed in the paper by Goebbels and Reiter [5]. For present purposes, it is required to derive a combination of measurement and fracture models which permit the determination of optimum accept/reject decision schemes. Hence, basic defect induced fracture mechanisms in ceramic materials are outlined in the initial sections of the paper. Thereafter, the acoustic techniques pertinent to the measurement of the significant fracture characteristics of the defects are described. Finally, probabilistic schemes for establishing accept/reject decision policies are derived and illustrated by several examples of reject decisions applied to hot pressed silicon nitride.

2. FRACTURE INITIATING DEFECTS

Three major failure categories can be distinguished in ceramic materials. At low or intermediate temperatures fracture is elastic, and generally, the strength is either invariant with temperature or increases with temperature (fig. 3). Fracture within this temperature range can occur either from flaws that exist prior to stress application [2,6] (materials with a fine scale microstructure), from the coalescence of stress induced microcracks, or from flaws smaller than the material grain size (materials with a coarse microstructure). Strong effects of grain size on strength obtain in the coarse grained regime [2]. By contrast, fracture at elevated temperatures is preceded by some nonlinear deformation and the fracture strength diminishes with increase in temperature or decrease in strain-rate (fig. 3). This fracture process involves the formation and coalescence of cavities; a phonomenon associated with either the diffusive transport of matter from the growing cavity or viscous hole growth within an amorphous second phase. The three failure classes are briefly examined in this section (detailed probabilistic failure relations for the important defect types can be located elsewhere) [1,7]. The explicit defect characterization requirements imposed on NDE measurements are then cursorily examined.

2.1 Pre-Existent Flaws

Typical pre-existent flaws include surface cracks, large voids (or void clusters), inclusions and (infrequently) large grains. The surface cracks are created during machining or surface

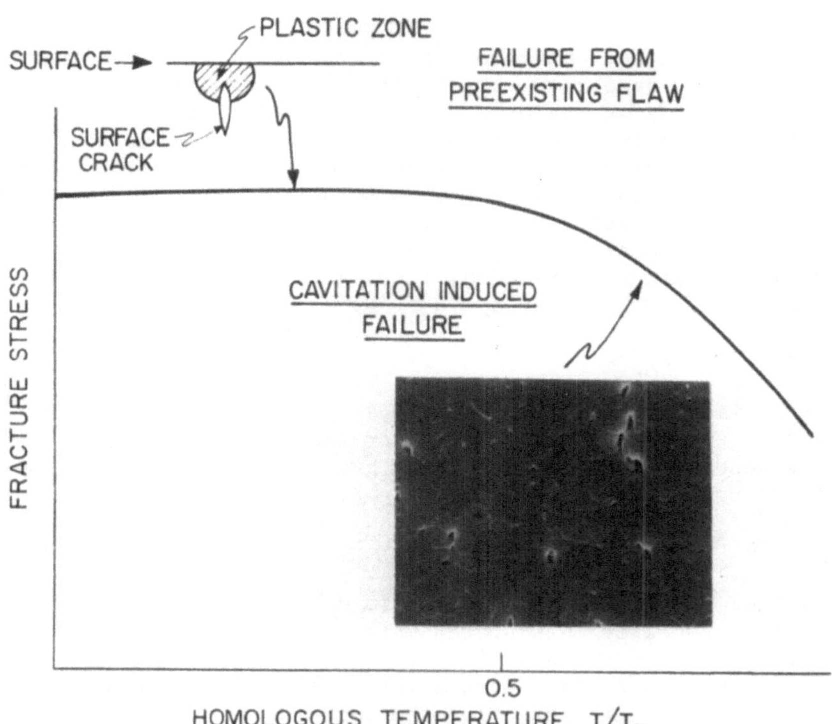

Fig. 3 The temperature dependence of the failure strength of a typical ceramic.

finishing. Most other flaws develop during fabrication. Surface crack and large grain induced fractures are primarily a consequence of residual stress. Void related failures are generally related to stress concentration effects. While inclusion initiated failure involves some combination of residual stress and stress concentrating influences.

2.1.1 Surface Cracks. Surface cracks form whenever a hard particle plastically penetrates the surface, as inevitably occurs during grinding, polishing or any other surface finishing process [8-10]. The cracks develop in response to a residual stress that results from the creation of a confined plastic zone [8] (fig. 4). These cracks extend upon application of an external stress σ_∞ and induce failure [9]. The initial crack extension is stable, because of the dominance of the local residual field (fig. 5). Criticality (as manifest in a maximum value of σ_∞) develops when the remote stress attains a level, σ^∞, given by [8];

$$\sigma_f^\infty = \beta K_{IC}/\sqrt{a_o} \tag{1}$$

where K_{IC} is the fracture toughness, a_o is the initial flaw depth and β is a parameter ≈ 0.56 for isolated cracks and ≈ 0.4 for crack beneath a uniform machining groove [11]. The distribution of failure strengths is primarily related to the spectrum of machining forces, which dictate the range of flaw radii, a_o, generated in the surface.

2.1.2 Voids. Large voids and clusters constitute a frequent failure origin, especially in sintered materials (fig. 6a) [12,13,14]. Voids are indirect sources of fracture (because a void does not usually create a sufficient stress intensification to permit the direct induction of fracture from the void surface). Fracture from voids is typically dictated by the presence of other defects located in their immediate vicinity [13,15]. The defects interact with the stress concentration around the void to produce the fracture (fig. 6b). The nature of these defects is not well comprehended at this juncture. Several possibilities exist. Voids located near the surface can interact with surface cracks [12]. Microcracking could be enhanced in the regions of stress concentration around the void in the presence of an appreciable localized residual stress (e.g. induced by thermal contraction anisotropy). The material around the void could be mechanically or chemically degraded as manifest in low levels of local toughness.

The existent models of void induced failure are not of sufficient generality to embrace all observed fracture characteristics. Nevertheless, it is apparent that, because of the statistical nature of the defects that must interact with the voids to produce failure, considerable variability in fracture stress exists even for voids of uniform size [1,13]. Also, the strength is often

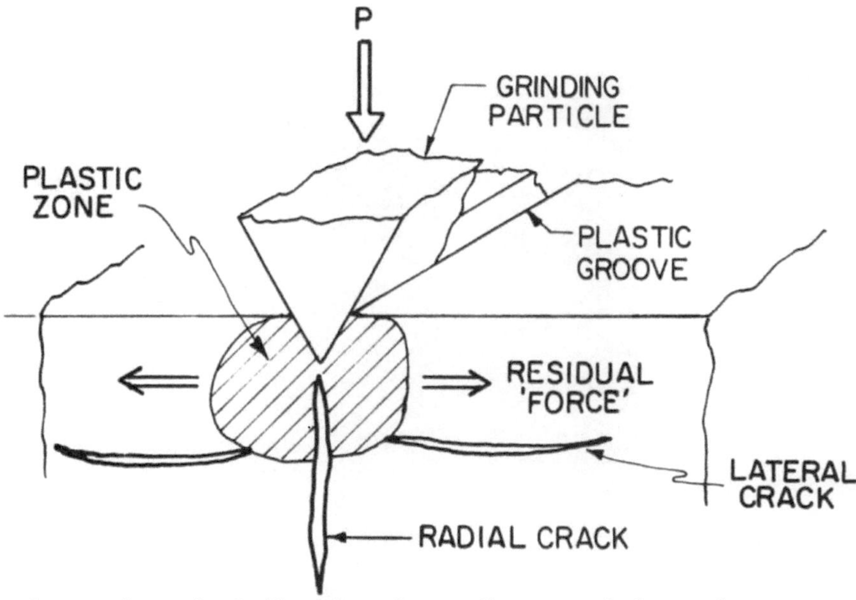

Fig. 4 A schematic indicating the surface crack formation process during machining.

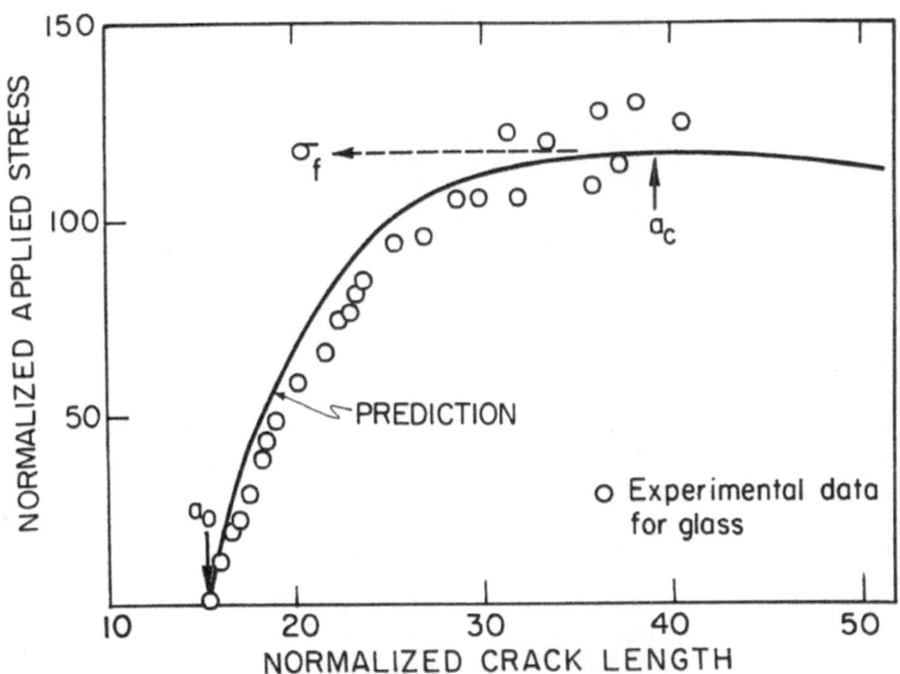

Fig. 5 The change in surface crack length with applied stress for a typical machining crack.

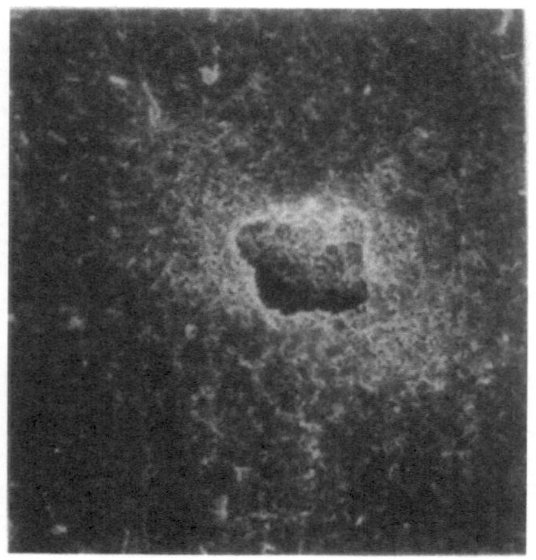

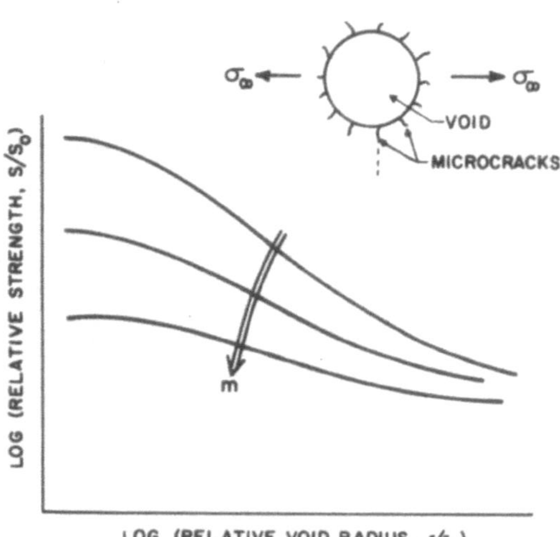

Fig. 6 (a) A scanning electron micrograph of a fracture initiating void in SiC.
(b) A schematic indicating the strength of a material containing an <u>internal</u> void as a function of the void size and of the shape parameter of the surrounding microcracks.

weakly dependent on void size (fig. 6b). This behavior contrasts with the limited fracture variability encountered with surface cracks of a specified size (eqn 1) [9]. Voids are thus particularly deleterious with regard to the predictability of failure, based upon non-destructive measurements (section 4.2.2).

An expression for failure from <u>internal</u> voids, appropriate for NDE analysis, derived in terms of a statistical distribution of small cracks at the void surface, has the form

$$- \ell n[1 - P(c{=}0|a)] = 8a^2(\sigma_{eff}/S_o)^m \exp[0.52m - 1.4] \qquad (2)$$

where

$$\sigma_{eff} = \sigma_\infty \left[0.3 + 0.7\left((1 + (\alpha/2)^2\right)^{-1}\right]$$

$$\alpha = (1/a)(K_c/\sigma_\infty)^2$$

$P(c{=}0|a)$ is the probability of failure, given the void radius a, m is the shape parameter and S_o the scale parameter associated with the microcrack distribution.

2.1.3 Inclusions. Inclusions are a major source of premature failure. Inclusion induced fracture exhibits a spectrum of possibilities [1], as illustrated in figs. 1 and 8. Failure from each inclusion/host combination should be examined on a separate basis. The first distinguishing feature is the tendency for <u>residual stress</u> associated cracking attributed to thermal contraction mismatch (fig. 7). If the thermal expansion coefficient of the inclusion is appreciably lower than that for the matrix, tensile stresses create radial matrix cracks (when the defect exceeds a critical size) [16]. This situation can produce severe strength degradation. But such behavior is unusual for structural materials (which must have an intrinsically low thermal expansion coefficient in order to resist thermal shock). Alternatively, if the expansion coefficient of the inclusion exceeds that for the matrix, several possibilities can result. Highly contracting, high modulus inclusions will tend to detach from the matrix, and produce a defect comparable in character to a <u>void</u>. Inclusions that are more compliant, or exhibit smaller relative contractions, remain attached to the matrix. The expected failure mode then depends upon the elastic modulus and the fracture toughness of the inclusion, vis-a-vis the matrix. For example, when the inclusion has a larger toughness than the matrix, stress concentration effects cause fracture to initiate within the matrix, usually from microflaws located within (or adjacent to) the interface. If the bulk modulus of the inclusion also exceeds that of the matrix, the tensile stresses (in a direction suitable for continued extension of the crack due to the applied stress) are confined to a relatively small zone near the poles of the inclusion (fig. 8). For this case, the fracture probability is anticipated to be relatively

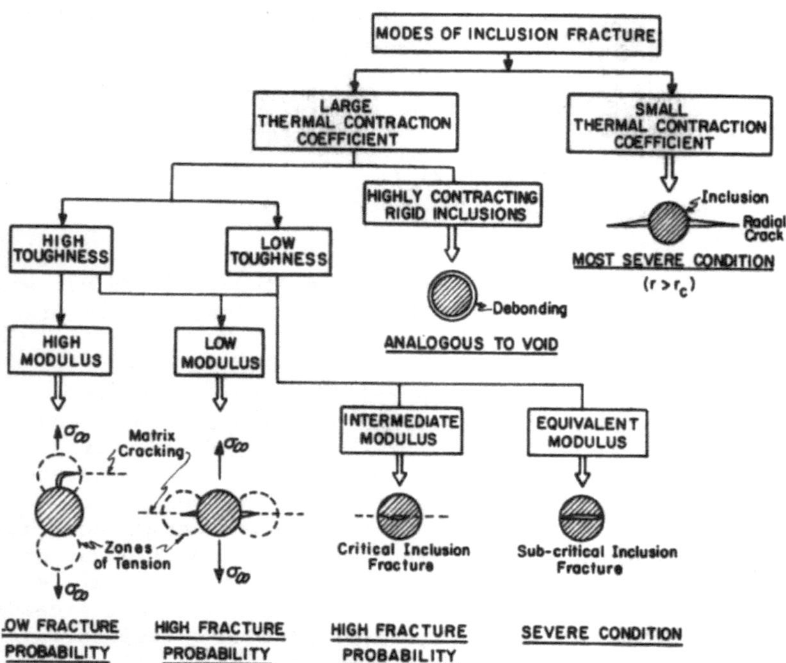

Fig. 7 A schematic indicating the principal modes of failure from inclusions.

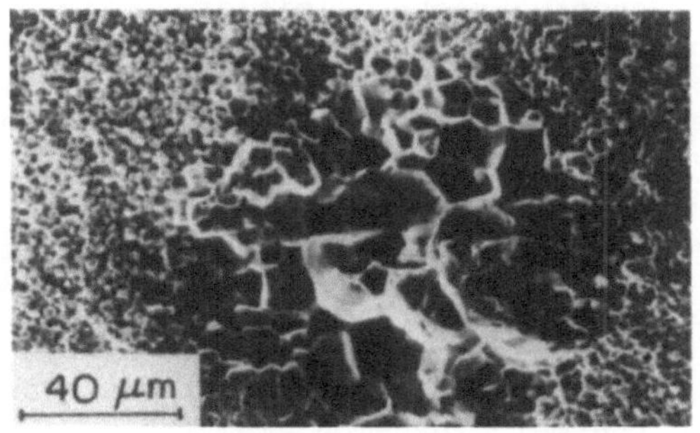

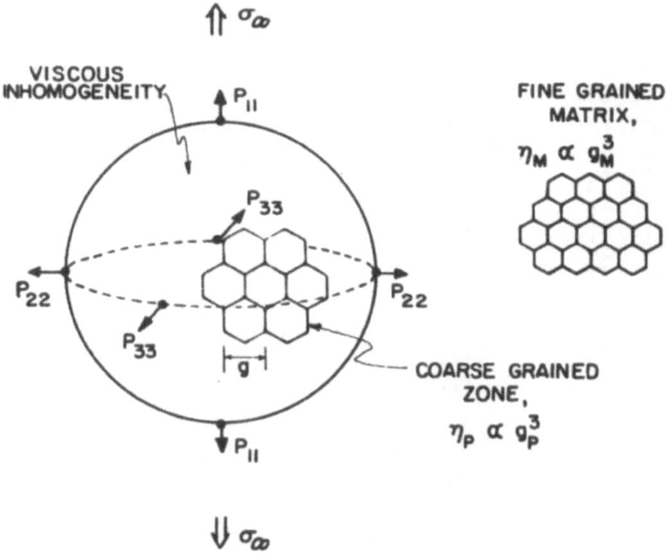

Fig. 8 A coarse grained high temperature failure origin in Al_2O_3, and a schematic indicating the stress enhancement that occurs in the large grained region.

small, as exemplified by WC inclusions in silicon nitride (fig. 1). Alternatively, when the modulus of the inclusion is small, the maximum tensile stress concentrations occur in the matrix near the equatorial plane. Fracture initiates within this region of the matrix, irrespective of the inclusion toughness. The fracture condition is thus comparable to that for a void. This case is expected to be an important one, because inclusions are often porous (following high temperature mass transport driven by thermal contraction anisotropy) and hence, of low effective modulus. Finally, low toughness (e.g. large grained or single crystal) inclusions with a similar modulus to the matrix can form sub-critical cracks. The cracks then dictate strength in accord with standard principles of fracture mechanics (e.g. Si or SiC inclusions in Si_3N_4), given appropriate recognition of the residual stress associated with thermal expansion mismatch.

Several models of fracture that accord with the general principles outlined above have been developed [1]. However, model development is at an elementary stage, and expressions relating fracture to inclusion size and composition have not been adequately correlated. Again, statistical variability poses concern with regard to fracture prediction based upon non-destructive measurements. A specific model that will be used in subsequent discussion concerns failure from low toughness inclusions, wherein a crack with dimensions equal to the inclusion size develops sub-critically. The failure condition for this process is given (section 4.2.1) by;

$$P(c=0|x) = \phi[c^{-1/2}(\sigma_\infty - \alpha - \beta\sigma_p(x))] \qquad (3)$$

where

$$\phi(u) = (1/\sqrt{2\pi}) \int_{-\infty}^{u} \exp(-t^2/2)\,dt$$

$$\sigma_p = K_c/z\,(c'/a')\sqrt{\pi c'}$$

$$z(u) = \left[\int_{o}^{\pi/2} d\Psi (1 - (1-u^2)\sin^2\Psi)^{1/2}\right]^{-1}$$

$$G(u,c) = (2\pi c)^{-1/2} \exp\,(-c^{-1/2}\,u^2/2)$$

G(.,.) is the Gaussian function, α and β are constants obtained from experimental fracture data and c' and a' are the minor and major axes of the ellipsoidal inclusion.

2.1.4 Relative Role of Dominant Flaws. Most brittle materials can be expected to contain at least two populations of flaws: surface cracks and fabrication defects. The different spatial character of these two flaw populations signifies a variable relative role of each population in the fracture response observed with different components or test configurations. Strength tests are frequently conducted in flexure; a test procedure which imposes the largest tensile stresses on the specimen surface. Hence, fractures that occur in flexure tests generally originate from surface cracks and flexural strengths often reflect the surface crack population. Conversely, tensile tests are more likely to induce failure from fabrication defects and hence, tensile strengths often reflect the presence of inclusions, voids, or large grains.

The different spatial character of the important flaw populations also impacts the failure modes experienced by components. For example, thermal stresses usually exhibit a sharp maximum at the surface and hence, thermal shock failures are most sensitive to the surface crack population. In other situations (e.g. rotational stresses in turbine discs) the fabrication flaws are more critical.

2.2 Flaw Creation By Microcrack Coalescence

Grain boundary located microcracks develop in certain ceramics. The cracks usually occur as a consequence of thermal expansion anisotropy, which leads to the development of large localized stresses [17]. Calculations of the stresses induced at grain boundaries indicate that failure must be initiated from defects that pre-exist at the triple junction [18]. It has thus been surmized that the pores which occur at certain junctions are of sufficient size that microcrack initiation can occur at these locations. This premise provides a good estimate of the microcracking condition, but has yet to be afforded a direct experimental validation. The basis for the analysis is the stress intensity factor, K , (as deduced from the residual stress) and its variation with crack length. A peak value of K is found to exist. This peak value is equated to the local fracture resistance in order to deduce a critical microfracture condition. The critical condition can be expressed in terms of a critical grain facet length, ℓ_c given by [19];

$$\ell_c = \frac{2(1+\nu^2)\gamma_{g.b.}}{E(\Delta\alpha\Delta T)^2} \exp\left[-\frac{8\sigma_\infty(1-\nu^2)}{5\Delta\alpha\Delta T}\right] \tag{4}$$

where $\gamma_{g.b.}$ is the grain boundary fracture energy, E is Young's modulus, $\Delta\alpha$ is the thermal anisotropy, ΔT is the cooling range and ν is Poisson's ratio.

Initial microcrack formation does not usually coincide with failure. Several contiguous microcracks generally need to accumulate before a crack of critical size is generated. This process is understood at the conceptual level, but adequate mathematical models have yet to emerge.

The strong influence of grain size upon the microcracking process suggests that zones of exceptionally large grains could be a source of premature failure, particularly if there exists a sufficient number of contiguous grains to provide a coalesced crack of critical size. This has not yet been a widely observed source of brittle fracture. Nevertheless, large grained regions should be regarded as a potential detriment to mechanical performance. Their detection by NDE should thus be considered desirable.

2.3 Flaw Formation by High Temperature Cavitation

At elevated temperatures, fracture from pre-existent flaws is frequently suppressed, as demonstrated by the observed stable opening of surface cracks [20]. Critical fractures initiate from other, preferred sites: by means of a process involving the nucleation, growth and coalescence of cavities (fig. 3) [21]. The cavity growth is dictated by diffusion and/or by viscous hole growth.

A superior (but still incomplete) comprehension of failure has emerged for predominantly single phase materials. In such materials cavitation typically occurs by a combination of surface and grain boundary diffusion [20,23]. Generally, cavities evolve from three grain junctions and extend across the intervening grain facets. The cavitation is <u>inhomogeneous</u>, and failure evolves from certain <u>preferred microstructural regions</u>. The dominant characteristics of the failure origins are still under investigation.

Recent studies indicate that large grained regions are a source of premature failure [20] (fig. 8). The enhanced failure rates derive from the stress concentrations that develop in the vicinity of coarse grained zones by virtue of their high viscosity (fig. 8).

2.4 NDE Requirements

The preceding discussion of fracture processes identified surface cracks, large voids, inclusions and large grains as prominent sources of failure in ceramics. The preference for a particular defect type depends upon material, stress state and temperature. A comprehensive non-destructive failure prediction capability should be able to both detect and determine the dimensions of each of these defects. Additionally, (fig. 1) the non-destructive measurement process must be capable of characterizing defects with sizes in the range 40-200 μm. Acoustic (elastic wave) techniques best satisfy these requirements (this does not, of course, discount the merit of other techniques for characterizing a limited range of defects more effectively than the acoustic methods). The versatility of acoustic techniques resides in the strong similarity of the response of defects both to the <u>elastic</u>

stress waves used for inspection and to the (subsequently imposed) elastic stresses associated with fracture. Specifically, the scattering of acoustic waves relies on a mismatch in acoustic impedence: a condition satisfied by all of the above defects.

3. ACOUSTIC MEASUREMENT TECHNIQUES

3.1 Surface Waves

The most directly successful acoustic method involves the use of surface acoustic waves to predict failure from machining induced surface cracks; in particular, the use of long wavelength, $\lambda >> a_o$, surface waves [24,25]. In the long wavelength limit, the scattering of an acoustic wave by a crack is closely analagous to the interaction of the crack with an applied stress field. Both the scattering coefficient S_1 and the strain energy release rate are related to the crack surface integral;

$$\int_{A_s} \sigma_{ij} \Delta u'_j n_i dA_s \tag{5}$$

where σ_{ij} is the stress across the crack plane in the absence of the crack, u'_j is the displacement of the crack surfaces and A_s is the crack surface area. Hence, it is straightforward to demonstrate that the scattering coefficient is directly related to the crack extension stress σ_c. The specific relationship depends upon the crack geometry and the residual field. For example, a small, open ellipsoidal crack can be characterized by;

$$\frac{K_c}{\sigma_c} = 2\left[\frac{6(1-\nu)\lambda^2 S_1 w}{\pi^5 f_z \eta}\right]^{1/6} \tag{6}$$

where w is the beam width, η is the transducer efficiency and $f_z \sim 0.4$. This result is strictly correct when both the acoustic wave and the applied stress are normal to the crack plane, as exemplified by failure prediction results for glass (fig. 9). Minor modifications to eqn (6) are required for inclined cracks.

The behavior of cracks induced by machining is modified by the presence both of residual stresses and of partial crack closure effects [24]. As-machined surfaces contain cracks that are essentially closed over the plastic zone, as well as being subject to a residual opening force [9] (imposed by the plastic zone). These effects modify both the acoustic scattering and the crack propagation behavior. The acoustic scattering coefficient can be expressed in terms of a relationship between $2\pi w|S_1|/\lambda$ and $2\pi(b+h)/\lambda$ where b and h are the crack dimensions defined in fig. 10 ($b+h = a_o$, the effective initial radius of the surface crack, fig. 10). The relation between a_o and

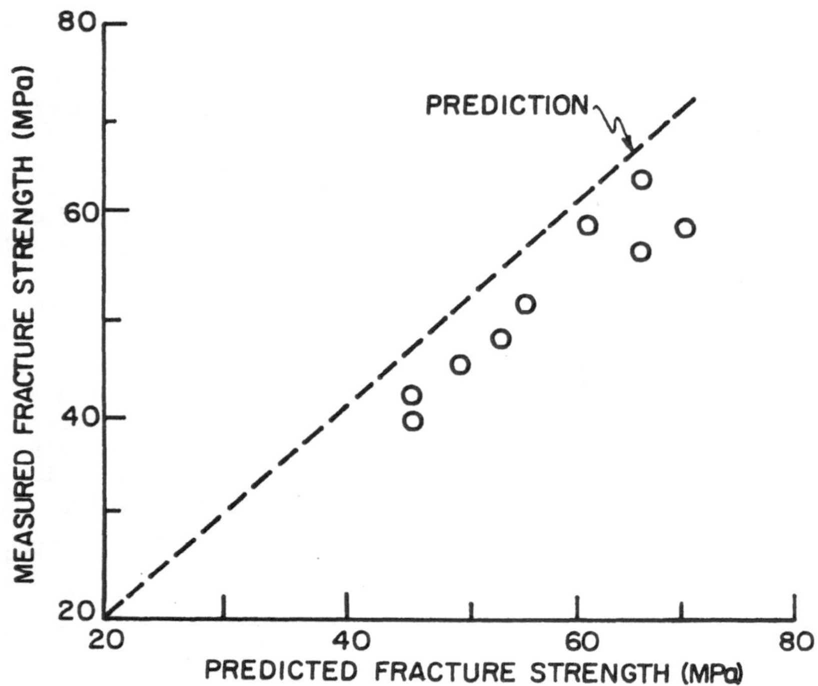

Fig. 9 A comparison of measured and predicted fracture strengths for surface cracks in glass.

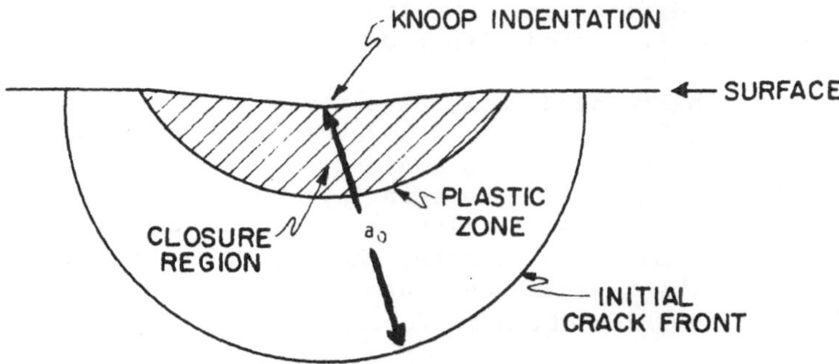

a) AS-INDENTED CRACK CONFIGURATION

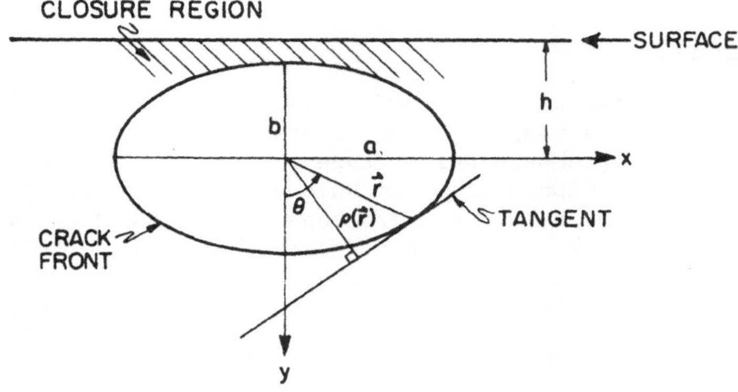

b) CRACK MODEL FOR ACOUSTIC SCATTERING ANALYSIS

Fig. 10 A schematic of the surface crack and of the sub-surface configuration used for the scattering analysis.

the failure stress in the presence of the machining induced residual stress has already been discussed, and given by eqn (1). Imposing these modifications, the fracture of machined Si_3N_4 components can be predicted with good accuracy [11] (table 1).

The machining related residual stresses can be relieved by annealing (at 1200°C for Si_3N_4). However, relief of the residual stresses causes 'asperities' to contact over a significant portion of the crack surface [24]. The asperity contact modifies the acoustic signal and the scattering measurements are difficult to interpret. It is more appropriate, therefore, to perform the measurements in the as-machined state. A safety margin may then be realized if the inspection is followed by an annealing treatment.

The surface acoustic wave method can be applied straightforwardly to components of complex shape, because surface waves propagate over curved surfaces. For example, a Si_3N_4 turbine blade can be inspected for surface cracks by exciting the waves at a flat seciton of the surface and allowing the wave to propagate over the highly stressed regions of principal interest.

3.2 Bulk Waves

The characterization of bulk defects is more complex. Information over a wide range of frequencies appears to be needed to obtain a highly probable defect type classification and hence, a size estimation. Appropriate techniques are available including: the scanning laser acoustic microscope, 200 MHz ZnO transducers and conventional (2-50 MHz) transducers. The most critical issue concerns the appropriate choice of algorithms to obtain the most reliable defect characterization. A typical set of algorithms and their interaction are illustrated in fig. 11 using low and high frequency information as well as acoustic microscopy. This set has not yet been fully evaluated. Redundancies may thus exist. Four algorithms are employed in this scheme: (i) long wavelength scattering [7], (ii) intermediate wavelength Born approximation [26], (iii) high frequency spectroscopy [4] and (iv) cross sectional information from acoustic microscopy [27]. Impulse response functions [28] (fig. 12) are firstly used to determine whether the defect is a void, an inclusion or an array of large grains. The void has an impulse response function (fig. 12a) characteristic of the transducer, while inclusions and large grains have more complex functions (fig. 12b). Thereafter, voids can be analyzed straightforwardly, using a variety of algorithms. For example, a long wavelength algorithm similar to that described for surface cracks may be employed. In the long wavelength limit the scattered amplitude is related to the void volume V and the frequency ω by [7];

TABLE I

A comparison of measured strengths attributed to surface crack controlled fracture with strengths predicted from surface acoustic wave measurements [11].

		Initial Flaw Sizes (μm)		Failure Strength (MPa)	
		OPTICAL	ACOUSTIC	MEASURED	PREDICTED
INDENTATION CRACKS	1	115	97	238	240
	2	118	94	239	243
	3	118	96	242	242
	4	114	111	233	224
MACHINING CRACKS	1	52	46	240	223
	2	35	42	306	233
SURFACE SCRATCH		42	36	239	251
ROW OF INDENTATIONS		140	98	140	152

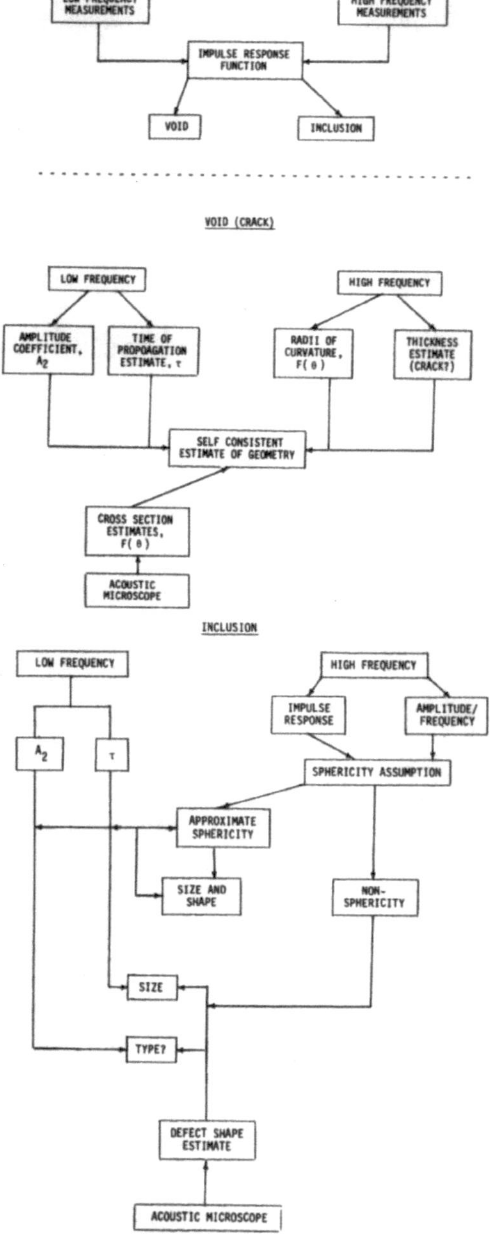

Fig. 11 A chart indicating the acoustic measurement algorithms used for characterizing bulk defects.

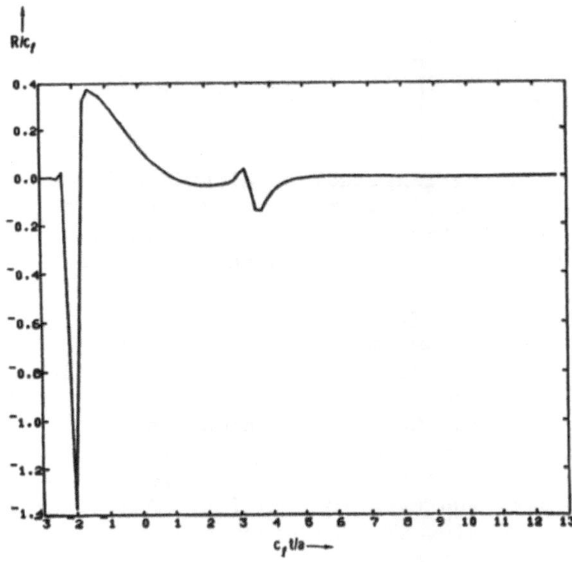

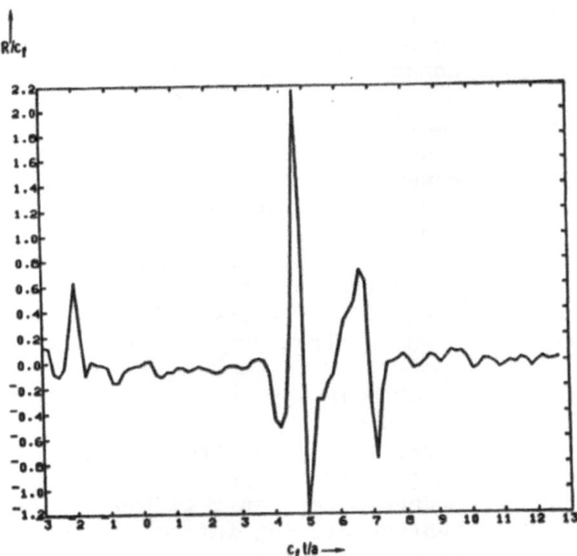

Fig. 12 (a) The impulse response function for a void in Si_3N_4.
(b) The impulse response function for a WC sphere in Si_3N_4.

$$S_1 = \frac{V\omega^2}{(4\pi c_\ell^2)^2} \left[1 + \frac{1+\nu}{2(1-2\nu)} + \frac{10(1-2\nu)}{7-5\nu}\right]^2 \tag{7}$$

where c_ℓ is the longitudinal elastic wave speed in the host. Alternately, acoustic microscopy [27] could be used to generate a diffraction pattern (fig. 13) which can then be reconstructed in order to determine the void radius.

Inclusions are more difficult to analyze. The combined use of several algorithms may be required. For nearly spherical inclusions, the interpretation is least complex. For example, a combination of the long wavelength algorithm (which contains coupled volume and type information) and the Born approximation (which provides an independent estimate of the distance from the geometric center to the back face of the inclusion) can yield the requisite size and type information. A typical result, obtained for a 100 μm radius Si inclusion in Si_3N_4, is illustrated in fig. 14; wherein the joint probability of the defect type and size is plotted as a function of the estimated size. In order to obtain this result, six possible inclusion types were permitted to exist within the material (selected on the basis of detailed failure analyses conducted on this material).

However, the smaller defects are difficult to detect at the relatively low frequencies associated with these measurements, especially in components of complex shape. High frequency methods thus seem to be required in order to assure a high level of reliability at large working stresses. These methods include determination of the time of flight from impulse response functions, as well as estimates of the front face radius of curvature from the scattered amplitude (using ZnO/sapphire transducers), imaging techniques (using transducer arrays), etc.

The application of bulk wave methods to components of complex geometry can pose an appreciable practical challenge. Information can be obtained at relatively low frequencies ($\lesssim$ 50 MHz) by using a water bath scanning (C scan) system. For planar surfaces, the scanning can be conducted very quickly (on a point-by-point basis) and yields consistent scattering information that can be uniquely related to the defect characteristics. Similar planar surface scanning methods can be employed at high frequencies (50-350 MHz) if the water is replaced by a thin viscous medium with an acoustic impedence comparable to that of the ceramic material (honey appears to be an ideal choice for Si_3N_4). The same basic scanning scheme can be extended to components of simple geometry, consisting of planar or cyclindrical surfaces. However, for components of complex geometry a computer controlled scanning device is required that allows the acoustic beam to enter the component surface at an approximately constant inclination at all locations.

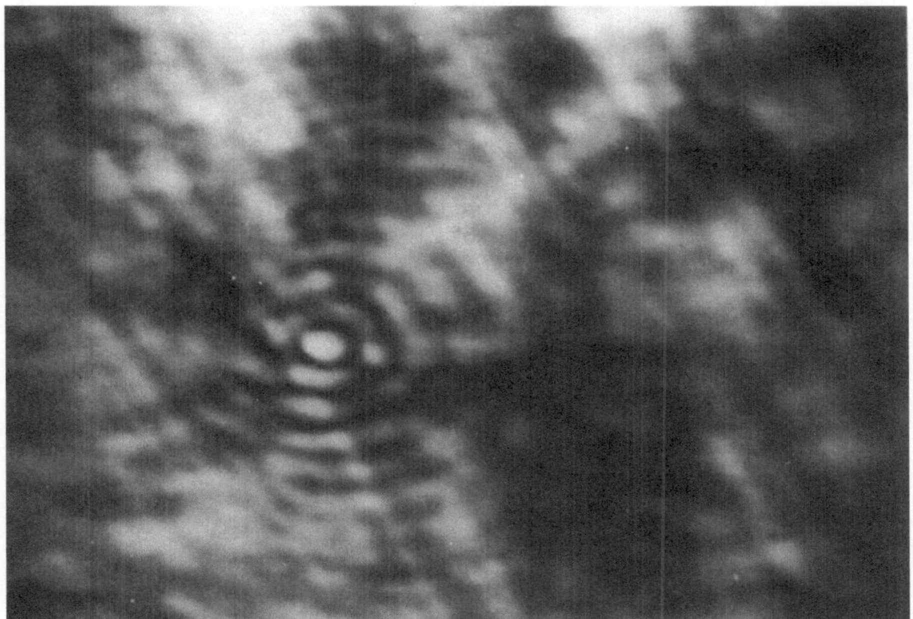

Fig. 13 A diffraction pattern from a defect in Si_3N_4 obtained using a scanning laser acoustic microscope.

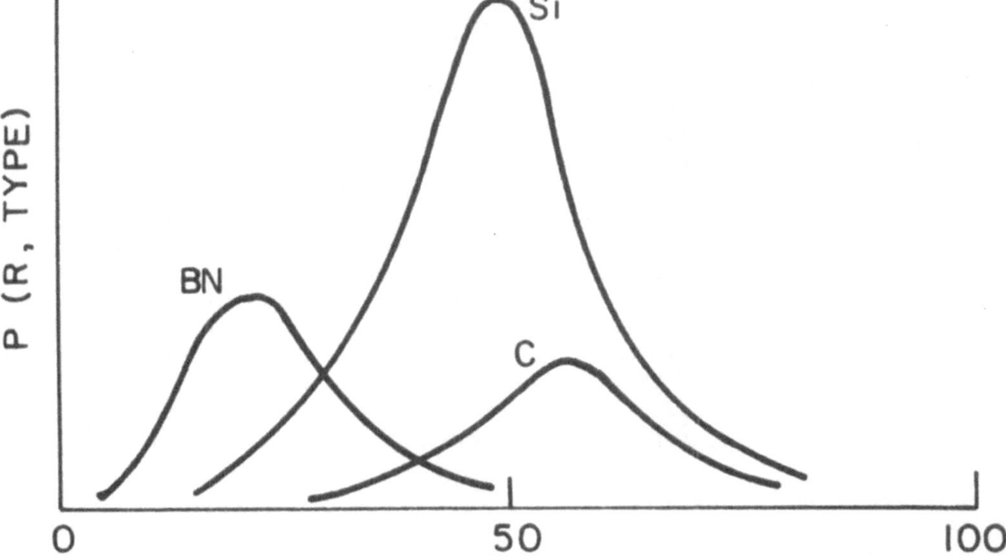

Fig. 14 The joint probability of defect shape and size deduced from a coupled wave length, Born approximation algorithm.

4. ACCEPT/REJECT DECISIONS

4.1 General Principles

Accept/reject decisions based on non-destructive measurements of defect scattering must be cognizant of the probabilistic character of the problem [7]. At least threee probabilities enter the analysis: the failure probability, given the defect dimensions $P(c|x)$; the joint probability of identifying the defect type and of estimating its size, $P(y|x)$; the apriori distribution of defect sizes, $P(x)$. These probabilities are combined and integrated to various inspection levels, y^*, to obtain two interrelated probabilities: the false-accept probability $P(c=1|y)$ and the false-reject probability, $P(c=0|y)$ (fig. 15), where

$$P(c|y) = \int P(c|x)\, P(x|y)\, P(x)dx \tag{8}$$

The inspection level y^* refers to the estimated defect dimension(s) selected for the rejection or acceptance of the component, e.g. all components with an estimated maximum dimension less than y^* are rejected. The false-accept probability is thus the probability that components accepted in accord with the specified inspection level will contain defects more severe than indicated by the estimate, and will actually fail in service (i.e., related to the failure probability). The false-reject probability is the (related) probability that rejected components would, in fact, have performed satisfactorily in service, because the defect severity has been overestimated. This probability increases as y^* decreases. However, it is crucial to recognize that these probabilities are interrelated. This interdependence is exemplified in fig. 15: a typical plot relating the false-accept and false-reject probabilities. Once one of these probabilities has been selected, the other probability, as well as the associated inspection level, are <u>necessarily</u> defined. It is now apparent from fig. 15 that the inspection technique (or combination of techniques) that would be preferred is that which yields a curve as close as possible to the probability axes. For example, technique B is preferred over technique A, because the rejection of satisfactory components required to satisfy the failure probability requirements is much lower. Such curves (referred to as the operating characteristic) thus represent a quantitative method for characterizing the failure prediction capabilities of various inspection technques, for a given material and service condition.

4.2 Examples

The accept/reject decision process is best illustrated by means of examples. The first example relates to failure from brittle inclusions, such as Si (or SiC) in Si_3N_4, using a combination of NDE measurements. In this instance, it will be demonstrated that NDE can greatly reduce the false rejection of ceramic components. A similar, satisfactory NDE operating characteristic obtains for failure from surface cracks (generated by machining) characterized using the long wavelength surface acoustic wave method. The second example selected concerns failure from isolated voids, using a long wavelength method to characterize the void size. For this situation, it will be demonstrated that the appreciable intrinsic variability of the fracture process minimizes the utility of the NDE measurements.

4.2.1 Sub-Critical Inclusions. Fully dense, brittle inclusions (such as Si or SiC in Si_3N_4) can fracture at applied stress levels below the final failure strength of the material. The failure process is then dictated by the propagation of the crack through the inclusion interface into the host.

A semi-infinite specimen with known host material is examined. The boundary of the specimen is parallel to the xy-plane and the outward pointing normal lies in the positive z-direction. The defect is regarded as an ellipsoidal inclusion (although the subsequent analysis is limited for the sake of brevity to the spheroidal case) with an included material determined unambiguously by a high frequency impulse response function method. The mode of inspection consists of a pulse-echo (i.e., backscatter) measurement with the incident wave in the negative z-direction. For a spheroidal inclusion of known composition the state vector x is four dimensional

$$x = \begin{pmatrix} a \\ b \\ \theta \\ \gamma_z \end{pmatrix} \tag{9}$$

where θ is the azimuthal angle (in the xy-plane) with respect to the symmetry axis (defined by the unit vector $\vec{w}$), γ_z is the direction cosine of $\vec{w}$ relative to the z-axis, and a and b are the major and minor semi-axis lengths of the inclusion.

The acoustic measurements consist of an arbitrary number of low-frequency longitudinal-to-longitudinal backscatter processes. These are collectively represented by a standard stochastic model of the generic form

$$y = f(x) + r \tag{10}$$

where y , $f(x)$ and r are N-dimensional vectors (but considerable attention will be devoted to the case N=1).

The crack is assumed to form on a plane intersecting the geometrical center of the spheroid and having an orientation perpendicular to the axis of the applied stress, i.e., the x-axis.

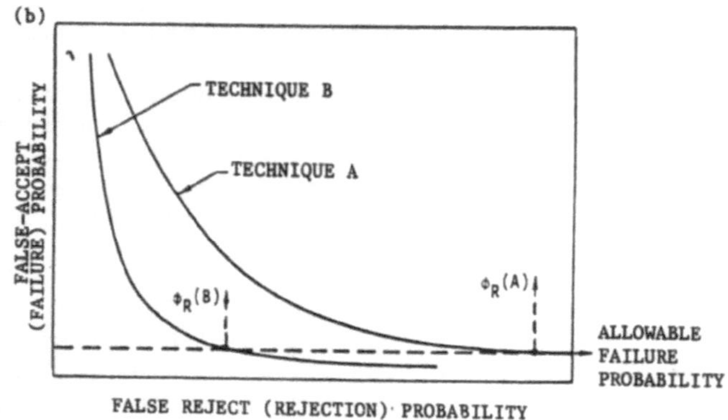

Fig. 15 False accept, false reject probabilities for two hypothetical measurement failure processes.

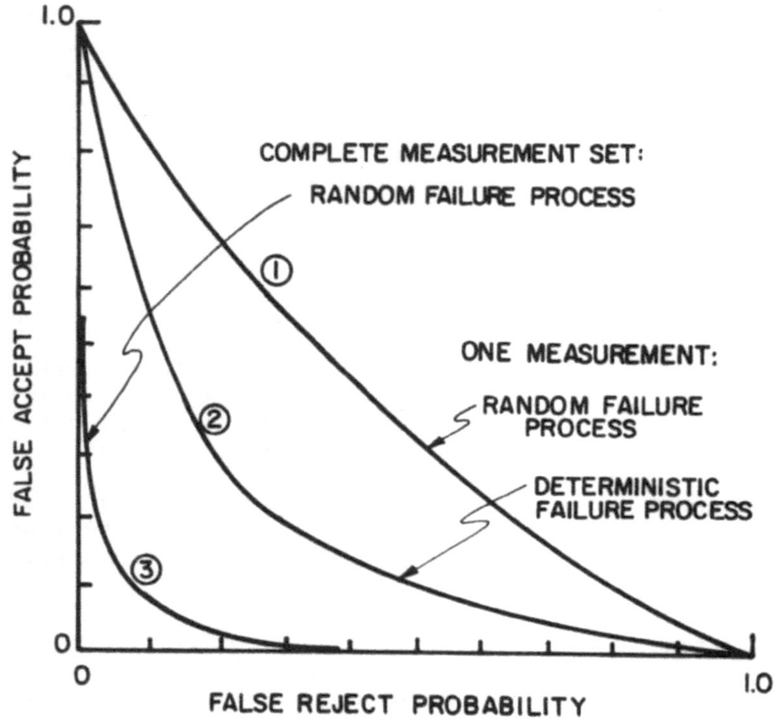

Fig. 16 NDE operating characteristics for long wavelength acoustic measurements pertinent to failure from Si inclusions in Si_3N_4.

Then, the dimensions, a' and c', of the elliptical crack can be related to the inclusion dimensions by;

$$a' = a$$
$$b' = [a^{-2} + (b^{-2} - a^{-2})\, w_{\ell}^{2}]^{-1/2} \tag{11}$$

where w_{ℓ} is the length of the projection of $\vec{w}$. For the fracture process involving extension of the inclusion crack into the host, the following result obtains. The performance variable c is given by

$$c = H(\sigma_F - \sigma_\infty) \tag{12}$$

where H(.) is the Heaviside unit step function, σ_∞ is the applied stress, and σ_F is the failure stress. The latter quantity is a random variable

$$\sigma_F = \alpha + \beta\sigma_p + s \tag{13}$$

where σ_p is the predicted failure stress, α and β are empirical recalibration constants, and s is a Gaussian random variable with zero mean and variance C_s. The computation of the fracture stress under the assumption that the crack is surrounded solely by host material gives

$$\sigma_p = \frac{K_c}{Z(b'/a')\sqrt{\pi b'}} \tag{14}$$

where K_c is the fracture toughness and Z(·) is a function

$$Z(u) = \left[\int_0^{\pi/2} d\Psi(1-(1-u^2)\,\sin^2\Psi)^{1/2}\right]^{-1}$$

The conditional probability density of σ_F is given by

$$P(\sigma_F|x) = G[\sigma_F-\alpha-\beta\sigma_p(x),\, C_s] \tag{15}$$

where G(.,.) is the Gaussian function which in the present case is specialized to a case of a scalar variable, i.e.,

$$G(u,C) = (2\pi c)^{-1/2}\,\exp(-C^{-1}\,u^2/2) \tag{16}$$

Hence, using eqn (12),

$$P(c=0|x) = \phi[C_s^{-1/2}(\sigma_\infty-\alpha-B\sigma_p(x)] \tag{17}$$

where the function $\phi(u)$ is the error integral

$$\phi(u) = \frac{1}{\sqrt{2\pi}} \int_{-\infty}^{u} \exp\,(-t^2/2)\,dt$$

The apriori probability density is assumed to have the form of an extreme value function given by;

$$P(a,c) = \beta\ g(a)\ h(c) \tag{18}$$

where

$$g(a) = \left(\frac{a}{a_c}\right)^{k-1} \left(\frac{k}{a_c}\right) \left[\exp - \left(\frac{a}{a_c}\right)^k\right]$$

$$h(b) = \left(\frac{b}{b_c}\right)^{k-1} \left(\frac{k}{b_c}\right) \left[\exp - \left(\frac{b}{b_c}\right)^k\right]$$

The resultant probabilities associated with the accept/reject decision process derived from the preceding results are;

$$P(y) = \int G(y-f(x),\ C_r)\ dx \tag{19a}$$

$$P(c=0|y) = \int \phi[C_z^{-1/2}(\sigma_\infty-\alpha-\beta\sigma_p(x))]'G[y-f(x),C_r]P(x)dx \tag{19b}$$

$$P(c=0|y) + P(c=1|y) = P(y) \tag{19c}$$

Case 1. One Measurement. Random Measurement and Failure Processes. In this case a single pulse-echo, longitudinal-to-longitudinal measurement is made with the incident propagation in the negative z-direction. The random experimental error corresponds to a signal-to-noise ratio of about 10. It is to be stressed that the single measurement assumed here represents a decidely incomplete set of measurements since the state vector is four-dimensional. In the failure process, σ_∞ = 250 MPa and $C_s^{1/2}$ = 37 MPa. A rather poor NDE performance emerges, as indicated by the operating characteristic shown in fig. 16. The poor performance is associated with three factors: incompleteness in the set of measurements, randomness in the measurement process, and randomness in the failure process. The two remaining cases will differentiate between these possibilities.

Case 2. One Measurement. Deterministic Measurement and Failure Processes. Here a single measurement is again used. However, the randomness is eliminated from the measurement and failure processes by setting the variances $C_r = C_s = 0$. The resultant NDE performance (hypothetical) is given by the plot in fig. 16. Although there is a marked improvement in the performance, i.e., the curve has moved closer to the horizontal and vertical axes, the performance is still poor. This is due, as one might expect, to the serious incompleteness of the measurement set.

Case 3. Complete Measurement Set. Deterministic Measurement Process but a Random Failure Process. In this case, a significantly large diversity of very accurate measurements is assumed, such that the measurement vector y implies a unique estimate of x, namely $\hat{x}(y)$, with a negligible a posteriori variance (more precisely, a covariance matrix $Cov(x|y)$, whose eigenvalues are small). The resultant plot of the NDE operating characteristic is presented in fig. 16. This highly satisfactory result demonstrates clearly that randomness in the present failure process is not a signficant contributor to the degradation of NDE performance.

4.2.2 Fracture From Voids. An analysis similar to that described in detail for sub-critical inclusion fracture has been conducted for fracture from internal voids [7]. The NDE operating characteristics described for void induced failure are plotted in fig. 17. The results indicate a rather poor NDE performance due to an excessive overlap of surviving and failing populations. This overlap is due almost entirely to inherent randomness in the failure process (remaining even when the state $x = a$ is known with precision). However, fracture from sub-surface voids, interacting with surface cracks, may provide a superior NDE performance, because of a reduced statistical variability of the fracture process.

5. CONCLUSION

The use of acoustic scattering methods for assuring the survival of ceramic components has been discussed and illustrated with examples. Considerable practical success has been achieved with regard to the prediction of failure from machining damage, based on a long wavelength surface acoustic wave measurement method and a residual stress fracture model. The surface acoustic wave method has the advantages that it can be readily applied to components of complex shape, and exhibits an ability to predict failure (in Si_3N_4) at strength levels up to $\sim$400 MPa (a safety factor may also be imposed by incorporating annealing treatments following the inspection process).

Bulk defects are more difficult to analyze and their detectability by acoustic methods in components of complex shape requires sophisticated equipment. Nevertheless, appreciable advances have been achieved using select combinations of acoustic measurement algorithms. The progress has been illustrated by two important examples: failure controlled by sub-critical inclusions (such as Si in Si_3N_4) and failure from large internal voids. For the sub-critical inclusion fracture problem good NDE operating characteristics were demonstrated, contingent upon a comprehensive

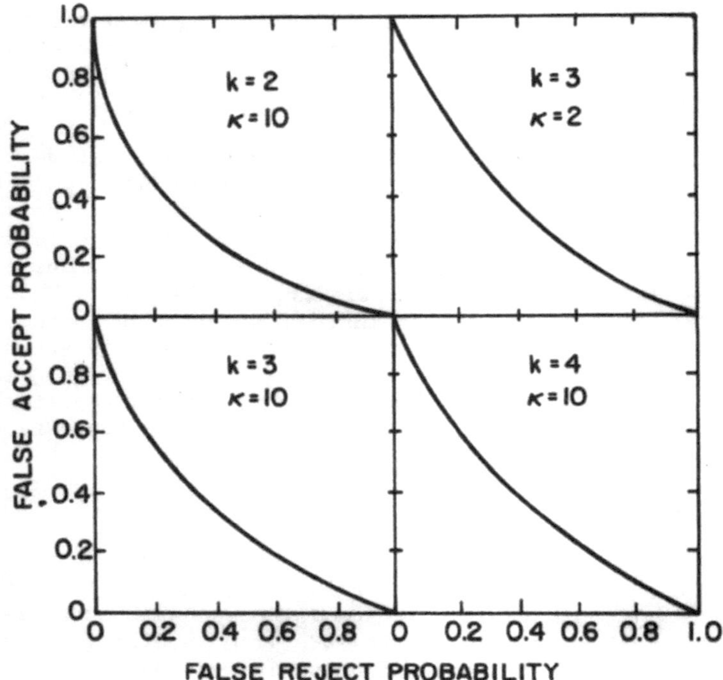

Fig. 17 NDE operating characteristics for long wavelength acoustic measurements pertinent to failure from internal voids.

set of acoustic measurements for each detected inclusion (i.e. scattering measurements obtained from several transducer locations). The basic utility of NDE for minimizing component rejection has thus been established for this important class of defect. Conversely, very poor NDE operating characteristics were obtained for failure from internal voids, even for a comprehensive set of NDE measurements. The minimal utility of NDE in this instance resides in the appreciable intrinsic variability of the fracture process associated with this class of defect. Proof testing, or NDE methods which detect the surrounding microscopic damage, may thus be required in order to achieve satisfactory operating characteristics with materials that fail predominantly from internal voids.

ACKNOWLEDGMENT

This work was supported by the U.S. Office of Naval Research, under contract No. N00014-79-C-0159.

REFERENCES

1. Evans, A.G., M.E. Meyer, K.W. Fertig and H.R. Baumgartner, Jnl. Non-Destructive Evaluation 1 (1980) 11.
2. Rice, R.W., Fracture Mechanics of Ceramics (Ed. R.C. Bradt, D.P.H. Hasselman and F.F. Lange), Plenum, N.Y. (1974) vol. 1, p. 323.
3. Evans, A.G., Jnl. Amer. Ceram. Soc., in press.
4. Thompson, R.B. and A.G. Evans, Sonics and Ultrasonics 23 (1976) 292.
5. Goebbels, K. and H. Reiter, this volume.
6. Rice, R.W., Processing of Crystalline Ceramics (Ed. H. Palmour and R.F. Davis) Plenum, N.Y. (1978) p. 303.
7. Richardson, J. and A.G. Evans, Jnl. Non-Destructive Evaluation 1 (1980) 37.
8. Lawn, B.R., A.G. Evans and D.B. Marshall, Jnl. Amer. Ceram. Soc. 63 (1980) 574.
9. Marshall, D.B., B.R. Lawn and P. Chantikul, Jnl. Mater. Sci. 14 (1979) 2225.
10. Rice, R.W. and J.J. Mecholsky, The Science of Ceramic Machining and Surface Finishing, NBS STP 562 (1979) 351.
11. Marshall, D.B., this volume.
12. Rice, R.W., Treatise On Materials Science and Technology (Ed. H. Herman), Academic Press 11 (1977) 189.
13. Evans, A.G., D.S. Biswas and R.M. Fulrath, Jnl. Amer. Ceram. Soc. 62 (1979) 101.
14. Seshadri, S.G.and M. Srinivasan, Jnl. Amer. Ceram. Soc. 64 (1981) C 69.
15. Baratta, F.I., Jnl. Amer. Ceram. Soc. 61 (1978) 490.
16. Lange, F.F., Fracture Mechanics of Ceramics, ibid., vol. 4 (1978) p. 799.
17. Kuzyk, J.A. and R.C. Bradt, Jnl. Amer. Ceram. Soc. 56 (1973) 420.
18. Evans, A.G., Acta Met 26 (1978) 1845.
19. Fu, Y., M.S. Thesis, Univ. Calif. Berkeley, August 1981.
20. Evans, A.G. and W. Blumenthal, Fracture Mechanics of Ceramics (Ed. R.C. Bradt, D.P.H. Hasselman, F.F. Lange and A.G. Evans), Plenum, N.Y., vol. 6 (1982).
21. Evans, A.G. and A.S. Rana, Acta Met 28 (1980) 129.
22. Chuang, T.J., K.I. Kagawa, J.R. Rice and L.R. Sills, Acta Met 27 (1979) 265.
23. Hsueh, C.H. and A.G. Evans, Acta Met, in press.
24. Tien, J.J.W., B.T. Khuri-Yakub, G.S. Kino, D.B. Marshall and A.G. Evans, Jnl. Appl. Phys., in press.
25. Khuri-Yakub, B.T., G.S. Kino and A.G. Evans, Jnl. Amer. Ceram. Soc. 63 (1980) 65.
26. Rose, J.H. and J.A. Krumhansl, Materials Science Center Report, 2846, Cornell Univ. (1977).
27. Kupperman, D.S., L. Pahi, D. Yuhas and T.E. McGraw, Bull. Amer. Ceram. Soc. 59 (1980) 814.
28. Chou, C.H., B.T. Khuri-Yakub, G.S. Kino and A.G. Evans, Jnl. Non-Destructive Evaluation, in press.

NON-DESTRUCTIVE EVALUATION OF CERAMIC GAS TURBINE COMPONENTS BY X-RAYS AND OTHER METHODS

K. Goebbels and H. Reiter

Fraunhofer-Institut für zerstörungsfreie Prüfverfahren, Universität, Gebäude 37, D-6600 Saarbrücken, W. Germany.

ABSTRACT. Quality assurance of ceramic gas turbine components needs special efforts. The material s brittleness is prohibitive against the release of stresses by plastic deformation at defects under load. Therefore, critical defect dimensions are 10 to 100µm, still smaller for heavily loaded surface regions. The contribution discusses the following techniques - excluding ultrasonics which were described in this volume by A.G. Evans - : Microfocus X-rays (microradiography). The analysis of macroscopic volumes (cm^3) with microscopic resolution (µm) shows its greatest advantage for complex shaped components like turbine blades, stator, rotor, inlet cone. The optimization process for a 15 µm - focus X-ray beam and projection technique results in an easy way to detect defects with dimensions ≥ 20 µm, to classify defects (lower/ higher X-ray absorption than the surrounding structure) and to describe structure homogeneity. Other techniques described under the above mentioned aspects are vibration analysis, optical holographical interferometry, microwaves and - but without success - acoustic emission.

1. INTRODUCTION

The brittleness of ceramic components is prohibitive against the release of stresses by plastic deformation at defects under load (compressive and tensile stresses, temperature gradients). Therefore critical defect dimensions are 10 to 100 µm depending upon the kind of defect. Still smaller values have to be avoided for heavily loaded surface regions. Non-destructive testing methods for the detection and evaluation of the types and sizes of defects do not belong to the state of the art. But they are under develop-

Riley, F.L. (ed.) Progress in Nitrogen Ceramics

ment where three goals should be reached:

- Detection and description (kind, shape, size) of single defects.
- Characterization of the structure (homogeneity, density variations). Especially in the defects' environment the structure analysis is as important for the judgement, e.g. following fracture mechanics, as the
- Stress situation. Fabrication of material and component cannot prevent that locally residual stresses occur, so the non-destructive stress analysis is of importance, too.

This contribution describes the optimization of given methods and the development of better or new non-destructive testing (NDT) techniques. Main work was done within the American (DOE, DARPA) and German (BMFT) gas turbine programs. Referring to the potential of NDT techniques, mostly ultrasonic methods (see the contribution of A.G. Evans in this volume) and high resolution X-rays will be applied (1). Other methods like vibration analysis, acoustic emission, microwaves or optical holographical interferometry were analyzed too. But they do not have the wide and high resolution application possibility (2).

The interaction between a defect and analyzing radiation (mechanical, electromagnetic waves) or load (impact, temperature shock) can be reduced to a few parameters. The essentials are

- the difference in the physical properties of matrix and defect (e.g. X-ray absorption, elastic and anelastic properties) and
- the relation defect dimension d to wavelength and defect cross section to analyzing beam parameters.

To obtain maxima for both parameter groups means optimized detectability. This includes too that not all of the possible defect kinds can be found best with one NDT technique alone: a Si-inclusion in Si_3N_4 is hard to find with X-rays (low difference in the absorption coefficients) but easy with microwaves (big difference in the electrical conductivities); the orientation of the defect relative to the analyzing radiation is of the same importance, because a crack perpendicular to the surface can be found better by X-rays than by ultrasonics while ultrasonics detect cracks parallel to the surface easier than X-rays; finally the component geometry influences strongly ultrasonics and microwaves but only to a minor extent X-rays.

2. X-RAY ANALYSIS

As well as single defects, extended structure heterogeneities (density variations, anisotropy of the structure, phase boundaries and bonding) can be imaged and analyzed with X-rays. The main question of the resolution and detectability (signal-to-noise ratio, contrast) has to be answered by the focal spot of the X-ray tube, the imaging technique and the receiver (film, image intensifier, fluorescent screen). The geometrical unsharpness, U_g, defined by [1]

$$U_g = f.b/a \qquad [1]$$

is proportional to the focal spot diameter f inside the tube (a - distance X-ray source to specimen, b - distance specimen to receiver). It follows from this relation that usual X-ray analysis with f ≈ mm cannot be used for the detection of defects ≥ 10 μm. Even the compensation of f by a small b/a value needs - besides other disadvantages - an expensive secondary enlargement, to make visible the microscopic defects. For the high definition X-ray analysis with fine focus tubes (f < 100 μm) and projection technique ("microradiography") (see Fig. 1), a secondary enlargement is unnecessary. The imaging technique reproduces on film or image intensifier specimen details with a primary magnification m = (a+b)/a. With usual values b = 2000 mm, a = 100 mm, this corresponds to a magnification x20. Focal spot sizes of 15 μm are commercially available* so that the geometrical unsharpness does not influence too much the resolution of defect details: a defect of 50 μm diameter will be imaged on the film as a 1 mm object with 0.3 mm penumbral unsharpness.

Microradiography has the following additional advantages:

- analysis of greater cross sections (≈ cm^2) and volumes (≈ cm^3) with microscopic resolution (≥ 20 μm);
- high depth of focus. Any detail inside a component will be imaged with the same geometrical unsharpness;
- the projection technique increases the contrast by reduced scattered radiation;
- the primary magnification allows the use of high sensitive film/screen combinations with short exposure times, otherwise applied only in the medical diagnosis.

The other advantages of X-ray analysis, like multiple angle through radiation and electronic image processing, are valid for microradiography too.

Outstanding developments can be characterized by the goal of X-ray topography for ceramic gas turbine parts. But the most important question to be answered at the moment is the electronic, automatic image recording with e.g. image intensifiers, X-ray sensitive cameras etc. Up to now the film is the most sensitive medium as shown in Figures 2-5. The loss of contrast and detail

---

* e.g. Wardray Company, Oxford, England. f ≈ 15 μm; beam diverging angle 18°, 80 kV, 0.5 mA.

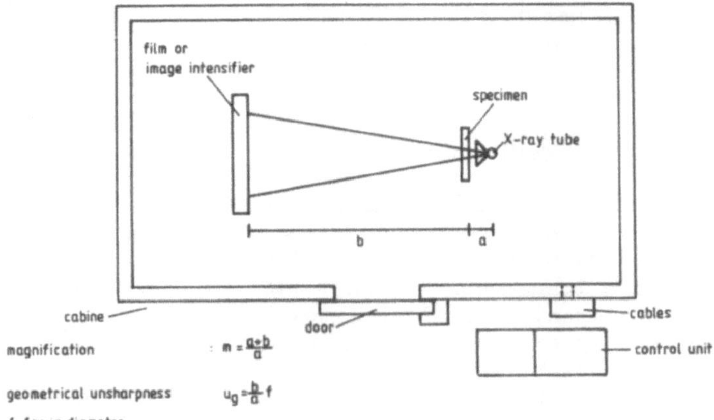

Fig. 1 Schematic Diagram of the Microradiographic Technique

Fig. 2 Surface Defect in a RBSN Turbine Blade

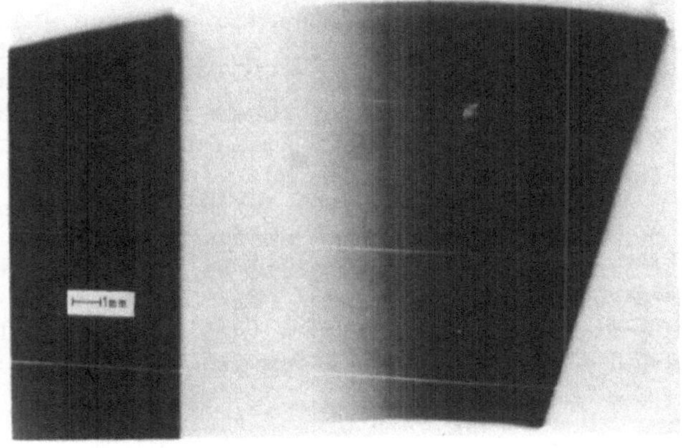

Fig. 3 Pores in a RBSN stator

detectability (film: details ≥ 20 μm; image intensifier: details ≥100 μm) could be balanced at least partly by microchannel plates, signal processing by averaging techniques, etc.

3. VIBRATION ANALYSIS

The excitation of mechanical long wavelength vibrations - where the wavelength corresponds to the dimensions of the specimen under test - leads to resonance vibrations which can be measured by piezoelectric transducers or just microphones. The simple solution* uses an impact excitation (broadband spectrum) and the analysis of the specimen-related main vibration mode (resonance frequency in kHz or period of vibration in μs). For simple geometries (bar, disc) the values are related to the elastic modulus, for complex shaped components like turbine blades or whole rotors they are characteristic parameters of the parts. More detailed information can be obtained exciting and analyzing with small bandwidth the spectrum of ground modes and higher harmonic vibration modes**, the so-called modal analysis. Caused by the overall excitation the usual vibration analysis allows only an integral component characterization. But for a series of equally fabricated parts this seems to be a simple and reliable judgement about the homogeneity of the fabrication and manufacturing process. Figure 6 shows the vibration time (two periods) against the weight of three series of reaction bonded silicon nitride (RBSN) turbine blades and the ultrasonic longitudinal wave velocity of two series of RBSN four point bending test specimens (3). The bandwidth of data scatter and tendencies are clearly seen. For a final judgement of this method there are still outstanding extended non-destructive and destructive tests.

4. ACOUSTIC EMISSION

Generally, the potential of acoustic emission analysis is very high for brittle materials. The elastic energy stored during a loading process cannot be released by plastic deformation but will be used almost totally for the generation of cracks. Especially for oxide ceramics (with grain sizes in the ten μm region and above) this was proved in the past (4). In agreement with the known literature, experiments together with MTU (Motoren-Turbinen-Union) have shown (2) that for RBSN specimens with and without Knoop indentations no correlation could be found with the strength or the damage. Until now the reason for this is assumed to be related to the fine grained structure (grain sizes smaller than 1 μm). It seems that too many bridges are broken instead of

---

*e.g. Grindosonic, Lemmens-Elektronika, Belgium

**e.g. Elastomat, Institut Dr. Förster, Reutlingen.

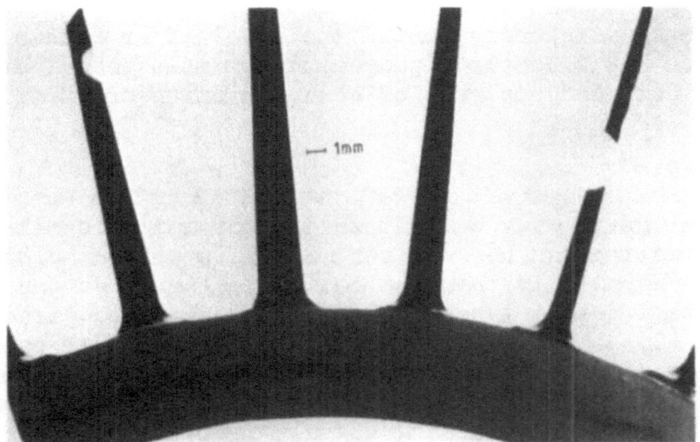

Fig. 4 Cracks in a RBSN Rotor Ring

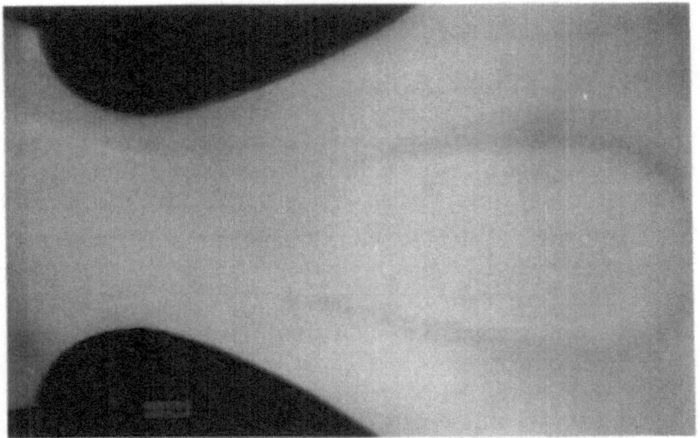

Fig. 5 Density Variations in a HPSiC Rotor Cut

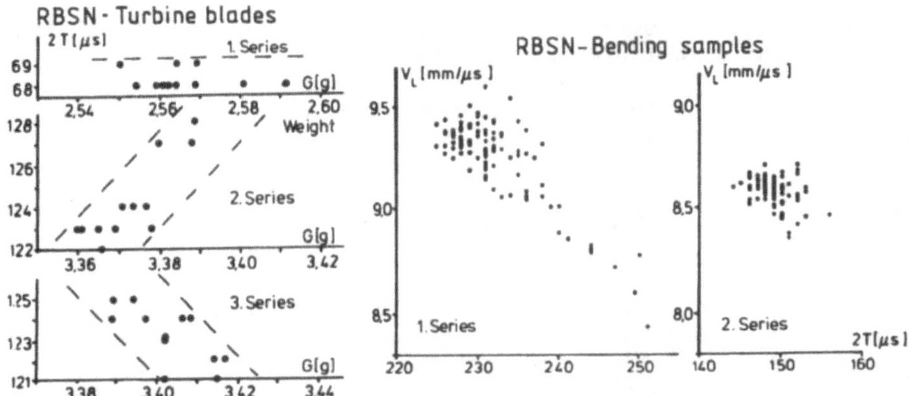

Fig. 6 Vibration Analysis of Series of Different RBSN Samples

storing the elastic energy until the level of an "audible" noise is reached. But a complete judgement for this behaviour needs still more detailed analysis of acoustic emission experiments.

5. MICROWAVES

In analogy to eddy current testing of metallic materials, microwave testing can be seen for materials without electrical conductivity. Especially then the semiconductor Si as one of the most important defect types in Si_3N_4 can easily be detected with microwaves (2,5). Detailed analysis has shown their capability to detect other defect types too, like WC, Fe, BN, pores and C. The necessary frequency range of more than 100 GHz at the moment includes unfortunately disturbing signals from the component geometry (curvature, edges etc.) which cannot be suppressed. So up to now it seems not to be possible to develop this technique without extensive work in the direction of practical application.

6. OPTICAL HOLOGRAPHICAL INTERFEROMETRY

With the optical holographical interferometry it is possible to image defects in specimens under load (mechanical, thermal loading). The ability was shown by different experiments (2,6), but compared to other techniques like ultrasonics and X-rays the lack of resolution especially for internal defects does not recommend at the moment this technique for practice.

7. DISCUSSION

From the point of view of NDT for ceramic gas turbine components, two relevant defect types have to be detected and described:

- surface defects with dimensions ≥ 5 µm, and
- internal defects with dimensions $\gtrsim 10$–100 µm, depending on the kind of the defect (inclusion, pore).

The variety of developments raises two methods with the highest application potential and which are nearest to practical application now:

- high resolution ultrasonics, and
- high definition X-ray analysis (micrgradiography).

They complement each other as well during detection and location as during judgement about kind, shape and size. The detection limit at the moment for both lies at about 20 µm with some deviations to lower or higher values depending upon the defect kind. Especially for the complex shaped turbine parts the microradiography seems to be the most suitable technique for practical use.

Future aspects for ultrasonic testing are developments to better detect and reconstruct surface flaws. Other techniques seem to be suited only for limited use. Additionally emphasis should be given to the point that still more research has to be done asking which component under which load and at which point will fail caused by which defect. But this question - important to concentrate NDT developments - cannot be answered by NDT, only by materials science, fracture mechanics and turbine design engineers.

REFERENCES

1. H. Reiter and K. Goebbels. Mikroradiographie. IzfP-Report No. 780106-TW (Saarbrücken, 1978).
2. H. Reiter et al. ZfP von Hochtemperatur-Keramik-Bauteilen für Kfz-Turbinen. IzfP Reports 790204 (1979), 800305 (1980), 810418 (1981), Saarbrücken.
3. K. Goebbels and H. Reiter. ZfP-Verfahren für Komponenten der Keramik-Gasturbine (BMFT-Status-Seminar, Bad Neuenahr, Nov. 1980).
4. L.J. Graham and G.A. Alers. Rockwell Science Center Reports AD-745000 (1972), AD-754839 (1972), AD-778015 (1974) (Thousand Oaks, CA).
5. A.J. Bahr. "Microwaves NDE of ceramics". Report AD/A-048582, AMMRC-CTR-77 (Stanford Research Institute, Stanford, CA, Nov. 1977);
   Proceedings ARPA/AFML; AFML-TR-78-209 (Jan. 1979), p.236-241.
6. D.S. Kupperman et al. "Preliminary evaluation of NDE techniques for structural ceramics". Proceedings ARPA/AFML; AFML-TR-78-209 (Jan. 1979), p.214-227.

ACKNOWLEDGMENT

This contribution is based on work performed with the support of the German Ministry for Research and Technology (ceramic gas turbine program).

DISCUSSION

Popper: What method is best for flaw detection in unfired bodies?

Goebbels: I see no obstacle against projection radiography. We do not know enough about the size, orientations and shapes of "defects" in unfired ceramics.

Katz: In the DARPA/Ford programme injection moulded silicon bodies were successfully screened prior to burn-out; slip cast preforms present handling difficulties.

SURFACE DAMAGE IN CERAMICS: IMPLICATIONS FOR STRENGTH DEGRADATION, EROSION AND WEAR

D. B. Marshall

Department of Materials Science and Mineral Enginnering, University of California, Berkeley, CA 94720

ABSTRACT

Recent advances in the analysis of elastic/plastic indentation fracture, as a model for contact induced damage, are reviewed. The central feature of the analysis is the dominant role played by by the residual stress in the evolution of the two major crack systems (one relating to strength and the other relating to material removal), and in the subsequent response of the cracks to an applied tension. The use of surface acoustic wave scattering as a method of non-destructive evaluation is discussed in the context of the indentation flaw model; the technique provides direct evidence that machining induced cracks are subjected to residual stress effects.

1. INTRODUCTION

Wear, erosion and strength degradation of ceramic surfaces all originate from localized cracking associated with small-scale contact events. Two fundamentally different types of contact can be distinguished [1,2]. One involves only elastic contact stresses, and results in relatively low material removal rates and strength loss. The more severe contact, which is the subject of this paper, involves elastic/plastic deformation, and occurs when the surface is penetrated by a hard, sharp object.

The general problem of predicting the damage resulting from contact of a flat surface by an irregular particle is complex. A large number of variables are expected to enter the description; as well as the variation of material properties (toughness,

Riley, F.L. (ed.) Progress in Nitrogen Ceramics

hardness, elastic modulus) and contact load (or velocity), there would appear to be an infinite number of geometrical variations. However, after the onset of plastic deformation 1), the contact site is surrounded by an approximately hemispherical plastic zone, which tends to reduce the sensitivity of the stress field to the details of contact geometry [3,4]. This holds true particularly for the residual field, which arises from mismatch of the plastic zone and the surrounding elastic matrix [3]. It will be shown in Section 2 that it is the residual field which dominates most of the ensuing fracture. Therefore the general features of contact induced cracking are insensitive to details of the contact geometry, and it becomes possible to model the general contact in terms of an ideal geometry. Indentation with a standard diamond indenter (Vickers or Knoop) provides a convenient model contact system.

In the first studies of elastic/plastic indentation fracture, by Lawn and co-workers [1,5-7], two primary crack systems were identified; the "median/radial" cracks which propagate normal to the surface and relate to strength properties, and the "lateral" cracks which form on planes closely parallel to the specimen surface and relate to the material removal. Analysis of median/radial crack growth within an elastic, point-contact field provided relations between indenter load, crack size and material toughness. Subsequently, Evans and Wilshaw [8] used a semi-empirical dimensional analysis to infer the dependence of the median/radial dimension on the elastic/plastic material properties. Neither analysis was applicable to the lateral cracks, a system which is complicated by the close proximity of the crack plane and the specimen free surface.

In this paper some recent advances in the analysis of indentation fracture are reviewed. Models, which provide predictions of the extent of cracking in terms of contact parameters and material properties, have been developed for both crack systems (Sections 3 and 5). The models are based on the observation that the primary crack driving force derives from the residual indentation stress field. Perhaps the most important implications of the radial crack model lie in the insight that it provides into the micromechanics of flaw response to applied tension; the model predicts contrasting crack growth behavior for contact-induced flaws and ideal Griffith flaws. The use of surface acoustic wave scattering to monitor the response of machining induced cracks to applied loading provides the most direct experimental support for the general application of the model, and is discussed in detail in Section 4.

1. Plastic deformation is taken to refer to any irreversible deformation process which accommodates the volume displaced by penetration of the surface, and may include densification, viscous flow and shear cracks, as well as the usual plastic deformation mechanisms.

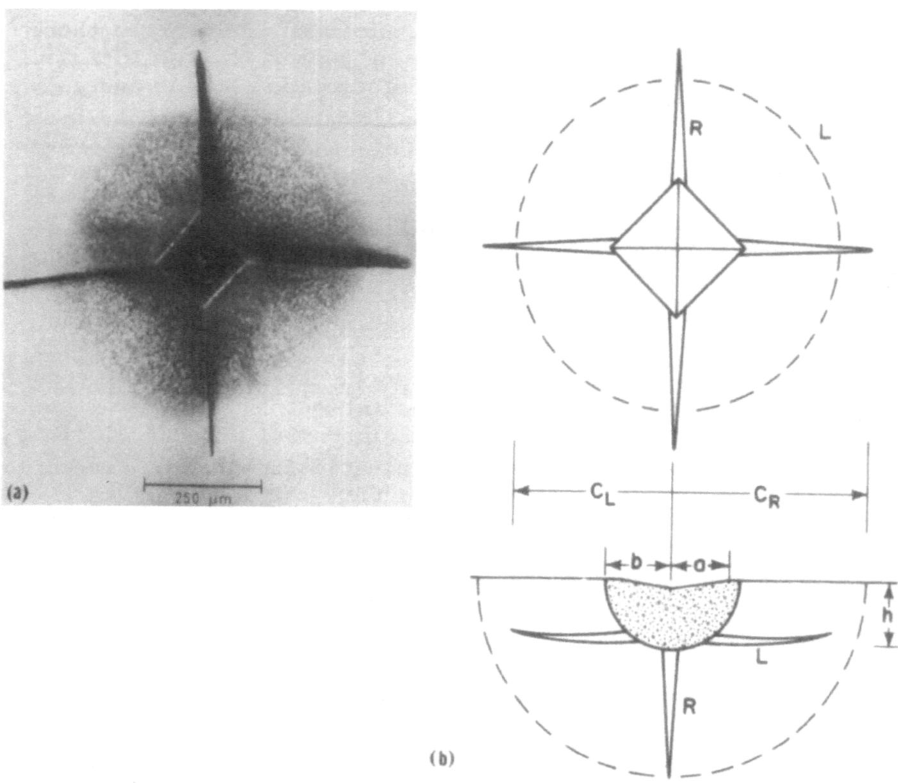

Fig. 1 Elastic/plastic contact damage. (a) Vickers indentation in ZnS. (b) Schematic surface and cross-section views of (a); R = radial crack, L = lateral crack, shading represents plastic zone.

2. INDENTATION FRACTURE AND RESIDUAL STRESSES

2.1 Fracture Geometry

The deformation/fracture pattern characteristic of elastic/plastic contact is typified in Fig. 1a by a Vickers indentation in ZnS, and illustrated schematically in Fig. 1b. Two crack systems are evident. The median/radial crack system, hereafter referred to as "radial" cracks, comprises two half-penny shaped cracks, each parallel to the load axis and one of the indenter diagonals. The lateral crack system comprises one or more penny-shaped cracks, parallel to the specimen surface and centered beneath the contact site. Surrounding the hardness impression is a zone of irreversible deformation, which is faintly visible in Fig. 1a as a brighter, out-of-focus, region. The detailed mechanisms of deformation within this zone are generally unknown for most brittle materials: in glass, KCl and LiF, bands of localized shear have been identified [9,10]; in ZnS, grain boundary cracking, which may be the result of slip within grains, is observed [11]; and in MgO, Al_2O_3 and SiC intense dislocation defomation has been observed by transmission electron microscopy [4,12].

Modifications of the indentation fracture pattern due to indenter shape are manifest mainly in the number of radial cracks and their locations. Cracking initiates near the elastic/plastic boundary, either from pre-existing nuclei, or from nuclei which are generated by inhomogeneity of the plastic deformation [13-15]. Initiation of radial cracks is enhanced in regions of local stress concentration near corners of the indentation; thus the Vickers indenter generates two orthogonal radial cracks, and the elongated Knoop indenter produces only one radial crack (Fig. 6).

2.2 The Role of Residual Stress

A residual stress field, due to mismatch between the plastic zone and the surrounding elastic matrix, remains after the indenter is removed [16,17]. The key step in analyzing indentation fracture is the recognition that the residual field dominates the evolution of both crack systems. This point is demonstrated in Fig. 2, which shows a sequence of optical observations of an indentation crack system during the load/unload cycle. Stress birefringence in the glass specimen is made visible by use of crossed polars in the illumination system; in the absence of stress birefringence the field would be completely dark. Two important observations can be made from Fig. 2. Firstly the final crack configuration is achieved as the indenter is removed from the surface; the radial cracks increase in size continuously as the load is reduced from the peak value in Fig. 2b to zero in Fig. 2d, and the lateral cracks initiate as the load approaches zero. Secondly, a high level of residual stress remains after removal of the indenter, as evidenced by the strong stress birefringence in Fig. 2d. Therefore, the residual field provides the driving force for propagation of both crack systems. Most importantly, the residual driving force continues to act after completion of the indentation cycle, and supplements the opening force of an applied tension during a subsequent breaking test.

2.3 Stress Analysis

Complete analysis of the residual indentation stress field is a complex elastic/plastic deformation problem. Therefore it is convenient to distinguish between near and far fields about the elastic/plastic boundary. Crack initiation is sensitive to the near field, and prediction of fracture thresholds requires detailed stress analysis. Such analysis has recently been

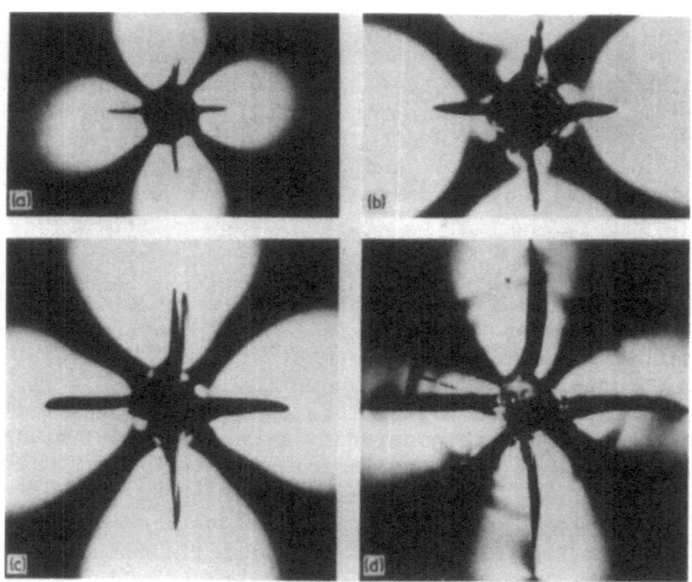

Fig. 2 In-situ observation of indentation crack evolution; soda-lime glass loaded with Vickers pyramid, while being viewed optically through the opposite face. (a) P = 50N, (b) P = 90N (peak load), (c) P = 30N, (d) P = 0 (complete unload). Width of field 720 μm.

performed [3,13], and good correlation between predicted and observed thresholds was achieved. Propagation of well developed cracks (i.e. crack length, c , >> plastic zone dimension, b) is governed by the far field. In this region the fracture mechanics analysis is simplified by an approximation which eliminates the complex stress analysis, but allows the important material parameters describing plasticity (hardness, H), elasticity (Young's modulus, E) and fracture (toughness, K_c), to be incorporated [18].

The residual stresses in the far field region are most conveniently evaluated in terms of a residual pressure acting at the boundary of the plastic zone [18]. The magnitude of the residual pressure, p , can be related to the indentation load, P , and the material properties by allowing the plastic zone to be created by a sequence of hypothetical cutting, deforming and healing operations [18], analagous to the procedure used by Eshelby to solve the problem of a transformed inclusion [19]. Then, in the limit of well developed cracks, the residual pressure may be regarded as concentrated at a point located at the origin of contact, and a residual crack opening force P_r is obtained by integrating the components of p normal to the crack plane over the zone cross-section within the plane 2);

$$P_r \sim pb^2 \sim (E/H)^{1/2} \ (\cot\Psi)^{2/3} \ P \ , \qquad (1)$$

where 2Ψ is the angle between opposite edges of the indenter. This relation provides the basis for the indentation fracture models of Sections 3 and 5.

3. STRENGTH DEGRADATION: RADIAL CRACKS.

3.1 Radial Crack Model

The evolution of radial cracks at all stages during the indentation cycle may be evaluated by regarding the indentation stress field as a superposition of the residual field and a reversible elastic field due to indenter loading [18]. However, since our aim in this section is to investigate strength degradation, the model will be concerned only with the final crack configuration, which is determined solely by the residual field.

The model is illustrated in Fig. 3; the half-penny shaped radial crack is subject to a residual force P_r located at the center of the surface trace. It is assumed that P_r is not relaxed significantly by radial crack growth; this assumption has been supported by stress birefringence measurements [9], and by theoretical analysis [18]. The stress intensity factor for a penny-like crack with constant force loading, P_r, at its center is [16]

$$K_r \sim P_r/c^{3/2} \qquad (2)$$

Equations 1 and 2 combine to give

$$K_r = \chi_r P/c^{3/2} \qquad (3a)$$

where

$$\chi_r = \xi_r (E/H)^{1/2} (\cot\Psi)^{2/3} \qquad (3b)$$

and ξ_r is a dimensionless constant, independent of material properties and indenter shape. The equilibrium crack length after completion of the indentation cycle is given by $c = c_o$ at $K_r = K_c$:

---

2. An expression for the plastic zone radius is obtained from the stress analysis of Chiang et al. [3]; $b \sim a(E/H)^{1/2}(\cot\Psi)^{1/3}$, where $a \sim (P/H)^{1/2}$ is the indentation dimension (Fig. 1).

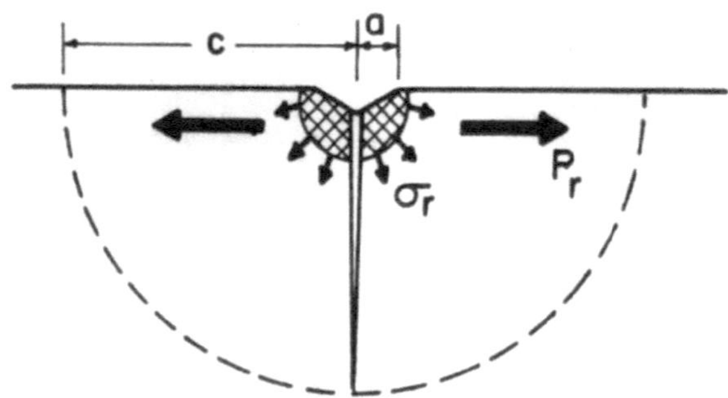

Fig. 3 Model for radial fracture; residual field σ_r is approximated as point loading P_r at the center of the surface trace of the semicircular radial crack.

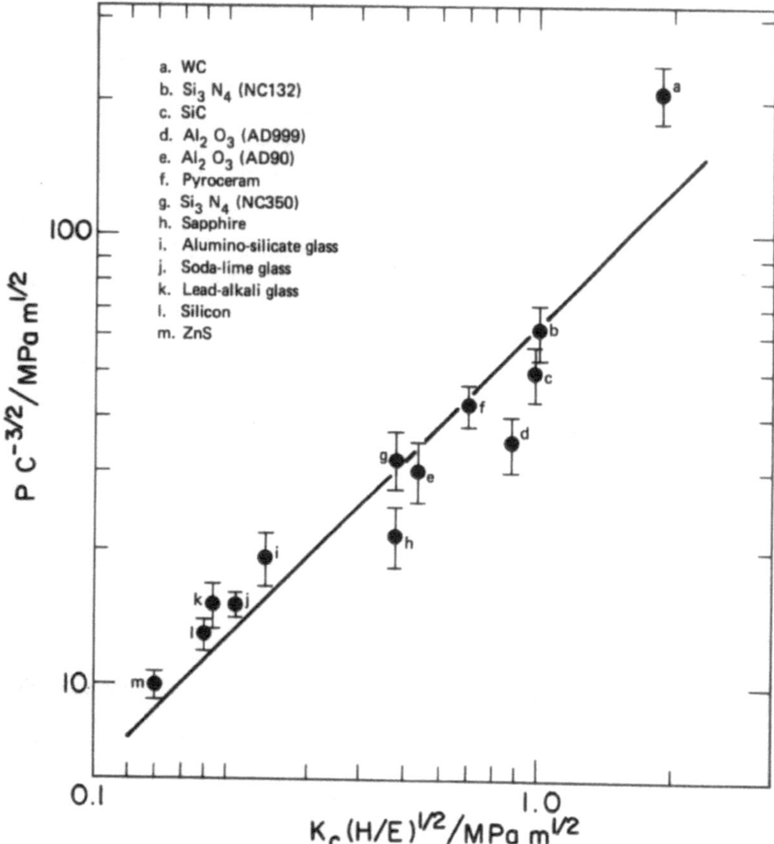

Fig. 4 Plot of radial crack measurements according to the prediction of Eqn. 4. Error bars represent standard deviation for at least 40 measurements, taken over a wide range of loads. Data from Ref. 20.

$$c_o = (\chi_r P/K_c)^{2/3} = \left\{\xi_r[(E/H)^{1/2}/K_c](\cot\Psi)^{2/3}\,P\right\}^{2/3} \quad (4)$$

The prediction of Eqn. 4 has been compared with measurements of radial cracks generated by Vickers indentation in ceramic materials representing a wide variation of material properties (E/H and K_c)[20]. The results are summarized in Fig. 4, where crack dimensions corresponding to a wide range of loads for each material are plotted as $P/c^{3/2}$ versus $(H/E)^{1/2}\,K_c$, as dictated by Eqn. 4. A fit of Eqn. 4 to the data provides calibration of the dimensionless constant; $\xi_r = 0.037$.

3.2 Response of Radial Cracks to Applied Tension

The residual force persists after the indenter is removed. In the absence of environmental stress corrosion, the crack length at the beginning of a breaking test is given by Eqn. 4 and the system is in a state of mechanical equilibrium, i.e. $K_r = K_c$. However, if the cracks are exposed to a non-inert environment between indentation and strength testing, they extend subcritically under the influence of K_r to some non-equilibrium size $c_o' > c_o$, and K_r at the beginning of the breaking test is given by Eqn. 3a with $c = c_o'$. Departure from the equilibrium configuration can also be caused by lateral crack chipping which relaxes the residual field and thereby reduces χ_r [21].

During a breaking test, the residual force supplements the applied tension in driving the radial cracks to failure. Assuming that one crack is oriented normal to the tensile direction, the stress intensity factor due to applied loading σ_a is of the standard form

$$K_a = \sigma_a(\pi\Omega c)^{1/2} \quad (5)$$

where Ω is a crack geometry parameter ($\Omega = 4/\pi^2$ for an ideal penny crack). The net stress intensity factor is the sum of K_a and K_r;

$$K = \chi_r P/c^{3/2} + \sigma_a(\pi\Omega c)^{1/2} \quad (6)$$

An applied stress/equilibrium-crack-size function follows from Eqn. 6 by putting $K = K_c$ and solving for σ_a;

$$\sigma_a = [K_c/(\pi\Omega c)^{1/2}][1 - \chi_r P/K_c c^{3/2}] \quad (7)$$

A plot of $\sigma_a(c)$ is shown in Fig. 5 for non-zero and zero χ_r; for non-zero χ_r the function exhibits a maximum at

$$c_m = (4\chi_r P/K_c)^{2/3} \quad (8a)$$

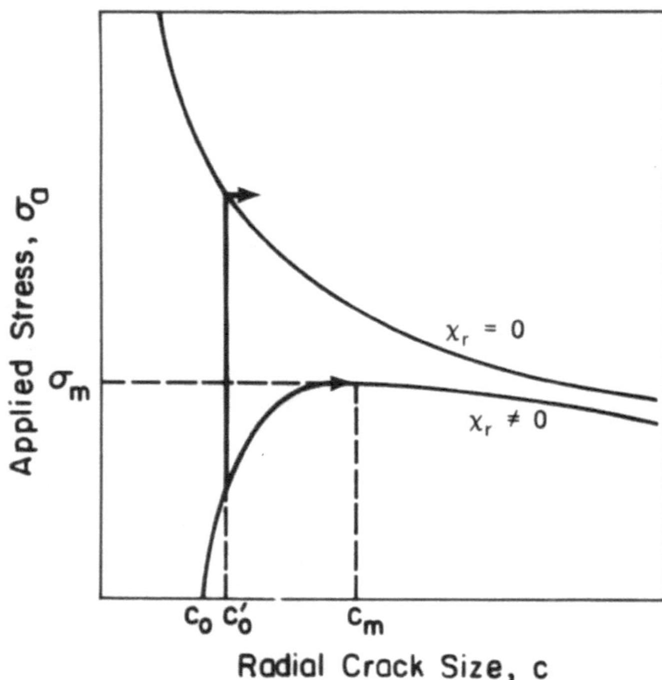

Fig. 5 Response of radial cracks to applied tension; applied-stress/equilibrium-crack-size relation from Eqn. 7. Curve for $\chi_r = 0$ (stress free flaws) represents unstable configuration. Curve for $\chi_r \neq 0$ (indentation flaws) represents stable configuration for $c < c_m$.

$$\sigma_m = 3K_c/4(\pi\Omega c_m)^{1/2} \tag{8b}$$

The two curves in Fig. 5 represent contrasting behavior. For $\chi_r = 0$ failure occurs spontaneously at $\sigma_a = K_c/(\pi\Omega c_o')^{1/2}$, according to the familiar Griffith description. For $\chi_r > 0$ (and $c_o' < c_m$) the crack extends stably from c_o' to c_m, prior to achieving instability at $\sigma_a = \sigma_m$. The resulting strength is given by Eqns. 8a and 8b

$$\sigma_a = \sigma_m = \left[(27/256)K_c{}^4/\chi_r(\pi\Omega)^{3/2}\right]^{1/3} P^{-1/3} \tag{9}$$

The strength is insensitive to the initial crack size, being controlled instead by the contact load P and the magnitude of the residual field, as defined by χ_r.

Direct observations of indentation cracks during failure testing have confirmed the existence of precursor, stable crack extension in a wide range of brittle materials (glass [17], silicon

[21], glass ceramics [22] and silicon nitride [23]). Representative data are shown in Figs. 6 and 7 3). In Fig. 6 the responses of cracks generated by two widely different indentation geometries (Vickers and Knoop) in silicon nitride are compared. The identical responses indicate that, in analyzing strength, the assumed point loading of the residual field provides a reasonable approximation, even for elongated contacts. In Fig. 7 two sets of data for Vickers indentations in soda-lime glass are plotted. In one set an inert environment was maintained throughout the indentation and subsequent failure test; the initial crack length was then c_o and the relative crack extension $c_m/c_o = 4^{2/3} = 2.5$ is observed. In the other set, subcritical crack extension was allowed between the indentation and failure tests (inert conditions being restored for the failure test), so that the initial crack size was $c_o' > c_o$; the extent of stable growth is reduced, but the final instability point is not influenced by the initial crack size.

3.3 Strength Degradation

According to Eqns. 7 and 8 the residual stress causes a severe reduction in strength. This prediction is consistent with results for comparative strength tests on indented materials with and without the residual stress removed [24]; typically the presence of the residual field reduces the strength by 30-40%.

The dependence of strength on material properties and contact parameters is obtained from Eqns. 3 and 9[25];

$$\sigma = A(H/E)^{1/6} K_c^{4/3}/P^{1/3} \tag{10}$$

where A is a dimensionless, material independent constant which includes the effects of indenter geometry. Strength degradation data obtained from Vickers indentation on ceramic materials covering a wide range of properties are summarized in Fig. 8 [25]. The data are plotted, according to Eqn. 10 as $(\sigma P^{1/3})^{3/4}$ versus $(H/E)^{1/8} K_c$; the dimensionless constant is calibrated as A = 2.02.

4. STRENGTH PREDICTION; ACOUSTIC WAVE SCATTERING

Techniques based on scattering of acoustic waves from defects are capable of providing non-destructive measurements of the defect dimensions [26]. One variant of this approach [27], designed for detection of surface cracks, is illustrated in Fig. 9a; transducer 1 excites surface (Raleigh) waves incident nearly normal to

---

3. The data are plotted in terms of normalized coordinates; putting $S_a = \sigma_a/\sigma_m$, $C = c/c_m$. Eqn. 7 becomes $S_a = C^{-1}[1-1/4C^{3/2}]$, and explicit dependence on material properties, indenter geometry and contact load is eliminated.

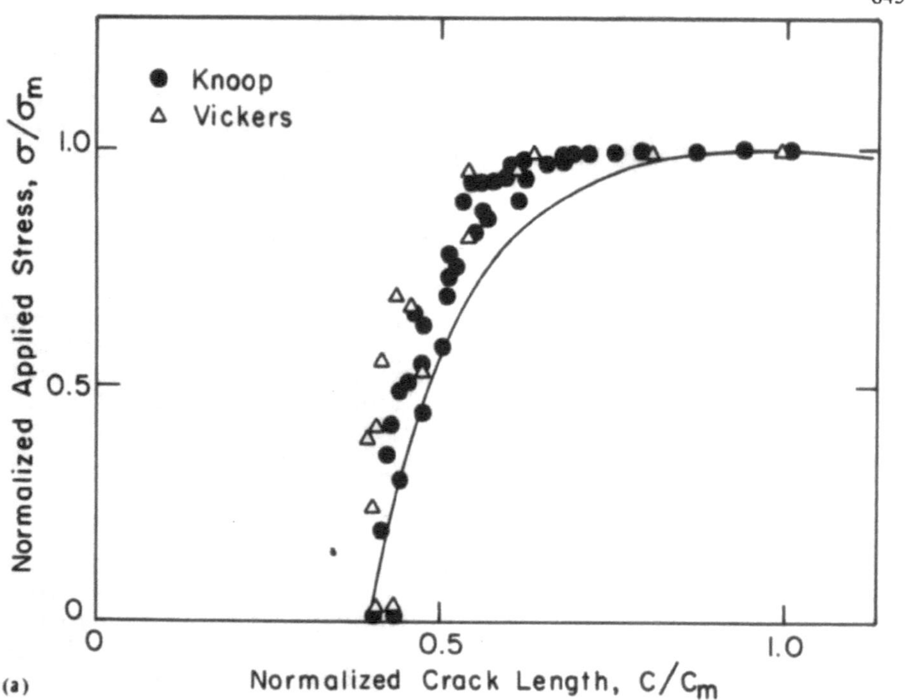

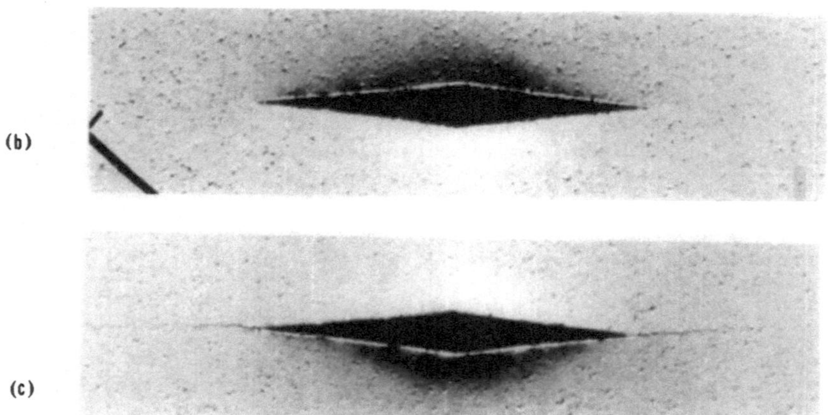

Fig. 6 Stable radial crack extension during breaking test; $HPSi_3N_4$ bars, indented at 50N load with Vickers or Knoop indenters, and broken in four-point bending. (a) Surface traces of radial cracks, measured optically as a function of applied loading, σ_a. (b) and (c) Optical micrographs of Knoop indentations; (b) at the beginning of breaking test, (c) at $\sigma_a = 0.9\sigma_m$.

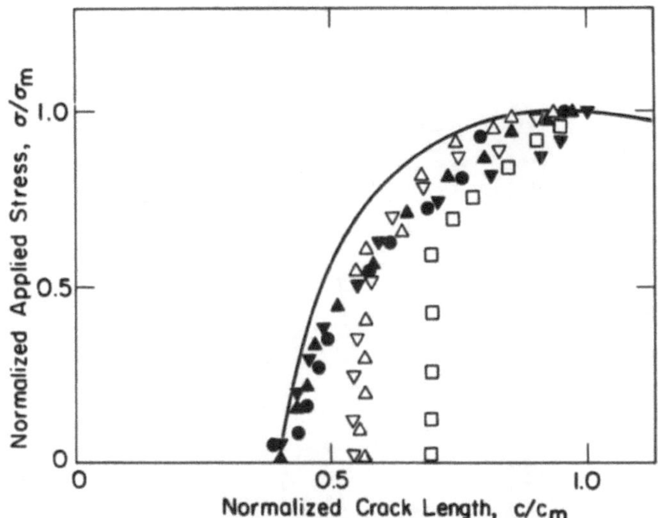

Fig. 7 Stable crack extension during failure test; soda-lime glass bars, indented at 5N (Vickers) and broken in four-point bending. Full symbols; inert environment maintained throughout indentation and failure tests. Open symbols; corrosive environment (air) admitted between indentation and failure tests, inert environment restored for failure test.

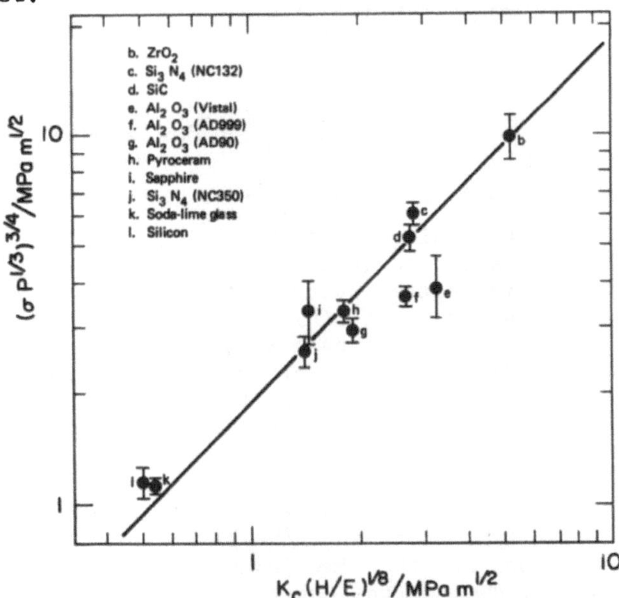

Fig. 8 Plot of strength measurements according to Eqn. 10. Error bars represent standard deviation of at least 15 measurements corresponding to a wide range of indentation loads. Data from Ref. 25.

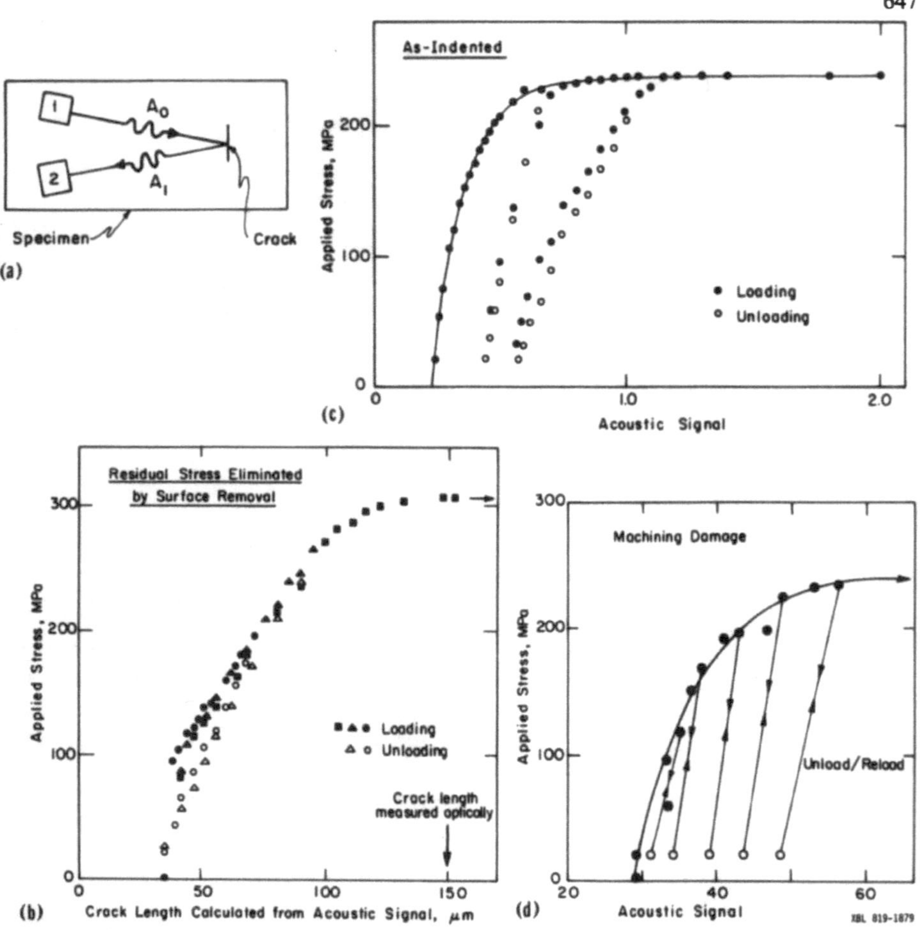

Fig. 9 (a) Schematic diagram of the method used to detect surface cracks by acoustic wave scattering. (b-d) Variation of acoustic scattering from strength-controlling cracks during breaking test. (b) Residual-stress-free crack. (Note reversible increase in scattering with applied load.) (c) Indentation crack. (Note irreversible increase in scattering.) (d) Machining-induced cracks. (Note irreversible increase in scattering.)

the crack surface, and transducer 2 detects the back-scattered waves. The relative amplitude of the back-scattered wave defines the crack dimension, and the time delay between generating and receiving the signal defines the crack position.

Strength prediction from acoustic scattering is a two-step procedure. In the first step the reflection coefficient, in conjunction with scattering analysis, provides the crack dimension. In the second step the crack dimension, in conjunction with fracture mechanics analysis, provides the strength. A knowledge of the flaw characteristics is essential to both steps.

4.1 Acoustic Scattering from Indentation Cracks

The sensitivity of the scattering analysis to the flaw model can be demonstrated by comparing the variation of acoustic scattering from indentation cracks and equivalent stress-free cracks 4) during failure tests [27] (Figs. 9b and c). The reflection coefficient for stress-free cracks in Fig. 9b is expressed in terms of the calculated crack radius, assuming an open, surface half-penny crack [27]. The reversible increase in apparent crack radius with applied loading can be interpreted in terms of reversible opening and closing of the crack surfaces. At zero applied stress, complete crack closure is prevented by contacts at asperities over the crack surface. The areas between the contacts scatter as small cracks, but, since the scattered amplitude is approximately proportional to a^3, the total scattered amplitude is considerably smaller than that of a fully open crack. Applied tension relieves the contacts continuously until, at the failure point, all contacts are relieved, and the true crack radius is measured. Scattering from cracks with residual stress (Fig. 9c) is not subject to the reversible opening and closing effects because, even in the absence of applied loading, the residual field holds the crack surfaces fully separated. However, an irreversible increase in acoustic signal with applied tension, corresponding to genuine crack extension is detected. Despite some complication in modelling the crack geometry 5), a true measure of the crack dimension is obtained at all stages during the failure test.

Non-destructive strength prediction requires a measurement of the initial crack length c_o. For indentation cracks a reliable measure of c_o is obtained from acoustic measurements. The strength is then obtained from Eqns. 8a, 8b and 4.

$$\sigma = A\, K_c/(\pi\Omega c_o)^{1/2} \tag{11}$$

where $A = 0.47$ for the ideal indentation crack, but can be as

---

4. Obtained by removing the plastic zone of an indentation by mechanical polishing.

5. The crack does not penetrate the plastic zone. Therefore the crack exhibits the geometry of a semi-annulus with inner radius dictated by the plastic zone radius. Calculations based on a subsurface elliptical crack have provided a good approximation [27].

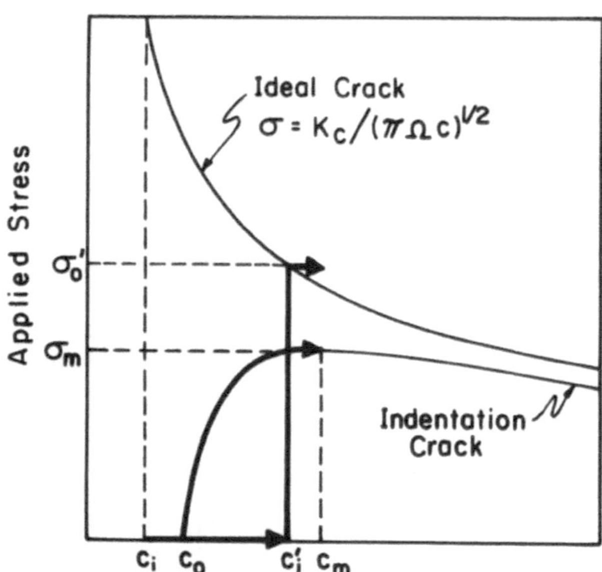

Fig. 10 Comparison of predicted strengths according to the two flaw models (indentation flaw and stress-free flaw) for a given acoustic scattering factor.

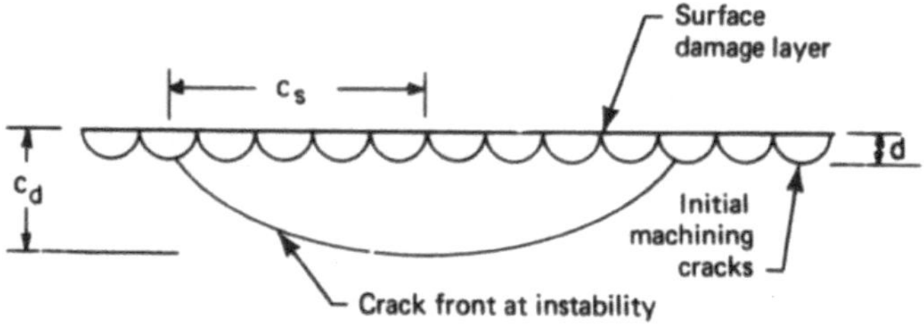

Fig. 11 Fracture surface resulting from failure from machining cracks.

high as 0.6 if relief of residual stress occurs [21] or if the initial crack length is $c_o' > c_o$. For stress free flaws, the true crack radius is measured immediately prior to failure, and is not related in a straightforward way to the apparent crack radius measured at zero applied stress. Therefore acoustic measurements provide a sound basis for strength prediction for flaws with residual stress, but not for stress free-flaws. Nevertheless it is of interest to compare the strength predictions for the two flaw types, for a given reflection coefficient (Fig. 10). With the ratio of apparent initial crack length to true crack length in Fig. 9b, the strength of a stress free crack is given by σ_o' in Fig. 10. The predicted strength of an indentation crack is $\sigma_m < \sigma_o'$. Therefore, if it is not known apriori whether or not residual stresses are present, the calculation based on the assumed existence of residual stress provides a conservative strength prediction.

4.3 Machining Damage

Acoustic scattering provides a unique method of monitoring the response of machining-induced cracks to an applied stress (the cracks are entirely subsurface and obscured from optical observation by a surface damage layer). The variation of acoustic reflection with applied loading is illustrated in Fig. 9d, for cracks induced in hot-pressed Si_3N_4 by surface grinding with a 240 grit diamond wheel. The irreversible increase in scattering amplitude provides a measure of stable extension preceding failure, and therefore provides direct evidence that machining induced cracks are subject to residual crack-opening forces.

The characteristic fracture surface markings produced by failure from machining damage in Si_3N_4 are illustrated schematically in Fig. 11. Similar features are produced by failure from a scratch and from a row of Knoop indentations. The initial machining damage is evident as a series of approximately penny-shaped cracks. Under the influence of an appied tension some of the cracks coalesce and extend in a stable manner (Fig. 11). The configuration at instability is related empirically to the initial crack dimension by $\sqrt{c_s c_d}/d = 5.5$, 6), where c_d, c_s and d are defined in Fig. 11.

5. EROSION AND WEAR; LATERAL CRACKS

5.1 Lateral Crack Model

A model for lateral crack evolution has been developed along similar lines to the radial crack model of Section 3, [28]. However a major difference between the two models arises because of the lack of bulk constraint of the material between the lateral crack plane and the specimen surface; the closeness of the lateral crack to the free surface allows significant relaxation of the residual driving force with crack extension. Accordingly, in the model for well developed lateral crack propagation, the material above the crack plane is regarded as a thin elastic plate clamped at its outer edges to a rigid substrate, and loaded in the center by the residual force P_r (Fig. 12). The residual force acts as a precompressed spring;

---

6. The dimension $\sqrt{c_s c_d}$ provides a useful characteristic crack dimension because the stress intensity factor for a semi-elliptical surface crack under uniform tension can be approximated by $K = \sigma(\pi\Omega\sqrt{c_s c_d})^{1/2}$.

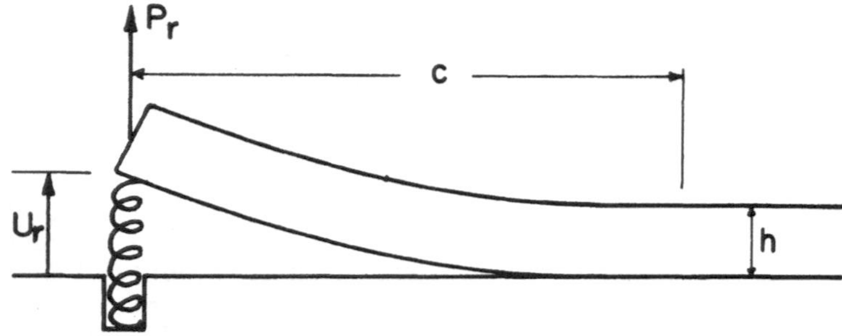

Fig. 12 Model for lateral fracture

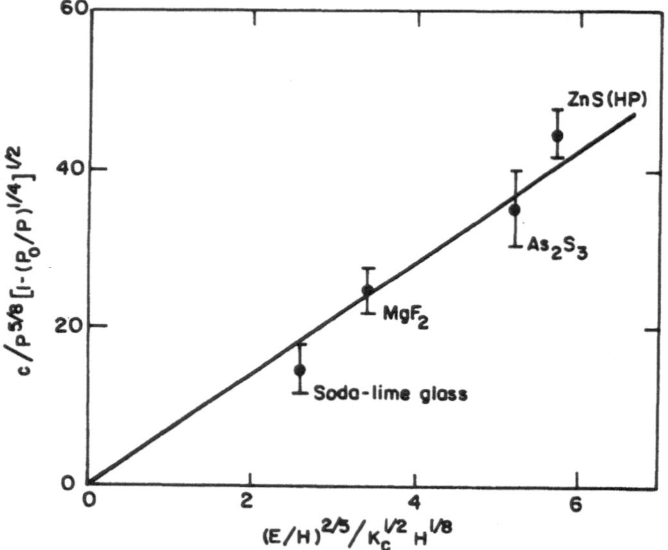

Fig. 13 Plot of lateral crack measurements according to Eqns. 14 and 15. Error bars represent standard deviations of measurements taken over a wide range of loads. Data from Ref. 28.

$$P_r = P_{r_o} \; (1 - U_r/U_{r_o}) \tag{12}$$

where U_r is the crack opening displacement at the crack center, P_{r_o} is the residual force for a fully closed crack (given by Eqn. 1), and U_{r_o} is the crack opening which completely relieves the residual force 7).

7. An expression for U_{r_o} can be derived from the analysis described in section 2.3; $U_{r_o} \sim [(H/E)/H^{1/2}]\; P^{1/2}$,[28].

Fracture mechanics analysis is based on evaluation of the crack extension force, G , in terms of the compliance, λ , of the elastic plate;

$$G = (P_r^2/4\pi c)(d\lambda/dc) \tag{13}$$

where $\lambda = Ac^2/Eh^3$, h is the depth of the crack below the surface (taken to be equal to the plastic zone dimension), and A is a dimensionless constant which is dependent on the relative dimensions of the lateral and radial cracks ($A = 3/4$ for lateral cracks smaller than radial cracks) [28]. Then, by calculating the stress intensity factor $K = [G\,E/(1-\nu^2)]^{1/2}$, and putting $K = K_c$ for equilibrium, an expression for the lateral crack length is obtained;

$$c = \eta\,[1 - (P_o/P)^{1/4}]^{\frac{1}{2}}\,P^{5/8} \tag{14}$$

where

$$\eta = \zeta_L (E/H)^{2/5}/K_c^{1/2}\,H^{1/8} \tag{15a}$$

$$P_o = \zeta_o\,K_c^4/H^3 \tag{15b}$$

and ζ_L and ζ_o are dimensionless,material independent constants.

Lateral crack measurements are compared with the predictions of Eqn. 14 in Fig. 13. Since the lateral cracks are entirely subsurface, and sectioning techniques are likely to modify the residual field and hence the lateral crack configuration, reliable crack measurements can only be obtained from materials that are at least partially transparent. Within that constraint the materials in Fig. 13 represent a wide variety of material properties (E/H and K_c). The data in Fig. 13 are plotted in terms of the load/crack-length, and material parameter combinations suggested by Eqn. 14 8). Reasonable agreement with the prediction is found, and a curve fit of Eqn. 14 provides a calibration of the dimensionless constant $\zeta_L = 0.096$.

5.2 Solid Particle Erosion

Equation 14 provides a basis for calculating the rate of erosion due to solid particle impact. The potential material

8. First, the parameter P_o was evaluated from Eqn. 15b for each material, after calibrating the dimensionless constant ζ_o by curve fitting $c(P)$ data for ZnS. Then the load/crack length combination, $c/[1-(P_o/P)^{1/4}]^{\frac{1}{2}}P^{5/8}$, was shown to be constant for each material under as wide a range of loads as possible (limited at the lower end by crack nucleation and at the upper end by chipping).

removal per impact event is defined by the volume between the lateral crack and the specimen free surface (Fig. 12);

$$V \sim c^2 h \tag{16}$$

Assuming that the mean contact pressure during impact is determined by the quasi-static hardness, H , the peak impulsive load is [29];

$$P \sim (H \tan^2\Psi)^{1/3}\ U_K^{2/3} \tag{17}$$

where U_K is the incident kinetic energy. Combination of Eqns. 14, 15, 16 and 17 then yields

$$V \sim \left\{(E/H)^{7/6}/K_c\ H^{1/6}\right\}\ U_K^{7/6} \tag{18}$$

for the erosion rate. This result is consistent with observed velocity and particle size dependence of erosion data, within the limits of experimental accuracy.

5.3 Machining

The extension of linear lateral cracks, appropriate to the morphology of plastic grooves produced by grinding and scratching, has been analyzed in terms of a model that is equivalent to the axisymmetric model of Section 5.1 [30]. The only difference between the two solutions for c appears as a small change in the exponent of E/H. Since less than an order of magnitude in the range of E/H covers most ceramic materials, the influence of grinding morphology on crack extension is considered to be of minor importance.

The volume removal removal corresponding to a groove of length ℓ is

$$V = 2\ h\ c\ \ell \tag{19}$$

In the limit of severe contact ($P_r \gg P_o$, where P_r is the normal grinding force and P_o is defined in Eqns. 14 and 15) Eqn. 19 becomes (with Eqns. 1, 14, 15)

$$V \sim [(E/H)^{4/5}/K_c^{1/2}\ H^{5/8}]\ P_r^{9/8}\ \ell \tag{20}$$

Equation 20 provides a reasonable correlation between experimentally measured grinding forces (at constant removal rate) and material properties [30].

6. DISCUSSION

The indentation fracture models provide quantitative predictions of the extent of cracking and the subsequent strength degradation and material removal rates which result from elastic/plastic contact. The predictions have been confirmed by extensive crack growth and strength data from Vickers indentations. The material parameters which provide optimum performance (small cracks, high strength, low material removal rate) are high K_c and, to a lesser extent, high H and H/E.

Real contact damage in ceramic surfaces can be divided into two groups; contacts with approximately axisymmetric geometry (e.g. solid particle impacts) and those with linear geometry (e.g. machining and scratching damage). Because of the dominant role played by the residual stress field, the indentation analysis applies directly to the general point contact damage (fig. 6), although minor modification can arise from relaxation of the residual stress by chipping associated with severe, or overlapping damage sites [21,31]. Analysis of linear contact damage requires some modification. The strength controlling radial cracks appear as a row of penny-shaped cracks beneath the damage zone, suggesting that the axisymmetric model may provide a good estimate of the extent of cracking (a linear grinding groove is generated by intermittant contact with an individual particle on the grinding wheel). However detailed analysis of the response of machining cracks to applied tension requires account to be taken of the linear distribution of residual forces along the surface trace of the dominant crack. The linear distribution of residual forces causes an increase in the extent of stable crack extension during a breaking test compared with the axisymmetric case. In the case of lateral cracks the influence of grinding morphology has been shown to be of minor importance.

An application of indentation fracture mechanics which has not been addressed here, is the evaluation of material properties. The analysis of Section 3 provides the basis for two independent methods of toughness measurement. The first involves measurement of the load and crack length for a controlled indentation, in conjunction with Eqn. 4. This method is extremely simple and economical in its use of material, and is therefore ideally suited for assessment of the effect of processing changes in the development of new materials [20]. The correlation with independent toughness measurements is demonstrated in Fig. 4. The second method involves the use of a controlled flaw as the origin of failure in a strength test, in conjunction with Eqn. 10 [25]. This technique holds the attraction that, because of the insensitivity of the failure condition to the initial crack length, no crack measurement is required; the only measurements entering Eqn. 10 are the indentation load and the strength. Another attractive feature of both of these techniques is that the cracks being tested are of the same dimensions and character as the flaws that control the strength of real components.

ACKNOWLEDGMENTS

This work was supported by the U.S. Office of Naval Research. Parts of the work were carried out in collaboration with B.R. Lawn at the University of New South Wales, Australia, A.G. Evans at the University of California at Berkeley, and B.T. Khuri-Yakub, G.S. Kino and J. Tien at Stanford University.

REFERENCES

1. Lawn, B.R. and T.R. Wilshaw. "Indentation fracture: principles and applications", J. Mat. Sci. 10 (1975) 1049-1081.
2. Lawn, B.R. and D.B. Marshall. Fracture Mechanics of Ceramics (R.C. Bradt, D.P.H. Hasselman and F.F. Lange, Plenum, 1978).
3. Chiang, S.S., D.B. Marshall and A.G. Evans. "The response of solids to elastic/plastic indentation: I. Stresses and residual stresses", J. Appl. Phys., to be published.
4. Hockey, B.J. and S.M. Wiederhorn. "Erosion of ceramic materials: the role of plastic flow", Proc. 5th Int. Conf. on Erosion by Solid and Liquid Impact, 1979, Ch. 26.
5. Lawn, B.R. and M.V. Swain. "Microfracture beneath point indentations in brittle solids", J. Mat. Sci. 10 (1975) 113-122.
6. Lawn, B.R. and E.R. Fuller. "Equilibrium penny-like cracks in indentation fracture", J. Mat. Sci. 10 (1975) 2016-2024.
7. Lawn, B.R., E.R. Fuller and S.M. Wiederhorn. "Strength degradation of brittle surfaces: sharp indenters", J. Amer. Ceram. Soc. 59 (1976) 193-197.
8. Evans, A.G. and T.R. Wilshaw. "Quasi-static solid particle damage in brittle solids: I. Observations, analysis and implications", Acta Metallurgica 24 (1976) 939-956.
9. Arora, A., D.B. Marshall, B.R. Lawn and M.V. Swain. "Indentation deformation/fracture of normal and anomalous glasses", J. Non-Cryst. Solids, 31 (1979) 415-428.
10. Hagan, J.T. "Micromechanics of crack nucleation during indentations", J. Mat. Sci. 14 (1979) 2975-2980.
11. Van Der Zwaag, S., J.T. Hagan and J.E. Field. "Studies of contact damage in polycrystalline zinc sulphide", J. Mat. Sci. 15 (1980) 2465-2972.
12. Hockey, B.J. and B.R. Lawn. "Electron microscopy of microcracking about indentations in Al_2O_3 and SiC", J. Mat. Sci. 10 (1975) 1275-1284.
13. Chiang, S.S., D.B. Marshall and A.G. Evans. "The response of solids to elastic/plastic indentation: II. Fracture initiation", J. Appl. Phys., to be published.
14. Hagan, J.T. "Micromechanics of crack nucleation during indentations", J. Mat. Sci. 14 (1979) 2975-2980.

15. Lawn, B.R. and A.G. Evans. "A model for crack initiation in elastic/plastic indentation fields", J. Mat. Sci. 12 (1977) 2195-2199.
16. Marshall, D.B. and B.R. Lawn. "Residual stress effects in sharp contact cracking: I. Indentation fracture mechanics", J. Mat. Sci. 14 (1979) 2001-2012.
17. Marshall, D.B., B.R. Lawn and P. Chantikul. "Residual stress effects in sharp contact cracking: II. Strength degradation", J. Mat. Sci. 14 (1979) 2225-2235.
18. Lawn, B.R., A.G. Evans and D.B. Marshall. "Elastic/plastic indentation damage in ceramics: The median/radial crack system", J. Amer. Ceram. Soc. 63 (1980) 574-581.
19. Eshelby, J.D. "The determination of the elastic field of an ellipsoidal inclusion, and related problems", Proc. Roy. Soc. Lond. 241A (1957), 376-396.
20. Anstis, G.R., P. Chantikul, B.R. Lawn and D.B. Marshall. "A critical evaluation of indentation techniques for measuring fracture toughness: I. Direct crack measurements", J. Amer. Ceram. Soc., in press.
21. Lawn, B., D.B. Marshall and P. Chantikul. "Mechanics of strength-degrading contact flaws in silicon", J. Mat. Sci., in press.
22. Lawn, B.R. and R. Cook. Unpublished work.
23. Marshall, D.B. Unpublished work.
24. Petrovic, J.J., R.A. Dirks, L.A. Jacobson and M.G. Mendiratta. "Effects of residual stresses on fracture from controlled surface flaws", J. Amer. Ceram. Soc. 59 (1976) 177-178.
25. Chantikul, P., G.R. Anstis, B.R. Lawn and D.B. Marshall. "A critical evaluation of indentation techniques for measuring fracture toughness: II. Strength method", J. Amer. Ceram. Soc., in press.
26. Khuri-Yakub, B.T., G.S. Kino and A.G. Evans. "Acoustic surface wave measurements of surface cracks in ceramics", J. Amer. Ceram. Soc. 63 (1980) 65-71.
27. Tien, J.J.W., B.T. Khuri-Yakub, G.S. Kino, A.G. Evans and D.B. Marshall. "Surface acoustic wave measurements of surface cracks in ceramics", J. Appl. Phys., in press.
28. Lawn, B.R., A.G. Evans and D.B. Marshall. "Elastic/plastic indentation damage in ceramics: the lateral crack system", in preparation.
29. Wiederhorn, S.M. and B.R. Lawn. "Strength degradation of glass impacted with sharp particles: I. Annealed surfaces", J. Amer. Ceram. Soc. 62 (1979) 66-70.
30. Evans, A.G. and D.B. Marshall. Fundamentals of Friction and Wear of Materials (ASM: 1980).
31. Marshall, D.B. and B.R. Lawn. "Residual stresses in dynamic fatigue of abraded glass", J. Amer. Ceram. Soc. 64 (1981) c6-c7.

CONTACT STRESSES AT CERAMIC INTERFACES

David W. Richerson

Garrett Turbine Engine Company
A Division of the Garrett Corporation.

ABSTRACT. Contact damage at ceramic-ceramic and ceramic-metal interfaces has been identified as an important failure mode for ceramic heat-engine components. Stress distributions determined by closed-form equations and finite element analysis will be reviewed for Hertzian and biaxial loading. Experimental results reviewing some effects of contact geometry, applied load, coefficient of friction, temperature and oxidation will be presented for silicon nitride and silicon carbide materials.

INTRODUCTION

Contact damage at ceramic-ceramic and ceramic-metal interfaces has been identified as an important failure mode in many applications. Components exposed to such loading include bearings, seals, wear surfaces, cutting tools, extrusion dies, heat-engine parts, and many more. To successfully design and utilize ceramics for these types of components requires an understanding of the stress distribution under uniaxial and biaxial loading and the effects of contact geometry, applied loads, coefficient of friction and, in some cases, temperature.

This paper briefly reviews the stress distribution for common contact geometries and presents experimental friction, contact geometry and retained-strength data for a reaction-bonded silicon nitride material and a sintered silicon carbide material for sliding contact at temperatures up to 1100°C.

Riley, F.L. (ed.) Progress in Nitrogen Ceramics

CONTACT STRESS DISTRIBUTIONS

Figure 1 compares the contact stress distributions for frequently encountered interface geometries for normal and biaxial loading. These stress distributions were determined analytically assuming elastic response of both contacting surfaces.

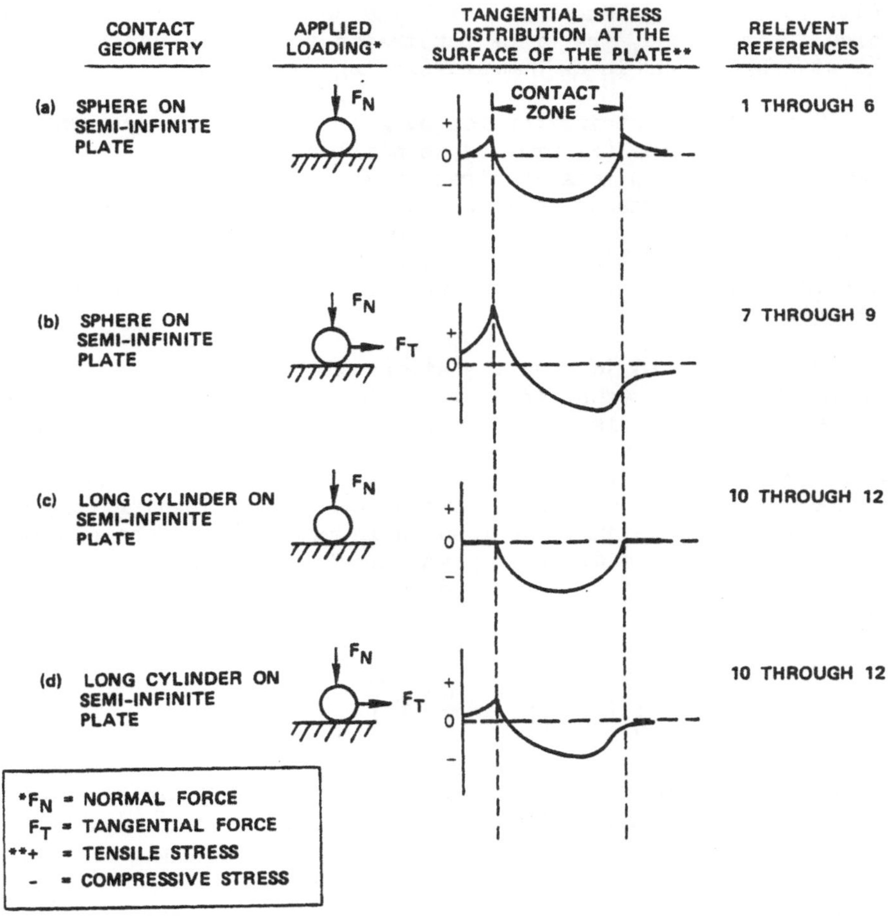

Figure 1. Contact Stress Distributions (Determined Analytically, Assuming Elastic Conditions) for Simple Contact Geometries for Normal and Biaxial Loading.

Figure 1a was initially analyzed by Hertz (1) and is commonly referred to as Hertzian loading. The stress field is symmetrical. The stress directly under the contact is compressive and adjacent to the contact is tensile. The tensile stress peaks at the edge of the circular contact zone and decreases with distance away from the contact zone(2). Fracture initiates when the tensile stress becomes high enough to cause an existing material flaw at or near the surface adjacent to the contact zone to propagate. The crack initially forms a ring concentric with the contact zone. As the normal force is increased, the crack penetrates to greater depth following the stress contour in an outward sloping cone(4-6).

The stress distribution for normal loading of a cylinder on a semi-infinite plane (Figure 1c) is similar to Hertzian loading, except that no hoop tensile stress is present adjacent to the contact zone. Thus, under ideal contact, the contact stresses for a long cylinder on a plate will be compressive. However, for a short cylinder or for realistic conditions where surface asperities prevent ideal contact, localized tensile stresses are likely to be present.

Figures 1b and 1d show the stress distributions when a tangential force is present in addition to a normal force(7-8). The stress distribution is skewed to produce a sharp tensile stress spike on the material surface at the trailing edge of the contact zone. The tensile spike for this biaxial loading is of much higher magnitude than the tensile stress in Hertzian loading (for a given normal force) and thus results in material surface damage at much lower normal load. Gilroy and Hirst(9) derived the following equation to estimate the relative damage tolerance of a material under normal versus biaxial loading:

$$\frac{P_N}{P_S} = \left(1 + \frac{3\pi(4+\nu)}{8(1-2\nu)} f \right)^3 \qquad (1)$$

where P_N is the load to cause Hertzian damage with normal loading of a spherical indenter, P_S is the load to cause surface damage with a sliding spherical indenter, ν is Poisson's ratio of the material and f is the coefficient of friction.

Although the derivation of this equation involved some simplifying assumptions, it does provide an indication of the effect of the tensile spike during biaxial loading on ceramic surface damage. For example, for a typical reaction bonded silicon nitride (Si_3N_4) material with a Poisson's ratio of 0.18 and a coefficient of friction of 0.5*, $P_S \approx P_N/114$. For a silicon carbide (SiC) material with a Poisson's ratio of 0.17 and a coefficient of friction

*Typical friction data included in Figure 5.

of 0.3*, $P_S \approx P_N/34$.

Fracture or surface damage initiates under the same criteria for biaxial loading as for Hertzian loading, i.e. cracks propagate from existing flaws when the local tensile stress (σ_f) reaches a critical value dependent on the flaw size (c), the elastic modulus (E) and the fracture surface energy (γ)[13].

$$\sigma_f = \frac{Z}{Y}\left(\frac{2E\gamma}{c}\right)^{1/2} \qquad (2)$$

where Z and Y are dimensionless parameters which depend on flaw shape, position and size and on the test configuration.

For sliding contact of a sphere on a plate, shallow arc-shaped cracks (with the concave side pointing in the direction of the tangential force) form[8]. These are shallow because the tensile stress spike is localized at the surface and decreases rapidly with depth. For most cases the stress is compressive within 100 µm of the surface. The cracks therefore stop when they reach the compressive zone and will not propagate further until the stress field at the crack tip becomes tensile.

The magnitude of the tensile spike is lower for biaxial loading of a cylinder on a semi-infinite plate. The normal load is distributed over a larger contact area. However, in real applications ideal line contact will not occur. Loading will undoubtedly be concentrated at asperities or regions of tolerance mismatch such that locally high tensile stress can occur. Due to the resulting variation in tensile stress along the contact line, surface damage will be intermittent and the crack morphology less well defined than for point contact.

The magnitude of the tensile spike for biaxial loading is strongly dependent on the coefficient of friction at the interfaces. This is illustrated in Figure 2 for biaxial contact of an aluminum cylinder against a hot pressed Si_3N_4 plate. The stress distributions were calculated for the various assumed coefficients of friction using closed-form equations[10,11].

The presence of a high tensile stress at the surface of a ceramic under sliding or biaxial contact loading has important implications and may explain the cause of failure or surface damage of ceramic components in many applications. Components exposed to such loading include bearings, seals, high-temperature engine parts, wear surfaces, cutting tools (and the workpieces

---

*Typical friction data included in Figure 5.

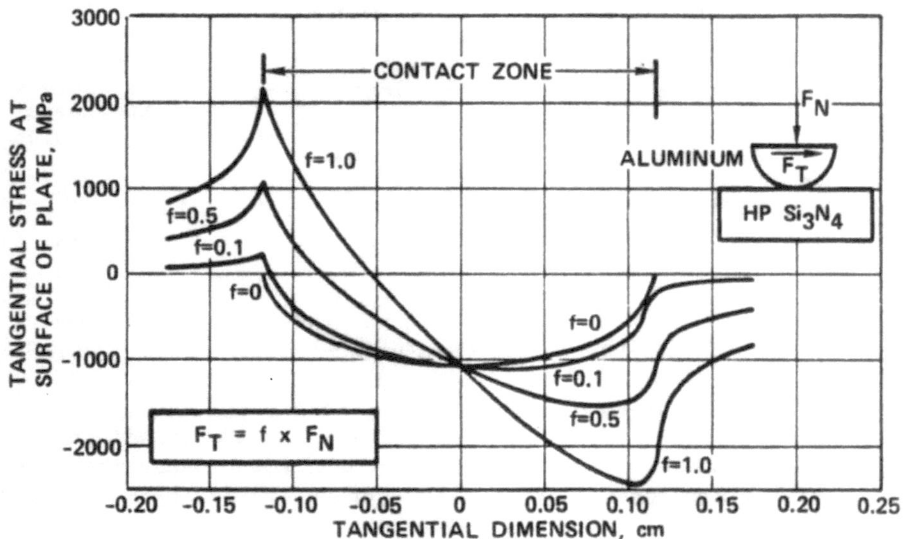

Figure 2. Contact Tangential Stress for Different Friction Values.

being machined), extrusion dies and many more. Ceramic component designers and users need to be aware that tensile stresses are present at contact interfaces in addition to compressive stresses.

EXPERIMENTAL RESULTS

Static and sliding contact tests have been conducted with reaction-bonded Si_3N_4* and sintered alpha-SiC** using contact geometries of a cylinder at 90 degrees to another cylinder ("point" contact) and of a cylinder on a flat surface ("line" contact). The test apparatus, which has been described previously(14), can provide measurements of normal force and tangential force at temperatures up to 1200°C. The test specimen (a rectangular bar 0.32 x 0.54 x 7.6 cm) then can be tested in four-point bending and the data compared with baseline data to determine if strength reduction has resulted from the contact test conditions.

---

*NC-350 from the Norton Company and RBN101 and RBN104 from the AiResearch Casting Company.

**Sintered alpha-SiC (Hexoloy SA SiC) from the Carborundum Company, Niagara Falls, N.Y.

Static point contact tests at normal loads up to 140 kg resulted in no surface damage or material strength reduction. Addition of a tangential force resulted in visible surface damage and strength reduction for normal loads of only 26 and 11.4 kg. Specifically for the 26 kg loading, the average four-point bend strength of the reaction-bonded Si_3N_4 was decreased from a baseline of 310 MPa to less than 140 MPa and sintered alpha-SiC from a baseline of 440 to 260 MPa. Line contact resulted in substantially less strength reduction. The data are summarized in Figure 3. Typical contact and fracture surfaces for reaction-bonded Si_3N_4 are shown in Figure 4.

Sliding contact tests have been conducted at temperatures up to 1100°C. The test specimen is held under normal load at temperature for 30 minutes prior to initiation of the sliding motion. The test apparatus produces a plot of the tangential force as a function of the relative motion. A series of plots for reaction-bonded Si_3N_4 and sintered alpha-SiC are compared in Figure 5. For reference, the apparent coefficient of friction is plotted on the right.

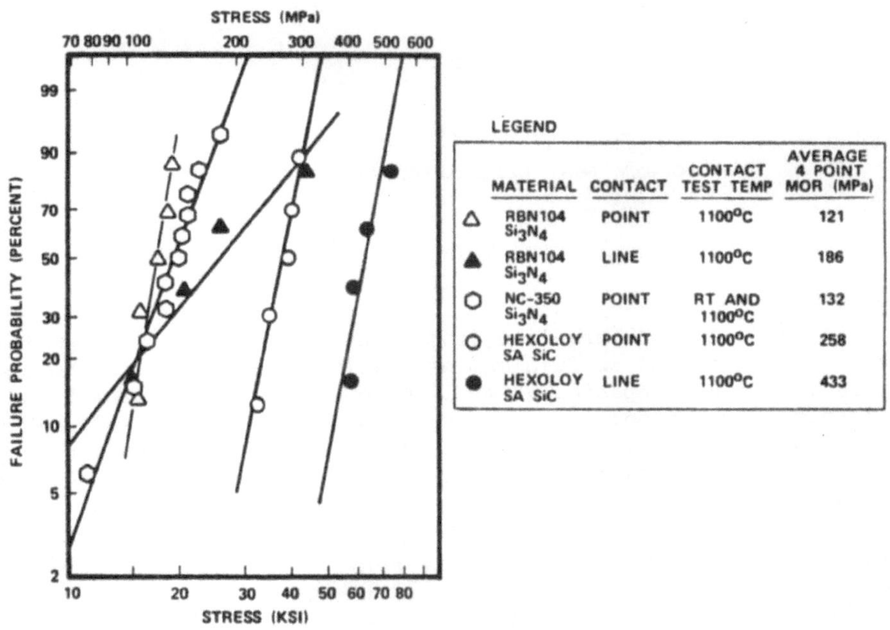

Figure 3. Room Temperature Strength of RBN104, NC-350 and Hexoloy SA SiC After Line and Point Contact Tests at 26 kg Normal Load.

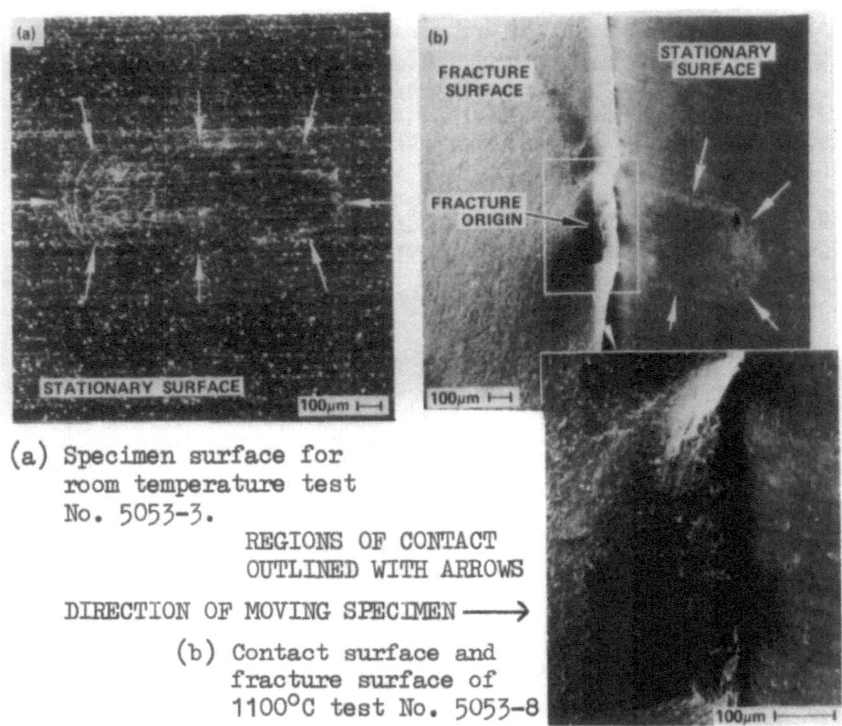

(a) Specimen surface for room temperature test No. 5053-3.

REGIONS OF CONTACT OUTLINED WITH ARROWS

DIRECTION OF MOVING SPECIMEN ⟶

(b) Contact surface and fracture surface of 1100°C test No. 5053-8

Figure 4. SEM photomicrographs of reaction-bonded Si_3N_4 stationary test bars after contact testing.

An important observation from Figure 5 is that the tangential force (and apparent coefficient of friction) increased as temperature increased, especially for line contact. Examination by scanning electron microscopy of the surface of the sintered alpha-SiC tested at 1100°C revealed a glassy-appearing surface layer adjacent to the initial contact zone and along the contact surface. One of the more pronounced examples is shown in Figure 6. Some of the reaction-bonded Si_3N_4 specimens tested at elevated temperatures also had a glassy appearance on the contact surface, but it was less obvious than for the sintered alpha-SiC. In some cases, such as the specimen shown in Figure 4(b), a glassy surface layer was not visible.

Energy dispersive X-ray analysis of the glassy region on the sintered alpha-SiC specimen shown in Figure 6 identified only silicon, suggesting (based upon prior experience) that the glass was

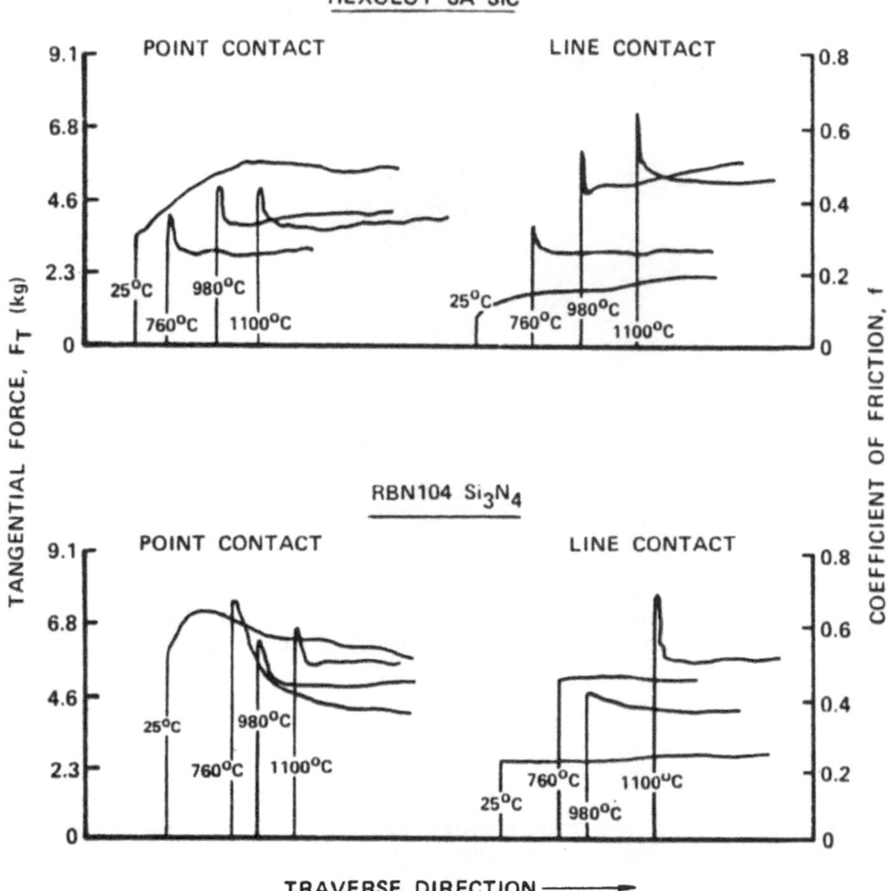

Figure 5. Typical Point and Line Contact Load-Relative Motion for Hexoloy SA SiC and RBN104.

SiO_2. SiO_2 is the natural oxidation product of SiC and Si_3N_4. The mechanism for the preferential build-up adjacent to the contact zone is not precisely known. It could be due to increased chemical activity associated with the localized high stress.

The increase in tangential force with increasing temperature is consistent with the formation of a glass at the interface*.

*It could also be consistent with other phenomena such as chemical reaction between the materials or localized deformation due to yielding, but no physical evidence of these was detected.

Figure 6. Scanning electron photomicrograph showing the contact surface features for Hexoloy SA SiC tested at 1100°C with a normal load of 11.4 kg.

At room temperature the glass would not be present (or, if present, would be solid). The only increment of the tangential force would be to overcome the interface friction. At progressively higher temperatures, more glass would form and provide adhesion at the interface. Now, an additional increment of tangential force would be necessary to overcome this adhesion in addition to the tangential force necessary to overcome interface friction.

The high-temperature contact test results have important implications on design of ceramics for high-temperature structures such as heat-engine components. The designer must be aware that the contact behaviour may be controlled by different mechanisms at elevated temperature and that analysis based upon room temperature data and current analysis models may not adequately predict the component behaviour during duty cycle operation.

REFERENCES

1. H. Hertz, J. Reine Angew, Math 92, 156 (1881); reprinted in English in "Hertz's Miscellaneous Papers", Chapter 5, 6, Macmillan & Co. Ltd., London, 1896.
2. F.C. Roesler, Proc. Phys. Soc. Lond. B69, 981 (1956).
3. F.C. Frank, and B.R. Lawn, Proc. Roy. Soc. Lond. A299, 191 (1967).
4. T.R. Wilshaw, J. Phys. D: Appl. Phys. Vol. 4, 1567-81 (1971).
5. B.R. Lawn and D.B. Marshall, in Fracture Mechanics of Ceramics Vol. 3 (ed. by R.C. Bradt, D.P.H. Hasselman and F.F. Lange), pp. 205-229 (1978), Plenum Press, New York.
6. B.R. Lawn, S.M. Wiederhorn and H.H. Johnson, J. Amer. Ceram. Soc. 58 (9-10), 428-32 (1975).
7. G.M. Hamilton and L.E. Goodman, J. Appl. Mech. 33, 371 (1966).
8. B.R. Lawn, Proc. Roy. Soc. Lond. A299, 307 (1967).
9. D.R. Gilroy and W. Hirst, J. Phys. D: Appl. Phys. 2, 1784-7 (1969).
10. J.O. Smith and C.K. Liu, "Stresses due to tangential and normal loads on an elastic solid with application to some contact stress problems", J. Appl. Mech., pp. 157-166, June 1953.
11. D.G. Finger, "Contact stress analysis of ceramic-to-metal interfaces", Final Report, ONR Contract N00014-78-C-0547, Sept. 1979.
12. D.W. Richerson, D.G. Finger and J.M. Wimmer, "Analytical and experimental evaluation of biaxial contact stress", to be published in Volume 6 of Fracture Mechanics of Ceramics (ed. by R.C. Bradt, D.P.H. Hasselman and F.F. Lange), Plenum Press.
13. A.G. Evans and G. Tappin, Proc. Brit. Ceram. Soc. 20 (1972) pp. 275-297.
14. D.W. Richerson, L.J. Lindberg, W.D. Carruthers and J. Dahn, "Contact stress effects on Si_3N_4 and SiC interfaces", to be published in Ceramic Engineering and Science Proceedings, July-August 1981.

Section I

APPLICATIONS

SINTERING, PROPERTIES AND FABRICATION OF Si_3N_4 + Y_2O_3 BASED CERAMICS

C.L. Quackenbush, J.T. Smith, J.T. Neil, K.W. French

GTE Laboratories, Inc.
40 Sylvan Road
Waltham, MA 02254

ABSTRACT

This paper will discuss powder materials consolidation in the Si_3N_4 + Y_2O_3 system where full density has been achieved by liquid phase sintering. The role of the sintering additives during densification and their influence on structural properties will be reviewed. Room temperature strength is controlled by residual porosity over a wide range of phase composition and chemical content. Strength maintenance to elevated temperature is increased by the reduction of flux content in the liquid phase. Fluxes increase fluidity of the liquid phase and suppress its crystallization. Oxidation performance depends on the in situ oxidation characteristics of the minor phase composition and the influence of fluxes from the substrate which dissolve into the surface layer altering its oxygen diffusion rate. Turbine blades have been fabricated from this family of ceramics by injection molding. Final sintered densities >99% theoretical have been achieved with close dimensional control.

INTRODUCTION

Without the use of densification additives pure silicon nitride shows a remarkable resistance to sintering. For example, a density of about 85% theoretical was all that could be achieved using reactive decomposition hot pressing (1) or under high pressure forming conditions (2) (1750°C, 60 kilobars pressure). In densifying Si_3N_4 to >98% theoretical density one to 20 w/o of a densification aid is used. MgO (3-6) and Y_2O_3 (7-13) have received the most widespread study. Both display high room temperature strength and excellent thermal shock resistance. One distinct

Riley, F.L. (ed.) Progress in Nitrogen Ceramics

advantage of the yttria additive is that it forms a more refractory silicate grain boundary phase than does MgO. Compositions containing Y_2O_3 have considerably improved elevated temperature strength as compared to those with MgO. The Y_2O_3 additive system meets the additional requirement that it is not too refractory to be sintered at temperatures where Si_3N_4 decomposition can be controlled. This is a virtual necessity if actual hardware is to be produced from the material.

Results will be presented which show the effect of chemical content (composition, raw material impurities, processing contaminants)on sinterability and resulting strength and oxidation properties. Turbine blades have been fabricated from Si_3N_4 + Y_2O_3 ceramics by injection molding. Final densities >99% theoretical have been achieved with close dimensional control.

SINTERING

In the microstructure of liquid phase sintered Si_3N_4 + Y_2O_3 a continuous yttrium silicate phase separates the Si_3N_4 grains (14,15). Silicon nitride has limited solid solubility so minor phases (e.g. Al_2O_3, Fe_2O_3) become concentrated in the grain boundary phase during sintering. In general these materials act as fluxes for the yttrium silicate liquid increasing its fluidity and facilitating sintering. Increased grain boundary fluidity however also degrades elevated temperature strength by promoting grain boundary deformation at reduced temperatures.

The maximum temperature required to reach sintered density of >98% theoretical in 2 hours decreases with increased amounts of flux added to the system, Table I. Alumina is particularly effective; a 2.0 w/o addition drops the sintering temperature 150°C, from 1975°C to 1825°C. The sintering temperature is also decreased by increasing Y_2O_3 content. A 16.5 w/o increase in Y_2O_3 decreases the sintering temperature 75°C, from 1825°C to 1750°C. In terms of sintering temperature Al_2O_3 has approximately 15 times the

Table I. Sintering Temperature vs. Composition

COMPOSITION (w/o)			SINTERING TEMPERATURE (°C)
Si_3N_4*	Y_2O_3	Al_2O_3	
94.0	6.0	0	1975
94.5	3.5	2.0	1825
78.0	20.0	2.0	1750

SN-502, GTE Towanda, PA 18848

fluxing power of Y_2O_3. During sintering Si_3N_4 + Y_2O_3 the Al_2O_3 is not completely removed from the grain boundary phase to form a SiAlON. Instead it is partitioned between the Si_3N_4 and grain boundary phases (18). Thus the lower the flux content the more refractory the grain boundary phases. This necessitates higher sintering temperatures.

STRENGTH

Room temperature strength is structure-sensitive and does not systematically depend on Y_2O_3 or Al_2O_3 content providing high density is achieved (Figure 1). At full density, room temperature strength is about 650 to 700 MN m^{-2}. There is a strong effect of porosity on room temperature strength. Figure 2 indicates approximately a 15% reduction in room temperature strength from 650 to 550 MN m^{-2} with 2.5% residual porosity. This reduction corresponds to an "n" value of about 7 in the Ryskewitsch relationship

$$\sigma = \sigma_o \exp(-np)$$

where p is fractional porosity and n is a constant typically between 4 and 7 for ceramics undergoing brittle fracture.

Strength at 1400°C is essentially the same for compositions containing 2 w/o Al_2O_3 and from 2 to 10 w/o Y_2O_3 (Figure 1). A reduction of the Al_2O_3 content to zero increases 1400°C strength by a factor of 3. Figure 3 expands this trend of strength at 1400°C vs. Al_2O_3 content for sintered and hot pressed Si_3N_4 + Y_2O_3 materials. A slow increase in strength is noted with reduced Al_2O_3

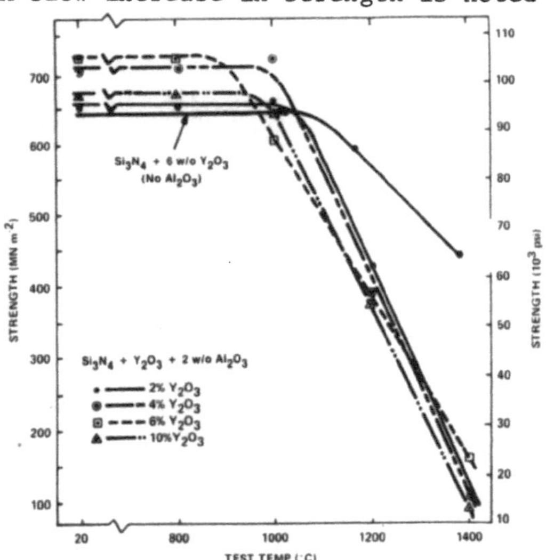

Figure 1. Strength of as-sintered Si_3N_4 specimens as a function of Y_2O_3 and Al_2O_3. All samples >99% theoretical density.

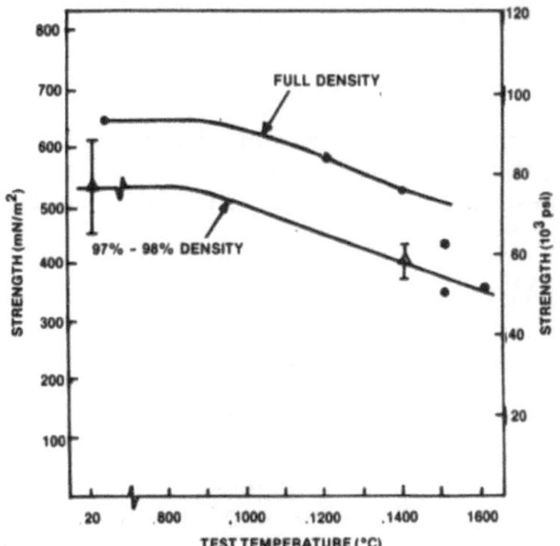

Figure 2. Strength vs. test temperature for sintered Si_3N_4 + 6 w/o Y_2O_3 (no Al_2O_3).

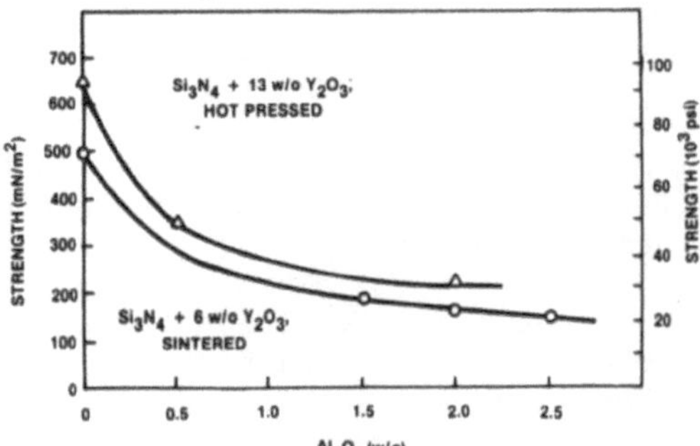

Figure 3. Strength at 1400°C of hot pressed and sintered Si_3N_4 + Y_2O_3 as a function of Al_2O_3 content.

to about 0.5 w/o Al_2O_3 where the curve steepens. Between 0.5 w/o and 0 w/o Al_2O_3 measured strength nearly doubles. This indicates the strong ability of even small amounts of Al_2O_3 to increase fluidity of the grain boundary phase at the test temperature. A fluid grain boundary phase allows sliding between Si_3N_4 grains resulting in integranular failure. The 25 to 30% higher strength for the hot pressed vs. sintered material is typical and related to an oriented microstructure in the hot pressed parts.

Fracture at room temperature is predominantly transgranular and strength is structure-sensitive. Fracture at elevated temperatures (e.g. 1400°C) is trans- or intergranular depending on composition. Materials containing >1.0 w/o Al_2O_3 generally deform at high loads, suffer slow crack growth, and have highly intergranular fracture surfaces. Materials with no Al_2O_3 added show predominantly transgranular fracture surfaces.

OXIDATION

Oxidation performance depends on the in situ oxidation characteristics of the minor phase composition and the influence of fluxes from the substrate introduced into the surface layer which alter its oxygen diffusion rate.

Oxidation characteristics will be discussed in terms of phase field (Figure 4), the presence of fluxes (Al_2O_3, iron) and impurities (carbon, WC) over the temperature range 750°C to 1350°C.

Phase field position is determined by the initial Y_2O_3/SiO_2 ratio. Specification of the Y_2O_3 content alone does not establish the likely identity of the final phases. Phase position is more sensitive to the absolute level of SiO_2 content than Y_2O_3 content. Silica is present as a surface layer on the Si_3N_4 raw material.

Phase Field I and II Materials

These materials have as their minor phases Si_2ON_2, $Y_2Si_2O_7$ and $Y_5Si_3O_{12}N$.

Phase field I materials show stable oxidation behavior from 1000°C to 1350°C (Figure 5). This is in agreement with literature reports (20) for these compositions.

Most Si_3N_4 + Y_2O_3 materials development work at GTE Labs has focused on phase field II materials with second phases of $Y_2Si_2O_7$ and/or $Y_5Si_3O_{12}N$. They too have stable and predictable oxidation

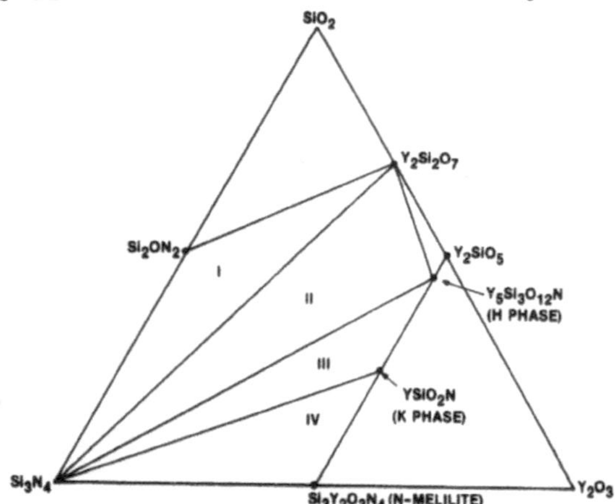

Figure 4. Si_3N_4 + Y_2O_3 + SiO_2 phase compatability diagram (after reference 19).

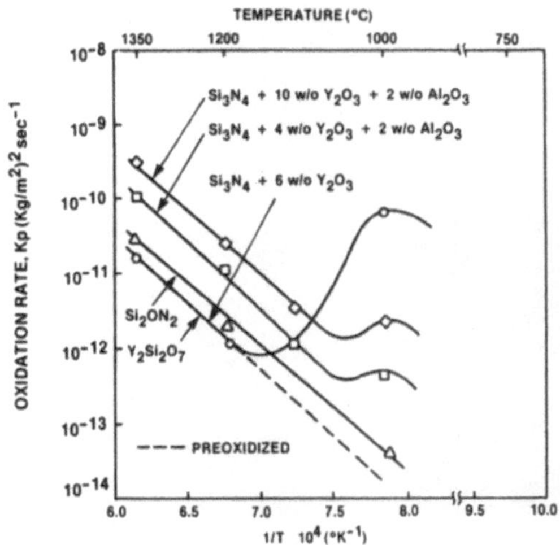

Figure 5. Arrhenius oxidation plot for Si_3N_4 - based samples with the indicated compositions and minor phases. All samples from phase fields I or II.

characteristics.

With phase field II materials two oxidation mechanisms have been identified separated by a critical transition temperature (T_c). Above T_c, (1200, 1350°C) the oxidation layer is dense and coherent; oxidation kinetics are parabolic with time. Below T_c, (1000°C) the surface oxidation layer contains connected porosity, it is nonprotective and oxidation kinetics are linear with time. Figure 6 shows a typical apperance of the two types of oxidation surface layers.

At temperatures $>T_c$ oxidation rate increases with additive content (21). Actually it is fluxing power of the additive which is important. As discussed in the sintering section Al_2O_3 is a stronger flux than Y_2O_3. Accordingly Si_3N_4 materials show increasing oxidation rates with the following ranking of additives: 6 w/o Y_2O_3 + 0 w/o Al_2O_3; 4 w/o Y_2O_3 + 2 w/o Al_2O_3; 10 w/o Y_2O_3 + 2 w/o Al_2O_3 (see Figure 5). Increasing amounts of flux decreases the glass transition temperature of the surface oxide glass. This increases the oxygen diffusion rate and thus the oxidation rate of the substrate (21).

Excessive fluxing of oxidation surface layers by Al_2O_3 should be avoided. The samples in Figure 7 are Si_3N_4 (SN-502 + 13 w/o Y_2O_3 containing 3.3 w/o and 4.7 w/o Al_2O_3. Blistering of the latter sample at 1350°C supports the conclusion that for this composition there is a threshold limit for Al_2O_3 which lies between 3.2 and 4.7 w/o.

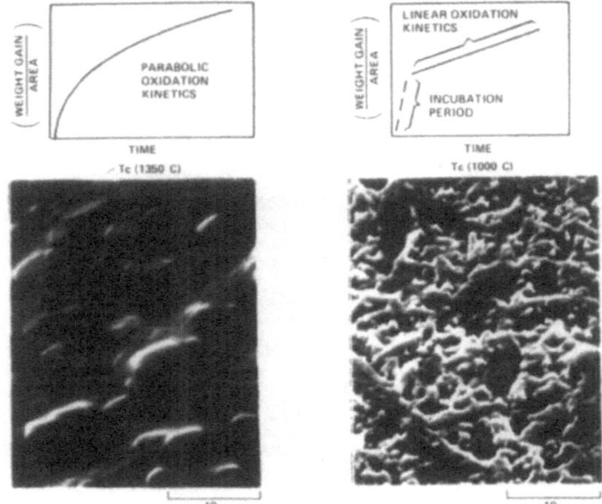

Figure 6. Oxidation surface layers and schematic weight gain curves for a Si_3N_4 + 6 w/o Y_2O_3 sample sintered free of Al_2O_3.

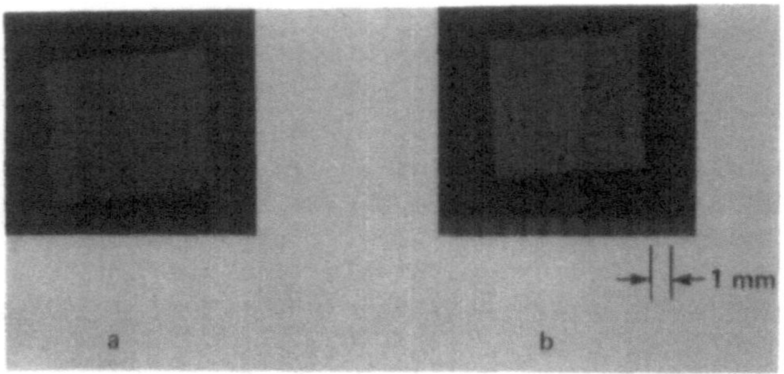

Figure 7. Samples made from Si_3N_4 (SN-502) + 13 w/o Y_2O_3 + 3.2 w/o Al_2O_3 (Fig. 7a) and 4.7 w/o Al_2O_3 (Fig. 7b). Oxidized 100 hours, 1350°C in air.

Excessive fluxing has also been observed in samples prepared from a commercially available high iron (1.3 w/o) Si_3N_4 powder [π]. The sample in Figure 8 contained 13 w/o Y_2O_3 and 4.0 w/o Al_2O_3. The combined fluxing action of the iron and alumina is evident as severe blistering of the surface glass at 1350°C. An otherwise identical sample was made from this same raw material powder whose iron level was chemically reduced from 1.3 w/o to 0.5 w/o. The low iron sample has a 1350°C oxidation rate nearly on order of magnitude lower than the high iron sample (5×10^{-10} vs. 3×10^{-9} $Kg^2\ m^{-4} sec^{-1}$). It is concluded that for best performance iron levels should be kept below 0.5 w/o.

At temperatures $<T_c$ samples from phase field II have porous

Figure 8. Oxidation surface of Si_3N_4 + 13% Y_2O_3 + 4% Al_2O_3 samples made from "standard product" Nippon Denko Si_3N_4 powder (1.3 w/o Fe) (250 hours at 1350°C).

surface oxidation layers which are nonprotective and thus oxidation rates are high. These layers can be compacted by heating to a temperature $>T_c$ forming a protective surface (21). Preoxidation at a temperature $>T_c$ will accomplish the same thing. Following such treatment the oxidation rate at 1000°C is reduced by more than 3 orders of magnitude (Figure 5). It should be emphasized that cracking is not observed in these samples whether or not they are preoxidized.

Samples from phase field II are also stable with respect to catastrophic oxidation aggravated by carbon in the raw materials. (This is not true for phase field III and IV materials.) Phase field II materials containing up to 0.6 w/o carbon do not show evidence of cracking.

Phase Field III and IV Materials

These materials have as their minor phases $Y_5Si_3O_{12}N$, $YSiO_2N$ and $Si_3Y_2O_3N_4$. Samples were prepared from high purity Si_3N_4 (SN-502) and had x-ray identified minor phases from phase fields III or IV. They demonstrated stable oxidation at 1200°C and 1350°C, Figure 9. In the temperature range 750° to 1000°C materials from phase fields III and IV are characterized by unpredictable oxidation behavior. This behavior has been reported (22).

Samples prepared with high purity SN-502 incorporating x-ray idenfified minor phases from phase fields III and/or IV were oxidized at 1000°C. Samples which contained $Y_5Si_3O_{12}N$ as their only minor phase exhibited no cracking and had a low oxidation rate, 2×10^{-12} $Kg^2m^{-4}sec^{-1}$ (Figure 9). This is not unexpected because the phase $Y_5Si_3O_{12}N$ is common to phase fields II and III and phase field II shows stable oxidation behavior. Samples which contained $YSiO_2N$ and/or $Si_3Y_2O_3N_4$ exhibited from minor to extreme cracking.

The presence of carbon in phase field III or IV materials ag-

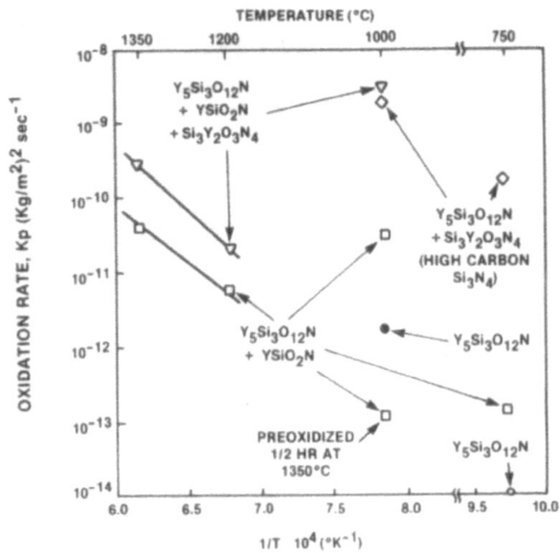

Figure 9. Arrhenius oxidation plot for Si_3N_4 - based samples with indicated minor phases. All samples from phase fields III and IV.

grevates this 1000°C instability (23) preoxidation (1/2 hour at 1350°C) successfully eliminated cracking in samples made from low carbon SN-502 (final carbon from 0.05 to 0.1 w/o) and lowered oxidation rates more than two orders of magnitude. The same preoxidation treatment had no effect on a sample made from a third Si_3N_4 raw material powder* containing 0.6 w/o carbon (Figure 10). Oxidation at 750°C of phase field III or IV samples containing 0.6 w/o carbon revealed cracking where none was observed for samples prepared with low carbon SN-502 having carbon levels from 0.05 to 0.1 w/o carbon.

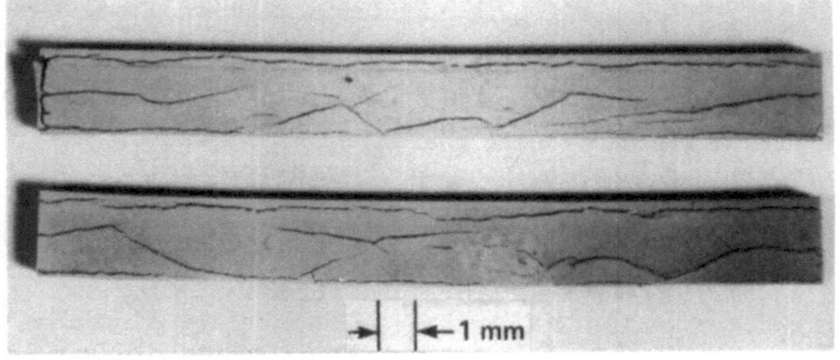

Figure 10. Sample made from carbon-containing T-186 Si_3N_4 powder + 13 w/o Y_2O_3. Oxidized 290 hours at 1000°C. Preoxidation for 1/2 hour at 1350°C had a similar appearance.

Tungsten Carbide and Oxidation at 600°C

Tungsten carbide milling balls are commonly used for milling of Si_3N_4 powders. Following one procedure (24) described in the literature, a sample of SN-502 + 12 w/o CeO_2 was wet milled in ethanol for 16 hours with WC media prior to hot pressing. The final WC level was about 14 w/o. This level is similar to that reported by other workers (25) who also wet milled Si_3N_4 with WC media. The hot pressed samples were >98% dense and had a room temperature strength of 135,000 psi. Oxidation at 600°C, however, completely disintegrated these hot pressed samples in less than 24 hours (Figure 11). It is presumed that this oxidation behavior is due to the nearly three-fold volumetric expansion of WC as it oxidizes to WO_3. The same composition hot pressed from powders milled with Si_3N_4 media showed no detectable weight change or sample deterioration after 600°C oxidation.

This behavior is an extreme effect because of the large amount of WC present. Another material milled with WC is NC-132 (<3 w/o WC). It does not show any measurable weight gain at 600°C nor any structural damage.

SHAPE FABRICATION BY INJECTION MOLDING

A primary advantage of sinterable vs. hot pressed materials is the potential of densifying complex parts to near net shape. Under a subcontract to the Detroit Diesel Allison, Div. of General Motors as part of their CATE program, technologies to injection mold turbine components are under development at GTE Laboratories.

Axial turbine blades (Figure 12) have been molded and sintered to >99% theoretical density. Airfoil tolerances on individual blades have been achieved which fall within a 0.004 inch blade contour envelope. Twist with respect to the blade attachment is correct to ±3/4°.

Three processing steps are involved in injection molding: compounding , molding and binder removal. In this study it has been

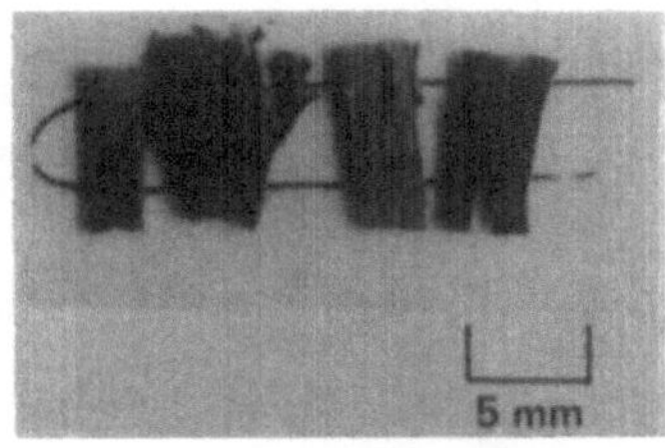

Figure 11. Delaminated samples of hot pressed Si_3N_4 (SN-502) +12 w/o CeO_2 milled with WC. Oxidized 16 hours at 600°C.

Figure 12. Axial turbine blade of injection molded Si_3N_4 (1.3 inches tall).

found that there is a wide overlap between those mixes which have enough binder to be moldable and those mixes which have enough ceramic to be sintered to >98% density. In fact, it is believed possible to achieve a higher volume percent ceramic solids level by injection molding then by dry pressing. This is true because in injection molding the hot liquid binder gives more efficient particle repacking to high density during the forming operation.

Attainment of adequate flow to fill the mold cavity has presented relatively little difficulty. Removal of the binder without disrupting the part geometry is considerably more difficult and is extremely sensitive to part cross section.

Sintering shrinkage is uniform and predictable. Importantly, warpage and slumping during sintering can be controlled despite the fact that the system is liquid phase sintered.

SUMMARY

In sintering Si_3N_4 + Y_2O_3 increased Y_2O_3 and particularly the addition of Al_2O_3 increases fluidity of the liquid (grain boundary) phase. This allows easy sintering to >98% theoretical density at a reduced temperature. Compositions not containing Al_2O_3 are considerably more difficult to sinter to >98% theoretical density. High density is responsible for high room temperature strength regardless of Y_2O_3 and Al_2O_3 levels. Strength at 1400°C is decreased sharply with as little as 0.5 w/o Al_2O_3. In general compositional changes which facilitate sintering degrade strength maintence to elevated temperatures.

Stable oxidation resistance over the temperature range 750° to 1350°C is observed for Si_3N_4 samples from phase fields I and II i.e. those containing Si_2ON_2, $Y_2Si_2O_7$ or $Y_5Si_3O_{12}N$ as minor phases. Oxidation rate in these phase fields increases with flux content. Excessive fluxing of these materials by iron and/or alumina causes bubbles and surface blisters at 1350°C. Iron levels should be kept below 0.5 w/o and alumina below 4.0 w/o for best oxidation performance at 1350°C.

Samples from phase fields III and IV with $YSiO_2N$ and $Si_3Y_2O_3N_4$ second phases show oxidation cracking at 1000°C but not at 1350, 1200, or 750°C. Cracking can be avoided by preoxidation. The presence of carbon aggrevates this instability. At 600°C in air, the presence of large amounts (14 w/o) of WC from self-abrasion of milling media can cause complete disintegration of an otherwise strong sample.

Complex shapes of $Si_3N_4 + Y_2O_3$ based ceramic can be formed by injection molding. Final densities >99% theoretical density are possible with close dimensional control.

Acknowledgement

The injection molding results are based on work performed for the Detroit Diesel Allision Div. of General Motors as part of their CATE program. The CATE program is conducted under NASA contract DEN 3-17 to the Lewis Research Center and is funded by the U.S. Department of Energy.

References

1. P.E.D. Morgan, "Research on Densification, Character and Properties of Dense Silicon Nitride," NTIS AD-757-748, March 1973.

2. C. Greskovich et al, "Ceramic Sintering," NTIS AD-A014-480 p 9, July 1975.

3. S. Wild et al, "The Role of Magnesia in Hot Pressed Silicon Nitride," p 377-84 in Special Ceramics, Vol. 5, P. Popper ed., Brit. Cer. Res. Assn. Stoke-on-Trent, U.K. (1972).

4. D.W. Richerson, "Effect of Impurities on the High Temperature Properties of Hot Pressed Silicon Nitride," Cer. Bull (7), 560-69, (1973).

5. G. Terwilliger, F.F. Lange, "Pressureless Sintering of Si_3N_4," J. Mater. Sci. 1169-1174, (1975).

6. M. Mitomo et al "Sintering of Si_3N_4," J. Am. Ceram. Soc., 55 (3) 313 (1976).

7. G.E. Gazza, "Hot Pressed Si_3N_4," J. Am. Ceram. Soc., 56 (12) 662 (1973).

8. G.E. Gazza, "Effect of Yttria Additions on Hot-Pressed Si_3N_4," Am. Ceram. Soc. Bull., 54 (9) 778-81 (1975).

9. A. Tsuge, H. Kudo, and K. Komeya, "Reaction of Si_3N_4 and Y_2O_3 in Hot-Pressing," J. Am. Ceram. Soc. 57 (6) 269-70 (1974)

10. A. Tsuge, K. Nishida, and M. Komatsu, "Effect of Crystallizing the Grain-Boundary Glass Phase on the High-Temperature Strength of Hot-Pressed Si_3N_4, Containing Y_2O_3," ibid, 58 (7-8) 323-26 (1975).

11. G.Q. Weaver and J.W. Lucek, "Optimization of Hot-Pressed Si_3N_4 + Y_2O_3 Materials," Am. Ceram. Soc. Bull., 57 (12) 1136-34, 1136 (1978).

12. M. Mitomo, "Sintering of Si_3N_4 with Al_2O_3 and Y_2O_3," Yogyo-Kyokai-Shi, 85, (8) 50-54 (1977).

13. J.T. Smith and C.L. Quackenbush, "Phase Effects in Si_3N_4 Containing Y_2O_3 or CeO_2: I Strength," Bull. Am. Ceramic Soc., 59, 5 529 (1980).

14. D.R. Clarke and G. Thomas, "Microstructure of Y_2O_3 Fluxed Hot Pressed Si_3N_4," J. Am. Ceramic Soc., 61 3-4 114-8 (1978).

15. O.L. Krivanek et al, "The Microstructure and Distribution of Impurities in Hot Pressed and Sintered Silicon Nitrides," J. Am. Ceramic Soc., 62, 11-12 585 (1979).

16. J.T. Smith, C.L. Quackenbush, "A Study of Sintered Si_3N_4 Compositions with Y_2O_3 and Al_2O_3 Densification Aids," pp. 426-42 in Proc. of Int'l. Symp. of Factors in Densification and Sintering of Oxide and Non-Oxide Ceramics, Hakone Japan (1978).

17. J.T. Smith et al, U.S. Patent 4,280,850 July 28, 1981.

18. D.J. Rowcliffe, "Microstructure Development in Sintered Si_3N_4," Annual Mtg. Am. Ceramic Soc., Cincinnati (1979) For abstract see Cer. Bull 3 347 (1979).

19. L. Gauckler et al, "The System Si_3N_4 - SiO_2 - Y_2O_3," J. Am. Ceram. Soc. 63, (1-2) 35-37 (1980).

20. F.F. Lange, S.C. Sighal, U.S. Patent 4,102,698, July 25, 1978.

21. C.L. Quackenbush and J.T. Smith, "Phase Effects in Si_3N_4 Containing Y_2O_3 or CeO_2: 11, Oxidation," Bull. Am. Ceramic Soc. 59 (5) 533-6 (1980).

22. F.F. Lange, S.C. Sighal, and R.K. Kyznicki, "Phase Relations and Stability Studies in the Si_3N_4 - SiO_2 - Y_2O_3 Pseudoternary System," J. Am. Ceramic Soc., 60 (5-6) 249-52 (1977).

23. H. Knoch and G.E. Gazza, "Effect of Carbon Impurity on the Thermal Degradation of an Si_3N_4 - Y_2O_3 Ceramic," J. Am. Ceramic Soc., 62, (11-12) 634-5 (1979).

24. H.F. Priest et al, "Sintering of Si_3N_4 under High Nitrogen Pressure," J. Am. Ceramic Soc., 60, (1-2) 81 (1979).

25. R.W. Rice and W.J. McDonough, "Hot-Pressed Si_3N_4 with Zr-Based Additions," J. Am. Ceramic Soc., 58 (5-6) 264 (1975).

π "Standard Product", Nippon Denko, Tokyo

* T-186, Starck Co., Berlin

DISCUSSION

Thompson: Do compositions containing $Y_2Si_2O_7$ also suffer catastrophic oxidation at 1000°C?

Quackenbush: The high oxidation rate at 1000°C did not result in a cracked sample, and so is not 'catastrophic'. The 1000°C oxidized material shows a duplex oxidation layer (Figure 6), with regions of thin uniform oxidation coverage and distribution similar to that of the Si_3N_4 matrix phase. This suggests that a more protective, uniform, layer may be achieveable with fine-grained material.

Jack: Are you trying to eliminate the glassy phase?

Quackenbush: We are looking at glass crystallization in a Si_3N_4-6 w/o Y_2O_3-2 w/o Al_2O_3 material. A Si_3N_4-6 w/o Y_2O_3 material has a crystalline secondary phase, as formed. For present rotating turbine components the properties of the glass-containing material are adequate because the highest stresses are in the hub at < 1000°C, and vice versa for the blade tips.

LUCAS SYALONS: COMPOSITION, STRUCTURE, PROPERTIES AND USES

R.J. Lumby and E. Butler

Lucas Industries Ltd., Group Research Centre, Solihull, U.K.

M.H. Lewis

Department of Physics, University of Warwick, Coventry, U.K.

1. INTRODUCTION

The successful development of a pressureless sintered material with properties, either as good as or superior to those of Hot Pressed Silicon Nitride (HPSN), has been the main objective of programmes at the Lucas Research Centre since 1972. Following the early pioneering work by Jack and co-workers at the University of Newcastle (1), investigations were started at LRC to prepare strong materials based on a matrix of expanded $\beta' Si_3N_4$. Identification of the correct composition of the expanded β' phase, first reported at "Special Ceramics 6" at Stoke-on-Trent in July 1974 (2), allowed a greater understanding of the control of the non-β' phases. It also allowed better control of the properties of predominantly β' materials (3,4). Structural studies using transmission electron microscopy at the University of Warwick (4) has allowed a programme of material optimisation to proceed. The use of yttrium oxide as a liquid phase sintering aid has led not only to the development of strong materials but also strong materials which are capable of retaining strength at temperatures up to 1300°C.

The facility that pressureless sintering offers has been demonstrated in that powders have not only been isopressed but, die pressed, extruded, slip cast, and injection moulded (Fig. 1). These shaping routes, followed by sintering, have led to strong materials, as strong as those proposed by isopressing.

The correct composition has been emphasised in a number of publications (3-6). Variations from the ideal result from lack of control in preparation or the use of reactive constituents which do not survive the effect of binders, the shaping environment, or

Riley, F.L. (ed.) Progress in Nitrogen Ceramics

Fig. 1. SYALON Ceramic Shapes.

Die Pressed — Extruded
Slip Cast — Injection Moulded

the debonding treatment. Aluminium nitride was found in particular to be a constituent which was a source of problems. Successive batches varied widely in oxygen content, and often contained an unacceptably high level of free aluminium. It was found that the polytype phases of AlN represented an improved unreacted starting constituent. Work in this area has been successfully protected (7). The use of the AlN polytypes in a starting powder leads to characteristic chemical balances and structures which are associated with the optimum properties developed. The use of polytype as a starting constituent is thus easy to detect. Lucas SYALON ceramic materials are therefore combinations of a high α-Si_3N_4 powder, a polytype, and a glass forming metal oxide, currently yttria. Having developed these materials an intensive effort has been mounted over the past four years to exploit them. Many engineering environments have been investigated. A summary of some of these environments is presented as an indication of the possible markets which may develop.

2. COMPOSITION - Effects of balance on property

Early work on aluminium silicon nitride oxide materials showed that although increasing additions of aluminium and oxygen increased the sinterability of a sialon powder, a gradual coarsening in the structure of the sintered product occurred and a reduction in strength and thermal conductivity was also obtained. The optimum composition was identified as one which was a compromise between sinterability, and strength. Additionally it was understood that a move in the sialon composition towards the SiO_2 or Al_2O_3 corners (Fig. 2), i.e. away from the β' sialon line, improved sinterability but at the expense of high temperature properties, resulting from an increase in the non-β' phases. Increases in the metal oxide addition also increased sinterability but again at the expense of increasing the non-β' phase content. The addition which would increase sinterability early in the sintering cycle without finally affecting high temperature properties was a combination of aluminium and nitrogen. Current compositions thus combine Si_3N_4 with an AlN polytype and Y_2O_3. The non-β' phase content is controlled by the balance of Si_3N_4 to AlN polytype, and Y_2O_3.

It was emphasised at the Canterbury NATO Advanced Study Institute (3) that the addition of Al and N to a sialon material controlled the glassy non-β' phase content and consequently the room and high temperature properties. The procedure for resolving the suitability of composition by deliberately preparing a range of materials and characterising the properties of that range is termed a balance exercise. The data presented at Canterbury merely demonstrated a principle in relation to hot-pressed materials. The characteristic range of properties obtained with the specific increase in aluminium and nitrogen content was found to be predictable and repetitive. This practice has been used with materials prepared by pressureless sintering and although the range of properties obtained differs, a characteristic range of properties has

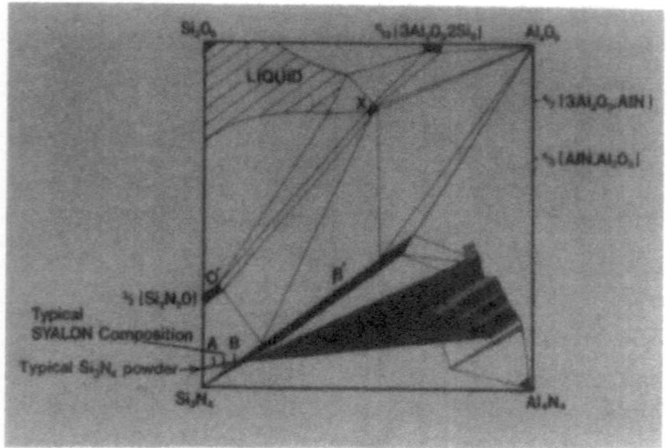

Fig. 2. A Schematic Representation of Lucas Syalon Compositions.

been obtained for a large number of starting materials. The data presented at Canterbury showed that as the increasing additions of aluminium and nitrogen caused a decrease in the intergranular phase content of hot pressed materials, a major improvement in creep occurred as a minor decrease in room temperature strength was obtained. In contrast, a similar increase in the Al and N content of yttria-containing materials prepared by pressureless sintering, results in the remnant intergranular phase approaching that of yttrium aluminium garnet (YAG).

The development of a pressureless sintered material relies on achieving compatibility between the β' and the residual intergranular phase. The dual objectives are again sinterability and strength (at room and high temperatures). Sinterability relies on the formation of a liquid phase allowing capillary-induced densification and transformation. It is essential that the residual non-β' phases at the completion of the reaction should be sufficiently refractory so as to not impair the superior properties conferred by the β'-matrix phases.

There are two ways of achieving this:

(1) by compositionally controlling the intergranular phase such that it solidifies as a glass with a high transition temperature, or

(2) by balancing the composition to allow a controlled glass-ceramic like transformation.

In practice these two materials have evolved from the balance exercise. By varying the Si_3N_4-polytype ratio the following data have been obtained from specific (Fig. 3) starting materials. It can be seen that an optimum β'/glass material is obtained at composition 348. This procedure has been followed for five Si_3N_4 starting powders differing in supplier and in nitriding technique. Optimum β'/glass material with a mean modulus of rupture value in excess of 800 MPa has been obtained. This value is typical of the hundreds of kilogrammes of β'/glass syalon ceramics prepared to date.

With the current development programme covering the complete spectrum from synthesis of starting materials to syalon chemistry, shaping and sintering, the simple objective is the further improvement of all aspects of material behaviour and properties. Six objectives with respect to properties are:

(1) Modulus of Rupture 25°C - 1500 MPa
(2) Modulus of Rupture 1300°C - 1000 MPa
(3) Thermal Conductivity 25°C - 25 $W^{-1}m^{-1}K^{-1}$
(4) Maintenance of the measured creep characteristics
(5) Retention of the freedom from slow crack growth at temperatures up to 1400°C
(6) Development of oxidation behaviour to allow the use of materials at 1400°C.

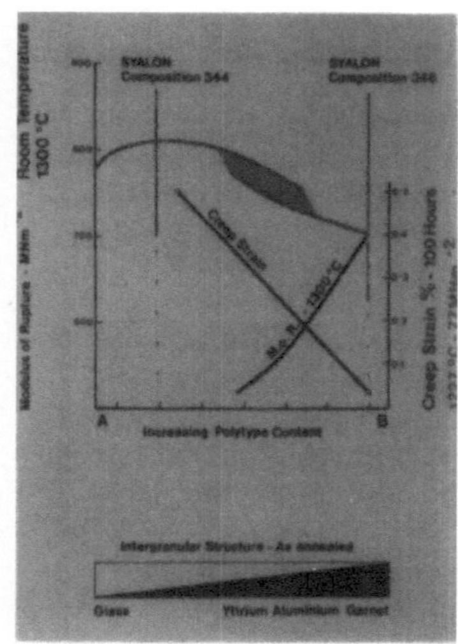

Fig. 3. Property Data for a Balance Exercise.
The effect of increasing polytype content on Modulus of Rupture and Creep.

3. STRUCTURE-PROPERTY RELATIONSHIPS

Fig. 4 shows a typical β'/glass material. This ion-beam thinned section shows 0.5 μm β' crystals embedded in a highly electron absorbing second phase. There is ample evidence of the growth of β'-phase into a liquid from their faceted hexagonal prism morphology. There is also evidence of a range of β'-crystal dimensions occurring as a result of its unconstrained growth which is permitted by the larger liquid volume in this pressureless sintered material. This morphological anisotropy and the interlocking form of the β' growth is probably the source of this material's remarkable toughness.

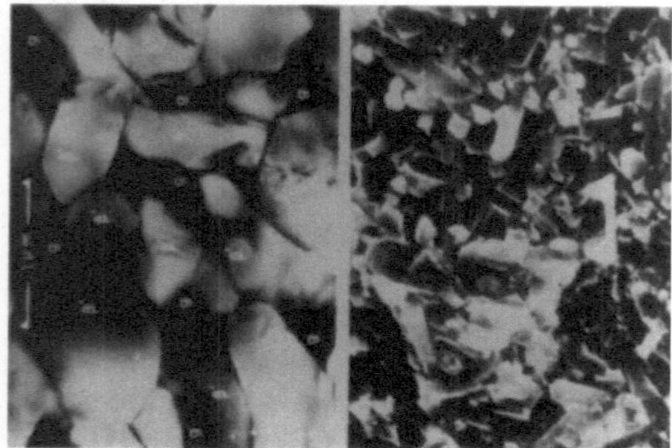

Fig. 4. β'/glass material. Transmission and Scanning Electron Micrographs.

The result of a recrystallising treatment on materials of higher aluminium and nitrogen content is to cause conversion of the intergranular phase to YAG. At a sufficiently high level of Al and N (i.e. at a sufficiently high polytype addition) the intergranular phase is totally converted. In contrast to β'/glass materials, β'-YAG structures have a smaller volume fraction of intergranular phase (Fig. 5). There is also evidence for the constancy of the orientation of YAG over distances of many grain diameters.

Examination of β'-YAG materials which have been deformed at high temperatures shows that the distribution of the YAG intergranular phase has been modified (Fig. 6).

The semicontinuous form of the YAG present in the as-recrystallised state has been modified to one of discrete islands at grain triple points. This form of the material is termed "morphologically stable".

Fig. 3 shows that the most significant difference between an unannealed β'/glass material of optimum composition and an optimum β'-YAG composition is in the creep behaviour. Studies of high temperature deformation have shown that as expected β'-glass material fails in creep as a direct result of slow crack growth. However the behaviour of β'-YAG materials is close to that of the best single phase hot pressed sialon materials reported elsewhere (8).

Fig. 7 shows data for measurements of crack velocity as a function of stress intensity. In contrast to HPSN, glassy hot

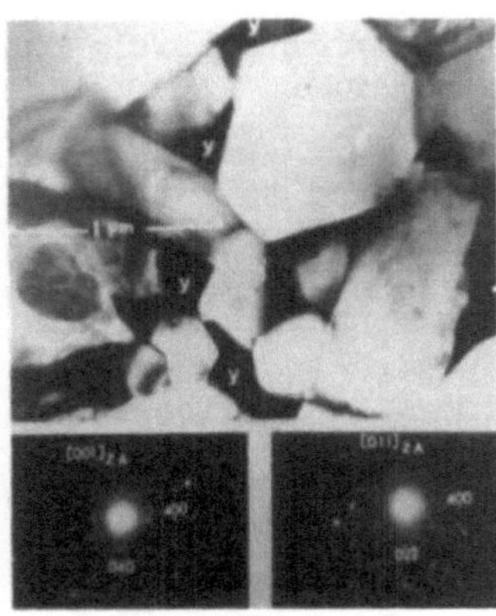

Fig. 5. Transmission electron micrograph of β'-YAG material.

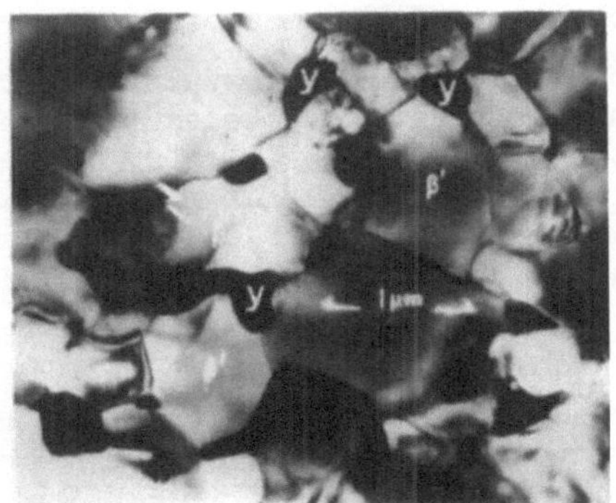

Fig. 6. Transmission electron micrograph of deformed β'-YAG.

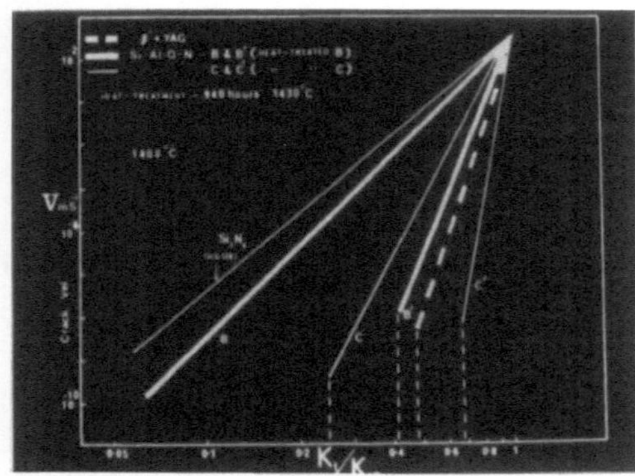

Fig. 7. High temperature deformation behaviour of various nitrogen ceramics; crack velocity plotted as a function of stress intensity factor.

pressed sialon and β'-glass materials, β'-YAG only undergoes slow crack growth at a very high proportion of its critical stress intensity factor (K_{IC}), even at temperatures as high as 1300°C. Again β'-YAG materials can sustain very high levels of creep strain.

Tensile face cracking occurs at relatively low levels of creep strain in a glass-containing syalon material. The more highly strained glass-free material below shows no tensile face cracking either by eye or down to an electron micrograph level of examination. Structurally, strain in a β'-YAG material is accommodated by the compliant YAG phase and by general grain deformation, both of which phenomena are readily observed in crept materials. The morphologically stable β'-YAG material has been shown to have creep performance at least one order of magnitude better than that of the β'-YAG material from which it is derived. This is probably due to the much higher incidence of β'-β' grain boundaries.

The oxidation of Lucas syalon materials is still the subject of continuing studies. The best performance obtained to date has been for the β'-YAG material up to temperatures of 1350°C. The morphologically stable β'-YAG material can survive at temperatures up to 1400°C. Isolation of the intergranular YAG phase causes a substantial reduction in oxidation rates.

4. APPLICATIONS

The study of the exploitation of Lucas syalon materials has been restricted exclusively to β'-glass materials, even though in some instances the choice of a β'-YAG material might have been a more wise choice.

Metal Cutting:

The first major use found for syalon materials was in metal cutting, as a result of which Lucas have licenced two of the world's major tungsten carbide insert manufacturers to manufacture and market on a world-wide basis. Kennametal launched their version of a syalon ceramic as Kyon 2000 in September 1981. Preliminary cutting trials were with difficult materials. One typical example is the cutting of hardened EN31 Bearing steel (hardness Rc65) at a surface speed of 100-150 m/min. The cutting of turbine discs with syalon ceramic at a surface speed of 300 m/min compares with a maximum cutting speed for tungsten carbide of 10-15 m/min and for Al_2O_3 inserts of 100 m/min. Cutting cast iron has illustrated the toughness of the β-glass syalon material. It is possible to face mill with syalon inserts at speeds of around 300m/min (optimum for a coated tungsten carbide insert). At this speed a tool life advantage was demonstrated under production conditions. Since the optimum speed for syalon inserts on cast iron is more than 600 m/min, the optimum use of syalon inserts will result in massive increases in productivity in cast iron face milling operations. This cutting performance data is now confirmed in the introductory publicity literature from Kennametal for their Kyon 2000 range of inserts.

Rock Cutting:

Fig. 8 shows typical rock cutting tools used in the evaluation of wear resistance. Syalon inserts have been evaluated on Darley Dale sandstone (an abrasive standard rock material) and have shown improved wear resistance over the conventional tungsten carbide material. Impact testing on a similar rock has shown syalon to have promise.

Fig. 8. Typical Rock Cutting Picks.

Welding:

Syalon materials have been evaluated as location devices and as weld shrouds.

Fig. 9 shows a Lucas starter motor component assembled on a syalon location pin. Syalon pins have shown lives of 100-200 times more than alternative materials. Fig. 10 shows location pins used for the welding of captive weld nuts. Pins of this type used in a tobotic operation have survived over 5,000,000 operations (in contrast to a conventional pin life of 7,000 operations). This operation involves severe thermal and mechanical stresses.

Syalon is also used in weld shrouds. This orbital welding operation required a shroud of 6 mm diameter with a wall thickness of 0.76 mm, and is an operation which could not be performed by other materials.

Metal Forming:

Hot pressed silicon nitride was used by Tube Investments Ltd. as a floating mandrel in the drawing of seamed stainless steel tubing. The absence of any metallic phase in HPSN avoided pick-up and cold welding. This resulted in scrap rate being reduced and made expensive lubrication system unnecessary. Syalon ceramic drawing plugs function with equal success but could be easily fabricated from a simple extrusion at much reduced cost.

Molten Metals:

HPSN coracles have been used with success by Metals Research

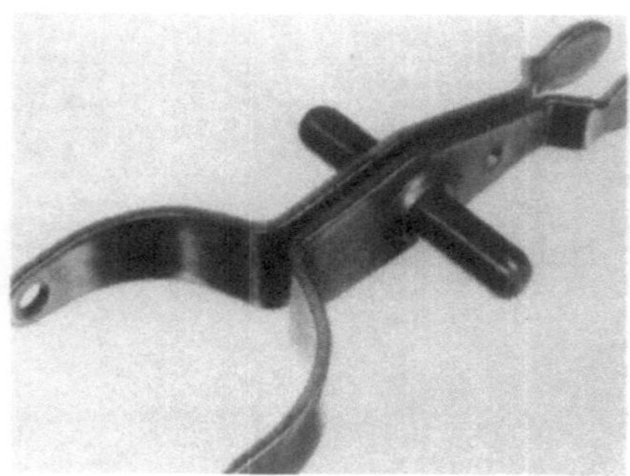

Fig. 9. A Lucas starter motor component assembled on a syalon ceramic location pin.

Fig. 10. Syalon ceramic location pins for the Captive Weld Nut Operation.

Fig. 11. A slip cast and sintered syalon ceramic coracle, and the final machined component.

in their crystal growing of Gallium Phosphide. Fig. 11 shows that the simple slip casting route for syalon coracles results in massive materials savings and a large reduction in machining cost.

Bearings:

A variety of bearing components (Fig. 12) are currently under investigation. These range from low friction bearings operating at ambient temperature to those at elevated temperatures in hostile environments. If the toughness differences revealed in metal cutting comparisons of syalon ceramics with HPSN can be assumed to be a good guide, then syalon elements would be expected to have excellent performance in rolling contact fatigue.

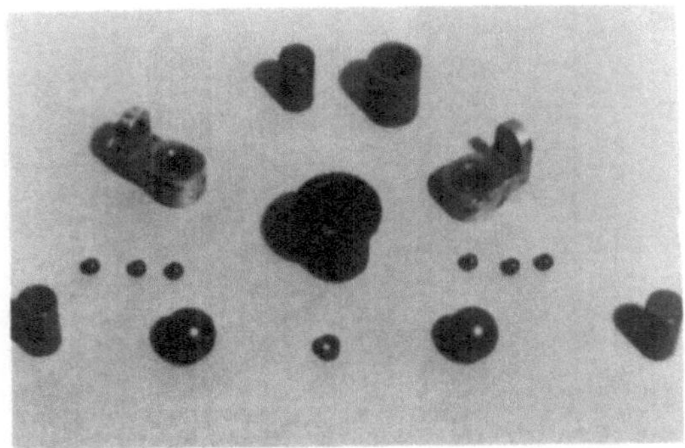

Fig. 12. A variety of syalon ceramic bearing components.

Seals:

The rotating shaft seal behaviour of syalon ceramics has been shown to be superior to most other conventional materials, especially when run against itself. Syalon on syalon has run at high speed and high face pressure; a PV value of 5 x 10 ft/min x psi has been achieved. More conventional shaft sealing operations are now being studied and edge seals are also being investigated (Fig. 13).

Engine Components:

The absence of any formal support for engine programmes in the United Kingdom has made engine component development difficult. Investigation in the area has had to be delayed in favour of the more short term potential markets. It has been possible however to demonstrate that a syalon ceramic tappet could offer a considerable advantage in terms of life. After 60,000 kms of running in a diesel engine, only 0.75 µm of wear could be detected in one such tappet.

The shaping capability of syalon ceramics makes the material a candidate in turbine programmes (Fig. 14). Further development in the chemistry and shaping capability will allow the properties and high temperature behaviour of β'-YAG materials to be fully exploited.

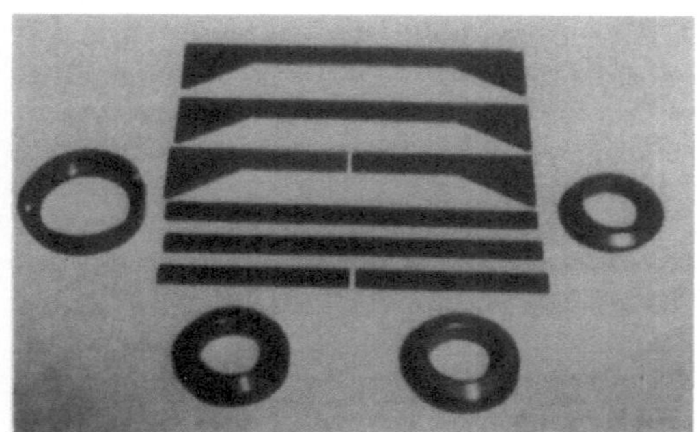

Fig. 13. Syalon ceramic seal elements.

Fig. 14. An injection moulded and sintered syalon ceramic nozzle guide vane.

REFERENCES

1. K.H. Jack. Trans. Brit. Ceram. Soc. 72 (1973) 376.
2. R.J. Lumby, B. North and A.J. Taylor. "Special Ceramics 6" edited by P. Popper, B.Ceram.R.A. (Stoke-on-Trent), 1975, p.283.
3. R.J. Lumby, B. North and A.J. Taylor. "Nitrogen Ceramics", NATO ASI Series E: Applied Science - No. 23, Noordhoff (Leyden), 1977, p.393.
4. M.H. Lewis, B.D. Powell, P. Drew, R.J. Lumby, B. North and A.J. Taylor. J. Mater. Sci. 12 (1977) 61.
5. R.J. Lumby, B. North and A.J. Taylor. "Ceramics for High Performance Applications II", edited by J.J. Burke, E.M. Lenoe and R.N. Katz, Brook Hill Publishing Co. (Chestnut Hill) 1978.
6. M.H. Lewis, A.R. Bhatti, R.J. Lumby and B. North. J. Mater. Sci. 15 (1980) 103-113.
7. U.K. Patent Number 1573299, U.S. Patent Number 4113503 and equivalents.
8. B.S.B. Karunaratue and M.H. Lewis. J. Mater.Sci. 15 (1980) 449-462.

DISCUSSION

Quackenbush: How does the strength of the S-Y-Al-ON materials vary with temperature?

Lumby: The bend strength of the β'-glass material falls from 800 MPa at 25°C to ~400 MPa at 400°C. As the polytype content of the material is increased the reduction diminishes. The β'-YAG (type '348') has a bend strength of 700 MPa at 25°C and 700 MPa at 1300°C.

Thümmler: What materials did you compare cutting rates with?

Lumby: Standard WC-Co hard metals on bearing steel (EN31) - maximum for the carbide 10 m min^{-1}. On Incalloy 901 the carbide maximum is 25 m min^{-1}. We also compared hot-pressed Al_2O_3/TiC composites for interrupted cutting of cast iron, when Syalon material performed much better.

THE NATURE OF SiC FOR USE IN HEAT ENGINES AS COMPARED TO Si_3N_4: AN OVERVIEW OF PROPERTY DIFFERENCES

D. C. Larsen, J. W. Adams
IIT Research Institute
Chicago, Illinois, USA

and

R. Ruh
Air Force Wright Aeronautical Laboratories
Dayton, Ohio, USA

ABSTRACT

An overview is presented of the thermal and mechanical properties of dense forms of silicon carbide. Hot-pressed and sintered SiC are included. Comparison is made to dense Si_3N_4. The properties discussed include flexural strength, elastic modulus, stress-strain behavior, creep strength, thermal shock resistance, thermal diffusivity, and thermal expansion. Fracture toughness and oxidation resistance are treated elsewhere in this volume, and are thus only briefly mentioned here. Tests were conducted up to 1500°C in static air, and the results are interpreted with respect to the influence of microstructure, secondary phases, etc.

INTRODUCTION

Silicon-base ceramics, silicon nitride and silicon carbide, have structural application in advanced heat engines such as gas turbines and diesel engines. Components include rotor blades, stator vanes, combustion chambers, piston caps, cylinder liners, etc. Such utilization of ceramics offers several advantages, including higher temperature operation leading to increased efficiency and decreased specific fuel consumption, decreased weight, lower potential life cycle cost, decreased complexity through the use of non-cooled components, and elimination of the need for the use of strategic materials (e.g., cobalt and chrome).

Riley, F.L. (ed.) Progress in Nitrogen Ceramics

At the first NATO Advanced Study Institute, held in 1976, the emphasis was on hot-pressed (HP) Si_3N_4. Many of the papers at this second ASI involve sintered Si_3N_4. This contribution to the meeting is an initial attempt to introduce silicon carbide to the ASI attendees. The properties of the dense forms of SiC (sintered and hot-pressed) are overviewed. The emphasis is placed on the general nature of SiC, as contrasted to hot-pressed Si_3N_4. It is assumed that the reader is familiar with the microstructure and properties of HP-Si_3N_4. The behavior of silicon-densified or siliconized SiC will not be treated in detail in this paper.

STRENGTH AND ELASTIC MODULUS

Silicon carbides have high elastic modulus compared to Si_3N_4 materials. The Young's modulus is typically 350-450 GPa (50-60 x 10^6 psi). This is ~50% higher than the elastic modulus of dense HP-Si_3N_4. The strength of hot-pressed and sintered SiC is intermediate to the strength of hot-pressed and reaction-sintered Si_3N_4. Figure 1 illustrates the 4-point flexure strength of various commercial hot-pressed and sintered SiC materials. Room temperature strength in SiC is usually controlled by the grain size. Hot-pressed SiC using Al_2O_3 as a densification aid has higher strength than when B_4C is used. Al_2O_3 inhibits grain growth; the

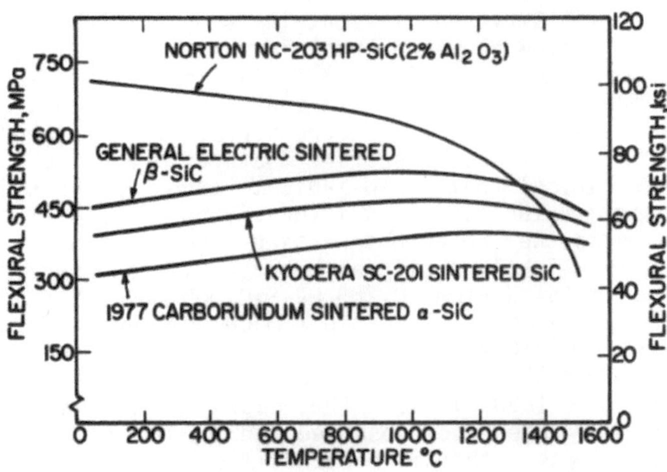

Figure 1. Flexural strength of silicon carbide materials.

grain size is typically 1-2 μm, and the strength is high. The converse is true for HP-SiC doped with B_4C, where the grain size ranges from 10-40 μm. Sintered SiC is processed in α- and β-SiC crystal structures (predominately hexagonal and cubic, respectively). Bend strengths are typically 300-450 MPa (45-65 ksi) as shown in Fig. 1. The β-SiC structure appears consistently stronger than α-SiC. Fracture origins in α-SiC are often large (∿50 μm) individual needles/platelets that have experienced exaggerated grain growth during processing. Single isolated surface-connected pores are also found in α-SiC. Both forms of SiC fracture predominately in the transgranular mode. It is for this reason that fracture phenomena in SiC are more often associated with grain size, rather than porosity or intergranular phases. The predominance of transgranular fracture also makes fracture surface analysis and the identification of fracture origins more difficult than in HP-Si_3N_4, where the fracture path is usually intergranular.

The behavior of SiC materials at elevated temperature is analagous in many respects to the high temperature behavior of Si_3N_4 materials. However, the effect of intergranular phases resulting from processing additives on subcritical crack growth and strength reduction is much less pronounced in silicon carbide. Figure 1 illustrates the strength-temperature behavior of sintered SiC, which contains few additives, if any, and that of hot-pressed SiC, which contains 1-2% Al_2O_3 additive to achieve densification, but which results in aluminosilicate grain boundary phases. The low temperature strength is maintained much better out to 1500°C for the sintered materials. Sintered SiC contains very little residual oxide impurity. A typical microstructure is shown in Fig. 2. Figure 3 shows, analagous to high purity RS-Si_3N_4, that linear stress-strain behavior is obtained in sintered SiC at 1500°C. Figure 4 illustrates a typical fracture surface for sintered SiC tested at 1500°C. No fracture surface features are visible (at this level of magnification) that would indicate an operative slow crack growth mechanism. Oxide additives in HP-SiC do not affect the strength as much as they do in hot-pressed Si_3N_4. Figure 5 illustrates that only slightly nonlinear behavior is observed in 1500°C stress-strain data for Norton NC-203 HP-SiC, which contains ∿2% Al_2O_3. Most HP-Si_3N_4 materials exhibit much more pronounced nonlinear stress-strain behavior at 1500°C. For the same reason, the temperature at which the strength begins to decrease precipitously is lower for Si_3N_4 (∿1250°C) (1). The reason for the less pronounced temperature dependence observed for SiC may be due to a combination of effects: the SiC aluminosilicate intergranular phase being relatively refractory, and the fracture mode in SiC being predominately transgranular.

Figure 2. Reflected light micrograph of 1977 Carborundum sintered α-SiC etched 10.3 min with boiling Murikami's reagent (60 g KOH + 60 g $K_3Fe(CN)_6$ + 120 ml H_2O).

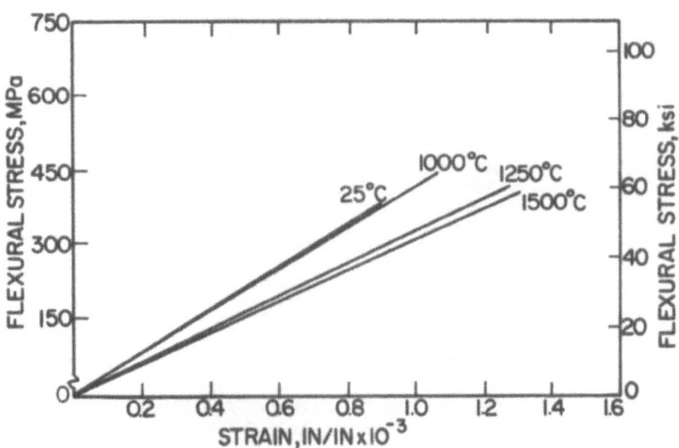

Figure 3. Flexural stress-strain behavior of Kyocera SC-201 sintered SiC.

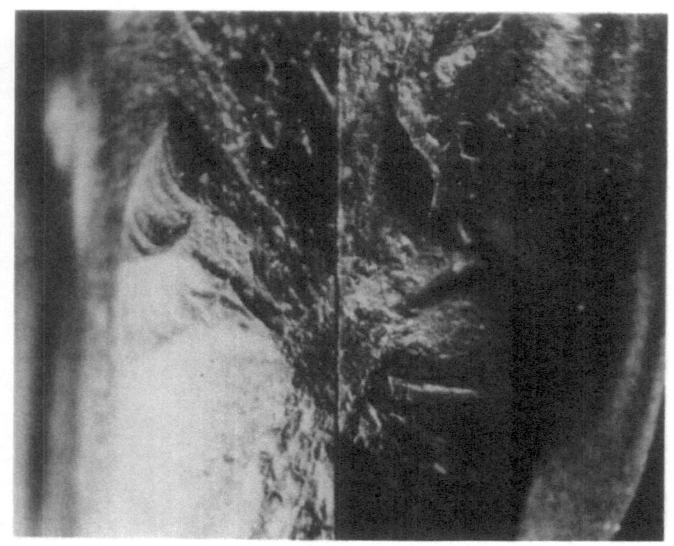

gure 4. Fracture surface (tensile surfaces together) of Kyocera SC-201 sintered SiC tested at 1500°C showing an absence of subcritical crack growth.

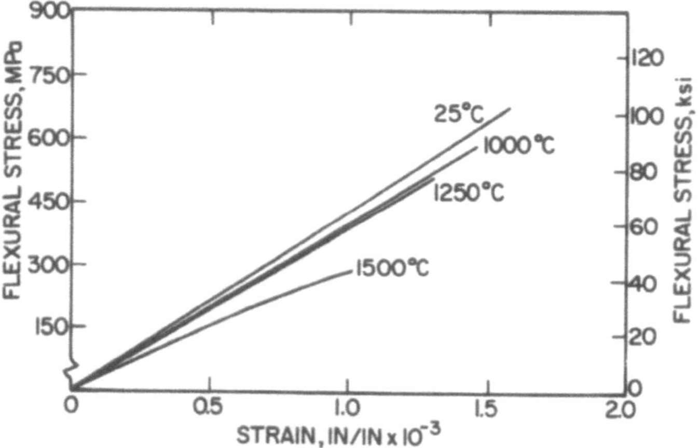

Figure 5. Stress-strain behavior for Norton NC-203 HP-SiC (2% Al_2O_3)

FRACTURE TOUGHNESS

The potential use of ceramics in gas turbines and other structural applications requires careful consideration of the brittle nature of these materials. Their fracture toughness or resistance to crack propagation is much lower than that of metals, and when failure occurs it does so in a rapid and catastrophic manner. The conditions controlling fracture may be considered in terms of fracture initiation or crack propagation. Fracture may occur if the applied stress is increased or intensified locally around a flaw such that it exceeds the theoretical strength of the material. Alternatively, cracks will propagate as long as the rate of strain energy release is greater or equal to the energy needed for forming new crack surface. The parameter that is most often used to characterize the fracture toughness of ceramics is K_{Ic}, the critical stress intensity factor.

The fracture toughness of dense SiC is inherently lower than the fracture toughness of dense Si_3N_4. The critical stress intensity factor of HP-Si_3N_4 is typically K_{Ic} = 4.5-6 $MNm^{-3/2}$, whereas the fracture toughness of hot-pressed and sintered SiC is nominally K_{Ic} = 3-4 $MNm^{-3/2}$. Based on data generated in our laboratory, there is no one SiC material that has fracture toughness significantly higher than others. There appears to be no manufacturing process for conventional silicon-base ceramics, either SiC or Si_3N_4, that offers any great clearcut advantage. All ceramics have fracture toughness roughly an order of magnitude lower than metals. Thus, to make major improvements in the fracture toughness of ceramics, it appears that innovative ceramic materials must be developed; for instance, ceramic-ceramic composites, where the addition of a second phase increases the work of fracture.

LONG TIME HIGH-TEMPERATURE EFFECTS

Silicon carbide materials are, in general, more stable in long term high-temperature applications than are Si_3N_4 materials (2). The strength distribution in SiC appears to be altered less by environmental effects such as oxidation. Weight gain by oxidation is lower in SiC, and the oxide scale thickness is less when compared to Si_3N_4. This is the case because in general, SiC materials are of higher purity and higher density than Si_3N_4 materials. Oxidation in HP-Si_3N_4 is mainly affected by the alkali impurities that segregate in grain boundaries, and migrate to and modify the oxide scale causing increased rates of oxidation. Reaction-sintered Si_3N_4 is admittedly very pure, but a trade-off is made; the porosity is high (10-20%) in RS-Si_3N_4, and thus surface area and total oxidation are large. Sintered SiC is both pure and dense, and thus has excellent oxidation resistance. Recent forms of HP-Si_3N_4, however, show much promise for good oxidation resistance. A particular example is a Westinghouse HP-Si_3N_4 that contains nominally 4% Y_2O_3, and an undetermined amount of SiO_2 (2).

As discussed above, another long term effect of concern in structural ceramics is strength reduction by slow or subcritical crack growth (SCG). Figure 6 illustrates the fracture stress vs. time-to-failure from dynamic fatigue tests (5). The n value in Fig. 6 is related to the slope of the strength-time relation, and is the exponent n in the relation $v = AK^n$, where v is the crack velocity and K is the stress intensity at the crack tip (A is a constant). A large value of n means little slow crack growth and time-invariant strength. As shown in Fig. 6, SiC exhibits much less SCG and strength degradation than does Si_3N_4.

The mechanisms of SCG appear to be different in Si_3N_4 and SiC. In Si_3N_4 the existence of slow crack growth is usually correlated with a grain boundary sliding mechanism. Accommodation for this deformation is provided by the nucleation of intergranular voids or the extension of pre-existing grain triple-point voids. The current speculation for SiC is that slow crack growth is an atomistic-level process occurring at the crack tip, being related to stress corrosion by an oxidation mechanism. For instance, Srinivasan et al. (6) report no detectable SCG in sintered α-SiC tested in argon at 1500°C. However, in air at 1500°C there was evidence of significant SCG, the mechanism being atmospheric attack of intergranular regions leading to grain separation. McHenry and Tressler (7) also indicate the mechanism of SCG in NC-203 HP-SiC to be stress-corrosion by oxidation

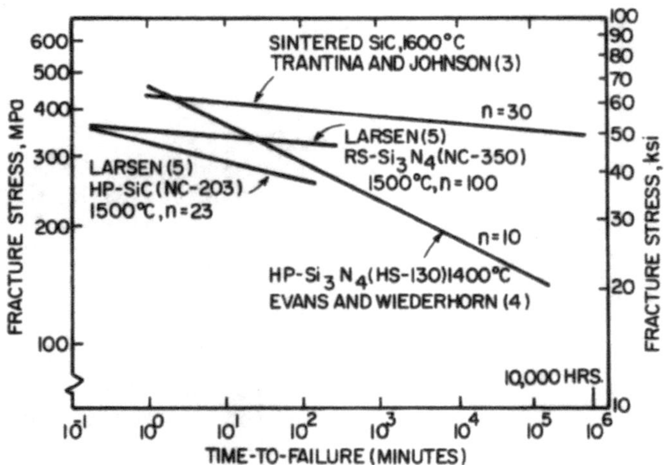

Figure 6. Strength degradation due to subcritical crack growth.

CREEP RESISTANCE

Creep resistance is of primary concern in the rotating components of a turbine engine. High creep rates can lead to both excessive deformations and uncontrolled stresses. We have investigated the creep behavior of various silicon-base ceramics at this laboratory. The results are presented in Fig. 7, expressed as bands of behavior on a creep strain rate vs. stress plot as a function of material type and processing method.

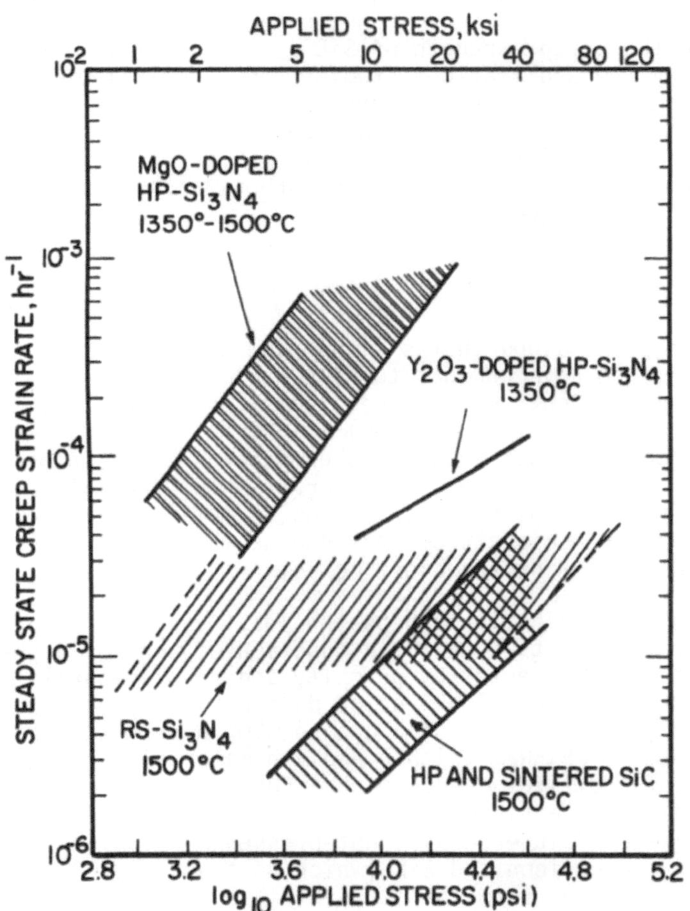

Figure 7. Flexural creep behavior of various Si_3N_4 and SiC materials.

For both sintered and hot-pressed forms of SiC, creep rates are extremely low, and a linear stress dependence is observed. Creep rates in HP-Si_3N_4 are high, primarily because of the existence of oxide intergranular phases. Mechanistic interpretation usually involves grain boundary sliding, cavitation, microcracking, etc. The creep rates in RS-Si_3N_4 are lower because of the absence of intergranular oxide phases. Current work in Si_3N_4 involves the use of Y_2O_3 as a densification aid. This results in oxide intergranular phases that can be crystallized by post-densification heat treatment. In this manner the creep rate for HP-Si_3N_4 is being decreased, as shown in Fig. 7.

Since the creep strength of SiC is so high (i.e., creep rate so low) very few applications studies have identified creep deformation to be a predominant failure mode for SiC. For this reason, only a few mechanistic studies have been undertaken. The linear stress dependence of the creep rate in SiC suggests that diffusion is the rate-controlling process. A carbon-vacancy diffusion mechanism has been proposed (8,9).

THERMAL EXPANSION

One of the reasons that silicon base ceramics are prime candidates for use in advanced gas turbine applications is their low expansion coefficient, which makes them less susceptible than many other ceramics (especially oxides) to thermal shock damage. However, we recognize that such low thermal expansion can be a disadvantage also; for instance, in the creation of thermal expansion mismatch situations with higher expansion metal engine components. As with many other aspects of heat engine materials selection and component design, many trade-offs exist.

Thermal expansion is perhaps the least variable property of silicon ceramics. It is mainly a function of the solid phase and thus not strongly affected by porosity and minor impurities. Figure 8 presents thermal expansion data bands for various silicon-based ceramics. All forms of SiC have ∿50% higher thermal expansion than all forms of Si_3N_4. Within a given material performance band, the effect of additives can be seen. For instance, within the thermal expansion data band for SiC shown in Fig. 8 exist curves for additive-free sintered SiC, hot-pressed SiC (which contains ∿2% Al_2O_3), and siliconized SiC (which contains ∿10-20% silicon metal). Siliconized forms of SiC exhibit the lowest expansion, sintered SiC has intermediate expansion, and Al_2O_3-doped hot-pressed SiC the highest expansion. The low expansion for Si/SiC is due to the low expansion of the silicon metal phase. The high expansion for hot-pressed SiC is caused by their slight aluminosilicate grain boundary phase (oxides having much higher expansion than pure SiC).

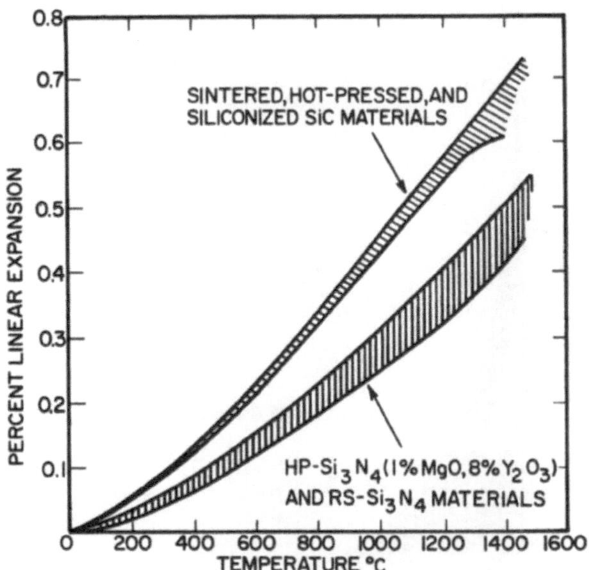

Figure 8. Thermal expansion data bands for Si_3N_4 and SiC materials.

THERMAL DIFFUSIVITY/CONDUCTIVITY

Along with thermal expansion, the other thermal properties that determine thermal stress are the thermal conductivity and thermal diffusivity. Thermal diffusivity is a derived property, being defined as the ratio of the thermal conductivity to the density-specific heat product. These thermophysical properties determine temperature distribution within a component and thus the thermal strain field. Specific heat is a volumetric property, not as variable with respect to impurities and microstructure as thermal conductivity and thermal diffusivity.

Figure 9 presents thermal diffusivity data bands for various types of silicon-base ceramics measured in this laboratory. Si_3N_4 materials have relatively low thermal diffusivity (and conductivity). Their properties are strongly affected by porosity, phase content (α/β ratio), and perhaps free silicon content. HP-Si_3N_4 generally has higher conductivity than RS-Si_3N_4 due to higher density. In fact, very consistent relations have been developed to describe the porosity dependence of thermal conductivity in RS-Si_3N_4 materials (5).

Silicon carbide has much higher thermal diffusivity (and therefore thermal conductivity) than Si_3N_4, especially at low temperatures. This is illustrated in Fig. 9. The temperature dependence

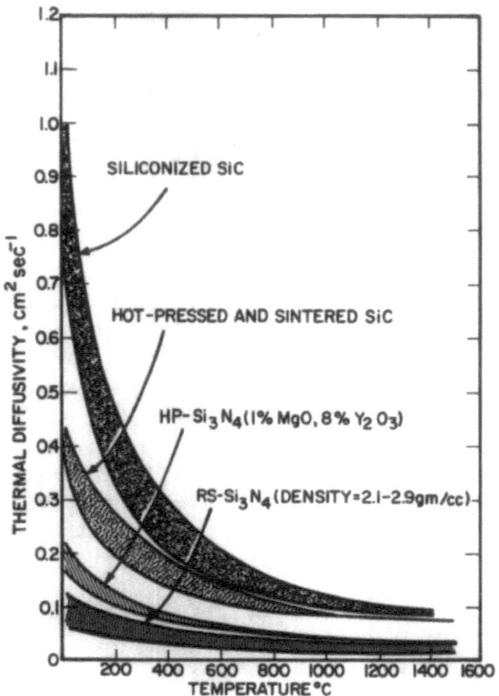

Figure 9. Thermal diffusivity of silicon ceramics.

is extremely strong near room temperature. Much data variability exists at 25°C where the exact role of grain size, purity, crystal phase, etc. are not well understood for SiC. The low temperature thermal diffusivity is very high for the siliconized forms of SiC. At elevated temperature very little scatter exists and the temperature dependence is relatively small.

It is noted that the difference in thermal diffusivity or thermal conductivity between SiC and Si_3N_4 is most apparent at low temperature. Thus, thermal transients in turbine components made from SiC vs. Si_3N_4 would be most critical during light-off rather than during a flame-out. This effect could be offset, however, by the fact that during light-off (severe thermal upshock) component surfaces are in compression, and thus are significantly stronger than the surfaces during the severe down-shock of a flame-out, where component surfaces are in tension.

THERMAL SHOCK RESISTANCE

The ability to withstand the thermal stresses generated during ignition, flame-out, and operating temperature excursions is an important consideration in evaluating potential ceramic heat engine materials. Thermally created stresses may initiate a fracture which can result in a catastrophic failure, or cause existing flaws to grow giving a gradual loss of strength and eventual loss of component integrity. However, the evaluation of thermal stress resistance is a complex task since performance is dependent not only on material thermal and mechanical properties, but is also influenced by heat transfer and geometric factors (i.e., heat transfer coefficient and component size).

The thermal shock resistance of various silicon ceramics has been determined in our laboratory by the water quench method, with the initiation of thermal shock damage being detected by internal friction measurement. This technique was chosen not in an attempt to simulate in-service engine conditions, but rather as a relative ranking of candidate materials in severe thermal down-shock. In conducting this test, internal friction is measured before and after water quench from successively higher temperatures using the flexural resonant frequency Zener bandwidth method. A marked change in internal friction (specific damping capacity) indicated the onset of thermal shock damage (i.e., thermal stress-induced crack initiation). This defined the critical quench temperature difference ΔT_c, which is compared to analytical thermal stress resistance parameters. The parameter which is most applicable to the experimental severe water quench is $R = \sigma(1-\mu)/\alpha E$, where σ is the strength, μ is Poisson's ratio, α is thermal expansion, and E is the elastic modulus (10).

The experimental and analytical results for various Si_3N_4 and SiC materials are directly compared in Fig. 10. All forms of silicon carbide (sintered, hot-pressed, or siliconized) exhibit poor thermal shock resistance when compared to Si_3N_4 (either hot-pressed or reaction-sintered). This is due to the extremely high thermal expansion and elastic modulus for SiC. It is interesting to note that despite the very high strength of some HP-SiC materials such as Norton NC-203 (>700 MPa, 100 ksi), their thermal shock resistance remains about the same as the other SiC materials. The reason for this is the overriding influence of the high αE product for SiC. For example, for NC-203 to exhibit an R value of 500°C (that is, approximately mid-range of the materials shown in Fig. 10), it would have to have a strength of 1120 MPa (160 ksi). For NC-203 HP-SiC to have an R value as high as the best HP-Si_3N_4, i.e., R = 750°C, as shown in Fig. 10, it would have to have a strength of almost 1700 MPa (∿240 ksi). Thus, SiC inherently has lower thermal shock resistance than Si_3N_4, and no conventional processing methods for SiC can overcome the high αE factor. In certain situations the higher thermal conductivity for SiC helps, but not enough to override the effect of high elastic modulus and high thermal expansion. It is for this reason that SiC has found application in

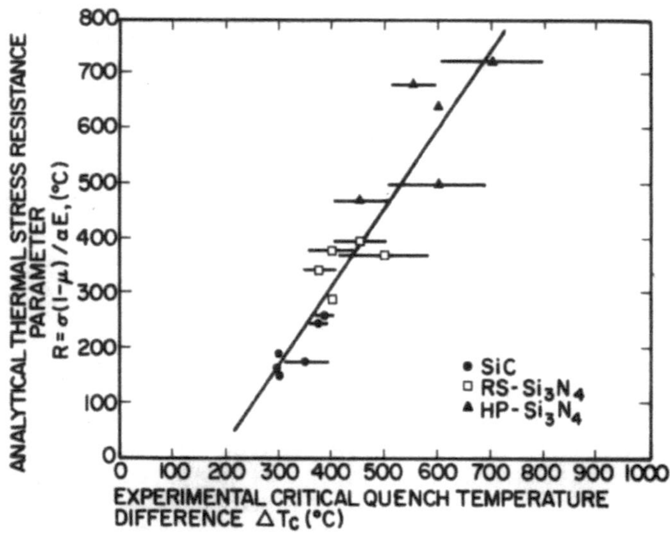

Figure 10. Analytical vs. experimental thermal shock results for various silicon ceramics.

combustors, where high thermal conductivity is useful to minimize hot spots, and the temperature is relatively static.

CONCLUSION

For ceramics to be successfully utilized as structural heat engine components, various requirements must be met including: the development of an overall life prediction methodology; the employment of a realistic design methodology; and the demonstration of fabrication process feasibility for the required ceramic component configurations. The biggest challenges facing designers in using structural ceramics include: (1) dealing with the statistical nature of these brittle materials; (2) the low fracture toughness and surface sensitivity; (3) the existence of subcritical crack growth; (4) the lack of data regarding the environmental effects of oxidation, corrosion, erosion, and deposition from the fuel combustion products; (5) batch-to-batch variability resulting from inadequate process control; and (6) the lack of an NDE technology capable of detecting critical strength-limiting flaws. These challenges are compounded by the fact that new candidate materials for advanced heat engine applications are continually emerging, and their properties are strongly dependent on microstructure, purity, and processing history. These comments apply generally to all structural ceramics, including silicon carbide and silicon nitride.

A comparison of the major characteristics of the dense forms of SiC and Si_3N_4 is summarized in Table 1. SiC is intermediate in

TABLE 1. A COMPARISON OF MAJOR CHARACTERISTICS OF DENSE FORMS OF SILICON CARBIDE AND SILICON NITRIDE

Silicon Carbide	Silicon Nitride
• Intermediate Strength	• High Strength
• High Thermal Expansion	• Low Thermal Expansion
• Poor Thermal Shock Resistance	• Good Thermal Shock Resistance
• Low Fracture Toughness	• Higher Fracture Toughness
• High Temperature Stability	• Strength Degradation by Slow-Crack Growth
• High Thermal Conductivity	• Low Thermal Conductivity
• Good Oxidation Resistance	• Oxidation Strong Function of Impurities
• High Creep Strength	• Low Creep Strength
• Low Influence of Impurities	• Second Phases Influence Behavior
• High Elastic Modulus	• Lower Elastic Modulus
• Transgranular Fracture	• Intergranular Fracture
• Little Subcritical Crack Growth	• Nonlinear Stress-Strain

strength to Si_3N_4. However, SiC experiences much less strength degradation by subcritical crack growth, and thus SiC is superior at temperatures >1400°C. Similarly, SiC is more stable and oxidation resistant at extremely high temperatures. It is, in general, a purer material and not as affected by impurities or secondary phases as is Si_3N_4. SiC has extremely good creep strength, which coupled with its good oxidation resistance and microstructural stability make SiC a likely candidate for very long time high temperature applications. SiC has much higher thermal conductivity and thermal diffusivity compared to Si_3N_4. This means that SiC would be a good candidate for a combustor in a gas turbine engine, but would not be a good candidate as a cast iron diesel engine cylinder liner. The major disadvantages of SiC when directly compared to Si_3N_4 are the lower fracture toughness and lower thermal shock resistance for SiC. The low toughness of SiC is due to its low critical stress intensity factor and low fracture surface energy. Poor thermal shock resistance is often cited as the most critical difference between silicon carbide and silicon nitride. The low thermal shock resistance of SiC is due to the combination of its higher thermal expansion and higher elastic modulus in comparison to Si_3N_4. These properties are more or less inherent, and cannot be modified to any meaningful extent by varying composition or processing method. Thus SiC and Si_3N_4 are unique as engineering materials as are the various heat engine component designs for which they are being used.

ACKNOWLEDGEMENT

This work was performed at IIT Research Institute for the Air Force under AFWAL contracts F33615-75-C-5196 and F33615-79-C-5100. We are indebted to all organizations who have supplied test samples to these programs.

REFERENCES

1. D. C. Larsen, J. W. Adams, S. A. Bortz, and R. Ruh, "Evidence of Strength Degradation by Subcritical Crack Growth in Si_3N_4 and SiC," International Symposium on Fracture Mechanics of Ceramics, July 15-17, 1981, Penn State University.

2. D. C. Larsen and J. W. Adams, "Property Screening and Evaluation of Ceramic Turbine Materials," IITRI Semiannual Interim Report No. 10 on AFWAL Contract No. F33615-79-C-5100 (April 1981).

3. G. G. Trantina and C. A. Johnson, "Subcritical Crack Growth in Boron-Doped SiC," J. Amer. Ceram. Soc. 58 (7-8), pp. 344-345 (1975).

4. A. G. Evans and S. M. Wiederhorn, "Crack Propagation and Failure Prediction in Silicon Nitride at Elevated Temperatures," J. Mater. Sci., 9 (2), pp. 270-278, (1974).

5. D. C. Larsen, "Property Screening and Evaluation of Ceramic Turbine Engine Materials," AFML-TR-79-4188 (October 1979).

6. M. Srinivasan, R. H. Smoak, and J. A. Coppola, "Static Fatigue Resistance of Sintered Alpha SiC," Paper 10-C-79C presented at the American Ceramic Society 3rd Annual Conference on Composites and Advanced Materials, Merritt Island, Florida, January 21-24, 1979.

7. K. D. McHenry and R. E. Tressler, "Subcritical Crack Growth in Silicon Carbide," J. Mater. Sci. 12, pp. 1272-1278 (1977).

8. T. L. Francis and R. L. Coble, "Creep of Polycrystalline Silicon Carbide," J. Amer. Ceram. Soc., 51 (2), pp. 115-116 (1968).

9. P. L. Farnsworth, R. L. Coble, "Deformation Behavior of Dense Polycrystalline SiC," J. Amer. Ceram. Soc., 49 (5), pp. 264-268 (1966).

10. D. P. H. Hasselman, "Thermal Stress Resistance Parameters for Brittle Refractory Ceramics: A Compendium," Bull. Amer. Ceram. Soc., 49, pp. 1033-1037 (1970).

Fabrication of Complex Shaped Ceramic Articles by Slip Casting and Injection Molding

J. A. Mangels

Ceramic Materials Department
Research
Ford Motor Company
Dearborn, Michigan 48121

Abstract

Complex shaped ceramic articles can routinely be formed to net shape by either slip casting or injection molding. Both of these processes are reviewed in general terms. The effect of both material and process variables are discussed. Case studies are presented to illustrate important features of each process.

Summary

Complex shaped ceramic articles, such as turbine rotors or stators, can be made to near net shape by either slip casting or injection molding.

Slip Casting

The general techniques for slip casting of metal powders (silicon) are essentially the same as for the slip casting of oxides. However, reactions between the metal powder and the casting vehicle (water) are possible and must be considered. Before slips can be used they must be "aged" until any surface reactions have been completed. Deflocculants must be compatible with the reacted powder surface.

The behavior of metal powder slips and the relation of slip parameters (pH, viscosity, specific gravity) to casting properties (casting time, green density, green shrinkage) are similar to what would be expected for oxide or clay slips. The particle size distribution of the powder can also effect the green density of the casting.

Complex shaped components, like those shown in Figure 1, can be produced using the Ford-patented fugitive wax slip casting process. This

Riley, F.L. (ed.) Progress in Nitrogen Ceramics

process begins with the fabrication of a positive model of the component using a water soluble wax. The wax slip casting mold is made using a water insoluble/organic soluble wax and is formed by repeated dipping of the positive model into liquid wax. The positive model is then removed by dissolving in water. The wax mold is placed on a plaster mold, which draws the water out of the slip, resulting in a undirectional casting. After the casting has solidified, the wax mold is removed by dissolving in an appropriate organic solvent. The principal advantage of the fugitive wax slip casting process, is the relatively low cost for tooling to produce the positive wax model, which makes this process ideal for prototype hardware development.

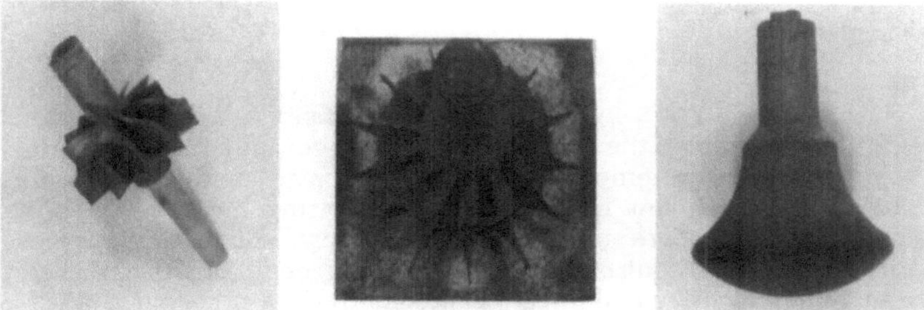

Figure 1 Components Fabricated Using the Fugitive Wax Slip Casting Method.

Injection Molding

Injection molding is a process intended for high volume production of complex shaped articles, such as those shown in Figure 2. Except for various material considerations, the injection molding of ceramics is generally the same as the conventional injection molding process used for plastics.

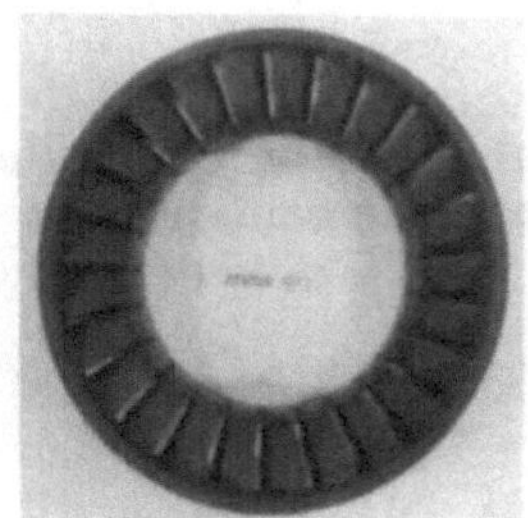

Figure 2 Components Fabricated Using Injection Molding.

The materials used in an injection molding composition (ceramic powder and polymer vehicle) are critical to the success of the process. The rheological properties of the injection molding material is governed by the ceramic powder characteristics as well as the volume fraction of powder (V_p) in the mix. The green density (ρ green) of a molded article can be expressed as

$$\rho \text{ green} = \left(\frac{V_P}{V_{Total}}\right) \times S.\,G\,P).$$

where S.G. (P) is the specific gravity of the ceramic powder.

The relative viscosity (η_r) of the molding material is also a function of V_p;

$$\eta_r = (1-V_p)^{-k}$$

where $k \simeq 5$. However the actual viscosity of ceramic molding materials is a function of both temperature and shear rate, and these materials often behave as Bingham fluids, that is exhibiting a yield stress which must be exceeded before the material will flow. Successful processing therefore generally requires a reduction in green density in order to reduce molding viscosity and improve moldability.

Binder removal, which must be accomplished before the part is fired, is governed by the polymer composition. Consequently the selection of the polymer vehicle depends not only on its compatability with the ceramic powder with respect to its injection moldability, but also on its "burnout" characteristics.

With plastics, component quality is related to material conditions existing in the die cavity. Four fundamental injection molding variables were identified; material temperature, flow rate (shear rate), cavity pressure and cooling rates. These variables can be affected by the various machine variables as well as the tooling design. It should be cautioned that changes in one machine variable can effect more than one of the fundamental variables simultaneously. Consequently measurement of conditions existing in the cavity, together with careful analysis are required to effect improvements in part quality.

Bibliography

Slip Casting:

R. E. Cowan, "Slip Casting," Treatise on Materials Science and Technology, F. F. Y. Wang, editor, Academic Press, N.Y., 1976, p. 253.

E. F. Adams, "Slip Cast Ceramics," High Temperature Oxides Part IV, Academic Press, N.Y., 1971, p. 145.

P. D. S. St. Pierre, "Slip Casting Non-Clay Ceramics," General Electric Laboratory Report No. 57-RL-1752, May, 1957.

P. D. S. St. Pierre, "Slip Casting Metal Powders: Molybdenum," General Electric Laboratory Report No. 58-RL-2058, Sept., 1958.

P. D. S. St. Pierre, "Slip Casting Metal Powders," General Electric Laboratory Report No. 59-RL-2281M, Oct., 1959.

A. Ezis, "The Fabrication and Properties of Slip Cast Silicon Nitride," *Ceramics for High Performance Applications*, J. J. Burke, R. N. Katz, A. E. Gorum, editors, Brook Hill Publishing Company, Chestnut Hill, Mass, 1974, p. 207.

A. Ezis, J. M. Nicholson, "Method of Manufacturing a Slip Cast Article," U.S. Patent 4,067,943, Jan., 1978.

E. A. Fisher, A. F. McLean, "Brittle Materials Design, High Temperature Gas Turbine," Contract No. DAAG 46-71-C-0162, Final Report, to be published in 1981.

Injection Molding:

I. I. Rubin, *Injection Molding - Theory and Practice*, John Wiley and Sons, N.Y., 1972.

K. Schwartzwalder, "Injection Molding of Ceramic Materials," Bull. Amer. Ceram. Soc. Vol. 28, No. 11, 1949, p. 459.

R. Westlake, "Injection Molding Complex Ceramics," Presented at the 71st Annual Meeting of the American Ceramic Society, 1969.

D. L. Mann, "Injection Molding of Sinterable Silicon-Base Nonoxide Ceramics," AFML-TR-78-200, 1978.

R. A. Giddings and C. A. Johnson, "Investigation of Sinterable Silicon Carbide for High Temperature Turbine Components," Report No. NADC-77096-30, 1980.

E. A. Fisher, A. F. McLean, "Brittle Materials Design, High Temperature Gas Turbine," Contract No. DAAG 46-71-C-0162, Final Report, to be published in 1981.

T. J. Whalen, C. F. Johnson, "Injection Molding of Ceramics," Bull. Amer. Ceram. Soc., Vol. 60, No. 2, 1981, p. 216.

J. A. Mangels, "Development of Injection Molded Reaction Bonded Si_3N_4," *Ceramics for High Performance Applications-II*, J. J. Burke, E. N. Lenoe, R. N. Katz, editors, Brook-Hill Publishing Company, Chestnut Hill, Mass, 1978, p. 113.

J. A. Mangels, "Development of a Moldable, High Density, Reaction Bonded Silicon Nitride," DOE/NASA Contract No. DEN 3-20, Final Report, to be published in 1981.

R. J. Farris, "Prediction of the Viscosity of Multimodal Suspensions from Unimodal Viscosity Data," Trans. of the Soc. Rheology, Vol. 12, No. 2, 1968, p. 281.

J. L. White, H. B. Dee, "Flow Visualization for Injection Molding of Polyethylene and Polystyrene Melts," Polymer Engineering and Science, Vol. 14, No. 3, 1974, p. 212.

DISCUSSION

Quackenbush: What is the maximum volume percentage of ceramic which can be molded into a complex shaped cavity?

Mangels: 73-76 $^{v}/o$ solids should be capable of processing into complex shape. The molding technology is transferable between shapes, but specific conditions must be experimentally optimized.

Gugel: Injection moulding is considered to be the most economical method for mass production of a low priced component - but is this true when we see the difficulties in removing the organic binder? Imagine what a furnace one would need for mass production when two weeks are needed for this production step!

Mangels: There are still problems, but it is the task of the production people to make this process economic.

HOT ISOSTATIC PRESSING OF CERAMICS

Hans T Larker
ASEA AB
High Pressure Laboratory
S-915 00 Robertsfors, Sweden

SUMMARY

The characteristic of hot isostatic pressing (HIP) to yield virtually theoretically dense products from various powders at a comparatively low temperature, often 50-70 % of their melting point, is very valuable in the manufacture of advanced ceramics. Strongly temperature dependent phenomena like grain growth and dissociation, or reactions in multi-phase systems, can thus be avoided. The ability to HIP parts to near net shape is particularly important for hard ceramics because of high machining costs. Different alternatives for applying HIP to make shaped ceramic parts are discussed.

1. INTRODUCTION

Hot Isostatic Pressing (HIP) is a process utilizing isostatic (triaxial) pressure to assist sintering (1). Products with virtually theoretical density can usually be made at a comparatively low temperature. This is particularly advantageous for materials for which undesired reactions or changes occur at higher temperatures, e g grain growth, growth of secondary phases, reactions between different components of dispersion strengthened or fibre reinforced materials and dissociation or decomposition.

For materials of very high melting point a lowered processing temperature can in itself be very advantageous. The commercially utilized ability of HIP to drastically reduce porosity and related types of defects in structural parts is particularly important for materials with low K_{IC}-values.

Riley, F.L. (ed.) Progress in Nitrogen Ceramics

The ability to make shaped parts, which is inherent in the process, has up to now found limited production use and only for shaped blanks. The utilization of this inherent feature is particularly attractive for difficult-to-machine materials like hard ceramics.

2. BASIC TECHNIQUE

The material to be treated is placed in what can be characterized as a high pressure furnace. The basic difference to the well-known vacuum furnace is the pressure of the gas in the work zone of the furnace. Typically vacuum furnaces operate at gas pressures seven to ten orders of magnitude lower than atmospheric and pressure furnaces for HIP three orders of magnitude above atmospheric pressure. In both cases the pressure difference to ambient is taken over a vessel wall cooled by water (or air). Heating is made inside an insulation tailored to be efficient under the conditions at the prevailing pressure level.

Even if the equipment used has basic similarities and the work-pieces in both the vacuum and the high pressure processes can be stacked to fully utilize the available volume in the work zone, there are fundamental differences in the processing. In contrast to vacuum sintering, in HIP the interior of a body to be densified must be sealed off and the high pressure gas, usually argon, not allowed to penetrate into it. Apart from making the isostatic pressing action on the body by the gas possible, this also means isolation of the interior of the body. Virtually no exchange of gases from the body to the high pressure gas takes place, an exemption being hydrogen isotopes through metal containment at high temperature. This is generally an advantage and particularly so for decomposing materials. However, processes like desoxydation of SiC by excess C forming volatile CO must be carried out before the body is isolated from the environment.

3. HIP METHODS FOR SHAPED CERAMIC PARTS

3.1 Part Shape Determined by HIP Containment

The conventional way to make shaped parts is to use a container of sheet metal or glass with an interior shape. Powder of the desired composition is filled into the cavity to as high fill density as possible, e g using vibration. Outgassing of the powder is done, usually at elevated temperature, through an evacuation tube which is then crimped and sealed by welding or melting. During the initial stages of the subsequent HIP cycle the shape of the powder mass is determined by the container. When the pressure acting on the outside of the container increases and forces the powder mass to shrink, the latter gradually becomes

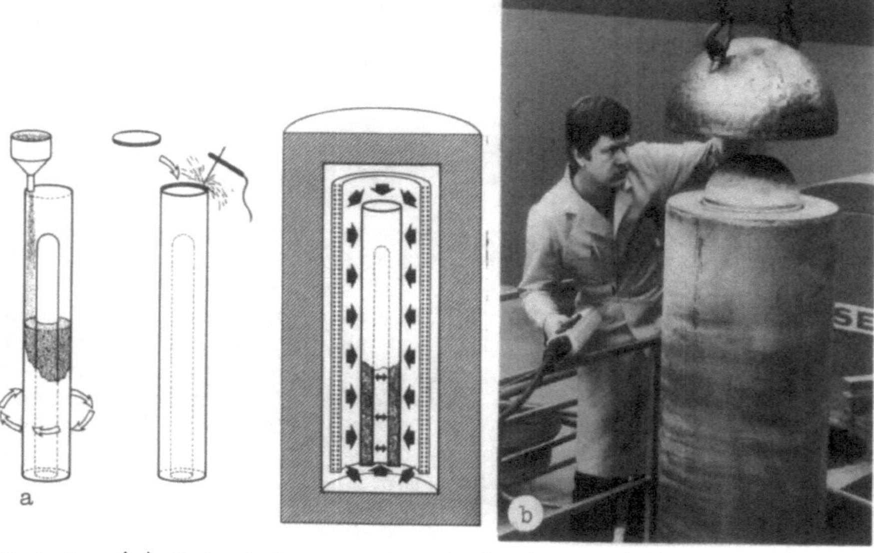

Fig. 1. (a) Principle for HIP of alumina containers.
(b) Assembling of HIPed alumina container and lid for joining by HIP bonding.

more rigid and takes over the control of the shape. Distorsion of the shape before this happens is however a problem, because the container wall, which is of dense material from the beginning, must be plastically deformed in such a way that its surface decreases and its thickness increases. The resistance of the container wall to this deformation results in distorsion such that the total surface of the part pressed from powder will be larger than if no distorsion took place. Uniform and high fill density of the powder (which is difficult to obtain with fine ceramic powders) and low flow stress of the container wall material during initial deformation can reduce this problem.

This method was e g used for the full scale feasibility study of making high purity, high density alumina canisters for safe long term containment of spent nuclear fuel which was carried out at ASEA a couple of years ago (2). Mechanical tamping (Fig. 1. a) was used to obtain a high and uniform fill density in the annular cavity of the mild steel container. Good shape control was obtained in spite of the reduction of height from 3 m to 2.5 m, OD from 600 mm to 500 mm and the increase of the mild steel container wall from 3.0 to 4.3 mm. Lids were made by a similar method and could (Fig. 1. b) be joined to the bottomed cylinder by a HIP bonding technique after cutting the surfaces to be joined clean by diamond grinding.

3.2 Part Shape Determined by the Green Powder Body

Processes based on this generic principle can give much better shape accuracy, particularly of complicated parts, than methods based on container geometry. Such a process was as well as HIP of silicon nitride pioneered (3) and has been further developed by ASEA (4, 5). Fig. 2 illustrates several alternative techniques.

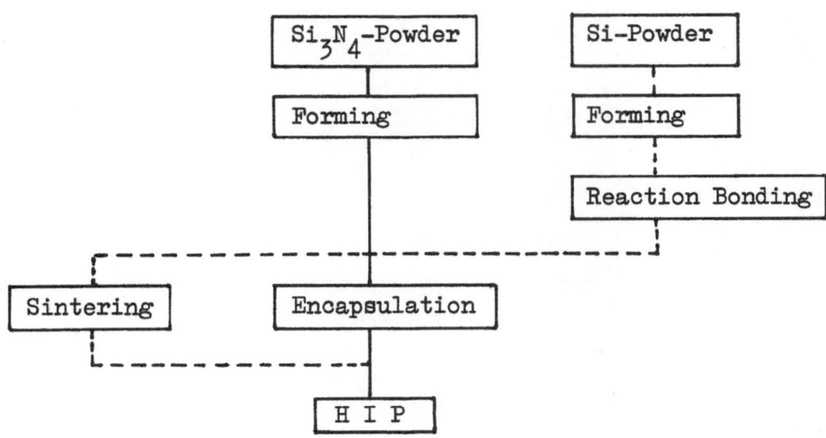

Fig. 2. Four alternative routes to manufacture dense silicon nitride parts from shaped green powder bodies.

The first method illustrated by the vertical line at the center of the figure has been particularly developed at ASEA. A green body is formed of silicon nitride powder with additives by any appropriate method e g cold isostatic pressing followed by machining, injection molding, extrusion, slip casting etc. Good shape precision and high repeatability of the density distribution in the green body is required for narrow tolerances in the finished part. Forming additives like plasticizers and binders are removed before encapsulation in glass is made.

An early method for encapsulation which can be useful for making material samples and small products of less complicated shape is illustrated in Fig. 3. A cold isostatically pressed blank of silicon nitride powder with additives, e g 1 % MgO or 5 % Y_2O_3, is placed in an over-sized ampoule of high silica glass which is hot evacuated and sealed. The ampoule is heated to a soft state before applying the pressure. It then folds over and conforms to the shape of the green powder body. After HIP processing at predetermined temperature and pressure the part is cooled and the glass which is not removed during the process is taken off by

Fig. 3. Glass ampoule encapsulation for silicon nitride.

Fig. 4. As-HIPed silicon nitride wheels (over and right) made by glass particle encapsulation of injection molded green body (left).

sand blasting. Good surface finish and accurate shape can be obtained with this simple technique but it is hardly suited for mass production. The stress on protruding parts of the green body by the viscous glass during the folding over the body limits the application to parts of simple shape. Filling the ampoule with a pressure transmitting powder around the green body can be used for some shapes but precision is adversely affected.

A further development of the same principle capable of handling weak protruding sections of the green body and being suitable for automated handling and mass production has however been developed (cf. Ref. 4. p 18-3f). The glass envelope around the green body is formed from glass particles applied on all surfaces of the body. Before sealing the glass envelope, hot evacuation is carried out. The temperature is then raised to seal the glass. This is at present made in the high pressure furnace but a production line could use a vacuum furnace until the encapsulation is sealed and use hot transfer of the workload to the high pressure furnace, thus minimizing the time in the high pressure equipment.

After a sustain time needed for full densification at selected pressure and temperature, these are decreased. A special processing schedule during the cooling facilitates removing the glass envelope but sand blasting is used for the final cleaning of the products.

Products with thin (0.3 mm) edges and sharp corners, internal cavities and protruding members on a thicker part e g integrated turbine wheels (Fig. 4) can be made with this technique.

A second alternative processing route starting from silicon nitride powder is sintering (Fig 2 left) until no interconnecting porosity is left. This is usually the case above 95 % of TD. The partly densified parts can then be HIPed to higher density without further encapsulation. This basic process has been used in production for over ten years for eliminating porosity in large cemented carbide parts (Ref. 1 p. 334). Pores in contact with the surface will however not be closed but have to be machined off.

A third method starts with forming a green body of silicon powder with sintering aids which is nitrided (Fig. 2 right). It is then encapsulated and HIPed as in the first method. Heinrich and Böhmer (6) have used this technique.

A fourth alternative as suggested by Giachello, Martinengo, Tommassini and Popper (Ref. 7 p. 1214) is to sinter RBSN to high density to avoid encapsulation during HIP.

The first alternative offers full possibilities to process material with low amounts or with compositions of sintering aids which do not allow good densification with other techniques. The gas impervious envelope which is sealed at a relatively low temperature prevents decomposition and weight loss. It also prevents undesirable reactions with the silicon nitride part of impurities in the gas used as pressure medium.

A typical magnitude of the driving force in sintering (the surface tension divided by the neck radius) is about 5 MPa. The applied isostatic pressure is typically more than one order of magnitude higher which is important not only for the densification of the material. It also substantially reduces the influence of gravity on distorsion during densification of intricately shaped parts which are difficult to support.

The second and fourth alternative both require amounts and compositions of sintering aids, usually mixtures of several oxides, which allow sintering to above about 95 % TD. Even with the use of nitrogen gas over-pressure and/or powder beds there is usually a significant weight loss which effects composition and tolerances of the part. Powder beds restrict the shrinkage of shaped parts (particularly with recesses) which leads to distorsion.

The third alternative (as well as the fourth) has the disadvantage of the long processing time needed for nitridation. It has the advantage of a very low shrinkage from green body to fully dense part, 5-10 % linearly, the higher values for thick parts which require lower density for complete nitridation. In work at ASEA according to the first alternative 12-13 % is typical for injection molded parts, while a linear shrinkage of 20 % or more is often reported in direct sintering.

4. PRODUCT PROPERTIES

The properties of ceramics produced by HIP are as well as every alternative process dependent on factors like raw material properties and the processing steps before and after HIP. Several authors (4, 6, 8) have given examples of very good material properties. It is generally correct to state that at least as good properties as are obtained in uniaxial hot pressing can be reached with HIP of the same starting material. The inherent effect of healing voids and fissures is particularly important in large structural components.

Tolerances similar to products made by investment casting seem to be obtainable by injection molding, encapsulation with glass particles and HIP. The surface finish is usually quite satisfactory.

5. PRODUCTION POSSIBILITIES AND ECONOMICS

Such critical processing steps, which differ from alternative processes, e g encapsulation and HIPing according to the first alternative in chapter 3.2, appear now to be technically feasible also in production. The process is comparatively easy to scale up for mass production. After the green forming process, e g injection molding, and placing the objects on a support structure, a line production with a minimum of manual handling can be utilized. Production cost estimates indicate that an acceptable level can be achieved for parts like integrated turbine wheels with blades for passenger car gas turbines.

6. CONCLUDING REMARKS

HIP is a very versatile process suitable also for some ceramic materials, particularly for making complicated and highly stressed parts. It can generally be applied as a post-process for reduction of defects in products made by other techniques. It can however also with advantage be used for direct production of accurately shaped parts starting from formed green bodies of silicon nitride or silicon powder.

REFERENCES

1. H.T. Larker. "Hot Isostatic Pressing - Characteristics and Prospects in Industrial Use", in High Pressure Science and Technology, Vol. 3, p. 329-337, ed. B. Vodar and Ph. Marteau, Pergamon Press (1980).
2. H.T.Larker, in Ceramics for Nuclear Waste Management, p. 169-173, ed. T.D. Chikalla and J.E. Mendel (1979), NTIS, Springfield, Va 22161, U.S.A.
3. J. Adlerborn, H.T. Larker. "Method of Manufacturing Bodies of Silicon Nitride", British Patent No. 1522 705, priority date 1974-11-11.
4. H.T. Larker. "Hot Isostatic Pressing of Silicon Nitride Parts", in High Pressure Science and Technology, p.651-655, ed. K.D. Timmerhaus and M.S. Barber, Plenum Publ. Corp., New York (1979).
5. H.T. Larker. "HIP Silicon Nitride", AGARD CP-276, p. 18-1 f.
6. J. Heinrich and M. Böhmer. "Microstructure and Mechanical Properties of Hot Isostatic Silicon Nitride", to be published in Science of Ceramics, Vol. 11.
7. A. Giachello, P.C. Martinengo, G. Tommasini and P. Popper. "Sintering and properties of silicon nitride containing Y_2O_3 and MgO", Cer. Bull. 59 No. 12, p. 1212-1215 (1980).
8. R.R. Wills and M.C. Brockway. "Hot Isostatic Pressing of Ceramics", Special Ceramics 7 (1981).

DISCUSSION

Popper: In cladless hipping and in the two-pressure method, nitrogen might be used in conjunction with a graphite heater. Do you foresee any problems?

Larker: Metal thermocouples do not behave well in high pressure nitrogen at high temperatures. We have not found cyanogen formation to be a problem.

Richerson: Is there any chemical interaction between the glass encapsulation and the silicon nitride?

Larker: No significant glass penetration into the Si_3N_4 body is found. The Si_2N_2O concentration is often higher close to the surface. No barrier layers between the glass and the Si_3N_4 are applied.

Katz: What kind of Weibull moduli do you obtain?

Larker: Hipping is probably an optimal densification method (controlled grain growth, defect healing) for obtaining high Weibull moduli. A modulus of 40 has been obtained but 15-20 is more typical for common starting materials.

Section J

NATIONAL RESEARCH PROGRAMMES

US NATIONAL PROGRAMS IN CERAMICS FOR ENERGY CONVERSION

R. Nathan Katz
Army Materials & Mechanics Research Center,
Watertown, MA 02172
and
Robert B. Schulz
Department of Energy, Washington, D.C. 20585.

INTRODUCTION

There are many potential high temperature applications for nitrogen ceramics as shown in Fig. 1. Our paper will primarily focus on ceramics for advanced heat engine applications. Other papers at this Advanced Study Institute (ASI) such as the one by Lumby will deal with other developing applications such as tool bits and high temperature dies. Clearly, the driving force for the major efforts in the United States over the past decade on the application of ceramics to energy conversion has been the need to lessen our dependence on imported oil and imported strategic metals. It has been previously pointed out that full implementation of ceramic configured gas turbines and industrial heat exchangers could save $17.5 billion in oil imports [1].

Fig. 2 shows the current major U.S. government and privately supported ceramic applications programs. In comparison to similar listings of U.S. programs five years ago [2], there are two important differences. First, the significant increases have occurred in programs aimed at civilian as opposed to military applications. This has been in response to the world energy crisis as represented by oil shortages and much higher fuel prices. Ceramics used in heat engine applications offer the potential of reduced petroleum consumption through improvements in conversion efficiency, and their use in alternative engines such as the gas turbine may facilitate switching to domestically available nonpetroleum fuels, such as coal-derived fuels. The greatest impact of ceramics on energy conservation can be made in the civilian area because of its much greater energy consumption compared to the military. The military can be expected to continue using strategic and other high-cost materials because of their higher priority for performance over

Riley, F.L. (ed.) Progress in Nitrogen Ceramics

POTENTIAL HIGH TEMPERATURE APPLICATIONS OF N_2 CERAMICS

HEAT ENGINES

- GAS TURBINE COMPONENTS
- DIESEL AND ADIABATIC DIESEL COMPONENTS
- TURBOCHARGERS
- BEARINGS
- HEAT EXCHANGERS

INDUSTRIAL ENERGY CONVERSION

- INDUSTRIAL HEAT EXCHANGERS
- BURNER VAPORIZORS
- COMBINED BURNER/HEAT EXCHANGERS
- HOT GAS VALVING FOR COAL GASIFICATION
- Si RIBBON DRAWING DIES

METAL WORKING

- TOOL BITS
- DIES

MILITARY

- RADOMES
- GUN BARREL LINERS

DEFINITION: HIGH TEMPERATURE = MATERIALS OPERATING TEMPERATURE IN EXCESS OF CURRENT PRACTICE.

Figure 1

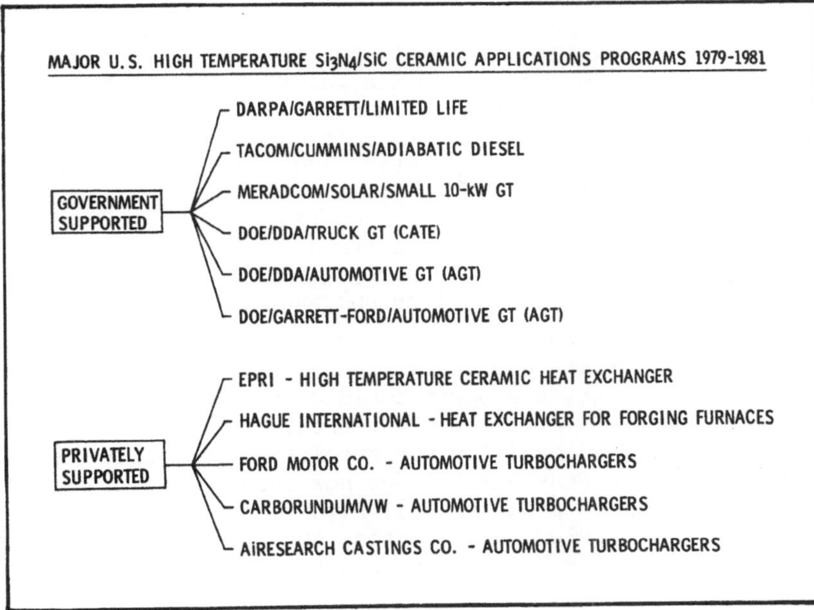

Figure 2

cost. However, military interest in ceramics continues to grow because of the demonstrated performance improvements, and military applications should follow as ceramic reliability questions are resolved. The second difference is that private industry is also funding more programs themselves, because they see the opportunities growing for commercial applications of ceramics. While the stimulus of government-funded programs has helped bring this about, the present trends do not eliminate the need for continued government support for long-term and high-risk, but high-payoff programs. Indeed the private programs are not addressing such high-risk areas.

The balance of this paper will look at the past experience in the application of ceramics to gas turbine engines, with particular emphasis on the lessons learned in various major programs. Next, we briefly review what is happening today in various gas turbine and diesel programs. And finally we will try to project into the future taking into account the economic policies of the new Administration.

Let us now briefly review the history of the major U.S. programs on gas turbine ceramics over the past decade.

U.S. CERAMIC CONFIGURED GAS TURBINE PROGRAMS, 1971-1980

As shown in Fig. 3, there has been a progression of four major U.S. government-funded ceramic configured gas turbine programs over the past ten years. Since these major programs were sponsored by different agencies (or groups of agencies), and since each subsequent program was initiated before the conclusion of the preceding one (or ones), it might have seemed to the casual observer that there was little coordination or technology transfer among these programs. In fact quite the opposite was the case. Each program, as indicated in Fig. 3, had its own major question to answer. In general, the answer to this major question was obtained or understood prior to the program completion and thus it was logical to initiate a program to address the next major issue as soon as possible. The DOE Advanced Gas Turbine (AGT) and Ceramic Applications in Turbine Engines (CATE) industrial contract team members and their ceramic suppliers certainly gained valuable experience and capabilities under the earlier DoD and DOE-funded programs.

The DARPA/Ford Program [3] was to establish that design with ceramics in gas turbine engines was feasible. Once this was sufficiently demonstrated, a program could be structured to address performance issues (i.e. the DARPA/Garrett Program [4]). Questions of reliability and suitability for automotive use (both performance and mass production feasibility) are now being addressed in the various DOE-supported programs [5]. Figs. 4 and 5 summarize the key accomplishments, problems highlighted, and issues left as open questions on several major U.S. government-sponsored programs through 1980.

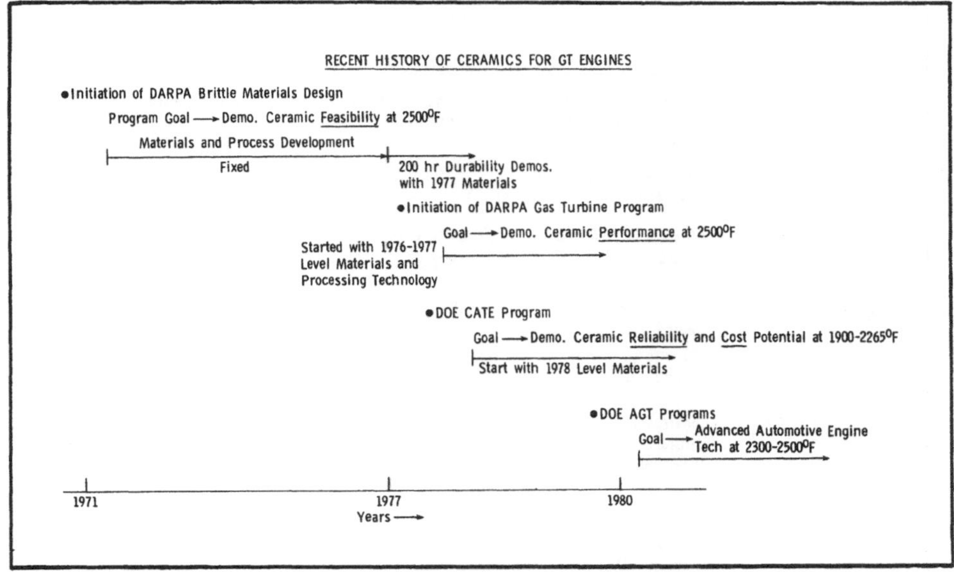

Figure 3

Coordination among government funding agencies was insured through a variety of means, among which the most prominent has been the Interagency Coordination Group for the Application of Ceramics to Turbine Engines [6]. Industry/government agency/university coordination and inter-industry technology transfer has been fostered by meetings held under Army [7], Navy [8], and DOE [9] auspices. Valuable international cooperation has occurred through the International Energy Agency (IEA) and NATO.

U.S. CERAMIC CONFIGURED HEAT ENGINE PROGRAMS - 1981

Gas Turbines:

In the spring of 1981, a major milestone in the progression of ceramic gas turbine technology was attained on an extension of the DARPA/Garrett Program. A ceramic bladed axial rotor successfully completed two 7-1/2-hour full-performance engine tests, with challenging duty cycle requirements. The rotors survived intact and the engine with the ceramic blades yielded higher power and lower specific fuel consumption than an engine with cooled metal blades operating at the same turbine inlet temperature (2200°F) [10]. This outstanding achievement validates that the performance benefits long promised by ceramics are indeed obtainable with components that can survive in an engine.

The results achieved this Spring in automotive ceramic rotor technology have been equally dramatic. Fig. 6 recounts some of the recent breakthroughs toward automotive and, in particular, small

CERAMIC STATOR TECHNOLOGY

Program	Accomplishment	Problems Highlighted	Open Questions
DARPA/Ford	• 200 hr One Piece RBSN + S SiC Stators ≈1930–2500+ °F • Significant Progress in Manufacturing	• Oxidation of RBSN (Internal)	• Reliability • Durability
DOE/NASA/Ford (cont. of above program)	• 500 hr and 30,000 Thermal Shocks ≈2200°F	• Some Cracking at ≈500 hr	• Reliability • Durability
DARPA/Garrett	~8-10 hr of Assembled Stator • Δ + 215 hp Δ - 0.04 lb/hp-hr	• Attachment and Contact Stresses	• Reliability • Durability
MERADCOM/Solar	250 hr ≈1700°F SiC Shrouds, HPSN Vanes 10 kW Power Output	• Need to Keep All Hot Section Components within Their Specified Envelopes	• Reliability • Cost
DOE/CATE	• 890 hr ≈1900°F Assembled SiC Stator in Engine Vanes Rig Qualified to 2070°F		• Reliability • Durability • Cost

Figure 4

CERAMIC ROTOR TECHNOLOGY

Program	Accomplishment	Problems Highlighted	Open Questions
DARPA/Ford	• 200 hr All Ceramic Rotor ≈2200°F • 36.5 hr All Ceramic Rotor ≈2500+ °F [Ceramics are Feasible]	• Ceramic/Metal Attachment • High Temperature Lube for Curvic Coupling • Need to Prevent Over-temperature on Attachment	• Reliability • Durability • Cost
DARPA Garrett	• ~8 hr Ceramic Bladed Rotor • Integration of 102 Ceramic Components in a Structure • Δ + 215 hp Δ - 0.04 lb/hp-hr [Ceramics can Deliver Increased Performance] [Inserted Blade Rotors with SOA Materials Work at 2200°F Average TIT]	• Contact Stresses in Stationary Multi-Component Structures	• Reliability • Durability • Cost

Figure 5

DOE/NASA AGT PROGRAM

BREAKING THROUGH IN CERAMICS

- ON MAY 20, 1981, FORD COLD-SPUN A SRBSN SIMULATED AGT ROTOR TO 95,000 RPM BEFORE FAILURE
- ON MAY 29, 1981, GARRETT COLD-SPUN AN ACC SSN SIMULATED AGT ROTOR TO 115,000 RPM WITH NO FAIULRE
- ON JUNE 1, 1981, DDA HOT-SPUN A CATE αSiC-BLADED METAL ROTOR TO 37,000 RPM, 100% SPEED, AT 1720°F WITH NO FAILURE
- ON JUNE 3, 1981, DDA COLD-SPUN A CBO αSiC FULLY-BLADED AGT ROTOR TO 109,000 RPM WITH NO FAILURE
- ON JULY 7, 1981, FORD COLD-SPUN A SRBSN SIMULATED AGT ROTOR TO 134,000 RPM WITH NO FAILURE - CENTERLINE STRESS LEVEL EQUIVALENT TO TURBINE OPERATING AT 3080 ft/sec TIP SPEED

Figure 6 (Courtesy of R. Ragsdale, LERC)

radial rotors, which are coming out of the DOE/NASA Programs. These spin tests represent a breakthrough for the AGT Program, because they demonstrate that ceramics fabricated by net shape techniques have the required strength levels in the thick hub cross sections of a radial rotor. The tests show the feasibility of the more highly stressed single-stage radial rotor in a small single-shaft engine. The AGT Program rotor accomplishments are also overtaking the CATE Program accomplishment of hot spinning ceramic bladed/metal hub wheels, because of the better cost and efficiency potential of the one-piece bladed radial rotors in single- and two-stage turbine engines. The encouraging results obtained for as-fabricated rotors spin tested at room temperature now need to be validated by high temperature spin and durability testing.

Diesel Engine Technology:

The U.S. Army Tank and Automotive Research and Development Command (TARADCOM), together with the Cummins Engine Company, have been developing an adiabatic-turbo-compound diesel engine for the past five years. Fig. 7 gives a summary of the status of that program. As far as we are aware, the 0.285 specific fuel consumption measured in a test at Cummins represents the highest level of fuel efficiency ever demonstrated for any experimental vehicular engine. This is a major accomplishment [11].

While the first demonstration of the adiabatic engine utilized HPSN piston caps, hot pressed to near net shape by Gazza and coworkers at AMMRC, focus on materials development is now shifting to transformation-toughened zirconia as noted in the introductory paper.

ARMY'S ADIABATIC TURBO COMPOUND DIESEL ENGINE PROGRAM

DEVELOPMENT TEAM:

- TACOM/CUMMINS
- CERAMIC MATERIALS CONSULTATION, AMMRC

GOAL:

- 0.280 lb/hp-hr, SFC

DEMONSTRATED ACCOMPLISHMENTS:

- 250 hr FULL PERFORMANCE DEMO OF CERAMIC PISTON CAP IN SINGLE CYLINDER ENGINE
- 250 hr OF SINGLE CYLINDER ADIABATIC OPERATION
- 0.285 lb/hp-hr, SFC IN MULTI-CYLINDER ENGINE TESTS

LOWEST SFC OF ANY VEHICULAR ENGINE

Figure 7

However, the use of the Si_3N_4 cannot be ruled out and indeed it is being actively developed for piston caps and other diesel components, particularly in Japan [12].

Turbochargers:

While most turbocharger work is being privately funded, recent publications by a Carborundum/VW team and Ford [13,14] indicate significant progress as summarized in Fig. 8. The outstanding feature of the ceramic turbocharger is the reduction of the rotating assembly response lag and the opportunity to reduce the weight of the housing due to reduced burst containment requirements.

With this summary of past and current progress, let us now turn to consideration of the future.

CERAMIC TURBOCHARGER TECHNOLOGY

INJECTION MOLDED α-SiC - Currently Under Road T&E in a 1.6L VW Dasher Diesel at ~950°C T_{max}

SLIP-CAST RBSN - Currently Under Road T&E in a 2.3L Ford Motor Co. Zephyr (Gasoline Engine) at ~860°C T_{max}

Figure 8

THE FUTURE OF U.S. NATIONAL PROGRAMS

The year 1981 has seen major changes in U.S. government policies as the result of our recent change in administration. There will be much less reliance on federal intervention in the private sector and more reliance on market forces to influence private investment decisions. The President's action to end oil price controls in the United States was a major step in implementing an energy policy focused on market forces. In keeping with the Administration's policy to significantly reduce federal spending, support has been withdrawn from programs at DOE where sufficient market incentives are believed to exist. Accordingly, in the DOE Automotive Technology Program there will be a transition in 1981/1982 from near-term commercialization and demonstration activities to emphasize long-term, high-risk but high-payoff generic research and development. However, at the same time, support exists within Congress for continuation of a revised AGT gas turbine "proof of concept" engine program with emphasis on the ceramics and component technologies. While the AGT and CATE programs may be cut back in 1982, the Administration does support a continued generic ceramics research program, with broader heat engine applications including diesel engines. Of course, U.S. government support of ceramics R & D for military applications will continue, and this support could increase in the future, especially if DOE funding is drastically reduced.

In conclusion, it is generally recognized that there are areas of high-risk, high-payoff research from which the U.S. economy will benefit, but at this stage they are still too risky for major commitments of private capital. Such areas will continue to receive federal R & D funding. It is widely acknowledged that ceramic heat engine components and materials is such an area. It has been decided that the demonstration of a ceramic configured engine in a civilian vehicle will not be supported. However, a significant program aimed at further development of ceramic materials and component technology will continue.

REFERENCES

1. Katz, R.N. "Science" 208, 23 May 1980, p. 841-847.
2. Katz, R.N. In Nitrogen Ceramics, ed. F.L. Riley, Noordhoff International Publishing, Leyden, 1977, p.646.
3. McLean, A.F. In Ceramics for High Performance Applications - II, eds. J.J. Burke, E.M. Lenoe and R.N. Katz, Brook Hill Publishing Co., Chestnut Hill, MA, 1978, p.1-34.
4. Wallace, F.B., Stone, A.J., and Nelson, N.R. Ibid p.593-624.
5. Automotive Technology Development, Contractors Coordination Meeting, Dept. of Energy, Office of Transportation Programs, DOE CONF 801182 (see p.61-193).
6. Report of the Committee on Structural Ceramics, National Materials Advisory Board, National Academy of Sciences, Report NMAB-32, 1975, p.27.

7. Ceramics for High Performance Applications - III, eds. E.M.Lenoe, R.N. Katz and J.J. Burke, Plenum Press, 1981.
8. Rice, R.W. and Fairbanks, J.W. Proceedings of the (1977) DARPA/NAVSEA Ceramic Gas Turbine Demonstration Engine Program Review MCIC Report 78-36, March 1978.
9. Workshop on Ceramics for Advanced Heat Engines, ERDA, January 1977, CONF-770110.
10. Personal Communication, Keith Johanson, Garrett Turbine Engine Co.
11. Brysik, W. See Ref. 5, p.269-277.
12. Hamano, Y. and Nakaharu, Y. 5th International Automotive Propulsion Systems Symposium, Proceedings DOE, Office of Transportation Programs, DOE CONF-8000419, April 1980, p.803-812.
13. Rottenkolber, P., Langer, M., Storm, R. and Frechette, F. "Design, Fabrication and Testing of an Experimental - SiC Turbocharter Rotor", SAE Technical Paper 810523, 1981.
14. DeBell, G.C. and Secord, J.R. "Development and Testing of a Ceramic Turbocharger Rotor", ASME Paper 81-GT-195, 1981.

STATUS REPORT 1981 ON THE GERMAN BMFT-SPONSORED PROGRAMME "CERAMIC COMPONENTS FOR VEHICULAR GAS TURBINES"

W. Bunk and M. Böhmer

DFVLR, German Aerospace Res. Est.,
Materials Laboratory, Köln.

"Ceramic components for vehicular gas turbines" - already many years ago, turbine manufacturers regarded this as a possible approach towards the economical use of uncooled small gas turbines in vehicles. But the first tests, performed with conventional ceramics (such as aluminium oxide), failed, mainly because of the ceramic's low thermal shock resistance. Only the advancement of non-oxide ceramics re-established the designers' optimism so that in the year 1974 the programme, a joint effort of the ceramics and automobile industries, was started. It took, however, a long time for the designers to get accustomed to the brittle materials "ceramics". But the ceramists, too, had to learn how to meet the requirements imposed by mechanical engineering.

Material-matched construction, improved testing methods, application of fracture mechanics and of statistical evaluation methods and the development of non-destructive test-methods have been very helpful in this process.

At a Status Seminar in the spring of 1978, the companies participating in the project had set themselves goals for the years until 1981. To what extent could these goals be realized, what difficulties were encountered in the past few years, what are the problems still existing today, what is the state-of-the-art in material and component development? These questions will be answered by the present Status Report.

The report is divided into the two key tasks of the programme, "material development" and "component development".

Riley, F.L. (ed.) Progress in Nitrogen Ceramics

1. Material development

When, two or three years ago, it became apparent that the material RBSN, then favoured for many turbine components, considerably lost its strength under long-term thermal loading (continuous and discontinuous), the effort as a consequence was more and more shifted to components production from silicon carbide. But a further consequence has also been the intensified development of the RBSN material in the ceramics industry leading to improved material characteristics and renewed RBSN component development. So the general statement may be placed at the beginning that the whole material family of the Si_3N_4 and SiC group continues to be of interest. In the following, an account will be given of the individual material advancements over the past few years. Fig. 1 gives an overview of which material has been developed by which company under the programme.

Company	Material	Forming Technology	Component
Annawerk	RBSN HPSN SiSiC SSiC	Slip Casting Injection Moulding Hot Pressing	Stator Inlet Cone Rotor Hubs
Degussa	RBSN	Injection Moulding	Blade Rings Single Blades
ESK	HPSiC	Hot Pressing	Blades, Hubs Rotor
Feldmühle	RBSN	Injection Moulding	Blades, Rings
Rosenthal	RBSN SiSiC	Foils Lamination Injection Moulding	Recuperator Blades Segment
Sigri	SiSiC	Injection Moulding Slip Casting	Combustor Inlet-Scroll
H.C. Starck	SiC Si_3N_4	Powder	-

Fig. 1: Overview about the tasks of the ceramic companies in the project "Ceramic Components for Vehicular Gas Turbines".

1.1 Silicon nitride

1.1.1 Reaction-bonded silicon nitride, RBSN

In the field of reaction-bonded Si_3N_4, materials development has been greatly advanced in the past few years by most of the companies involved, due to the reasons described above. As will be shown in the following, significant material improvements could be achieved.

Bend test specimens (3.5 x 4.5 x 45 mm) made from injection-molded RBSN with a room-temperature flexural strength of $\sigma_B \geq 300$ MPa at a density of 2580 kg m^{-3} did not show any significant decrease in strength even after 3000 hours of annealing at 900°C in air. This is a decisive step in the development of this material as, for the first time, realistic prospects exist for the applicability of the porous reaction-bonded silicon nitride as a long-term stable turbine blade material.

On bend test specimens, injection-molded simultaneously with rotor blades, σ_B-mean values at room temperature - as fired - of up to 330 MPa were found by another manufacturer. Scatter of the individual values of the test series yielded Weibull values of m = 10-20. Room temperature spin tests conducted on the rotor blades obtained the specific values for a number of blade series.

The oxidation resistance of the RBSN materials could also be improved. After 1000 h annealing at 1260°C, a decrease in strength of 50-100 MPa was measured, while after 1000 h of annealing at 900°C even an increase in strength was identified.

Previous RBSN-qualities showed an extremely unsatisfactory creep behaviour. Through appropriate raw material selection in terms of composition and impurities, as well as through suitable grain structure leading to slightly higher density of about 2500 kg m^{-3}, this disadvantage could be eliminated. Further improvement was brought about by different steps leading to even higher densities around 2700 kg m^{-3}, so that these RBSN materials showed a more favourable creep behaviour at comparable thermal and mechanical loads. With these measures, it was at the same time possible to eliminate and/or minimize other disadvantageous properties as well. So the drop in strength at room temperature after oxidation treatment could be considerably reduced, so that now the strengths after annealing are about the same as with untreated specimens. The progress is to be seen in connection with the oxidation phenomenon and with the measures for its elimination. Internal oxidation plays a special role which, in turn, is very much dependent on the pore structure.

The strength level could also be increased. So values around 300 MPa and more at densities of 2500 kg m^{-3} have been achieved rather frequently. Even higher strengths are anticipated for the denser qualities if appropriate microstructure optimization is performed.

These material improvements incited renewed component developments from RBSN in the individual manufacturing companies. In particular not only have injection-molded rotor blades proved their worth, but also stator vanes of complex shape.

1.1.2 Hot-pressed silicon nitride HPSN

The "conventional" hot-pressed silicon nitride with MgO as sintering aid could be further optimized with respect to the material and to the technology for component fabrication. Previous grades showed unsatisfactory properties for gas turbine applications, due to the poor powder qualities and the hot pressing technology not adjusted to the individual Si_3N_4 powders. Successive adaption and optimization have given rise to a material which, at favourable flexural strengths of 700 MPa at room temperature and of 500 MPa at 1200°C, is characterized by satisfactory creep and oxidation resistance as well as by excellent homogeneity with Weibull factors of $m > 30$. This guarantees a tensile strength of 400 MPa for this material.

A newly developed material with Y_2O_3 as sintering aid has as yet not shown the negative oxidation effects known from the literature. The improved characteristics compared to the MgO-containing material show themselves in addition to the mean room temperature strength (800 MPa) especially in the strength at high temperatures (1200°C: 680 MPa; 1400°C: 590 MPa), in the resistance to cyclic temperature changes (after 96 h shock annealing according to a cycle defined by the automobile industry still 760 MPa) and in the creep rate (1250°C, 80 MPa: 2×10^{-6} h^{-1}; 1300°C, 120 MPa: 5×10^{-6} h^{-1}). This applies also to resistance against slow crack growth as well as to oxidation resistance. After a hardly detectable weight increase following 100 h annealing at temperatures in the temperature range of 500 to 1300°C, practically no decrease in room temperature strength was observed at the annealed specimens.

In the future, we have to improve further the HPSN qualities, but placing principal emphasis on other technologies, such as sintering and hot-isostatic pressing, which promise more economical fabrication of gas turbine components.

1.1.3 Sintered silicon nitride SSN and hot-isostatic pressed Si_3N_4 HIPSN

The materials SSN and HIPSN are only in the initial stage of development. Namely for the SSN development (Fig. 2), a sinteractive powder had first to be developed which allows dense sintering to be achieved with as small an amount of sintering additives as possible. By optimizing the powder composition and preparation, where the sintering aids are homogeneously distributed in the Si_3N_4 powder in dissolved form by means of spray drying, sinter densities of ≥ 95% of theoretical density could be obtained by adding only 2 wt % MgO. Such a low concentration of sintering

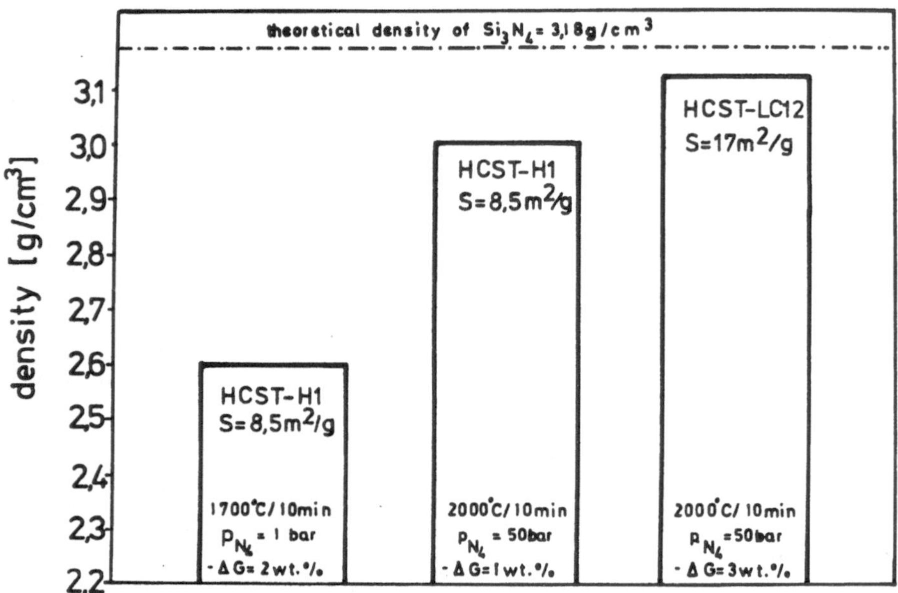

Fig. 2: Density of sintered silicon nitride SSN (sintering additive 2% MgO) at different starting conditions with different starting materials.

aid is to be strived for, to reach the best mechanical properties at high temperatures.

These special powders have been given selective development in the past few years. For SSN, densities of 95-98% of the theoretical density can be achieved today. One possibility of increasing the density to 100% is offered by the HIP technique which, as a combination of reaction sintering with good molding capability and hot-pressing with its high strength values, furthermore seems to allow fabrication of complex-shaped components of high strength. A prerequisite for this is, however, a suitable encapsulation technology, which separates the porous RBSN from the pressure-transmitting medium. Such encapsulation techniques are under development. Since a suitable facility is available, it is hoped that it will soon be possible to test appropriate materials and first components.

1.2 Silicon Carbide

Shortly after the German Federal Ministry of Research and Technology had begun to sponsor the development of ceramic materials and gas turbine components for application in automobiles, various companies started improving SiC materials. Therefore extensive knowhow is available covering all SiC material variations. Following the above mentioned temporary setback with RBSN, it was mainly the material Si-SiC, containing free silicon, which first took the place of the RBSN.

1.2.1 Silicon carbide with free silicon, Si-SiC

The materials developed exhibit excellent high-temperature behaviour (strength, creep, oxidation). The room temperature strength is 300 to 400 MPa. For the fabrication of components a special slip casting method has been developed which, in comparison with the injection-molding method attempted for mass production, allows a relatively economical manufacturing of even rather complex components. So a workable approach has been adopted for preliminary testing of both material suitability and design (Fig.3).

In terms of mechanical properties and especially their scatter, Si-SiC still can be improved. After suitability tests with positive results, economically valid processes for component fabrication must now be developed.

1.2.2 Pressureless sintered silicon carbide, SSiC

Silicon carbide can be pressureless sintered to high density. But the strengths are still rather unsatisfactory. The technology for the production of prototype components is not yet at a very advanced stage.

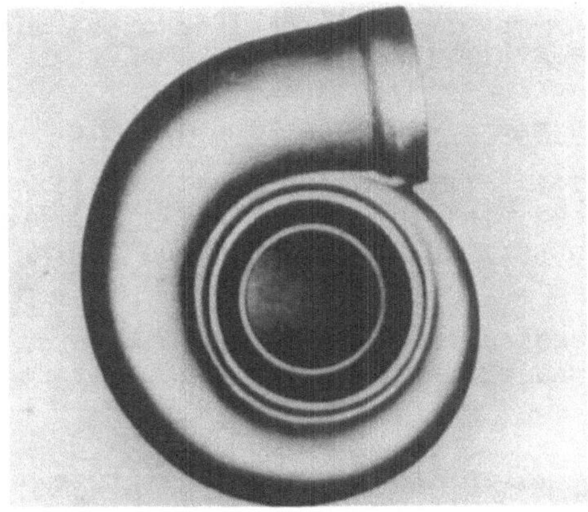

Fig. 3: Inlet scroll (Si-SiC, slip cast).

Primary attention was given to material advancement with the objective of further increasing the density, maintaining at the same time a fine-grained structure. As a result of these activities materials with 97 to 99% theoretical density were accomplished. The average strengths are 320-350 MPa, maximum 400 MPa. Parallel to the materials development, special sinterable powders had to be developed. Type and quantity of the sintering additives are to be matched optimally to the molding and sintering process. The quality of the undoped powders - in spite of being fine-grained - distinguishes itself by a high green density of typically 65% theoretical density, and recently by reduced oxygen contents. Both factors have a positive effect on the shrinkage, the weight loss, and the achieveable sintering density. For pressureless sintering and hot-pressing α- or β-SiC powders of 13-17 m^2/g surface have turned out to be of favourable use. Even finer powders can be fabricated, but then the advantage of a higher sinter-active surface is off-set by the disadvantage of higher oxygen contents.

1.2.3 Hot-pressed silicon carbide, HPSiC

Similarly as with Si_3N_4, the HPSiC shows the highest strength potential in the SiC family. But here too, we have the disadvantage of expensive machining, so that the goal for the near future must be development of HIPSiC components. In connection with HP-SiC it was found that SiC powders of finest particle size can be hot-pressed to maximum densities and strengths with small amounts of additives. It was also recognized that the strength and the K_{IC} values of the SiC specimen with intercrystalline fracture mechanism decrease with increasing temperature, while the same characteristics are largely independent of the temperature, if a transcrystalline fracture mechanism is observed. The fracture mechanism depends on the type and quantity of the sintering additive and, on the other hand, on the modification of the selected SiC starting powder. Hence the development of HP-SiC with transcrystalline fracture behaviour may be of special interest with respect to automobile gas turbine application.

1.2.4 Hot-isostatic pressed silicon carbide, HIPSiC

Application of hot-isostatic pressing for post-densification of hot-pressed and pressureless sintered SiC slightly improved the mechanical properties and, in particular, reduced scattering of these values.

For hot-isostatic compression of SiC powders, the development of suitable encapsulation methods is also given prior attention at present.

1.3 Supporting activities at materials research institutes

Besides the material manufacturers and the users, a number of institutes are involved in the programme carrying out basic materials research and determining, at a neutral place, characteristics of all materials developed by industry and making these values available to all participants in the programme.

In the past few years, the Max Planck Institute for Materials Science has investigated the possibilities of optimizing mixed ceramics on the basis of SiC and Si_3N_4. Primary objective of the activities has above all been improvement of the mechanical high-temperature characteristics. One example is the five-phase system with ZrO_2 embedded in a SiAlON matrix (Fig. 4). Such relatively complex investigations are supported by computer programs for thermodynamic calculations.

The work carried out at Berlin Technical University concentrates on basic investigations of powder preparation and of the sintering behaviour of SiC and Si_3N_4. The major results are presented in other papers.

The activities at Karlsruhe University and at DFVLR in the framework of "assessment" have as their objective to determine on the materials manufactured by the German ceramics industry high-temperature tensile and bending parameters, creep characteristics as well as long-term and oxidation behaviour and to provide the

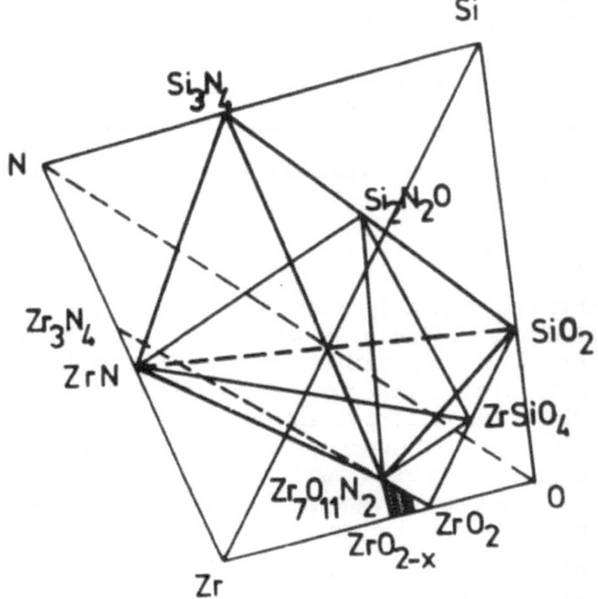

Fig. 4: System with ZrO_2-inclusions in SiAlON.

users with comparative results for their optimization activities. The HIP tests performed by DFVLR towards the development of HIPSN are run outside the programme, but on facilities funded by BMFT, so that all companies and institutes involved in the programme have access to the HIP capacity.

2. Component Development

In the first phase of the programme, activities were focussed on the area "material development", including at the most testing of preliminary parts or, e.g., non-shaped discs. The objectives of the second phase are more and more directed towards components and first aerodynamically shaped blades and blade assemblies. Fig. 5 shows the goals of the work carried out by the three gas turbine companies participating in the programme.

Company	Power of GT	Components	Rotor - Concept
Volkswagen	100 KW	- Rotor - Stator - (Combustor)	- Hybridrotor - Monolith, RBSN-Rotor
Daimler-Benz	150 KW	- Rotor - Inlet-Scroll - Recuperator	- Monolith, HPSN-Rotor
MTU	300 KW	- Rotor - Stator - Combustor	- Hybridrotor

Fig. 5: Goals of the gas turbine companies participating in the project.

2.1 Static components

The component which is probably most advanced in development and nearing realization most closely is the combustor. In the first phase, combustors made from different materials had already achieved good results in stationary operation without any pressure: a running period of 22.3 hours without any damage could be observed for a combustor made from Si-SiC under pressure conditions at cyclic loading, while in atmospheric tests and cyclic loading 250 hours were sustained without any damage. Related tests were performed with inlet cones. No cracks were discovered.

Far greater problems are encountered with the inlet scroll, mainly for purely constructional reasons, but above all with respect to the thermal stresses occurring in operation.

The stator development was frequently leading to revised designs, due to rather different results obtained during the past few years. No other component is available in so many modifications which presented sometimes positive, sometimes negative results (Fig. 6). The spectrum of construction ranges from inserted and slipped-in individual blades, bonded stator segments via monolithic cast vane assemblies as far as construction types with slotted internal and external rings and most recently even hollow blades, and includes a great variety of materials.

Different types of stators have been tested up to 1375°C. In stator segments, no damage occurred. Integral stators had cracks in the shroud.

After the testing of first-generation stators with typical fractures in the centre of the blade rear edge and in the outer ring, the stators were redesigned by means of FEM calculations. A type with straight heat flow from the thin vanes into the tip plates showed that tensile stresses of < 100 MPa are to be expected. The preferred material is RBSN of a recent grade. First hot tests at temperatures up to 1225°C (continuous operation) did not show any failures after 20 hours. With turbine vane assemblies of the second generation, the intermediate goal of 100 hours was achieved in combined test cycles. At maximum gas temperatures of 1325°C only one vane failed as a result of insufficient nitridation. Further tests under more severe operating conditions in the demonstrator are envisaged.

2.2 Turbine rotor

Development of the rotor covers different concepts.

Fig. 6: Segments and single-vanes for stators.

2.2.1 Hybrid rotor

The hybrid turbine rotor concept - metallic disc with inserted ceramic blades - of the first programme phase was modified on the basis of revised design data. HPSN blade dummies passed cold spin endurance testing with 5000 load cycles from 0 to > 500 m/s circumferential speed, related to the blade tip, without any damage. Rotor blades made from HPSN and RBSN were subjected to hot spin testing (maximum gas temperature 2200°C). In the automobile turbine, using latest RBSN material grades, the measured circumferential speeds, related to the blade cross section, are above the design value of 382 m/s. Preparations for hot runs with complete hybrid rotors have been made. First, the rotors are planned to be proof tested up to ~300 m/s.

2.2.2 Monolithic RBSN axial wheel

The only moderate progress achieved in the development of duo-density rotors, essentially due to manufacturing problems, led to new considerations directed towards integral RBSN rotors. According to FEM calculations, a RBSN rotor without central bore could be realized if today's material grades with more than 350 MPa flexural strength could be realized in the component. Manufacturing difficulties are anticipated as soon as larger volumes would have to be processed to homogeneous components in injection-molding technique.

2.2.3 Monolithic HPSN rotor

Experience acquired so far shows that monolithic turbine rotors made from HPSN have proved to be successful (Fig. 7). These rotors are fabricated through ultrasonic processing of blanks supplied by the ceramics industry. Maximum speeds achieved so far are ~ 1000 Hz at a gas temperature of 1250°C and 833 Hz at 1350°C.

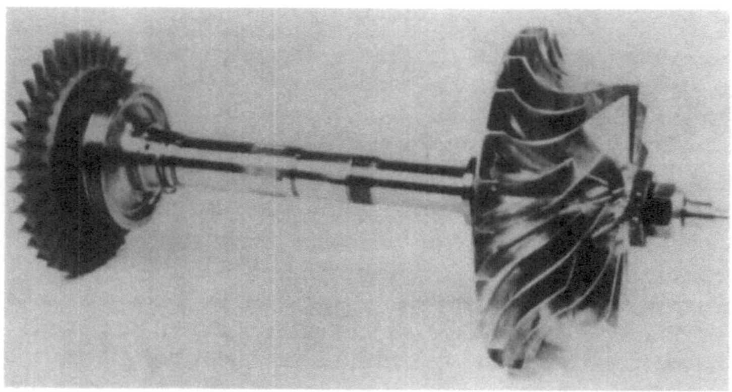

Fig. 7: Monolithic HPSN-rotor.

2.2.4 Radial wheel

Besides the further development of axial wheels, development of radial turbine wheels was started. The size of the wheel is designed for an exhaust gas turbine. The objective is to possibly develop a radial wheel type - if the outcome of the activities is successful - which is designed for a vehicular gas turbine of ~100 kW engine power. FEM calculations have shown that under the given boundary conditions of an exhaust gas turbine, SiSiC material is better suited to meet the requirements than RBSN. After several iteration steps, the intermediate goal of 400 m/s, related to the blade external diameter, was accomplished in cold testing. In first hot runs, acceleration tests under realistic operating conditions of an exhaust gas turbine achieved ~340 m/s.

2.3 Heat exchanger

In contrast to other ceramic gas turbine projects, the objective of our programme is to develop a recuperative heat exchanger which promises some advantages over the regenerator, especially because of low leakage and a greater variety of arrangement possibilities.

It became apparent that very thorough detailed tests on specimens were necessary (leakage, bursting strength, oxidation tests). Although progress has been made, it has not yet been possible to manufacture and test a complete recuperator element. Further activities will be concentrated on manufacturing development. At present, it cannot yet be stated when and whether it can be considered for use in a turbine. If necessary, recourse must be made to a regenerator concept.

2.4 Supplementary activities at research institutes

In addition to the institutes already named in connection with the material development description, mention should be made of further institutes supporting the project, which are primarily working on the development of testing methods.

One activity under the programme includes the review of existing non-destructive testing methods for their applicability and the further development as well as new development of non-destructive testing procedures for quality control of ceramic materials and components (Fig. 8).

For quality control and quality improvement of ceramic materials and components, efficient non-destructive testing is necessary. The results achieved support the capability of the various non-destructive procedures of verifying the failure parameters required with respect to material characteristics and load. The state of development of the individual methods may in part be regarded as ready for application, e.g. microradiography. Other procedures still require considerable effort before the method

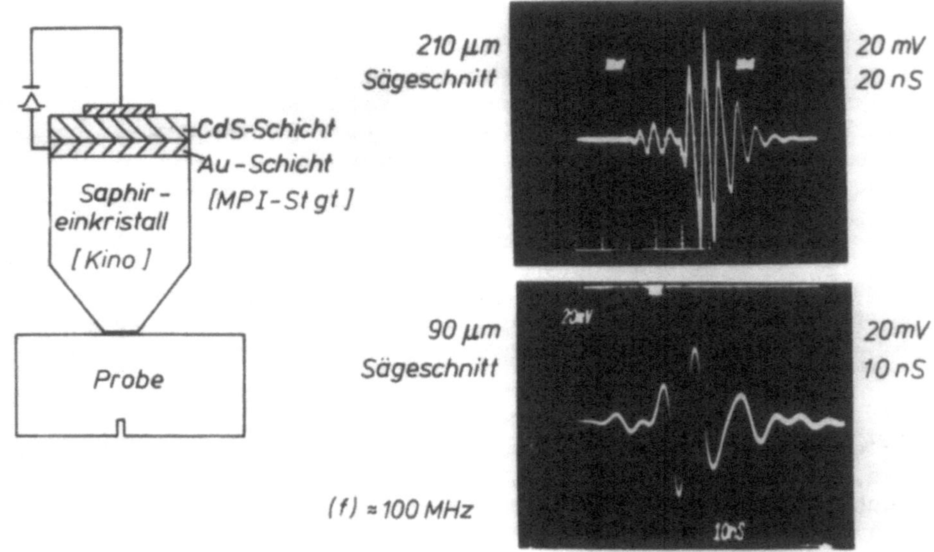

Fig. 8: Testing of a HPSN bend-test specimen with seeded defects by HF ultrasonics.

can be applied in situ in production control, e.g. high-frequency ultrasonic testing. In particular objectivization of the results of non-destructive testing (nature, size, form and orientation of defects as well as their relevance under fracture-mechanical aspects) is a major task to be given attention in the further development.

Another institute is concerned with the fracture-mechanical characterization of hot-pressed silicon nitride in the temperature range up to 1400°C, with interest being focussed on the relationships between microstructure and slow crack propagation.

With a view to the planned application of HPSN as a material for automobile gas turbines, the results achieved as yet may be assessed as follows:

- In the case of the MgO-doped material, regard must be paid to the fact that at operating temperatures above ~900–1000°C – in spite of relatively high $K_{I_{max}}$ values – crack growth may already start at considerably lower K_I values.
- The Y_2O_3-doped material, as compared to the MgO-doped material, shows a better fracture behaviour at temperatures up to 1400°C; according to the manufacturer, the scatter of the material properties may be reduced by optimizing the method for incorporation of the sintering aid.

An essential step in the realization of a vehicular gas turbine is to solve the problem of joining the ceramic and metal materials, e.g. the interface of rotor disc and shaft. Problems are also encountered in attaching ceramic parts to each other.

The following joining techniques were studied by one institute: 1. mechanical connection, 2. cement bonding, 3. slip and nitridation bonding, 4. hot press welding, 5. diffusion welding, 6. direct soldering.

3. Outlook

If one looks at the numerous results obtained under the programme, one cannot see a breakthrough yet, in spite of many successful test data, neither on the material side nor on the gas turbine component side.

The development goals set for 1981, as mentioned above, could not yet be fully achieved. This was essentially due to the time factor. The technical goals needed only slight modification (Fig. 9). This fact, i.e. that the right way has been followed for several years, encouraged us to start the last phase of the programme, running to 1983. This means that both the Federal Ministry of Research and Technology (BMFT), and industry, are funding partners. All participants have learned that the new materials and processing techniques have a large potential for high-temperature applications. They intend to concentrate their efforts, in the third phase of the programme, on optimization of the qualities, procedures and components. Then the time will hopefully have come where BMFT may retire, for industry to continue the remaining way using its own resources.

During the first six years of the project (from 1974 until 1980) about 64 million DM have been spent on the project. With an additional 33 million DM for the period 1980 until 1983 at the end of the project, about 100 million DM will be spent in industry and research laboratories. Because of a 50% funding of industry and 100% funding of research laboratories, the German government (BMFT) will pay 53 million DM, whereas industry will spend about 44 million DM.

As a result of the exchange of information under the Implementing Agreement with the USA, we know that there too, considerable funds are spent to enhance realization of a ceramic gas turbine. Japan has great plans as well and has made considerable progress in the development of ceramic materials and components. Furthermore, intensive work is also performed in Sweden on a ceramic gas turbine.

Finally, a Project Support Status Report cannot leave unmentioned that the experience meanwhile acquired by the material manufacturers and users with these new ceramic materials has led to developments and applications which are going far beyond the gas turbine. This was to be anticipated, but is not included in the subject of this paper and therefore is just indicated briefly at the end.

DEVELOPMENT GOALS UNTIL 1983

1. Stationary Components (Combustor, nose, cone, stator, rotor shroud).

   Production of original components from various materials and 200 H-tests in a simulated duty cycle with a maximum combustor outlet temperature of 2500 F and an inlet pressure of 73 psi.

2. Heat Exchanger

   Production of Si_3N_4-recuperator with a wall thickness of 2 mm, a pressure ratio of 5, an allowable leak rate of 5% and a max. temperature of 2200 F inlet temperature, 10 H-test.

3. Rotor

   - Metal-Ceramic-Rotor (cars and trucks)
     200 H-test in duty-cycle
     (inlet temperature 2285 F)
   - All Ceramic Rotor (car)
     50 H-test in duty-cycle
     (inlet temperature 2500 F).

Fig. 9.

NITROGEN CERAMICS IN FRANCE

M. Billy, J. Desmaison and P. Goursat

Laboratoire de Céramiques Nouvelles (L.A. C.N.R.S. No.320), Université de Limoges, 87060 Limoges Cedex, France.

Since the 1976 NATO Advanced Study Institute on 'Nitrogen Ceramics', increasing attention has been directed in France to the properties of ceramic materials based either on silicon carbide or on silicon nitride and related systems (in particular 'Sialon' systems).

Almost all the research programmes on the development of silicon-based ceramics for structural applications at high temperature have been sponsored by several State Departments, such as Research and Technology (CNRS, DGRST), Education, Industry, Defence (DRET), and conducted both by private and public research centres.

On the industrial side, in the main, only two companies are interested at present in the production of such ceramics: CGE (Ceraver) and Lafarge (SEPR, CEC, Desmarquest). For example, Ceraver is able to produce several ceramic components for gas turbines (6 wt% Al RBSN, pressureless sintered Sialon, and hot-pressed or pressureless sintered silicon carbide). For its part, Lafarge is more interested in the production of refractories for metallurgical applications (carbonitrides, "nitride bonded-silicon carbide", or sialons obtained by carbothermal reduction of clays). However this company is also able to produce hot-pressed or pressureless sintered silicon carbide.

In France, the main potential users are involved with research development programmes in new energy conversion devices: Renault and PSA (automobile); SNIAS, SNECMA, and Turbomeca (aeronautics); SNIAS and SEP (aerospace). But the actual tendency (in particular in aeronautics and aerospace) is to give more emphasis to the

Riley, F.L. (ed.) Progress in Nitrogen Ceramics

development of composite materials rather than to monolithic ceramics.

A great part of the basic research in this area is undertaken in public research centres (CNRS, Universities, School of Engineers, ONERA, CEA ....) in close connection with industrial research centres which are generally taking charge of the more applied aspects of the programmes. Special interest is paid to these materials by the CNRS through the so-called "Nitrogen Ceramics" Laboratory (LA no. 320) at the Ceramic Centre of Limoges. In addition to the other research teams participating at this Advanced Study Institute (Ecole des Mines de Paris, INSA de Lyon, Université de Rennes), there are different University Laboratories that have just started working in this promising field (e.g. Ecole des Mines de St-Etienne, Ecole Centrale de Lyon, Universities of Bordeaux, Caen, Orleans, Lyon, CNRS at Odeillo).

In conclusion, we may say that despite the lack of important private or public programmes on nitrogen ceramics such as exist in U.S.A., Germany or Japan, France is making a positive contribution to the development of these materials.

CURRENT JAPANESE RESEARCH PROGRAMMES INTO NITROGEN CERAMICS

Hiroshige Suzuki

Tokyo Institute of Technology, Japan
(Oral presentation by K. Suzuki, Asahi Glass Co. Ltd., Japan).

In this report, an outline of the present status of research and development on nitrogen ceramics in Japan will be described. Although silicon carbide ceramics are also regarded as of great importance as silicon nitride or sialons, only a brief mention of silicon carbide ceramics is contained herein.

1. HISTORICAL BACKGROUND

Silicon nitride was studied at the outset as a refractory material in Japan. For example, H. Suzuki has studied since 1958 effective catalysts, and the resulting phases, in the nitridation reaction of silicon, concluding that some transition metals, or their compounds, are the most effective catalysts, and that the proper choice between them can realize the formation of silicon nitrides of mainly α- or β-phase (1). Subsequently Oyama and Kamigaito (2) at the Laboratory of Toyota Motors Co. discovered the wide solid-solubility range between silicon nitride and alumina.

This author believes that many people in the world know that this was at roughly the same time that Jack reported the identical fact in England (3). While the study of highly densified bodies through the hot pressing process (HPSN) started later in Japan than in England, Komeya and Tsuge at the Toshiba Corporation Laboratory, in extending their study on $AlN-Y_2O_3-SiO_2$ system (4), succeeded in obtaining silicon nitride test pieces with the highest strength in the world (see Fig. 1) by crystallization of grain boundary phase in the $Si_3N_4-Y_2O_3-Al_2O_3$ component system (5).

On the other hand, the manufacture of more intricate shapes with sufficiently high density through pressureless sintering has been pursued by several ceramic makers (6) such as Asahi Glass Co.,

Riley, F.L. (ed.) Progress in Nitrogen Ceramics

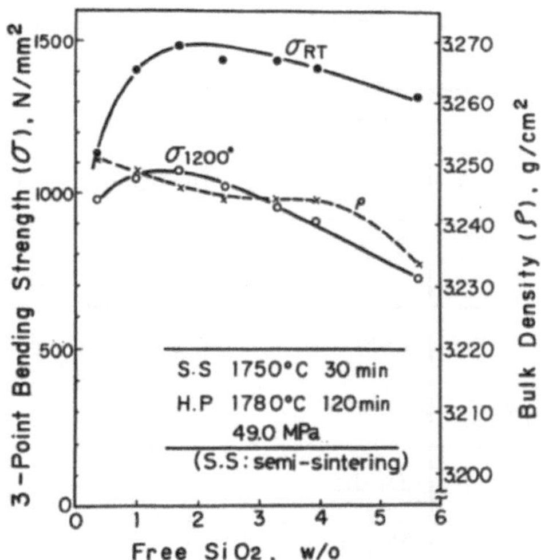

Fig. 1. Effect of free SiO_2 on the strength at 25°C and 1200°C(5b)

N.G.K. Spark Plug, N.G.K. Insulators, and Toyota, and has reached a stage where some of the products match the best in the world.

More recently, trials of sintering of silicon nitride in higher nitrogen pressure, and of sialons in controlled p_{SiO} atmospheres, have begun. In this field, the contribution of Mitomo et al. (7) in the National Institute for the Research of Inorganic Materials (NIRIM) is noteworthy.

A means of densification of RBSN (reaction bonded silicon nitride) by reheating in high-pressure nitrogen has been the subject of a patent application by Azuma (8) of the Government Industrial Technology Research Institute, Nagoya, based on his research work, and this concept is now seen to be expanding its application into practice. It is also being continuously studied in the experiments forming part of the "Moonlight" project, which will be mentioned later.

Other fundamental work on nitrogen ceramics is listed, together with the above-mentioned studies, in Table 1.

2. RECENT EXPANSION OF RESEARCH AND DEVELOPMENT PROGRAMMES

In the context of fundamental studies, and manufacturing technology, of silicon nitride and sialons, continued from the above-mentioned work, the following subjects are being energetically pursued in various Japanese universities, Government laboratories, and private company laboratories. These are listed in Table 2.

Table 1. Research and Development in Nitrogen Ceramics in Japan (to ~ 1977).

SUBJECT	REPRESENTATIVE OF AUTHORS
Si Nitridation Catalysts (1958 ~)	Tokyo Inst. Tech. (H. Suzuki)
SIALON (1971 ~)	Toyota Centr. R & D Lab. (Oyama)
HP-Si_3N_4 (Y_2O_3-Al_2O_3) Grain Boundary Crystallization (1974 ~ 1977)	Toshiba Corp. (Tsuge, Komeya)
PLS-Si_3N_4	Asahi Glass NGK Insulators NGK Spark Plug Toyota Centr. R & D Lab. etc.
Sintering of Si_3N_4 under High N_2 Pressure (1976) (SiO Pressure for SIALON)	Natl. Inst. Res. Inorg. Mat. (Mitomo)
Post-Sintering of RB-Si_3N_4 (1976)	Govt. Ind. Res. Inst., Nagoya (Azuma)
AlN (Fundamental Study) (1973 ~ 1977) Si_3N_4 (1974 1977) Formation of Single Crystal Synthesis of α-Phase (1974 ~ 1977) Etc.	Natl. Inst. Res. Inorg. Mat. (Sakai (9)) (Kijima (10))

Table 2. Recent Research and Development of Nitrogen Ceramics in Japan (1978 ~)

SUBJECT		REPRESENTATIVE OF AUTHORS
Powder	Si_3N_4, $(Si + N_2)$	Denki Kagaku Kogyo
high-purity	$(SiO_2 + C + N_2)$	Toshiba Corp. - Toshiba Ceramics (11)
	$(Si(NH_2)_2)$	Toyota Central R & D Lab. (12)
easier-sintering	SIALONS	NIRIM* (13)
Whisker,	Si_3N_4	GIRI** - Osaka (14)
Single Crystal	SiC	Tateho Kagaku, Asahi Kasei
Coating, N-Glass	$La_4Si_2O_7N_2$	NIRIM (15), Toshiba Ceramic (16)
Sintering	Si_3N_4 (PLS, HP, RB) Additive-Phase Relation Microstructure, G-B Structure Mechanism	Many groups (17)
	Si_3N_4 (CVD)	Tohoku Univ. (18), Toshiba Ceramic
	SIALON (HP, PLS) (β)	GIRI - Kyusyu (19)
	(β and α)	NIRIM (20)
	$AlN\text{-}Al_2O_3$ (HP)	NIRIM (Sakai) (21)
	Super High Pressure Sintering	
	(Impact Treatment)	Tokyo Inst. Tech. (22)
	(Non-Additive)	Osaka Univ. (23), NIRIM (24)
	HIP	Kobe Steel (25), GIRI-Osaka
	Composite (Si_3N_4-SiC. Whisker, Fibre)	GIRI - Osaka (26)
Joining	Si_3N_4, SiC	GIRI - Osaka (27), Tokyo Inst. Tech. (28)

*National Institute for Research in Inorganic Materials

**Government Industrial Technology Research Institute.

(Cont.)

Table 2 (Cont.)

SUBJECT		REPRESENTATIVE OF AUTHORS
Properties	Mechanical Properties	GIRI - Nagoya (29)
	Strength measurement	Natl. Aerospace Lab. (30)
	Statistical Treatment	Asahi Glass (31)
	Fracture Mechanics	GIRI - Nagoya (32)
	Life Prediction	NGK - Spark Plug (32)
	Fatigue Tests	NGK - Insulator (32)
	Proof Tests	Tokyo Inst. Tech. (32)
	NDE	
	Crystal Structure	NIRIM (33)
	Oxidation Behaviour	NIRIM (34), Tohoku Univ.
Design Technology (Ceramic Tile Heat Seal)		Centr. Res. Inst. Elect. Power Industry (35).

3. PRACTICAL APPLICATIONS

Some of the results of these research and development activities have led to the acceptance of the practical application of silicon nitride or sialon ceramics (and also some kinds of silicon carbide). Many ceramic manufacturers in Japan are examining their own trial products for applicability to various branches of industry in place of identical metallic materials. Asahi Glass Co. promptly determined to commercialise their pressureless sintered silicon nitride and silicon carbide products as heat-radiation tubes, conveyor rollers, high-temperature fans and valves etc., mainly for metallurgical industries (see Fig. 2). Large and rather complicated shapes can be made, the maximum length of which to date has reached some 2 m.

The market for these industrial applications is expected to be established, and to develop steadily.

Relating to the expectancy for future heat-engine parts in nitrogen ceramics, there have been rather few published data among ceramic manufacturers and automobile manufacturers. Toyota is believed to have a joint development programme with Toshiba, while Isuzu Motors formed a combined development plan with Kyocera concerning ceramic diesel engines in 1978, and which has progressed recently to actual 4-cylinder car engine tests. However, in each case details are unknown. Toshiba, Kyocera and NGK are also playing a part in the Advanced Gas Turbine programmes, supported by United States Government funds, and according to this programme

Fig. 2. High-strength ceramics "CERAROI" (17c)
(courtesy Asahi Glass Co. Ltd.).

the first ceramic automobile engine is expected to run in 1985. However, it must undoubtedly be said that there are still a great many things to be done before a practical ceramic-engined automobile can be operated with sufficient reliability. The commercialization of ceramics for this field cannot be expected in the short-term future.

4. GOVERNMENT POLICY AND CIVILIAN PROMOTION SCHEMES

4.1 Plans and Proceedings

As mentioned so far, research and development in the area of nitrogen ceramics (together with silicon carbide and/or oxide ceramics) in Japan has been expanding in the last 10 years. The Japanese Government (MITI and Science and Technology Agency) is always positive in promoting this trend. Along with the utilization of these materials as a number of mechanical parts for the purposes of enhancement of industrial resources and of energy saving, an ambition to establish a more generalized-base technology for the coming decade (probably '90's) has arisen. Recently two national projects for each of these objectives have been started, and which will be explained later. There is additionally a certain amount of Government official grant money paid every year to several companies as an encouragement for their own high risk development work.

The Ministry of Education provides scientific research funds for universities and their attached laboratories of which, however, the share for ceramics technology is not yet so significant compared with those connected with nuclear fusion, space development, natural disasters, the environment, or cancer.

The 124th Committee (Chairman: Prof. H. Suzuki) of the Japanese Society for the Promotion of Science has paid special attention to silicon nitride and sialon materials together with silicon carbide and zirconia, and this committee is playing its role as a centre of the co-operative, or information-exchanging, sphere between academic and industrial circles, and occasionally preparing suggestions for the Government. As another mission of the Society lies in promotive work for international academic cooperation, it may be expected that plans concerning personnel exchange and cooperative schedules will be discussed through this organisation.

There are also several private foundations starting to make increasingly large contributions for the subsidy of researches relating to high-temperature ceramics.

4.2 <u>The "Moonlight" Project</u> (a large-scale, energy-saving technology development project).

This is a national project sponsored by the Agency of Industrial Science and Technology (MITI) of Japan, for 7 years, from 1978 through 1984. It includes five sub-projects as follows:

Waste Heat Utilizing Technical System
MHD
Advanced Gas Turbine
New-Type Battery and Power-Storage System
Fuel Cells

Nitrogen ceramics technology is concerned mainly in the third sub-project, namely the Advanced Gas Turbine. The total budget of this sub-project is about 21,000 M yen (~ 91M \$ US) and its object is the accomplishment of a prototype gas turbine adequate to provide a total thermal efficiency (steam-combined cycle generation system, including low-heat utilization) as high as 55% and of a 100-MW scale (per one gas turbine). In the approach, five main items are to be studied. They are:

1) Super heat-resistant materials
   a. Metallic
   b. Ceramic
2) Gas turbine component technology
3) Manufacture and tests of a pilot and a prototype turbine
4) Assessment for the environment
5) Optimum system specification.

The total research and development budget for ceramic materials in these items was about 2000M yen (~ 8.6M \$ US) in all, through seven years. Three government laboratories, namely Nagoya, Osaka and

Kyushu Industrial Technology Research Institutes, are joining to work with private company laboratories, the latter organizing an Engineering Research Association. 14 companies constitute this association, in which Asahi Glass, Toshiba, NGK and Kyocera are the main members for research and development of ceramics. In addition to these, turbine makers, namely Ishikawajima-Harima Heavy Industries, Kawasaki H.I. and Mitsubishi H.I., and also as the user, Central Research Institute of the Electric Power Industry, are participating in the development of design technology using ceramic parts. The Central Research Institute of the Electric Power Industry is also concerned with developing a metal-ceramic dual-mosaic type heat-insulation structure ("Ceramic Tile Heat Seal") and is very hopeful about its applicability as a combustion chamber. Combustion chambers, shrouds, stator vanes, and possibly rotors are the subjects of the fabrication technology development, with which the development of testing technologies capable of evaluating performances of the ceramic parts with sufficient reliability are included in the task. As possible materials, sialon, fibre reinforced silicon nitride and silicon carbide, RBSN, sintered silicon nitride, sintered silicon carbide, and HPSN are all being studied.

4.3 Industrial Base Technology Development Project

This new project sponsored by MITI is just concluding its master-planning and is probably going to start in the autumn of 1981. The overall objective is the establishment of industrial base technologies for the decade of 1990. It consists of three different fields of study, namely: (i) new materials, (ii) biotechnology, (iii) new-functional units. In turn the high performance ceramics are placed as one of four sub-projects belonging to the first field of 'new materials'. There are 12 sub-projects in all, covering the 3 fields. The whole budget for this project is estimated at 104,000M yen (~ 450M $ US), with full sponsorship of MITI, and the amount which will be shared for the high performance ceramics is assumed to be 13,000M yen (~ 56M $ US) in total. The periods of time of the development differ slightly among the sub-projects, and it will be 10 years in the case of the ceramics. The budget for the first year (1981) for the new materials field will add up to about 2700M yen (~ 12M $ US), the division of which among the four sub-projects constituting this field is not yet fixed. The sub-projects are (a) high-performance ceramics (called "Fine Ceramics" in Japan), (b) polymers with special functions, (c) special metallic materials, (d) special composite materials.

The outline of the plan of the research and development of "Fine Ceramics" is as follows: The substances are silicon nitride, sialons, silicon carbide and several kinds of oxide. Large, intricately-shaped sintered parts of sufficiently high strength, corrosion resistance or wear resistance, according to fields of application, with high dimensional accuracy and material reliability (the objective numerical data in Table 3) are the requisite

Table 3. Objectives for Performance Properties of Fine Ceramics

CLASSIFICATION	OBJECTIVE	
High Strength Material	1) ≥1200°C in air, after 1000 hrs holding:	
	Reliability (Weibull Modulus)	m ≥ 20
	Strength (Average tensile strength)	$\bar{\sigma}$ ≥ 294 MPa
	2) 1200°C in air, 1000 hrs continuous loading:	
	Resistance (creep rupture strength)	$\bar{\sigma}$ ≥ 98 MPa
High Corrosion Resistant Material	1) ≥1300°C in air, after 1000 hrs holding:	
	Reliability (Weibull Modulus)	m ≥ 20
	Corrosion resistance (oxidation weight gain)	≤ 10^{-2} kgm^{-2}
	Strength (Average tensile strength)	$\bar{\sigma}$ ≥ 196 MPa
High Wear Resistant Material	1) 800°C in air, after 1000 hrs holding:	
	Reliability (Weibull Modulus)	m ≥ 22
	Strength (Average tensile strength)	$\bar{\sigma}$ ≥ 490 MPa
	2) RT	
	Wear Resistance (Specific wear amount)	≤ 10^{-14} m^2kg^{-1}
	Tolerance (surface flatness)	R_{MAX} ≤ 2 μm

objects, and the corresponding manufacturing process, the evaluation, as well as the application technologies, have to be developed.

The subject matter in each technology is as follows:

1. Manufacturing technology
   Synthesis of powder materials
   Moulding and sintering
   Processing (finishing and joining)

2. Evaluation technology
   Strength evaluation corresponding basically to practical conditions.

Non-destructive testing
Proof-testing on account of life prediction.

3. Application technology
Design technology (of brittle materials)
Verification of performance with test rigs.

The theme is to be carried out mainly by the efforts of private companies, though several government laboratories will also join and contribute on the more fundamental aspects (for example, analyses of sintering mechanisms, abrasion mechanisms, etc., and also the standardization of property-measurement manuals). In order to achieve effective progress, a unique method of management will be adopted which introduces a competitive principle, by committing similar items to more than one member (company). The executive management commission and the evaluation commission will be appointed respectively by the Government. However, at present neither the names of private companies in the commission, nor the names of members of the commission, are fixed yet.

5. OUTLOOK FOR THE FUTURE

The background of the research and development of nitrogen ceramics in Japan rests largely on three standpoints:

1) Energy saving (heat engines with high thermal efficiency and heat exchangers etc.).

2) Resources saving (of strategic metals such as Ni, Cr, Co, W).

3) The establishment of an Advanced Materials Technology as an Industrial Base Technology for the Coming Age.

Most recently, the latter is regarded as being most important; thus the primary objective is the attainment of the novel and systematic material science and technology itself, rather than that of a certain "definite" system, having a limited specification. The author would like to stress that this practice derives from our intention of achieving a wider field of application for these high-performance ceramics to the Japanese industrial world. Men of intelligence agree that it is necessary and most profitable to co-operate with advanced foreign countries. Therefore it may be said that there will be some distinct movement to this direction shortly.

REFERENCES

1a) H. Suzuki and T. Yamauchi "Effects of various additions on the synthesis of silicon nitride and its polymorphism", 18th IUPAC Congress, Montreal, Canada, August 6-12 (1961), pp. 213-215.

1b) H. Suzuki, "The synthesis and properties of silicon nitride", Bull. of Tokyo Institute of Technology No. 54, 159-161 (1963).

2a) Y. Oyama and O. Kamigaito, "Solid solubility of some oxide in Si_3N_4", Short Notes, Japan. J. Appl. Phys. 10 1637 (1971).

2b) Y. Oyama and O. Kamigaito, "Hot pressing of Si_3N_4-Al_2O_3", Yogyo-Kyokai-Shi 80 (8), 327 (1972).

2c) Y. Oyama, "Solid solution in the system Si_3N_4-AlN-Al_2O_3", Short Notes, Japan. J. Appl. Phys. 11 760 (1972).

2d) Y. Oyama, "Solid solution in the ternary system Si_3N_4-Al_2O_3-Ga_2O_3", Short Notes, ibid, 11 1572 (1972).

2e) Y. Oyama, "Solid solution in the system Si_3N_4-Ga_2O_3-Al_2O_3", ibid, 12 500 (1973).

3) K.H. Jack and W.I. Wilson. Nat. Phys. Sci. 283 28 (1972).

4a) K. Komeya, A. Tsuge, H. Inoue and M. Murata, "Textural determination in sintered AlN-Y_2O_3 with fibrous structure", Note, Yogyo-Kyokai-Shi, 79 (12), 470 (1971).

4b) K. Komeya, H. Inoue and A. Tsuge, "Role of Y_2O_3 and SiO_2 additions in sintering of AlN", J. Amer. Ceram. Soc. 57 (9), 414 (1974).

5a) K. Nishida, M. Komatsu, T. Ochiai and A. Tsuge, "Hot pressing of silicon nitride articles with high volume", Proc. Int. Symp. on Factors in Densification and Sintering of Oxide and Non-Oxide Ceramics, Japan, 1978, pp. 557-579, Eds. S. Somiya and S. Saito (Tokyo Institute of Technology, 1979).

5b) A. Tsuge, K. Komeya and H. Inoue, "Effect of powder characteristic on hot pressing of silicon nitride with Y_2O_3 and Al_2O_3", presented at 81st Annual Meeting of Am. Ceram. Soc., Cincinnati (1979).

6a) Y. Hattori and Y. Matsuo (NGK Spark Plug), "High strength silicon nitride sintered materials", U.S. Patent 4,216,021.

6b) Y. Hattori and T. Miyachi (NGK Spark Plug), "Process for producing compact silicon nitride ceramics", U.S. Patent 4,205,033.

6c) I. Oda, M. Kaneno and N. Yamamoto (NGK Insulators), "Pressureless sintering silicon nitride", in 'Nitrogen Ceramics', Ed. F.L. Riley, NATO ASI Applied Science Series (E23), Noordhoff, Leyden, 1977.

6d) H. Masaki and O. Kamigaito (Toyota Central R & D Lab. Inc.), "Pressureless sintering of silicon nitride with addition of MgO, Al_2O_3 and/or spinel", Yogyo-Kyokai-Shi, 84 (10), 508-512 (1976).

6e) O. Kamigaito and Y. Oyama (Toyota), "Method for producing ceramics of silicon nitride (AlN-Si_3N_4-Al_2O_3)", U.S. Patent 3,903,230, Brit. Patent 139,216, German Patent 2262785, Canada Patent 967175.

7a) M. Mitomo, M. Tsutsumi, E. Bannai and T. Tanaka, "Sintering of Si_3N_4", Am. Ceram. Soc. Bull. 55 313 (1976).

7b) M. Mitomo, "Pressure sintering of Si_3N_4", J. Mat. Sci. 11, 1103 (1976).

7c) M. Mitomo, N. Kuramoto and Y. Inomata, "The fabrication of high strength β-sialon by reaction sintering", J. Mat. Sci. 14 2309 (1979).

7d) M. Mitomo, N. Kuramot and Y. Yajima, "Thermal decomposition reaction of sialon", Yogyo-Kyokai-Shi, 88 42 (1980).

8) N. Azuma, T. Yamada and S. Matsuno, "Method for producing high-density Si_3N_4 materials", Japan, laid open Pat. Applic. No.33028/'78.

9a) T. Tanaka, T. Sakai and M. Iwata, "The effect of oxygen impurity on high temperature thermal conductivity of AlN", Yogyo-Kyokai-Shi, 81 399-400 (1973).

9b) T. Sakai and M. Iwata, "Effect of oxygen on sintering of AlN", J. Mat. Sci. 12 1659-65 (1977).

10a) K. Kijima, N. Setaka and H. Tanaka, "Preparation of Si_3N_4 single crystals by chemical vapour deposition", J. Crystal Growth, 24/25 183-187 (1974).

10b) K. Kato, Z. Inoue, K. Kijima et al. "Structural approach to the problem of oxygen content in α-Si_3N_4", J. Amer. Ceram. Soc. 58 (3-4), 90 (1975).

10c) H. Yamamura, K. Kijima, S. Shirasaki, Y. Inomata and H. Suzuki, "Mössbauer effect of Fe^{57}-doped silicon nitride", J. Mat. Sci. 11 1754 (1976).

10d) K. Kijima et al. "Nitrogen self-diffusion in Si_3N_4", J. Chem. Phys. 15 (7), 2688 (1976).

11) M. Mori, H. Inoue and T. Ochiai, this Advanced Study Institute.

12a) O. Kamigaito, "Method for producing Si_3N_4 powder", Japan. Pat. Application Pub. No. 14520/'78.

12b) O. Kamigaito and H. Doi, "Synthesis of α-Si_3N_4", ibid, No. 40199/'78.

12c) O. Kamigaito et al. "Method for producing the powder of Si_3N_4 solid solution", ibid., No. 43486/'77.

13) M. Mitomo et al., "Sialon formation by shock compression", Yogyo-Kyokai-Shi, 89 (7), 390 (1981).

14) T. Ogura et al., "Growth of SiC whiskers", Bull. Govt. Ind. Res. Inst., Osaka, 3 (31), 199 (1980).

15a) N. Ii, M. Mitomo and Z. Inoue, "Single-crystal growth of $La_4Si_2O_7N_2$ by the floating zone method", J. Mat. Sci. 15, 1961 (1980).

15b) A. Makishima, M. Mitomo, H. Tanaka, N. Ii and M. Tsutsumi, "Preparation of La-Si-O-M oxynitride glass of high nitrogen content", Yogyo-Kyokai-Shi, 88 (11), 701 (1980).

16) S. Matsuo, M. Watanabe et al. (Toshiba Ceramic & Toshiba R & D Centre), "Oxygen-free silicon single crystal grown from Si_3N_4 crucible", presented at Annual Meeting of the Electro-Chemical Soc., Minneapolis, May 10-16 (1981).

17a) A. Tsuge and K. Nishida, "High strength hot pressed Si_3N_4 with concurrent Y_2O_3 and Al_2O_3 additions", J. Amer. Ceram. Soc. 57 (4), 424 (1978).

17b) O. Kamigaito and Y. Oyama, "Method for producing ceramic sintered material of Si_3N_4-AlN-Ga_2O_3 system", Japan, Pat. Applic. Pub. No. 1764/'78.

(Cont.)

17b) (Cont.) "Method for producing ceramic sintered material of Si_3N_4-Al_2O_3-Ga_2O_3-AlN system", ibid. No. 1765/'78.

"Method for producing ceramic sintered material of Si_3N_4-spinel-Y_2O_3 system", ibid. No. 27471/'81.

17c) T. Ono and H. Abe, "High strength ceramics CERAROI, their properties and applications", Engineering Materials, 29 (1), 114 (1981).

17d) S. Horiuchi and M. Mitomo, "Crystal structure of Si_3N_4-Y_2O_3 examined by 1 MV high-resolution electron microscope", J. Mat. Sci. 14 2543 (1979).

18a) F. Itoh, T. Honda, K. Niihara, T. Hirai and K. Suzuki (Tohoku Univ.), "Chemical bond of CVD-Si_3N_4 by Compton scattering measurement", J. Phys. Soc. Japan, 48 (2), 561-66 (1980).

18b) K. Niihara and T. Hirai, "Growth morphology and slip system of α-Si_3N_4 single crystal", J. Mat. Sci. 14 1952-60 (1979).

19) S. Umebayashi et al. (Govt. Ind. Res. Inst., Kyusyu), "Relation between composition and bending strength of β-sialon ceramics", presented at the Annual Meeting of Ceram. Soc., Japan (1981).

20a) M. Mitomo, N. Kuramoto, M. Tsutsumi and H. Suzuki, "The formation of single phase Si-Al-O-N ceramics", Yogyo-Kyokai-Shi, 88 (11), 526-531 (1978).

20b) M. Mitomo, "Sialon formation by shock compression", Yogyo-Kyokai-Shi, 89 (7), 390 (1981).

20c) M. Mitomo, H. Tanaka, K. Muramatsu, N. Ii and Y. Fujii, "The strength of α-sialon ceramics", J. Mat. Sci. 15 2661 (1981).

20d) M. Mitomo, Y. Moriyoshi, T. Sakai, T. Ohsaka and M.M. Kobayashi, "β-sialon ceramics", J. Mat. Sci., to be published.

21a) T. Sakai, "Hot-pressing of the AlN-Al_2O_3 system", Yogyo-Kyokai-Shi, 86 125-30 (1978).

21b) T. Sakai and A. Watanabe, "Thermal expansion of sintered oxynitride in AlN-Al_2O_3 system", Am. Ceram. Soc. Bull. 59 (8), 853 (1980).

22a) A. Sawaoka et al. "Very high pressure sintering of covalent materials", Proc. of International Symposium on Factors in Densification and Sintering of Oxide and Non-Oxide Ceramics, Hakone, Japan (1978), pp. 339-344.

23) A. Tanaka, M. Shimada and M. Koizumi (Osaka Univ.), "High pressure sintering of α and β type of Si_3N_4", presented at Annual Meeting of Ceram. Soc. Japan (1981).

24) M. Mitono and N. Setaka, "Consolidation of Si_3N_4 by shock compression", J. Mat. Sci. 16 851 (1981).

25) K. Homma and T. Fujikawa, "HIP treatment of Si_3N_4 sintered materials", JSPS, 124th Committee Report No. 4-12 (1980).

26) Y. Toibana, N. Tamari and M. Kinoshita, "Research on SiC whisker-Si_3N_4 composite materials", Bull. Govt. Ind. Tech. Res. Inst. 24 (7), 7 (1980).

27) Y.Y. Ehata, S. Oori, M. Kinoshita and Y. Toibana, "Method for joining Si_3N_4 materials", Japan laid open Pat. Applic. No. 131683/'81.

28) T. Iseki, K. Yamashita and H. Suzuki, "Joining of self-bonded SiC by Ge metal", J. Amer. Ceram. Soc. 64 (1) C-13 (1981).

29a) K. Ozuka, T. Yamamoto and E. Oota, "Ultrasonic non-destructive testing for sintered Si_3N_4 materials", Bull. Govt. Ind. Tech. Res. Inst., Nagoya, to be published.

29b) S. Itoh (Govt. Ind. Tech. Res. Inst., Nagoya), "Cutting, finishing and bend strength measurement of HP-Si_3N_4 materials", JSPS, the 124th Committee Report, 3-6 (1980).

30) K. Takahara, "High temperature ceramics materials for engine parts", JSPS, the 124th Committee Report No. 1-1 (1980).

31a) M. Kawai, H. Abe and J. Nakayama, "The effect of surface roughness on the strength of Si_3N_4", Proc. International Symposium on Factors in Densification and Sintering of Oxide and Non-Oxide Ceramics, 1978, Japan, pp. 545-56.

31b) H. Abe, "Instrumented Charpy Impact Testing of SiC", Rept. of Res. Lab. of Asahi Glass Co. Ltd., 28 (1), 35 (1978).

31c) M. Kawai and H. Abe, "Weible statistics for brittle materials comparison of estimation method", Rept. of Res. Lab. Asahi Glass Co. Ltd., 30 (2) 111-120 (1980).

31d) M. Kawai, H. Abe and J. Nakayama, "Indentation-induced flow method for measuring crack velocity in sintered Si_3N_4", International Symposium on Fracture Mechanics of Ceramics, Penn. State Univ. U.S.A., July 1981.

32a) Y. Matsuo, "Properties of pressureless sintered Si_3N_4", presented at 82nd Annual Meeting of Amer. Ceram. Soc., 1981.

32b) T. Matsuo, "Cyclic fatigue of ceramics", this Advanced Study Institute.

32c) I. Oda, M. Matsui and T. Soma, "High-temperature fatigue failure in PLS silicon nitride", this Advanced Study Institute.

32d) T. Iseki and H. Suzuki, "Fracture toughness measurement by indentation", JSPS, the 124th Committee Rept. No. 3-9 (1980).

33) Z. Inoue, M. Mitomo and N. Ii, "A crystallographic study of a new compound of lanthanum silicon nitride, $LaSi_3N_4$", J. Mat. Sci. 15 2915-2920 (1980).

34a) Y. Hasegawa, H. Tanaka, M. Tsutsumi and H. Suzuki, "Oxidation behaviour of hot-pressed Si_3N_4 with addition of Y_2O_3 and Al_2O_3", Yogyo-Kyokai-Shi 88 (5), 292-297 (1980).

34b) Y. Hasegawa, "Oxidation behaviour of hot-pressed Si-M-O-N materials", JSPS, the 124th Committee Rept. No. 3-10 (1980).

35) H. Ishikawa, S. Suhara and T. Abe (Central Res. Inst. of E.P.I.), "An application of ceramics to high temperature gas turbine components", JSPS, the 124th Committee Rept. No. 5-15 (1981).

SURVEY AND CONCLUSIONS

CONCLUDING REMARKS

K.H. Jack

Crystallography Laboratory,
The University of Newcastle upon Tyne, U.K.

These final few words merely mark the formal end of the second NATO Advanced Study Institute on "Nitrogen Ceramics" and can in no way be regarded as a summary of our proceedings.

I have said on other occasions that the 1976 meeting in Canterbury was one of the best conferences I had attended and I thought it would be impossible for the present one to reach the same standard. On the whole, I think it has. We have followed the same pattern and have had nine sections covering widely different aspects of materials science and engineering. Indeed, these are so wide that it is quite impossible for me to assess everything that has been achieved. I can give only personal impressions and these are certainly not objective. You will know from some of our discussions that I am likely to be in a minority of one!

We tend as individuals or groups to be somewhat specialised and this has been evident in the past two weeks. However, this is not always such a bad thing because if we try to do everything we sometimes end up by achieving nothing. Nevertheless, this tendency to specialisation is exactly why it is so essential to meet and exchange views and to assess whether what we are doing is worthwhile. Dr. Bunk has told us that it was necessary for his scientists, engineers and designers to meet three times a year before they started to talk to each other. Once every five years must be inadequate, particularly if, - as Dr. Lumby claims - we do not read each other's papers.

A glance at the 1976 Proceedings shows how far we have progressed, and if we have not looked at it since receiving it in

Riley, F.L. (ed.) Progress in Nitrogen Ceramics

1977 the better appreciation and understanding of the whole field of nitrogen ceramics that now emerges is astonishing.

There is almost agreement that β-silicon nitride is thermodynamically more stable than the α-phase, even though kinetics and not thermodynamics usually determines which variety is formed under any set of circumstances. With the recent determination of the crystal structure of sialon-X-phase, the compositions, the inter-relationships and the structures of all the phases in the Si-Al-O-N system have been established - but still only over a limited temperature range. I do not think that any of Dr. Katz's engine trials were delayed by not knowing the X-phase structure, but if they <u>are</u> in the future, Dr. Morgan can predict any crystal structure with a fifty per cent probability of success. Seriously, there are now structural data, coefficients of expansion, densities, and limited information on properties for most phases likely to occur in nitrogen ceramics. The behaviour diagrams for metal-Si-O-N and metal-Si-Al-O-N systems are still incomplete, or are in dispute, and considering how necessary they are in the processing of ceramic compositions, this is a need that must be met. Dr. McCauley gave a fantastic example of the painstaking, detailed work that is required to correlate phase diagrams and microstructures, and incidentally to give us AlON, a new oxynitride ceramic.

Nor have techniques stood still. In 1976 Dr. Clarke presented, for the first time, direct observations of the crystal lattice planes up to and on either side of grain boundaries in silicon nitride, and showed that there was no glassy phase at the boundaries he investigated. Electron microscopy lattice imaging must have improved because he now cannot find a grain boundary without glass! The array of techniques available for the non-destructive evaluation of ceramic components is formidable and, as has been emphasized, is absolutely essential to the success of the high-temperature gas turbine. The method of controlled indentation fracture coupled with acoustic-wave scattering, as described by Dr. Marshall, is very elegant and appears not too difficult to apply in the assessment of wear, erosion and strength degradation.

In 1976 there were four materials: HPSN was the most developed and best understood; RBSN was a close second, while SSN and sialons were in their infancy. Now, HPSN is rejected because of cost and the difficulty of fabricating complex shapes. In its place there are the Ford and Fiat powder-bed processes for producing dense SRBSN. These, when coupled with slip-casting and injection-moulding methods of fabricating silicon powder shapes, give a complete facility for mass-production of high density silicon nitride components. These magnificent achievements are most stimulating and offer optimistic prospects for the future commercial success of silicon nitride. It is further claimed that the RBSN and SRBSN processes might be combined with cladless hot isostatic pressing to give further improvements in properties.

The nitriding of silicon to produce RBSN is still imperfectly understood, and although Dr. Mangels has the exotherm of the reaction well under control by using N_2:H_2:He gas mixtures, Dr. Popper pointed out that the pre-treatment of the silicon powder, as described by Dr. Evans, increased the rate of nitriding by more than two orders of magnitude. Dr. Schwier described the first-class development carried out by H.C. Starck Berlin to produce good silicon nitride powder on a commercial scale. An alternative route, by carbothermal reduction of silica in molecular nitrogen, is attractive because pure silica is cheaper than pure silicon and Mr. Mori was sufficiently confident about Toshiba's product, made in this way, to distribute free samples.

The oxynitride liquid that is necessary for the high-temperature densification of all nitrogen ceramics cools to give an intergranular glass that degrades properties. Much time was spent in discussing the effects of this glass and also the secondary crystalline phases that are sometimes formed with it. In some systems, the choice of the appropriate initial composition and a post-densification heat-treatment can react the glass with the matrix to give virtually a single-phase product with a minimum of glass. This is the principle by which Dr. Greskovich at G.E. produces a $Be_{0.1}Si_{2.0}O_{0.2}N_{3.8}$ β-phase. Alternatively, the glass can be reacted to give a refractory intergranular crystalline phase compatible with the matrix; this happens in the formation of Lucas SYALON, a β'-sialon with intergranular yttrium-aluminium garnet. Both this and the Be-β-phase oxynitride have outstanding properties and it would seem that similar principles might be applied in other systems to produce good engineering ceramics.

Although bulk glasses have been made with nitrogen contents as high as 15 atomic %, it is a little disappointing that the increases in hardness, viscosity, strength and refractive index compared with corresponding oxide glasses are not greater. However, Dr. Katz predicts that these nitrogen-containing glasses have a future, and in any case their characterization is important because, as previously stated, they occur as grain boundary phases in all nitrogen ceramics. The paper by Professor Raj, presented by Dr. Shaw, is significant because it suggests, on thermodynamic grounds, that grain-boundary glasses can never be fully devitrified.

These are some aspects of the meeting that have interested me, but quite different selections will appeal to others and I am sure that there is no paper that has not been of value to someone. In his admirable opening address, Dr. Katz said that one major commercial success was needed urgently. Going back to the 1976 Proceedings, three applications were considered in detail: (i) gas turbines; (ii) diesel engine components; and (iii) ball and roller bearings. They are still being considered, but using modified materials and modified processing methods. It is hoped that one application that was not considered in 1976 - Syalon cutting tools for machining metals - will be the commercial success that meets our need.

Dr. Bunk has called for more international cooperation. I agree with whoever said that this is best achieved on a personal basis. For Britain, in its present economic climate, there is no chance of it being on any other basis! The great thing about both NATO meetings has been the delightful extent of the social as well as technical cooperation between all members from all countries.

AUTHOR INDEX

SUBJECT INDEX

INDEX OF CRYSTAL STRUCTURES